PRINCIPLES OF

ELECTRONIC
COMMUNICATION
SYSTEMS

Louis E. Frenzel

Austin Community College
Riverside Campus
Austin, Texas

McGraw-Hill **Glencoe McGraw-Hill**

New York, New York   Columbus, Ohio   Mission Hills, California   Peoria, Illinois

Library of Congress Cataloging–in–Publication Data

Frenzel, Louis E.
    Principles of electronic communication systems / Louis E. Frenzel.
       p.   cm.
    Includes index.
    ISBN 0-02-800409-4
    1. Telecommunication. I. Title.
  TK5101.F664   1998
621.382—dc20
                             96-30979
                                CIP
                                AC

Glencoe/McGraw-Hill

*A Division of The **McGraw·Hill** Companies*

Principles of Electronic Communication Systems

Send all inquiries to:
Glencoe/McGraw-Hill
936 Eastwind Drive
Westerville, Ohio 43081

ISBN 0-02-800409-4

  3 4 5 6 7 8 9 10   027/043   04 03 02 01 00 99

CONTENTS

CONTENTS

C O N T E N T S

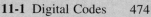

CONTENTS

PREFACE

This book is written for the introductory course in electronic communications at the technology level. It can be used in community colleges, technical institutes, and colleges of technology. It is assumed that students have had a good background in algebra and trigonometry fundamentals and have taken a course in AC principles and circuits.

ADVANTAGES OF THIS BOOK

This new book on electronic communications reflects the author's experience as both a teacher at the community college level and technician, engineer, and manager in industry.

1. *Four-color printing.* This book takes full advantage of color in its illustrations, photographs, layout, and design.

2. *Broad coverage of advanced communications technologies.* Most electronic communications texts focus on two-way radio technology and offer merely a selection of advanced applications. This book covers all major segments of the communications field in a balanced way. Examples of this expanded coverage are the cellular telephone system, satellites, paging, and facsimile.

3. *Introduction to advanced circuits and processes.* This text includes advanced circuits such as spread spectrum, digital signal processing (DSP), and switched capacitor filters.

4. *Emphasis on systems, not circuits.* The technician must understand the overall objectives and concepts of the system. The technician must also know what the major components of a piece of equipment are, how each works, and how the components relate to one another. The technician needs to know about signal flow at the block diagram level: What are the inputs and outputs, and what processing steps and circuits are involved? This book emphasizes concepts and explains "how things work."

5. *Broad coverage of data communications.* The technology of electronic communications has reached the point at which digital communications cannot be completely separated from traditional techniques. All technicians must be familiar with these concepts, because all communications systems and techniques now involve digital methods. This book addresses this need by offering extensive coverage of digital techniques and data communications methods.

6. *Coverage of test equipment and troubleshooting.* A major omission from many communications books is coverage of the test instruments and measurements used by technicians in the communications field. Communications "techs" install, operate, test, troubleshoot, maintain, service, and repair communications equipment and systems. Chapter 20 summarizes the types of test instruments used in communications and provides an overview of the basic tests and measurements that technicians carry out.

7. *Flexible, modern organization.* The material in this book is organized in a way that makes it easy for the instructor to present, and the student to understand. The difference between this text and others is the placement of subjects in context with related subjects. For example, the topic of noise is included in the same chapter with receivers, where it is relevant. Modulation theory and circuits are separated from transmitters and receivers. Television is presented at the end of the book, rather than at the beginning. Television coverage includes advanced circuits used in combination with VCRs and signal reception via cable or dish antenna..

OBJECTIVES OF THE BOOK

1. *Make the book easy to read.* The easier a book is to read, the easier it is for students to understand. This book contains more than 1000

line drawings and many color photographs, designed to aid comprehension.

2. *Make this a technician-level book.* Many texts include engineering-level material that technicians will never need. Technicians install, troubleshoot, maintain, and repair equipment. This text contains the material that technicians will really need, such as the theory of operation and signal flow in real equipment.

3. *Emphasize communications systems rather than specific circuits.* This text has moved away from the standard "chasing of electrons" and gives attention to signal flow analysis in block diagrams. The emphasis is on the universal model of input—process—output, describing what the inputs are, how they are processed, and what the new outputs are. Most circuits are now ICs, and the technician will do little analysis or troubleshooting at the component level.

A QUICK OVERVIEW

The following summary gives the highlights of various chapters, to illustrate the focus of the book and its advantages.

Chapter 1: This chapter is designed to introduce the student to the field of communications. It includes a section designed to give students some ideas about job opportunities and levels of education needed to qualify for jobs.

Chapter 2: This chapter reviews many concepts that students should know before they start a course in communications. The sections on decibels and filters are particularly pertinent to modern communications techniques.

Chapters 3 and 4: These chapters cover amplitude modulation. The theory and specific circuits are dealt with separately because students seem to grasp complex subjects like this more readily when the material is presented in discrete segments. SSB and DSB are also covered in these chapters.

Chapters 5 and 6: In these chapters, angle modulation, both FM and PM, are covered. Again, theory and the actual circuits are discussed separately, to simplify the presentation.

Chapter 7: This chapter, on transmitters, deals with the special circuits in transmitters, with emphasis on RF power amplifiers. It contains a unique coverage of class D and class E switching power amplifiers. Impedance-matching networks are also covered.

Chapter 8: This chapter is devoted to receivers. IC receiver circuits are covered extensively, and frequency synthesis is also included.

Chapter 9: The digital techniques and circuits commonly used in communications are introduced. Students need this background before more advanced communications techniques are presented. Among the subjects in this chapter are analog-to-digital and digital-to-analog conversion methods, basic digital communications methods, digital signal processing, and pulse communications methods.

Chapter 10: This chapter deals with multiplexing techniques, including complete coverage of frequency- and time-division multiplexing concepts and circuits.

Chapter 11: In this basic data communications chapter, all codes, transmission techniques, and theories are covered, including baseband signaling, encoding, and media. Modulation techniques such as PSK and QAM, modems, and spread spectrum are presented.

Chapter 12: The subject of data communications is continued in this chapter, with the addition of coverage of LANs, topologies, protocols, equipment, and specific systems such as Ethernet, Token Ring, and FDDI.

Chapter 13: Expanded coverage is given of the practical aspects of transmission lines (popular types and their specifications and connectors). Since most current communications equipment operates in the ultrahigh-frequency and microwave regions, it is essential to include this topic. The chapter also includes an explanation of Smith charts.

Chapter 14: The theory and basic types of antennas are given here. Impedance matching, antenna tuners, and the fundamentals of radio wave propagation are included.

Chapter 15: Microwaves are the core of most modern communications systems. Both components and circuits are covered here. Among the other related subjects are microstrip and stripline, as used in microwave transmitters and receivers; typical semiconductor components and circuits; waveguides; horns; parabolic dishes; specialty antennas such as slots, patch antennas, and arrays; and microwave applications including microwave relay links and radar.

Chapter 16: In this chapter, on satellite operation and applications, in-depth coverage of the Global Positioning System (GPS) for navigation and a look at the forthcoming satellite-based cellular telephone systems are included.

Chapter 17: What communications book would be complete without a chapter on the telephone system? Details on telephone circuits and operation are given. Standard, electronic, and cordless telephones are included, as is complete coverage of the cellular telephone system. Other telephone-related technologies such as paging, facsimile, and ISDN, are discussed in detail.

Chapter 18: The emphasis is on fiber-optic communications. The chapter also covers the other common optical (infrared) communications, such as TV wireless remote control and wireless LAN.

Chapter 19: Cable TV, direct broadcast TV, satellite TV, and high-definition TV are discussed in detail in this chapter.

Chapter 20: Test equipment and troubleshooting related to communications equipment and systems are the topic of this chapter.

FLEXIBILITY IN THE USE OF THIS BOOK

Each instructor has a preference as to the sequence of presentation of material. Also, textbooks differ in approaching and sequencing. The organization of this text permits the instructor to make assignments based upon personal preference, the school's program, the curriculum, and the content of the course. All material needed for a comprehensive communications course is contained in this book. For a shorter course, the instructor may assign the following chapters. This list contains brief suggestions for using selected chapters in a shorter course.

Chapter 1: This is the overview—"the big picture." Consider beginning the first class with a discussion on Section 1–8, "Jobs and Careers in the Communications Industry."

Chapter 2: This chapter is optional; the amount of class time devoted to this chapter will depend upon the needs of the students.

Chapters 3 to 8: Most of the material in these chapters should be covered in a communications course—even an abbreviated one.

Chapters 9 to 12: Ideally, some coverage of all four chapters (on digital and data communi-

cations) should be included, with the emphasis on Chapter 9.

Chapters 13 and 14: The topics of antennas and transmission lines are essential to communications.

Chapter 15: Do not neglect the subject of microwaves, for it is important.

Chapters 16 to 19: These chapters can be considered optional in an abbreviated course. The decision about whether to use any or all of them should be based on factors such as students' interests and local employers' needs.

Chapter 20: Be sure to include this chapter, for it discusses what technicians do. Encourage individual or group discussions to stimulate critical thinking.

The author is grateful to Dr. Foster Chin, Tulsa Junior College, Tulsa, Oklahoma; Walter Brock, McHenry County College, Crystal Lake, Illinois; Anthony Hearn, Community College of Philadelphia, Philadelphia, Pennsylvania; David Heiserman, formerly of DeVry Institute of Technology, Columbus, Ohio; Dean A. Honadle, State Technical Institute of Memphis, Memphis, Tennessee; Richard A. Honeycutt, Davidson County Community College, Lexington, North Carolina; Donald Markel, DeVry Institute of Technology, Woodbridge, New Jersey; Charles A. Schuler, California University of Pennsylvania, California, Pennsylvania; Tseng-Yuan Woo, Durham Technical Community College; and Robert Wooten, Director of Education, RETS Electronics Institute, Louisville, Kentucky, for their detailed technical and practical suggestions. The author is especially grateful to Dr. Abraham Pallas, Dean of Applied Sciences, Columbia State Community College, Columbia, Tennessee, who critiqued the original manuscript and the proofs.

And finally, thanks to my wife, Joan, for her constant loving support and keyboarding expertise. A special thank-you also to Olive Collen, of Publishing Advisory Service, and John J. Beck and Rob Ciccotelli of Glencoe/McGraw-Hill.

I welcome suggestions from readers, who can communicate with me through the publisher. All ideas, needs, and comments will be considered for inclusion in subsequent editions of this book.

Louis E. Frenzel
Austin, Texas

INTRODUCTION TO ELECTRONIC COMMUNICATIONS

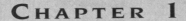 Objectives

After completing this chapter, you will be able to:

◆ *Explain* the functions of the three main parts of an electronic communications system.

◆ *Describe* the system used to classify different types of electronic communications and *list* examples of each type.

◆ *Discuss* the role of modulation and multiplexing in facilitating signal transmission.

◆ *Define* the electromagnetic spectrum and *explain* why the nature of electronic communications makes it necessary to regulate the electromagnetic spectrum.

◆ *Explain* the relationship between frequency range and bandwidth and *give* the frequency ranges for spectrum uses ranging from voice to ultra-high-frequency television.

◆ *List* the major branches of the field of electronic communications and *describe* the qualifications necessary for different jobs.

Communications is the process of exchanging information. People communicate in order to convey their thoughts, ideas, and feelings to others. The process of communications is inherent to all human life. Early in human history, a large component of communications was nonverbal. Body movements and facial expressions were effective ways to communicate. Later, languages were invented, and still later, written communications was developed. Our early ancestors learned to draw pictures to depict their thoughts. In some parts of the world, alphabets were developed; in others, systems consisting of symbols representing whole objects or having complex meanings were created. Humans wrote letters to one another and recorded history using these systems. The printing press was invented in 1440, and eventually newspapers and books were printed rather than handwritten. Although the bulk of human communication today is still oral, a huge volume of information is exchanged by means of the written word. Today, although there is a glut of printed information of every conceivable variety, most of our communication is still verbal as we talk to one another face to face or over the telephone.

Two of the main barriers to human communication are language and distance. Humans from different tribes, nations, or races usually do not speak the same language. Language barriers can, however, be overcome. People who speak one language can learn to speak other languages, or can use interpreters.

Communicating over long distances is another problem. Communication between early human beings was limited to face-to-face encounters. Long-distance communication was probably first accomplished by sending simple signals such as drumbeats, horn blasts, and smoke signals and later by waving signal flags (semaphores). With these methods, transmission distances were limited. Signals launched from a hill, mountain, or chain of high towers might cover distances of several miles. By relaying messages from one location to another, even greater distances could be covered.

The distance over which communication could be sent was extended by the written word. Messages and letters were transported from one place to another. For many years, long-distance communication was limited to the sending of verbal or written messages by human runner, horseback, ship, and, later, trains.

Human communication took a dramatic leap forward in the late nineteenth century when electricity was discovered and its many applications were explored. The telegraph was invented in 1844 and the telephone in 1876. Radio was discovered in 1887 and demonstrated in 1895. Figure 1-1 is a timetable listing important milestones in the history of communications.

Well-known forms of electronic communications, such as the telephone, radio, and TV, have increased our ability to share information. Today, it is hard to imagine what our lives would be like without the knowledge and information that come our way from around the world by the various means of electronic communications. The way that we do things and the success of our work and personal lives are directly related to how well we communicate. It has been said that the emphasis in our society has now shifted from that of manufacturing and mass production of goods to the accumulation, packaging, and exchange of information. Ours is an information society, and a key part of it is communications. Without electronic communications, we could not access and apply the

DID YOU KNOW?

Fax machines have been in use since the 1930s. These early machines were primarily utilized by news services to transmit photographs using free space or radio waves rather than telephone lines. (World Book, vol. 7, 1995)

WHEN?	WHERE OR WHO?	WHAT?
ca. 100,000 B.C.	Early humans	Development of language.
ca. 35,000	Africa	Counting devices used.
25,000 B.C.	Africa, Australia	Rock carvings, cave paintings.
6000 B.C.	South America	Pictographs and repeated symbols.
3000 B.C.	Mesopotamia	Pictographs used in a system of writing.
ca. 1500 B.C.	Syria, Lebanon, Israel	The first phonetic alphabet adapted.
300 B.C.	Egypt	First library built.
105 A.D.	China	Paper invented.
300		Use of the codex, a precursor of the book, where both sides of the page could be written on.
780	China	Wood block printing used to mass produce books.
868	China	First known printed book with both text and illustrations.
1041–1048	China	The invention of movable type.
1157	Spain	The first European paper mill built.
1440	Johann Gutenberg	Invention of the printing press.
1500	Countries, states, cities	Messenger services using men on horseback.
1794	Claude Chappe in France and later Depillion in Britain	Semaphore towers used to signal ships and to send messages across country by way of lights or flags.
1823–1832	Charles Babbage	The "difference engine," the first mechanical computer, invented
1837	Samuel Morse	Invention of the telegraph (patented in 1844).
1843	Alexander Bain	Invention of facsimile.
1866	U.S. and England	The first trans-Atlantic telegraph cable laid
1876	Alexander Bell	Invention of the telephone.
1877	Thomas Edison	Invention of the phonograph.
1879	George Eastman	Invention of photography.
1887	Heinrich Hertz (German)	Discovery of radio waves.
1887	Guglielmo Marconi (Italian)	Demonstration of "wireless" communications by radio waves.
1901	Marconi (Italian)	First trans-Atlantic radio contact made.
1903	John Fleming	Invention of the two-electrode vacuum tube rectifier.
1906	Reginald Fessenden	Invention of amplitude modulation; first electronic voice communications demonstrated.

(continues on next page)

FIG. 1-1   Milestones in the history of communications.

When?	Where or Who?	What?
1906	Lee de Forest	Invention of the triode vacuum tube.
1914	Hiram P. Maxim	Founding of American Radio Relay League, the first amateur radio organization.
1920	KDKA Pittsburgh	First radio broadcast.
1923	Vladimir Zworykin	Invention and demonstration of television.
1933–1939	Edwin Armstrong	Invention of the superheterodyne receiver and frequency modulation.
1939	United States	First use of two-way radio (walkie talkies).
1940–1945	Britain, United States	Invention and perfection of radar (World War II).
1948	John von Neumann and others.	Creation of the first stored program electronic digital computer.
1948	Bell Laboratories	Invention of transistor.
1953	RCA/NBC	First color TV broadcast.
1958–1959	Jack Kilby (Texas Instruments) and Robert Noyce (Fairchild)	Invention of integrated circuits.
1958–1962	United States	First communications satellite tested.
1961	United States	Citizen's band radio first used.
1975	United States	First personal computers.
1977	United States	First use of fiber-optic cable.
1983	United States	Cellular telephone networks.
1990s	United States	Major adoption and growth of computer networking, including local area returns (LANS). Use of the Global Positioning System (GPS) for satellite navigation. The Internet and World Wide Web.

FIG. 1-1 (continued)

available information in a timely way. The so-called information superhighway of the future is the epitome of electronic communications technology.

This book is about electronic communications, and how electrical and electronic principles, components, circuits, equipment, and systems facilitate and improve our ability to communicate. Rapid communication is critical in our very fast-paced world. It is also addictive. Once we adopt and get used to any form of electronic communications, we become hooked on its benefits. In fact, we cannot imagine conducting our lives or our businesses without it. Just imagine our world without the telephone, radio, or television. The fax is a good example of how quickly modern technology makes us dependent on the benefits of communications convenience and speed. Today, people wonder how they managed, just a few years ago, without their fax machines.

1-2 COMMUNICATIONS SYSTEMS

All electronic communications systems have the basic components shown in Fig. 1-2: a transmitter, a communications channel or medium, and a receiver. The process of communications begins when a human generates some kind of message, data, or other intelligence that must be received by others. In electronic communications systems, the message is referred to as *information,* or an intelligence signal. This message, in the form of an electronic signal, is fed to the transmitter, which then transmits the message over the communications channel. The message is picked up by the receiver and relayed to another human. Along the way, noise is picked up in the communications channel and in the receiver. *Noise* is the general term applied to any phenomenon that degrades or interferes with the transmitted information. Let's take a closer look at each of these basic elements.

TRANSMITTER

The first step in sending a message is to convert it into electronic form suitable for transmission. For voice messages, a microphone is used. The microphone is a transducer that translates the sound into an electronic *audio* signal. For TV, a camera converts the light information in the scene into a video signal. In computer systems, the message is typed on a keyboard and converted into binary codes that can be stored in memory or transmitted serially. Regardless of the kind of information to be sent, it must first be put into the form of an electrical signal.

The transmitter itself is a collection of electronic components and circuits designed to convert the electrical signal into a signal suitable for transmission over a given communications medium. Transmitters are made up of oscillators, amplifiers, tuned circuits and filters, modulators, frequency mixers, frequency synthesizers, and other circuits. The original intelligence signal usually modulates a higher-frequency carrier sine wave generated by the transmitter, and the combination is raised in amplitude by power amplifiers resulting in a signal that is compatible with the selected transmission medium.

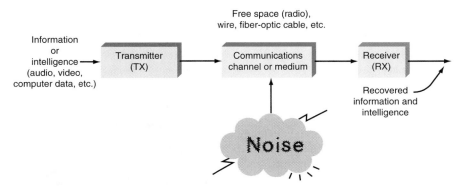

FIG. 1-2 A general model of all communications systems.

COMMUNICATIONS CHANNEL

The communications channel is the medium by which the electronic signal is sent from one place to another. Many different types of media are used in communications systems, including wire conductors, fiber-optic cable, and free space.

ELECTRICAL CONDUCTORS. In its simplest form, the medium may simply be a pair of wires that carry a voice signal from a microphone to a headset. It may be a coaxial cable such as that used to carry cable TV signals. Or it may be a twisted pair cable used in a local area network (LAN) for personal computers.

OPTICAL MEDIA. The communications medium may also be a fiber-optic cable or "light pipe" that carries the message on a light wave. These are widely used today by telephone companies to carry long distance calls. The information is converted into digital form that can be used to turn a light emitting diode (LED) or laser diode off and on at high speeds. Alternatively, audio or video analog signals can be used to vary the amplitude of the light.

FREE SPACE. When free space is the medium, the resulting system is known as radio. *Radio* is the broad general term applied to any form of wireless communication from one point to another. Radio makes use of the electromagnetic spectrum. Intelligence signals are converted into electric and magnetic fields that propagate readily through space over long distances.

OTHER TYPES OF MEDIA. While the most widely used media are conducting cables and free space (radio), other types of media are used in special communications systems. For example, in sonar, water is used as the medium. Passive sonar "listens" for underwater sounds with sensitive hydrophones. Active sonar uses an echo-reflecting technique similar to that used in radar for determining how far away objects under water are and in what direction they are moving.

The earth itself can be used as a communications medium, because it conducts electricity and can also carry low-frequency sound waves.

Alternating current power lines, the electrical conductors that carry the power to operate virtually all of our electrical and electronic devices, can also be used as communications channels. The signals to be transmitted are simply superimposed on or added to the power-line voltage. This is known as *carrier current transmission*. It is used for some types of voice intercoms, remote control of electrical equipment, and in some LANs.

RECEIVERS

A *receiver* is a collection of electronic components and circuits that accepts the transmitted message from the channel and converts it back into a form understandable by humans. Receivers contain amplifiers, oscillators, mixers, tuned circuits and filters, and a demodulator or detector that recovers the original intelligence signal from the modulated carrier. The output is the original signal, which is then read out or displayed. It may be a voice signal sent to a speaker, a video signal that is fed to a cathode-ray tube for display, or binary data that is received by a computer and then printed out or displayed on a video monitor.

TRANSCEIVERS

Most electronic communications is two way, and so both parties participating in the communications must have both a transmitter and a receiver. As a result, most communications equipment incorporates circuits that both send and receive. These units are commonly referred to as *transceivers*. All the transmitter and receiver circuits are packaged within a single housing and usually share some common circuits such as the power supply. Telephones, fax machines, handheld CB radios, cellular telephones, and computer modems are examples of transceivers.

ATTENUATION

Signal *attenuation,* or degradation, is inevitable no matter what the medium of transmission. Media are usually frequency-selective, in that a given medium will act as a low-pass filter to a transmitted signal, distorting digital pulses in addition to greatly reducing signal amplitude over long distances. Thus considerable signal amplification, in both the transmitter and the receiver, is required for successful transmission. Any medium also slows signal propagation to a speed slower than the speed of light.

NOISE

All communications systems are also subject to noise both in the communications channel and in the receiver. *Noise* is random, undesirable electronic energy that enters the communications system via the communicating medium and interferes with the transmitted message. Some noise is also produced in the receiver. Noise comes from the atmosphere (e.g., from lightning, which produces static); from outer space, where the sun and other stars emit various kinds of radiation that can interfere with communications; and from manufactured equipment, such as the electric ignition systems of cars, electric motors, fluorescent lights, and other types of equipment which generate signals that can interfere with transmission.

Finally, many electronic components generate noise internally due to thermal agitation of the atoms. Resistors and transistors are common noise-producing components. Although such noise signals are low-level, they can often seriously interfere with the extremely low-level signals that appear greatly attenuated at the receiver after being transmitted over a long distance. In some cases, noise completely obliterates the message. In others, it results in partial interference. Noise is one of the more serious problems of electronic communications. Although it probably cannot be completely eliminated, there are ways to deal with it, as later sections will discuss.

1-3 TYPES OF ELECTRONIC COMMUNICATIONS

Electronic communications are classified according to whether they are (1) one-way (simplex) or two-way (full duplex or half duplex) transmissions and (2) analog or digital signals.

SIMPLEX

The simplest way in which electronic communications is conducted is one-way communications, normally referred to as *simplex*. In simplex communications, the information travels in one direction only. Examples are shown in Fig. 1-3. The most common form of simplex communications is radio and TV broadcasting. The listeners do not respond. Another example of one-way communication is transmission via a paging system to a personal receiver (beeper).

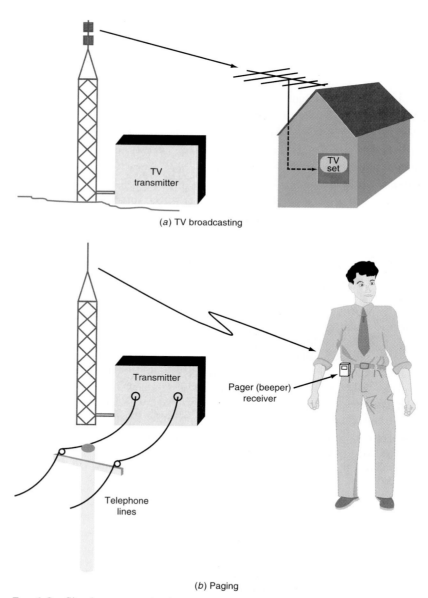

(a) TV broadcasting

(b) Paging

Fig. 1-3 Simplex communications.

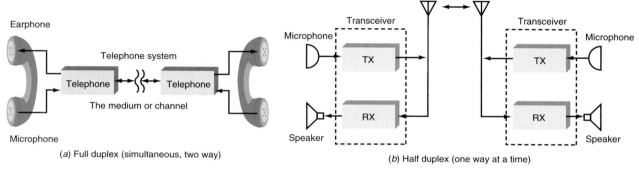

(a) Full duplex (simultaneous, two way)

(b) Half duplex (one way at a time)

FIG. 1-4   Duplex communications. (a) Full duplex (simultaneous two way). (b) Half duplex (one way at a time).

FULL DUPLEX

The bulk of electronic communications is two-way, or *duplex.* Typical duplex applications are shown in Fig. 1-4. For example, people communicating with one another over the telephone can talk and listen simultaneously, as Fig. 1-4(a) illustrates. Such simultaneous send-and-receive two-way communications is called *full duplex.*

HALF DUPLEX

The form of two-way communications in which only one party transmits at a time is known as *half duplex* [see Fig. 1-4(b)]. The communications is two-way, but the direction alternates: the communicating parties take turns transmitting and receiving. Most radio transmissions, such as those used in the military, fire, police, aircraft, marine, and other services, are half-duplex communications. Citizens band (CB) and amateur radio communications are also typically half duplex. Most wired intercom systems permit only one party to transmit at a time.

ANALOG SIGNALS

An *analog* signal is a smoothly and continuously varying voltage or current. Some typical analog signals are shown in Fig. 1-5. A sine wave is a single-frequency analog signal. Voice and video voltages are analog signals which vary in accordance with the sound or light variations that are analogous to the information being transmitted.

DIGITAL SIGNALS

Digital signals, in contrast to analog signals, do not vary continuously, but change in steps or in discrete increments. Most digital signals use binary or two-state codes. Some examples are shown in Fig. 1-6. The earliest forms of both wire and radio communi-

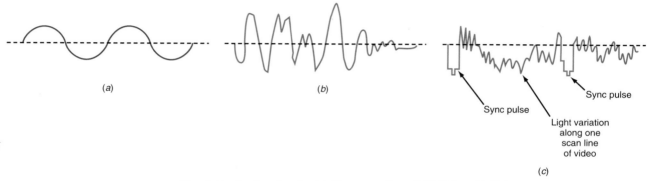

Sync pulse
Sync pulse
Light variation along one scan line of video

(c)

Fig. 1-5 Analog signals. (*a*) Sine wave "tone." (*b*) Voice. (*c*) Video (TV) signal.

cations used a type of on-off digital code. The telegraph used Morse code, with its system of short and long signals (dots and dashes) to designate letters and numbers. See Fig. 1-6(*a*). In radio telegraphy, also known as continuous wave (CW) transmission, a sine wave signal is turned off and on for short or long durations to represent the dots and dashes. Refer to Fig. 1-6(*b*).

Data used in computers is also digital. Binary codes representing numbers, letters, and special symbols are transmitted serially by wire, radio, or optical medium. The most commonly used digital code in communications is the *American Standard Code for Information Interchange* (*ASCII*, pronounced ask key). Figure 1-6(*c*) shows a serial binary code.

Many transmissions are of signals which originate in digital form, e.g., telegraphy messages or computer data, but which must be converted to analog form to match the transmission medium. An example is the transmission of digital data over the telephone network, which was designed to handle analog voice signals only. If the digital data is converted into analog signals, such as tones in the audio frequency range, it can be transmitted over the telephone network. Two common examples are given in Fig. 1-7. In Fig. 1-7(*a*), the data is converted to frequency varying tones. This is called *frequency-shift keying (FSK)*. In Fig. 1-7(*b*), the data introduces a 180-degree phase shift. This is called *phase-shift keying (PSK)*. Devices called *modems* (*mod*ulator-*dem*odulator) translate the data from digital to analog and back again.

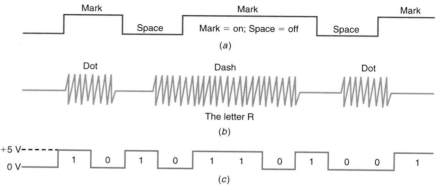

Fig. 1-6 Digital signals. (*a*) Telegraph (Morse code). (*b*) Continuous wave (CW) code. (*c*) Serial binary code.

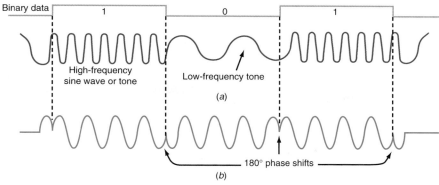

FIG. 1-7 Transmitting binary data in analog form. (*a*) FSK. (*b*) PSK.

Analog signals can also be transmitted digitally. It is very common today to take voice or video analog signals and digitize them with an analog-to-digital (A/D) converter. The data can then be transmitted efficiently in digital form and processed by computers and other digital circuits.

1-4 MODULATION AND MULTIPLEXING

Modulation and multiplexing are electronic techniques for transmitting information efficiently from one place to another. *Modulation* makes the information signal more compatible with the medium, and *multiplexing* allows more than one signal to be transmitted concurrently over a single medium. Modulation and multiplexing techniques are basic to electronic communications. Once you have mastered the fundamentals of these techniques, you will easily understand how most modern communications systems work.

BASEBAND TRANSMISSION

Before it can be transmitted, the information or intelligence must be converted into an electronic signal compatible with the medium. For example, a microphone changes voice signals (sound waves) into an analog voltage of varying frequency and amplitude. This signal is then passed over wires to a speaker or headphones. This is the way the telephone system works.

A video camera generates an analog signal that represents the light variations along one scan line of the picture. This analog signal is usually transmitted over a coaxial cable. Binary data is generated by a keyboard attached to a computer. The computer stores the data and processes it in some way. The data is then transmitted on cables to peripherals such as a printer or to other computers over a LAN. Regardless of whether the original information or intelli-

DID YOU KNOW?

Multiplexing has been used in the music industry to create stereo sound. In stereo radio, two signals are transmitted and received—one for the right and one for the left channel of sound. (For more information on multiplexing, see Chap. 10.)

gence signals are analog or digital, they are all referred to as baseband signals.

In a communications system, baseband information signals can be sent directly and unmodified over the medium or can be used to modulate a carrier for transmission over the medium. Putting the original voice, video, or digital signals directly into the medium is referred to as *baseband transmission.* For example, in many telephone and intercom systems, it is the voice itself that is placed on the wires and transmitted over some distance to the receiver. In some computer networks, the digital signals are applied directly to coaxial cables for transmission to another computer.

In many instances, baseband signals are incompatible with the medium. Although it is theoretically possible to transmit voice signals directly by radio, realistically it is impractical. Voice signals occur in the 300- to 3000-hertz (Hz) frequency range. After being increased in amplitude by a power amplifier such as that used in a common stereo system, the voice signals could be sent to a long wire antenna instead of a speaker. The resulting electromagnetic waves would be propagated through space to a receiver, an audio amplifier attached to a long wire antenna. In order for a system like this to work effectively, the antenna would have to be enormous. Antenna length is usually one-quarter or one-half wavelength of the signal it is set up to transmit. Thus the antenna for audio signals would have to be many miles long, a practical impossibility.

Second, simultaneously transmitted audio signals would interfere with one another since they occupy the same frequency range. The audio portion of the spectrum (300 to 3000 Hz) would be nothing but a jumble of simultaneous voice communications. The antenna would pick them all up at the same time, and the receiver would amplify them concurrently. There would be no way to separate them and select the desired signal.

For these reasons, the baseband information signal, be it audio, video, or data, is normally used to modulate a high-frequency signal called a *carrier.* The higher-frequency carriers radiate into space more efficiently than the baseband signals themselves. Such wireless signals consist of both electric and magnetic fields. These electromagnetic signals, which are able to travel through space for long distances, are also referred to as *radio-frequency (RF) waves,* or just radio waves.

BROADBAND TRANSMISSION

Modulation is the process of having a baseband voice, video, or digital signal modify another, higher-frequency signal, the carrier. The process is illustrated in Fig. 1-8. The information or intelligence to be sent is said to be *impressed* upon the carrier. The carrier is usually a sine wave generated by an oscillator. The carrier is fed to a circuit called a modulator along with the baseband intelligence signal. The intelligence signal changes the carrier in a unique way. The modulated carrier is amplified and sent to the antenna for transmission. This process is called *broadband transmission.*

At the receiver, an antenna picks up the signal which is then amplified and processed in other ways. It is then applied to a demodulator or detector, where the original baseband signal is recovered. See Fig. 1-9.

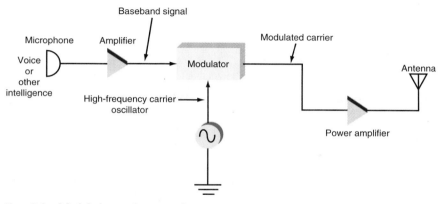

Fig. 1-8 Modulation at the transmitter.

Consider the common mathematical expression for a sine wave:

$$v = V_p \sin(2\pi ft + \theta) \qquad \text{or} \qquad v = V_p \sin(\omega t + \theta)$$

where v = instantaneous value of sine-wave voltage

$\quad V_p$ = peak value of sine wave

$\quad f$ = frequency, Hz

$\quad \omega$ = angular velocity = $2\pi f$

$\quad t$ = time, s

$\quad \omega t = 2\pi ft$ = angle, rads ($360° = 2\pi$ rad)

$\quad \theta$ = phase angle

There are three ways to make the baseband signal change the carrier sine wave: vary its amplitude, vary its frequency, or vary its phase angle. The two most common methods of modulation are *amplitude modulation (AM)* and *frequency modulation (FM)*. In AM, the baseband information signal called the modulating signal varies the amplitude of the higher-frequency carrier signal, as shown in Fig. 1-10(*a*). It changes the V_p part of the equation. In FM, the information signal varies the frequency of the carrier, as shown in Fig. 1-10(*b*). The carrier amplitude remains constant. FM varies the value of *f* in the first angle term inside the parentheses. Varying the phase angle produces *phase modulation (PM)*. Here, the second term inside the parentheses (θ) is made to vary by the intelligence signal. Phase modulation produces frequency modulation; therefore, the PM signal is similar in appearance to a frequency-modulated carrier. Both FM and PM are forms of *angle modulation*.

At the receiver, the carrier with the intelligence signal is amplified and then demodulated to extract the original baseband signal. Another name for the demodulation process is *detection*.

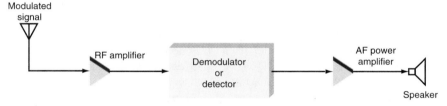

Fig. 1-9 Recovering the intelligence signal at the receiver.

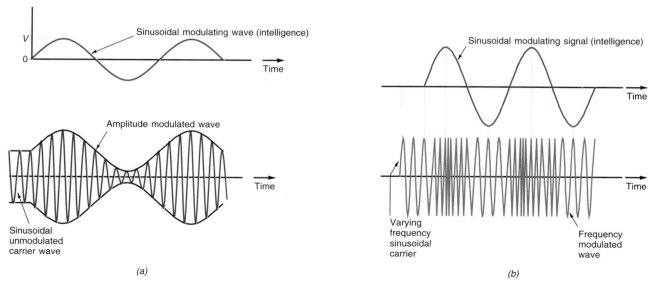

(a)

(b)

FIG. 1-10 Types of modulation. (*a*) Amplitude modulation. (*b*) Frequency modulation.

MULTIPLEXING

The use of modulation also permits another technique, known as multiplexing, to be used. *Multiplexing* is the process of allowing two or more signals to share the same medium or channel; see Fig. 1-11. A multiplexer converts the individual baseband signals into a composite signal that is used to modulate a carrier in the transmitter. At the receiver, the composite signal is recovered at the demodulator, then sent to a demultiplexer where the individual baseband signals are regenerated (see Fig. 1-12).

There are two types of multiplexing: frequency division and time division. In *frequency-division multiplexing,* the intelligence signals modulate subcarriers that are then added together and the composite signal is used to modulate the carrier. In *time-division multiplexing,* the multiple intelligence signals are sequentially sampled and a small piece of each is used to modulate the carrier. If the information signals are sam-

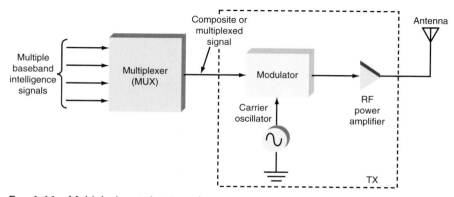

FIG. 1-11 Multiplexing at the transmitter.

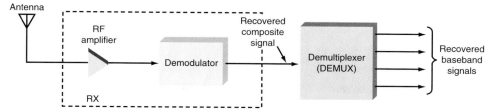

FIG. 1-12   Demultiplexing at the receiver.

pled fast enough, a sufficient amount of detail is transmitted so that at the receiving end the signal can be reconstructed with great accuracy.

1-5 THE ELECTROMAGNETIC SPECTRUM

Electromagnetic waves are signals that oscillate; that is, the amplitudes of the electric and magnetic fields vary at a specific rate. The field intensities fluctuate up and down, and the polarity reverses a given number of times per second. The electromagnetic waves vary sinusoidally. Their frequency is measured in cycles per second (cps) or hertz (Hz). These oscillations may occur at a very low frequency or at an extremely high frequency. The range of electromagnetic signals encompassing all frequencies is referred to as the *electromagnetic spectrum.*

All electric and electronic signals that radiate into free space fall into the electromagnetic spectrum. Not included are signals carried by cables. Signals carried by cable may share the same frequencies of similar signals in the spectrum, but they are not radio signals. Figure 1-13 shows the entire electromagnetic spectrum, giving both frequency and wavelength. In the middle ranges are the most commonly used radio frequencies for two-way communications, television, and other applications. At the upper

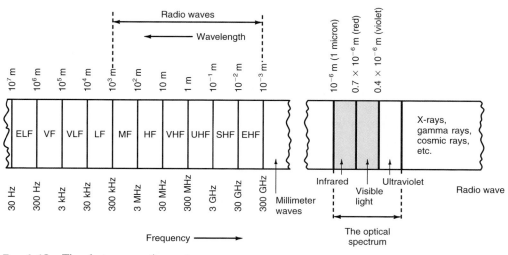

FIG. 1-13   The electromagnetic spectrum.

Name	Frequency	Wavelength
Extremely low frequencies (ELF)	30–300 Hz	10^7–10^6 m
Voice frequencies (VF)	300–3000 Hz	10^6–10^5 m
Very low frequencies (VLF)	3–30 kHz	10^5–10^4 m
Low frequencies (LF)	30–300 kHz	10^4–10^3 m
Medium frequencies (MF)	300 kHz–3 MHz	10^3–10^2 m
High frequencies (HF)	3–30 MHz	10^2–10^1 m
Very high frequencies (VHF)	30–300 MHz	10^1–1 m
Ultra high frequencies (UHF)	300 MHz–3 GHz	1–10^{-1} m
Super high frequencies (SHF)	3–30 GHz	10^{-1}–10^{-2} m
Extremely high frequencies (EHF)	30–300 GHz	10^{-2}–10^{-3} m
Infrared	—	0.7–10 μm
The visible spectrum (light)	—	0.4×10^{-6}–0.8×10^{-6} m

Units of Measure and Abbreviations:
KHz = 1000 Hz
MHz = 1000 kHz = 1×10^6 = 1,000,000 Hz
GHz = 1000 MHz = 1×10^6 = 1,000,000 kHz
    = 1×10^9 = 1,000,000,000 Hz
m = meter
μm = micron = $\dfrac{1}{1,000,000}$ m = 1×10^{-6} m

FIG. 1-14   The electromagnetic spectrum used in electronic communications.

end of the spectrum are infrared and visible light. Figure 1-14 is a listing of the generally recognized segments in the spectrum used for electronic communications.

FREQUENCY AND WAVELENGTH

A given signal is located on the frequency spectrum according to its frequency and wavelength.

FREQUENCY.   Frequency is the number of times a particular phenomenon occurs in a given period of time. In electronics, frequency is the number of cycles of a repetitive wave that occur in a given time period. A cycle consists of two voltage polarity reversals, current reversals, or electromagnetic field oscillations per one second. The cycles repeat, forming a continuous but repetitive wave. Frequency is measured in cycles per second (cps). In electronics, the unit of frequency is the hertz, named for the German physicist Heinrich Hertz, who was a pioneer in the field of electromagnetics. One cycle per second is equal to one hertz, abbreviated (Hz). Therefore, 440 cps = 440 Hz.

Figure 1-15(*a*) shows a sine wave variation of voltage. One positive and one negative alternation form a cycle. If 2500 cycles occur in 1 s, the frequency is 2500 Hz.

Prefixes representing powers of 10 are often used to express frequencies. The most frequently used prefixes are:

$$k = \text{kilo} = 1,000 = 10^3$$
$$M = \text{mega} = 1,000,000 = 10^6$$
$$G = \text{giga} = 1,000,000,000 = 10^9$$
$$T = \text{tera} = 1,000,000,000,000 = 10^{12}$$

Thus, 1000 Hz = 1 kHz (kilohertz). A frequency of 9,000,000 Hz is more commonly expressed as 9 MHz (megahertz). A signal with a frequency of 15,700,000,000 Hz is written as 15.7 GHz (gigahertz).

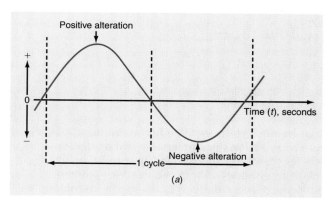
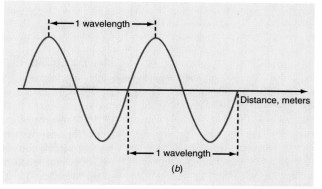

FIG. 1-15 Frequency and wavelength. (*a*) One cycle. (*b*) One wavelength.

Example 1-1

Use the following general rules to convert from one frequency to another.

To Convert	To	Perform This Operation
Hz	kHz	divide by 10^3
Hz	MHz	divide by 10^6
Hz	GHz	divide by 10^9
kHz	Hz	multiply by 10^3
MHz	Hz	multiply by 10^6
GHz	Hz	multiply by 10^9
kHz	MHz	divide by 10^3
kHz	GHz	divide by 10^6
MHz	kHz	multiply by 10^3
MHz	GHz	divide by 10^3
GHz	kHz	multiply by 10^6
GHz	MHz	multiply by 10^3

a. Convert 31.5 kHz to Hz.
$$31.5 \text{ kHz} \times 10^3 = 31,500 \text{ Hz}$$

b. Convert 980 kHz to MHz.
$$\frac{980 \text{ kHz}}{10^3} = 0.980 \text{ MHz}$$

c. Convert 2.45 GHz to MHz.
$$2.45 \times 10^3 = 2450 \text{ MHz}$$

d. Convert 17,030 MHz to GHz.
$$\frac{17,030}{10^3} = 17.03 \text{ GHz}$$

e. Convert 1.9 GHz to Hz.
$$1.9 \times 10^9 = 1,900,000,000 \text{ Hz}$$

In 1887, German physicist Heinrich Hertz was the first to demonstrate the effect of electromagnetic radiation through space. The distance of transmission was only a few feet, but this proved that radio waves could travel from one place to another without the need for any connecting wires. Hertz also proved that radio waves, although invisible, travel at the same velocity as light waves. (Grob, *Basic Electronics,* 8th ed., Glencoe/McGraw-Hill, 1997, p. 2)

WAVELENGTH. Wavelength is the distance occupied by one cycle of a wave, and is usually expressed in meters. One meter (m) is equal to 39.37 in. (just over 3 ft, or 1 yd). Wavelength is measured between identical points on succeeding cycles of a wave, as Fig. 1-15(*b*) shows. If the signal is an electromagnetic wave, one wavelength is the distance that one cycle occupies in free space. It is the distance between adjacent peaks or valleys of the electric and magnetic fields making up the wave.

Wavelength is also the distance traveled by an electromagnetic wave during the time of one cycle. Electromagnetic waves travel at the speed of light, or 299,792,800 m/s. The speed of light and radio waves in a vacuum or in air is usually rounded off to 300,000,000 m/s (3×10^8 m/s), or 186,000 mi/s. The speed of transmission in other media is less.

The wavelength of a signal, which is represented by the Greek letter lambda (λ), is computed by dividing the speed of light by the frequency (f) of the wave in hertz: $\lambda = 300,000,000/f$. For example, the wavelength of a 4,000,000-Hz signal is

$$\lambda = 300,000,000/4,000,000 = 75 \text{ m}$$

If the frequency is expressed in megahertz, the formula can be simplified to $\lambda = 300/f$.

The 4,000,000-Hz signal can be expressed as 4 MHz. Therefore $\lambda = 300/4 = 75$ m.

A wavelength of 0.697 m, as in the second equation in Example 1-2, is what is known as a *very high frequency signal wavelength*. Very high frequency wavelengths are usually expressed in centimeters (cm). One meter equals 100 centimeters, so we would have 0.697 m = 69.7 (about 70) cm.

Brokers and commodity traders using multiple telephone lines transmit worldwide buy-sell orders.

Example 1-2

Find the wavelengths of *(a)* a 150-MHz, *(b)* a 430-MHz, *(c)* an 8-MHz, and *(d)* a 750-kHz signal.

a. $\lambda = \dfrac{300}{150} = 2$ m

b. $\lambda = \dfrac{300}{430} = 0.697$ m

c. $\lambda = \dfrac{300}{8} = 37.5$ m

d. For Hz (750 kHz = 750,000 Hz):

$$\lambda = \frac{300,000,000}{750,000} = 400 \text{ m}$$

For MHz (750 KHz = 0.75 MHz):

$$\lambda = \frac{300}{0.75} = 400 \text{ m}$$

If the wavelength of a signal is known or can be measured, the frequency of the signal can be calculated by rearranging the basic formula $f = 300/\lambda$. Here, f is in megahertz and λ is in meters. As an example, a signal with a wavelength of 14.29 m has a frequency of $f = 300/14.29 = 21$ MHz.

Example 1-3

A signal with a wavelength of 1.5 m has a frequency of

$$f = \frac{300}{1.5} = 200 \text{ MHz}$$

Example 1-4

A signal travels a distance of 75 ft in the time it takes to complete one cycle. What is its frequency?

$$1 \text{ m} = 3.28 \text{ ft}$$

$$\frac{75 \text{ ft}}{3.28} = 22.86 \text{ m}$$

$$f = \frac{300}{22.86} = 13.12 \text{ MHz}$$

Example 1-5

The maximum peaks of an electromagnetic wave are separated by a distance of 8 in. What is the frequency in megahertz? In gigahertz?

$$1 \text{ m} = 39.37 \text{ in.}$$

$$8 \text{ in.} = \frac{8}{39.37} = 0.203 \text{ m}$$

$$f = \frac{300}{0.203} = 1477.8 \text{ MHz}$$

$$\frac{1476.375}{10^3} = 1.4778 \text{ GHz}$$

Frequency Ranges from 30 Hz to 300 GHz

For the purpose of classification, the electromagnetic frequency spectrum is divided into segments as shown in Fig. 1-13. Below are the signal characteristics and applications for each segment.

EXTREMELY LOW FREQUENCIES. Extremely low frequencies (ELFs) are those in the 30- to 300-Hz range. These include ac power-line frequencies (50 and 60 Hz are common), as well as those frequencies in the low end of the human audio range.

VOICE FREQUENCIES. Voice frequencies (VFs) are those in the range of 300 to 3000 Hz. This is the normal range of human speech. Although human hearing extends from approximately 20 to 20,000 Hz, most intelligible sound occurs in the VF range.

VERY LOW FREQUENCIES. Very low frequencies (VLFs) include the higher end of the human hearing range up to about 15 or 20 kHz. Many musical instruments make sounds in this range, as well as in the ELF and VF ranges. The VLF range is also used in some government and military communications. For example, VLF radio transmission is used by the navy to communicate with submarines.

LOW FREQUENCIES. Low frequencies (LFs) are those in the 30- to 300-kHz range. The primary communications services using this range are in aeronautical and marine navigation. Frequencies in this range are also used as *subcarriers,* signals which are modulated by the baseband information. Usually, two or more subcarriers are added together and the combination is used to modulate the final high-frequency carrier.

MEDIUM FREQUENCIES. Medium frequencies (MFs) are in the 300- to 3000-kHz (0.3- to 3.0-MHz) range. The major application of frequencies in this range is AM radio broadcasting (535 to 1605 kHz). Other applications in this range include various marine and aeronautical communications.

HIGH FREQUENCIES. High frequencies (HFs) are those in the 3- to 30-MHz range. These are the frequencies generally known as short waves. All kinds of simplex broadcasting and half-duplex two-way radio communications take place in this range. Broadcasts from Voice of America and the British Broadcasting Company occur in this range. Government and military services use these frequencies for two-way communications. An example is diplomatic communications between embassies. Amateur radio and CB communications also occur in this part of the spectrum.

VERY HIGH FREQUENCIES. Very high frequencies (VHFs) encompass the 30- to 300-MHz range. This popular frequency range is used by many services, including mobile radio, marine and aeronautical communications, FM radio broadcasting (88 to 108 MHz), and television channels 2 through 13. Radio amateurs also have numerous bands in this frequency range.

ULTRAHIGH FREQUENCIES. Ultrahigh frequencies (UHFs) encompass the 300- to 3000-MHz range. This, too, is a widely used portion of the frequency spectrum. It includes the UHF TV channels 14 through 67, and is used for land mobile communications and services such as cellular telephones, as well as for military communications. Some radar and navigation services occupy this portion of the frequency spectrum, and radio amateurs also have bands in this range.

MICROWAVES AND SHFs. Frequencies between the 1000-MHz (1-GHz) and 30-GHz range are called *microwaves*. Microwave ovens usually operate at 2.45 GHz. Superhigh frequencies (SHFs) are those in the 3- to 30-GHz range. These microwave frequencies are widely used for satellite communications and radar. Some specialized forms of two-way radio communications, e.g., wireless networks, also occupy this region.

EXTREMELY HIGH FREQUENCIES. Extremely high frequencies (EHFs) extend from 30 to 300 GHz. Equipment used to generate and receive signals in this range is extremely complex and expensive. Presently there is only a limited amount of activity in this range, but it does include satellite communications and some specialized radar. As technological developments permit equipment advances, this frequency range will be more widely used.

FREQUENCIES BETWEEN 300 GHz AND THE OPTICAL SPECTRUM. Those electromagnetic signals whose frequencies are higher than 300 GHz are referred to as millimeter waves. This portion of the spectrum is just now being developed. As hardware techniques advance, use of millimeter wave frequencies will no doubt increase, primarily for military, radar, and other specialized applications.

With three sons serving abroad during World War II, the Rubis family of Muse, Pennsylvania, listened to President Roosevelt's Sunday radio address in 1943.

THE OPTICAL SPECTRUM

Right above the millimeter wave region is what is called the *optical spectrum,* the region occupied by light waves. There are three different types of light waves, infrared, visible, and ultraviolet.

INFRARED. The infrared region is sandwiched between the highest radio frequencies (i.e., millimeter waves) and the visible portion of the electromagnetic spectrum. Infrared occupies the range between approximately 0.1 millimeter (mm) and 700 nanometers (nm), or 0.7 to 100 microns, where 1 micron is one millionth of a meter, called a micrometer (μm). Infrared wavelengths are often given in microns.

Infrared radiation is generally associated with heat. Infrared is produced by light bulbs, our bodies, and any physical equipment that generates heat. Infrared signals can also be generated by special types of light-emitting diodes and lasers.

Infrared signals are used for various special kinds of communications. For example, infrared is used in astronomy to detect stars and other physical bodies in the universe, and for guidance in weapons systems, where the heat radiated from airplanes or missiles can be picked up by infrared detectors and used to guide missiles to targets. Infrared is also used in most new TV remote-control units where special coded signals are transmitted by an infrared LED to the TV receiver for the purpose of changing channels, setting the volume, and other functions. Infrared is the basis for some of the newer wireless LANs.

Infrared signals have many of the same properties as signals in the visible spectrum. Optical devices such as lenses and mirrors are often used to process and manipulate infrared signals, and infrared light is the signal usually propagated over fiber-optical cables.

THE VISIBLE SPECTRUM. Just above the infrared region is the visible spectrum we ordinarily refer to as light. Light is a special type of electromagnetic radiation that has a wavelength in the 0.4- to 0.8-μm range. Light wavelengths are usually expressed in terms of angstroms (Å). An angstrom is one ten-thousandth of a μm; for example, $1 \text{ Å} = 10^{-10}$ m. The visible range is approximately 8000 Å (red) to 4000 Å (violet). Red is low-frequency or long-wavelength light, while violet is high-frequency or short-wavelength light.

Light is used for various kinds of communications. Light waves can be modulated and transmitted through glass fibers, just as electric signals can be transmitted over wires. Fiber optics is one of the fastest growing specialties of communications electronics. The great advantage of light wave signals is that their very high frequency gives them the ability to handle a tremendous amount of information. That is, the bandwidth of the baseband signals can be very wide.

Light signals can also be transmitted through free space. Various types of communications systems have been created using a laser that generates a light beam at a specific visible frequency. Lasers generate an extremely narrow beam of light which is easily modulated with voice, video, and data information.

ULTRAVIOLET. Ultraviolet light (UV) covers the range from about 4 to 400 nm. UV generated by the sun is what causes sunburn; repeated exposure to UV can result in skin cancer. UV is also generated by mercury vapor lights and some other types of lights such as fluorescent lamps and sun lamps. Ultraviolet is not used for communications; its primary use is medical.

Beyond the visible region are the x-rays, gamma rays, and cosmic rays. These are all forms of electromagnetic radiation, but they do not figure into communications systems and are not covered here.

1-6 BANDWIDTH

Bandwidth is that portion of the electromagnetic spectrum occupied by a signal. It is also the frequency range over which a receiver or other electronic circuit operates. More specifically, bandwidth (BW) is the difference between the upper and lower frequency limits of the signal or the equipment operation range. Figure 1-16 shows the bandwidth of the voice frequency range from 300 to 3000 Hz. The upper frequency is f_2 and the lower frequency is f_1. The bandwidth, then is

$$BW = f_2 - f_1$$
$$= 3000 - 300$$
$$= 2700 \text{ Hz}$$

Example 1-6

A commonly used frequency range is 902 to 928 MHz. What is the width of this band?

$$f_1 = 902 \text{ MHz} \qquad f_2 = 928 \text{ MHz}$$
$$BW = f_2 - f_1 = 928 - 902 = 26 \text{ MHz}$$

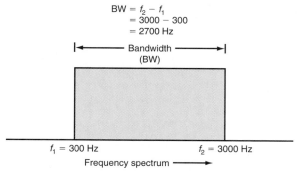

FIG. 1-16 Bandwidth is the frequency range over which equipment operates or that portion of the spectrum occupied by the signal. This is the voice frequency bandwidth.

Example 1-7

A television signal occupies a 6-MHz bandwidth. If the low frequency limit of channel 2 is 54 MHz, what is the upper frequency limit?

$$BW = 54 \text{ MHz} \qquad f_1 = 6 \text{ MHz}$$
$$BW = f_1 - f_2$$
$$f_2 = BW + f_1 = 6 + 54 = 60 \text{ MHz}$$

CHANNEL BANDWIDTH

When information is modulated onto a carrier somewhere in the electromagnetic spectrum, the resulting signal occupies a small portion of the spectrum surrounding the carrier frequency. The modulation process causes other signals, called *sidebands,* to be generated at frequencies above and below the carrier frequency by an amount equal to the modulating frequency. For example, in AM broadcasting, audio signals up to 5 kHz can be transmitted. If the carrier frequency is 1000 kHz, or 1 MHz, and the modulating frequency is 5 kHz, sidebands will be produced at $1000 - 5 = 995$ kHz and at $1000 + 5 = 1005$ kHz. In other words, the modulation process generates other signals which take up spectrum space. It is not just the carrier at 1000 kHz that is transmitted. Thus the term *bandwidth* refers to the range of frequencies that contain the information. The term *channel bandwidth* refers to the range of frequencies required to transmit the desired information.

The bandwidth of the AM signal described above is the difference between the highest and lowest transmitting frequencies: $BW = 1005 \text{ kHz} - 995 \text{ kHz} = 10 \text{ kHz}$. In this case, the channel bandwidth is 10 kHz. An AM broadcast signal, therefore, takes up a 10-kHz piece of the spectrum.

Signals transmitting on the same frequency or on overlapping frequencies do, of course, interfere with one another. Thus a limited number of signals can be transmitted in the frequency spectrum. As communications activities have grown over the years, there has been a continuous demand for more frequency channels over which communications can be transmitted. This has caused a push for the development of equipment that operates at the higher frequencies. Prior to World War II, frequencies above 1 GHz were virtually unused, since there were no electronic components suitable for generating signals at those frequencies. But technological developments over the years have given us many microwave components such as klystrons, magnetrons, and traveling-wave tubes, and today transistors and other semiconductor devices that work in the microwave range.

MORE ROOM AT THE TOP

The benefit of using the higher frequencies for communications carriers is that a signal of a given bandwidth represents a smaller percentage of the spectrum at the higher frequencies than at the lower frequencies. For example, at 1000 kHz, the 10-kHz-wide AM signal discussed earlier represents 1 percent of the spectrum:

$$\% \text{ of spectrum} = \frac{10 \text{ kHz}}{1000 \text{ kHz}} \times 100$$
$$= 1\%$$

But at 1 GHz, or 1,000,000 kHz, it only represents a thousandth of 1 percent:

$$\% \text{ of spectrum} = \frac{10 \text{ kHz}}{1,000,000 \text{ kHz}} \times 100$$
$$= 0.001\%$$

In practice, this means that there are many more 10-kHz channels at the higher frequencies than at the lower frequencies. In other words, there is more spectrum space for information signals at the higher frequencies.

The higher frequencies also permit wider-bandwidth signals to be used. A TV signal, for example, occupies a bandwidth of 6 MHz. Such a signal cannot be used to modulate a carrier in the MF or HF ranges because it would use up all the available spectrum space. Television signals are transmitted in the VHF and UHF portions of the spectrum, where sufficient space is available.

Today, virtually the entire frequency spectrum between approximately 30 kHz and 300 MHz has been spoken for. Some open areas and portions of the spectrum are not heavily used, but for the most part, the spectrum is filled with communications activities of all kinds generated from all over the world. There is tremendous competition for these frequencies, not only between companies, individuals, and government services in individual carriers but also between the different nations of the world. The electromagnetic spectrum is one of our most precious natural resources. Because of this, communications engineering is devoted to making the best use of that finite spectrum. A considerable amount of effort goes into developing communications techniques that will minimize the bandwidth required to transmit given information and thus conserve spectrum space. This provides more room for additional communications channels and gives other services or users an opportunity to take advantage of it. Many of the techniques discussed later in this book evolved in an effort to minimize transmission bandwidth.

SPECTRUM MANAGEMENT

Governments of the United States and other countries recognized early on that the frequency spectrum was a valuable and finite natural resource and set up agencies to control spectrum use. In the United States, Congress passed the Communications Act of 1934. This act and its various amendments established regulations for the use of spectrum space. It also established the Federal Communications Commission (FCC), a regulatory body whose function is to allocate spectrum space, issue licenses, set standards, and police the airwaves. The FCC controls all telephone and radio communications in this country and, in general, regulates all electromagnetic emissions. The National Telecommunications and Information Administration (NTIA) performs a similar function for government and military services. Other countries have similar organizations.

The International Telecommunications Union (ITU), an agency of the United Nations that is headquartered in Geneva, Switzerland, has 154 member countries that meet at regular intervals to promote cooperation and negotiate national interests. Typical of these meetings are the World Administrative Radio Conferences, held approximately every two years. Various committees of the ITU set standards for various areas within the communications field. The two most important committees are the Radio Consultative Committee (CCIR) and the International Telegraph and Telephone Consultative Committee (CCITT). The CCITT has been renamed ITU-T (where the T suffix stands for *telecommunications*). The ITU brings together the various countries to discuss how the frequency spectrum is to be divided up and shared. Because many of the signals generated in the spectrum do not carry for long distances, countries can use these frequencies simultaneously without interference. On the other hand, some ranges of the frequency spectrum can, literally, carry signals around the world. High-frequency shortwave signals are an example. As a result, countries must negotiate with one another to coordinate usage of various portions of the high-frequency spectrum.

1-7 A Survey of Communications Applications

The applications of electronic techniques to communications are so common and pervasive that you are already familiar with most of them. You use the telephone, listen to the radio, and watch television. You may also use or be aware of other forms of electronic communications such as cellular telephones, ham radios, CB radios, pagers, electronic mail, and remote-control garage door openers. There are many more. Figure 1-17 lists all the various major applications of electronic communications. The list is divided into simplex and duplex categories. Read through the list to refresh your memory or possibly to discover some new applications you may not know about.

1-8 Jobs and Careers in the Communications Industry

The electronics industry is roughly divided into four major specializations. The largest in terms of people employed and the dollar value of equipment purchased is the computer field, closely followed by the communications field. The industrial control and instrumentation fields are considerably smaller. There are hundreds of thousands of employees in the communications field, and billions of dollars of equipment is purchased each year. The growth rate varies from year to year depending upon the economy, technological developments, and other factors. But, like most areas in electronics, the communications field has grown steadily over the years, creating a relatively constant opportunity for employment. If your interests lie in communications, you will be glad to know that there are many opportunities for long-term jobs and careers. The next section of the chapter outlines the types of jobs available and the major kinds of employers.

SIMPLEX (ONE WAY)

1. AM and FM radio broadcasting. Commercial stations broadcast music, news, weather reports, as well as other programs for entertainment and information.

2. TV broadcasting. Commercial stations broadcast a wide variety of entertainment, informational, and educational programs.

3. Cable television. The distribution of movies, sports events, and other programs by local cable companies to subscribers by coax cable. Cable stations originate some programming, but they primarily distribute "packaged" programming received by satellite from services such as HBO and CNN.

4. Facsimile. The transmission of printed visual material over the telephone lines. A facsimile, or fax, machine scans a photo or other document and converts it into electronic signals that are sent over the telephone system for reproduction in original printed form by another fax machine at the receiving end.

5. Wireless remote control. A mechanism that controls missiles, satellites, robots, and other vehicles or remote plants or stations. A garage door opener is a special form of remote control which uses a tiny battery-operated radio transmitter in the car to operate a receiver in the garage that activates a motor to open or close the door. The remote control on your TV set uses digital data modulated on an infrared light beam to control volume, channel and other functions.

6. Paging services. A radio system for paging individuals, usually in connection with their work. Persons carry tiny battery-powered receivers or "beepers" that can pick up signals from a local paging station that receives telephone requests to locate and page individuals when they are needed.

7. Navigation and direction-finding services. Transmission by special stations of signals that can be picked up by receivers with highly directional antennas for the purpose of identifying exact location (latitude and longitude) or determining direction and/or distance from a station. Such systems employ both land-based and satellite stations. The services are used primarily by boats and ships or airplanes, although systems for cars and trucks are being developed.

8. Telemetry. The transmission of measurements over a long distance. Telemetry sys-

(continues on next page)

FIG. 1-17 Applications of electronic communications.

tems use sensors to determine physical conditions (temperature, pressure, flow rate, voltages, frequency, etc.) at a remote location. The sensors modulate a carrier signal that is sent by wire or radio to a remote receiver which stores and/or displays the data for analysis. Telemetry systems permit remote equipment and systems to be monitored for the purpose of determining their status and condition. Examples are satellites, rockets, pipelines, plants, and factories.

9. *Radio astronomy.* Radio signals, including infrared, are emitted by virtually all heavenly bodies such as stars and planets. By using large directional antennas and sensitive high-gain receivers, these signals may be picked up and used to plot star locations and study the universe. Radio astronomy is an alternative and supplement to traditional optical astronomy.

10. *Surveillance.* Surveillance means discreet monitoring or "spying." Electronic techniques are widely used by police forces, governments, the military, business and industry, and others to gather information for the purpose of gaining some competitive advantage. Techniques include phone taps, tiny wireless "bugs," clandestine listening stations, as well as reconnaissance airplanes and satellites.

11. *Teletext and Viewdata.* The transmission of text and graphical data to be displayed on a TV receiver with an appropriate adapter. Teletext data is transmitted by telephone lines, whereas Viewdata or Videotext is transmitted during the vertical picture blanking intervals on standard TV broadcast signals.

12. *Music services.* The transmission of continuous background music for doctors' offices, stores, elevators, and so on by local FM radio stations on special high-frequency subcarriers that cannot be picked up by conventional FM receivers.

DUPLEX (TWO WAY)

13. *Telephones.* One-on-one verbal communications over the vast worldwide telephone networks employing wire, radio relay stations, and satellites. Cordless telephones provide short-distance convenience communication without restrictive wires. Cellular radio systems provide telephone service in vehicles and for portable use. Digital data may also be transmitted over the telephone system.

14. *Two-way radio.* Commercial, industrial, and governmental communications between vehicles or between vehicles and a base station. Examples: police, fire, taxi,

FIG. 1-17 *(continued)*

forestry service, trucking companies, etc. Other forms of two-way radios are used in aircraft, marine, military, and space applications. Government applications are broad and diverse and include embassy communications, Treasury, Secret Service, and CIA.

15. *Radar.* A special form of communications that makes use of reflected microwave signals for the purpose of detecting ships, planes, and missiles and for determining their range, direction, and speed. Most radar is used in military applications, but civilian aircraft and marine services also use it. Police use radar in speed detection and enforcement.

16. *Sonar.* Underwater communications in which audible baseband signals use water as the transmission medium. Submarines and ships use sonar to detect the presence of enemy submarines. Passive sonar uses audio receivers to pick up water, prop, and other noises. Active sonar is like an underwater radar where reflections from a transmitted ultrasonic pulse are used to determine direction, range, and speed of an underwater target.

17. *Amateur ratio.* A hobby for individuals interested in radio communications. Individuals may become licensed

"hams" to build and operate two-way radio equipment for personal communication with other hams.

18. *Citizen's radio.* Citizen's band (CB) radio is a special service that any individual may use for personal communication with others. Most CB radios are used in trucks and cars for exchanging information about traffic conditions, speed traps, and for emergencies.

19. *Data communications.* The transmission of binary data between computers. Computers frequently use the telephone system as the medium. Devices, called modems, make the computers and telephone networks compatible. Data communications also take place over twisted pair and coax cables, fiber-optic cables as well as terrestrial microwave relay links and satellites. Data communications techniques make possible on-line services and information sharing via the Internet.

20. *Local area networks (LANs).* Wired (or wireless) interconnections of personal computers (PCs) or PCs and mini or mainframe computers within an office or building for the purpose of sharing mass storage, peripherals, and data.

Fɪɢ. 1-17 (*continued*)

The two major types of technical positions available in the communications field are engineer and technician. A third type, technologist, may perform either engineer's or technician's work or some combination.

ENGINEERS. *Engineers* design communications equipment and systems. They have bachelor's (B.S.E.E.), masters (M.S.E.E.), or doctoral (Ph.D.) degrees in electrical engineering, giving them a strong science and mathematics background combined with specialized education in electronic circuits and equipment. Engineers work from specifications and create new equipment or systems which are then manufactured.

Some engineers specialize in design; others work in manufacturing, testing, quality control, and management, among other areas. Engineers may also serve as field service personnel, installing and maintaining complex equipment and systems. If your interest lies in the design of communications equipment, then an engineering position may be for you.

Although a degree in electrical engineering is generally the minimum entrance requirement for engineers' jobs in most organizations, people with other educational backgrounds (e.g., physics and math) do become engineers. Technicians who obtain sufficient additional education and appropriate experience may go on to become engineers.

Modern electronics stores carry a great variety of stereo equipment, television sets, video cameras, cellular telephones, and other communications equipment. Many electronic technicians seek career technical sales opportunities.

TECHNICIANS. *Technicians* have some kind of postsecondary education in electronics, from a vocational or technical school, a community college, or a technical institute. Many technicians are educated in military training programs. Most technicians have an average of two years of formal post high school education and an associate degree. Common degrees are associate in arts (A.A.), associate in science (A.S.) or associate of science in engineering technology or electronic engineering (A.S.E.T. or A.S.E.E.), and associate in applied science (A.A.S.). The A.A.S. degrees tend to cover more occupational and job-related subjects; the A.A. and A.S. degrees are more general, and are designed to provide a foundation for transfer to a bachelor's degree program. Technicians with an associate degree from a community college can usually transfer to a bachelor of technology program (see below) and complete the bachelor's degree in another two years. However, associate degree holders are usually not able to transfer to an engineering degree program, but must literally start over if the engineering career path is chosen.

Technicians are most often employed in service jobs. The work typically involves equipment installation, troubleshooting and repair, maintenance and adjustment, or operation. Technicians in such positions are sometimes called field service technicians, field service engineers, or customer representatives.

Technicians can also be involved in engineering. Engineers may use one or more technicians to assist in the design of equipment. They build and troubleshoot prototypes and in many cases actually participate in equipment design. A great deal of the work involves testing and measuring. In this capacity, the technician is known as an engineering technician, lab technician, engineering assistant, or associate engineer.

Technicians are also employed in manufacturing. They may be involved in the actual building and assembling of the equipment, but more typically are concerned with final testing and measurement of the finished products. Other positions involve quality control or repair of defective units.

TECHNOLOGISTS. A less well known technical position is that of technologist. A *technologist* has a bachelor's degree in electronics technology from a technical college or university. Some typical degree titles are bachelor of technology (B.T.), bachelor of engineering technology (B.E.T.), and bachelor of science in engineering technology (B.S.E.T.).

Bachelor of technology programs are generally extensions of two-year associate degree programs. In the two additional years required for a bachelor of technology degree, the student takes more complex electronics courses along with additional science, math, and humanities courses. The main difference between the B.T. graduate and the engineering graduate is that the technologist usually gets less math and science, and the electronics courses are more practical and hands-on than engineering courses. Whereas few engineering schools have labs associated with their courses, all technology courses have strong lab segments. Holders of B.T. degrees can generally design electronic equipment and systems but do not typically have the depth of background in analytical mathematics or science that is required for complex design jobs. However, technologists are generally employed as engineers. While many do design work, others are employed in engineering positions in manufacturing and field service rather than design.

OTHER POSITIONS. There are many jobs in the communications industry other than those of engineer or technician. For example, there are many outstanding jobs in technical sales. Selling complex electronic communications equipment often requires a strong technical education and background. The work may involve determining customer needs and related equipment specifications, writing technical proposals, making sales presentations to customers, and attending shows and exhibits where equipment is sold. The pay potential in sales is generally much higher than in the engineering or service positions.

Another position is that of technical writer. Technical writers generate the technical documentation for communications equipment and systems, producing installation and service manuals, maintenance procedures, and customer operations manuals. This important task requires considerable depth of education and experience.

Finally, there is the position of trainer. Engineers and technicians are often used to train other engineers and technicians or customers. With the high degree of complexity that exists in communications equipment today, there is a major need for training. Many individuals find the education and training positions to be very desirable and satisfying. The work typically involves developing curriculum and programs, generating the necessary training manuals and presentation materials, and conducting classroom training sessions in-house or at a customer site.

MAJOR EMPLOYERS

The overall structure of the communications electronics industry is shown in Fig. 1-18. The four major segments of the industry are manufacturers, resellers, service organizations, and end users.

MANUFACTURERS. It all begins, of course, with customer needs. Manufacturers translate customer needs into products, purchasing components and materials from other electronics companies to use in creating the products. Engineers design the products and manufacturing produces them. There are jobs for engineers, technicians, salespeople, field service personnel, technical writers, and trainers.

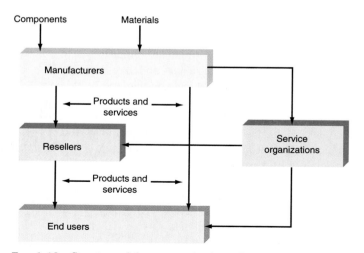

FIG. 1-18 Structure of the communications electronics industry.

RESELLERS. Manufacturers who do not sell products directly to end users sell the products to reselling organizations, which, in turn, sell them to the end user. For example, a manufacturer of marine communications equipment may not sell directly to the boat owner, but instead to a regional distributor or marine electronics store or shop. This shop not only sells the equipment but also takes care of installation, service, and repairs. A cellular telephone or fax machine manufacturer also typically sells to a distributor or dealer who takes care of sales and service. Most of the jobs available in the reselling segment of the industry are in sales, service, and training.

SERVICE ORGANIZATIONS. These companies usually perform some kind of service, such as repair, installation, or maintenance. One example is an avionics company that does installation or service work on electronic equipment for private planes. Another is a systems integrator, a company that designs and assembles a piece of communications equipment or more often an entire system using the products of other companies. Systems integrators put together systems to meet special needs and customize existing systems for particular jobs.

END USERS. The end user is the ultimate customer—and a major employer. Today, almost every person and organization is an end user of communications equipment. The major categories of end users in the communications field are

> Telephone companies
> Radio users—mobile, marine, aircraft, etc.
> Radio and TV broadcast stations and cable TV companies
> Business and industry users of satellites, networks, etc.
> Transportation companies (airlines, shipping, railroads)
> Government and military
> Personal and hobby

There are an enormous number of communications jobs with end users. Most of them are of the service type: installation, repair, maintenance, and operation of equipment.

Most communications technicians perform installation, maintenance, and trouble-shooting on equipment like this.

SUMMARY

All electronic communications devices consist of three basic components: a transmitter, a communications channel (medium), and a receiver. Messages are converted into electrical signals and sent over electrical or fiber-optic cable or free space to a receiver. Attenuation (weakening) and noise can interfere with transmission.

Electronic communications are classified as (1) one-way (simplex) or two-way (full duplex or half duplex) transmissions and (2) analog or digital signals. Analog signals are smoothly varying, continuous signals. Digital signals are discrete, two-state (on/off) codes. Electronic signals are often changed from analog to digital and vice versa. Before transmission, electronic signals are known as baseband signals.

Amplitude and frequency modulation make an information signal compatible with the channel over which it is to be sent, modifying the carrier wave by changing its amplitude, frequency, or phase angle and sending it to an antenna for transmission, a process known as broadband communications. Frequency-division and time-division multiplexing allow more than one signal at a time to be transmitted over the same medium.

All electronic signals that radiate into space are part of the electromagnetic spectrum; their location on the spectrum is determined by frequency. Most information signals to be transmitted occur at lower frequencies and modulate a carrier wave of a higher frequency.

How much information a given signal can carry depends in part on its bandwidth. Available space for transmitting signals is limited, and signals transmitting on the same frequency or on overlapping frequencies interfere with one another. Research efforts are being devoted to developing use of higher-frequency signals and minimizing the bandwidth required.

There are many job opportunities in the field of electronic communications. The four major specialties are computers, communications, industrial control, and instrumentation.

KEY TERMS

Analog signals
Attenuation
Bandwidth
Baseband transmission
Broadband transmission
Carrier
Channel bandwidth
Digital signals
Duplex communication
Electromagnetic spectrum
Electronic communication systems
Engineers
Extremely high frequency (EHF)
Extremely low frequency (ELF)

Frequency
High frequency (HF)
Infrared region
International Telecommunications Union (ITU)
International Telegraph and Telephone Consultative Committee (CCITT)
Low frequency (LF)
Medium frequency (MF)
Modem
Modulation
Multiplexing
National Telecommunications and Information Administration (NTIA)
Noise

Optical spectrum
Radio
Radio Consultative Committee (CCIR)
Receiver
Simplex communication
Super high frequency (SHF)
Technicians
Technologists
Transceivers
Transmitters
Ultrahigh frequency (UHF)
Very high frequency (VHF)
Very low frequency (VLF)
Visible spectrum
Voice frequency
Wavelength

REVIEW

QUESTIONS

1. In what century did electronic communications begin?
2. Name the four main elements of a communications system and draw a diagram that shows their relationship.
3. List five types of media used for communications and state which three are the most commonly used.
4. Name the device used to convert an information signal into a signal compatible with the medium over which it is being transmitted.
5. What piece of equipment acquires a signal from a communications medium and recovers the original information signal?
6. What is a transceiver?
7. What are two ways in which a communications medium can affect a signal?
8. What is another name for *communications medium?*
9. What is the name given to undesirable interference that is added to a signal being transmitted?
10. Name three common sources of interference.
11. What is the name given to the original information or intelligence signals that are transmitted directly via a communications medium?
12. Name the two forms in which intelligence signals can exist.
13. What is the name given to one-way communications? Give three examples.
14. What is the name given to simultaneous two-way communications? Give three examples.
15. What is the term used to describe two-way communications in which each party takes turns transmitting? Give three examples.
16. What type of electronic signals are continuously varying voice and video signals?
17. What are on/off intelligence signals called?
18. How are voice and video signals transmitted digitally?
19. What terms are often used to refer to original voice, video, or data signals?
20. What technique must sometimes be used to make an information signal compatible with the medium over which it is being transmitted?
21. What is the process of recovering an original signal called?
22. What is a broadband signal?
23. Name the process used to transmit two or more baseband signals simultaneously over a common medium.
24. Name the technique used to extract multiple intelligence signals that have been transmitted simultaneously over a single communications channel.
25. What is the name given to signals that travel through free space for long distances?
26. What does a radio wave consist of?
27. Calculate the wavelength of signals with frequencies of 1.5kHz, 18 MHz, and 22 GHz.
28. Why aren't audio signals transmitted directly by electromagnetic waves?
29. What is the human hearing frequency range?
30. What is the approximate frequency range of the human voice?
31. Do radio transmissions occur in the VLF and LF ranges?
32. What is the frequency range of AM radio broadcast stations?

R
E
V
I
E
W

33. What is the name given to radio signals in the HF range?
34. In what segment of the spectrum do television channels 2 to 13, and FM broadcasting, appear?
35. List five major uses of the UHF band.
36. What are frequencies above 1 GHz called?
37. What are the frequencies just above the EHF range called?
38. What is a micron and what is it used to measure?
39. Name the three segments of the optical frequency spectrum.
40. What is a common source of infrared signals?
41. What is the approximate spectrum range of infrared signals?
42. Define the term angstrom and explain how it used.
43. What is the wavelength range of visible light?
44. Light signals use which two channels or media for electronic communications?
45. Name two methods of transmitting visual data over a telephone network.
46. What is the name that is given to the signaling of individuals at remote locations by radio?
47. What term is used to describe the process of making measurements at a distance?
48. List four ways radio is used in the telephone system.
49. What principle is used in radar?
50. What is underwater radar called? Give two examples.
51. What is the name of a popular radio communications hobby?
52. What device enables computers to exchange digital data over the telephone network?
53. What do you call the systems of interconnections of PCs and other computers in offices or buildings?
54. What is a generic synonym for radio?
55. Name the three main types of technical positions available in the communications field.
56. What is the main job of an engineer?
57. What is the primary degree for an engineer?
58. What is the primary degree for a technician?
59. Name a type of technical degree in engineering other than engineer or technician.
60. Can the holder of an associate of technology degree transfer the credits to an engineering degree program?
61. What types of work does a technician ordinarily do?
62. List three other types of jobs in the field of electronic communications that do not involve engineering or technician's work.
63. What are the four main segments of the communications industry? Explain briefly what the function of each is.

PROBLEMS

1. Calculate the frequency of signals with wavelengths of 40 m, 5 m, and 8 cm. ◆
2. In what frequency range does the common AC power-line frequency fall?
3. Convert the following frequencies to different units as indicated: 2.2 GHz to MHz, 8333 kHz to MHz, 27875 kHz to MHz, 17,500 MHz to GHz. ◆
4. What is the primary use of the SHF and EHF ranges?

Answers to selected problems are given at the end of the book. The ◆ symbol appearing at the end of a problem indicates that the answer is provided.

36 ◆ *Chapter 1* INTRODUCTION TO ELECTRONIC COMMUNICATIONS

CRITICAL THINKING

1. Name three ways that a higher-frequency signal called the carrier can be varied to transmit the intelligence.
2. Name two common household remote-control units and state the type of media and frequency ranges used for each.
3. How is radio astronomy used to locate and map stars and other heavenly bodies?
4. In what segment of the communications field are you interested in working and why?
5. Assume that all of the electromagnetic spectrum from ELF through microwaves were fully occupied. Explain some ways that communications capability could be added.
6. What is the speed of light in feet per microsecond? In inches per nanosecond?
7. Make a general statement comparing the speed of light with the speed of sound. Give an example of how the principles mentioned might be demonstrated.
8. List five real-life communications applications not specifically mentioned in this chapter.
9. "Invent" five new communications, wired or wireless, that you think would be practical.
10. Assume that you have a wireless application you would like to design, build, and sell as a commercial product. You have selected a target frequency in the UHF range. How would you decide what frequency to use and how would you get permission to use it?
11. Make an exhaustive list of all of the electronic communications products that you own, have access to at home or in the office, and/or use on a regular basis.
12. You have probably seen or heard of a simple communications system made of two paper cups and a long piece of string. How could such a simple system work?

THE FUNDAMENTALS OF ELECTRONICS: A REVIEW

Objectives

After completing this chapter, you will be able to:

- *Calculate* voltage, current, gain, and attenuation in decibels and *apply* these formulas in applications involving cascaded circuits.

- *Explain* the relationship between Q, resonant frequency, and bandwidth.

- *Describe* the basic configuration of the different types of filters that are used in communications networks and *compare and contrast* active filters with passive filters.

- *Explain* how using switched capacitor filters enhances selectivity.

- *Describe* how transformers are used in electrical isolation, voltage step up and step down, impedance transformation, and phase inversion.

- *Calculate* bandwidth using Fourier analysis.

To understand communications electronics as presented in this book, you need a knowledge of certain basic principles of electronics, including the fundamentals of alternating and direct current, AC and DC circuits, semiconductor operation and characteristics, and basic electronic circuit operation (amplifiers, oscillators, power supplies, and digital logic circuits). Some of the basics are particularly critical to understanding the chapters to follow. These include the expression of gain and loss in decibels; *LC* tuned circuits, resonance and filters, and the Fourier theory. The purpose of this chapter is to review briefly all of these subjects. If you have had the material before, it will simply serve as a review and reference. If, because of your own schedule or the school's curriculum, you have not previously covered this material, use this chapter to learn the necessary information before continuing.

2-1 GAIN, ATTENUATION, AND DECIBELS

Most electronic circuits in communications are used to process signals, that is, to manipulate signals to produce a desired result. All signal-processing circuits involve either gain or attenuation.

GAIN

Gain means amplification. If a signal is applied to a circuit such as the amplifier shown in Fig. 2-1 and the output of the circuit has a greater amplitude than the input signal, the circuit has gain. Gain is simply the ratio of the output to the input. For input (V_{in}) and output (V_{out}) voltages, voltage gain A_V is expressed as follows:

$$A_V = \frac{\text{output}}{\text{input}} = \frac{V_{out}}{V_{in}}$$

The number obtained by dividing the output by the input shows how much larger the output is than the input. For example, if the input is 150 μV and the output is 75 mV, the gain is $A_V = (75 \times 10^{-3})/(150 \times 10^{-6}) = 500$.

The formula can be rearranged to obtain the input or the output given the other two variables: $V_{out} = V_{in} \times A_V$ and $V_{in} = V_{out}/A_V$.

If the output is 0.6 V and the gain is 240, the input is $V_{in} = 0.6/240 = 2.5 \times 10^{-3} = 2.5$ mV.

Example 2-1

What is the voltage gain of an amplifier that produces an output of 750 mV for a 30-μV input?

$$A_V = \frac{V_{out}}{V_{in}} = \frac{750 \times 10^{-3}}{30 \times 10^{-6}} = 25000$$

V gain

$10 \log \frac{V_0}{V_I}$

P gain

$20 \log \frac{V_6}{V_I}$

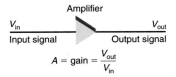

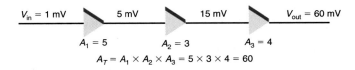

Fig. 2-1   An amplifier has gain.

Fig. 2-2   Total gain of cascaded circuits is the product of individual stage gains.

Since most amplifiers are also power amplifiers, the same procedure can be used to calculate power gain A_P:

$$A_P = P_{out}/P_{in},$$

where P_{in} is is the power input and P_{out} is the power output.

Example 2-2

The power output of an amplifier is 6 watts (W). The power gain is 80. What is the input power?

$$A_P = \frac{P_{out}}{P_{in}} \quad \text{therefore} \quad P_{in} = \frac{P_{out}}{A_P}$$

$$P_{in} = \frac{6}{80} = 0.075 \text{ W} = 75 \text{ mW}$$

When two or more stages of amplification or other forms of signal processing are cascaded, the overall gain of the combination is the product of the individual circuit gains. Figure 2-2 shows three amplifiers connected one after the other so that the output of one is the input to the next. The voltage gains of the individual circuits are as marked. To find the total gain of this circuit, simply multiply the individual circuit gains: $A_T = A_1 \times A_2 \times A_3 = 5 \times 3 \times 4 = 60$.

If an input signal of 1 mV is applied to the first amplifier, the output of the third amplifier will be 60 mV. The outputs of the individual amplifiers depend upon their individual gains. The output voltage from each amplifier is shown in Fig. 2-2.

Example 2-3

Three cascaded amplifiers have power gains of 5, 2, and 17. The input power is 40 mW. What is the output power?

$$A_P = A_1 \times A_2 \times A_3 = 5 \times 2 \times 17 = 170$$

$$A_P = \frac{P_{out}}{P_{in}} \quad \text{therefore} \quad P_{out} = A_P P_{in}$$

$$P_{out} = 170 \, (40 \times 10^{-3}) = 6.8 \text{ W}$$

Example 2-4

A two-stage amplifier has an input power of 25 μW and an output power of 1.5 mW. One stage has a gain of 3. What is the gain of the second stage?

$$A_P = \frac{P_{\text{out}}}{P_{\text{in}}} = \frac{1.5 \times 10^{-3}}{25 \times 10^{-6}} = 60 \qquad A_p = A_1 \times A_2$$

If $A_1 = 3$, then $60 = 3 \times A_2$ and $A_2 = 60/3 = 20$.

ATTENUATION

Attenuation refers to a loss introduced by a circuit. Many electronic circuits reduce the amplitude of a signal rather than increase it. If the output signal is lower in amplitude than the input, the circuit has loss, or attenuation. Like gain, attenuation is simply the ratio of the output to the input. The letter A is used to represent attenuation as well as gain:

$$\text{Attenuation } A = \frac{\text{output}}{\text{input}} = \frac{V_{\text{out}}}{V_{\text{in}}}$$

Circuits that introduce attenuation have a gain that is less than 1. In other words, the output is some fraction of the input.

An example of a simple circuit with attenuation is a voltage divider like that shown in Fig. 2-3. The output voltage is the input voltage multiplied by a ratio based on the resistor values. With the resistor values shown, the gain or attenuation factor of the circuit is $A = R_2/(R_1 + R_2) = 100/(200 + 100) = 100/300 = .3333$. If a signal of 10 V is applied to the attenuator, the output is $V_{\text{out}} = V_{\text{in}}A = 10 \ (0.3333) = 3.333$ V.

When several circuits with attenuation are cascaded, the total attenuation is, again, the product of the individual attenuations. The circuit in Fig. 2-4 is an example. The attenuation factors for each circuit are shown. The overall attenuation is:

$$A_T = A_1 \times A_2 \times A_3$$

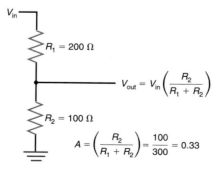

$$A = \left(\frac{R_2}{R_1 + R_2}\right) = \frac{100}{300} = 0.33$$

FIG. 2-3 A voltage divider introduces attenuation.

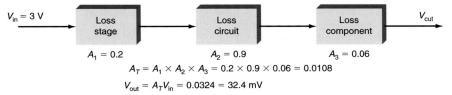

$$A_1 = 0.2 \qquad A_2 = 0.9 \qquad A_3 = 0.06$$
$$A_T = A_1 \times A_2 \times A_3 = 0.2 \times 0.9 \times 0.06 = 0.0108$$
$$V_{out} = A_T V_{in} = 0.0324 = 32.4 \text{ mV}$$

FIG. 2-4   Total attenuation is the product of individual attenuations of each cascaded circuit.

With the values shown in Fig. 2-4, the overall attenuation is:

$$A_T = 0.2 \times 0.9 \times 0.06 = .0108$$

Given an input of 3 V, the output voltage is:

$$V_{out} = A_T V_{in} = 0.0108(3) = 0.0324 = 32.4 \text{ mV}$$

It is common in communications systems and equipment to cascade circuits and components that have gain and attenuation. For example, loss introduced by a circuit can be compensated for by adding a stage of amplification that offsets it. An example of this is shown in Fig. 2-5. Here the voltage divider introduces a 4-to-1 voltage loss, or an attenuation of 0.25. To offset this, it is followed with an amplifier whose gain is 4. The overall gain or attenuation of the circuit is simply the product of the attenuation and gain factors. In this case, the overall gain is $A_T = A_1 A_2 = 0.25(4) = 1$.

Another example is shown in Fig. 2-6, which shows two attenuation circuits and two amplifier circuits. The individual gain and attenuation factors are given. The overall circuit gain is $A_T = A_1 A_2 A_3 A_4 = (0.1)(10)(0.3)(15) = 4.5$.

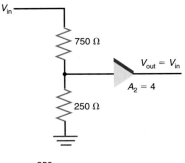

$$A_1 = \frac{250}{(750 + 250)} \qquad A_T = A_1 A_2 = 0.25(4) = 1$$
$$A_1 = \frac{250}{1000} = 0.25$$

FIG. 2-5   Gain exactly offsets the attenuation.

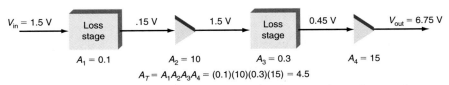

$$A_1 = 0.1 \qquad A_2 = 10 \qquad A_3 = 0.3 \qquad A_4 = 15$$
$$A_T = A_1 A_2 A_3 A_4 = (0.1)(10)(0.3)(15) = 4.5$$

FIG. 2-6   The total gain is the product of the individual stage gains and attenuations.

For an input voltage of 1.5 V, the output voltage at each circuit is shown in Fig. 2-6.

In this example, the overall circuit has a net gain. But in some instances, the overall circuit or system may have a net loss. In any case, the overall gain or loss is obtained by multiplying the individual gain and attenuation factors.

Example 2-5

A voltage divider like that shown in Fig. 2-5 has values of $R_1 = 10\text{k}\Omega$ and R_2 470 Ω.

a. What is the attenuation?
$$A_1 = \frac{R_2}{R_1 + R_2} = \frac{470}{10,470} \qquad A_1 = 0.045$$

b. What amplifier gain would you need to offset the loss for an overall gain of 1?
$$A_T = A_1 A_2$$
where A_1 is the attenuation and A_2 is the amplifier gain.
$$1 = 0.045 A_2 \qquad A_2 = \frac{1}{0.045} = 22.3$$
Note: To find the gain that will offset the loss for unity gain, just take the reciprocal of attenuation: $A_2 = 1/A_1$.

Example 2-6

An amplifier has a gain of 45,000, which is too much for the application. With an input voltage of 20 μV, what attenuation factor is needed to keep the output voltage from exceeding 100 mV? Let A_1 = amplifier gain = 45,000; A_2 = attenuation factor; A_T = total gain.
$$A_T = \frac{V_{\text{out}}}{V_{\text{in}}} = \frac{100 \times 10^{-3}}{20 \times 10^{-6}} = 5000$$

$$A_T = A_1 A_2 \qquad \text{therefore} \qquad A_2 = \frac{A_T}{A_1} = \frac{5000}{45000} = 0.1111$$

DECIBELS

The gain or loss of a circuit is expressed in *decibels,* a unit of measurement that was originally created as a way of expressing the hearing response of the human ear to various sound levels. A *decibel* is one-tenth of a bel.

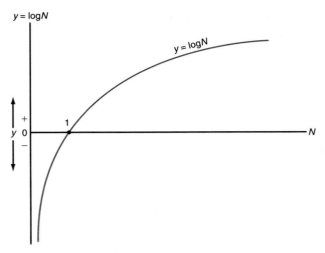

$y = \log N$

$y = \log N$

Fig. 2-7 Logarithmic curve.

When gain and attenuation are both converted into decibels, the overall gain or attenuation of an electronic circuit can be computed by simply adding the individual gains or attenuations, expressed in decibels.

It is common for electronic circuits and systems to have extremely high gains or attenuations, often in excess of 1 million. Converting these factors into decibels and using logarithms results in smaller gain and attenuation figures, which are easier to use.

LOGARITHMS. Figure 2-7 shows the *logarithmic curve.* It is a plot of the expression

$$y = \log N$$

The horizontal axis is the number N whose log is to be taken and the vertical axis is the logarithm y. Note that as the number gets bigger, its logarithm also gets bigger but not in the same proportion. The log curve is *flattened,* meaning that the log is a smaller number.

Note also that for numbers greater than 1, the log is positive and for numbers less than 1, the log is negative. When $N = 1$, the log is zero.

Remember that the logarithm y is the exponent to which some base or radix number B must be raised to get the number N; $N = B^y$. Rewriting this in log form,

$$y = \log_B N$$

The bases used are 10 and e; $e = 2.71828$. Logarithms to the base 10 are called *common logarithms;* logs to the base e are called *natural logarithms.* In decibel calculations, common logarithms are used.

Although log tables appear in many scientific reference books, the easiest way to find the logarithm of a number is to use a scientific calculator. Simply enter the number and press the $\boxed{\log}$ key to get the common log or the $\boxed{\ln}$ key to get the natural log.

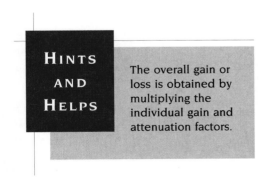

HINTS AND HELPS

The overall gain or loss is obtained by multiplying the individual gain and attenuation factors.

DECIBEL CALCULATIONS. The formulas for computing the decibel gain or loss of a circuit are

$$dB = 20 \log \frac{V_{out}}{V_{in}} \tag{1}$$

$$dB = 20 \log \frac{I_{out}}{I_{in}} \tag{2}$$

$$dB = 10 \log \frac{P_{out}}{P_{in}} \tag{3}$$

Formula (1) is used for expressing the voltage gain or attenuation of a circuit; formula (2) for current gain or attenuation. The ratio of the output voltage or current to the input voltage or current is determined as usual. The base 10 or common log of the input/output ratio is then obtained and multiplied by 20. The resulting number is the gain or attenuation in decibels.

Formula (3) is used to compute power gain or attenuation. The ratio of the power output to the power input is computed, then its logarithm is multiplied by 10.

Example 2-7

a. An amplifier has an input of 3 mV and an output of 5 V. What is the gain in decibels?

$$dB = 20 \log \frac{5}{0.003} = 20 \log 1666.67 = 20 \,(3.22) = 64.4$$

b. A filter has a power input of 50 mW and an output of 2 mW. What is the gain or attenuation?

$$dB = 10 \log \left(\frac{2}{50} \right) = 10 \log \,(0.04) = 10 \,(-1.398) = -13.98$$

Note that when the circuit has gain, the decibel figure is positive. If the gain is less than 1, meaning that there is an attenuation, the decibel figure is negative.

Now, to calculate the overall gain or attenuation of a circuit or system, you simply add together the decibel gain and attenuation factors of each circuit. An example of this is shown in Fig. 2-8, where there are two gain stages and an attenuation block. The overall gain of this circuit is

$$A_T = A_1 + A_1 + A_2 = 15 - 20 + 35 = 30 \text{ dB}$$

Decibels are widely used in the expression of gain and attenuation in communications circuits. Table 2-1 shows some common gain and attenuation factors and their corresponding decibel figures.

Ratios less than 1 give negative decibel values, indicating attenuation. Note that a 2:1 ratio represents a 3-dB power gain or a 6-dB voltage gain.

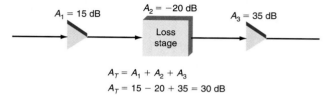

$$A_T = A_1 + A_2 + A_3$$
$$A_T = 15 - 20 + 35 = 30 \text{ dB}$$

FIG. 2-8 Total gain or attenuation is the algebraic sum of the individual stage gains in decibels.

ANTILOGS. To calculate the input or output voltage or power given the decibel gain or attenuation and the output or input, the *antilog* is used. The antilog is the number obtained when the base is raised to the logarithm which is the exponent:

$$dB = 10 \log \frac{P_{\text{out}}}{P_{\text{in}}} \quad \text{and} \quad \frac{dB}{10} = \log \frac{P_{\text{out}}}{P_{\text{in}}}$$

and

$$\frac{P_{\text{out}}}{P_{\text{in}}} = \text{antilog} \frac{dB}{10}$$

The antilog is simply the base 10 raised to the dB/10 power.

TABLE 2-1

Ratio (power or voltage)	dB Gain or Attenuation	
	Power	**Voltage**
0.000001	−60	−120
0.00001	−50	−100
0.0001	−40	−80
0.001	−30	−60
0.01	−20	−40
0.1	−10	−20
0.5	−3	−6
1	0	0
2	3	6
10	10	20
100	20	40
1000	30	60
10000	40	80
100000	50	100

Remember that the logarithm y of a number N is the power to which the base 10 must be raised to get the number.

$$N = 10^y \quad \text{and} \quad y = \log_{10} N$$

Therefore

$$\frac{P_{out}}{P_{in}} = 10^{dB/10}$$

The antilog is readily calculated on a scientific calculator. To find the antilog for a common or base 10 logarithm, you normally press the (Inv) or (2nd) function key on the calculator, then the (log) key. Sometimes the log key is marked with 10^x, which is the antilog. The antilog with base e is found in a similar way, using the (Inv) or (2nd) function on the (ln) key. It is sometimes marked e^x, which is the same as the antilog.

Example 2-8

A power amplifier with a 40-dB gain has an output power of 100 W. What is the input power?

$$dB = 10 \log \frac{P_{out}}{P_{in}}$$

$$\frac{dB}{10} = \log \frac{P_{out}}{P_{in}}$$

$$\frac{40}{10} = \log \frac{P_{out}}{P_{in}}$$

$$\frac{P_{out}}{P_{in}} = \text{antilog } 4$$

$$\frac{P_{out}}{P_{in}} = 10^4 = 10000$$

$$P_{in} = \frac{P_{out}}{10000} = \frac{100}{10000} = 0.01 \text{ W} = 10 \text{ mW}$$

Example 2-9

An amplifier has a gain of 60 dB. If the input voltage is 50 μV, what is the output voltage? (For voltage calculations use the formula $V_{out}/V_{in} = 10^{dB/20}$.)

$$\frac{V_{out}}{V_{in}} = 10^{dB/20}$$

$$\frac{V_{out}}{50 \times 10^{-6}} = 10^{60/20} = 10^3$$

$$\frac{V_{out}}{V_{in}} = 10^3 = 1000$$

$$V_{out} = 1000 V_{in} = 1000 (50 \times 10^{-6}) = 0.05 \text{ V} = 50 \text{ mV}$$

dBm. When the gain or attenuation of a circuit is expressed in decibels, implicit is a comparison between two values, the output and the input. In computing the ratio, the units of voltage or power are canceled, making the ratio a dimensionless, or relative, figure. When you see a decibel value, you really do not know the actual voltage or power values. In some cases this is not a problem; in others, it is useful or necessary to know the actual values involved. When an absolute value is needed, you can use a *reference value* to compare any other value.

An often-used reference level in communications is 1 mW. When a decibel value is computed by comparing a power value to 1 mW, the result is a value called the *dBm*. It is computed with the standard power decibel formula with 1 mW as the denominator of the ratio:

$$dBm = 10 \log \frac{P_{out}}{0.001}$$

Here P_{out} is the output power, or some power value you want to compare to 1 mW and 0.001 is 1 mW expressed in watts.

The output of a 1-W amplifier expressed in dBm is, for example,

$$dBm = 10 \log \frac{1}{0.001} = 10 \log 1000 = 10(3) = 30 \text{ dBm}$$

Sometimes the output of a circuit or device is given in dBm. For example, if a microphone has an output of -50 dBm, the actual output power can be computed as follows:

$$\frac{dBm}{10} = \frac{-50}{10} = -5$$
$$\frac{P_{out}}{0.001} = 10^{dBm/10} = 10^{-5} = 0.00001$$
$$P_{out} = 10^{-5} \times 0.001 = 10^{-8} \text{ W} = 10 \times 10^{-9} = 10 \text{ nW}$$

The boy enjoys the "music" as he plays his trumpet in Grandpa's ear. Sound is measured in decibels.

Example 2-10

A power amplifier has an input of 90 mV across 10 kΩ. The output is 7.8 V across an 8-Ω speaker. What is the power gain, in decibels?

$$P = \frac{V^2}{R}$$

$$P_{in} = \frac{(90 \times 10^{-3})^2}{10^4} = 8.1 \times 10^{-7} \text{ W}$$

$$P_{out} = \frac{(7.8)^2}{8} = 7.605 \text{ W}$$

$$A_P = \frac{P_{out}}{P_{in}} = \frac{7.605}{8.1 \times 10^{-7}} = 9.39 \times 10^6$$

$$A_P \text{ (dB)} = 10 \log A_P = 10 \log 9.39 \times 10^6 = 69.7 \text{ dB}$$

Example 2-11

An amplifier has a power gain of 28 dB. The input power is 36 mW. What is the output power?

$$\frac{P_{out}}{P_{in}} = 10^{dB/10} = 10^{2.8} = 630.96$$

$$P_{out} = 630.96 P_{in} = 630.96 (36 \times 10^{-3}) = 22.71 \text{ W}$$

Example 2-12

A circuit consists of two amplifiers with gains of 6.8 and 14.3 dB and two filters with attenuations of −16.4 and −2.9 dB. If the output voltage is 800 mV, what is the input voltage?

$$A_T = A_1 + A_2 + A_3 + A_4 = 6.8 + 14.3 - 16.4 - 2.9 = 1.8 \text{ dB}$$

$$A_T = \frac{V_{out}}{V_{in}} = 10^{dB/20} = 10^{1.8/20} = 10^{0.09}$$

$$\frac{V_{out}}{V_{in}} = 10^{0.09} = 1.23$$

$$V_{in} = \frac{V_{out}}{1.23} = \frac{800}{1.23} = 650.4 \text{ mV}$$

Example 2-13

Express $P_{\text{out}} = 123$ dBm in watts.

$$\frac{P_{\text{out}}}{0.001} = 10^{\text{dBm}/10} = 10^{123/10} = 10^{12.3} = 2 \times 10^{12}$$

$$P_{\text{out}} = 0.001 \times 2 \times 10^{12} = 2 \times 10^9 = 2 \text{ GW}$$

2-2 TUNED CIRCUITS

Virtually all communications equipment contains *tuned circuits,* circuits made up of inductors and capacitors that resonate at specific frequencies. In this section you will review how to calculate the reactance, resonant frequency, impedance, *Q,* and bandwidth of series and parallel resonance circuits.

REACTIVE COMPONENTS

All tuned circuits and many filters are made up of inductive and capacitive elements, including discrete components such as coils and capacitors and the stray and distributed inductance and capacitance that appear in all electronic circuits. Both coils and capacitors offer an opposition to alternating current flow known as *reactance,* which is expressed in ohms. Like resistance, reactance is an opposition that directly affects the amount of current in a circuit. In addition, reactive effects produce a phase shift between the currents and voltages in a circuit. Capacitance causes the current to lead the applied voltage, while inductance causes the current to lag the applied voltage. Coils and capacitors used together form tuned, or resonant, circuits.

CAPACITORS. A *capacitor* is, essentially, two parallel conductors separated by an insulating medium known as the *dielectric.* Capacitors are capable of storing energy in the form of a charge and an electric field. The basic action of a capacitor is charging and discharging as energy is alternately stored and released.

The capacitance is dependent on the physical characteristics of the capacitor such as plate size, plate spacing, number of plates, and type of dielectric material. Capacitance is expressed in farads (F). The farad is an extremely large unit, and most capacitors used in communications circuits have values in the microfarad (μF = 10^{-6} F) and picofarad (pF = 10^{-12} F) range. The unit nanofarad (nF = 10^{-9}) is also used in some cases.

A capacitor used in an AC circuit continually charges and discharges. A capacitor tends to oppose voltage changes

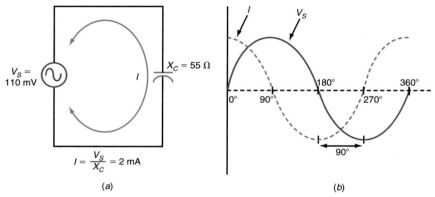

$$I = \frac{V_s}{X_C} = 2 \text{ mA}$$

(a)

(b)

Fig. 2-9 A basic capacitive circuit.

across it. This translates into an opposition to alternating current known as capacitive reactance X_C. In the basic capacitive circuit shown in Fig. 2-9, the current in the circuit is inversely proportional to the capacitive reactance. This is expressed by the simple Ohm's law expression

$$I = \frac{V_s}{X_C}$$

If the capacitive reactance is 55 Ω and the applied voltage (V_s) is 110 mV, the circuit current by Ohm's law is

$$I = \frac{110 \times 10^{-3}}{55} = 2 \times 10^{-3} \text{ A or 2 mA}$$

The reactance of a capacitor is inversely proportional to the value of capacitance C and operating frequency f. It is given by the familiar expression

$$X_C = \frac{1}{2\pi f C}$$

The reactance of a 100 pF capacitor at 2 MHz is

$$X_C = \frac{1}{6.28 \, (2 \times 10^6) \, (100 \times 10^{-12})} = 796.2 \ \Omega$$

The formula can also be used to calculate either frequency or capacitance depending on the application. These formulas are

$$f = \frac{1}{2\pi X_C C} \qquad \text{and} \qquad C = \frac{1}{2\pi f X_C}$$

When the circuit consists of capacitance only, the current in the circuit will lead the voltage by 90° as indicated by the sine waves in Fig. 2-9(b). When a capacitor is combined with a resistor in either series or parallel form (Fig. 2-10), the current will still lead the voltage but by some phase angle Θ between 0 and 90° depending on the values of the resistance and capacitance or the frequency.

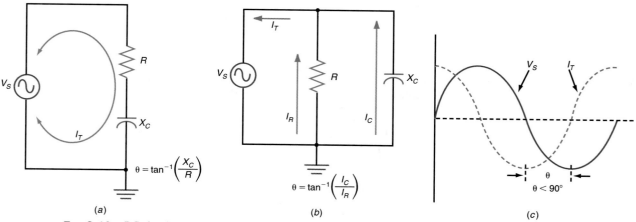

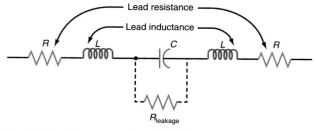

Fig. 2-10 *RC* circuits. (*a*) Series. (*b*) Parallel. (*c*) Phase shift.

Combining resistance and capacitance in series or parallel also produces a total opposition of the combined components known as *impedance* (*Z*). The impedance of a series *RC* circuit like that shown in Fig. 2-10(*a*) is given by the simple expression

$$Z = \sqrt{(R)^2 + (X_C)^2}$$

For a parallel *RC* circuit like that in Fig. 2-10(*b*), the total impedance can be computed by the expression

$$Z = \frac{RX_C}{\sqrt{(R)^2 + (X_C)^2}}$$

The wire leads of a capacitor have resistance and inductance, and the dielectric has leakage which appears as a resistance value in parallel with the capacitor. These characteristics, which are illustrated in Fig. 2-11, are sometimes referred to as *residuals*. The series resistance and inductance are very small and the leakage resistance is very high, so these factors can be ignored at low frequencies. At radio frequencies, however, these residuals become noticeable, and the capacitor functions as a complex *RLC* circuit. Most of these effects can be greatly minimized by keeping the capacitor leads very short.

Capacitance is generally added to a circuit by a capacitor of a specific value, but capacitance can occur between any two conductors separated by an insulator. For example, there is capacitance between the parallel wires in a cable; between a wire and a metal chassis; and between parallel adjacent copper patterns on a printed circuit board.

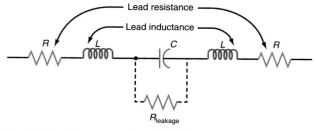

Fig. 2-11 What a capacitor looks like at high frequencies.

These are known as *stray,* or *distributed,* capacitances. Stray capacitances are typically small, but they cannot be ignored, especially at the high frequencies used in communications. Stray and distributed capacitances can significantly affect the performance of a circuit.

INDUCTORS.   An *inductor,* also called a *coil* or *choke,* is simply a winding of multiple turns of wire. When current is passed through a coil, a magnetic field is produced around the coil. If the applied voltage and current are varying, the magnetic field alternately expands and collapses. This causes a voltage to be self-induced into the coil winding, which has the effect of opposing current changes in the coil. This effect is known as inductance.

The basic unit of inductance is the henry (H). Inductance is directly affected by the physical characteristics of the coil, including number of turns of wire in the inductor, the spacing of the turns, the length of the coil, the diameter of the coil, and the type of magnetic core material. Practical inductance values are in the millihenry (mH = 10^{-3} H) and microhenry (μH-10^{-6} H) region.

Figure 2-12 shows several different types of inductor coils.

- Figure 2-12(*a*) is an inductor made of a heavy, self-supporting wire coil.
- In Fig. 2-12(*b*) the inductor is formed of a copper spiral that is etched right on to the board itself.
- In Fig. 2-12(*c*) the coil is wound on an insulating form containing a powdered iron or ferrite core in the center, to increase its inductance.
- Figure 2-12(*d*) shows another common type of inductor, one using turns of wire on a torroidal or donut-shaped form.
- Figure 2-12(*e*) shows an inductor made by placing a small ferrite bead over a wire; the bead effectively increases the wire's small inductance.

In a DC circuit, an inductor will have little or no effect. Only the ohmic resistance of the wire affects current flow. However, when the current changes, such as during the time the power is turned off or on, the coil will oppose these changes in current.

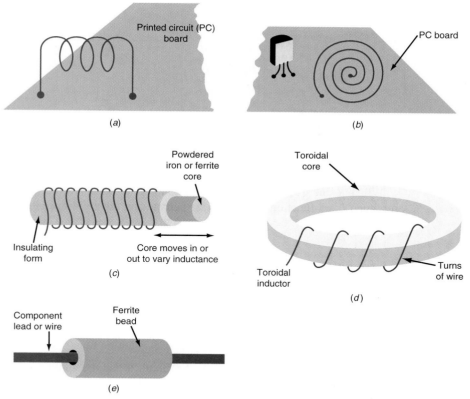

FIG. 2-12 Types of inductors. (*a*) Heavy self-supporting wire coil. (*b*) Inductor made as copper pattern. (*c*) Insulating form. (*d*) Torroidal inductor. (*e*) Ferrite bead inductor.

When an inductor is used in an AC circuit, this opposition becomes continuous and constant and is known as *inductive reactance*. Inductive reactance X_L is expressed in ohms and is calculated using the expression

$$X_L = 2\pi fL$$

For example, the inductive reactance of a 40-μH coil at 18 MHz is

$$X_L = 6.28 \ (18 \times 10^6) \ (40 \times 10^{-6}) = 4522 \ \Omega$$

The inductor has resistance, and so its equivalent circuit has the form shown in Fig. 2-13(*a*). The combined reactance and resistance form a total opposition known as *impedance* (*Z*), computed using the same expression for computing impedance in an *RC* circuit:

$$Z = \sqrt{(R)^2 + (X_L)^2}$$

In circuits in which the reactance is much greater than the circuit resistance, the resistance may essentially be ignored. In such cases, the impedance is approximately equal to the inductive reactance. As you will see, in most practical circuits, the resistance is large enough to be a factor not only in controlling the current but also in affecting other characteristics of the circuit.

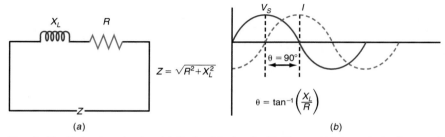

(a)

$$Z = \sqrt{R^2 + X_L^2}$$

(b)

$$\theta = \tan^{-1}\left(\frac{X_L}{R}\right)$$

FIG. 2-13 Inductive circuits. (*a*) Series *RL* circuit. (*b*) Phase shift in an inductive conductor.

When a sine wave of voltage is applied to the inductor, current flows. The current amplitude *I* is a function of the applied voltage and the coil impedance:

$$I = \frac{V_s}{Z}$$

If the resistance is much much smaller than the reactance, the impedance will be approximately the same as the reactance. Therefore the current in the coil will be

$$I = \frac{V_s}{X_L}$$

An inductor causes the current in the circuit to lag the applied voltage. In a pure inductive circuit, the current lags the applied voltage by 90°. With some resistance in the circuit, the current lags the applied voltage by an angle less than 90°, the exact value being dependent on the values of reactance and resistance. See Fig. 2-13(*b*).

In addition to the resistance of the wire in an inductor, there is stray capacitance between the turns of the coil. See Fig. 2-14(*a*). The overall effect is as if a small capacitor were connected in parallel with the coil, as shown in Fig. 2-14(*b*). This is the equivalent circuit of an inductor at high frequencies. At low frequencies, capacitance may be ignored, but at radio frequencies, it is sufficiently large to affect circuit operation. The coil then functions not as a pure inductor, but as a complex *RLC* circuit with a self-resonating frequency.

Any wire or conductor exhibits a characteristic inductance. The longer the wire, the greater the inductance. While the inductance of a straight wire is only a fraction of a microhenry, at very high frequencies, the reactance can be significant. For this rea-

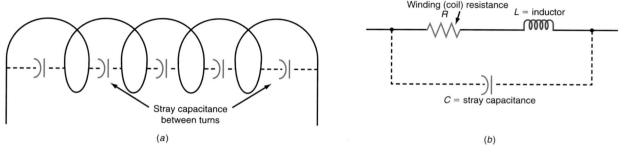

FIG. 2-14 Equivalent circuit of an inductor at high frequencies. (*a*) Stray capacitance between turns. (*b*) Equivalent circuit of an inductor at high frequencies.

son, it is important to keep all lead lengths short in interconnecting components in *RF* circuits. This is especially true of capacitor and transistor leads, since stray or distributed inductance can significantly affect the performance and characteristics of a circuit.

Another important characteristic of an inductor is its quality factor Q, the ratio of inductive power to resistive power:

$$Q = \frac{I^2 X_L}{I^2 R} = \frac{X_L}{R}$$

This is the ratio of the power returned to the circuit versus the power actually dissipated by the coil resistance. For example, the Q of a 3-μH inductor with a total resistance of 45 Ω at 90 MHz is calculated as follows:

$$Q = \frac{2\pi f L}{R} = \frac{6.28(90 \times 10^6)(3 \times 10^{-6})}{45} = \frac{1695.6}{45} = 37.68$$

RESISTORS. At low frequencies, a standard low-wattage color-coded resistor offers nearly pure resistance, but at high frequencies its leads have considerable inductance and stray capacitance between the leads causes the resistor to act like a complex *RLC* circuit, as shown in Fig. 2-15. To minimize the inductive and capacitive effects, the leads are kept very short in radio applications.

The tiny resistor chips used in surface mount construction of the electronic circuits preferred for radio equipment have practically no leads except for the metallic end pieces soldered to the printed circuit board. They have virtually no lead inductance and little stray capacitance.

Many resistors are made from a carbon-composition material in powdered form sealed inside a tiny housing to which leads are attached. The type and amount of carbon material determine such resistors' value. These resistors contribute noise to the circuit in which they are used caused by thermal effects and the granular nature of the resistance material. The noise contributed by such resistors in an amplifier used to amplify very-low-level radio signals may be so high as to obliterate the desired signal.

To overcome this problem, film resistors were developed. These are made by depositing a carbon or metal film in spiral form on a ceramic form. The size of the spiral and the kind of metal film determine the resistance value. Carbon film resistors are quieter than carbon-composition resistors, and metal film resistors are quieter than carbon film resistors. Metal film resistors should be used in amplifier circuits that must deal with very-low-level RF signals. Most surface mount resistors are of the metallic film type.

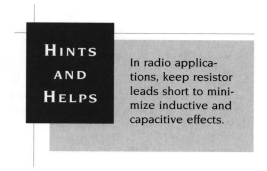

HINTS AND HELPS

In radio applications, keep resistor leads short to minimize inductive and capacitive effects.

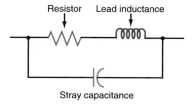

FIG. 2-15 Equivalent circuit of a resistor at high (radio) frequencies.

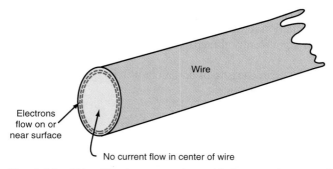

FIG. 2-16 Skin effect increases wire and inductor resistance at high frequencies.

SKIN EFFECT. The resistance of any wire conductor, whether it is a resistor or capacitor lead or the wire in an inductor, is primarily determined by the ohmic resistance of the wire itself. However, other factors influence it. The most significant one is *skin effect,* the tendency of electrons flowing in a conductor to flow near and on the outer surface of the conductor frequencies in the VHF, UHF, and microwave regions (Fig. 2-16). This has the effect of greatly decreasing the total cross-sectional area of the conductor, thus increasing its resistance and significantly affecting the performance of the circuit in which the conductor is used. For example, skin effect lowers the Q of an inductor at the higher frequencies, causing unexpected and undesirable effects. Thus many high-frequency coils, particularly those in high-powered transmitters, are made with copper tubing. Since current does not flow in the center of the conductor, but only on the surface, tubing provides the most efficient conductor. Very thin conductors, such as a copper pattern on a printed circuit board, are also used. Often these conductors are silver- or gold-plated to further reduce their resistance.

TUNED CIRCUITS AND RESONANCE

A tuned circuit is made up of inductance and capacitance and resonates at a specific frequency, the resonant frequency. In general, the terms *tuned circuit* and *resonant circuit* are used interchangeably. Because tuned circuits are frequency-selective, they respond best at their resonant frequency and at a narrow range of frequencies around the resonant frequency.

SERIES RESONANT CIRCUITS. A series resonant circuit is made up of inductance, capacitance, and resistance as shown in Fig. 2-17. Such circuits are often referred to as *LCR* circuits or *RLC* circuits. The inductive and capacitive reactances depend upon the frequency of the applied voltage. Resonance occurs when the inductive and capacitive reactances are equal. A plot of reactance versus frequency is shown in Fig. 2-18, where f_r is the resonant frequency.

As stated previously, the total impedance of the circuit is given by the expression

$$Z = \sqrt{R^2 + (X_L - X_C)^2}$$

When X_L equals X_C, they cancel one another, leaving only the resistance of the circuit to oppose the current. At resonance, the total circuit impedance is simply the value of

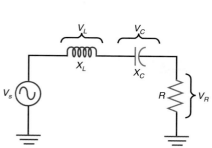

FIG. 2-17   Series *RLC* circuit.

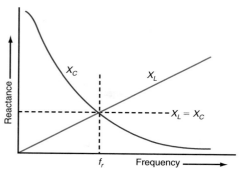

FIG. 2-18   Variation of reactance with frequency.

all series resistances in the circuit. This includes the resistance of the coil and the resistance of the component leads, as well as any physical resistor in the circuit.

The resonant frequency can be expressed in terms of inductance and capacitance. A formula for resonant frequency can be easily derived. First, express X_L and X_C as an equivalence: $X_L = X_C$. Rearranging the equation $X_L = 2\pi fL$ to solve for f, we have

$$f_r = \frac{1}{2\pi\sqrt{LC}}$$

In this formula, the frequency is in Hertz, the inductance is in henries, and the capacitance is in farads.

Example 2-14

What is the resonant frequency of a 2.7-pF capacitor and a 33-nH inductor?

$$f_r = \frac{1}{2\pi\sqrt{LC}} = \frac{1}{6.28\sqrt{33 \times 10^{-9} \times 2.7 \times 10^{-12}}}$$
$$= 5.33 \times 10^8 \text{ Hz or } 533 \text{ MHz}$$

It is often necessary to calculate capacitance or inductance given one of those values and the resonant frequency. The basic resonant frequency formula can be rearranged to solve for either inductance and capacitance as follows;

$$L = \frac{1}{4\pi^2 f^2 C} \qquad \text{and} \qquad C = \frac{1}{4\pi^2 f^2 L}$$

For example, the capacitance that will resonate at a frequency of 18 MHz with a 12-μH inductor is determined as follows:

$$C = \frac{1}{4\pi^2 f_r^2 L} = \frac{1}{39.478(18 \times 10^6)^2(12 \times 10^{-6})}$$

$$= \frac{1}{39.478(3.24 \times 10^{14})(12 \times 10^{-6})} = 6.5 \times 10^{-12} \text{ F or 6.5 pF}$$

Example 2-15

What value of inductance will resonate with a 12-pF capacitor at 49 MHz?

$$L = \frac{1}{4\pi^2 f_r^2 C} = \frac{1}{39.478\ (49 \times 10^6)^2(12 \times 10^{-12})}$$
$$= 8.79 \times 10^{-7}\ \text{H or 879 nH}$$

As indicated earlier, the basic definition of resonance in a series tuned circuit is the point at which X_L equals X_C. With this condition, only the resistance of the circuit impedes the current. The total circuit impedance at resonance is $Z = R$. For this reason, resonance in a series tuned circuit can also be defined as the point at which the circuit impedance is lowest and the circuit current is highest. Since the circuit is resistive at resonance, the current is in phase with the applied voltage. Above the resonant frequency, the inductive reactance is higher than the capacitive reactance and the inductor voltage drop is greater than the capacitor voltage drop. Therefore, the circuit is inductive and the current will lag the applied voltage. Below resonance, the capacitive reactance is higher than the inductive reactance; the net reactance is capacitive, thereby producing a leading current in the circuit. The capacitor voltage drop is higher than the inductor voltage drop.

The response of a series resonant circuit is illustrated in Fig. 2-19, which is a plot of the frequency and phase shift of the current in the circuit with respect to frequency.

At very low frequencies, the capacitive reactance is much greater than the inductive reactance; therefore the current in the circuit is very low because of the high impedance. In addition, because the circuit is predominantly capacitive, the current leads the voltage by nearly 90°. As the frequency increases, X_C goes down and X_L goes up. The amount of leading phase shift decreases. As the values of reactances approach one another, the current begins to rise. When X_L equals X_C, their effects cancel and the im-

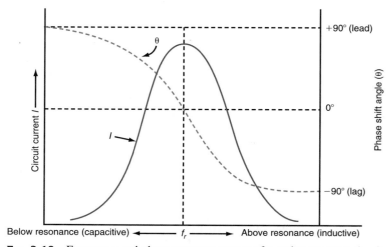

FIG. 2-19 Frequency and phase response curves of a series resonant circuit.

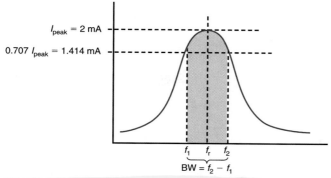

Fig. 2-20 Bandwidth of a series resonant circuit.

pedance in the circuit is just that of the resistance. This produces a current peak, where the current is in phase with the voltage (0°). As the frequency continues to rise, X_L becomes greater than X_C. The impedance of the circuit increases and the current decreases. With the circuit predominantly inductive, the current lags the applied voltage. If the output voltage were being taken from across the resistor in Fig. 2-17, the response curve and phase angle of the voltage would correspond to those in Fig. 2-19. As Fig. 2-19 shows, the current is highest in a region centered around the resonant frequency. The narrow frequency range over which the current is highest is called the *bandwidth*. This area is illustrated in Fig. 2-20.

The upper and lower boundaries of the bandwidth are defined by two cutoff frequencies designated f_1 and f_2. These cutoff frequencies occur where the current amplitude is 70.7 percent of the peak current. In the figure, the peak circuit current is 2 mA and the current at both the lower (f_1) and upper (f_2) cutoff frequencies is 0.707 of 2 mA, or 1.414 mA.

Current levels at which the response is down 70.7 percent are called the *half-power points* because the power at the cutoff frequencies is one-half the power peak of the curve.

$$P = I^2 R = (0.707 \ I_{peak})^2 R = 0.5 \ T_{peak}{}^2 R$$

The bandwidth of the tuned circuit is defined as the difference between the upper and lower cutoff frequencies:

$$BW = f_2 - f_1$$

For example, assuming a resonant frequency of 75 kHz and upper and lower cutoff frequencies of 76.5 and 73.5 kHz, respectively, the bandwidth is BW = 76.5 − 73.5 = 3 kHz.

The bandwidth of a resonant circuit is determined by the Q of the circuit. Recall that the Q of an inductor is the ratio of the inductive reactance to the circuit resistance. This holds true for a series resonant circuit, where Q is the ratio of the inductive reactance to the total circuit resistance, which includes the resistance of the inductor plus any additional series resistance:

$$Q = \frac{X_L}{R_T}$$

Bandwidth is then computed as

$$\text{BW} = \frac{f_r}{Q}$$

If the Q of a circuit resonant at 18 MHz is 50, then the bandwidth is BW = 18/50 = 0.36 MHz = 360 kHz.

Example 2-16

What is the bandwidth of a resonant circuit with a frequency of 28 MHz and a Q of 70?

$$\text{BW} = \frac{f_r}{Q} = \frac{28 \times 10^6}{70} = 400,000 \text{ Hz} = 400 \text{ kHz}$$

The formula can be rearranged to compute Q given the frequency and the bandwidth:

$$Q = \frac{f_r}{\text{BW}}$$

Thus the Q of the circuit whose bandwidth was computed above is Q = 75 kHz/ 3 kHz = 25.

Since the bandwidth is approximately centered around the resonant frequency, f_1 is the same distance from f_r as f_2 is from f_r. Knowing this allows you to calculate the resonant frequency by knowing only the cutoff frequencies:

$$f_r = \sqrt{f_1 \times f_2}$$

For example, if f_1 = 175 kHz and f_2 = 178 kHz, the resonant frequency is

$$f_r = \sqrt{175 \times 10^3 \times 178 \times 10^3} = 176.5 \text{ kHz}$$

For a linear frequency scale, you can calculate the center or resonant frequency using an average of the cutoff frequencies.

$$f_r = \frac{f_1 + f_2}{2}$$

If the current Q is very high (> 100), then the response curve is approximately symmetrical around the resonant frequency. The cutoff frequencies will then be roughly equidistant from the resonant frequency by an amount of BW/2. Thus the cutoff frequencies can be calculated if the bandwidth and the resonant frequency are known:

$$f_1 = f_r - \frac{\text{BW}}{2} \qquad \text{and} \qquad f_2 = f_r + \frac{\text{BW}}{2}$$

For instance, if the resonant frequency is 49 MHz (49,000 kHz) and the bandwidth is 10 kHz, then the cutoff frequencies will be

$$f_1 = 49,000 \text{ k}\Omega - \frac{10k}{2} = 49,000 \text{ k}\Omega - 5 \text{ k}\Omega = 48995 \text{ kHz}$$

$$f_2 = 49,000 \text{ k}\Omega + 5 \text{ k}\Omega = 49,005 \text{ kHz}$$

Keep in mind that although this procedure is an approximation, it is useful in many applications.

The bandwidth of a resonant circuit defines its *selectivity,* that is, how the circuit responds to varying frequencies. If the response is to produce a high current only over a narrow range of frequencies, a narrow bandwidth, the circuit is said to be highly selective. If the current is high over a broader range of frequencies, that is, the bandwidth is wider, the circuit is less selective. In general, circuits with high selectivity and narrow bandwidths are more desirable. However, the actual selectivity and bandwidth of a circuit must be optimized for each application.

The relationship between circuit resistance, Q, and bandwidth is extremely important. The bandwidth of a circuit is inversely proportional to Q. The higher Q is, the smaller the bandwidth. Low Q's produce wide bandwidths or less selectivity. In turn, Q is a function of the circuit resistance. A low resistance produces a high Q, a narrow bandwidth, and a highly selective circuit. A high circuit resistance produces a low Q, wide bandwidth, and poor selectivity. In most communication circuits, circuit Q's are at least 10 and typically higher. In most cases, Q is controlled directly by the resistance of the inductor. Figure 2-21 shows the effect of different values of Q on bandwidth.

Example 2-17

The upper and lower cutoff frequencies of a resonant circuit are found to be 8.07 and 7.93 MHz. Calculate (*a*) bandwidth, (*b*) approximate resonant frequency, and (*c*) Q.

a. $BW = f_2 - f_1 = 8.07 \text{ MHz} - 7.93 \text{ MHz} = 0.14 \text{ MHz} = 140 \text{ kHz}$

b. $f_r = \sqrt{f_1 f_2} = \sqrt{(8.07 \times 10^6)(7.93 \times 10^6)} = 8 \text{ MHz}$

c. $Q = \dfrac{f_r}{BW} = \dfrac{8 \times 10^6}{140 \times 10^3} = 57.14$

Example 2-18

What are the approximate 3-dB down frequencies of a resonant circuit with a Q of 200 at 16 MHz?

$$BW = \frac{f_r}{Q} = \frac{16 \times 10^6}{200} = 80,000 \text{ Hz} = 80 \text{ kHz}$$

$$f_1 = f_r - \frac{BW}{2} = 16,000,000 - \frac{80,000}{2} = 15.96 \text{ MHz}$$

$$f_2 = f_r + \frac{BW}{2} = 16,000,000 + \frac{80,000}{2} = 16.04 \text{ MHz}$$

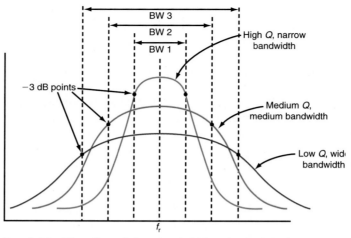

Fig. 2-21 The effect of Q on bandwidth and selectivity in a resonant circuit.

Resonance produces an interesting but useful phenomenon in a series *RLC* circuit. Consider the circuit in Fig. 2-22(*a*). At resonance, assume $X_L = X_C = 500 \ \Omega$. The total circuit resistance is 10 Ω. The Q of the circuit is then

$$Q = \frac{X_L}{R} = \frac{500}{10} = 50$$

If the applied or source voltage V_s is 2 V, the circuit current at resonance will be

$$I = \frac{V_s}{R} = \frac{2}{10} = 0.2 \ \text{A}$$

Knowing the reactances, the resistances, and the current, the voltage drops across each component can be computed:

$$V_L = IX_L = 0.2(500) = 100 \ \text{V}$$
$$V_C = IX_C = 0.2(500) = 100 \ \text{V}$$
$$V_R = IR = 0.2(10) = 2 \ \text{V}$$

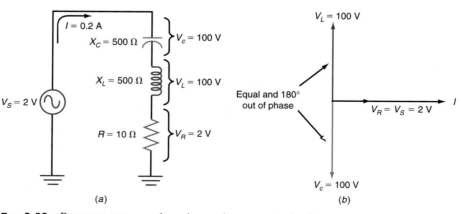

Fig. 2-22 Resonant step-up voltage in a series resonant circuit.

As you can see, the voltage drops across the inductor and capacitor are significantly higher than the applied voltage. This is known as the *resonant step-up voltage*. Although the sum of the voltage drops around the series circuit is still equal to the source voltage, at resonance, the voltage across the inductor leads the current by 90° and the voltage across the capacitor lags the current by 90° [see Fig. 2-22(*b*)]. Therefore the inductive and reactive voltages are equal but 180° out of phase. As a result, when added, they cancel one another, leaving a total reactive voltage of 0. This means that the entire applied voltage appears across the circuit resistance.

The resonant step-up voltage across the coil or capacitor can be easily computed by multiplying the input or source voltage by Q:

$$V_L = V_C = QV_s$$

In the example in Figure 2-22, $V_L = 50(2) = 100$ V.

This interesting and useful phenomenon means that small applied voltages can essentially be stepped up to a higher voltage—a form of simple amplification without active circuits that is widely applied in communications circuits.

Example 2-19

A series resonant circuit has a Q of 150 at 3.5 MHz. The applied voltage is 3 μV. What is the voltage across the capacitor?
$$V_C = QV_s = 150(3 \times 10^{-6}) = 450 \times 10^{-6} = 450 \ \mu V$$

PARALLEL RESONANT CIRCUITS. A *parallel resonant circuit* is formed when the inductor and capacitor are connected in parallel with the applied voltage, as shown in Fig. 2-23(*a*). In general, resonance in a parallel tuned circuit can also be defined as the point at which the inductive and capacitive reactances are equal. The resonant frequency is therefore calculated by the resonant-frequency formula given earlier. If we assume lossless components in the circuit (no resistance), then the current in the inductor equals the current in the capacitor:

$$I_L = I_C$$

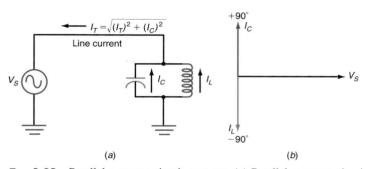

(a)   (b)

FIG. 2-23 Parallel resonant circuit currents. (*a*) Parallel resonant circuit. (*b*) Current relationships in parallel resonant circuit.

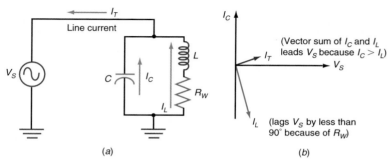

Fig. 2-24 A practical parallel resonant circuit. (*a*) Practical parallel resonant circuit with coil resistance R_w. (*b*) Phase relationships.

While the currents are equal, they are 180° out of phase, as the phasor diagram in Fig. 2-23(*b*) shows. The current in the inductor lags the applied voltage by 90°, and the current in the capacitor leads the applied voltage by 90°, for a total of 180°.

Now, applying Kirchhoff's current law to the circuit, the sum of the individual branch currents equals the total current drawn from the source. With the inductive and capacitive currents equal and out of phase, their sum is 0. Thus, at resonance, a parallel tuned circuit appears to have infinite resistance, draws no current from the source and thus has infinite impedance, and acts like an open circuit. However, there is a high circulating current between the inductor and capacitor. Energy is being stored and transferred between the inductor and capacitor. Because such a circuit acts as a kind of storage vessel for electric energy, it is often referred to as a *tank circuit* and the circulating current is referred to as the *tank current*.

In a practical resonant circuit where the components do have losses (resistance), the circuit still behaves as described above. Typically we can assume that the capacitor has practically zero losses and the inductor contains a resistance, as illustrated in Fig. 2-24(*a*). At resonance, where $X_L = X_C$, the impedance of the inductive branch of the circuit is higher than the impedance of the capacitive branch because of the coil resistance. The capacitive current is slightly higher than the inductive current. Even if the reactances are equal, the branch currents will be unequal and therefore there will be some net current flow in the supply line. The source current will lead the supply voltage, as shown in Fig. 2-24(*b*). Nevertheless, the inductive and capacitive currents will in most cases cancel because they are approximately equal and of opposite phase, and consequently the line or source current will be significantly lower than the individual branch currents. The result is a very high resistive impedance, approximately equal to

$$Z = \frac{V_s}{I_T}$$

The circuit in Fig. 2-24(*a*) is not easy to analyze. One way to simplify the mathematics involved is to convert the circuit to an equivalent circuit in which the coil resistance is translated to a parallel resistance that gives the same overall results, as shown in Fig. 2-25. The equivalent inductance L_{eq} and resistance R_{eq} are calculated with the formulas

$$L_{eq} = \frac{L(Q^2 + 1)}{Q^2} \qquad \text{and} \qquad R_{eq} = R_w(Q^2 + 1)$$

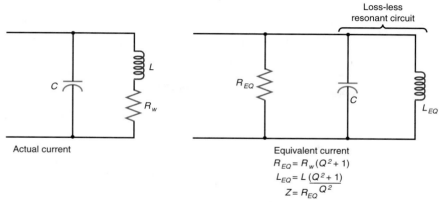

Actual current

Equivalent current
$R_{EQ} = R_w (Q^2 + 1)$
$L_{EQ} = L \, \underline{(Q^2 + 1)}$
$Z = R_{EQ} \, Q^2$

FIG. 2-25 An equivalent circuit makes parallel resonant circuits easier to analyze.

Q is determined by the formula

$$Q = \frac{X_L}{R_W}$$

where R_w is the coil winding resistance.

If Q is high, usually more than 10, L_{eq} is approximately equal to the actual inductance value L. The total impedance of the circuit at resonance is equal to the equivalent parallel resistance:

$$Z = R_{eq}$$

Example 2-20

What is the impedance of a parallel LC circuit with a resonant frequency of 52 MHz and a Q of 12? $L = .15 \ \mu H$.

$$Q = \frac{X_L}{R_w}$$

$$X_L = 2\pi f L = 6.28 \ (52 \times 10^6)(.15 \times 10^{-6}) = 49\Omega$$

$$R_w = \frac{X_L}{Q} = \frac{49}{12} = 4.1 \ \Omega$$

$$Z = R_{eq} = R_w \ (Q^2 + 1) = 4.1(12^2 + 1) = 4.1 \ (145) = 592 \ \Omega$$

If the Q of the parallel resonant circuit is greater than 10, the following simplified formula can be used to calculate the resistive impedance at resonance:

$$Z = \frac{L}{CR_w}$$

The value of R is the winding resistance of the coil.

Example 2-21

Calculate the impedance of the circuit given in Example 2-20 using the formula $Z = L/CR_w$.

$$f_r = 52 \text{ MHz} \qquad R_w = 4.1 \ \Omega \qquad L = 0.15 \ \mu\text{H}$$

$$C = \frac{1}{4\pi^2 f_r^2 L} = \frac{1}{[39.478(52 \times 10^6)^2(0.15 \times 10^{-6})]}$$

$$= 6.245 \times 10^{-11}$$

$$Z = L/CR_w = \frac{0.15 \times 10^{-6}}{(62.35 \times 10^{-12})(4.1)} = 586 \ \Omega$$

This is close to the previously computed value of 592 Ω. The formula $Z = L/CR_w$ is an approximation.

A frequency and phase response curve of a parallel resonant circuit is shown in Fig. 2-26. Below the resonant frequency, X_L is less than X_C; thus the inductive current is greater than the capacitive current and the circuit appears inductive. The line current lags the applied voltage. Above the resonant frequency, X_c is less than X_L; thus the capacitive current is less than the inductive current and the circuit appears capacitive. Therefore, the line current leads the applied voltage.

At the resonant frequency, the impedance of the circuit peaks. This means that the line current at that time is at its minimum. At resonance, the circuit appears to have a very high resistance and the small line current is in phase with the applied voltage.

Note that the Q of a parallel circuit, which was previously expressed as $Q = X_L/R_w$, can also be computed with the expression

$$Q = \frac{R_P}{X_L}$$

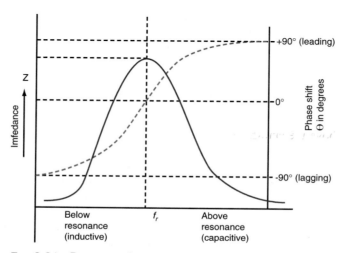

FIG. 2-26 Response of a parallel resonant circuit.

where R_P is the equivalent parallel resistance R_{eq} is parallel with any other parallel resistance and X_L is the inductive reactance of the equivalent inductance L_{eq}.

2-3 FILTERS

A *filter* is a frequency-selective circuit. Filters are designed to pass some frequencies and reject others. The series and parallel resonant circuits reviewed in the previous section are examples of filters.

There are numerous ways to implement filter circuits. Simple filters created by using resistors and capacitors or inductors and capacitors are called *passive filters* because they use passive components that do not amplify. In communications work, most filters are of the passive *LC* variety, although many other types are used. Some special types of filters are active filters which use *RC* networks with feedback in op amp circuits, switched capacitor filters, crystal and ceramic filters, surface acoustic wave (SAW) filters, and digital filters implemented with digital signal processing (DSP) techniques.

The five basic kinds of filter circuits are:

Low-pass filter. Passes frequencies below a critical frequency called the cutoff frequency and greatly attenuates those above the cutoff frequency.
High-pass filter. Passes frequencies above the cutoff but rejects those below it.
Bandpass filter. Passes frequencies over a narrow range between lower and upper cutoff frequencies.
Band-reject filter. Rejects or stops frequencies over a narrow range but allows frequencies above and below to pass. *notch filter*
All-pass filter. Passes all frequencies equally well over its design range but has a fixed or predictable phase shift characteristic.

RC FILTERS

RC filters use combinations of resistors and capacitors to achieve the desired response. Most *RC* filters are of the low-pass or high-pass type. Some band-reject or notch filters are also made with *RC* circuits. Bandpass filters can be made by combining low-pass and high-pass *RC* sections, but this is rarely done.

LOW-PASS FILTER. A low-pass filter is a circuit that introduces no attenuation at frequencies below the cutoff frequency but completely eliminates all signals with frequencies above the cutoff. Low-pass filters are sometimes referred to as *high cut filters.*

The ideal response curve for a low-pass filter is shown in Fig. 2-27. This response curve cannot be realized in practice. In practical circuits, instead of a sharp transition at the cutoff frequency, there is a more gradual transition between little or no attenuation and maximum attenuation.

The simplest form of low-pass filter is the *RC* circuit shown in Fig. 2-28(*a*). The circuit forms a simple voltage divider with one frequency-sensitive component, in this case the capacitor. At very low frequencies, the capacitor has very high reactance com-

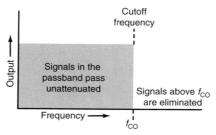

FIG. 2-27   Ideal response curve of a low-pass filter.

pared to the resistance and therefore the attenuation is minimum. As the frequency increases, the capacitive reactance decreases. When the reactance becomes smaller than the resistance, the attenuation increases rapidly. The frequency response of the basic circuit is illustrated in Fig. 2-28(b). The cutoff frequency of this filter is that point where R and X_C are equal. The cutoff frequency, also known as the critical frequency, is determined by the expression

$$f_{co} = \frac{1}{2\pi RC}$$

For example, if $R = 4.7k\ \Omega$ and $C = 560$ pF, the cutoff frequency is

$$f_{co} = \frac{1}{2\pi(4700)(560 \times 10^{-12})} = 60469 \text{ Hz} \quad \text{or} \quad 60.5 \text{ kHz}$$

Example 2-22

What is the cutoff frequency of a single-section RC low-pass filter with $R = 8.2$ kΩ and $C = 0.0033$ μF?

$$f_{co} = \frac{1}{2\pi RC} = \frac{1}{2\pi(8.2 \times 10^3)(0.0033 \times 10^{-6})}$$

$$= 5884.54 \text{ Hz or } 5.88 \text{ kHz}$$

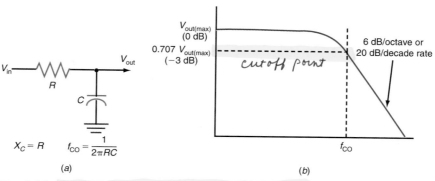

FIG. 2-28   *RC* low-pass filter. (*a*) Circuit. (*b*) Low-pass filter.

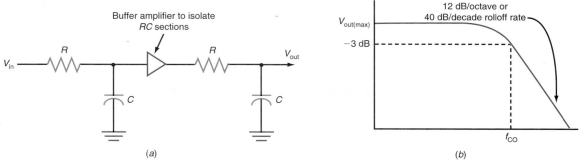

(a)

(b)

FIG. 2-29 Two stages of *RC* filter improves the response but increases signal loss. (*a*) Circuit. (*b*) Response curve.

At the cutoff frequency, the output amplitude is 70.7 percent of the input amplitude at lower frequencies. This is the so-called 3-dB down point. In other words, this filter has a voltage gain of -3 dB at the cutoff frequency. At frequencies above the cutoff frequency, the amplitude decreases at a linear rate of 6 dB per octave or 20 dB per decade. An *octave* is defined as a doubling or halving of frequency, while a *decade* represents a one-tenth or *times 10* relationship. Assume that a filter has a cutoff of 600 Hz. If the frequency doubles to 1200 Hz, the attenuation will increase by 6 dB, or from 3 dB at cutoff to 9 dB at 1200 Hz. If the frequency increased by a factor of 10 from 600 Hz to 6 kHz, the attenuation would increase by a factor of 20 dB from 3 dB at cutoff to 23 dB at 6 kHz.

If a faster rate of attenuation is required, two *RC* sections set to the same cutoff frequency can be used. Such a circuit is shown in Fig. 2-29(*a*). With this circuit, the rate of attenuation is 12 dB per octave or 40 dB per decade. Two identical *RC* circuits are used, but an isolation or buffer amplifier such as an emitter-follower (gain $\approx$ 1) is used between them to prevent the second section from loading the first. Cascading two *RC* sections without the isolation will give an attenuation rate less than the theoretically ideal 12-dB octave because of the loading effects.

With a steeper attenuation curve, the circuit is said to be more selective. The disadvantage of cascading such sections is that higher attenuation makes the output signal considerably smaller. This signal attenuation in the passband of the filter is called *insertion loss*.

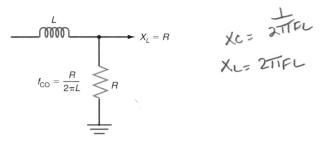

FIG. 2-30 A low-pass filter implemented with an inductor.

A low-pass filter can also be implemented with an inductor and a resistor, as shown in Fig. 2-30. The response curve for this *RL* filter is the same as that shown in Fig. 2-28(*b*). The cutoff frequency is determined using the formula

$$f_{co} = \frac{R}{2\pi L}$$

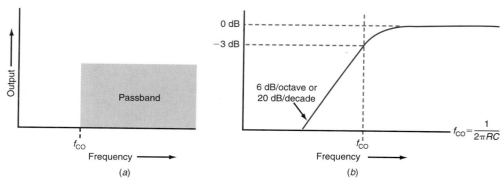

FIG. 2-31 Frequency response curve of a high-pass filter. (*a*) Ideal. (*b*) Practical.

RL low-pass filters are not as widely used as *RC* filters because inductors are usually larger, heavier, and more expensive than capacitors. Inductors also have greater loss than capacitors because of their inherent winding resistance.

HIGH-PASS FILTER. A high-pass filter passes frequencies above the cutoff frequency with little or no attenuation but greatly attenuates those signals below the cutoff. The ideal high-pass response curve is shown in Fig. 2-31(*a*). Approximations to the ideal response curve shown in Fig. 2-31(*b*) can be obtained with a variety of *RC* and *LC* filters.

The basic *RC* high-pass filter is shown in Fig. 2-32(*a*). Again, it is nothing more than a voltage divider with the capacitor serving as the frequency-sensitive component in a voltage divider. At low frequencies, X_C is very high. When X_C is much higher than *R,* the voltage divider effect provides high attenuation of the low-frequency signals. As the frequency increases, the capacitive reactance decreases. When the capacitive reactance is equal to or less than the resistance, the voltage divider gives very little attenuation. Therefore, high frequencies pass relatively unattenuated.

The cutoff frequency for this filter is the same as that for the low-pass circuit and is derived from setting X_C equal to *R* and solving for frequency:

$$f_{\text{co}} = \frac{1}{2\pi RC}$$

The roll-off rate is 6 dB per octave or 20 dB per decade.

A high-pass filter can also be implemented with a coil and a resistor, as shown in Fig. 2-32(*b*). The cutoff frequency is

$$f_{\text{co}} = \frac{R}{2\pi L}$$

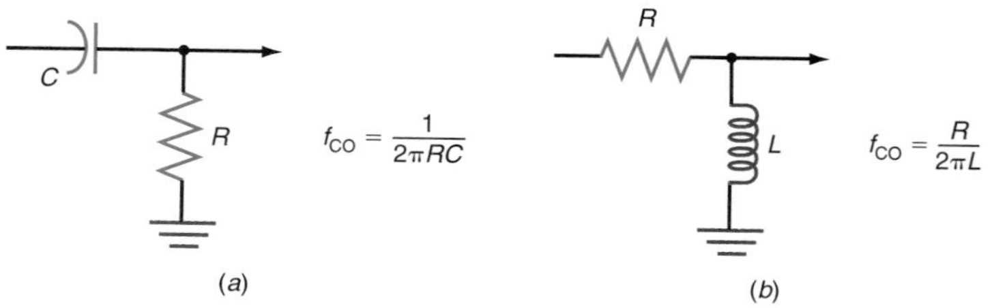

$$f_{\text{co}} = \frac{1}{2\pi RC} \qquad\qquad f_{\text{co}} = \frac{R}{2\pi L}$$

(*a*) (*b*)

FIG. 2-32 (*a*) *RC* high-pass filter. (*b*) *RL* high-pass filter.

The response curve for this filter is the same as that shown in Fig. 2-31(b). The rate of attenuation is 6 dB per octave or 20 dB per decade, as was the case with the low-pass filter. Again, improved attenuation can be obtained by cascading filter sections.

Example 2-23

What is the closest standard EIA resistor value that will produce a cutoff frequency of 3.4 kHz with a 0.047-μF capacitor in a high-pass RC filter?

$$f_{co} = \frac{1}{2\pi RC}$$

$$R = \frac{1}{2\pi f_{co}C} = \frac{1}{2\pi(3.4 \times 10^3)(0.047 \times 10^{-6})} = 996 \ \Omega$$

The closest standard values are 910 Ω and 1000 Ω, with 1000 being the closest.

RC NOTCH FILTER. Notch filters are also referred to as *bandstop* or *band-reject* filters. Band-reject filters are used to greatly attenuate a narrow range of frequencies around a center point. Notch filters accomplish the same purpose, but for a single frequency.

A simple notch filter that is implemented with resistors and capacitors as shown in Fig. 2-33(a) is called a *parallel-T* or *twin-T* notch filter. This filter is a variation of a bridge circuit. Recall that in a bridge circuit the output is zero if the bridge is balanced. If the component values are precisely matched, the circuit will be in balance and produce an attenuation of an input signal at the design frequency as high as 30 to 40 dB. A typical response curve is shown in Fig. 2-33(b).

The center notch frequency is computed with the formula

$$f_{notch} = \frac{1}{2\pi RC}$$

DID YOU KNOW?

Twin-T notch filters are used at low frequencies to eliminate power line hum from audio circuits and medical equipment amplifiers.

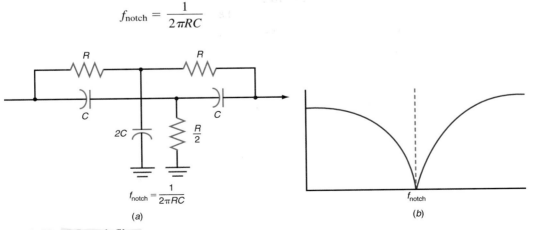

$$f_{notch} = \frac{1}{2\pi RC}$$

(a)

f_{notch}

(b)

FIG. 2-33 *RC* notch filter.

For example, if the values of resistance and capacitance are 100 kΩ and 0.02 μF, the notch frequency is

$$f_{notch} = \frac{1}{6.28(10^5)(0.02 \times 10^{-6})} = 79.6 \text{ Hz}$$

Twin-T notch filters are used primarily at low frequencies, audio and below. A common use is to eliminate 60-Hz power-line hum from audio circuits and low-frequency medical equipment amplifiers. The key to high attenuation at the notch frequency is precise component values. The resistors and capacitor values must be matched to achieve high attenuation.

Example 2-24

What values of capacitors would you use in an *RC* twin-T notch filter to remove 120 Hz if $R = 220$ kΩ?

$$f_{notch} = \frac{1}{2\pi RC}$$

$$C = \frac{1}{2\pi f_{notch} R} = \frac{1}{6.28(120)(220 \times 10^3)}$$
$$= 6.03 \times 10^{-9} = 6.03 \text{ nF or } 0.006 \ \mu\text{F}$$
$$2C = 0.012 \ \mu\text{F}$$

LC FILTERS

RC filters are used primarily at the lower frequencies. They are very common at audio frequencies but rarely used above about 100 kHz. At radio frequencies, their passband attenuation is just too great, and the cutoff slope is too gradual. It is more common to see *LC* filters made with inductors and capacitors. Inductors for lower frequencies are large, bulky, and expensive, but those used at higher frequencies are very small, light, and inexpensive. Over the years, a multitude of filter types have been developed. These include the basic constant-*k* and *m*-derived filters as well as many special variations that are designed to improve some particular characteristic of the filter.

FILTER TERMINOLOGY. When working with filters, you will hear a variety of terms to describe the operation and characteristics of filters. The following definitions will help you understand filter specifications and operation.

1. *Passband.* This is the frequency range over which the filter passes signals. It is the frequency range between the cutoff frequencies or between the cutoff frequency and zero (for low pass) or between the cutoff frequency and infinity (for high pass).
2. *Stop band.* This is the range of frequencies outside the passband, that is, the range of frequencies that is greatly attenuated by the filter. Frequencies in this range are rejected.

Q — quality

3. **Attenuation.** This is the amount by which undesired frequencies in the stop band are reduced. It can be expressed as a power ratio or voltage ratio of the output to the input. Attenuation is usually given in decibels.

4. **Insertion loss.** Insertion loss is the loss the filter introduces to the signals in the passband. Passive filters introduce attenuation because of the resistive losses in the components. Insertion loss is typically given in decibels.

5. **Impedance.** Impedance is the resistive value of the load and source terminations of the filter. Filters are usually designed for specific driving source and load impedances that must be present for proper operation.

6. **Ripple.** Amplitude variation with frequency in the passband, or the repetitive rise and fall of the signal level in the passband of some types of filters, is known as ripple. It is usually measured in decibels. There may also be ripple in the stop bandwidth in some types of filters.

7. **Shape factor.** Shape factor, also known as bandwidth ratio, is the ratio of the pass bandwidth to the stop bandwidth of a bandpass filter. It compares the bandwidth at minimum attenuation, usually at the −3-dB points or cutoff frequencies, to that of maximum attenuation and thus gives a relative indication of attenuation rate or selectivity. The smaller the ratio, the greater the selectivity. The ideal is a ratio of 1, which in general cannot be obtained with practical filters. The filter in Fig. 2-34 has a bandwidth of 6 kHz at the 3-dB attenuation point and a bandwidth of 14 kHz at the −40-dB attenuation point. The shape factor then is 14 kHz/6 kHz = 2.333. The points of comparison vary with different filters and manufacturers. The points of comparison may be at the 6-dB down and 60-dB down points or at any other designated two levels.

8. **Pole.** A pole is a frequency at which there is a high impedance in the circuit. It is also used to describe one *RC* or *LC* section of a filter. A simple low-pass *RC* filter like that in Fig. 2-28(*a*) has one pole. The two-section filter in Fig. 2-29 has two poles.

9. **Zero.** This term is used to refer to a frequency at which there is zero impedance in the circuit.

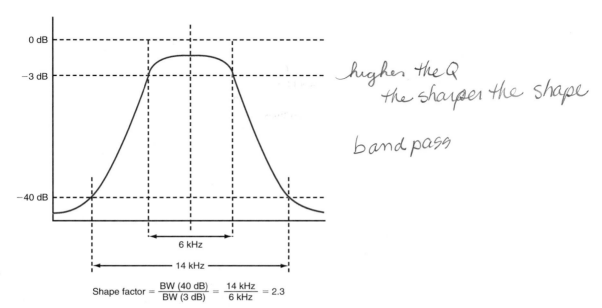

higher the Q the sharper the shape

band pass

$$\text{Shape factor} = \frac{\text{BW (40 dB)}}{\text{BW (3 dB)}} = \frac{14 \text{ kHz}}{6 \text{ kHz}} = 2.3$$

Fig. 2-34 Shape factor.

10. *Envelope delay.* Also known as *time delay,* envelope delay is the time it takes for a specific point on an input waveform to pass through the filter.

11. *Roll-off.* Also called the attenuation rate, roll-off is the rate of change of amplitude with frequency in a filter. The faster the roll-off, or the higher the attenuation rate, the more selective the filter is, that is, the better able it is to differentiate between two closely spaced signals, one desired and the other not.

LC LOW-PASS FILTERS. There are two basic types of *LC* filters: constant-k and m-derived filters. Constant-k filters make the product of the capacitive and inductive reactances a constant value k. This kind of filter has constant resistive input and output impedances. *M*-derived filters use a tuned circuit in the filter to introduce a point of infinite attenuation for the purpose of making the roll-off or attenuation rate faster. The rate of attenuation is a function of the ratio of the filter cutoff frequency to the frequency of infinite attenuation, or m.

Constant-k low-pass filters can be implemented in three basic ways, as shown in Fig. 2-35. The circuit in Fig. 2-35(a) is called a half, or L-section, filter. T- and π-section versions are shown in Figs. 2-35(b) and 2-35(c), respectively. Note the values of L and C. The filters must be driven by a generator with an internal impedance equal to R_L and terminated by its characteristic or load impedance R_L, which is determined by the expression

$$R_L = k = \sqrt{\frac{L}{C}}$$

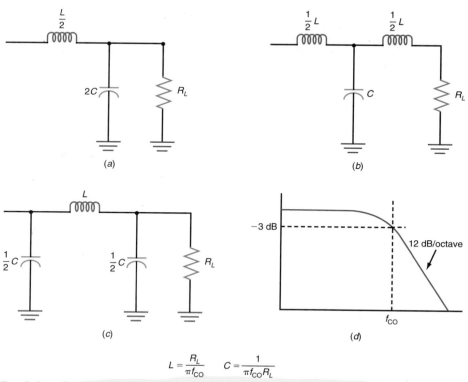

$$L = \frac{R_L}{\pi f_{CO}} \quad C = \frac{1}{\pi f_{CO} R_L}$$

FIG. 2-35 Constant-k low-pass filter. (a) L section. (b) T section. (c) π section. (d) Response curve.

The -3-dB down cutoff frequency of these circuits is

$$f_{co} = \frac{1}{4\pi\sqrt{LC}}$$

The roll-off rate of the filters in Fig. 2-35 is 12 dB per octave, as shown in Fig. 2-35(d).

Additional sections of these filters may be cascaded to obtain higher attenuation of the frequencies above cutoff and a steeper roll-off curve for better selectivity.

Example 2-25

Calculate the values for inductance and capacitance needed for a π-section low-pass constant-k filter using a 50-Ω load at 40 MHz. Refer to the design equations in Fig. 2-35.

$$L = \frac{R_L}{\pi f_{co}} = \frac{50}{3.14(40 \times 10^6)} = 3.98 \times 10^{-7} = 398 \times 10^{-9}$$
$$= 398 \text{ nH}$$
$$C = \frac{1}{\pi f_{co}R_L} = \frac{1}{3.14(40 \times 10^6)(50)} = 1.592 \times 10^{-10}$$
$$= 1.592 \times 10^{-12} = 159.2 \text{ pF}$$

Since a π-section filter is specified, each capacitor is $C/2 = 159.2/2 = 79.61$ pF.

When good selectivity near the cutoff point is required, the low-pass filter can be modified by including a tuned or resonant circuit. Two examples of such m-derived filters are shown in Fig. 2-36. In Fig. 2-36(a), a parallel LC circuit forms a high impedance at a frequency near the cutoff. This provides a deep attenuation notch near the

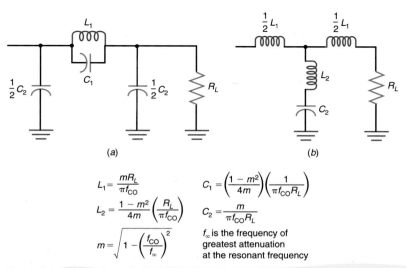

$$L_1 = \frac{mR_L}{\pi f_{co}}$$

$$L_2 = \frac{1-m^2}{4m}\left(\frac{R_L}{\pi f_{co}}\right)$$

$$m = \sqrt{1 - \left(\frac{f_{co}}{f_\infty}\right)^2}$$

$$C_1 = \left(\frac{1-m^2}{4m}\right)\left(\frac{1}{\pi f_{co}R_L}\right)$$

$$C_2 = \frac{m}{\pi f_{co}R_L}$$

f_∞ is the frequency of greatest attenuation at the resonant frequency

FIG. 2-36 Low-pass m-derived filters. (a) π-type network. (b) T-type network.

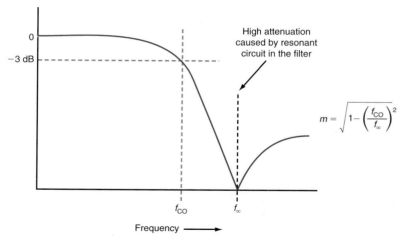

FIG. 2-37 Frequency response of an *m*-derived low-pass filter.

cutoff frequency, as shown in Fig. 2-37. The same effect can be obtained with a series resonant circuit placed in shunt with the output as shown in Fig. 2-36(*b*). *M*-derived filters provide extra steepness of the response curve and thereby improve selectivity. The main disadvantage is that there is less attenuation beyond the cutoff point because of the effect of the resonant circuit.

The value of *m*, typically in the range from 0.5 to 0.9, is determined by the ratio of the cutoff frequency to the frequency of infinite attenuation:

$$m = \sqrt{1 - \left(\frac{f_{co}}{f_\infty}\right)^2}$$

The higher the value of *m*, the greater the selectivity but the greater the output beyond the cut-off frequency. A common value of *m* is 0.6, which is a good balance between selectivity and attenuation above cutoff.

LC High-Pass Filters. Three constant-*k* *LC* high-pass filters are shown in Fig. 2-38. The formulas for calculating these filters are given in the illustration. Some *m*-derived high-pass filter circuits are shown in Fig. 2-39. The response curve

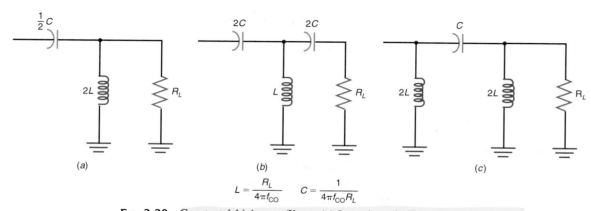

$$L = \frac{R_L}{4\pi f_{co}} \qquad C = \frac{1}{4\pi f_{co} R_L}$$

FIG. 2-38 Constant-*k* high-pass filters. (*a*) L section. (*b*) T section. (*c*) π section.

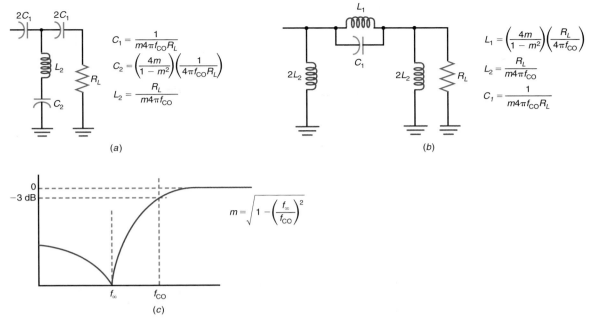

$$C_1 = \frac{1}{m4\pi f_{co}R_L}$$

$$C_2 = \left(\frac{4m}{1-m^2}\right)\left(\frac{1}{4\pi f_{co}R_L}\right)$$

$$L_2 = \frac{R_L}{m4\pi f_{co}}$$

$$L_1 = \left(\frac{4m}{1-m^2}\right)\left(\frac{R_L}{4\pi f_{co}}\right)$$

$$L_2 = \frac{R_L}{m4\pi f_{co}}$$

$$C_1 = \frac{1}{m4\pi f_{co}R_L}$$

(a)

(b)

$$m = \sqrt{1 - \left(\frac{f_\infty}{f_{co}}\right)^2}$$

(c)

FIG. 2-39 High-pass m-derived filters and their response. (a) T-type network. (b) π-type network. (c) Response curve.

is shown in Fig. 2-39(c). Both parallel and series LC sections are combined with the other components to produce a deep attenuation point to improve selectivity.

There are two important things to remember about filter circuits such as these. First, LC circuits are designed to work with specific input and output impedances, and if the correct input amount impedances are not provided, the correct filtering results

Example 2-26

Calculate the values of L and C for a T-type m-derived high-pass filter with $m = 0.6$, $f_{co} = 28$ MHz, and $R_L = 75$ Ω.

$$L_2 = \frac{R_L}{m4\pi f_{co}} = \frac{75}{(0.6)(4)(3.14)(28 \times 10^6)} = 3.55 \times 10^{-7}$$
$$= 0.355 \times 10^{-6} = 0.355 \ \mu\text{H or } 355 \text{ nH}$$

$$C_1 = \frac{1}{m4\pi f_{co}R_L} = \frac{1}{(0.6)(4)(3.14)(28 \times 10^6)(75)}$$
$$= 6.32 \times 10^{-12} = 63.2 \text{ pF}$$
$$2C_1 = 2(63.2) = 126 \text{ pF}$$

$$C_2 = \frac{4m}{1 - m^2}\frac{1}{4\pi f_{co}R_C}$$

$$= \frac{4(0.6)}{1 - (0.6)^2}\frac{1}{4(3.14)(28 \times 10^6)(75)}$$
$$= (3.75)(3.79 \times 10^{-11}) = 1.42 \times 10^{-10}$$
$$= 142 \times 10^{-12} \text{ F or } 142 \text{ pF}$$

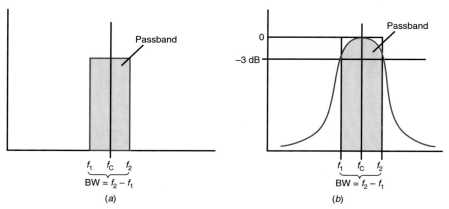

Fig. 2-40 Response curves of a bandpass filter. (*a*) Practical. (*b*) Ideal.

will not be obtained. Second, the amount of attenuation and selectivity is dependent not only upon the filter configuration but also upon the number of different *L*, *T*, or *π* sections cascaded. The desired level of selectivity and attenuation can be obtained by cascading sections until the desired level is reached. However, overall signal attenuation in the passband is increased. It is often necessary to add gain to the circuit to offset the high attenuation introduced by the circuit in the passband. This attenuation is referred to as *insertion loss*.

BANDPASS FILTERS. A bandpass filter is one that allows a narrow range of frequencies around a center frequency f_c to pass with minimum attenuation but rejects frequencies above and below this range. The ideal response curve of a bandpass filter is shown in Fig. 2-40(*a*). It has both upper and lower cutoff frequencies, f_2 and f_1, as indicated. The bandwidth of this filter is the difference between the upper and lower cutoff frequencies, or BW = $f_2 - f_1$. Frequencies above and below the cutoff frequencies are eliminated.

The ideal response curve is not obtainable with practical circuits, but close approximations can be obtained. A practical bandpass filter response curve is shown in Fig. 2-40(*b*). The simple series and parallel resonant circuits described in the previous section have a response curve like that in the figure and make good bandpass filters. The cutoff frequencies are those at which the output voltage is down 0.707 percent from the peak output value. These are the 3-dB attenuation points.

Two types of bandpass filters are shown in Fig. 2-41. In Fig. 2-41(*a*), a series resonant circuit is connected in series with an output resistor, forming a voltage divider. At frequencies above and below the resonant frequency, either the inductive or capacitive reactance will be high compared to the output resistance. Therefore, the output

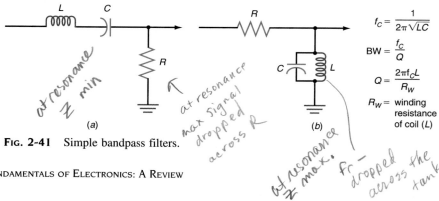

Fig. 2-41 Simple bandpass filters.

$$f_c = \frac{1}{2\pi\sqrt{LC}}$$

$$BW = \frac{f_c}{Q}$$

$$Q = \frac{2\pi f_c L}{R_W}$$

R_W = winding resistance of coil (*L*)

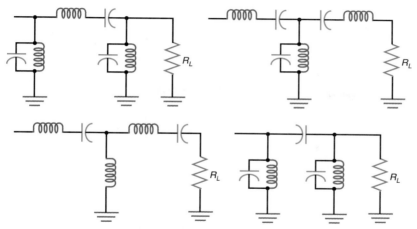

$$f_R = \frac{1}{2\pi\sqrt{LC}}$$

Fig. 2-42 Some common bandpass filter circuits.

amplitude will be very low. However, at the resonant frequency, the inductive and capacitive reactances cancel, leaving only the small resistance of the inductor. Thus most of the input voltage appears across the larger output resistance. The response curve for this circuit is shown in Fig. 2-40(*b*). Remember that the bandwidth of such a circuit is a function of the resonant frequency and *Q*: BW = f_c/Q.

A parallel resonant bandpass filter is shown in Fig. 2-41(*b*). Again, a voltage divider is formed with resistor *R* and the tuned circuit. This time the output is taken from across the parallel resonant circuit. At frequencies above and below the center resonant frequency, the impedance of the parallel tuned circuit is low compared to that of the resistance. Therefore, the output voltage is very low. Frequencies above and below the center frequency are greatly attenuated. At the resonant frequency, the reactances are equal and the impedance of the parallel tuned circuit is very high compared to that of the resistance. Therefore, most of the input voltage appears across the tuned circuit. The response curve is similar to that shown in Fig. 2-40(*b*).

Improved selectivity with steeper "skirts" on the curve can be obtained by cascading several bandpass sections. Several ways to do this are shown in Fig. 2-42. As sections are cascaded, the bandwidth becomes narrower and the response curve becomes steeper. An example is shown in Fig. 2-43. As indicated earlier, using multiple

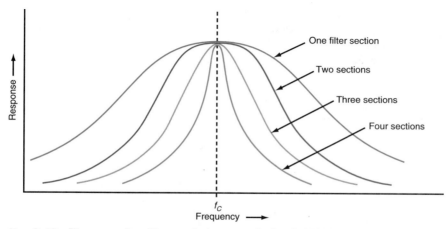

Fig. 2-43 How cascading filter sections narrow the bandwidth and improve selectivity.

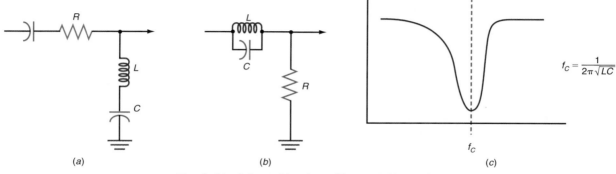

$$f_c = \frac{1}{2\pi\sqrt{LC}}$$

(a)   (b)   (c)

FIG. 2-44 *LC* tuned bandstop filters. (*a*) Shunt. (*b*) Series. (*c*) Response curve.

filter sections greatly improves the selectivity but increases the passband attenuation (insertion loss) which must be offset by added gain.

BAND-REJECT FILTERS. Band-reject filters, also known as bandstop filters, reject a narrow band of frequencies around a center of notch frequency. Two typical *LC* bandstop filters are shown in Fig. 2-44. In Fig. 2-44(*a*), the series *LC* resonant circuit forms a voltage divider with input resistor *R*. At frequencies above and below the center rejection or notch frequency, the *LC* circuit impedance is high compared to that of the resistance. Therefore, signals at frequencies above and below center frequency are passed with minimum attenuation. At the center frequency, the tuned circuit resonates, leaving only the small resistance of the inductor. This forms a voltage divider with the input resistor. Since the impedance at resonance is very low compared to the resistor, the output signal is very low in amplitude. A typical response curve is shown in Fig. 2-44(*c*).

A parallel version of this circuit is shown in Fig. 2-44(*b*), where the parallel resonant circuit is connected in series with a resistor from which the output is taken. At frequencies above and below the resonant frequency, the impedance of the parallel circuit is very low; there is, therefore, little signal attenuation, and most of the input voltage appears across the output resistor. At the resonant frequency, the parallel *LC* circuit has an extremely high resistive impedance compared to the output resistance and so minimum voltage appears at the center frequency. *LC* filters used in this way are often referred to as *traps.*

Another bridge-type notch filter is the bridge-T filter shown in Fig. 2-45(*c*). This filter, which is widely used in RF circuits, uses inductors and capacitors and thus has a steeper response curve than the *RC* twin-T notch filter. Since *L* is variable, the notch is tunable.

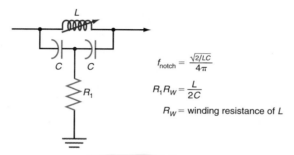

$$f_{notch} = \frac{\sqrt{2/LC}}{4\pi}$$

$$R_1 R_w = \frac{L}{2C}$$

R_w = winding resistance of *L*

FIG. 2-45 Bridge-T notch filter.

Filters are one of the key building blocks of communications systems. The major types of *LC* filters in use are named after the person who discovered and developed the analysis and design method for each filter. The most widely used filters are Butterworth, Chebyshev, Cauer (elliptical), and Bessel. Each can be implemented in constant-*k* and *m*-derived forms.

Butterworth: This type of filter has maximum flatness in response in the passband and a uniform attenuation with frequency. The attenuation just outside the passband is not as great as can be achieved with other types of filters. See Fig. 2-46 for an example of a low-pass Butterworth filter.

Chebyshev: Chebyshev (or Tchebyschev) filters have extremely good selectivity; that is, their attenuation rate or roll-off is high, much higher than that of the Butterworth filter (see Fig. 2-46). The attenuation just outside the passband is also very high—again, better than the Butterworth. The main problem with the Chebyshev filter is that it has ripple in the passband, as is evident from the figure. The response is not flat or constant, as it is with the Butterworth filter. This may be a disadvantage in some applications.

Cauer (Elliptical): Cauer filters produce an even greater attenuation or roll-off rate than Chebyshev filters and greater attenuation out of the passband. However, they do this with an even higher ripple in the passband as well as outside of the passband.

Bessel: Also called *Thomson* filters, these circuits provide the desired frequency response (i.e., low-pass, bandpass, etc.) but have a constant time delay in the passband. Bessel filters have what is known as a *flat group delay:* as the signal frequency varies in the passband, the phase shift or time delay it introduces is constant. In some applications constant group delay is necessary to prevent distortion of the signals in the passband due to varying phase shifts with frequency. Filters that must pass pulses are an example. To achieve this desired response, the Bessel filter has lower attenuation just outside the passband.

Whatever the type, passive filters are usually designed and built with discrete components. It is not practical to put them into IC form. A number of filter design software packages are available to simplify and speed up the design process. However, filters can also be purchased as components. These filters are predesigned and packaged in small sealed housings with only input, output, and ground terminals and can be used just like integrated circuits. A wide range of frequencies, response characteristics, and attenuation rates can be obtained.

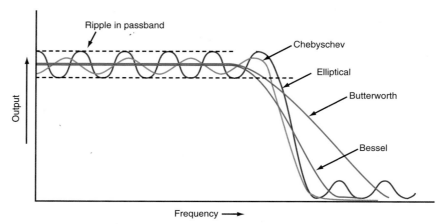

FIG. 2-46 Butterworth, elliptical, Bessel, and Chebyschev response curves.

ACTIVE FILTERS

Active filters are frequency-selective circuits that incorporate *RC* networks and amplifiers with feedback to produce low-pass, high-pass, bandpass, and bandstop performance. These filters can replace standard passive *LC* filters in many applications. They offer the following advantages over standard passive *LC* filters.

1. *Gain.* Because active filters use amplifiers, they can be designed to amplify as well as filter, thus offsetting any insertion loss.
2. *No inductors.* Inductors are usually larger, heavier, and more expensive than capacitors and have greater losses. Active filters use only resistors and capacitors.
3. *Easy to tune.* Because selected resistors can be made variable, the filter cutoff frequency, center frequency, gain, *Q*, and bandwidth are adjustable.
4. *Isolation.* The amplifiers provide very high isolation between cascaded circuits because of the amplifier circuitry, thereby decreasing interaction between filter sections.
5. *Easier impedance matching.* Impedance matching is not as critical as with *LC* filters.

Figure 2-47 shows two types of low-pass active filters and two types of high-pass active filters. Note that these active filters use op amps to provide the gain. The voltage divider, made up of R_1 and R_2, sets the circuit gain in the circuits of Figs. 2-47(*A*) and (*C*) as in any noninverting op amp. The gain is set by R_3 and/or R_1 in Fig. 2-47(*b*) and by C_3 and/or C_1 in Fig. 2-47(*d*). All circuits have what is called a *second-order response,* which means that they provide the same filtering action as an L, T, or π constant-*k LC* filter. The roll-off rate is 12 dB per octave, or 40 dB per decade. Multiple filters can be cascaded to provide faster roll-off rates.

Two active band-pass filters and a notch filter are shown in Fig. 2-48. In Fig. 2-48(*a*), both *RC* low-pass and high-pass sections are combined with feedback to give a bandpass result. In Fig. 2-48(*b*), a twin-T *RC* notch filter is used with negative feedback to provide a bandpass result. A notch filter using a twin-T is illustrated in Fig. 2-48(*c*). The feedback makes the response sharper than with a standard passive twin-T.

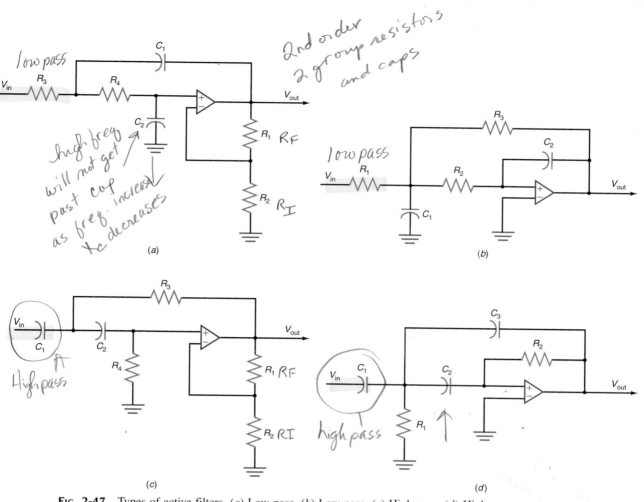

FIG. 2-47 Types of active filters. (a) Low-pass. (b) Low-pass. (c) High-pass. (d) High-pass.

A special form of active filter is the variable-state filter, which can simultaneously provide low-pass, high-pass, and bandpass operation from one circuit. The basic circuit is shown in Fig. 2-49(a). It uses op amps and RC networks and a feedback arrangement. Op amps 2 and 3 are connected as integrators or low-pass filters. Op amp 1 is connected as a summing amplifier that adds the input signal to the feedback signals from op amps 2 and 3. Note the outputs at each op amp. The center and cutoff frequencies are set by the integrator feedback capacitors and the value of R_f. R_q and R_g set the Q and gain of the circuit. The circuit can be made tunable by simultaneously varying the values of R_f.

A variation of the variable-state filter is the biquad filter shown in Fig. 49(b). It, too, uses two op amp integrators and a summing amplifier. Again, low-pass, high-pass, and bandpass characteristics are obtained simultaneously. However, the primary use of the biquad filter is bandpass filtering. Again, the center or cutoff frequencies are set by the value of the integrator feedback capacitors and the value of R_f. R_b sets the filter bandwidth, and R_g sets the circuit gain.

Active filters can be made with IC op amps and discrete RC networks. They can be designed to have any of the responses discussed earlier, such as Butterworth and Chebyshev, and they are easily cascaded to provide even greater selectivity. Active filters are also available as complete packaged components. The primary disadvantage of

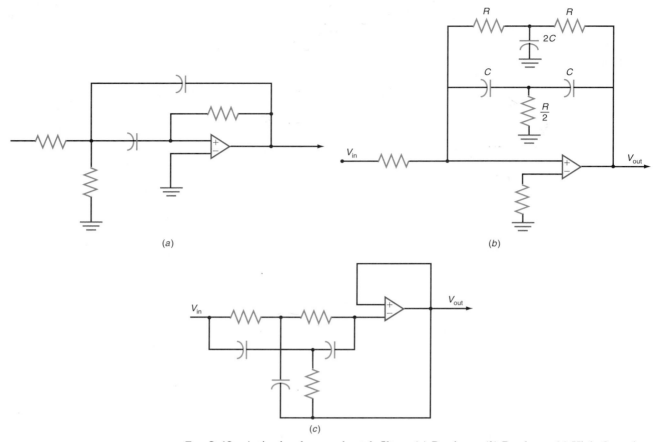

Fig. 2-48 Active bandpass and notch filters. (*a*) Bandpass. (*b*) Bandpass. (*c*) High-*Q* notch.

active filters is that their upper frequency of operation is limited by the frequency response of the op amps and the practical sizes of resistors and capacitors. Most active filters are restricted to frequencies below 1 MHz, and most active circuits operate in the audio range and slightly above.

CRYSTAL AND CERAMIC FILTERS

The selectivity of a filter is limited primarily by the *Q* of the circuits, which is generally the *Q* of the inductors used. With *LC* circuits, it is difficult to achieve *Q* values over 200. In fact, most *LC* circuit *Q*'s are in the range 10 to 100 and as a result, the roll-off rate is limited. In some applications, however, it is necessary to select one desired signal, distinguishing it from a nearby undesired signal (see Fig. 2-50). A conventional filter has a slow roll-off rate, and the undesired signal is not, therefore, fully attenuated. The way to gain greater selectivity and higher *Q,* so that the undesirable signal will be almost completely rejected, is to use filters that are made of thin slivers of quartz crystal or certain types of ceramic materials. These materials exhibit what is called *piezoelectricity.* When they are physically bent or otherwise distorted, they develop a voltage across the faces of the crystal. Alternatively, if an AC voltage is applied across the crystal or ceramic, the material vibrates at a very precise frequency, a

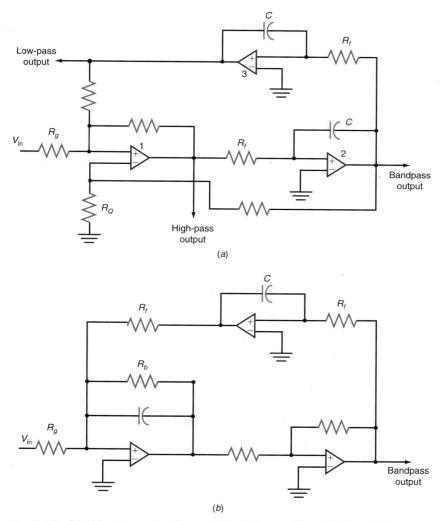

FIG. 2-49 Multifunction active filters. (*a*) Variable-state filter. (*b*) Biquad filter.

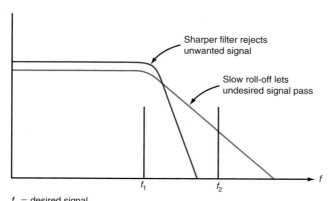

f_1 = desired signal
f_2 = undesired signal

FIG. 2-50 How selectivity affects the ability to discriminate between signals.

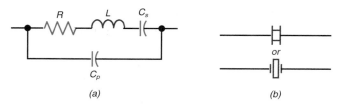

(a)   (b)

FIG. 2-51   Quartz crystal. (*a*) Equivalent circuit. (*b*) Schematic symbol.

frequency which is determined by the thickness, shape, and size of the crystal, as well as the angle of cut of the crystal faces. In general, the thinner the crystal or ceramic element, the higher the frequency of oscillation.

Crystals and ceramic elements are widely used in oscillators to set the frequency of operation to some precise value, which is held despite temperature and voltage variations that may occur in the circuit.

Crystals and ceramic elements can also be used as circuit elements to form filters, specifically bandpass filters. The equivalent circuit of a crystal or ceramic device is a tuned circuit with a *Q* of 10,000 to 1,000,000, permitting highly selective filters to be built.

CRYSTAL FILTERS.   Crystal filters are made from the same type of quartz crystals normally used in crystal oscillators. When a voltage is applied across a crystal, it vibrates at a specific resonant frequency, which is a function of the size, thickness, and direction of cut of the crystal. Crystals can be cut and ground for almost any frequency in the 100-kHz to 100-MHz range. The frequency of vibration of crystal is extremely stable, and crystals are therefore widely used to supply signals on exact frequencies with good stability.

The equivalent circuit and schematic symbol of a quartz crystal are shown in Fig. 2-51. The crystal acts as a resonant *LC* circuit. The series *LCR* part of the equivalent circuit represents the crystal itself, whereas the parallel capacitance C_P is the capacitance of the metal mounting plates with the crystal as the dielectric.

Figure 2-52 shows the impedance variations of the crystal as a function of frequency. At frequencies below the crystal's resonant frequency, the circuit appears capacitive and has a high impedance. However, at some frequency, the reactances of the equivalent inductance *L* and the series capacitance C_S are equal, and the circuit resonates. The series circuit is resonant when $X_L = X_C$. At this series resonant frequency

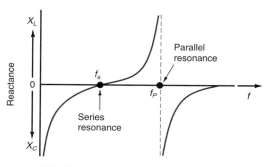

FIG. 2-52   Impedance variation with frequency of a quartz crystal.

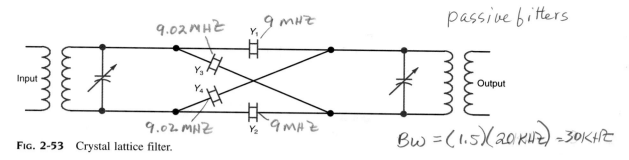

FIG. 2-53 Crystal lattice filter.

handwritten annotations on figure: 9.02 MHZ, 9 MHZ, Y_1, Y_3, Y_4, 9.02 MHZ, Y_2, 9 MHZ, passive filters

$$Bw = (1.5)(20\,KHz) = 30\,KHz$$

handwritten: -3db, 15K 30K 15K, (1.5) (1.5)

f_S, the circuit is resistive. The resistance of the crystal is extremely low, giving the circuit an extremely high Q. Values of Q in the 10,000 to 1,000,000 range are common. This makes the crystal a highly selective series resonant circuit.

If the frequency of the signal applied to the crystal is above f_S, the crystal appears inductive. At some higher frequency, the reactance of the parallel capacitance C_P equals the reactance of the net inductance. When this occurs, a parallel resonant circuit is formed. At this parallel resonant frequency f_P, the impedance of the circuit is resistive but extremely high.

Because the crystal has both series and parallel resonant frequencies which are close together, it makes an ideal component for use in filters. By combining crystals with selected series and parallel resonant points, highly selective filters with any desired bandpass can be constructed.

The most commonly used crystal filter is the full crystal lattice shown in Fig. 2-53. It is a bandpass filter. Note that transformers are used to provide the input to the filter and to extract the output. Crystals Y_1 and Y_2 resonate at one frequency, while crystals Y_3 and Y_4 resonate at another frequency. The difference between the two crystal frequencies determines the bandwidth of the filter. The 3-dB down bandwidth is approximately 1.5 times the crystal frequency spacing. For example, if the Y_1 to Y_2 frequency is 9 MHz and the Y_3 to Y_4 frequency is 9.002 MHz, the difference is 9.002 − 9.000 = 0.002 MHz = 2 kHz. The 3-dB bandwidth is, then, 1.5 × 2 kHz = 3 kHz.

The crystals are also chosen so that the parallel resonant frequency of Y_3 to Y_4 equals the series resonant frequency of Y_1 to Y_2. The series resonant frequency of Y_3 to Y_4 is equal to the parallel resonant frequency of Y_1 to Y_2. The result is a passband with extremely steep attenuation. Signals outside the passband are rejected as much as 50 to 60 dB below those inside the passband. Such a filter can easily discriminate between very closely spaced desired and undesired signals.

Another type of crystal filter is the ladder filter shown in Fig. 2-54, which is also a bandpass filter. All the crystals in this filter are cut for exactly the same frequency. The number of crystals used and the values of the shunt capacitors set the bandwidth. At least six crystals must usually be cascaded to achieve the kind of selectivity needed in communications applications.

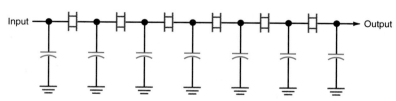

FIG. 2-54 Crystal ladder filter.

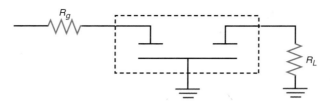

FIG. 2-55 Schematic symbol for a ceramic filter.

CERAMIC FILTERS. Ceramic is a manufactured crystal-like compound that has the same piezoelectric qualities as quartz. Ceramic disks can be made so that they vibrate at a fixed frequency, thereby providing filtering actions. Ceramic filters are very small and inexpensive and are, therefore, widely used in transmitters and receivers. Although the Q of ceramic does not have as high an upper limit as that of quartz, it is typically several thousand, which is very high compared to the Q obtainable with LC filters. Typical ceramic filters are of the bandpass type with center frequencies of 455 kHz and 10.7 MHz. These are available in different bandwidths depending upon the application. Such ceramic filters are widely used in communications receivers.

A schematic diagram of a ceramic filter is shown in Fig. 2-55. For proper operation, the filter must be driven from a generator with an output impedance of R_g and be terminated with a load of R_L. The values of R_g and R_L are usually 1.5 or 2 kΩ.

circuit is active
filter passive device
amplified before and after

SURFACE ACOUSTIC WAVE FILTERS. A special form of a crystal filter is the surface acoustic wave (SAW) filter. This fixed tuned bandpass filter is designed to provide the exact selectivity required by a given application. Figure 2-56 shows the schematic design of an SAW filter. SAW filters are made on a piezoelectric ceramic substrate such as lithium niobate. A pattern of interdigital fingers on the surface converts the signals into acoustic waves that travel across the filter surface. By controlling the shapes, sizes, and spacings of the interdigital fingers, the response can be tailored to any application. Interdigital fingers at the output convert the acoustic waves back into electrical signals.

SAW filters are normally bandpass filters used at very high radio frequencies where selectivity is difficult to obtain. Their common useful range is from 10 MHz to 1 GHz. They have a low shape factor, giving them exceedingly good selectivity at such high frequencies. They do have a significant insertion loss, usually in the 10- to 35-dB range, which must be overcome with an accompanying amplifier. SAW filters are widely used in modern TV receivers and in radar receivers.

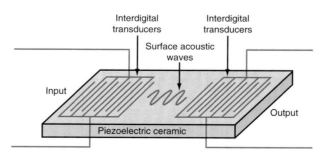

FIG. 2-56 A surface acoustic wave (SAW) filter.

SWITCHED CAPACITOR FILTERS

Switched capacitor filters (SCFs) are active IC filters made of op amps, capacitors, and transistor switches. Also known as *analog sampled data filters* or *commutating filters,* these devices are usually implemented with MOS or CMOS circuits. They can be designed to operate as high-pass, low-pass, bandpass, or bandstop filters. The primary advantage of SCFs is that they provide a way to make tuned or selective circuits in an IC without the use of discrete inductors, capacitors, or resistors.

Switched capacitor filters are made of op amps, MOSFET switches, and capacitors. All components are fully integrated on a single chip, making external discrete components unnecessary. The secret to the SCF is that all resistors are replaced by capacitors that are switched by MOSFET switches. Resistors are more difficult to make in IC form and take up far more space on the chip than transistors and capacitors. With switched capacitors, it is possible to make complex active filters on a single chip. Other advantages are selectibility of filter type, full adjustability of the cutoff or center frequency, and full adjustability of bandwidth. One filter circuit can be used for many different applications and set to a wide range of frequencies and bandwidths.

SWITCHED INTEGRATORS. The basic building block of SCFs is the classic op amp integrator, as shown in Fig. 2-57(*a*). The input is applied through a resistor and the feedback is provided by a capacitor. With this arrangement, the output is a function of the integral of the input:

$$V_{\text{out}} = \int \frac{-1}{RC} \, V_{\text{in}} \, dt$$

With AC signals, the circuit essentially functions as a low-pass filter with a gain of $1/RC$.

To work over a wide range of frequencies, the integrator RC values must be changed. Making low and high resistor and capacitor values in IC form is difficult. However, this problem can be solved by replacing the input resistor with a switched capacitor as shown in Fig. 2-57(b). The MOSFET switches are driven by a clock generator whose frequency is typically 50 to 100 times the maximum frequency of the AC signal to be filtered. The resistance of a MOSFET switch when on is usually less than 1000 Ω. When the switch is off, its resistance is many megohms.

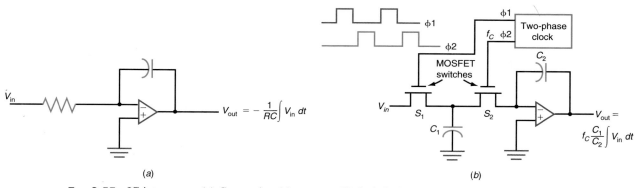

FIG. 2-57 IC integrators. (*a*) Conventional integrator. (*b*) Switched capacitor integrator.

The clock puts out two phases, designated ϕ_1 and ϕ_2, that drive the MOSFET switches. When S_1 is on, S_2 is off and vice versa. The switches are of the break-before-make type, meaning that one switch opens before the other is closed. When S_1 is closed, the charge on the capacitor follows the input signal. Since the clock period and time duration that the switch is on are very short compared to the input signal variation, a brief "sample" of the input voltage remains stored on C_1 and S_1 turns off.

Now, S_2 turns on. The charge on the capacitor C_1 is applied to the summing junction of the op amp. It discharges, causing a current to flow in the feedback capacitor C_2. The resulting output voltage is proportional to the integral of the input. But this time, the gain of the integrator is

$$ f\left(\frac{C_1}{C_2}\right) $$

where f is the clock frequency. The capacitor C_1 which is switched at a clock frequency of f with period T is equivalent to a resistor value of $R = T/C_1$.

The beauty of this arrangement is that it is not necessary to make resistors on the IC chip. Instead, capacitors and MOSFET switches, which are smaller than resistors, are used. Further, since the gain is a function of the ratio of C_1 to C_2, the exact capacitor values are less important than their ratio. It is much easier to control the ratio of matched pairs of capacitors than it is to make precise values of capacitance.

By combining several such switching integrators it is possible to create low-pass, high-pass, bandpass, and band-reject filters of the Butterworth, Chebyshev, elliptical, and Bessel type with almost any desired selectivity. The center frequency or cutoff frequency of the filter is set by the value of the clock frequency. This means that the filter can be tuned on the fly by varying the clock frequency.

A unique but sometimes undesirable characteristic of an SCF is that the output signal is really a stepped approximation of the input signal. Because of the switching action of the MOSFETs and the charging and discharging of the capacitors, the signal takes on a stepped digital form. The higher the clock frequency compared to the frequency of the input signal, the smaller this effect. The signal can be smoothed back into its original state by passing it through a simple RC low-pass filter whose cutoff frequency is set to just pass the signal.

A variety of SCF is available in IC form. Both dedicated single-purpose or universal SCFs can be purchased for less than $2 in bulk. One of the most popular is the MF10 made by National Semiconductor. It is a universal SCF that can be set for low-pass, high-pass, bandpass, or band-reject operation. It can be used for center or cutoff frequencies up to about 20 kHz. The clock frequency is about 50 to 100 times the operating frequency.

COMMUTATING FILTERS. An interesting variation of a switched capacitor filter is the commutating filter shown in Fig. 2-58. It is made of discrete resistors and capacitors with MOSFET switches driven by a counter and decoder. The circuit appears to be a low-pass RC filter, but the switching action makes the circuit function as a bandpass filter. The operating frequency f_{out} is related to the clock frequency f_c and the number N of switches and capacitors used.

$$ f_c = N f_{out} \qquad \text{and} \qquad f_{out} = \frac{f_c}{N} $$

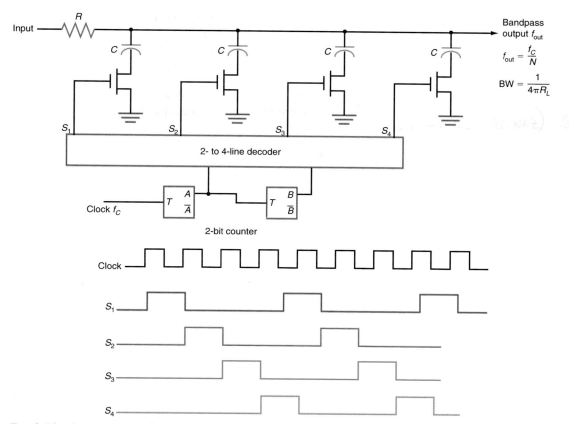

Fig. 2-58 A commutating SCF.

The bandwidth of the circuit is related to the RC values and number of capacitors and switches used as follows:

$$BW = \frac{1}{2\pi NRC}$$

For the filter in Fig. 2-58, the bandwidth is $BW = 1/8\pi RC$.

Very high Q and narrow bandwidth can be obtained, and varying the resistor value makes the bandwidth adjustable.

The operating waveforms in Fig. 2-58 show that each capacitor is switched on and off sequentially such that only one capacitor is connected to the circuit at a time. A sample of the input voltage is stored as a charge on each capacitor as it is connected to the input. The capacitor voltage is the average of the voltage variation during the time the switch connects the capacitor to the circuit.

Figure 2-59(a) shows typical input and output waveforms assuming a sine-wave input. The output is a stepped approximation of the input because of the sampling action of the switched capacitors. The steps are large, but their size can be reduced by simply using a greater number of switches and capacitors. Increasing the number of capacitors from four to eight, as in Fig. 2-54(b), makes the steps smaller, and thus the output more closely approximates the input. The steps can be eliminated or greatly

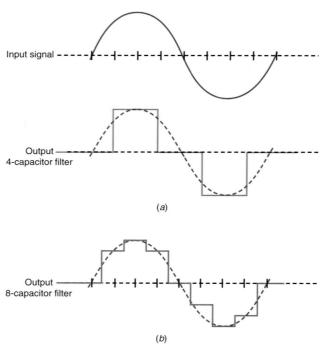

Input signal

Output
4-capacitor filter

(a)

Output
8-capacitor filter

(b)

Fig. 2-59   Input and output for commutating filter. (*a*) Four-capacitor filter. (*b*) Eight-capacitor filter.

minimized by passing the output through a simple *RC* low-pass filter whose cutoff is set to the center frequency value or slightly higher.

One characteristic of the commutating filter is that it is sensitive to the harmonics of the center frequency for which it is designed. Signals whose frequency is some integer multiple of the center frequency of the filter are also passed by the filter, although at a somewhat lower amplitude. The response of the filter, called a *comb* response, is shown in Fig. 2-60. If such performance is undesirable, the higher frequencies can be eliminated with a conventional *RC* or *LC* low-pass filter connected to the output.

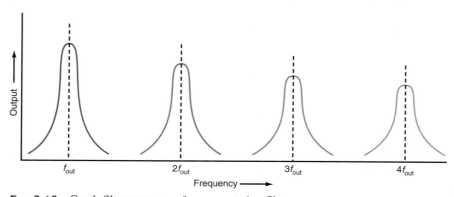

Fig. 2-60   Comb filter response of a commutating filter.

2-4 Transformers and Inductively Coupled Circuits

Inductively coupled circuits are those circuits that are connected to one another via electromagnetic induction. Inductively coupled circuits are not connected electrically but are linked by way of a magnetic field through a transformer. Many RF and communications circuits use transformer coupling for isolation and impedance matching. Both iron-core and air-core transformers are used. This section briefly reviews the principles of mutual induction and transformer operation and application.

Mutual Inductance

If two inductors are placed near one another, a current in one creates a magnetic field that cuts the turns of the other, inducing a voltage into it. Such inductors are said to be *coupled* because they are linked by a magnetic field. The amount of voltage induced into one inductor by a change in current in the other inductor is a function of the mutual inductance L_M, which is related to the inductances of the coils and the degree of coupling between them by the expression:

$$L_M = \frac{k}{L_1 L_2}$$

where L_1 and L_2 are the inductances of the two coils in henries and k is the *coefficient of coupling* between the coils (the percentage of the total number of magnetic lines of force produced by the first coil that cuts the turns of the second coil).

If the two coils are at right angles to one another, none of the lines of force generated by the first coil cut the turns of the second coil, so k is zero. If the coils are parallel to one another or wound on the same core, then a high percentage of the lines produced by the first coil will cut the turns of the second coil, producing a higher value of k. If a common iron core is used, the coefficient of coupling can approach 1.

Recall that if two or more inductors are connected in series, as shown in Fig. 2-61, the total inductance is simply the sum of the individual inductances:

$$L_T = L_1 + L_2$$

This assumes that there is no mutual inductance between the coils. If the coils are in series and there is mutual inductance between them, the total inductance is

$$L_T = L_1 + L_2 + 2L_M$$

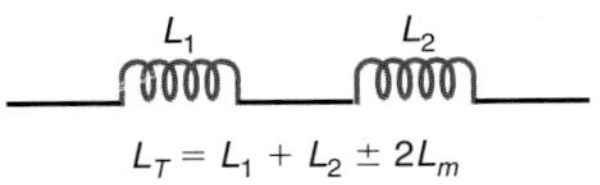

$$L_T = L_1 + L_2 \pm 2L_m$$

Fig. 2-61   Inductors in series with coupling.

Fig. 2-62 Transformer. (*a*) Air-core. (*b*) Iron-core.

This assumes that the magnetic fields are in the same direction and reinforcing one another.

If the magnetic fields oppose one another, the total inductance is

$$L_T = L_1 + L_2 - 2L_M$$

In most cases, the coils are not connected to form a single inductance. Instead, they are used separately to form a component known as a transformer.

A *transformer* is made up of two coils with mutual inductance. The coil to which an input voltage is applied is called the *primary winding*, and the winding from which the output is taken is called the *secondary winding*. Although the primary and secondary windings are not usually connected to one another electrically, they are generally wound on a common form or core. The core can simply be air, but can also be a magnetic core made of iron laminations, powdered iron, or a ferrite. The schematic symbols for such transformers are shown in Fig. 2-62. Transformers with laminated iron cores are used for power transformers in power supplied and in audio amplifier applications. Air, powdered iron, and ferrite cores are used in RF and communications applications.

IRON-CORE TRANSFORMERS

When a voltage is applied to the primary winding, current flows, producing a magnetic field in the core. This magnetic field cuts the turns of the secondary winding, inducing a voltage into them. If a load is connected to the secondary winding, current flows. Thus electric energy is transferred from a source to a load without a direct electrical connection. Energy is transferred by way of the magnetic field. With a magnetic core, most of the lines of force produced by the primary cut the turns of the secondary. There is some flux leakage, meaning that not all of the lines of force produced by the primary cut the secondary. Nevertheless, the coefficient of coupling *k* is very nearly equal to 1.

Iron-core transformers are widely used in electronics. They are found primarily in power supplies, audio equipment, telephone equipment, and in transmitters and receivers. In these applications, they perform four primary functions:

1. Electrical isolation
2. Voltage step up or step down
3. Impedance transformation
4. Phase inversion

ELECTRICAL ISOLATION. As mentioned above, since the primary and secondary windings of a transformer are not electrically connected, circuits connected to the primary are also electrically insulated from the circuits connected to the secondary. The transfer of energy takes place through the magnetic field. This makes it possible to interconnect circuits and equipment that must for some reason be electrically isolated from one another. For example, a high DC voltage in the primary circuit may need to be isolated from lower-voltage circuits in the secondary circuit. Another common application is in those cases where signals must be transferred

Radio disk jockeys provide us with easy listening music, rock music, and talk radio as we commute to school or to work.

between circuits or equipment with different grounds. In power applications, AC isolation transformers are used for safety purposes to prevent accidental cross-grounding of the AC power-line equipment.

VOLTAGE TRANSFORMATION. Transformers can be used to step up or step down AC voltage levels. The voltage applied to the primary winding can be transformed to a higher or lower voltage across the secondary winding. This transformation depends on the *turns ratio* of the transformer, the ratio of the number of turns of wire in the secondary to the number of turns in the primary:

$$N = \frac{N_s}{N_p}$$

If there are more turns of wire on the secondary than on the primary, the transformer is a step-up transformer. The output voltage is higher than the input voltage by a factor equal to the turns ratio. If the number of turns on the primary is less than the turns on the secondary, the transformer steps down the input voltage. The relationship between the input and output voltages and turns ratio is

$$\frac{V_s}{V_p} = \frac{N_s}{N_p}$$

Knowing the input voltage and the turns ratio, the output voltage can be computed:

$$V_s = V_p \frac{N_s}{N_p}$$

Knowing the output voltage and the turns ratio, the input or primary voltage can be computed:

$$V_p = V_s \frac{N_p}{N_s}$$

For example, if there are 550 turns on a secondary and 75 turns on a primary and 12 V rms is applied to the primary, the secondary voltage is $V_s = 12(550/75) = 12(7.333) = 88$ V.

While it is possible to step up or step down voltages by using a transformer, it should be kept in mind that the total power transferred between primary and secondary is constant. Assuming 100 percent efficiency, the power in the secondary equals the power in the primary:

$$P_s = P_p \quad \text{or} \quad I_s V_s = I_p V_p$$

While iron-core transformers are not 100 percent efficient, the efficiency of most such devices is greater than 95 percent, and it can be assumed for most applications.

As can be seen from the above relationships, the input and output currents are inversely proportional to the voltage ratio and the turns ratio:

$$\frac{I_p}{I_s} = \frac{V_s}{V_p} = \frac{N_s}{N_p}$$

In practice, this means that a voltage step-up transformer is a current step-down transformer, and vice versa.

IMPEDANCE TRANSFORMATION. It is often necessary in electronics circuits to match the impedance of a load to the impedance of the driving generator or circuit. Recall that one of the basic theories of electronics is that maximum power transfer takes place between a generator and a load when the load resistance is equal to the generator resistance. In applications requiring the development and transfer of power from one circuit or piece of equipment to another, this is critical.

In many applications, the value of the load impedance and the generator impedance are vastly different. In some cases, electronic circuits such as amplifiers can be inserted to correct the problem. However, one of the simplest and easiest ways to match the load and driving impedances is to use a transformer to cause a particular load impedance to appear to match that of the generator. For example, assume that you wish to match a transistor's amplifier output impedance of 1000 Ω to an 8-Ω speaker as shown in Fig. 2-63. The relationship between the input and output impedances and the transformer turns ratio is

$$\frac{N_s}{N_p} = \sqrt{\frac{Z_s}{Z_p}}$$

Thus the transformer must have a turns ratio of

$$\frac{N_s}{N_p} = \sqrt{\frac{8}{1000}} = \sqrt{0.008} = 0.089$$

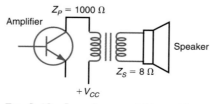

FIG. 2-63 Impedance matching with a transformer.

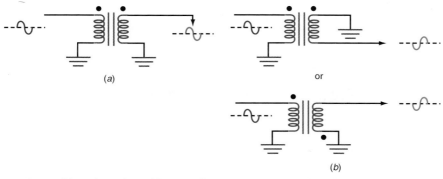

FIG. 2-64 Phase inversion with a transformer. (*a*) In phase (0°). (*b*) Phase inversion (180°).

As you can see, by simply selecting a transformer with the correct turns ratio, input and output impedances can be matched to ensure that maximum power transfer takes place between the generator and the load.

PHASE INVERSION. One of the most useful applications for a transformer is *phase inversion,* reversing the polarity of an AC signal. Transformers can be used to introduce a 180° phase inversion if desired. This is easily accomplished by simply making the correct connections to the transformer windings.

If the primary and secondary windings are both wound in the same direction on the core and the corresponding ends of the windings are both connected to ground reference, the output voltage from the secondary will be in phase with the input voltage to the primary. However, if the leads to the secondary winding are reversed, the output signal will be 180° out of phase with the input signal. The dots on the transformer leads in Fig. 2-64 indicate the phasing of the transformers. In Fig. 2-64(*a*), the primary and secondary signals are in phase, as indicated by the two dots at the top; in Fig. 2-64(*b*), there is a phase reversal, indicated by a dot at the top of the primary and a dot at the bottom of the secondary. This phase reversal can also be accomplished by reversing the connections to the primary winding rather than the turns of the secondary winding.

Transformers can also be used to obtain signals of both polarities simultaneously. This is done by providing for a center tap on the secondary winding, as shown in Fig. 2-65. With the center tap connected to ground, the voltage at the upper end of the secondary winding is one-half the total voltage produced by the secondary. The voltage at the bottom end of the secondary is also one-half the total secondary voltage with respect to ground. In addition, the phase of the voltage with respect to ground at the upper terminal is 180° out of phase with the signal at the bottom connection. Many applications in electronics require signals of equal voltage but opposite phase.

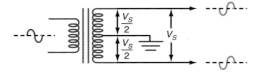

FIG. 2-65 Using a center-tapped transformer to obtain signals of equal amplitude and opposite phase.

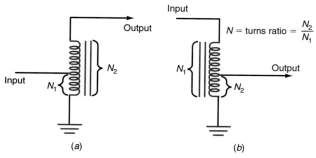

Fig. 2-66 The autotransformer. (*a*) Step up. (*b*) Step down.

AUTOTRANSFORMERS. Autotransformers are iron-core transformers that work just like standard two-winding transformers except that a single tapped winding is used to provide both input and output connections. Figure 2-66 shows the basic autotransformer connections for both step-up and step-down applications. All of the previously given formulas related to turns ratio, power, and impedance matching fully apply.

The primary advantage of autotransformers is their low cost relative to other types. Their primary disadvantage is that because only a single winding is used, no electrical isolation is produced.

Example 2-27

An iron core transformer had 22 turns on the primary and 100 turns on the secondary. The primary voltage is 37 mV and the secondary or load impedance is 93 Ω. Find (*a*) the turns ratio, (*b*) the secondary voltage, (*c*) the primary impedance, and (*d*) the primary current.

a. $N = N_s/N_p = 100/22 = 4.545$

b. $V_s/V_p = N_s/N_p$. Therefore,

$$V_s = V_p\left(\frac{N_s}{N_p}\right) = V_p N = (37 \times 10^{-3})(4.545)$$
$$= 0.168 \text{ V or } 168 \text{ mV} \qquad \text{(step-up)}$$

c. $\dfrac{N_s}{N_p} = \sqrt{\dfrac{Z_s}{Z_p}}$ and $\dfrac{Z_s}{Z_p} = \left(\dfrac{N_s}{N_p}\right)^2 = N^2$. Therefore,

$$Z_p = \frac{Z_s}{N^2} = \frac{93}{(4.545)^2} = \frac{93}{20.66} = 4.5 \text{ }\Omega$$

d. $I_p/I_s = N_s/N_p = N$.

$$I_s = \frac{V_s}{R_L} = \frac{168 \times 10^{-3}}{93} = 1.8 \times 10^{-3} = 1.8 \text{ mA}$$
$$I_p = NI_s = 4.545(1.8 \times 10^{-3}) = 8.21 \times 10^{-3} = 8.21 \text{ mA}$$

Because of the very high losses in the cores of iron-core transformers, they are not often used in high-frequency radio signal applications. At frequencies up to about 50 MHz, some powdered iron and ferrite transformer cores are used. Air-core transformers, however, are widely used. Such transformers use cordless coils of wire for primary and secondary windings. The coils can be made of heavy wire and self-supporting and placed adjacent to one another. In most cases, however, they share a common coil form which is usually tubular and made of plastic, coated cardboard, or some other insulating material.

With no magnetic core, the coefficient of coupling of air-core transformers is significantly less than 1. In fact, values of less than $k = 0.1$ are typical. Since all of the lines of force produced by the primary do not cut the turns of the secondary, the transfer of power from primary to secondary is less efficient.

Despite the lower efficiency of power transfer, air-core transformers do provide impedance matching and circuit isolation. In addition, in most RF applications, the primary and secondary windings serve as inductances to form tuned circuits along with capacitors connected in series or in parallel with them. Thus the transformers become tuned circuits, providing bandpass selectivity over a narrow range of frequencies depending upon the Q of the circuit.

FIG. 2-67   Double-tuned air-core transformer.

A common arrangement in receivers and transmitters is the use of coupled tuned circuits between amplifier stages. An example is shown in Fig. 2-67. Both the primary and secondary windings are resonated with capacitors. The output voltage versus frequency for such a double tuned coupled circuit is strictly dependent upon the amount of coupling or mutual inductance between the primary and secondary windings. That is, the spacing between the windings determines how much of the magnetic field produced by the primary will cut the turns of the secondary. This affects not only the amplitude of the output voltage, but also the bandwidth.

Figure 2-68 shows the effect of different amounts of coupling between the primary and secondary windings. When the windings are spaced far apart, the coils are said to be *undercoupled*. The result of undercoupling is low amplitude and a relatively narrow bandwidth.

At some specific level of coupling, the output reaches a peak value. This is known as the *critical coupling* point. In most applications, critical coupling provides the best gain if the bandwidth provided is adequate.

Moving the coils closer together and thus increasing the coupling causes an increase in bandwidth. The output signal amplitude is already at a maximum value and will not increase beyond that obtained at critical coupling, but the bandwidth can be widened further. This point is usually known as *optimum coupling*. Increasing the amount of coupling beyond that point produces an effect known as *overcoupling*. The result is a doubled peak output response curve with a considerably wider bandwidth. By setting the amount of coupling between the windings in the coupling transformers, the desired amount of bandwidth can be obtained.

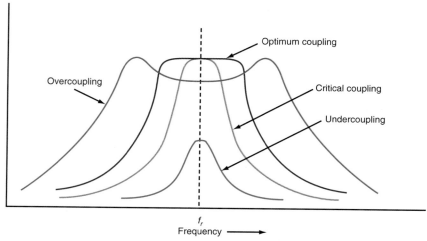

Fig. 2-68 Response curves of a double-tuned air-core transformer for various degrees of coupling.

2-5 FOURIER THEORY

The mathematical analysis of the modulation and multiplexing methods used in communications systems assume sine-wave carriers and information signals. This simplifies the analysis and makes operation predictable. However, in the real world, not all information signals are sinusoidal. Information signals are typically more complex voice and video signals which are essentially composites of sine waves of many frequencies and amplitudes. Information signals can take on an infinite number of shapes, including rectangular waves (i.e., digital pulses), triangular waves, sawtooth waves, and other nonsinusoidal forms. Such signals require that a non-sine-wave approach be taken to determine the characteristics and performance of any communications circuit or system. One of the methods used to do this is Fourier analysis, which provides a means of accurately analyzing the content of most complex nonsinusoidal signals. Although Fourier analysis requires the use of calculus and advanced mathematical techniques beyond the scope of this text, its practical applications to communications electronics are relatively straightforward.

BASIC CONCEPTS

Figure 2-69(*a*) shows a basic sine wave with its most important dimensions and the equation expressing it. A basic cosine wave is illustrated in Fig. 69(*b*). Note that the cosine wave has the same shape as a sine wave but leads the sine wave by 90°. A *harmonic* is a sine wave whose frequency is some integer multiple of a fundamental sine wave. For example, the third harmonic of a 2-kHz sine wave is a sine wave of 6 kHz. Figure 2-70 shows the first four harmonics of a fundamental sine wave.

What the Fourier theory tells us is that we can take a nonsinusoidal waveform and break it down into individual harmonically related sine wave or cosine wave compo-

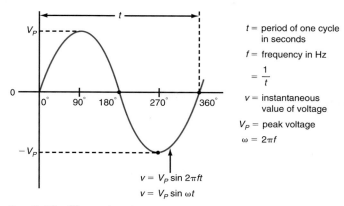

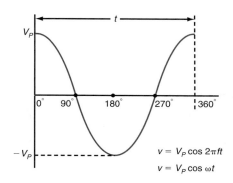

t = period of one cycle in seconds

f = frequency in Hz

$= \dfrac{1}{t}$

v = instantaneous value of voltage

V_P = peak voltage

$\omega = 2\pi f$

$v = V_P \sin 2\pi ft$

$v = V_P \sin \omega t$

$v = V_P \cos 2\pi ft$

$v = V_P \cos \omega t$

FIG. 2-69 Sine and cosine waves.

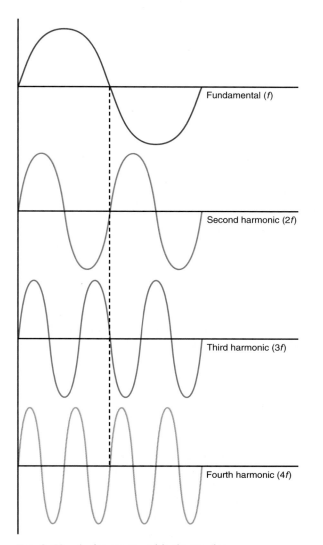

Fundamental (f)

Second harmonic ($2f$)

Third harmonic ($3f$)

Fourth harmonic ($4f$)

FIG. 2-70 A sine wave and its harmonics.

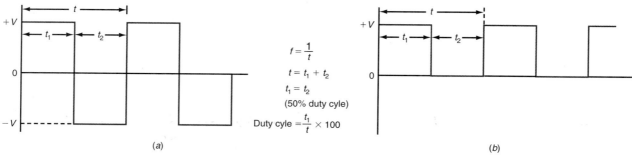

$$f = \frac{1}{t}$$
$$t = t_1 + t_2$$
$$t_1 = t_2$$
(50% duty cycle)
$$\text{Duty cyle} = \frac{t_1}{t} \times 100$$

FIG. 2-71   A square wave.

nents. The classic example of this is a *square wave,* which is a rectangular signal with equal-duration positive and negative alternations. In the AC square wave in Fig. 2-71, this means that t_1 is equal to t_2. Another way of saying this is that the square wave has a 50 percent *duty cycle D,* the ratio of the duration of the positive alteration t_1 to the period t expressed as a percentage:

$$D = \frac{t_1}{t} \times 100$$

Fourier analysis tells us that a square wave is made up of a sine wave at the fundamental frequency of the square wave plus an infinite number of odd harmonics. For example, if the fundamental frequency of the square wave is 1 kHz, the square wave can be synthesized by adding the 1-kHz sine wave and harmonic sine waves of 3 kHz, 5 kHz, 7 kHz, 9 kHz, etc.

Figure 2-72 shows how this is done. The sine waves must be of the correct amplitude and phase with relationship to one another. The fundamental sine wave in this case has a value of 20 V peak to peak (a 10-V peak). When the sine wave values are added instantaneously, the result approaches a square wave. In Fig. 2-72(*a*), the fundamental and third harmonic are added. Note the shape of the composite wave with the third and fifth harmonics added, as in Fig. 2-72(*b*). The more higher harmonics that are added, the more the composite wave looks like a perfect square wave. Figure 2-73 shows how the composite wave would look with 20 odd harmonics added to the fundamental. The results very closely approximate a square wave.

The implication of this is that a square wave should be analyzed as a collection of harmonically related sine waves rather than a single square-wave entity. This is confirmed by performing a Fourier mathematical analysis on the square wave. The result is the following equation, which expresses voltage as a function of time:

$$f(t) = \frac{4V}{\pi}\left[\sin 2\pi\left(\frac{1}{T}\right)t + \frac{1}{3}\sin 2\pi\left(\frac{3}{T}\right)t + \frac{1}{5}\sin 2\pi\left(\frac{5}{T}\right)t + \frac{1}{7}\sin 2\pi\left(\frac{7}{T}\right)t + \ldots \right]$$

where the factor $4V/\pi$ is a multiplier for all sine terms and V is the square-wave peak voltage. The first term is the fundamental sine wave, and the succeeding terms are the third, fifth, seventh, etc., harmonics. Note that the terms also have an amplitude factor. In this case, the amplitude is also a function of the harmonic. For example, the third harmonic has an amplitude that is one-third of the fundamental amplitude, and so on. The expression could also be rewritten with $f = 1/T$. If the square wave is direct current rather than alternating current, as shown in Fig. 2-71(*b*), the Fourier expression has a DC component:

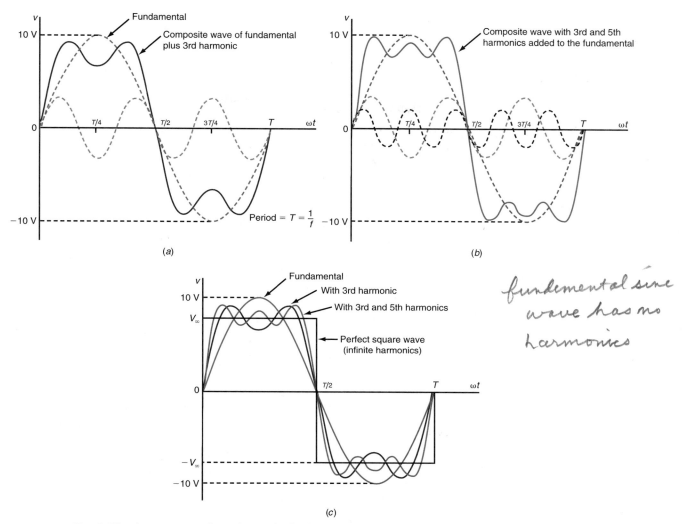

fundamental sine wave has no harmonics

Fig. 2-72 A square wave is made up of a fundamental sine wave and an infinite number of odd harmonics.

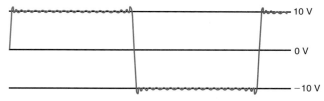

Fig. 2-73 Square wave made up of 20 odd harmonics added to the fundamental.

$$f(t) = \frac{V}{2} + \frac{4V}{\pi}\left(\sin 2\pi ft + \frac{1}{3}\sin 2\pi 3ft + \frac{1}{5}\sin 2\pi 5ft + \frac{1}{7}\sin 2\pi 7ft + \ldots\right)$$

In this equation, $V/2$ is the DC component, the average value of the square wave. It is also the baseline upon which the fundamental and harmonic sine waves ride.

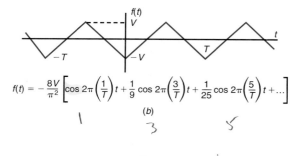

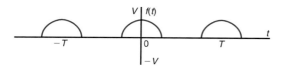

$$f(t) = \frac{4V}{\pi}\left[\sin 2\pi\left(\frac{1}{T}\right)t + \frac{1}{3}\sin 2\pi\left(\frac{3}{T}\right)t + \frac{1}{5}\sin 2\pi\left(\frac{5}{T}\right)t + ...\right]$$

F

(a)   3rd harmonics   5th harmonic

$$f(t) = -\frac{8V}{\pi^2}\left[\cos 2\pi\left(\frac{1}{T}\right)t + \frac{1}{9}\cos 2\pi\left(\frac{3}{T}\right)t + \frac{1}{25}\cos 2\pi\left(\frac{5}{T}\right)t + ...\right]$$

(b)

1   3   5

$T = \frac{1}{f}$

all even harmonics

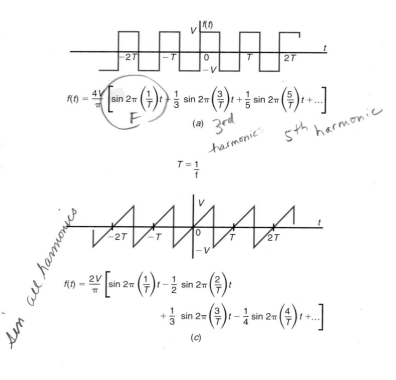

$$f(t) = \frac{2V}{\pi}\left[\sin 2\pi\left(\frac{1}{T}\right)t - \frac{1}{2}\sin 2\pi\left(\frac{2}{T}\right)t \right.$$
$$\left. + \frac{1}{3}\sin 2\pi\left(\frac{3}{T}\right)t - \frac{1}{4}\sin 2\pi\left(\frac{4}{T}\right)t + ...\right]$$

(c)

sin all harmonics

$$f(t) = \frac{V}{\pi} + \frac{V}{\pi}\left[\frac{\pi}{2}\cos 2\pi\left(\frac{1}{T}\right)t + \frac{2}{3}\cos 2\pi\left(\frac{2}{T}\right)t \right.$$
$$\left. - \frac{2}{15}\cos 2\pi\left(\frac{4}{T}\right)t + \frac{2}{35}\cos 2\pi\left(\frac{6}{T}\right)t + ...\right]$$

(d)

cos all harmonics

$$f(t) = \frac{2V}{\pi} + \frac{2V}{\pi}\left[\frac{2}{3}\cos 2\pi\left(\frac{1}{T}\right) - \frac{2}{15}\cos 2\pi\left(\frac{2}{T}\right)t + \frac{2}{35}\cos 2\pi\left(\frac{3}{T}\right)t + ...\right]$$

(e)

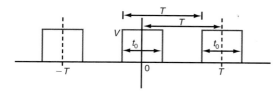

$$f(t) = \frac{Vt_0}{T} + \frac{2Vt_0}{T}\left[\frac{\sin\frac{\pi t_0}{T}}{\frac{\pi t_0}{T}}\cos\frac{\pi t_0}{T} \right.$$
$$\left. + \frac{\sin\frac{2\pi t_0}{T}}{\frac{2\pi t_0}{T}}\cos\frac{2\pi t_0}{T} + \frac{\sin\frac{3\pi t_0}{T}}{\frac{3\pi t_0}{T}}\cos\frac{3\pi t_0}{T} + ...\right]$$

(f)

FIG. 2-74   Common nonsinusoidal waves and their Fourier equations. (*a*) Square wave. (*b*) Triangle wave. (*c*) Sawtooth. (*d*) Half cosine wave. (*e*) Full cosine wave. (*f*) Rectangular pulse.

A general formula for the Fourier equation of a waveform is

$$f(t) = \frac{V}{2} + \frac{4V}{\pi n}\sum_{n=1}^{\infty}\left(\sin 2\pi nft\right)$$

where *n* is odd.

By using calculus and other mathematical techniques, the waveform is defined, analyzed, and expressed as a summation of sine and/or cosine terms as illustrated by the expression for the square wave above. Figure 2-74 gives the Fourier expressions for some of the most common nonsinusoidal waveforms.

Example 2-28

An AC square wave has a peak voltage of 3 V and a frequency of 48 kHz. Find (a) the frequency of the fifth harmonic and (b) the rms value of the fifth harmonic. Use the formula in Fig. 2-74(a).

a. 5×48 kHz $= 240$ kHz

b. Isolate the expression for the fifth harmonic in the formula, which is $\frac{1}{5}\sin 2\pi(5/T)t$. Multiply by the amplitude factor $4V/\pi$. The peak value of the fifth harmonic (V_p) is

$$V_p = \frac{4V}{\pi}\left(\tfrac{1}{5}\right) = \frac{4(3)}{5\pi} = 0.76$$

rms $= 0.707 \times$ peak value

$V_{rms} = 0.707 V_p = 0.707\,(0.76) = 0.537$ V

The triangular wave in Fig. 2-74(b) exhibits the fundamental and odd harmonics, but it is made up of cosine waves rather than sine waves. The sawtooth wave in Fig. 2-74(c) contains the fundamental plus all odd and even harmonics. Figure 2-74(d) and (e) shows half sine pulses like those seen at the output of half- and full-wave rectifiers. Both have an average DC component, as would be expected. The half-wave signal is made up of even harmonics only, whereas the full-wave signal has both odd and even harmonics. Figure 2-74(f) shows the Fourier expression for a DC square wave where the average DC component is Vt_o/T.

TIME DOMAIN VERSUS FREQUENCY DOMAIN

Most of the signals and waveforms that we discuss and analyze are expressed in the time domain. That is, they are variations of voltage, current, or power with respect to time. All of the signals shown in the previous illustrations are examples of time-domain waveforms. Their mathematical expressions contain the variable time (t) indicating that they are a time-variant quantity.

Fourier theory gives us a new and different way to express and illustrate complex signals. Here, complex signals containing many sine and/or cosine components are expressed as sine or cosine wave amplitudes at different frequencies. In other words, a graph of a particular signal is a plot of sine and/or cosine component amplitudes with respect to frequency.

A typical frequency-domain plot of the square wave is shown in Fig. 2-75(a). Note that the straight lines represent the sine-wave amplitudes of the fundamental and harmonics and these are plotted on a horizontal frequency axis. Such a frequency-domain plot can be made directly from the Fourier expression by simply using the frequencies of the fundamentals and harmonics and their amplitudes. Frequency-domain plots for some of the other common nonsinusoidal waves are also shown in Fig. 2-75. Note that the triangle wave in Fig. 2-75(c) is made up of the fundamental and odd harmonics. The third harmonic is shown as a line below the axis, which indicates a 180° phase shift in the cosine wave making it up.

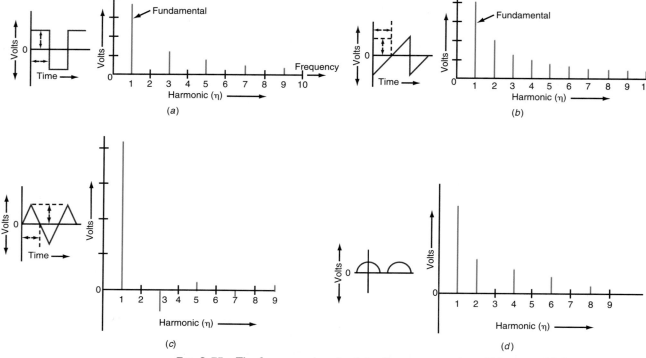

Fig. 2-75 The frequency-domain plots of common nonsinusoidal waves. (*a*) Square wave. (*b*) Sawtooth. (*c*) Triangle. (*d*) Half sine wave.

Figure 2-76 shows how the time and frequency domains are related. The square wave discussed earlier is used as an example. The result is a three-axis three-dimensional view.

Signals and waveforms in communications applications are expressed using both time-domain and frequency-domain plots, but in many cases the frequency-domain plot is far more useful. This is particularly true in the analysis of complex signal waveforms as well as the many modulation and multiplexing methods used in communications.

Test instruments for displaying signals in both time and frequency domain are readily available. You are already familiar with the oscilloscope, which displays the voltage amplitude of a signal with respect to a horizontal time axis.

The test instrument for producing a frequency-domain display is the *spectrum analyzer*. Like the oscilloscope, the spectrum analyzer uses a cathode ray tube for display, but the horizontal sweep axis is calibrated in Hertz and the vertical axis is calibrated in volts or power units or decibels.

The Importance of Fourier Theory

Fourier analysis allows us not only to determine the sine-wave components in any complex signal but also how much bandwidth a particular signal occupies. While a sine or cosine wave at a single frequency theoretically occupies no bandwidth, complex signals obviously take up more spectrum space. For example, a 1-MHz square wave with

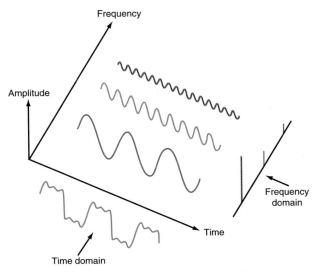

Fig. 2-76 The relationship between time and frequency domain.

harmonics up to the eleventh occupies a bandwidth of 11 MHz. If this signal is to pass unattenuated and undistorted, then all harmonics must be passed.

An example of this is shown in Fig. 2-77. If a 1-kHz square wave is passed through a low-pass filter with a cutoff frequency just above 1 kHz, all of the harmonics beyond the third harmonic are greatly attenuated or, for the most part, filtered out completely. The result is that the output of the low-pass filter is simply the fundamental sine wave at the square-wave frequency.

If the low-pass filter were set to cut off at a frequency above the third harmonic, then the output of the filter would consist of a fundamental sine wave and the third harmonic. Such a waveshape is shown in Fig. 2-72(*a*). As you can see, when the higher harmonics are not all passed, the original signal is greatly distorted. This is why it is important for communications circuits and systems to have a bandwidth wide enough to accommodate all of the harmonic components within the signal waveform to be processed.

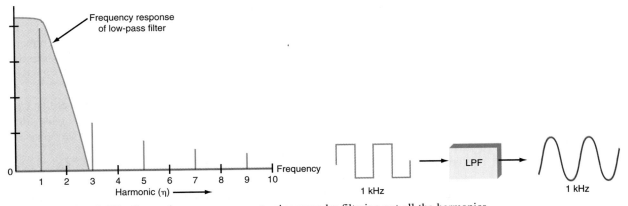

Fig. 2-77 Converting a square wave to sine wave by filtering out all the harmonics.

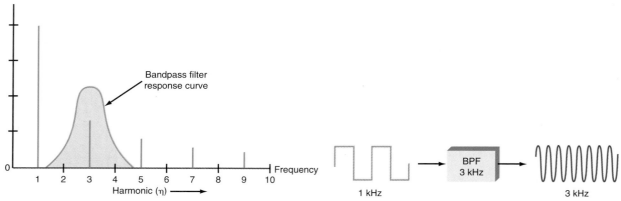

FIG. 2-78 Selecting the third harmonic with a bandpass filter.

Figure 2-78 shows an example in which a 1-kHz square wave is passed through a bandpass filter set to the third harmonic, resulting in a 3-kHz sine-wave output. In this case the filter used is sharp enough to select out the desired component.

PULSE SPECTRUM

The Fourier analysis of binary pulses is especially useful in communications, for it gives a way to analyze the bandwidth needed to transmit such pulses. Although, theoretically, the system must pass all of the harmonics in the pulses, in reality relatively few must be passed to preserve the shape of the pulse. In addition, the pulse train in data communications rarely consists of square waves with a 50 percent duty cycle. Instead, the pulses are rectangular in shape and exhibit varying duty cycles, from very low to very high. [The Fourier response of such pulses is given in Fig. 2-74(f).]

Look back at Fig. 2-74(f). The period of the pulse train is T and the pulse width is t_0. The duty cycle is t_0/T. The pulse train consists of DC pulses with an average DC value of Vt_0/T. In terms of Fourier analysis, the pulse train is made up of a fundamental and all even and odd harmonics. The special case of this waveform is where the duty cycle is 50 percent; in that case all of the even harmonics drop out. But with any other duty cycle the waveform is made up of both odd and even harmonics. Since this is a series of DC pulses, the average DC value is Vt_0/T.

A frequency-domain graph of harmonic amplitudes plotted with respect to frequency is shown in Fig. 2-79. The horizontal axis is frequency plotted in increments of ω (lowercase Greek omega), where $\omega = 2\pi f$ or $2\pi/T$, and T is the period. The first component is the average DC component at zero frequency, Vt_0/T, where V is the peak voltage value of the pulse.

Now, note the amplitudes of the fundamental and harmonics. Remember that each vertical line represents the peak value of the sine-wave components of the pulse train. Some of the higher harmonics are negative; that simply means that their phase is reversed.

The dashed line in Fig. 2-79, the outline of the peaks of the individual components, is what is known as the *envelope* of the frequency spectrum. The equation for the envelope curve has the general form sin x/x, where $x = n2\pi ft_0/2$, and t_0 is the pulse

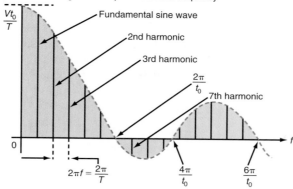

Amplitude of average DC component at zero frequency

$\frac{Vt_0}{T}$

Fundamental sine wave

2nd harmonic

3rd harmonic

$\frac{2\pi}{t_0}$

7th harmonic

$2\pi f = \frac{2\pi}{T}$

$\frac{4\pi}{t_0}$

$\frac{6\pi}{t_0}$

FIG. 2-79 Frequency domain of a rectangular pulse train.

width. This is known as the *sinc* function. In Fig. 2-79, the sinc function crosses the horizontal axis several times. These times can be computed and are indicated in the figure. Note that they are some multiple of $2\pi/t_0$.

The sinc function drawn on a frequency-domain curve is used in predicting the harmonic content of a pulse train and thus the bandwidth necessary to pass the wave. For example, in Fig. 2-79, as the frequency of the pulse train gets higher, the period T gets shorter and the spacing between the harmonics gets wider. This moves the curve out to the right. And as the pulse duration t_0 gets shorter, meaning that the duty cycle

Example 2-29

A DC pulse train like that in Fig. 2-74(f) has a peak voltage value of 5 V, a frequency of 4 MHz, and a duty cycle of 30 percent.

a. What is the average dc value? [$V_{avg} = Vt_0/T$. Use the formula given in Fig. 2-74(f).]

Duty cycle = $\frac{t_0}{T}$ = 30% or 0.30

$$T = \frac{1}{f} = \frac{1}{4 \times 10^6} = 2.5 \times 10^{-7} \text{ s} = 250 \times 10^{-9} \text{ s}$$
$$= 250 \text{ ns}$$
$$t_0 = \text{duty cycle} \times T = 0.3 \times 250 = 75 \text{ ns}$$
$$V_{avg} = \frac{Vt_0}{T} = V \times \text{duty cycle} = 5 \times 0.3 = 1.5 \text{ V}$$

b. What is the minimum bandwidth necessary to pass this signal without excessive distortions?

$$\text{Minimum bandwidth BW} = \frac{2\pi}{t_0} = \frac{6.28}{75 \times 10^{-9}}$$
$$= 8.373 \times 10^7 \text{ Hz or } 83.73 \times 10^6$$
$$= 83.73 \text{ MHz}$$

gets shorter, the first zero crossing of the envelope moves farther to the right. The practical significance of this is that higher-frequency pulses with shorter pulse durations have more harmonics with greater amplitudes and thus a wider bandwidth is needed to pass the wave with minimum distortion. For data communications applications, it is generally assumed that a bandwidth equal to the first zero crossing of the envelope is the minimum that is sufficient to pass enough harmonics for reasonable waveshape:

$$BW = \frac{2\pi}{t_0}$$

Most of the higher-amplitude harmonics and thus the most significant part of the signal power is contained within the larger area between zero frequency and the $2\pi/t_o$ point on the curve.

THE RELATIONSHIP BETWEEN RISE TIME AND BANDWIDTH

Because a rectangular wave such as a square wave theoretically contains an infinite number of harmonics, we can use a square wave as the basis for determining the bandwidth of a signal. If the processing circuit should pass all or an infinite number of harmonics, the rise and fall times of the square wave will be zero. As the bandwidth is decreased by rolling off or filtering out the higher frequencies, the higher harmonics are greatly attenuated. The effect this has on the square wave is that the rise and fall times of the waveform become finite and increase as more and more of the higher harmonics are filtered out. The more restricted the bandwidth, the fewer the harmonics passed and the greater the rise and fall times. The ultimate restriction is where all of the harmonics are filtered out, leaving only the fundamental sine wave (Fig. 2-77).

The concept of rise and fall times is illustrated in Fig. 2-80. The rise time t_r is the time it takes the pulse voltage to rise from its 10 percent value to its 90 percent value.

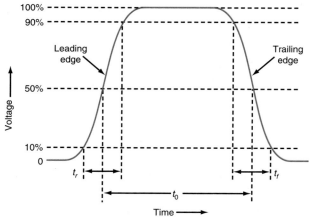

t_r = rise time
t_f = fall time
t_0 = pulse width (duration)

FIG. 2-80   Rise and fall times of a pulse.

The fall time t_f is the time it takes the voltage to drop from the 90 percent value to the 10 percent value. Pulse width t_o is usually measured at the 50 percent amplitude points on the leading (rise) and trailing (fall) edges of the pulse.

A simple mathematical expression relating the rise time of a rectangular wave and the bandwidth of a circuit required to pass the wave without distortion is

$$BW = \frac{0.35}{t_r}$$

Example 2-30

A pulse train has a rise time of 6 ns. What is the minimum bandwidth to pass this pulse train faithfully?

$$BW = \frac{0.35}{t_r} \qquad t_r = 6 \text{ ns} = 0.006 \ \mu s$$

$$\text{minimum BW} = \frac{0.35}{0.006} = 55.333 \text{ MHz}$$

This is the bandwidth of the circuit required to pass a signal containing the highest-frequency component contained in a square wave with a rise time of t_r. In this expression, the bandwidth is really the upper 3-dB down cutoff frequency of the circuit given in megahertz. The rise time of the output square wave is given in microseconds. For example, if the square-wave output of an amplifier has a rise time of 10 ns (0.01 μs), the bandwidth of the circuit must be at least BW = 0.35/0.01 = 35 MHz.

Rearranging the formula, you can calculate the rise time of an output signal from the circuit whose bandwidth is given: $t_r = 0.35/BW$. For example, a circuit with a 50-MHz bandwidth will pass a square wave with a minimum rise time of $t_r = 0.35/50 = 0.007 \ \mu s = 7$ ns.

This simple relationship permits you to quickly determine the approximate bandwidth of a circuit needed to pass a rectangular waveform with a given rise time. This relationship is widely used to express the frequency response of the vertical amplifier in an oscilloscope. Oscilloscope specifications often give only a rise-time figure for the vertical amplifier. An oscilloscope with a 60-MHz bandwidth would pass rectangular waveforms with rise times as short as $t_r = 0.35/60 = 0.041666 \ \mu s = 4.1666$ ns.

Example 2-31

A circuit has a bandwidth of 200 kHz. What is the fastest rise time this circuit will pass?

$$t_r = \frac{0.35}{f} \text{ MHz} \qquad \text{and} \qquad 200 \text{ kHz} = 0.2 \text{ MHz}$$

$$t_r = \frac{0.35}{0.2} = 1.75 \ \mu s$$

Example 2-32

An oscilloscope has a bandwidth of 60 MHz. The input square wave has a rise time of 15 ns. What is the rise time of the displayed square wave?

$$t_{ra} \text{ (oscilloscope)} = \frac{0.35}{60} = 0.005833 \ \mu s = 5.833 \text{ ns}$$

$$t_{ri} = 15 \text{ ns}$$

$$t_{ra} \text{ (composite)} = 1.1 \ \sqrt{t_{ri}^2 + t_{ra}^2} = 1.1 \ \sqrt{(15)^2 + (5.833)^2}$$
$$= 1.1 \ \sqrt{259} = 17.7 \text{ ns}$$

Similarly, an oscilloscope whose vertical amplifier is rated at 2 ns (0.002 μs) has a bandwidth or upper cutoff frequency of BW = 0.35/0.002 = 175 MHz. What this means is that the vertical amplifier of the oscilloscope has a bandwidth adequate enough to pass a sufficient number of harmonics so that the resulting rectangular wave has a rise time of 2 ns. This does not indicate the rise time of the input square wave itself. To take this into account, you use the formula

$$t_r = 1.1\sqrt{t_{ri}^2 + t_{ra}^2}$$

where t_{ri} = rise time of input square wave
t_{ra} = rise time of amplifier
t_r = composite rise time of amplifier output.

The expression can be expanded to include the effect of additional stages of amplification by simply adding the squares of the individual rise times to the above expression before taking the square root of it.

The spectrum analyzer shows a frequency domain plot of electronic signals. It is the key test instrument in designing, analyzing, and troubleshooting communications equipment.

SUMMARY

It is common in communications systems to cascade components that have gain and loss such that loss can be offset by adding a gain stage, and vice versa. The formulas for voltage, current, and power gain and loss are commonly expressed in terms of decibels, or in dBm. Tuned parallel or series circuits are made up of inductors and capacitors that resonate at specific frequencies. Both coils and capacitors offer an opposition to alternating current known as reactance. Like resistance, it is an opposition that directly affects the amount of current in the circuit. Another reactive effect is capacitance. Combining resistance, inductance, and capacitance produces a total opposition of the combined components known as impedance.

When current is passed through an inductor, the induced voltage has the effect, known as inductance, of opposing current changes in the coil. Two other circuit characteristics are Q, the ratio of the inductive power to the resistive power, and the resonant frequency f_r. A circuit's bandwidth determines its selectivity.

A filter is a frequency-selective circuit designed to pass some frequencies and reject others. There are both active and passive filters. The five basic kinds of filter circuits are low-pass, high-pass, bandpass, band-reject, and all-pass. The type of filter material (e.g., crystal or ceramic) affects selectivity.

To work over a wide range of frequencies, integrator RC values must be changed. This can be done by replacing the input resistor with a switched capacitor. By combining several such switching integrators it is possible to create filters with almost any desired selectivity.

One of the simplest and easiest ways to match the load and driving impedances in many applications is to use a transformer. Transformers can also be used to obtain phase inversion and signals of both polarities simultaneously.

Fourier analysis gives a way to analyze the bandwidth needed to transmit binary pulses by relating the rise time of a rectangular wave and the bandwidth of a circuit needed to pass the wave.

KEY TERMS

Air-core transformer
All-pass filter
Antilog
Attenuation
Autotransformer
Bandpass filter
Band-reject filter
Bandstop filter
Bandwidth
Bessel filter
Butterworth filter
Capacitor
Cauer filter
Ceramic filter
Chebyshev filter
Commutating filter
Constant-k filter
Coupled circuit
Crystal filter
dBm

Decibel
Dielectric
Electrical isolation
Envelope
Filter
Fourier theory
Gain
Harmonic
High-pass filter
Impedance
Impedance transformation
Inductor
Insertion loss
Iron-core transformer
LC filter
Logarithmic curve
Mutual inductance
Notch filter
Parallel resonant circuit
Passband

Phase inversion
Pulse spectrum
RC filter
Reactance
Resistor
Ripple
Rise time
Series resonant circuit
Shape factor
Skin effect
Step down
Step up
Stop band
Surface acoustic wave
(SAW) filter
Switched capacitor filter
Switched integrator
Transformer
Tuned circuit
Turns ratio

REVIEW

QUESTIONS

1. What happens to capacitive reactance as the frequency of operation increases?
2. As frequency decreases, how does the reactance of a coil vary?
3. What is skin effect and how does it affect the Q of a coil?
4. What happens to a wire when a ferrite bead is placed around it?
5. What is the name given to the widely used coil form that is shaped like a donut?
6. Describe the current and impedance in a series RLC circuit at resonance.
7. Describe the current and impedance in a parallel RLC circuit at resonance.
8. State in your own words the relationship between Q and the bandwidth of a tuned circuit.
9. What kind of filter is used to select a single signal frequency from many signals?
10. What kind of filter would you use to get rid of an annoying 120-Hz hum?
11. What does selectivity mean?
12. State the Fourier theory in your own words.
13. Define the terms time domain and frequency domain.
14. Write the first four odd harmonics of 800 Hz.
15. What waveform is made up of even harmonics only? What waveform is made up of odd harmonics only?
16. Why is a nonsinusoidal signal distorted when it passes through a filter?

PROBLEMS

1. What is the gain of an amplifier with an output of 1.5 V and an input of 30 μV?
2. What is the attenuation of a voltage divider like that in Fig. 2-3, where R_1 is 3.3 kΩ and R_2 is 5.1 kΩ?
3. What is the overall gain or attenuation of the combination formed by cascading the circuits described in Questions 1 and 2?
4. Three amplifiers with gains of 15, 22, and 7 are cascaded; the input voltage is 120 μV. What are the overall gain and the output voltages of each stage?
5. A piece of communications equipment has two stages of amplification with gains of 40 and 60 and two loss stages with attenuation factors of 0.03 and 0.075. The output voltage is 2.2 V. What are the overall gain (or attenuation) and the input voltage?
6. Find the voltage gain or attenuation, in decibels, for each of the circuits described in Questions 1 through 5.
7. A power amplifier has an output of 200 W and an input of 8 W. What is the power gain in decibels?
8. A power amplifier has a gain of 55 dB. The input power is 600 mW. What is the output power?
9. An amplifier has an output of 5 W. What is its gain in dBm?
10. A communications system has five stages, with gains and attenuations of 12, −45, 68, −31, and 9 dB. What is the overall gain?
11. What is the reactance of a 7-pF capacitor at 2 GHz?

12. What value of capacitance is required to produce 50 ohms of reactance at 450 MHz?
13. Calculate the inductive reactance of a 0.9-μH coil at 800 MHz.
14. At what frequency will a 2-μH inductor have a reactance of 300 Ω?
15. A 2.5-μH inductor has a resistance of 23 Ω. At a frequency of 35 MHz what is its Q?
16. What is the resonant frequency of a 0.55-μH coil with a capacitance of 22 pF?
17. What is the value of inductance that will resonate with an 80-pF capacitor at 18 MHz?
18. What is the bandwidth of a parallel resonant circuit that has an inductance of 33 μH with a resistance of 14 Ω and a capacitance of 48 pF?
19. A series resonant circuit has upper and lower cutoff frequencies of 72.9 and 70.5 MHz. What is its bandwidth?
20. A resonant circuit has a peak output voltage of 4.5 mV. What is the voltage output at the upper and lower cutoff frequencies?
21. What circuit Q is required to give a bandwidth of 36 MHz at a frequency of 4 GHz?
22. Find the impedance of a parallel resonant circuit with $L = 60$ μH, $R_w = 7$ Ω, and $C = 22$ pF.
23. Write the first four terms of the Fourier equation of a sawtooth wave that has a peak-to-peak amplitude of 5 V and a frequency of 100 kHz.
24. An oscilloscope has a rise time of 8 ns. What is the highest-frequency sine wave that the scope can display?
25. A low-pass filter has a cutoff frequency of 24 MHz. What is the fastest rise time that a rectangular wave that will pass through the filter can have?

CRITICAL THINKING

1. Explain how capacitance and inductance can exist in a circuit without lumped capacitors and inductor components being present.
2. How can the voltage across the coil or capacitor in a series resonant circuit be greater than the source voltage at resonance?
3. What type of filter would you use to prevent the harmonics generated by a transmitter from reaching the antenna?
4. What kind of filter would you use on a TV set to prevent a signal from a CB radio on 27 MHz from interfering with a TV signal on channel 2 at 54 MHz?
5. Explain why it is possible to reduce the effective Q of a parallel resonant circuit by connecting a resistor in parallel with it.
6. A parallel resonant circuit has an inductance of 800 nH, a winding resistance of 3 Ω, and a capacitance of 15 pF. Calculate (a) resonant frequency, (b) Q, (c) bandwidth, (d) impedance at resonance.
7. For the previous circuit, what would the bandwidth be if you connected a 33-kΩ resistor in parallel with the tuned circuit?
8. Calculate the values of L and C for a constant-k T-network high-pass filter with a load impedance of 300 Ω and a cutoff frequency of 500 MHz.
9. What is the minimum bandwidth needed to pass a periodic pulse train whose frequency is 28.8 KHz and duty cycle is 20 percent? 50 percent?
10. Refer to Fig. 2-74. Examine the various waveforms and Fourier expressions. What circuit do you think might make a good but simple frequency doubler?

CHAPTER 3

AMPLITUDE
MODULATION
FUNDAMENTALS

Objectives

After completing this chapter, you will be able to:

◆ *Calculate* the modulation index and percentage of modulation of an AM signal given the amplitudes of the carrier and modulating signals.

◆ *Define* overmodulation and *explain* how to alleviate its effects.

◆ *Explain* how the power in an AM signal is distributed between the carrier and sideband, and then *compute* the carrier and sideband powers given the percentage of modulation.

◆ *Compute* sideband frequencies given carrier and modulating signal frequencies.

◆ *Compare* time-domain, frequency-domain, and phasor representations of an AM signal.

◆ *Explain* what is meant by the terms DSB and SSB and *state* the main advantages of an SSB signal over a conventional AM signal.

◆ *Calculate* peak envelope power (PEP) given signal voltages and load impedances.

Information or intelligence signals such as voice, video, or binary data are sometimes transmitted from one point to another over a communications medium. However, when the distances involved are long, radio transmission is used. Because interference between signals would result if information signals were transmitted at their original frequency, it is necessary to modulate the signals. In the modulation process, the baseband voice, video, or digital signal modifies another, higher-frequency signal called the carrier, which is usually a sine wave. A sine-wave carrier can be modified by the intelligence signal through amplitude modulation, frequency modulation, or phase modulation. The focus of this chapter is amplitude modulation (AM).

3-1 AM CONCEPTS

As the name suggests, in AM, the information signal varies the amplitude of the carrier sine wave. The instantaneous value of the carrier amplitude changes in accordance with the amplitude and frequency variations of the modulating signal. Fig. 3-1 shows a single-frequency sine-wave intelligence signal modulating a higher-frequency carrier. The carrier frequency remains constant during the modulation process, but its amplitude varies in accordance with the modulating signal. An increase in the amplitude of the modulating signal causes the amplitude of the carrier to increase. Both the positive and negative peaks of the carrier wave vary with the modulating signal. An increase or decrease in the amplitude of the modulating signal causes a corresponding increase or decrease in both the positive and negative peaks of the carrier amplitude.

An imaginary line connecting the positive peaks and negative peaks of the carrier waveform (the dashed line in Fig. 3-1), gives the exact shape of the modulating information signal. This imaginary line on the carrier waveform is known as the *envelope*.

Because complex waveforms like that shown in Fig. 3-1 are difficult to draw, they are often simplified by representing the high-frequency carrier wave as many equally spaced vertical lines whose amplitudes vary in accordance with a modulating signal, as in Fig. 3-2. This method of representation is used throughout this book.

The signals illustrated in Figs. 3-1 and 3-2 show the variation of the carrier amplitude with respect to time and are said to be in the time domain. Time-domain sig-

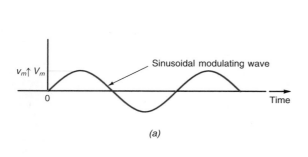

(a)

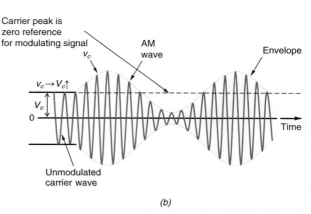

(b)

FIG. 3-1 Amplitude modulation. (*a*) The modulating or information signal. (*b*) The modulated carrier.

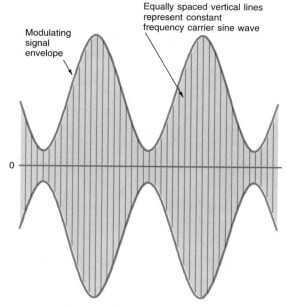

Modulating signal envelope

Equally spaced vertical lines represent constant frequency carrier sine wave

0

FIG. 3-2   A simplified method of representing an AM high-frequency sine wave.

nals, voltage or current variations that occur over time, are what is displayed on the screen of an oscilloscope.

Using trigonometric functions, we can express the sine-wave carrier with the simple expression

$$v_c = V_c \sin 2\pi f_c t$$

In this expression, v_c represents the instantaneous value of the carrier sine-wave voltage at some specific time in the cycle; v_c represents the peak value of the constant unmodulated carrier sine wave as measured between zero and the maximum amplitude of either the positive-going or negative-going alternations (Fig. 3-1); f_c is the frequency of the carrier sine wave; and t is a particular point in time during the carrier cycle.

A sine-wave modulating signal can be expressed with a similar formula:

$$v_m = V_m \sin 2\pi f_m t$$

where v_m = instantaneous value of intelligence signal
V_m = peak amplitude of intelligence signal
f_m = frequency of modulating signal

In Fig. 3-1, the modulating signal uses the peak value of the carrier rather than zero as its reference point. The envelope of the modulating signal varies above and below the peak carrier amplitude. That is, the zero reference line of the modulating signal coincides with the peak value of the unmodulated carrier. Because of this, the relative amplitudes of the carrier and modulating signal are important. In general, the amplitude of the modulating signal should be less than the amplitude of the carrier. When the amplitude of the

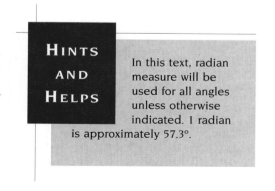

HINTS AND HELPS

In this text, radian measure will be used for all angles unless otherwise indicated. 1 radian is approximately 57.3°.

modulating signal is greater than the amplitude of the carrier, distortion will occur, causing incorrect information to be transmitted. In amplitude modulation, it is particularly important that the peak value of the modulating signal be less than the peak value of the carrier. Mathematically,

$$V_m < V_c$$

TRIGONOMETRIC REPRESENTATION OF SINE WAVES

A sine wave can be represented by a mathematical expression using trigonometry. If the sine wave is a voltage, as in Fig. A, then the instantaneous voltage v is given by the expression

$$v = V_p \sin \theta$$

where V_p is the peak value of the sine wave and θ (theta) is the angle along the horizontal axis. One cycle of a sine wave is spread out over 360°. To find the instantaneous value of the voltage, the peak amplitude is multiplied by the sine of the angle. For example, assume a sine wave with a peak amplitude of 15 V and an angle of 60°. The instantaneous voltage at 60° is (see Fig. A)

$$v = 15 \sin 60° = 15(0.866) = 13 \text{ V}$$

The angle can also be expressed in radians. A radian (rad) is equal to approximately 57.3°. There are 2π rad per cycle of a sine wave, as shown in Fig. B. Thus there are $2\pi ft$ rad in a sine wave, where f is the frequency of the sine wave and t is a time value at some point in the cycle; $2\pi f$ is called the angular velocity and is usually represented

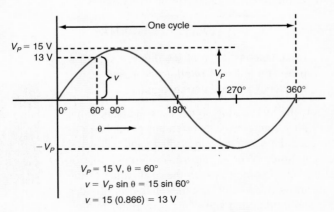

$$V_p = 15 \text{ V}, \theta = 60°$$
$$v = V_p \sin \theta = 15 \sin 60°$$
$$v = 15 \ (0.866) = 13 \text{ V}$$

Fig. A

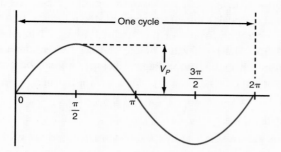

Fig. B

by the lowercase Greek letter omega (ω). Thus the angle is equal to ωt rad, and the sine-wave expression becomes

$$v = V_p \sin 2\pi f t = V_p \sin \omega t$$

While the horizontal axis of a sine wave is sometimes marked off in degrees or radians to calculate the instantaneous value of voltage or current, usually the horizontal axis shows time increments, as in Fig. C.

The instantaneous voltage at a specific time in the cycle can be computed as follows. Assume a sine wave with a peak voltage of 20 V and a frequency of 5 MHz (Fig. C). The time from the beginning of the cycle t is 70 ns. The instantaneous voltage at that time is

$$v = 20 \sin 2\pi (5 \times 10^6) \, (70 \times 10^{-9}) = 20 \sin 2.198$$

where 2.198 is the angle in radians. (Note that most calculators can find the sine of an angle in radians directly.) Then

$$v = 20 \sin 2.198 = 20 \, (0.80967) = 16.19 \text{ V}$$

Converting radians to degrees yields the same instantaneous value. An angle of 2.198 rad is equal to $2.198 \times 57.3 = 125.9°$. Then

$$v = 20 \sin 125.9° = 20(0.80967) = 16.19 \text{ V}$$

In this text we will use radian measure for all angles unless otherwise indicated.

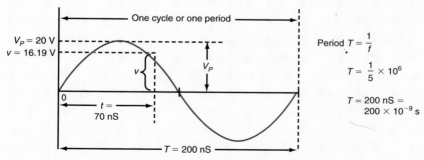

Fig. C

harmonic distortion

sum difference
2 original freq
double carrier freq
tri "intelligence"
double "
triple

non linear to do
amplifier to
mixing or modulation
algebraic sum

Values for the carrier signal and the modulating signal can be used in a formula to express the complete modulated wave. First, keep in mind that the peak value of the carrier is the reference point for the modulating signal; the value of the modulating signal is added to or subtracted from the peak value of the carrier. The instantaneous value of either the top or bottom voltage envelope, v_1, can be computed using the expression

$$v_1 = V_c + v_m = V_c + V_m \sin 2\pi f_m t$$

which expresses the fact that the instantaneous value of the modulating signal algebraically adds to the peak value of the carrier. Thus we can write the instantaneous value of the complete modulated wave v_2 by substituting v_1 for the peak value of carrier voltage V_c as follows:

$$v_2 = v_1 \sin 2\pi f_c t$$

Now substituting the previously derived expression for v_1 and expanding, we get the following:

$$v_2 = (V_c + V_m \sin 2\pi f_m t) \sin 2\pi f_c t = V_c \sin 2\pi f_c t + (V_m \sin 2\pi f_m t)(\sin 2\pi f_c t)$$

where v_2 is the instantaneous value of the AM wave (or v_{AM}), $V_c \sin 2\pi f_c t$ is the carrier waveform, and $(V_m \sin 2\pi f_m t)(\sin 2\pi f_c t)$ is the carrier waveform multiplied by the modulating signal waveform. It is the second part of the expression that is characteristic of AM. A circuit must be able to produce mathematical multiplication of the carrier and modulating signals in order for AM to occur. The AM wave is the product of the carrier and modulating signals.

The circuit used for producing AM is called a *modulator.* Its two inputs, the carrier and the modulating signal, and the resulting outputs, are shown in Fig. 3-3. Amplitude modulators compute the product of the carrier and modulating signals. Circuits that compute the product of two analog signals are also known as analog multipliers, mixers, converters, product detectors, and phase detectors. A circuit that changes a lower-frequency baseband or intelligence signal to a higher-frequency signal is usually called a modulator. A circuit that changes a higher-frequency signal down to a lower frequency is generally referred to as a mixer. A circuit used to recover the original intelligence signal from an AM wave is known as a detector or demodulator. Mixing and detection applications are discussed in detail in later chapters.

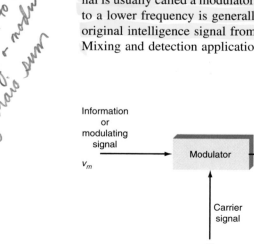

FIG. 3-3 Amplitude modulator showing input and output signals.

3-2 MODULATION INDEX AND
PERCENTAGE OF MODULATION

As stated previously, in order for undistorted AM to occur, the modulating signal voltage V_m must be less than the carrier voltage V_c. Therefore the relationship between the amplitude of the modulating signal and the amplitude of the carrier signal is important. This relationship, known as the *modulation index m* (also called the modulating factor or coefficient, or the degree of modulation), is the ratio

$$m = \frac{V_m}{V_c}$$

These are the peak values of the signals, and the carrier voltage is the unmodulated value.

Multiplying the modulation index by 100 gives the *percentage of modulation*. For example, if the carrier voltage is 9 V and the modulating signal voltage is 7.5 V, the modulation factor is 0.8333 and the percentage of modulation is $0.833 \times 100 = 83.33$.

OVERMODULATION AND DISTORTION

The modulation index should be a number between 0 and 1. If the amplitude of the modulating voltage is higher than the carrier voltage, m will be greater than 1, causing distortion of the modulated waveform. If the distortion is great enough, the intelligence signal becomes unintelligible. Distortion of voice transmissions produces garbled, harsh, or unnatural sounds in the speaker. Distortion of video signals produces a scrambled and inaccurate picture on a TV screen.

Simple distortion is illustrated in Fig. 3-4. Here a sine-wave information signal is modulating a sine-wave carrier, but the modulating voltage is much greater than the

Envelope is no longer the same shape as
original modulating signal

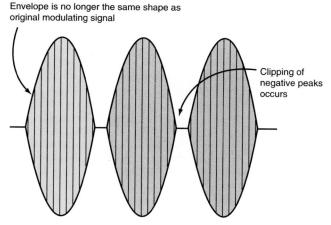

Clipping of
negative peaks
occurs

FIG. 3-4 Distortion of the envelope caused by overmodulation where the modulating signal amplitude V_m is greater than the carrier signal V_c.

carrier voltage, resulting in a condition called *overmodulation*. As you can see, the waveform is flattened at the zero line. The received signal will produce an output waveform in the shape of the envelope, which in this case is a sine wave whose negative peaks have been clipped off. If the amplitude of the modulating signal is less than the carrier amplitude, no distortion will occur. The ideal condition for AM is where $V_m = V_c$, or $m = 1$, which gives 100 percent modulation. This results in the greatest output power at the transmitter and the greatest output voltage at the receiver, with no distortion.

Preventing overmodulation is tricky. For example, at different times during voice transmission voices will go from low amplitude to high amplitude. Normally, the amplitude of the modulating signal is adjusted so that only the voice peaks produce 100 percent modulation. This prevents overmodulation and distortion. Automatic circuits called *compression circuits* solve this problem by amplifying the lower-level signals and suppressing or compressing the higher-level signals. The result is a higher average power output level without overmodulation.

Distortion caused by overmodulation also produces adjacent channel interference. Distortion produces a nonsinusoidal information signal. According to Fourier theory, any nonsinusoidal signal can be treated as a fundamental sine wave at the frequency of the information signal plus harmonics. As stated in Chap. 2, harmonics are sine waves whose frequencies are integer multiples of the fundamental frequency. A distorted 500-Hz sine wave might contain the second, third, fourth, etc., harmonics of 1000 Hz, 1500 Hz, 2000 Hz, etc. Obviously, these harmonics also modulate the carrier, and can cause interference with other signals on channels adjacent to the carrier.

PERCENTAGE OF MODULATION

The modulation index can be determined by measuring the actual values of the modulation voltage and the carrier voltage and computing the ratio. However, it is more common to compute the modulation index from measurements taken on the composite modulated wave itself. When the AM signal is displayed on an oscilloscope, the modulation index can be computed from V_{max} and V_{min} as shown in Fig. 3-5. The peak value of the modulating signal V_m is one-half the difference of the peak and trough values:

$$V_m = \frac{V_{max} - V_{min}}{2}$$

As shown in Fig. 3-5, V_{max} is the peak value of the signal during modulation, and V_{min} is the lowest value, or trough, of the modulated wave. The V_{max} is one-half the peak-to-peak value of the AM signal, or $V_{max(p-p)}/2$. Subtracting V_{min} from V_{max} produces the peak-to-peak value of the modulating signal. One-half of that, of course, is simply the peak value.

The peak value of the carrier signal V_c is the average of the V_{max} and V_{min} values:

$$V_c = \frac{V_{max} + V_{min}}{2}$$

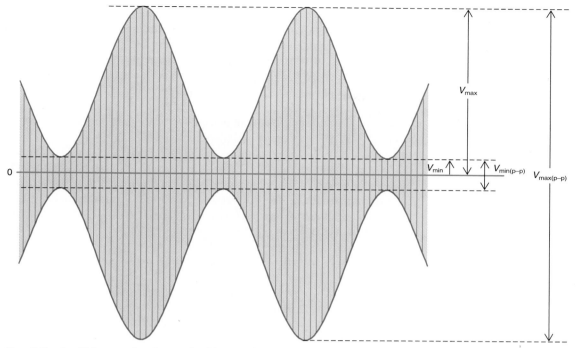

Fig. 3-5 An AM wave showing peaks (V_{max}) and troughs (V_{min}).

The modulation index is:

$$m = \frac{V_{max} - V_{min}}{V_{max} + V_{min}}$$

The values for $V_{max(p-p)}$ and $V_{min(p-p)}$ can be read directly from an oscilloscope screen and plugged directly into the formula to compute the modulation index.

In Carson, California, the Los Angeles County sheriff's deputies work closely with the communication staff to coordinate police activities.

The amount, or depth, of AM is more commonly expressed as the percentage of modulation rather than as a fractional value. In Example 3-1, the percentage of modulation is $100 \times m$, or 66.2 percent. The maximum amount of modulation without signal distortion, of course, is 100 percent, where V_c and V_m are equal. At this time, $V_{min} = 0$ and $V_{max} = 2V_m$, where V_m is the peak value of the modulating signal.

Example 3-1

Suppose that on an AM signal, the $V_{max(p-p)}$ value read from the graticule on the oscilloscope screen is 5.9 divisions and $V_{min(p-p)}$ is 1.2 divisions.

a. What is the modulation index?
$$m\frac{V_{max} - V_{min}}{V_{max} + V_{min}} = \frac{5.9 - 1.2}{5.9 + 1.2} = \frac{4.7}{7.1} = 0.662$$

b. Calculate V_c, V_m, and m if the vertical scale is 2 V per division. (*Hint:* Sketch the signal.)
$$V_c = \frac{V_{max} + V_{min}}{2} = \frac{5.9 + 1.2}{2} = \frac{7.1}{2} = 3.55 \ @ \ \frac{2 \ V}{div}$$
$$V_c = 3.55 \times 2 \ V = 7.1 \ V$$
$$V_m = \frac{V_{max} - V_{min}}{2} = \frac{(5.9 - 1.2)}{2} = \frac{4.7}{2}$$
$$= 2.35 \ @ \ \frac{2 \ V}{div}$$
$$V_m = 2.35 \times 2 \ V = 4.7 \ V$$
$$m = \frac{V_m}{V_c} = \frac{4.7}{7.1} = 0.662$$

3-3 SIDEBANDS AND THE FREQUENCY DOMAIN

Whenever a carrier is modulated by an information signal, new signals at different frequencies are generated as part of the process. These new frequencies, which are called *side frequencies,* or *sidebands,* occur in the frequency spectrum directly above and directly below the carrier frequency. More specifically, the sidebands occur at frequencies that are the sum and difference of the carrier and modulating frequencies. When signals of more than one frequency make up a waveform, it is often better to show the AM signal in the frequency domain rather than the time domain.

SIDEBAND CALCULATIONS

When only a single-frequency sine-wave modulating signal is used, the modulation process generates two sidebands. If the modulating signal is a complex wave, such as voice or video, a whole range of frequencies modulates the carrier, and thus a whole range of sidebands is generated.

The upper sideband f_{USB} and lower-sideband f_{LSB} are computed as

$$f_{USB} = f_c + f_m \qquad \text{and} \qquad f_{LSB} = f_c - f_m$$

where f_c is the carrier frequency and f_m is the modulating frequency.

The existence of sidebands can be demonstrated mathematically, starting with the equation for an AM signal described previously:

$$v_{AM} = V_c \sin 2\pi f_c t + (V_m \sin 2\pi f_m t)(\sin 2\pi f_c t)$$

Using the trigonometric identity that says that the product of two sine waves is

$$\sin A \sin B = \frac{\cos (A - B)}{2} - \frac{\cos (A + B)}{2}$$

and substituting this identity into the expression a modulated wave, the instantaneous amplitude of the signal becomes

$$V_{AM} = V_c \sin 2\pi f_c t + \frac{V_m}{2} \cos 2\pi t (f_c - f_m) - \frac{V_m}{2} \cos 2\pi t (f_c + f_m)$$

where the first term is the carrier; the second term, containing the difference $f_c - f_m$, is the lower sideband; and the third term, containing the sum $f_c + f_m$, is the upper sideband.

For example, assume that a 400-Hz tone modulates a 300-kHz carrier. The upper and lower sidebands are

$$f_{USB} = 300,000 + 400 = 300,400 \text{ Hz or } 300.4 \text{ kHz}$$
$$f_{LSB} = 300,000 - 400 = 299,600 \text{ Hz or } 299.6 \text{ kHz}$$

Observing an AM signal on an oscilloscope, you can see the amplitude variations of the carrier with respect to time. This time-domain display gives no obvious or outward indication of the existence of the sidebands, although the modulation process does indeed produce them, as the equation above shows. An AM signal is really a composite signal formed from several components: the carrier sine wave is added to the upper and lower sidebands, as the equation indicates. This is illustrated graphically in Fig. 3-6. Adding these signals together algebraically at every instantaneous point along the time axis and plotting the result yields the AM wave shown in the figure. It is a sine wave at the carrier frequency whose amplitude varies as determined by the modulating signal.

FREQUENCY-DOMAIN REPRESENTATION OF AM

Another method of showing the sideband signals is to plot the carrier and sideband amplitudes with respect to frequency, as in Fig. 3-7. Here the horizontal axis represents frequency and the vertical axis represents the amplitudes of the signals. The signals may be voltage, current, or power amplitudes and may be given in peak or rms values. A plot of signal amplitude versus frequency is referred to as a *frequency-*

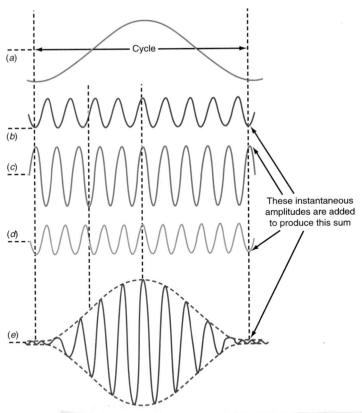

FIG. 3-6 The AM wave is the algebraic sum of the carrier and upper and lower sideband sine waves. (*a*) Intelligence or modulating signal. (*b*) Lower sideband. (*c*) Carrier. (*d*) Upper sideband. (e) Composite AM wave.

domain display. A test instrument known as a *spectrum analyzer* is used to display the frequency domain of a signal.

Figure 3-8 shows the relationship between the time- and frequency-domain displays of an AM signal. The time and frequency axes are perpendicular to one another. The amplitudes shown in the frequency domain display are the peak values of the carrier and sideband sine waves.

Whenever the modulating signal is more complex than a single sine-wave tone, multiple upper and lower sidebands are produced by the AM process. For example, a voice signal consists of many sine wave components of different frequencies mixed

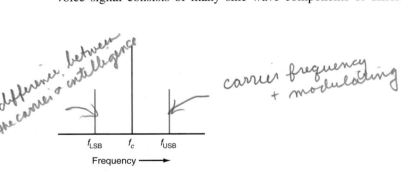

FIG. 3-7 A frequency-domain display of an AM signal.

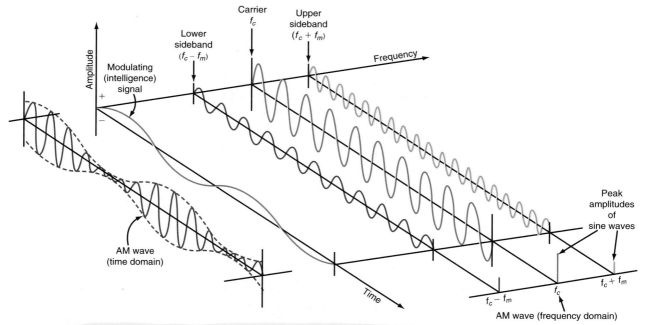

Fig. 3-8 The relationship between the time and frequency domains.

together. Recall that voice frequencies occur in the 300- to 3000-Hz range. Therefore, voice signals produce a range of frequencies above and below the carrier frequency, as shown in Fig. 3-9. These sidebands take up spectrum space. The total bandwidth of an AM signal is calculated by computing the maximum and minimum sideband frequencies. This is done by finding the sum and difference of the carrier frequency and maximum modulating frequency (3000 Hz, or 3 kHz, in Fig. 3-9). For example, if the carrier frequency is 2.8 MHz (2800 kHz), then the maximum and minimum sideband frequencies are

$$f_{USB} = 2800 + 3 = 2803 \text{ kHz} \qquad \text{and} \qquad f_{LSB} = 2800 - 3 = 2797 \text{ kHz}$$

The total bandwidth is simply the difference between the upper and lower sideband frequencies:

$$BW = f_{USB} - f_{LSB} = 2803 - 2797 = 6 \text{ kHz}$$

carrier freq.
53 MHz
voice 300 - 3 kHz
5.3 × 10⁷
5.2 ×

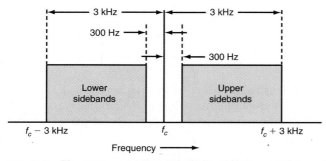

Fig. 3-9 The upper and lower sidebands of a voice modulator AM signal.

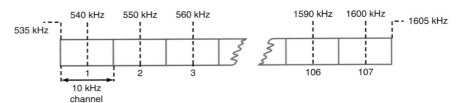

FIG. 3-10 Frequency spectrum of AM broadcast band.

As it turns out, the bandwidth of an AM signal is twice the highest frequency in the modulating signal: BW = $2f_m$, where f_m is the maximum modulating frequency. In the case of a voice signal whose maximum frequency is 3 kHz, the total bandwidth is simply

$$BW = 2(3 \text{ kHz}) = 6 \text{ kHz}$$

Example 3-2

A standard AM broadcast station is allowed to transmit modulating frequencies up to 5 kHz. If the AM station is transmitting on a frequency of 980 kHz, compute the maximum and minimum upper and lower sidebands and the total bandwidth occupied by the AM station.

$f_{USB} = 980 + 5 = 985 \text{ kHz}$
$f_{LSB} = 980 - 5 = 975 \text{ kHz}$
$BW = f_{USB} - f_{LSB} = 985 - 975 = 10 \text{ kHz}$ or
$BW = 2(5 \text{ kHz}) = 10 \text{ kHz}$

As Example 3-2 indicates, an AM broadcast station has a total bandwidth of 10 kHz. In addition, AM broadcast stations are spaced every 10 kHz across the spectrum from 540 kHz to 1600 kHz. This is illustrated in Fig. 3-10. The sidebands from the first AM broadcast frequency extend down to 535 kHz and up to 545 kHz, forming a 10-kHz channel for the signal. The highest channel frequency is 1600 kHz, with sidebands extending from 1595 Khz up to 1605 kHz. There are a total of 107 10-kHz-wide channels for AM radio stations.

PULSE MODULATION

When complex signals such as pulses or rectangular waves modulate a carrier, a broad spectrum of sidebands is produced. According to Fourier theory, complex signals such as square waves, triangular waves, sawtooth waves, and distorted sine waves are simply made up of a fundamental sine wave and numerous harmonic signals at different amplitudes. Assume that a carrier is amplitude-modulated by a square wave which is

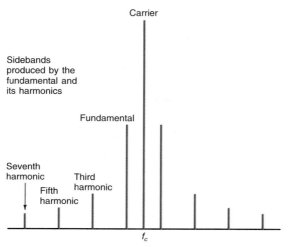

FIG. 3-11   Frequency spectrum of an AM signal modulated by a square wave.

made up of a fundamental sine wave and all odd harmonics. A modulating square wave will produce sidebands at frequencies based upon the fundamental sine wave as well as at the third, fifth, seventh, etc., harmonics, resulting in a frequency-domain plot like that shown in Fig. 3-11. As can be seen, pulses generate extremely wide bandwidth signals. In order for a square wave to be transmitted and faithfully received without distortion or degradation, all of the most significant sidebands must be passed by the antennas and the transmitting and receiving circuits.

Figure 3-12 shows the AM wave resulting when a square wave modulates a sine-wave carrier. In Fig. 3-12(*a*), the percentage of modulation is 50; in Fig. 3-12(*b*), it is 100. In this case, when the square wave goes negative, it drives the carrier amplitude to zero. Amplitude modulation by square waves or rectangular binary pulses is referred to as *amplitude shift keying (ASK)*. ASK is used in some types of data communications where binary information is to be transmitted.

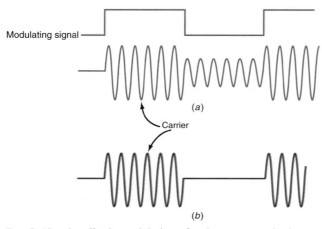

FIG. 3-12   Amplitude modulation of a sine-wave carrier by a pulse or rectangular wave is called amplitude shift keying (ASK). (*a*) Fifty percent modulation. (*b*) One hundred percent modulation.

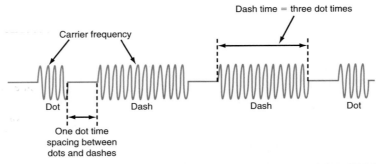

FIG. 3-13 Sending the letter P by Morse code. An example of ON–OFF keying (OOK).

Another crude type of amplitude modulation can be achieved by simply turning the carrier off and on. An example is transmitting Morse code using dots and dashes. A dot is a short burst of carrier while a dash is a longer burst of carrier. Figure 3-13 shows the transmission of the letter P, which is dot-dash-dash-dot (pronounced "dit-dah-dah-dit"). The time duration of a dash is three times the length of a dot, and the spacing between dots and dashes is one dot time. Code transmissions like this are usually called continuous-wave (CW) transmissions. This kind of transmission is also referred to as ON–OFF keying (OOK). Despite the fact that only the carrier is being transmitted, sidebands are generated by such ON–OFF signals. The sidebands result from the frequency or repetition rate of the pulses themselves plus their harmonics.

As indicated earlier, the distortion of an analog signal by overmodulation also generates harmonics. For example, the spectrum produced by a 500-Hz sine wave modulating a carrier of 1 MHz is shown in Fig. 3-14(*a*). The total bandwidth of the signal is 1 kHz. However, if the modulating signal is distorted, second, third, fourth, and higher harmonics are generated. These harmonics also modulate the carrier, producing many more sidebands, as illustrated in Fig. 3-14(*b*). Assume that the distortion is such that the harmonic amplitudes beyond the fourth harmonic are insignificant (usually less than 1 percent); then the total bandwidth of the resulting signal is about 4 kHz instead of the 1-kHz bandwidth that would result without overmodulation and distortion. The harmonics can overlap into adjacent channels, where other signals may be present and interfere with them. Such harmonic sideband interference is sometimes called *splatter* because of the way it sounds at the receiver. Overmodulation and splatter are easily eliminated simply by reducing the level of the modulating signal using gain control or in some cases by using amplitude-limiting or compression circuits.

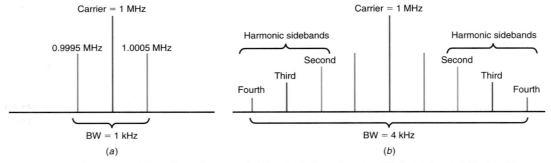

FIG. 3-14 The effect of overmodulation and distortion on AM signal bandwidth. (*a*) Sine-wave of 500 Hz modulating a 1-MHz carrier. (*b*) Distorted 500-Hz sine wave with significant second, third, and fourth harmonics.

Phasors are sometimes used to represent sine waves in circuit analysis. A phasor is a line or arrow whose length is proportional to the peak value of the sine wave being represented. The phasor is assumed to be rotating in the counterclockwise direction on an axis, as shown in Fig. 3-15(a). The angle between the phasor and the horizontal axis represents the angle during the cycle at some given time.

The phasor forms a right triangle with the axis, the height of which is proportional to the sine of the angle. Plotting the height values for the angles from zero to 360° or 2π rad yields the curve shown in Fig. 3-15(b), which traces out one cycle of the sine wave for one complete 360° rotation of the phasor.

Phasor addition is illustrated in Fig. 3-15(c). This is the equivalent of adding two sine waves of the same frequency but of different amplitudes and phase (time) positions. The phasors have different lengths due to the different peak amplitudes of the sine waves they represent. The addition is accomplished by forming a parallelogram with the phasors and drawing the diagonal whose length is the sum of the phasors.

To represent an AM signal with phasors, separate phasors for the carrier and for each sideband are used. Their lengths represent the peak values of the carrier and upper and lower sideband voltages. Assume a simple AM signal that results when a sine-wave carrier is modulated 100 percent by a sine-wave intelligence signal. Adding these three sine waves produces an AM wave like that shown in Fig. 3-6. A phasor representation of this AM signal is shown in Fig. 3-15(d). The sideband phasors are shown at the end of the carrier phasor. As the carrier phasor rotates, tracing out the carrier, the two sideband phasors also rotate around the point of the carrier phasor. The lower sideband phasor rotates at a speed slightly lower than that of the carrier because it is at a lower frequency. The upper sideband phasor rotates at a frequency slightly higher than the carrier phasor. To indicate this, the direction arrows on the sideband phasors in Fig. 3-15(d) are shown rotating in opposite directions. The resultant, or phasor sum of the sideband phasors, is added to the carrier phasor. For 100 percent modulation, the sideband phasor lengths are one-half the carrier length.

Figure 3-15(e) shows the amplitudes resulting when the phasors are at various positions in their cycles. At position 1, when the two sideband phasors coincide, with both pointing upward, they add, producing a resultant twice the sideband amplitude, which adds to the carrier phasor. This is the result for 100 percent modulation. At position 2, when the phasors are 180° apart, they effectively cancel one another, resulting in a sum that is just the carrier amplitude alone. At position 3, when two sideband phasors coincide, with both pointing downward, they again add, forming a resultant that is twice the sideband amplitude. This time, however, they subtract from the carrier, producing a zero resultant.

It is perhaps difficult to visualize three phasors all rotating at different speeds, and therefore adding and subtracting, producing the AM wave. Figure 3-16 shows one way to go about conceptualizing this process. First, assume that the carrier phasor remains fixed, pointing upward, while the upper and lower sideband phasors are allowed to rotate. The resulting wave plotted by the algebraic summation of all the phasors is the envelope of the AM wave which is the intelligence signal or modulating sine wave. Next, imagine the carrier phasor rotating as well; the signal traced out is the complete AM wave.

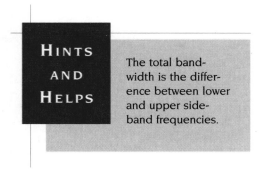

HINTS AND HELPS

The total bandwidth is the difference between lower and upper sideband frequencies.

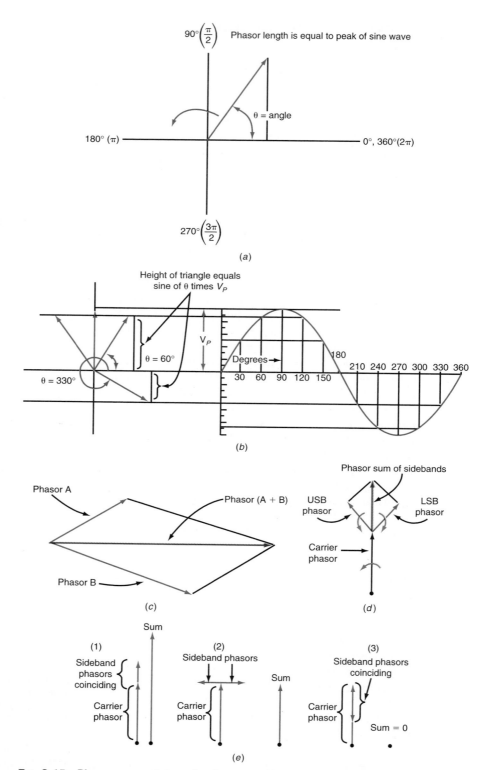

Fig. 3-15 Phasor representation of a sine wave. (*a*) Rotating vector or phasor is another way to represent a sine wave. (*b*) Relating the phasor and time-domain presentations of a sine wave. (*c*) Adding to phasors by drawing a parallelogram and the diagonal. (*d*) AM phasors. (*e*) Phasor additions at various positions.

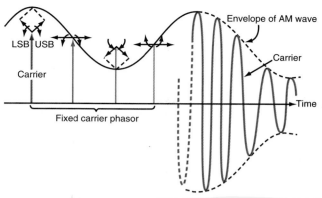

Fig. 3-16 Assuming a fixed vertical carrier phasor, as the sideband phasors rotate, their algebraic sum traces out the envelope of the carrier.

Time- or frequency-domain displays of AM signals are more often used than phasor diagrams. However, phasor plots can help give a visual representation of some types of circuits.

3-4 AM POWER

In radio transmission, the AM signal is amplified by a power amplifier and fed to the antenna with a characteristic impedance which is ideally, but not necessarily almost pure resistance. The AM signal is really a composite of several signal voltages, namely the carrier and the two sidebands, and each of these signals produces power in the antenna. The total transmitted power P_T is simply the sum of the carrier power P_c and the power in the two sidebands, P_{USB} and P_{LSB}:

$$P_T = P_c + P_{LSB} + P_{USB}$$

You can see how the power in an AM signal is distributed and calculated by going back to the original AM equation:

$$V_{AM} = V_c \sin 2\pi f_c t + \frac{V_m}{2} \cos 2\pi t(f_c - f_m) - \frac{V_m}{2} \cos 2\pi t(f_c + f_m)$$

where the first term is the carrier, the second term is the lower sideband, and the third term is the upper sideband.

Now, remember that V_c and V_m are peak values of the carrier and modulating sine waves. For power calculations, rms values must be used for the voltages. We can convert from peak to rms by dividing the peak value by $\sqrt{2}$ or multiplying by 0.707. The rms carrier and sideband voltages are then

$$V_{AM} = \frac{V_c}{\sqrt{2}} \sin 2\pi f_c t + \frac{V_m}{2\sqrt{2}} \cos 2\pi t(f_c - f_m) - \frac{V_m}{2\sqrt{2}} \cos 2\pi t(f_c + f_m)$$

The power in the carrier and sidebands can be calculated by using the power formula $P = V^2/R$, where P is the output power, V is the rms output voltage, and R is the resistive part of the load impedance, which is usually an antenna. We just need to use the coefficients on the sine and cosine terms above in the power formula:

$$P_T = \frac{(V_c/\sqrt{2})^2}{R} + \frac{(V_m/2\sqrt{2})^2}{R} + \frac{(V_m/2\sqrt{2})^2}{R} = \frac{(V_c)^2}{2R} + \frac{(V_m)^2}{8R} + \frac{(V_m)^2}{8R}$$

Remembering that we can express the modulating signal V_m in terms of the carrier V_c by using the expression given earlier for the modulation index, $m = V_m/V_c$, we can write

$$V_m = mV_c$$

Expressing the sideband powers in terms of the carrier power, the total power becomes

$$P_T = \frac{(V_c)^2}{2R} + \frac{(mV_c)^2}{8R} + \frac{(mV_c)^2}{8R} = \frac{(V_c)^2}{2R} + \frac{(m^2V_c^2)}{8R} + \frac{(m^2V_c^2)}{8R}$$

Since the term $(V_c)^2/2R$ is equal to the rms carrier power P_c, it can be factored out, giving

$$P_T = \frac{(V_c)^2}{2R}\left(1 + \frac{m^2}{4} + \frac{m^2}{4}\right)$$

Finally, we get a handy formula for computing the total power in an AM signal when the carrier power and the percentage of modulation are known:

$$P_T = P_c\left(1 + \frac{m^2}{2}\right)$$

For example, if the carrier of an AM transmitter is 1000 W, and it is modulated 100 percent ($m = 1$), the total AM power is

$$P_T = 1000\left(1 + \frac{1^2}{2}\right) = 1500 \text{ W}$$

Of the total power, 1000 W of it is in the carrier. That leaves 500 W in both sidebands. Since the sidebands are equal in size, each sideband has 250 W.

For a 100 percent modulated AM transmitter, the total sideband power is always one-half that of the carrier power. A 50-kW transmitter carrier that is 100 percent modulated will have a sideband power of 25 kW, 12.5 kW in each sideband. The total power for the AM signal is the sum of the carrier and sideband power, or 75 kW.

When the percentage of modulation is less than the optimum 100, there is much less power in the sidebands. For example, for a 70 percent modulated 250-W carrier, the total power in the composite AM signal is

$$P_T = 250\left(1 + \frac{0.7^2}{2}\right) = 250\,(1 + 0.245) = 311.25 \text{ W}$$

Of the total, 250 W is in the carrier, leaving $311.25 - 250 = 61.25$ W in the sidebands. There is 61.25/2 or 30.625 W in each sideband.

In the real world, it is difficult to determine AM power by measuring the output voltage and calculating power with the expression $P = V^2/R$. However, it is easy to measure the current in the load. It is very common to see an RF ammeter connected

single sideband

6:1 ratio
6x better than AM

$$m = \frac{V_m}{V_c}$$

% of modulation

$$m = \frac{V_{max} - V_{min}}{V_{max} + V_{min}}$$

(p 127)
modulating index

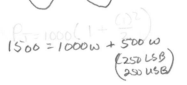

$$P_T = 1000\left(1 + \frac{(1)^2}{2}\right)$$

$1500 = 1000w + 500\,w$

(250 LSB)
(250 USB)

½ power of carrier is in sidebands

Example 3-3

An AM transmitter has a carrier power of 30 W. The percentage of modulation is 85 percent. Calculate (a) the total power, (b) the power in one sideband.

a. $P_T = P_c \left(1 + \dfrac{m^2}{2}\right) = 30 \left[1 + \dfrac{(0.85)^2}{2}\right] = 30\left(1 + \dfrac{0.7225}{2}\right)$

$= 30(1.36125) = 40.8 \text{ W}$

b. $P_{SB} \text{ (both)} = P_T - P_c = 40.8 - 30 = 10.8 \text{ W}$

$P_{SB} \text{ (one)} = \dfrac{P_{SB}}{2} = \dfrac{10.8}{2} = 5.4 \text{ W}$

in series with an antenna to observe antenna current. When the antenna impedance is known, the output power is easily calculated using the formula

$$P_T = (I_T)^2 R$$

where $I_T = I_c \sqrt{(1 + m^2/2)}$. I_c is the unmodulated carrier current in the load and m is the modulation index. For example, the total output power of an 85 percent modulated AM transmitter, whose unmodulated carrier current into a 50-Ω antenna load impedance is 10A, is

$$I_T = 10\sqrt{\left(1 + \frac{0.85^2}{2}\right)} = 10\sqrt{(1.36125)} = 11.67 \text{ A}$$
$$P_T = 11.67^2 \, (50) = 136.2 \, (50) = 6809 \text{ W}$$

One way to find the percentage of modulation is to measure both the modulated and unmodulated antenna currents. Then, by algebraically rearranging the formula above, m can be calculated directly:

$$m = \sqrt{2\left[\left(\frac{I_T}{I_c}\right)^2 - 1\right]}$$

Suppose that the unmodulated antenna current is 2.2 A. That is the current produced by the carrier only, or I_c. Now, if the modulated antenna current is 2.6 A, the modulation index is

$$m = \sqrt{2\left[\left(\frac{2.6}{2.2}\right)^2 - 1\right]} = \sqrt{2[(1.18)^2 - 1]} = \sqrt{0.7934} = 0.89$$

The percentage of modulation is 89.

As you can see, the power in the sidebands depends on the value of the modulation index. The greater the percentage of modulation, the higher the sideband power and the higher the total power transmitted. Of course, maximum power appears in the sidebands when the carrier is 100 percent modulated. The power in each sideband P_{SB} is given by the following expression:

$$P_{SB} = P_{LSB} = P_{USB} = \frac{P_c m^2}{4}$$

Assuming 100 percent modulation where the modulation factor $m = 1$, the power in each sideband is one-fourth, or 25 percent, of the carrier power. Since there are two sidebands, their power together represents 50 percent of the carrier power. For example, if the carrier power is 100 W, then at 100 percent modulation, 50 W will appear in the sidebands, 25 W in each. The total transmitted power, then, is the sum of the carrier and sideband powers, or 150 W. The goal in AM is to keep the percentage of modulation as high as possible without overmodulation so that maximum sideband power is transmitted.

The carrier power represents two-thirds of the total transmitted power. Assuming 100 W and a total power of 150 W, the carrier power percentage is 100/150 = 0.667, or 66.7 percent. The sideband power percentage is thus 50/150 = 0.333, or 33.3 percent.

The carrier itself conveys no information. The carrier can be transmitted and received, but unless modulation occurs, no information will be transmitted. When modulation occurs, sidebands are produced. It is easy to conclude, therefore, that all the transmitted information is contained within the sidebands. Only one-third of the total transmitted power is allotted to the sidebands, while the remaining two-thirds of it is literally wasted on the carrier.

At lower percentages of modulation, the power in the sidebands is even less. For example, assuming a carrier power of 500 W and a modulation of 70 percent, the power in each sideband is

$$P_{SB} = \frac{P_c m^2}{4} = \frac{500(0.7)^2}{4} = \frac{500(0.49)}{4} = 61.25 \text{ W}$$

and the total sideband power is 122.5 W. The carrier power, of course, remains unchanged at 500 W.

As stated previously, complex voice and video signals vary over a wide amplitude and frequency range, and 100 percent modulation only occurs on the peaks of

Example 3-4

An antenna has an impedance of 40 Ω. An unmodulated AM signal produces a current of 4.8 A. The modulation is 90 percent. Calculate (*a*) the carrier power, (*b*) the total power, (*c*) the sideband power.

a. $P_c = I^2 R = (4.8)^2 (40) = (23.04)(40) = 921.6 \text{ W}$

b. $I_T = I_c \sqrt{1 + \frac{m^2}{2}} = 4.8 \sqrt{1 + \frac{(0.9)^2}{2}} = 4.8 \sqrt{1 + \frac{0.81}{2}}$

$= 4.8 \sqrt{1.405} = 5.7 \text{A}$

$P_T = I_T^2 R = (5.7)^2 (40) = 32.49(40) = 1295 \text{ W}$

c. $P_{SB} = P_T - P_c = 1295 - 921.6 = 373.4 \text{ W}$ (186.7 W each sideband)

Example 3-5

The transmitter in Example 3-4 experiences an antenna current change from 4.8 A unmodulated to 5.1 A. What is the percentage of modulation?

$$m = \sqrt{2\left[\left(\frac{I_T}{I_c}\right)^2 - 1\right]}$$

$$= \sqrt{2\left[\left(\frac{5.1}{4.8}\right)^2 - 1\right]}$$

$$= \sqrt{2[(1.0625)^2 - 1]}$$

$$= \sqrt{2(1.13 - 1)}$$

$$= \sqrt{2(0.13)}$$

$$= \sqrt{0.26}$$

$$= 0.51$$

The percentage of modulation is 51.

Example 3-6

What is the power in one sideband of the transmitter in Example 3-4?

$$P_{SB} = m^2\,\frac{P_c}{4} = \frac{(0.9)^2\,(921.6)}{4} = \frac{746.5}{4} = 186.6 \text{ W}$$

the modulating signal. For this reason, the average sideband power is considerably lower than the ideal 50 percent that would be produced by 100 percent modulation. With less sideband power transmitted, the received signal is weaker and communication is less reliable.

Despite its inefficiency, AM is still widely used because it is simple and effective. It is used in AM radio broadcasting, CB radio, TV broadcasting, and in some aircraft and marine communications.

3-5 SINGLE-SIDEBAND MODULATION

In amplitude modulation, two-thirds of the transmitted power is in the carrier, which itself conveys no information. The real information is contained within the sidebands. One way to improve the efficiency of amplitude modulation is to suppress the carrier and eliminate one sideband. The result is a single-sideband (SSB) signal. SSB is a form of AM that offers unique benefits in some types of electronic communications.

DSB Signals

The first step in generating an SSB signal is suppressing the carrier, leaving the upper and lower sidebands. This type of signal is referred to as a *double-sideband suppressed carrier (DSSC or DSB)* signal. The benefit, of course, is that no power is wasted on the carrier. Double-sideband suppressed carrier modulation is simply a special case of AM with no carrier.

A typical DSB signal is shown in Fig. 3-17. This signal, the algebraic sum of the two sinusoidal sidebands, is the signal produced when a carrier is modulated by a single-tone sine-wave information signal. The carrier is suppressed, and the time-domain DSB signal is a sine wave at the carrier frequency, varying in amplitude as shown. Note that the envelope of this waveform is not the same as the modulating signal, as it is in a pure AM signal with carrier. A unique characteristic of the DSB signal is the phase transitions that occur at the lower-amplitude portions of the wave. In Fig. 3-17, note that there are two adjacent positive-going half cycles at the null points in the wave. That is one way to tell from an oscilloscope display whether the signal shown is a true DSB signal.

A frequency-domain display of a DSB signal is given in Fig. 3-18. As shown, the spectrum space occupied by a DSB signal is the same as that for a conventional AM signal.

Double-sideband suppressed carrier signals are generated by a circuit called a *balanced modulator.* The purpose of the balanced modulator is to produce the sum and difference frequencies but to cancel or balance out the carrier. Balanced modulators are covered in detail in Chap. 4.

Despite the fact that elimination of the carrier in DSB AM saves considerable power, DSB is not widely used because the signal is difficult to demodulate (recover) at the receiver. One important application for DSB, however, is the transmission of the color information in a TV signal.

SSB Signals

In DSB transmission, since the sidebands are the sum and difference of the carrier and modulating signals, the information is contained in both sidebands. As it turns out, there is no reason to transmit both sidebands in order to convey the information.

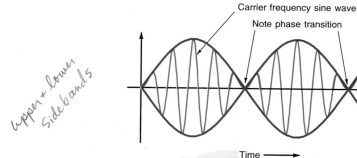

FIG. 3-17 A time-domain display of a DSB AM signal.

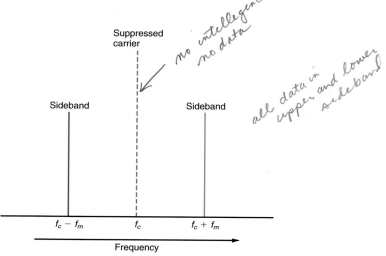

no intelligence no data

all data in upper and lower sideband

Fig. 3-18 A frequency-domain display of DSB signal.

One sideband can be suppressed; the remaining sideband is called a *single-sideband suppressed carrier (SSSC or SSB)* signal. SSB signals offer four major benefits.

1. The primary benefit of an SSB signal is that the spectrum space it occupies is only half that of AM and DSB signals. This greatly conserves spectrum space and allows more signals to be transmitted in the same frequency range.
2. All the power previously devoted to the carrier and the other sideband can be channeled into the single sideband, producing a stronger signal that should carry farther and be more reliably received at greater distances. Alternatively, SSB transmitters can be made smaller and lighter than an equivalent AM or DSB transmitter because less circuitry and power are used.
3. Because SSB signals occupy a narrower bandwidth, the amount of noise in the signal is reduced.
4. There is less selective fading of an SSB signal over long distances. An AM signal is really multiple signals, at least a carrier and two sidebands. These are on different frequencies, so they are affected in slightly different ways by the ionosphere and upper atmosphere, which have a great influence on radio signals of less than about 50 MHz. The carrier and sidebands may arrive at the receiver at slightly different times, causing a phase shift that can, in turn, cause them to add in such a way as to cancel one another rather than add up to the original AM signal. Such cancellation, or *selective fading*, is not a problem with SSB since only one sideband is being transmitted.

not talking no signal transmitted not wasting energy

An SSB signal has some unusual characteristics. First, when no information or modulating signal is present, no RF signal is transmitted. In a standard AM transmitter, the carrier is still transmitted even though it may not be modulated. This is the condition that might occur during a voice pause on an AM broadcast. But since there is no carrier transmitted in an SSB system, no signals are present if the information signal is zero. Sidebands are generated only during

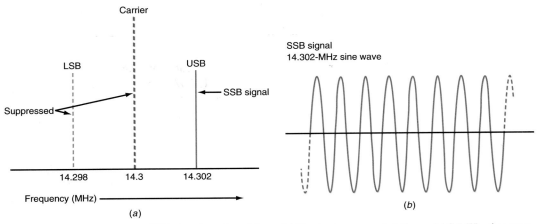

FIG. 3-19 An SSB signal produced by a 2-kHz sine wave modulating a 14.3-MHz sine-wave carrier.

the modulation process, for example, when someone speaks into a microphone. This explains why SSB is so much more efficient than AM.

Figure 3-19 shows the frequency- and time-domain displays of an SSB signal produced when a steady 2-kHz sine-wave tone modulates a 14.3-MHz carrier. Amplitude modulation would produce sidebands of 14.298 and 14.302 MHz. In SSB, only one sideband is used. Figure 3-19(a) shows that only the upper sideband is generated. The RF signal is simply a constant-power 14.302-MHz sine wave. A time-domain display of this SSB signal is shown in Fig. 3-19(b).

Of course, most information signals transmitted by SSB are not pure sine waves. A more common modulation signal is voice, with its varying frequency and amplitude content. The voice signal creates a complex RF SSB signal that varies in frequency and amplitude over the narrow spectrum defined by the voice signal bandwidth. The waveform at the output of the SSB modulator has the same shape as the baseband waveform, but it is shifted in frequency.

DISADVANTAGES OF DSB AND SSB

The main disadvantage of DSB and SSB signals is that they are harder to recover, or demodulate, at the receiver. Demodulation depends upon the carrier being present. If the carrier is not present, then it must be regenerated at the receiver and reinserted back into the signal. To faithfully recover the intelligence signal, the reinserted carrier must have the same phase and frequency as the original carrier. This is a difficult requirement. When SSB is used for voice transmission, the reinserted carrier can be made variable in frequency so that it can be adjusted manually while listening in order to recover an intelligible signal. This is not possible with some kinds of data signals.

To solve this problem, a low-level carrier signal is sometimes transmitted along with the two sidebands in DSB or a single sideband in SSB. Because the carrier has a low power level, the essential benefits of SSB are retained, but a weak carrier is received so that it can be amplified and reinserted to recover the original information. Such a low-level carrier is referred to as a *pilot carrier*. This technique is used in FM stereo transmissions as well as in the transmission of the color information in a TV picture.

SIGNAL POWER CONSIDERATIONS

In conventional AM, the transmitted power is distributed among the carrier and two sidebands. For example, given a carrier power of 400 W with 100 percent modulation, each sideband will contain 100 W of power and the total power transmitted will be 600 W. The effective transmission power is the combined power in the sidebands, or 200 W.

An SSB transmitter sends no carrier, so the carrier power is zero. A given SSB transmitter will have the same communications effectiveness as a conventional AM unit running much more power. For example, a 10-W SSB transmitter offers the performance capabilities of an AM transmitter running a total of 40 W, since they both show 10 W of power in one sideband. The power advantage of SSB over AM is 4:1. 6:1

In SSB, the transmitter output is expressed in terms of *peak envelope power (PEP)*, the maximum power produced on voice amplitude peaks. The PEP is computed by the equation $P = V^2/R$. For example, assume that a voice signal produces a 360-V peak-to-peak signal across a 50 Ω load. The rms voltage is 0.707 times the peak value and the peak value is one-half the peak-to-peak voltage. In this example, the rms voltage is $0.707(360/2) = 127.26$ V.

The peak envelope power is then:

$$PEP = V^2/R = \frac{(127.26)^2}{50} = 324 \text{ W}$$

The PEP input power is simply the DC input power of the transmitter's final amplifier stage at the instant of the voice envelope peak. It is the final amplifier stage DC supply voltage multiplied by the maximum amplifier current that occurs at the peak, or

$$PEP = V_s I_{max}$$

where V_s = amplifier supply voltage
I_{max} = current peak

For example, a 450-V supply with a peak current of 0.8 A produces a PEP of $450(0.8) = 360$ W.

It should be noted that voice amplitude peaks are produced only when very loud sounds are generated during certain speech patterns or when some word or sound is emphasized. During normal speech levels, the input and output power levels are much less than the PEP level. The average power is typically only one-fourth to one-third of the PEP value with typical human speech:

$$P_{avg} = \frac{PEP}{3} \quad \text{or} \quad P_{avg} = \frac{PEP}{4}$$

With a PEP of 240 W, the average power is only 60 to 80 W. Typical SSB transmitters are designed to handle only the average power level on a continuous basis, not the PEP.

HINTS AND HELPS

Because DSB and SSB signals are difficult to demodulate, a low-level carrier signal is sometimes transmitted along with the sideband(s). Because the carrier has a low power level, the benefits of SSB and DSB are retained. The carrier is then amplifed and reinserted to recover the information.

The transmitted sideband will, of course, change in frequency and amplitude as a complex voice signal is applied. This sideband will occupy the same bandwidth as one sideband in a fully modulated AM signal with carrier.

Incidentally, it does not matter whether the upper or lower sideband is used, since the information is contained in either. A filter is typically used to remove the unwanted sideband.

Example 3-7

An SSB transmitter produces a peak-to-peak voltage of 178 V across a 75-Ω antenna load. What is the PEP?

$$V_p = \frac{V_{p-p}}{2} = \frac{178}{2} = 89 \text{ V}$$

$$V_{rms} = 0.707 \qquad V_p = 0.707 \ (89) = 62.9 \text{ V}$$

$$P = \frac{V^2}{R} = \frac{(62.9)^2}{75} = 52.8 \text{ W}$$

$$\text{PEP} = 52.8 \text{ W}$$

Example 3-8

An SSB transmitter has a 24 Vdc power supply. On voice peaks the current achieves a maximum of 9.3 A.

a. What is the PEP?

$$\text{PEP} = V_s I_m = 24 \ (9.3) = 223.2 \text{ W}$$

b. What is the average power of the transmitter?

$$P_{avg} = \frac{\text{PEP}}{3} = \frac{223.2}{3} = 74.4 \text{ W}$$

$$P_{avg} = \frac{\text{PEP}}{4} = \frac{223.2}{4} = 55.8 \text{ W}$$

$$P_{avg} = 55.8 \text{ to } 74.4 \text{ W}$$

APPLICATIONS OF DSB AND SSB

Both DSB and SSB techniques are widely used in communications. Pure SSB signals are used in telephone systems as well as in two-way radio. Two-way SSB communications is used in marine applications, in the military, and by hobbyists known as radio amateurs (hams). DSB signals are used in FM and TV broadcasting to transmit two-channel stereo signals and to transmit the color information for a TV picture. They are also used in some types of phase-shift keying which is used for transmitting binary data.

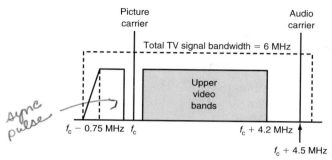

sync pulse

Fig. 3-20 Vestigial sideband transmission of a TV picture signal.

An unusual form of AM is that used in television broadcasting. A TV signal consists of the picture (video) signal and the audio signal, which have different carrier frequencies. The audio carrier is frequency-modulated, but the video information amplitude-modulates the picture carrier. The picture carrier is transmitted, but one sideband is partially suppressed.

Video information typically contains frequencies as high as 4.2 MHz. A fully amplitude-modulated television signal would then occupy 2(4.2) = 8.4 MHz. This is an excessive amount of bandwidth that is wasteful of spectrum space because not all of it is required to reliably transmit a TV signal. To reduce the bandwidth to the 6-MHz maximum allowed by the FCC for TV signals, a portion of the lower sideband of the TV signal is suppressed, leaving only a small part, or vestige, of the lower sideband. This arrangement, known as a *vestigial sideband signal,* is illustrated in Fig. 3-20. Video signals above 0.75 MHz (750 kHz) are suppressed in the lower (vestigial) sideband, while all video frequencies are transmitted in the upper sideband.

> **HINTS AND HELPS**
>
> For SSB transmissions, it does not matter whether the upper or lower sideband is used, since the information is contained in both.

3-6 CLASSIFICATION OF RADIO EMISSIONS

Figure 3-21 shows the codes used to designate the many types of signals that can be transmitted by radio and wire. The basic code is made up of a capital letter and a number, and lowercase subscript letters are used for more specific definition. For example, a basic AM voice signal such as that heard on the AM broadcast band or on a CB or aircraft radio has the code A3. All of the variations of AM using voice or video intelligence have the A3 designation, but subscript letters are used to distinguish them. Examples of codes designating signals described in this chapter are:

DSB two sidebands, full carrier = A3
DSB two sidebands, suppressed carrier = $A3_b$
SSB single sideband, suppressed carrier = $A3_j$
SSB single sideband, 10 percent pilot carrier = $A3_a$
Vestigial sideband TV = $A3_c$
OOK and ASK = A1

Letter	A	Amplitude modulation
	F	Frequency modulation
	P	Phase modulation
Number	0	Carrier ON only, no message (radio beacon)
	1	Carrier ON/OFF, no message (Morse code, radar)
	2	Carrier ON, keyed tone ON–OFF (code)
	3	Telephony, message as voice or music
	4	Fax, nonmoving graphics (slow-scan TV)
	5	Vestigial sideband (commercial TV)
	6	Four-frequency diplex telegraphy
	7	Multiple sideband, each with different message
	8	
	9	General (all others)
Subscripts	None	Double sideband, full carrier
	a	Single sideband, reduced carrier
	b	Double sideband, no carrier
	c	Vestigial sideband
	d	Carrier pulses only, pulse amplitude modulation (PAM)
	e	Carrier pulses only, pulse width modulation (PWM)
	f	Carrier pulses only, pulse position modulation (PPM)
	g	Quantized pulses, digital video
	h	Single sideband, full carrier
	j	Single sideband, no carrier

FIG. 3-21 Radio emission code designations.

Note that there are special designations for fax and pulse transmissions, and that the number 9 covers any special modulation or techniques not covered elsewhere. When there is a number preceding the letter code, the number refers to bandwidth in kHz. For example, the designation 10A3 refers to a 10-kHz-bandwidth voice AM signal. The designation 20A3 h refers to an AM SSB signal with full carrier and message frequency to 20 kHz.

Another system used to describe a signal is given in Fig. 3-22. It is similar to the method just described, but with some variations. This is the definition used by the standards organization International Telecommunications Union (ITU). Some examples are:

A3F   amplitude-modulated analog TV
J3E   SSB voice
F2D   FSK data
G7E   phase-modulated voice, multiple signals

The Los Angeles Fire Department Emergency Command Center requires advanced communications systems to handle the high volume of emergency calls.

Type of Modulation
  N   Unmodulated carrier
  A   Amplitude modulation
  J   Single sideband
  F   Frequency modulation
  G   Phase modulation
  P   Series of pulses, no modulation

Type of Modulating Signals
  0   None
  1   Digital, single channel, no modulation
  2   Digital, single channel, with modulation
  3   Analog, single channel
  7   Digital, two or more channels
  8   Analog, two or more channels
  9   Analog plus digital

Type of Intelligence Signal
  N   None
  A   Telegraphy, human
  B   Telegraphy, machine
  C   Fax
  D   Data, telemetry, control signals
  E   Telephony (human voice)
  F   Video, TV
  W   Some combination of any of the above

FIG. 3-22 ITU emissions designations.

SUMMARY

In amplitude modulation, an increase or decrease in the amplitude of the modulating signal causes a corresponding increase or decrease in both the positive and negative peaks of the carrier amplitude. Interconnecting the adjacent positive or negative peaks of the carrier waveform yields the shape of the modulating information signal, known as the envelope.

Using trigonometric functions, we can form mathematical expressions for the carrier and the modulating signal, and combine these to create a formula for the complete modulated wave. Modulators (circuits that produce amplitude modulation) compute the product of the carrier and modulating signals.

The relationship between the amplitudes of the modulating signal and carrier is expressed as the modulation index (m), a number between 0 and 1. If the amplitude of the modulating voltage is higher than the carrier voltage, $m > 1$, distortion, or overmodulation, will result.

When a carrier is modulated by an information signal, new signals at different frequencies are generated. These side frequencies, or sidebands, occur in the frequency spectrum directly above and below the carrier frequency. An AM signal is a composite of several signal voltages, the carrier, and the two sidebands, each of which produces power in the antenna. Total transmitted power is the sum of the carrier power and the power in the two sidebands.

AM signals can be expressed through time-domain displays or frequency-domain displays, or by using phasor representation.

In AM transmission, two-thirds of the transmitted power appears in the carrier, which itself conveys no information. One way to overcome this wasteful effect is to suppress the carrier. When the carrier is initially suppressed, both the upper and lower sidebands are left, leaving a double-sideband suppressed (DSSC or DSB) signal. Because both sidebands are not necessary to transmit the desired information, one of the remaining sidebands can then be suppressed, leaving a single-sideband (SSB) signal. SSB signals offer important benefits; they conserve spectrum space, produce strong signals, reduce noise, and result in less fading over long distances.

In SSB, the transmitter output is expressed as peak envelope power (PEP), the maximum power produced on voice amplitude peaks.

Both DSB and SSB techniques are widely used in communications. Pure SSB signals are used in telephone systems as well as in two-way radio. Two-way SSB communications are used in marine applications, in the military, and by hams. In some TV applications, to reduce the signal bandwidth to the 6-MHz maximum allowed by the FCC for TV signals, a vestigial sideband signal is used to suppress the lower sideband of the TV signal.

KEY TERMS

Amplitude shift keying (ASK)
Continuous-wave transmission
Distortion
Double-sideband modulation
Double-sideband suppressed carrier (DSSC or DSB)

Envelope
Frequency-domain display
Modulation
Modulation index
ON–OFF keying (OOK)
Overmodulation
Peak envelope power (PEP)
Percentage of modulation

Phasor
Pulse modulation
Sideband
Single-sideband suppressed carrier (SSSC or SSB)
Single-sideband modulation
Time-domain display
Vestigial sideband signal

REVIEW

QUESTIONS

1. Define modulation.
2. Explain why modulation is necessary or desirable.
3. Name the circuit that causes one signal to modulate another and give the names of the two signals applied to this circuit.
4. In AM, how does the carrier vary in accordance with the information signal?
5. True or false. The carrier frequency is usually lower than the modulating frequency.
6. What is the outline of the peaks of the carrier signal called and what shape does it have?
7. What are voltages that vary over time called?
8. Write the trigonometric expression for a sine-wave carrier signal.
9. True or false. The carrier frequency remains constant during AM.
10. What mathematical operation does an amplitude modulator perform?
11. What is the ideal relationship between the modulating signal voltage (V_m) and the carrier voltage (V_c)?
12. What is the modulation index called when it is expressed as a percentage?
13. Explain the effects of a modulation percentage greater than 100.
14. What is the name given to the new signals generated by the modulation process?
15. What is the name of the type of signal that is displayed on an oscilloscope?
16. What is the type of signal whose amplitude components are displayed with respect to frequency called and on what instrument is this signal displayed?
17. Explain why complex nonsinusoidal and distorted signals produce a greater bandwidth AM signal than a simple sine-wave signal of the same frequency.
18. What three signals can be added together to give an AM wave?
19. What is the name given to an AM signal whose carrier is modulated by binary pulses?
20. What is the value of phasor representation of AM signals?
21. True or false. The modulating signal appears in the output spectrum of an AM signal.
22. What percentage of the total power in an AM signal is in the carrier? One sideband? Both sidebands?
23. What type of load is used to dissipate an AM signal?
24. Does the carrier of an AM signal contain any information? Explain.
25. What is the name of a signal that has both sidebands but no carrier?
26. What is the name of the circuit that is used to eliminate the carrier in DSB and SSB transmissions?
27. What is the minimum AM signal that can be transmitted and still convey all of the necessary intelligence?
28. State the four main benefits of SSB over conventional AM.
29. Name two applications for SSB and two applications for DSB.
30. Name the type of AM used in TV picture transmission. Why is it used? Draw the frequency-domain spectrum of the TV signal.
31. Using Figs. 3-21 and 3-22, write the designations for a frequency-modulated radio station signal and an amplitude modulated (V_{SB}) analog fax signal.

1. Give the formula for modulation index and explain its terms.
2. An AM wave displayed on an oscilloscope has values of $V_{max} = 4.8$ and $V_{min} = 2.5$ as read from the graticule. What is the percentage of modulation?
3. What is the ideal percentage of modulation for maximum amplitude of information transmission?
4. To achieve 75 percent modulation of a carrier of $V_c = 50$ V, what amplitude of the modulating signal V_m is needed?
5. The maximum peak-to-peak value of an AM wave is 45 V. The peak-to-peak value of the modulating signal is 20 V. What is the percentage of modulation?
6. What is the mathematical relationship of the carrier and modulating signal voltages when over modulation occurs?
7. An AM radio transmitter operating on 3.9 MHz is modulated by frequencies up to 4 kHz. What are the maximum upper and lower side frequencies? What is the total bandwidth of the AM signal?
8. What is the bandwidth of an AM signal whose carrier is 2.1 MHz modulated by a 1.5-kHz square wave with significant harmonics up to the fifth? Calculate all of the upper and lower sidebands produced.
9. How much power appears in one sideband of an AM signal of a 5-kW transmitter modulated by 80 percent?
10. What is the total power supplied by an AM transmitter with a carrier power of 2500 W and modulation of 77 percent?
11. An AM signal has a 12-W carrier and 1.5 W in each sideband. What is the percentage of modulation?
12. An AM transmitter puts a carrier of 6 A into an antenna whose resistance is 52 Ω. The transmitter is modulated by 60 percent. What is the total output power?
13. The antenna current produced by an unmodulated carrier is 2.4 A into an antenna with a resistance of 75 Ω. When amplitude-modulated, the antenna current rises to 2.7 A. What is the percentage of modulation?
14. A ham transmitter has a carrier power of 750 W. How much power is added to the signal when the transmitter is 100 percent modulated?
15. An SSB transmitter has a power supply voltage of 250 V. On voice peaks, the final amplifier draws a current of 3.3 A. What is the PEP input power?
16. The peak-to-peak output voltage of 675 V appears across a 52-Ω antenna on voice peaks in an SSB transmitter. What is the PEP output power?
17. What is the average output power of an SSB transmitter rated at 100 W PEP?
18. An SSB transmitter with a carrier of 2.3 MHz is modulated by an intelligence signal in the 150-Hz to 4.2-kHz range. Calculate the frequency range of the lower sideband.

CRITICAL THINKING

1. Can intelligence be sent without a carrier? If so, how?
2. How is the output power of an SSB transmitter expressed?
3. A subcarrier of 70 kHz is amplitude-modulated by tones of 2.1 kHz and 6.8 kHz. The resulting AM signal is then used to amplitude-modulate a carrier of 12.5 MHz. Calculate all sideband frequencies in the composite signal and draw a frequency-domain display of the signal. Assume 100 percent modulation. What is the bandwidth occupied by the complete signal?

4. Explain how you could transmit two independent baseband information signals using SSB on a common carrier frequency.
5. An AM signal with 100 percent modulation has an upper-sideband power of 32 W. What is the carrier power?
6. Can an information signal have a higher frequency than the carrier signal? What would happen if a 1-kHz signal amplitude-modulated a 1-kHz carrier signal?

This technician is testing radio frequency circuits on semiconductor chips.

AMPLITUDE MODULATOR AND DEMODULATOR CIRCUITS

Objectives

After completing this chapter, you will be able to:

- *Explain* the relationship of the basic equation for an AM signal to the production of amplitude modulation, mixing, and frequency conversion by a diode or other nonlinear frequency component or circuit.

- *Describe* the operation of diode modulator circuits and diode detector circuits.

- *Compare* the advantages and disadvantages of low- and high-level modulation.

- *Explain* how the performance of a basic diode detector is enhanced by using full-wave rectifier circuits.

- *Define* synchronous detection and *explain* the role of clippers in synchronous detector circuits.

- *State* the function of balanced modulators and *describe* the differences between lattice modulators and IC modulator circuits.

- *Draw* the basic components of both filter-type and phase-shift-type circuits for generation of SSB signals.

Dozens of modulator circuits have been developed that cause carrier amplitude to be varied in accordance with the modulating information signal. There are circuits to produce AM, DSB, and SSB at low or high power levels. This chapter examines some of the more common and widely used discrete-component and integrated-circuit (IC) amplitude modulators. Also covered are demodulator circuits for AM, DSB, and SSB.

4-1 BASIC PRINCIPLES OF AMPLITUDE MODULATION

Examining the basic equation for an AM signal, introduced in the last chapter, gives us several clues as to how AM can be generated. The equation is

$$v_{AM} = V_c \sin 2\pi f_c t + (V_m \sin 2\pi f_m t)(\sin 2\pi f_c t)$$

where the first term is the sine-wave carrier and second term is the product of sine-wave carrier and modulating signals. (Remember that v_{AM} is the instantaneous value of the amplitude modulation voltage.) The modulation index m is the ratio of the modulating signal amplitude to the carrier amplitude, or $m = V_m/V_c$, and so $V_m = mV_c$. Then, substituting this for V_m in the basic equation yields $v_{AM} = V_c \sin 2\pi f_c t + (mV_c \sin 2\pi f_m t)(\sin 2\pi f_c t)$. Factoring, $v_{AM} = V_c \sin 2\pi f_c t(1 + m \sin 2\pi f_m t)$.

AM IN THE TIME DOMAIN

Looking at the expression for v_{AM}, it is clear that we need a circuit that can multiply the carrier by the modulating signal, then add the carrier. A block diagram of such a circuit is shown in Fig. 4-1. One way to do this is to develop a circuit whose gain (or attenuation) is a function of $(1 + m \sin 2\pi f_m t)$. If we call that gain A, the expression for the AM signal becomes

$$v_{AM} = A(v_c)$$

where A is the gain or attenuation factor. Figure 4-2 shows simple circuits based on this expression. In Fig. 4-2(a), A is a gain greater than 1 provided by an amplifier. In Fig. 4-2(b), the carrier is attenuated by a voltage divider. The gain in this case is less than 1 and is therefore an attenuation factor. The carrier is multiplied by a fixed fraction A.

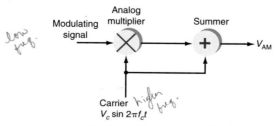

FIG. 4-1   Block diagram of a circuit to produce AM.

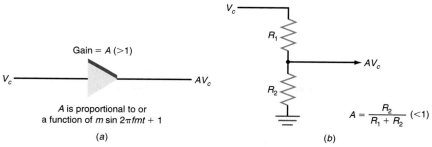

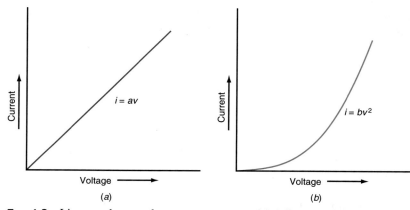

FIG. 4-2 Multiplying the carrier by a fixed gain A.

Now, if the gain of the amplifier or the attenuation of the voltage divider can be varied in accordance with the modulating signal plus 1, AM will be produced. In Fig. 4-2(a) the modulating signal would be used to increase or decrease the gain of the amplifier as the amplitude of the intelligence changed. In Fig. 4-2(b), the modulating signal could be made to vary one of the resistances in the voltage divider, creating a varying attenuation factor. A variety of popular circuits permit gain or attenuation to be varied dynamically with another signal, producing AM.

AM IN THE FREQUENCY DOMAIN

Another way to generate the product of the carrier and modulating signal is to apply both signals to a nonlinear component or circuit, ideally one that generates a square-law function. A linear component or circuit is one in which the current is a linear function of the voltage [see Fig. 4-3(a)]. A resistor or linearly biased transistor is an example of a linear device. The current in the device increases in direct proportion to increases in voltage. The steepness or slope of the line is determined by the coefficient a in the expression $i = av$.

A nonlinear circuit is one in which the current is not directly proportional to the voltage. A common nonlinear component is a diode which has the nonlinear response shown in Fig. 4-3(b), where increasing the voltage increases the current but not in a

FIG. 4-3 Linear and square-law response curves. (a) A linear voltage-current relationship. (b) A nonlinear or square-wave response.

straight line. Instead, the current variation is a square-law function. A square-law function is one that varies in proportion to the square of the input signals. A diode gives a good approximation of a square-law response. Bipolar and field-effect transistors (FETs) can also be biased to give a square-law response. An FET gives a near perfect square-law response, whereas diodes and bipolar transistors, which contain higher-order components, only approximate the square-law function.

The current variation in a typical semiconductor diode can be approximated by the equation

$$i = av + bv^2$$

where av is a linear component of the current equal to the applied voltage multiplied by the coefficient a (usually a DC bias current) and bv^2 is second-order or square-law component of the current. Diodes and transistors also have higher-order terms, such as cv^3, dv^4, etc.; however, these are smaller and often negligible, and so are neglected in an analysis.

To produce AM, the carrier and modulating signals are added and applied to the nonlinear device. A simple way to do this is to connect the carrier and modulating sources in series and apply them to the diode circuit, as in Fig. 4-4. The voltage applied to the diode is then

$$v = v_c + v_m$$

The diode current in the resistor is

$$i = a(v_c + v_m) + b(v_c + v_m)^2$$

Expanding, we get

$$i = a(v_c + v_m) + b(v_c^2 + 2v_c v_m + v_m^2)$$

Substituting the trigonometric expressions for the carrier and modulating signals, let $v_c \sin 2\pi f_c t = v_c \sin \omega_c t$, where $\omega = 2\pi f_c$, and let $v_m = \sin 2\pi f_m t = v_m \sin \omega_m t$, where $\omega_m = 2\pi f_m$. Then

$$i = aV_c \sin \omega_c t + aV_m \sin \omega_m t + bV_c^2 \sin^2 \omega_c t + 2bV_c V_m \sin \omega_c t \sin \omega_m t + bv_m^2 \sin^2 \omega_m t$$

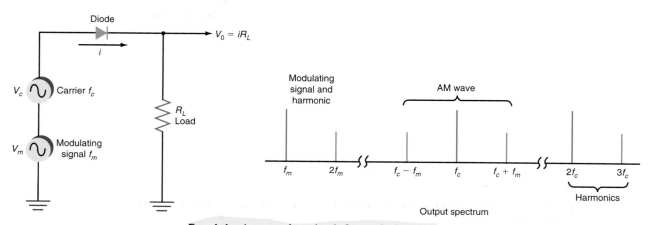

Fig. 4-4 A square-law circuit for producing AM.

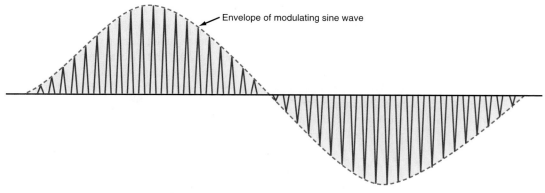

Envelope of modulating sine wave

Fig. 4-5 AM signal containing not only the carrier and sidebands but also the modulating signal.

Next, substituting the trigonometric identity $\sin^2 A = 0.5 (1 - \cos 2A)$ into the preceding expression gives the expression for the current in the load resistor in Fig. 4-4:

$$i = av_c \sin \omega_c t + av_m \sin \omega_m t + 0.5bv_c^2 (1 - \cos 2 \omega_c t)$$
$$+ 2bv_c v_m \sin \omega_c t \sin \omega_m t + 0.5bv_m^2 (1 - \cos \omega_m t)$$

The first term is the carrier sine wave, which is a key part of the AM wave; the second term is the modulating-signal sine wave. Normally, this is not part of the AM wave. It is substantially lower in frequency than the carrier, so it is easily filtered out. The third term, the product of the carrier and modulating signal sine waves, defines the AM wave. If we make the trigonometric substitutions explained in Chap. 3, we obtain two additional terms—the sum and difference frequency sine waves, which are, of course, the upper and lower sidebands. The term $\cos 2\omega_c t$ is a sine wave at two times the frequency of the carrier, that is, the second harmonic of the carrier. The term $\cos 2\omega_m t$ is the second harmonic of the modulating sine wave. These components are undesirable, but are relatively easy to filter out. Diodes and transistors whose function is not a pure-square law function produce third-, fourth-, and higher-order harmonics, which are sometimes referred to as *intermodulation products* and which are also easy to filter out.

Figure 4-4 shows both the circuit and the output spectrum for a simple diode modulator. The output waveform is shown in Fig. 4-5. This waveform is a normal AM wave to which the modulating signal has been added.

If a parallel resonant circuit is substituted for the resistor in Fig. 4-4, the modulator circuit shown in Fig. 4-6 results. This circuit is resonant at the carrier frequency

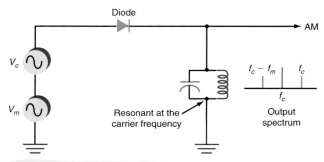

Diode

V_c

V_m

Resonant at the carrier frequency

AM

$f_c - f_m$ f_c

f_c

Output spectrum

Fig. 4-6 The tuned circuit filters out the modulating signal and carrier harmonics, leaving only the carrier and sidebands.

and has a bandwidth wide enough to pass the sidebands but narrow enough to filter out the modulating signal as well as the second- and higher-order harmonics of the carrier. The result is an AM wave across the tuned circuit.

This analysis applies not only to AM, but also to frequency translation devices such as mixers, product detectors, phase detectors, balanced modulators, and other heterodyning circuits. In fact, it applies to any device or circuit that has a square-law function. It explains how sum and difference frequencies are formed and also explains why most mixing and modulation is accompanied by undesirable components such as harmonics and intermodulation products.

4-2 AMPLITUDE MODULATORS

Amplitude modulators are generally one of two types: low level or high level. Low-level modulators generate AM with small signals and thus must be amplified considerably if they are to be transmitted. High-level modulators produce AM at high power levels, usually in the final amplifier stage of a transmitter. Although the discrete component circuits to be discussed in the following sections are still used to a limited extent, keep in mind that today most amplitude modulators and demodulators are in integrated circuit form.

LOW-LEVEL AM

DIODE MODULATORS.   One of the simplest amplitude modulators is the diode modulator described in the previous section. The practical implementation shown in Fig. 4-7 consists of a resistive mixing network, a diode rectifier, and an *LC* tuned circuit. The carrier is applied to one input resistor and the modulating signal to the other. The mixed signals appear across R_3. This network causes the two signals to be linearly mixed, that is, algebraically added. If both the carrier and modulating signal are sine waves, the waveform resulting at the junction of the two resistors will be like that shown in Fig. 4-8(*c*), where the carrier wave is riding on the modulating signal. This signal is not AM. Modulation is a multiplication process, not an addition process.

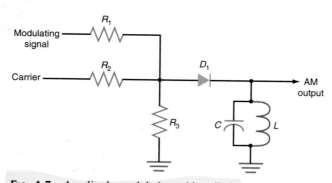

FIG. 4-7   Amplitude modulation with a diode.

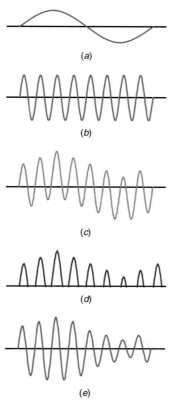

Fig. 4-8 Waveforms in the diode modulator.

The composite waveform is applied to a diode rectifier. The diode is connected so that it is forward-biased by the positive-going half cycles of the input wave. During the negative portions of the wave, the diode is cut off and no signal passes. The current through the diode is a series of positive-going pulses whose amplitude varies in proportion with the amplitude of the modulating signal (see Fig. 4-8(d)).

These positive-going pulses are applied to the parallel tuned circuit made up of L and C, which are resonant at the carrier frequency. Each time the diode conducts, a pulse of current flows through the tuned circuit. The coil and capacitor repeatedly exchange energy, causing an oscillation, or "ringing," at the resonant frequency. The oscillation of the tuned circuit creates one negative half cycle for every positive input pulse. High-amplitude positive pulses cause the tuned circuit to produce high-amplitude negative pulses. Low-amplitude positive pulses produce corresponding low-amplitude negative pulses. The resulting waveform across the tuned circuit is an AM signal, as Fig. 4-8(e) illustrates. The Q of the tuned circuit should be high enough to eliminate the harmonics and produce a clean sine wave and to filter out the modulating signal, and low enough so that its bandwidth accommodates the sidebands generated.

This signal produces high-quality AM, but the amplitudes of the signals are critical to proper operation. Because the nonlinear portion of the diode's characteristic curve occurs only at low voltage levels, signal levels must be low, less than a volt, to produce AM. At higher voltages, the diode current response is nearly linear. The circuit works best with millivolt-level signals.

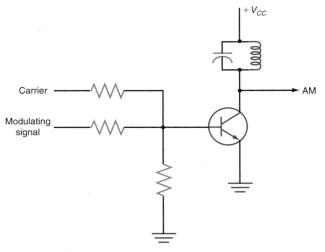

Fig. 4-9   Simple transistor modulator.

TRANSISTOR MODULATORS.   An improved version of the circuit just described is shown in Fig. 4-9. Because it uses a transistor instead of the diode, the circuit has gain. The emitter-base junction is a diode and a nonlinear device. Modulation occurs as described previously, except that the base current controls a larger collector current, and therefore the circuit amplifies. Rectification occurs because of the emitter-base junction. This causes larger half sine pulses of current in the tuned circuit. The tuned circuit oscillates (rings) to generate the missing half cycle. The output is a classical AM wave.

Another transistor modulator circuit is shown in Fig. 4-10. This circuit uses the principles of resistance variation to produce AM. Here the carrier is transformer-coupled to the base of the class A transistor amplifier. Bias comes from the voltage divider R_1–R_2 as it would in any single-stage common emitter amplifier. C_1 has a low reactance at the carrier frequency so that no carrier is developed across R_2. The modulating signal is capacitively coupled through C_3 and applied across the emitter resis-

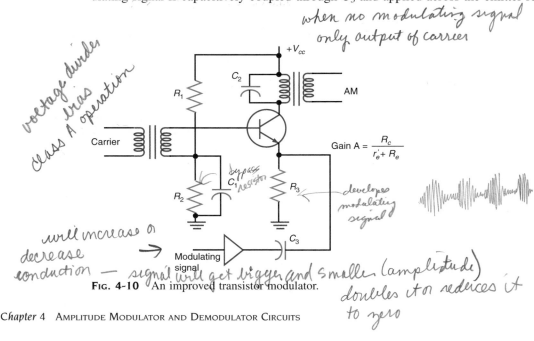

Gain $A = \dfrac{R_c}{r_e' + R_e}$

Fig. 4-10   An improved transistor modulator.

tor R_3. Therefore, it rides on the DC bias voltage developed across R_3 by the emitter current. This arrangement permits both the carrier and modulating signal to control the collector current.

According to basic transistor amplifier theory, the gain of a common emitter amplifier is approximately

$$A = \frac{R_C}{R_e + r'_e}$$

where R_C is the AC load impedance in the collector, which for this circuit is the resistive impedance of the parallel tuned circuit at resonance, R_e is the external emitter resistance in this circuit R_3, and r'_e is the AC resistance of the conducting emitter-base diode. The value of r'_e is usually fixed by the value of the emitter current (I_E), which is approximately $0.025/I_E$. R_e is usually much greater than r'_e. Varying the voltage across R_e in accordance with the modulating signal causes the effective bias to change, and the resistance of r'_e varies in the same way. Thus the gain of the circuit is changed in proportion to the modulating signal, and AM is produced across the output tuned circuit.

FET MODULATORS. The simple amplitude modulator shown in Fig. 4-11 consists of an operational amplifier (op amp) and an FET used as a variable resistor. The op amp is connected as a noninverting amplifier for the carrier signal. The gain A of the circuit is given by the expression $A = 1 + (R_f/R_i)$, where the feedback resistance R_f is a fixed value and R_i is the resistance of an N-channel junction FET. A negative DC bias keeps the gate source junction reverse-biased. A modulating signal is applied to the gate through capacitor C_1.

The carrier signal is applied to the noninverting (+) input of the op amp. Changing the gain of the amplifier in accordance with the modulating signal produces AM. With zero modulating signal, the FET resistance is a fixed value, and therefore the carrier output amplitude is constant. As a sine-wave modulating signal is applied, the resistance of the FET varies. A positive-going modulating input signal causes the FET resistance to decrease; a negative-going modulating signal causes it to increase. Increasing the FET resistance causes the op-amp circuit gain to decrease and vice versa. The result is an AM signal at the op-amp output.

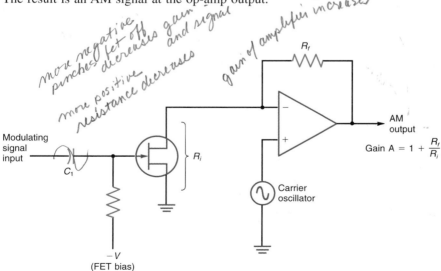

FIG. 4-11   Using an FET to vary the gain of an op amp to produce AM.

The critical component in this circuit is the FET, which should be biased such that its source-to-drain resistance is as linear as possible over a wide range. The best point of linearity can be predicted by examining the operating curves of the FET and setting the bias on the gate for the center of that linear range.

PIN Diode Modulators. Variable attenuator circuits for producing AM are shown in Fig. 4-12. These circuits use PIN diodes to produce AM at VHF, UHF, and microwave frequencies. The PIN diodes are a special type of silicon junction diode designed for use at frequencies above approximately 100 MHz. When forward-biased, these diodes act as variable resistors. The resistance of the diode varies linearly with the amount of current flowing through it. A high current produces a low resistance, whereas a low current produces a high resistance. As the modulating signal varies the forward-bias current through the PIN diode, AM is produced.

In Fig. 4-12(a), two PIN diodes are connected back to back and are forward-biased by a fixed negative DC voltage. The modulating signal is applied to the diodes through capacitor C_1. This AC modulating signal rides on the DC bias, adding to and subtracting from it and thus varying the resistance of the PIN diodes. These diodes appear in series with the carrier oscillator and the load. A positive-going modulating signal reduces the bias on the PIN diodes, causing their resistance to go up. This causes a reduction in amplitude of the carrier across the load. A negative-going modulating signal adds to the forward bias, causing the resistance of the diodes to go down, thereby increasing the carrier amplitude.

A variation of the PIN diode modulator circuit is shown in Fig. 4-12(b), where the diodes are arranged in a π network. This configuration is used when it is necessary to maintain a constant circuit impedance even under modulation.

In both circuits of Fig. 4-12, the PIN diodes form a variable attenuator circuit whose attenuation varies with the amplitude of the modulating signal. Such modulator circuits introduce a considerable amount of loss and must therefore be followed by am-

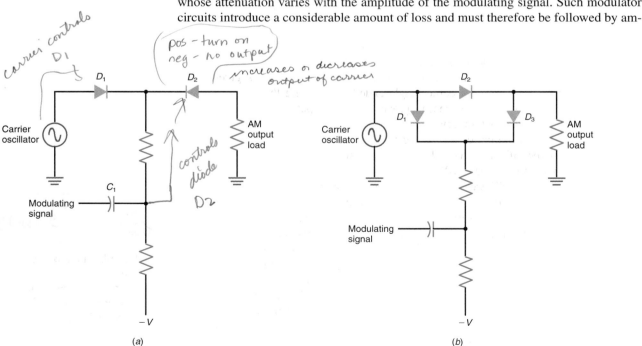

Fig. 4-12 High-frequency amplitude modulators using PIN diodes.

plifiers to increase the AM signal to a usable level. Despite this disadvantage, PIN modulators are widely used because they are one of the few methods available for producing AM at microwave frequencies.

DIFFERENTIAL AMPLIFIERS. A differential amplifier makes an excellent amplitude modulator. A typical circuit is shown in Fig. 4-13(a). Transistors Q_1 and Q_2 form the differential pair, and Q_3 is a constant-current source. Q_3 supplies a fixed emitter current I_E to Q_1 and Q_2, half of which flows in each transistor. The output is developed across the collector resistors R_1 and R_2.

The output is a function of the difference between inputs V_1 and V_2; that is, $V_{out} = A(V_2 - V_1)$, where A is the circuit gain. The amplifier can also be operated with a single input. When this is done, the other input is grounded or set to zero. In Fig. 4-13(a), if V_1 is zero, the output is $V_{out} = A(V_2)$. If V_2 is zero, the output is $V_{out} = A(-V_1) = -AV_1$. This means that the circuit inverts V_1.

The output voltage can be taken between the two collectors, producing a *balanced,* or *differential,* output. The output can also be taken from the output of either collector to ground, producing a *single-ended output.* The two outputs are 180° out of phase with one another. If the balanced output is used, the output voltage across the load is twice the single-ended output voltage.

No special biasing circuits are needed, since the correct value of collector current is supplied directly by the constant-current source Q_3 in Fig. 4-13(a). Resistors R_3, R_4, and R_5, along with V_{EE}, bias the constant-current source Q_3. With no inputs applied, the current in Q_1 equals the current in Q_2, which is $I_E/2$. The balanced output at this time is zero. The circuit formed by R_1 and Q_1 and R_2 and Q_2 is a *bridge circuit.* When no inputs are applied, R_1 equals R_2, and Q_1 and Q_2 conduct equally. Therefore, the bridge is balanced and the output between the collectors is zero.

Now if an input signal V_1 is applied to Q_1, the conduction of Q_1 and Q_2 is affected. Increasing the voltage at the base of Q_1 increases the collector current in Q_1 and decreases the collector current in Q_2 by an equal amount, so that the two currents sum to I_E. Decreasing the input voltage on the base of Q_1 decreases the collector current in Q_1 but increases it in Q_2. The sum of the emitter currents is always equal to the current supplied by Q_3.

The gain of a differential amplifier is a function of the emitter current and the value of the collector resistors. An approximation of the gain is given by the expression $A = (R_C I_E)/50$. This is the single-ended gain, where the output is taken from one of the collectors with respect to ground. If the output is taken from between the collectors, the gain is two times the above value.

R_C is the collector resistor value in ohms and I_E is the emitter current in milliamperes. If $R_C = R_1 = R_2 = 4.7$ kΩ and $I_E = 1.5$ mA, the gain will be about $A = 4700(1.5)/50 = 7050/50 = 141$.

In most differential amplifiers, both R_C and I_E are fixed, providing a constant gain. But as the formula above shows, the gain is directly proportional to the emitter current. Thus if the emitter current can be varied in accordance with the modulating signal, the circuit will produce AM. This is easily done by changing the circuit only slightly, as in Fig. 4-13(b). The carrier is applied to the base of Q_1 and the base of Q_2 is grounded. The output, taken from the collector of Q_2, is single-ended. Since the output from Q_1 is not used, its collector resistor can be omitted with no effect on the circuit. The modulating signal is applied to the base of the constant-current source Q_3. As the intelligence signal varies, it

DID YOU KNOW?

Differential amplifiers make excellent amplitude modulators because they have a high gain, good linearity, and can be 100 percent modulated.

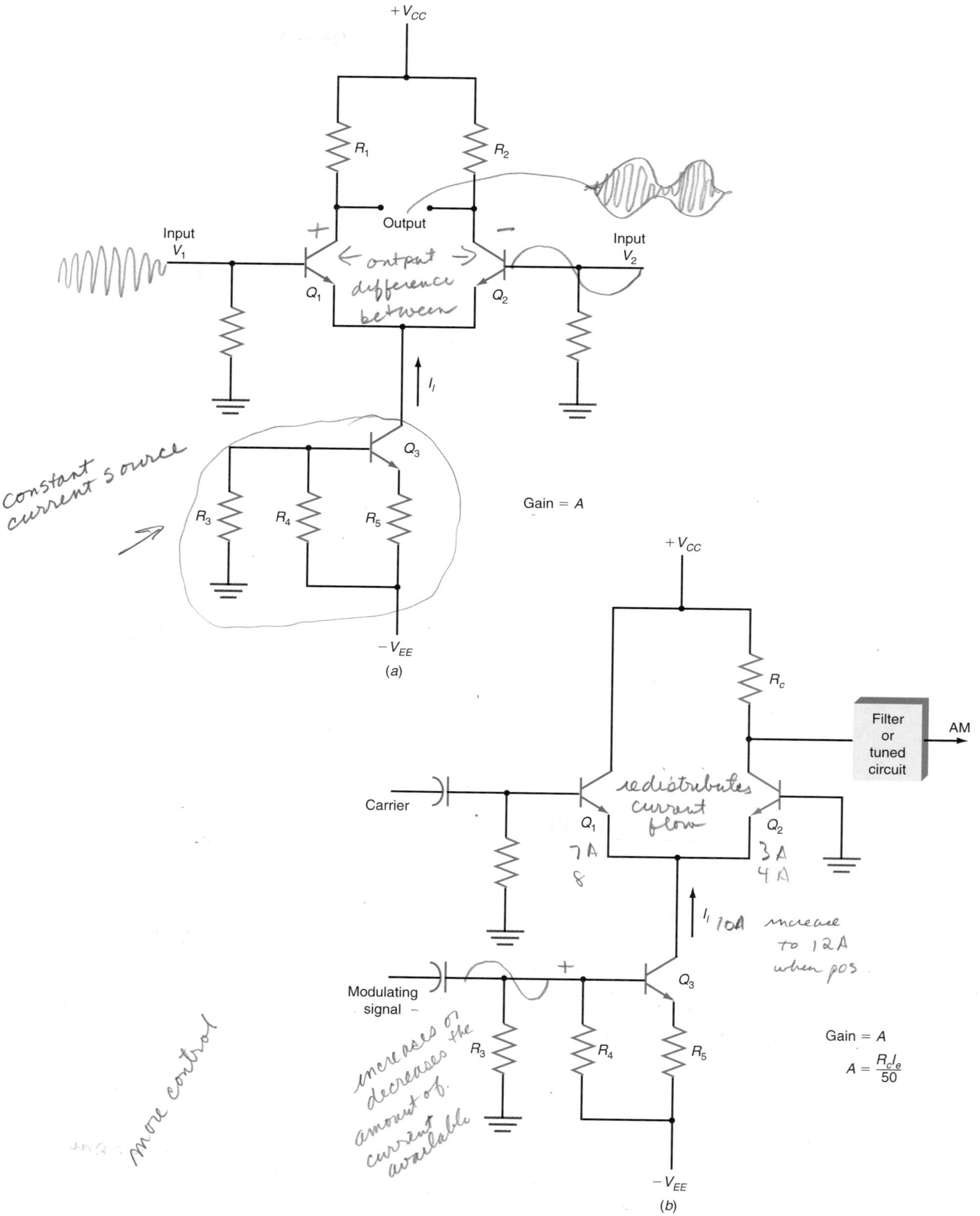

+V_{CC}
R_1
R_2
Output
Input V_1
Input V_2
Q_1
Q_2
I_I
Q_3
R_3
R_4
R_5
Gain = A
$-V_{EE}$
(a)
+V_{CC}
R_c
Filter or tuned circuit
AM
Carrier
Q_1
Q_2
I_I
Modulating signal
Q_3
R_3
R_4
R_5
Gain = A
$A = \frac{R_c I_e}{50}$
$-V_{EE}$
(b)

FIG. 4-13 (*a*) Basic differential amplifier. (*b*) Differential amplifier modulator.

varies the emitter current. This changes the gain of the circuit, amplifying the carrier by an amount determined by the modulating signal amplitude. The result is AM in the output.

This circuit, like the basic diode modulator, has the modulating signal in the output in addition to the carrier and sidebands. The modulating signal can be removed by using a simple high-pass filter on the output, since the carrier and sideband frequencies are usually much higher than the modulating signal. A bandpass filter centered on the carrier with sufficient bandwidth to pass the sidebands can also be used. A parallel tuned circuit in the collector of Q_2 replacing R_C can also be used.

The differential amplifier makes an excellent amplitude modulator. It has a high gain and good linearity, and can be modulated 100 percent. And if high-frequency transistors or a high-frequency IC differential amplifier is used, this circuit can be used to produce low-level modulation at frequencies well into the tens of megahertz region.

OP-AMP MODULATORS. Since most op amps use differential amplifiers, theoretically they can be used to produce AM. However, most op amps do not provide a way to vary the emitter current or the gain of an intermediate differential stage. However, in specialized op amps known as *programmable op amps,* or *operational transconductance amplifiers (OTAs),* an external resistor is used to set the current in one of the differential stages and thus set the circuit gain. OTAs can therefore be used to produce AM.

An example of an OTA is shown in Fig. 4-14. The IC in the figure is similar to a conventional op amp, but its gain can be varied by an external voltage. In addition, its output acts like a current source rather than a voltage source. The carrier is attenuated by R_1 and R_2 and applied to the inverting input, while the noninverting input is connected to ground through R_3. A DC bias voltage from R_4 is applied to this input via R_5 to balance out any DC offset in the circuit. The modulating signal is applied across

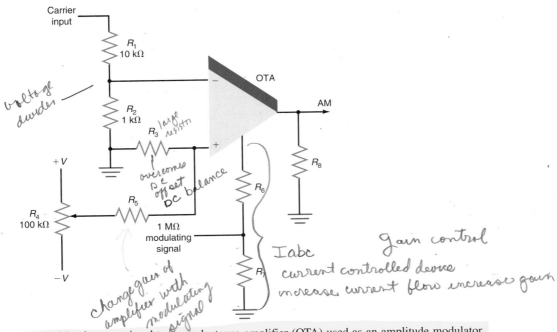

FIG. 4-14 An operational transconductance amplifier (OTA) used as an amplitude modulator.

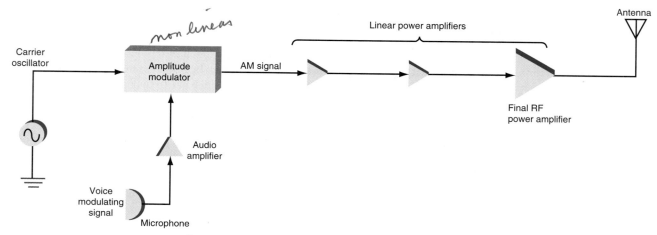

non linear

Fig. 4-15 Low-level modulation systems use linear power amplifiers to increase the AM signal level before transmission.

R_7 and through R_6 to the circuit source in the amplifier. The output is a current proportional to the carrier amplitude and the input current produced by the modulating signal. This current is passed through load resistor R_8 to convert it into a voltage which is an AM signal.

Examples of commercial OTAs are the CA3060, the CA3080, and the NE5517. Because of the frequency-response limitations of these amplifiers, the upper carrier frequency is less than 500 kHz.

AMPLIFYING LOW-LEVEL AM SIGNALS. In low-level modulator circuits such as those discussed above, the signals are generated at very low voltage and power amplitudes. The voltage is typically less than 1 V, and the power is in milliwatts. In systems using low-level modulation, the AM signal is applied to one or more linear amplifiers, as shown in Fig. 4-15, to increase its power level without distorting the signal. These amplifier circuits—class A, class AB, or class B—raise the level of the signal to the desired power level before the AM signal is fed to the antenna.

HIGH-LEVEL AM

In high-level modulation, the modulator varies the voltage and power in the final RF amplifier stage of the transmitter. The result is high efficiency in the RF amplifier and overall high-quality performance.

COLLECTOR MODULATORS. One example of a high-level modulator circuit is the collector modulator shown in Fig. 4-16. The output stage of the transmitter is a high-power class C amplifier. Class C amplifiers conduct for only a portion of the positive half cycle of their input signal. The collector current pulses cause the tuned circuit to oscillate (ring) at the desired output frequency. The tuned circuit, therefore, reproduces the negative portion of the carrier signal (see Chap. 7 for more detail).

The modulator is a linear power amplifier that takes the low-level modulating signal and amplifies it to a high-power level. The modulating output signal is coupled through modulation transformer T_1 to the class C amplifier. The secondary winding of

Class A
Conducting 100%

Class B   AB
*   50%   78%*
biased at cut off

Class C —
*   less than 50%*
biased beyond
cut off

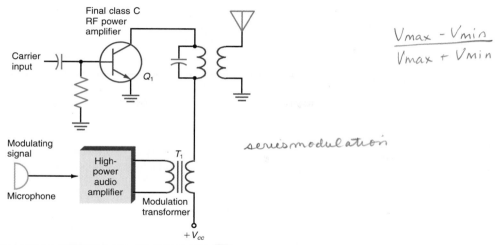

series modulation

FIG. 4-16 A high-level collector modulator.

the modulation transformer is connected in series with the collector supply voltage V_{CC} of the class C amplifier.

With a zero modulation input signal, there is zero modulation voltage across the secondary of T_1, the collector supply voltage is applied directly to the class C amplifier, and the output carrier is a steady sine wave.

When the modulating signal occurs, the AC voltage of the modulating signal across the secondary of the modulation transformer is added to and subtracted from the DC collector supply voltage. This varying supply voltage is then applied to the class C amplifier, causing the amplitude of the current pulses through transistor Q_1 to vary. As a result, the amplitude of the carrier sine wave varies in accordance with the modulated signal. When the modulation signal goes positive, it adds to the collector supply voltage, thereby increasing its value and causing higher current pulses and a higher-amplitude carrier. When the modulating signal goes negative, it subtracts from the collector supply voltage, decreasing it. For that reason, the class C amplifier current pulses are smaller, resulting in a lower-amplitude carrier output.

For 100 percent modulation, the peak of the modulating signal across the secondary of T_1 must be equal to the supply voltage. When the positive peak occurs, the voltage applied to the collector is twice the collector supply voltage. When the modulating signal goes negative, it subtracts from the collector supply voltage. When the negative peak is equal to the supply voltage, the effective voltage applied to the collector of Q_1 is zero, producing zero carrier output. This is illustrated in Fig. 4-17.

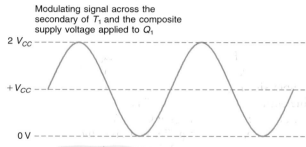

Modulating signal across the secondary of T_1 and the composite supply voltage applied to Q_1

FIG. 4-17 For 100 percent modulation the peak of the modulating signal must be equal to V_{CC}.

In practice, 100 percent modulation cannot be achieved with the high-level collector modulator circuit shown in Fig. 4-16 because of the transistor's nonlinear response to small signals. To overcome this problem, the driver amplifier driving the final class C amplifier is collector-modulated simultaneously, as shown in Fig. 4-18. The output of the modulation transformer is connected in series with the collector supply voltage both to the driver transistor Q_1 and to the final amplifier Q_2. Both are class C amplifiers. The inductors labeled RFC are RF chokes which provide a low-impedance path for DC and a very high impedance for AC. This technique, widely used in low-power CB transmitters, permits solid 100 percent modulation.

High-level modulation produces the best type of AM, but it requires an extremely high-power modulator circuit. In fact, for 100 percent modulation, the power supplied by the modulator must be equal to one-half the total class C amplifier input power. If the class C amplifier has an input power of 1000 W, the modulator must be able to deliver one-half this amount, or 500 W.

Example 4-1

An AM transmitter uses high-level modulation of the final RF power amplifier, which has a DC supply voltage V_{cc} of 48 V with a total current I of 3.5 A. The efficiency is 70 percent.

a. What is the RF input power to the final stage?

DC input power $= P_i = V_{cc}I$ $P = 48 \times 3.5 = 168$ W

b. How much AF power is required for 100 percent modulation? (*Hint:* For 100 percent modulation, AF modulating power P_m is one half the input power.)

$$P_m = \frac{P_i}{2} = \frac{168}{2} = 84 \text{ W}$$

c. What is the carrier output power?

$$\% \text{ efficiency} = \frac{P_{out}}{P_{in}} \times 100$$

$$P_{out} = \frac{\% \text{ efficiency} \times P_{in}}{100} = \frac{70(168)}{100} = 117.6 \text{ W}$$

d. What is the power in one sideband for 67 percent modulation?

$$P_s = \text{sideband power}$$
$$= \frac{P_c(m^2)}{4}$$
$$m = \text{modulation percentage } (\%) = 0.67$$
$$P_c = 168$$
$$P_s = \frac{168(0.67)^2}{4} = 18.85 \text{ W}$$

e. What is the maximum and minimum DC supply voltage swing with 100 percent modulation? (See Fig. 4-17.)

Minimum swing $= 0$

Supply voltage $v_{CC} = 48$ V

Maximum swing $2 \times V_{cc} = 2 \times 48 = 96$ V

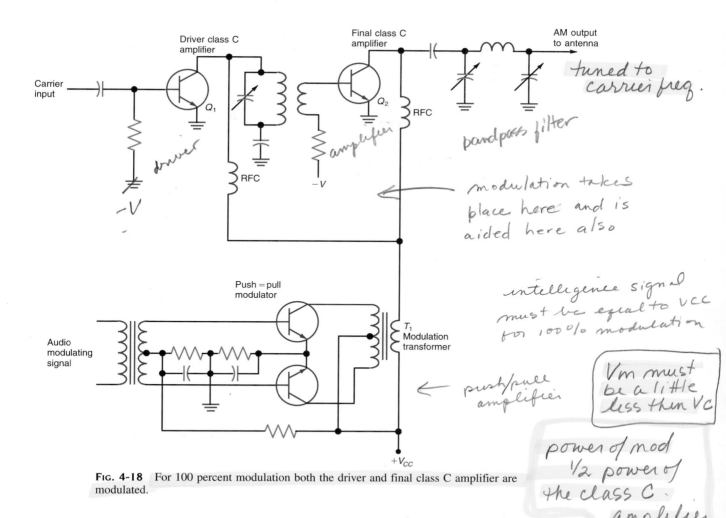

FIG. 4-18 For 100 percent modulation both the driver and final class C amplifier are modulated.

[Handwritten annotations:]
tuned to carrier freq.
bandpass filter
modulation takes place here and is aided here also
intelligence signal must be equal to vcc for 100% modulation
Vm must be a little less than Vc
power of mod 1/2 power of the class C amplifier
driver
amplifier
push/pull amplifier

SERIES MODULATORS. A major disadvantage of collector modulators is the need for the modulation transformer that connects the audio amplifier to the class C amplifier in the transmitter. The higher the power, the larger and more expensive the transformer. For very high-power applications, the transformer is eliminated and the modulation is accomplished at a lower level with one of the many modulator circuits described in the previous sections. The resulting AM signal is amplified by a high-powered linear amplifier. This arrangement is not preferred because linear RF amplifiers are less efficient than class C amplifiers.

One approach is to use a transistorized version of a collector modulator in which a transistor is used to replace the transformer, as in Fig. 4-19. This series modulator replaces the transformer with an emitter follower. The modulating signal is applied to the emitter follower Q_2, which is an audio power amplifier. Note that the emitter follower appears in series with the collector supply voltage $+V_{CC}$. This causes the amplified audio modulating signal to vary the collector supply voltage to the class C amplifier Q_1, as illustrated in Fig. 4-17. Q_2 simply varies the supply voltage to Q_1. If the modulating signal goes positive, the supply voltage to Q_1 increases; thus the carrier amplitude increases in proportion to the modulating signal. If the modulating signal goes negative, the supply voltage to Q_1 decreases, thereby decreasing the carrier am-

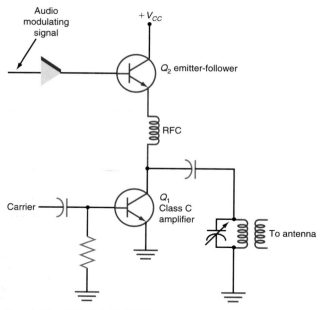

Fig. 4-19 Series modulation.

plitude in proportion to the modulating signal. For 100 percent modulation, the emitter follower can reduce the supply voltage to zero on maximum negative peaks.

Using this high-level modulating scheme eliminates the need for a large, heavy, and expensive transformer, and considerably improves frequency response. However, it is very inefficient. The emitter-follower modulator must dissipate as much power as the class C RF amplifier. For example, assume a collector supply voltage of 24 V and a collector current of 0.5 A. With no modulating signal applied, the percentage of modulation is 0. The emitter follower is biased so that the base and the emitter are at a DC voltage of about one-half the supply voltage, or in this example 12 V. The collector supply voltage on the class C amplifier is 12 V and the input power is therefore as follows:

$$P_{\text{in}} = V_{cc}I_c = 12\ (0.5) = 6\ \text{W}$$

To produce 100 percent modulation, the collector voltage on Q_1 must double, as must the collector current. This occurs on positive peaks of the audio input, as described above. At this time most of the audio signal appears at the emitter of Q_1; very little of the signal appears between the emitter and collector of Q_2, and so at 100 percent modulation, Q_2 dissipates very little power.

When the audio input is at its negative peak, the voltage at the emitter of Q_2 is reduced to 12 V. This means that the rest of the supply voltage, or another 12 V, appears between the emitter and collector of Q_2. Since Q_2 must also be able to dissipate 6 W, it has to be a very large power transistor. The efficiency drops to less than 50 percent. With a modulation transformer, the efficiency is much greater, in some cases as high as 80 percent.

This arrangement is not practical for very high-power AM, but it does make an effective higher-level modulator for power levels below about 100 W.

Demodulators, or *detectors,* are circuits that accept modulated signals and recover the original modulating information. The demodulator circuit is the key circuit in any radio receiver. In fact, demodulator circuits can be used alone as simple radio receivers.

Diode Detectors

The simplest and most widely used amplitude demodulator is the *diode detector* (see Fig. 4-20). As shown, the AM signal is usually transformer-coupled and applied to a basic half-wave rectifier circuit consisting of D_1 and R_1. The diode conducts when the positive half cycles of the AM signals occur. During the negative half cycles, the diode is reverse-biased and no current flows through it. As a result, the voltage across R_1 is a series of positive pulses whose amplitude varies with the modulating signal. A capacitor is connected across resistor R_1, effectively filtering out the carrier and thus recovering the original modulating signal.

One way to look at the operation of a diode detector is to analyze its operation in the time domain. The waveforms in Fig. 4-21 illustrate this. On each positive alternation of the AM signal, the capacitor charges quickly to the peak value of the pulses passed by the diode. When the pulse voltage drops to zero, the capacitor discharges into resistor R_1. The time constant of C_1 and R_1 is chosen to be long compared to the period of the carrier. As a result, the capacitor discharges only slightly during the time that the diode is not conducting. When the next pulse comes along, the capacitor again charges to its peak value. When the diode cuts off, the capacitor again discharges a small amount into the resistor. The resulting waveform across the capacitor is a close approximation to the original modulating signal.

Because the capacitor charges and discharges, the recovered signal has a small amount of ripple on it, causing distortion of the modulating signal. However, because the carrier frequency is usually many times higher than the modulating frequency, these ripple variations are barely noticeable.

Because the diode detector recovers the envelope of the AM signal, which is the original modulating signal, the circuit is sometimes referred to as an *envelope detector.* Distortion of the original signal can occur if the time constant of the load resistor R_1 and the shunt filter capacitor C_1 is too long or too short. If the time constant is too long, the capacitor discharge will be too slow to follow the faster changes in the mod-

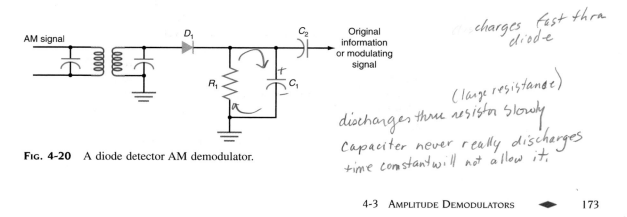

FIG. 4-20   A diode detector AM demodulator.

charges fast thru diode

(large resistance)
discharges thru resistor slowly
Capaciter never really discharges
time constant will not allow it.

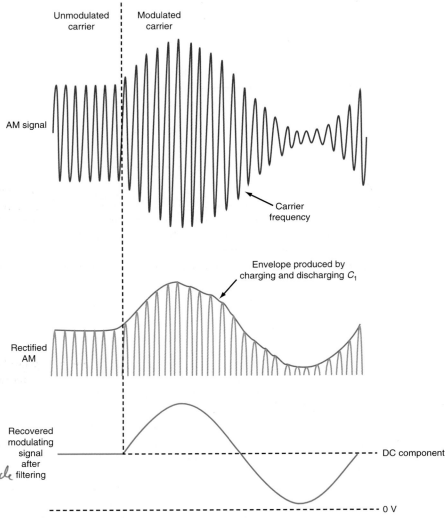

Unmodulated carrier   Modulated carrier

AM signal

Carrier frequency

Envelope produced by charging and discharging C_1

Rectified AM

Recovered modulating signal after filtering

DC component

0 V

If you reverse diode — will be negative half cycle negative voltage signal

FIG. 4-21   Diode detector waveforms.

ulating signal. This is referred to as *diagonal distortion*. If the time constant is too short, the capacitor will discharge too fast and the carrier will not be sufficiently filtered out. The DC component in the output is removed with a series coupling or blocking capacitor, C_2 in Fig. 4-20, which is connected to an amplifier.

Another way to view the operation of the diode detector is in the frequency domain. In this case, the diode is regarded as a nonlinear device to which is applied multiple signals where modulation will take place. The multiple signals are the carrier and sidebands, which make up the input AM signal to be demodulated. The components of the AM signal are the carrier f_c, the upper sideband $f_c + f_m$, and the lower sideband $f_c - f_m$. The diode detector circuit combines these signals, creating the sum and difference signals:

$$f_c + (f_c + f_m) = 2f_c + f_m$$
$$f_c - (f_c + f_m) = -f_m$$
$$f_c + (f_c - f_m) = 2f_c - f_m$$
$$f_c - (f_c - f_m) = f_m$$

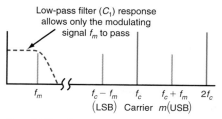

FIG. 4-22 Output spectrum of a diode detector.

All these components appear in the output. Since the carrier frequency is very much higher than that of the modulating signal, the carrier signal can easily be filtered out with a simple low-pass filter. In a diode detector, this low-pass filter is just capacitor C_1 across load resistor R_1. Removing the carrier leaves only the original modulating signal. The frequency spectrum of a diode detector is illustrated in Fig. 4-22. The low-pass filter, C_1 in Fig. 4-20, removes all but the desired original modulating signal.

CRYSTAL RADIO RECEIVERS

The crystal component of the crystal radio receivers that were widely used in the past is simply a diode. In Fig. 4-23 the diode detector circuit of Fig. 4-20 is redrawn, showing an antenna connection and headphones. A long wire antenna picks up the radio signal which is inductively coupled to the secondary winding of T_1, which forms a series resonant circuit with C_1. Note that the secondary is not a parallel circuit, because the voltage induced into the secondary winding appears as a voltage source in series with the coil and capacitor. The variable capacitor C_1 is used to select a station. At resonance, the voltage across the capacitor is stepped up by a factor equal to the Q of the tuned circuit. This resonant voltage rise is a form of amplification. This higher-voltage signal is applied to the diode. The diode detector D_1 and its filter C_2 recover the original modulating information, which causes current flow in the headphones. The headphones serve as the load resistance, and capacitor C_2 removes the carrier. The result is a simple radio receiver; reception is very weak because no active amplification is provided. Typically a germanium diode is used because its voltage threshold is lower than that of a silicon diode and permits reception of weaker signals. Crystal radio receivers can easily be built to receive standard AM broadcasts.

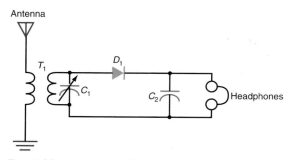

FIG. 4-23 A crystal radio receiver.

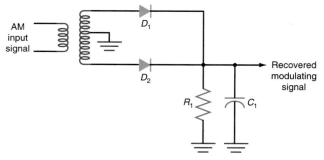

Fig. 4-24 A full-wave diode detector.

Full-Wave Diode Detectors

The performance of a basic diode detector can be improved by using a full-wave rectifier circuit as shown in Fig. 4-24 Here, two diodes and a center-tapped secondary on the RF transformer are used to form a standard full-wave rectifier circuit. With this arrangement, diode D_1 conducts on the positive half cycle and D_2 conducts on the negative half cycle. This diode detector produces a higher average output voltage, which is much easier to filter. The ripple is twice the carrier frequency. The capacitor value necessary to remove the carrier can be half the size of the capacitor value used in a half-wave diode detector. The primary benefit of this circuit is that the higher modulating frequencies are not distorted as much by ripple or attenuated as much by filtering as in the half-wave detector circuit. Thus the output amplitude is greater and the filtering is not as critical.

Synchronous Detection

Synchronous detectors use an internal clock signal at the carrier frequency in the receiver to switch the AM signal off and on, producing rectification similar to that in a standard diode detector. (See Fig. 4-25.) The AM signal is applied to a series switch that is opened and closed synchronously with the carrier signal. The switch is usually a diode or transistor that is turned ON or OFF by an internally generated clock signal equal in frequency to and in phase with the carrier frequency. The switch in Fig. 4-25 is turned ON by the clock signal during the positive half cycles of the AM signal, which therefore appears across the load resistor. During the negative half cycles of the AM signal, the clock turns the switch OFF, so no signal reaches the load or filter capacitor. The capacitor filters out the carrier.

A full-wave synchronous detector is shown in Fig. 4-26. The AM signal is applied to both inverting and noninverting amplifiers. The internally generated carrier signal operates two switches A and B. The clock turns A ON and B OFF or turns B ON and A OFF. This arrangement simulates an electronic single-pole double-throw (SPDT) switch. During positive half cycles of the AM signal, the A switch feeds the non-inverted AM output of positive half cycles to the load. During the

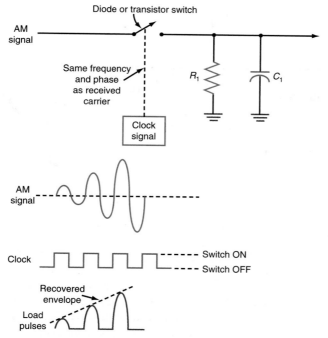

FIG. 4-25 Concept of a synchronous detector.

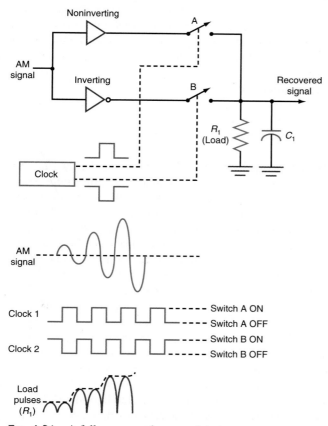

FIG. 4-26 A full-wave synchronous detector.

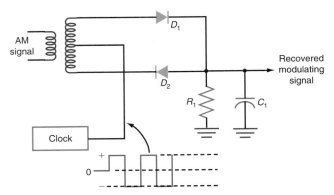

Fig. 4-27 A practical synchronous detector.

negative half cycles of the input, the B switch connects the output of the inverter to the load. The negative half cycles are inverted, becoming positive, and the signal appears across the load. The result is full-wave rectification of the signal.

The key to making the synchronous detector work is ensuring that the signal producing the switching action is perfectly in phase with the received AM carrier. An internally generated carrier signal from, say, an oscillator will not work. Even though the frequency and phase of the switching signal might be close to that of the carrier, they would not be perfectly equal. However, there are a number of techniques, collectively referred to as *carrier recovery circuits,* that can be used to generate a switching signal that has the correct frequency and phase relationship to the carrier.

A practical synchronous detector is shown in Fig. 4-27. A center-tapped transformer provides the two equal but inverted signals. The carrier signal is applied to the center tap. Note that one diode is connected oppositely from the way it would be if used in a full-wave rectifier. These diodes are used as switches, which are turned OFF and ON by the clock, which is used as the bias voltage. The carrier is usually a square wave rather than a sine wave. When the clock is positive, diode D_1 is forward-biased. It acts like a short and connects the AM signal to the load resistor. Positive half cycles appear across the load.

When the clock goes negative, D_2 is forward-biased. During this time, the negative cycles of the AM signal are occurring, which makes the lower output of the secondary winding positive. With D_2 conducting, the positive half cycles are passed to the load and the circuit performs full-wave rectification. As before, the capacitor across the load filters out the carrier, leaving the original modulating signal across the load.

The circuit shown in Fig. 4-28 is one way to supply the carrier to the synchronous detector. The AM signal to be demodulated is applied to a highly selective bandpass filter, which picks out the carrier and suppresses the sidebands, thus removing most of the amplitude variations. This signal is amplified and applied to a clipper or limiter, which removes any remaining amplitude variations from the signal, leaving only the carrier. The clipper circuit typically converts the sine-wave carrier into a square wave, which is amplified and thus becomes the clock signal. In some synchronous detectors, the clipped carrier is put through another bandpass filter to get rid of the square-wave harmonics and generate a pure sine-wave carrier. This signal is then amplified and used as the clock. A small phase shifter may be introduced to correct for any phase differences that occur during the carrier recovery process. The resulting carrier signal is exactly

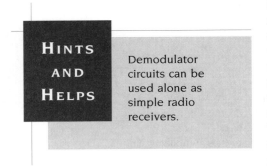

HINTS AND HELPS

Demodulator circuits can be used alone as simple radio receivers.

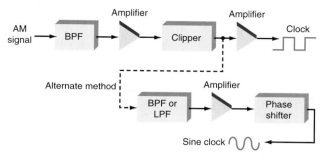

Fig. 4-28 A simple carrier recovery circuit.

the same frequency and phase of the original carrier, as it is indeed derived from it. The output of this circuit is applied to the synchronous detector. Some synchronous detectors use a phase-locked loop to generate the clock, which is locked to the incoming carrier.

Synchronous detectors are also referred to as *coherent detectors,* and were known in the past as homodyne detectors. Their main advantage over standard diode detectors is that they have less distortion and have a better signal-to-noise ratio. They are also less prone to *selective fading,* a phenomenon in which distortion is caused by the weakening of a sideband on the carrier during transmission.

4-4 BALANCED MODULATORS

A *balanced modulator* is a circuit that generates a DSB signal, suppressing the carrier and leaving only the sum and difference frequencies at the output. The output of a balanced modulator can be further processed by filters or phase-shifting circuitry to eliminate one of the sidebands, resulting in an SSB signal.

LATTICE MODULATORS

One of the most popular and widely used balanced modulators is the diode ring or *lattice modulator* illustrated in Fig. 4-29, consisting of an input transformer T_1, an output transformer T_2, and four diodes connected in a bridge circuit. The carrier signal is applied to the center taps of the input and output transformers and the modulating signal is applied to the input transformer T_1. The output appears across the secondary of the output transformer T_2. The connections in Fig. 4-29(a) are the same as those in Fig. 4-29(b), but the operation of the circuit is perhaps more easily visualized as represented in part (b).

The operation of the lattice modulator is relatively simple. The carrier sine wave, which is usually considerably higher in frequency and amplitude than the modulating signal, is used as a source of forward and reverse bias for the diodes. The carrier turns the diodes OFF and ON at a high rate of speed, and the diodes act like switches which connect the modulating signal at the secondary of T_1 to the primary of T_2.

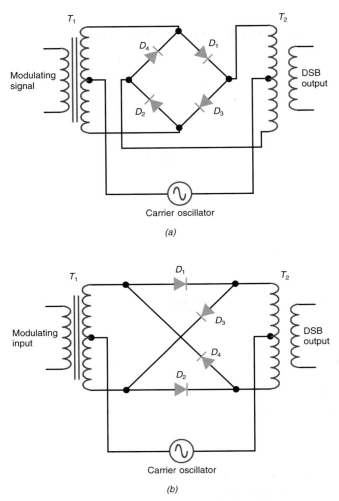

Fig. 4-29 Lattice-type balanced modulator.

Figures 4-30 and 4-31 show how lattice modulators operate. Assume that the modulating input is zero. When the polarity of the carrier is positive, as illustrated in Fig. 4-31(a), diodes D_1 and D_2 are forward-biased. At this time, D_3 and D_4 are reverse-biased and act like open circuits. As you can see, current divides equally in the upper and lower portions of the primary winding of T_2. The current in the upper part of the winding produces a magnetic field that is equal and opposite to the magnetic field produced by the current in the lower half of the secondary. The magnetic fields thus cancel each other out. No output is induced in the secondary and the carrier is effectively suppressed.

When the polarity of the carrier reverses, as shown in Fig. 4-31(b), diodes D_1 and D_2 are reverse-biased and diodes D_3 and D_4 conduct. Again, the current flows in the secondary winding of T_1 and the primary winding of T_2. The equal and opposite magnetic fields produced in T_2 cancel each other out. The carrier is effectively balanced out and its output is zero. The degree of carrier suppression depends on the degree of precision with which the transformers are made and the placement of the center tap: the goal is exactly equal upper and lower currents and perfect magnetic field cancellation. The degree of carrier attenuation also depends upon the diodes. The greatest carrier suppression occurs when the diode characteristics are perfectly matched. A carrier suppression of 40 dB is achievable with well-balanced components.

Now assume that a low-frequency sine wave is applied to the primary of T_1 as the modulating signal. The modulating signal appears across the secondary of T_1. The diode switches connect the secondary of T_1 to the primary of T_2 at different times depending upon the carrier polarity. When the carrier polarity is as shown in Fig. 4-31(a), diodes D_1 and D_2 conduct and act as closed switches. At this time, D_3 and D_4 are reverse-biased and are effectively not in the circuit. As a result, the modulating signal at the secondary of T_1 is applied to the primary of T_2 through D_1 and D_2.

When the carrier polarity reverses, D_1 and D_2 cut off and D_3 and D_4 conduct. Again, a portion of the modulating signal at the secondary of T_1 is applied to the primary of T_2, but this time the leads have been effectively reversed because of the connections of D_3 and D_4. The result is a 180° phase reversal. With this connection, if the modulating signal is positive, the output will be negative, and vice versa.

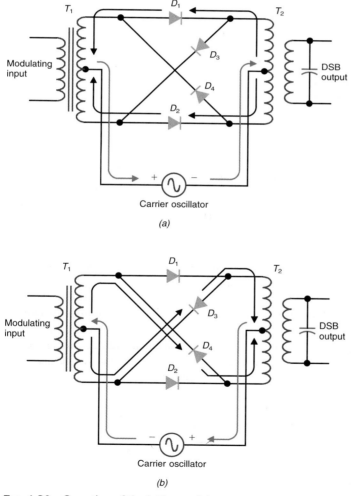

(a)

(b)

Fig. 4-30 Operation of the lattice modulator.

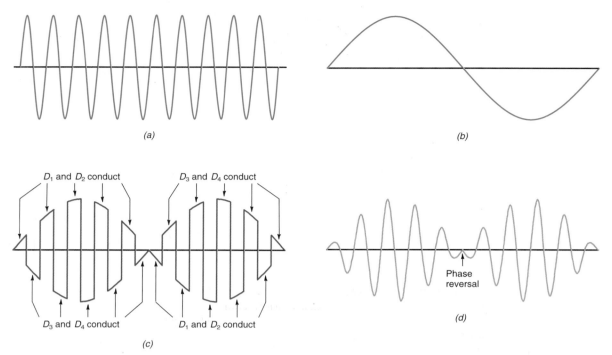

Fig. 4-31 Waveforms in the lattice-type balanced modulator. (*a*) Carrier. (*b*) Modulating signal. (*c*) DSB signal—primary T_2. (*d*) DSB output.

In Fig. 4-31, the carrier is operating at a considerably higher frequency than the modulating signal. Therefore, the diodes switch OFF and ON at a high rate of speed, causing portions of the modulating signal to be passed through the diodes at different times. The DSB signal appearing across the primary of T_2 is illustrated in Fig. 4-31(*c*). The steep rise and fall of the waveform is caused by the rapid switching of the diodes. Because of the switching action the waveform contains harmonics of the carrier. Ordinarily, the secondary of T_2 is a resonant circuit as shown, and therefore, the high-frequency harmonic content is filtered out, leaving a DSB signal like that shown in Fig. 4-31(*d*).

There are several important things to notice about this signal. First, the output waveform occurs at the carrier frequency. This is true even though the carrier has been removed. If two sine waves occurring at the sideband frequencies are added algebraically, the result is a sine-wave signal at the carrier frequency with the amplitude variation shown in Fig. 4-31(*c*) or (*d*). Observe that the envelope of the output signal is *not* the shape of the modulating signal. Note also the phase reversal of the signal in the very center of the waveform, which is one indication that the signal being observed is a true DSB signal.

Although lattice modulators can be constructed of discrete components, they are usually available in a single module containing the transformers and diodes in a sealed package. The unit can be used as an individual component. The transformers are carefully balanced, and matched hot-carrier diodes are used to provide a wide operating frequency range and superior carrier suppression.

The diode lattice modulator shown in Fig. 4-30 uses one low-frequency iron-core transformer for the modulating signal and an air-core

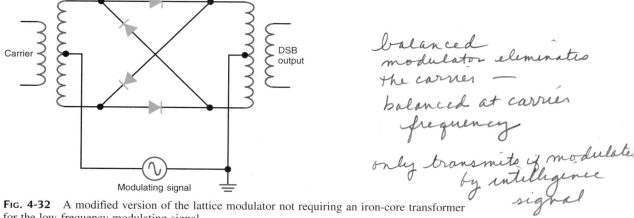

FIG. 4-32 A modified version of the lattice modulator not requiring an iron-core transformer for the low-frequency modulating signal.

balanced modulator eliminates the carrier — balanced at carrier frequency

only transmits if modulated by intelligence signal

transformer for the RF output. This is an inconvenient arrangement because the low-frequency transformer is large and expensive. More commonly, two RF transformers are used, as shown in Fig. 4-32, where the modulating signal is applied to the center taps of the RF transformers. The operation of this circuit is similar to that of other lattice modulators.

IC BALANCED MODULATORS

Another widely used balanced modulator circuit uses differential amplifiers. A typical example, the popular 1496/1596 IC balanced modulator, is shown in Fig. 4-33. This circuit can work at carrier frequencies up to approximately 100 MHz and can achieve a carrier suppression of 50 to 65 dB. The pin numbers shown on the inputs and outputs of the IC are those for a standard 14-pin dual in-line package (DIP) IC. The device is also available in a 10-lead metal can.

In Fig. 4-33, transistors Q_7 and Q_8 are constant-current sources that are biased with a single external resistor and the negative supply. They supply equal values of current to the two differential amplifiers. One differential amplifier is made up of Q_1, Q_2, and Q_5, and the other of Q_3, Q_4, and Q_6. The modulating signal is applied to the bases of Q_5 and Q_6. These transistors are connected in the current paths to the differential transistors and vary the amplitude of the current in accordance with the modulating signal. The current in Q_5 is 180° out of phase with the current in Q_6. As the current in Q_5 increases, the current through Q_6 decreases, and vice versa.

The differential transistors Q_1 through Q_4, which are controlled by the carrier, operate as switches. When the carrier input is such that the lower input terminal is positive with respect to the upper input terminal, transistors Q_1 and Q_4 conduct and act as closed switches and Q_2 and Q_3 are cut off. When the polarity of the carrier signal reverses, Q_1 and Q_4 are cut off and Q_2 and Q_3 conduct, acting as closed switches. These differential transistors, therefore, serve the same switching purpose as the diodes in the lattice modulator circuit discussed previously. They switch the modulating signal OFF and ON at the carrier rate.

Assume that a high-frequency carrier wave is applied to switching transistors Q_1 and Q_4 and that a low-frequency sine wave is applied to the modulating signal input

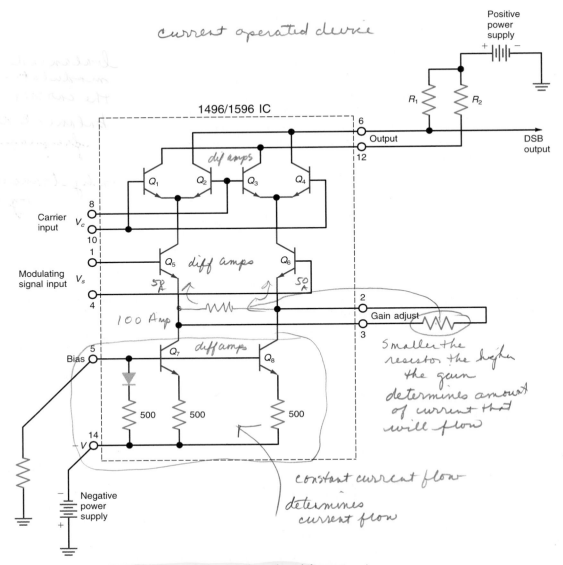

current operated device

dif amps

diff amps

diff amps

100 Amp

constant current flow determines current flow

Smaller the resistor the higher the gain determines amount of current that will flow

FIG. 4-33 Integrated-circuit balanced modulator.

at Q_5 and Q_6. Assume that the modulating signal is positive-going so that the current through Q_5 increases while the current through Q_6 decreases. When the carrier polarity is positive, Q_1 and Q_4 conduct. As the current through Q_5 increases, the current through Q_1 and R_2 increases proportionately; therefore, the output voltage at the collector of Q_1 goes in a negative direction. As the current through Q_6 decreases, the current through Q_4 and R_1 decreases. Thus the output voltage at the collector of Q_4 increases. When the carrier polarity reverses, Q_2 and Q_3 conduct. The increasing current of Q_5 is passed through Q_2 and R_1 and therefore the output voltage begins to decrease. The decreasing current through Q_6 is now passed through Q_3 and R_2, causing the output voltage to increase. The result of the carrier switching OFF and ON and the modulating signal varying as indicated produces the classical DSB output signal described before [see Fig. 4-31(c)]. The signal at R_1 is the same as the signal at R_2, but the two are 180° out of phase.

Figure 4-34 shows the 1496 connected as a balanced modulator. The additional components are added to the circuit in Fig. 4-33 to provide for single-ended rather than balanced inputs to the carrier, modulating signal inputs, and a way to fine-tune the carrier balance. The potentiometer on pins 1 and 4 allows tuning for minimum carrier output, compensates for minor imbalances in the internal balanced modulator circuits, and corrects for parts tolerances in the resistors, thus giving maximum carrier suppression. The carrier suppression can be adjusted to at least 50 dB under most conditions and as high as 65 dB at low frequencies.

APPLICATIONS FOR 1496/1596 ICs.

The 1496 IC is one of the most versatile circuits available for communications applications. In addition to its use as balanced modulator, it can be reconfigured to perform as an amplitude modulator or as a synchronous detector.

Figure 4-34 shows the 1496 connected as an amplitude modulator. The 1-kΩ resistors bias the differential amplifiers into the linear region so they amplify the input carrier. The modulating signal is applied to the series emitter transistors Q_5 and Q_6. An adjustable network using a 50-kΩ potentiometer allows control of the amount

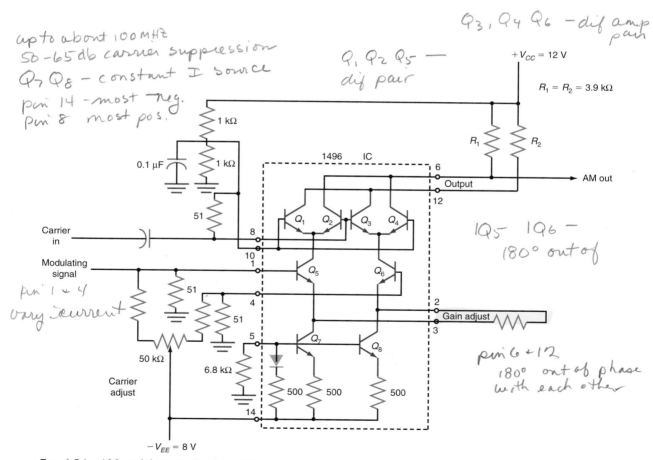

FIG. 4-34 AM modulator made with 1496 IC.

of modulating signal that is applied to each internal pair of differential amplifiers. If the potentiometer is set near the center, the carrier balances out and the circuit functions as a balanced modulator. When the potentiometer is fine-tuned to the center position, the carrier is suppressed and the output is DSB AM.

If the potentiometer is offset one way or another, one pair of differential amplifiers receives little or no carrier amplification and the other pair gets all or most of the carrier. The circuit becomes a version of the differential amplifier modulator shown in Fig. 4-13(b). This circuit works quite nicely, but has very low input impedances. The carrier and modulating signal input impedances are equal to the input resistor values of 51 Ω. This means that the carrier and modulating signal sources must come from circuits with low output impedances, such as emitter followers or op amps.

Figure 4-35 shows the 1496 connected as a synchronous detector for AM. The AM signal is applied to the series emitter transistors Q_5 and Q_6, thus varying the emitter currents in the differential amplifiers, which in this case are used as switches to turn the AM signal OFF and ON at the right time. The carrier must be in phase with the AM signal.

In this circuit, the carrier can be derived from the AM signal itself. In fact, connecting the AM signal to both inputs works if the AM signal is high enough in amplitude. When the amplitude is high enough, the AM signal drives the differential amplifier transistors Q_1 through Q_4 into cutoff and saturation, thereby removing any

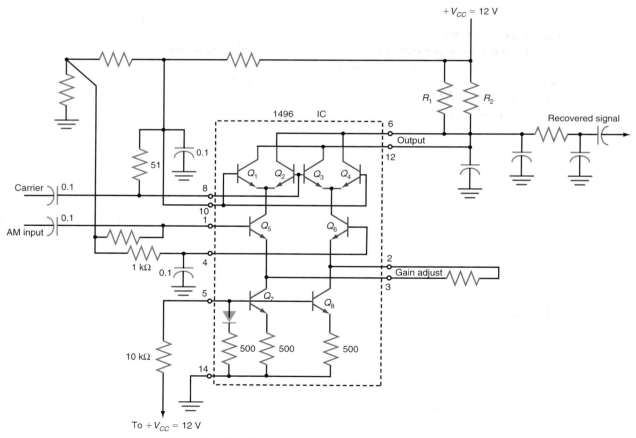

Fig. 4-35 Synchronous AM detector using a 1496.

amplitude variations. Since the carrier is derived from the AM signal, it is in perfect phase to provide high-quality demodulation. The carrier variations are filtered from the output by an *RC* low-pass filter, leaving the recovered intelligence signal.

ANALOG MULTIPLIERS.   Another type of IC that can be used as a balanced modulator is the *analog multiplier*. Analog multipliers are often used to generate DSB signals. The primary difference between an IC balanced modulator and an analog multiplier is that the balanced modulator is a switching circuit. The carrier, which may be a rectangular wave, causes the differential amplifier transistors to turn OFF and ON to switch the modulating signal. The analog multiplier uses differential amplifiers, but they operate in the linear mode. The carrier must be a sine wave, and the analog multiplier produces the true product of two analog inputs.

4-5 SSB CIRCUITS

GENERATING SSB SIGNALS: THE FILTER METHOD

The simplest and most widely used method of generating SSB signals is the filter method. Figure 4-36 shows a general block diagram of an SSB transmitter using the filter method. The modulating signal, usually voice from a microphone, is applied to the audio amplifier whose output is fed to one input of a balanced modulator. A crystal oscillator provides the carrier signal, which is also applied to the balanced modulator. The output of the balanced modulator is a DSB signal. An SSB signal is produced by passing the DSB signal through a highly selective bandpass filter which selects either the upper or lower sideband.

The primary requirement of the filter is, of course, that it pass only the desired sideband. Filters are usually designed with a bandwidth of approximately 2.5 to 3 kHz,

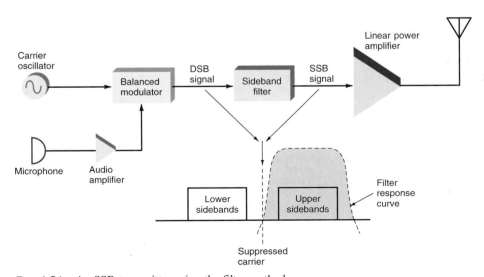

FIG. 4-36   An SSB transmitter using the filter method.

making them wide enough only to pass standard voice frequencies. The sides of the filter response curve are extremely steep, providing for excellent selectivity. Filters are *fixed* tuned devices; that is, the frequencies they can pass are not alterable. Therefore, the carrier oscillator frequency must be chosen so that the sidebands fall within the filter bandpass. Many commercially available filters are tuned to the 455-kHz, 3.35-MHz, or 9-MHz frequency ranges, although other frequencies are also used.

With the filter method, it is necessary to select either the upper or lower sideband. Since the same information is contained in both sidebands, it generally makes no difference which one is selected, provided that the same sideband is used in both transmitter and receiver. However, the choice of the upper or lower sideband as a standard varies from service to service, and it is necessary to know which has been used to properly receive an SSB signal.

There are two methods of sideband selection. Many transmitters simply contain two filters, one that will pass the upper sideband and another that will pass the lower sideband, and a switch is used to select the desired sideband [see Fig. 4-37(a)]. An alternative method is to provide two carrier oscillator frequencies. Two crystals change the carrier oscillator frequency to force either the upper sideband or the lower sideband to appear in the filter bandpass (see Fig. 4-37(b)).

As an example, assume that a bandpass filter is fixed at 1000 kHz and the modulating signal f_m is 2 kHz. The balanced modulator generates the sum and difference frequencies. Therefore, the carrier frequency f_c must be chosen so that the USB or LSB is at 1000 kHz. The balanced modulator outputs are USB $= f_c + f_m$ and LSB $= f_c - f_m$.

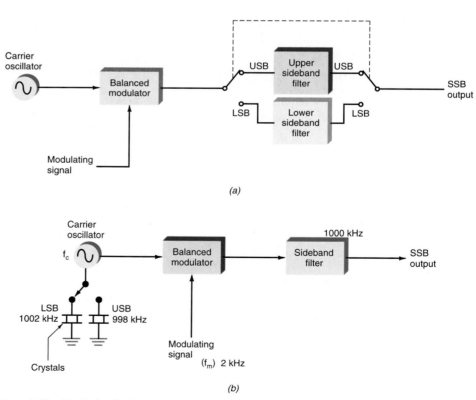

FIG. 4-37 Methods of selecting the upper or lower sideband. (*a*) Two filters. (*b*) Two carrier frequencies.

To set the USB at 1000 kHz, the carrier must be $f_c + f_m = 1000$, $f_c + 2 = 1000$, and $f_c = 1000 - 2 = 998$ kHz. To set the LSB at 1000 kHz, the carrier must be $f_c - f_m = 1000$, $f_c - 2 = 1000$, and $f_c = 1000 + 2 = 1002$ kHz.

Crystal filters, which are low in cost and relatively simple to design, are by far the most commonly used filters in SSB transmitters. Their very high Q provides extremely good selectivity. Ceramic filters are used in some designs. Typical center frequencies are 455 kHz and 10.7 MHz.

Mechanical filters are also used in SSB-generating equipment. These mechanical filters consist of small metal disks coupled together with rods to form an assembly that vibrates or resonates over a narrow frequency range. The diameter and thickness of the disks determine the resonant frequency, whereas the number of disks and their spacing and method of coupling determine the bandwidth. The AC signal to be filtered is applied to a coil, which creates a magnetic field. This magnetic field works against a permanent magnet to produce mechanical motion in the disks. If the input signal is within the bandpass resonant frequency range of the disks, they vibrate freely. This vibration is mechanically coupled to a coil. The moving coil cuts the field of a permanent magnet, inducing a voltage in the coil. This is the output signal. If the input signal is outside of the resonant frequency range of the disks, they do not vibrate and little or no output is produced. Such mechanical assemblies are extremely effective bandpass filters. Most are designed to operate over the 50- to 500-kHz range. A 455-kHz mechanical filter is commonly used.

Example 4-2

An SSB transmitter using the filter method of Fig. 4-36 operates at a frequency of 4.2 MHz. The voice frequency range is 300 to 3400 Hz.

a. Calculate the upper and lower sideband ranges.
Upper sideband
Lower limit $f_{LL} = f_c + 300 = 4,200,000 + 300 = 4,200,300$ Hz
Upper limit $f_{UL} = f_c + 3400 = 4,200,000 + 3400$
$$= 4,203,400 \text{ Hz}$$
Range, USB = 4,200,300 to 4,203,400 Hz
Lower sideband
Lower limit $f_{LL} = f_c - 300 = 4,200,000 - 300 = 4,199,700$ Hz
Upper limit $f_{UL} = f_c - 3400 = 4,200,000 - 3400$
$$= 4,196,600 \text{ Hz}$$
Range, LSB = 4,196,000 to 4,199,700 Hz

b. What should be the approximate center frequency of a bandpass filter to select the lower sideband? The equation for the center frequency of the lower sideband, f_{LSB}, is:
$$f_{LSB} = \sqrt{f_{LL} f_{UL}} = \sqrt{4,196,660 \times 4,199,700} = 4,198,149.7 \text{ Hz}$$
An approximation is:
$$f_{LSB} = \frac{f_{LL} + f_{UL}}{2} = \frac{4,196,600 + 4,199,700}{2} = 4,198,150 \text{Hz}$$

GENERATING SSB SIGNALS: PHASING

The phasing method of SSB generation uses a phase-shift technique that causes one of the sidebands to be canceled out. A block diagram of a phasing-type SSB generator is shown in Fig. 4-38. It uses two balanced modulators, which effectively eliminate the carrier. The carrier oscillator is applied directly to the upper balanced modulator along with the audio modulating signal. The carrier and modulating signal are then both shifted in phase by 90° and applied to the second, lower, balanced modulator. The

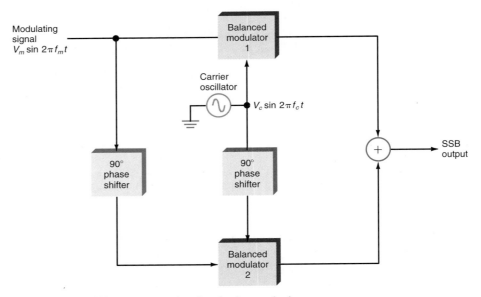

FIG. 4-38 An SSB generator using the phasing method.

phase-shifting action causes one sideband to be canceled out when the two balanced modulator outputs are added together to produce the output.

The carrier signal is $V_c \sin 2\pi f_c t$. The modulating signal is $V_m \sin 2\pi f_m t$. Balanced modulator 1 produces the product of these two signals; $(V_m \sin 2\pi f_m t)(V_c \sin 2\pi f_c t)$. Applying a common trigonometric identity,

$$\sin A \sin B = 0.5[\cos(A - B) - \cos(A + B)]$$

we have

$$(V_m \sin 2\pi f_m t)(V_c \sin 2\pi f_c t) = 0.5 V_m V_c[\cos(2\pi f_c - 2\pi f_m)t - \cos(2\pi f_c + 2\pi f_m)t]$$

Note that these are the sum and difference frequencies or the upper and lower sidebands.

It is important to remember that a cosine wave is simply a sine wave shifted by 90°; that is, it has exactly the same shape as a sine wave, but it occurs 90° earlier in time. A cosine wave *leads* a sine wave by 90° and a sine wave *lags* a cosine wave by 90°.

The 90° phase shifters in Fig. 4-38 create cosine waves of the carrier and modulating signals which are multiplied in balanced modulator 2 to produce $(V_m \cos 2\pi f_m t) \times (V_c \cos 2\pi f_c t)$. Applying another common trigonometric identity,

$$\cos A \cos B = 0.5[\cos(A - B) + \cos(A - B)]$$

we have

$$(V_m \cos 2\pi f_m t)(V_c \cos 2\pi f_c t) = 0.5 V_m V_c[\cos(2\pi f_c - 2\pi f_m)t + \cos(2\pi f_c + 2\pi f_m)t]$$

When these two expressions are added, the sum frequencies cancel while the difference frequencies add, producing only the lower sideband, $\cos[(2\pi f_c - 2\pi f_m)t]$.

CARRIER PHASE SHIFT. A phase shifter is usually an RC network that causes the output to either lead or lag the input by 90°. Many different kinds of circuits have been devised for producing this phase shift. A simple RF phase shifter consisting of two RC sections, each set to produce a phase shift of 45°, is shown in Fig. 4-39. The section made up of R_1 and C_1 produces an output that lags the input by 45°. The section made up of C_2 and R_2 produces a phase shift that leads the input by 45°. The total phase shift between the two outputs is 90°. One output goes to balanced modulator 1 and the other goes to balanced modulator 2.

Since a phasing-type SSB generator can be made with IC balanced modulators like the 1496 and since these can be driven by a square-wave carrier frequency signal, a digital phase shifter can be used to provide the two carrier signals that are 90° out of phase. Figure 4-40 shows two D-type flip-flops connected as a simple shift register with feedback from the complement output of the B flip-flop to the D input of the A flip-flop. JK flip-flops could also be used. It is assumed that the flip-flops trigger or change state on the negative-going edge of the clock signal. The clock signal is set to a frequency exactly four times higher than

HINTS AND HELPS

When the filter method is used to produce SSB signals, either the upper or lower sideband is selected. The choice of upper or lower sideband varies from service to service, and must be known in order to properly receive an SSB signal.

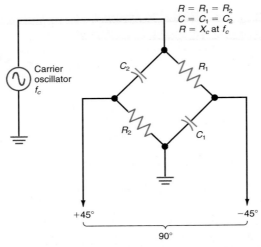

$R = R_1 = R_2$
$C = C_1 = C_2$
$R = X_c$ at f_c

FIG. 4-39 A single-frequency 90° phase shifter.

the carrier frequency. With this arrangement, each flip-flop produces a 50 percent duty cycle square wave at the carrier frequency and the two signals are exactly 90° out of phase with one another. These signals drive the differential amplifier switches in the 1496 balanced modulators, and this phase relationship is maintained regardless of the clock or carrier frequency. TTL flip-flops can be used at frequencies up to about 50 MHz. For higher frequencies, in excess of 100 MHz, emitter coupled logic (ECL) flip-flops can be used.

AUDIO PHASE SHIFT. The most difficult part of creating a phasing-type SSB generator is designing a circuit that maintains a constant 90° phase shift over a wide range of audio modulating frequencies. (Keep in mind that a phase shift is simply a time shift

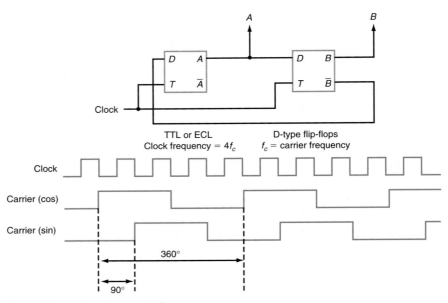

FIG. 4-40 A digital phase shifter.

Guglielmo Marconi (1874–1937), the Italian-born physicist and inventor, filed his famous patent No. 7777 for improvements in wireless telegraphy in 1900. He received the Nobel Prize in physics in 1909. He sent the first radio message from England to Australia in 1918.

between sine waves of the same frequency.) An *RC* network produces a specific amount of phase shift at only one frequency because the capacitive reactance varies with frequency. In the carrier phase shifter, this is not a problem, since the carrier is maintained at a constant frequency. However, the modulating signal is usually a band of frequencies, typically in the audio range from 300 to 3000 Hz.

One of the circuits commonly used to produce a 90° phase shift over a wide bandwidth is shown in Fig. 4-41. The phase-shift difference between the output to modulator 1 and the output to modulator 2 is 90° ± 1.5° over the 300- to 3000-Hz range. Resistor and capacitor values must be carefully selected to ensure phase-shift accuracy, since inaccuracies cause incomplete cancellation of the undesired sideband.

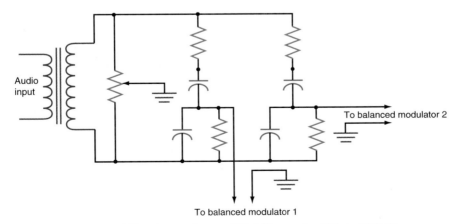

FIG. 4-41   A phase shifter that produces a 90° shift over the 300- to 3000-Hz range.

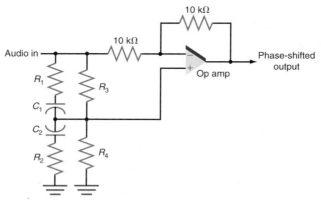

FIG. 4-42 An active phase shifter.

A wideband audio phase shifter that uses an op amp in an active filter arrangement is shown in Fig. 4-42. Careful selection of components will ensure that the phase shift of the output will be close to 90° over the audio frequency range of 300 to 3000 Hz. Greater precision of phase shift can be obtained by using multiple stages, with each stage having different component values and therefore a different phase-shift value. The phase shifts in the multiple stages will produce a total shift of 90°.

The phasing method can be used to select either the upper or lower sideband. This is done by changing the phase shift of either the audio or carrier signals to the balanced modulator inputs. For example, applying the direct audio signal to balanced modulator 2 in Fig. 4-38 and the 90° phase-shifted signal to balanced modulator 1 will cause the upper sideband to be selected instead of the lower sideband. The phase relationship of the carrier can also be switched to make this change.

The output of the phasing generator is a low-level SSB signal. The degree of suppression of the carrier depends on the configuration and precision of the balanced modulators, and the precision of the phase shifting determines the degree of suppression of the unwanted sideband. The design of phasing-type SSB generators is critical if complete suppression of the undesired sideband is to be achieved. The SSB output is then applied to linear RF amplifiers, where its power level is increased before being applied to the transmitting antenna.

DSB AND SSB DEMODULATION

To recover the intelligence in a DSB or SSB signal, the carrier that was suppressed at the receiver must be reinserted. Assume, for example, that a 3-kHz sine-wave tone is transmitted by modulating a 1000-kHz carrier. With SSB transmission of the upper sideband, the transmitted signal is 1000 + 3 = 1003 kHz. Now at the receiver, the SSB signal (the 1003-kHz USB) is used to modulate a carrier of 1000 kHz. See Fig. 4-43(a). If a balanced modulator is used, the 1000-kHz carrier is suppressed, but the sum and difference signals are generated. The balanced modulator is called a *product detector* because it is used to recover the modulating signal rather than generate a carrier that will transmit it. The sum and difference frequencies produced are

$$\text{Sum:} \quad 1003 + 1000 = 2003 \text{ kHz}$$
$$\text{Difference:} \quad 1003 - 1000 = 3 \text{ kHz}$$

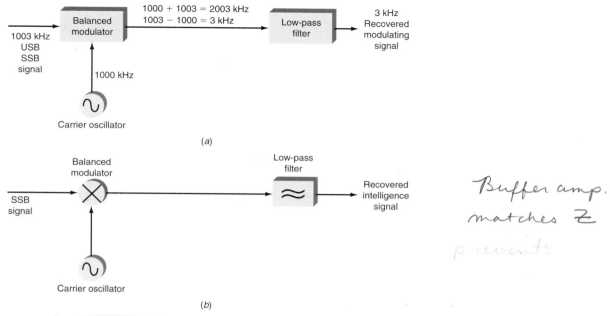

FIG. 4-43 A balanced modulator used as a product detector to demodulate an SSB signal.

The difference is, of course, the original intelligence or modulating signal. The sum, the 2003-kHz signal, has no importance or meaning. Since the two output frequencies of the balanced modulator are so far apart, the higher undesired frequency is easily filtered out by a low-pass filter that keeps the 3-kHz signal but suppresses everything above it.

Any balanced modulator can be used as a product detector to demodulate SSB signals. Many special product detector circuits have been developed over the years. Lattice modulators or ICs like the 1496 both make good product detectors. All that needs to be done is connect a low-pass filter on the output to get rid of the undesired high-frequency signal while passing the desired difference signal. Figure 4-43(b) shows a widely accepted convention for representing balanced modulator circuits. Note the special symbols used for the balanced modulator and low-pass filter.

Individuals now enjoy a variety of "Walkpersons" to listen to their cassettes and AM and FM radio programs.

SUMMARY

One type of AM circuit varies the gain of the amplifier or the attenuation of the voltage divider according to the modulating signal plus 1. Another applies the product of the carrier and modulating signals to a nonlinear component or circuit. A parallel tuned circuit resonant at the carrier frequency, with a bandwidth wide enough to filter out the modulating signal as well as the second and higher harmonics of the carrier, can be used to produce an AM wave.

Low-level AM can be produced by many types of circuits. In high-level modulation, the modulator varies the voltage and power in the transmitter's final RF amplifier stage.

Demodulator circuits accept a modulated signal and recover the original modulating information. Performance of the basic diode detector can be improved by using a full-wave rectifier circuit. Synchronous detectors use an internal clock signal to switch off the AM signal, producing rectification.

A balanced modulator is a circuit that generates a DSB signal. The diode ring or lattice modulator is a widely used balanced modulator.

Filters used to generate SSB signals must have high selectivity. Both crystal and mechanical filters are used.

Product detectors, which are circuits for demodulating or detecting DSB or SSB signals, generate the mathematical product of the SSB signal and the carrier.

KEY TERMS

1496/1596 circuits
Analog multiplier
Audio phase shift
Balanced modulator
Carrier phase shift
Collector modulator
Crystal radio
Demodulator (detector)
Differential amplifier
   modulator

Diode ring (lattice modulator)
Double sideband (DSB)
FET modulator
Full-wave diode detector
High-level AM
IC balanced modulator
Intermodulation products
Lattice modulator
Low-level AM
Modulation index m

Op-amp modulator
Operational transductance
   amplifier (OTA)
PIN diode modulator
Product detector
Programmable op amp
Series modulator
SSB circuit
Synchronous detection
Transistor modulator

QUESTIONS

1. What mathematical operation does an amplitude modulator perform?
2. A device that produces amplitude modulation must have what type of response curve?
3. Describe the two basic ways that amplitude modulator circuits generate AM.
4. What type of semiconductor device gives a near-perfect square-law response?

REVIEW

5. Which four signals and frequencies appear at the output of a low-level diode modulator?
6. What component does a PIN diode appear to be when it is used in an AM modulator?
7. Name the primary application of PIN diodes as amplitude modulators.
8. What kind of amplifier must be used to boost the power of a low-level AM signal?
9. How does a differential amplifier modulator work?
10. To what stage of a transmitter does the modulator connect in a high-level AM transmitter?
11. What is the simplest and most common technique for demodulating an AM signal?
12. What is the most critical component value in a diode detector circuit? Explain.
13. What are the advantages of a full rectifier diode detector over a half-wave diode detector?
14. What is the basic component in a synchronous detector? What operates this component?
15. What signals does a balanced modulator generate? Eliminate?
16. What type of balanced modulator uses transformers and diodes?
17. What is the most commonly used filter in a filter-type SSB generator?
18. What is the most difficult part of producing SSB for voice signals using the phasing methods?
19. Which type of balanced modulator gives the greatest carrier suppression?
20. What is the name of the circuit used to demodulate an SSB signal?
21. What signal must be present in an SSB demodulator besides the signal to be detected?

PROBLEMS

1. A collector modulated transmitter has a supply voltage of 48 V and an average collector current of 600 mA. What is the input power to the transmitter? How much modulating signal power is needed to produce 100 percent modulation? ◆
2. An SSB generator has a 9-MHz carrier and is used to pass voice frequencies in the 300- to 3300-Hz range. The lower sideband is selected. What is the approximate center frequency of the filter needed to pass the lower sideband?
3. A 1496 IC balanced modulator has a carrier-level input of 200 mV. The amount of suppression achieved is 60 dB. How much carrier voltage appears at the output? ◆

CRITICAL THINKING

1. State the relative advantages and disadvantages of synchronous detectors versus other types of amplitude demodulators.
2. Could a balanced modulator be used as a synchronous detector? Why or why not?
3. An SSB signal is generated by modulating a 5-MHz carrier with a 400-Hz sine tone. At the receiver, the carrier is reinserted during demodulation, but its frequency is 5.00015 MHz rather than exactly 5 MHz. How does this affect the recovered signal? How would a voice signal be affected by a carrier that is not exactly the same as the original?

FUNDAMENTALS OF FREQUENCY MODULATION

Objectives

After completing this chapter, you will be able to:

- *Compare* and *contrast* frequency modulation and phase modulation.

- *Calculate* the modulation index given the maximum deviation and the maximum modulating frequency and *use* the modulation index and Bessel coefficients to determine the number of significant sidebands in an FM signal.

- *Calculate* the bandwidth of an FM signal using (1) the modulation index and Bessel functions and (2) Carson's rule, and *explain* the practical significance of the difference between the two methods.

- *Explain* how pre-emphasis is used to solve the problem of the interference of high frequency components by noise.

- *List* the advantages of and disadvantages of FM as compared to AM.

- *Give* the reasons for FM's superior immunity to noise.

A sine wave carrier can be modified for the purpose of transmitting information from one place to another by varying its amplitude, frequency, or phase shift. The basic equation for a sine wave is

$$v = V_c \sin (2\pi ft \pm \theta)$$

where V_c = peak amplitude
  f = frequency
  θ = phase angle

Varying the amplitude of a carrier signal in accordance with the intelligence signal produces AM. Impressing an information signal on a carrier by changing its frequency produces FM. Varying the amount of phase shift that a carrier experiences in order to impress information on the carrier is known as phase modulation (PM). As it turns out, varying the phase shift of a carrier also produces FM, and thus FM and PM are closely related. They are collectively referred to as *angle modulation.* Since FM is generally superior in performance to AM, it is widely used in many areas of communications electronics.

5-1 BASIC PRINCIPLES OF FREQUENCY MODULATION

In FM, the carrier amplitude remains constant while the carrier frequency is changed by the modulating signal. As the amplitude of the information signal varies, the carrier frequency shifts proportionately. As the modulating signal amplitude increases, the carrier frequency increases. If the amplitude of the modulating signal decreases, the carrier frequency decreases. The reverse relationship can also be implemented. A decreasing modulating signal increases the carrier frequency above its center value, whereas an increasing modulating signal decreases the carrier frequency below its center value. As the modulating signal amplitude varies, the carrier frequency varies above and below its normal center, or *resting,* frequency with no modulation. The amount of change in carrier frequency produced by the modulating signal is known as the frequency deviation f_d. Maximum frequency deviation occurs at the maximum amplitude of the modulating signal.

The frequency of the modulating signal determines the *frequency deviation rate,* or how many times per second the carrier frequency deviates above and below its center frequency. If the modulating signal is a 500-Hz sine wave, the carrier frequency shifts above and below the center frequency 500 times per second.

An FM signal is illustrated in Fig. 5-1(*c*). Normally the carrier [Fig. 5-1(*a*)] is a sine wave, but it is shown as a triangular wave here to simplify the illustration. With no modulating signal applied, the carrier frequency is a constant-amplitude sine wave at its normal resting frequency.

The modulating information signal [Fig. 5-1(*b*)] is a low-frequency sine wave. As the sine wave goes positive, the frequency of the carrier increases proportionately. The highest frequency occurs at the peak amplitude of the modulating signal. As the modulating signal amplitude decreases, the carrier frequency decreases. When the modulating signal is at zero amplitude, the carrier is at its center frequency point.

DID YOU KNOW?

The frequency of the modulating signal determines the frequency deviation rate, or how many times per second the carrier frequency deviates above and below its center frequency.

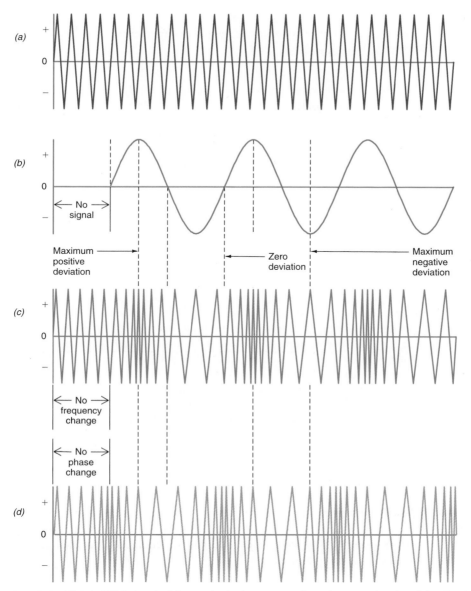

FIG. 5-1 FM and PM signals. The carrier is drawn as a triangular wave for simplicity, but in practice it is a sine wave. (*a*) Carrier. (*b*) Modulating signal. (*c*) FM signal. (*d*) PM signal.

When the modulating signal goes negative, the carrier frequency decreases. It continues to decrease until the peak of the negative half cycle of the modulating sine wave is reached. Then, as the modulating signal increases toward zero, the carrier frequency again increases. This phenomenon is illustrated in Fig. 5-1(*c*), where the carrier sine waves seem to be first compressed and then stretched by the modulating signal.

Assume a carrier frequency of 150 MHz. If the peak amplitude of the modulating signal causes a maximum frequency shift of 30 kHz, the carrier frequency will deviate up to 150.03 MHz and down to 149.97 MHz. The total frequency deviation is 150.03 − 149.97 = 0.06 MHz = 60 kHz. In practice, however, the frequency deviation is expressed as the amount of frequency shift of the carrier above or below the center frequency. Thus the frequency deviation for the 150-MHz carrier frequency is

Example 5-1

A transmitter operates on a frequency of 915 MHz. The maximum FM deviation is ±12.5 kHz. What are the maximum and minimum frequencies that occur during modulation?

$$915 \text{ MHz} = 915,000 \text{ kHz}$$
$$\text{Maximum deviation} = 915,000 + 12.5 = 915,012.5 \text{ kHz}$$
$$\text{Minimum deviation} = 915,000 - 12.5 = 914,987.5 \text{ kHz}$$

represented as ±30 kHz. This means that the modulating signal varies the carrier above and below its center frequency by 30 kHz. Note that the frequency of the modulating signal has no effect on the *amount* of deviation, which is strictly a function of the amplitude of the modulating signal.

Frequently, a modulating signal is a pulse train or series of rectangular waves, for example, serial binary data. When the modulating signal has only two amplitudes, the carrier frequency, instead of an infinite number of values, as it would have with a continuously varying (analog) signal, has only two values. This phenomenon is illustrated in Fig. 5-2. For example, when the modulating signal is a binary 0, the carrier frequency is the center frequency value. When the modulating signal is a binary 1, the carrier frequency abruptly changes to a higher frequency level. The amount of the shift depends on the amplitude of the binary signal. This kind of modulation, called *frequency-shift keying (FSK),* is widely used in the transmission of binary data, for example, when computer files are to be transmitted over the analog voice telephone network using a modem.

5-2 PRINCIPLES OF PHASE MODULATION

When the amount of phase shift of a constant-frequency carrier is varied in accordance with a modulating signal, the resulting output is a phase-modulation (PM) signal [see Fig. 5-1(*d*)]. Imagine a modulator circuit whose basic function is to produce a *phase shift*, that is, a time separation between two sine waves of the same frequency. Assume that a phase shifter can be built that will cause the amount of phase shift to vary with the amplitude of the modulating signal. The greater the amplitude of the modulating

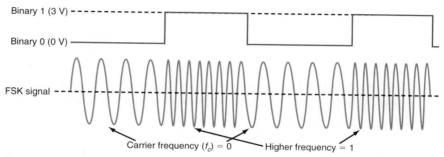

FIG. 5-2   Frequency-modulating of a carrier with binary data produces FSK.

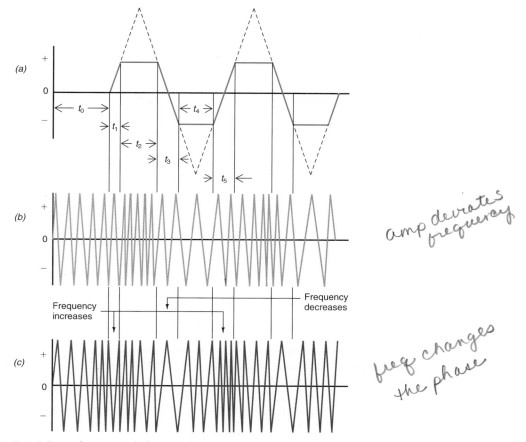

amp deviates frequency

freq changes the phase

FIG. 5-3 A frequency shift occurs in PM only when the modulating-signal amplitude varies. (a) Modulating signal. (b) FM signal. (c) PM signal.

signal, the greater the phase shift. Assume further that positive alternations of the modulating signal produce a lagging phase shift and negative signals produce a leading phase shift.

If a constant-amplitude–constant-frequency carrier sine wave is applied to the phase shifter whose phase shift is varied by the intelligence signal, the output of the phase shifter is a PM wave. As the modulating signal goes positive, the amount of phase lag, and thus the delay of the carrier output, increases with the amplitude of the modulating signal. The result at the output is the same as if the constant-frequency carrier signal had been stretched out, or had its frequency lowered. When the modulating signal goes negative, the phase shift becomes leading. This causes the carrier sine wave to be effectively speeded up, or compressed. The result is the same as if the carrier frequency had been increased.

Note that it is the dynamic nature of the modulating signal that causes the frequency variation at the output of the phase shifter: FM is produced only as long as the phase shift is varying. To understand this better, look at the modulating signal shown in Fig. 5-3(a), which is a triangular wave whose positive and negative peaks have been clipped off at a fixed amplitude. During time t_0, the signal is zero, so the carrier is at its center frequency.

Applying this modulating signal to a frequency modulator produces the FM signal shown in Fig. 5-3(b). During the time the waveform is rising (t_1), the frequency increases. During the time the positive amplitude is constant (t_2), the FM output frequency is constant. During the time the amplitude decreases and goes negative (t_3), the frequency decreases. During the constant-amplitude negative alternation (t_4), the frequency remains constant, at a lower frequency. During t_5, the frequency increases.

Now, refer to the PM signal in Fig. 5-3(c). During increases or decreases in amplitude (t_1, t_3, and t_5), a varying frequency is produced. However, during the constant-amplitude positive and negative peaks, no frequency change takes place. The output of the phase modulator is simply the carrier frequency which has been shifted in phase. This clearly illustrates that when a modulating signal is applied to a phase modulator, the output frequency changes only during the time that the amplitude of the modulating signal is varying.

The maximum frequency deviation produced by a phase modulator occurs during the time that the modulating signal is changing at its most rapid rate. For a sine-wave modulating signal, the rate of change of the modulating signal is greatest when the modulating wave changes from plus to minus or from minus to plus. As Fig. 5-3(c) shows, the maximum rate of change of modulating voltage occurs exactly at the zero crossing points. In contrast, note that in an FM wave the maximum deviation occurs at the peak positive and negative amplitude of the modulating voltage. Thus, although a phase modulator does indeed produce FM, maximum deviation occurs at different points of the modulating signal.

In PM, the amount of carrier deviation is proportional to the rate of change of the modulating signal, that is, the calculus derivative. With a sine-wave modulating signal, the PM carrier appears to be frequency-modulated by the cosine of the modulating signal. Remember that the cosine occurs 90° earlier (leads) than the sine.

Since the frequency deviation in PM is proportional to the rate of change in the modulating signal, the frequency deviation is proportional to the modulating signal frequency as well as its amplitude. As you will see later, this effect is compensated for prior to modulation.

As Fig. 5-4 shows, PM is also used with binary signals. When the binary modulating signal is 0 V, or binary 0, the PM signal is simply the carrier frequency. When a binary 1 voltage level occurs, the modulator, which is a phase shifter, simply changes the phase of the carrier, not its frequency. In Fig. 5-4, the phase shift is 180°. Each time the signal changes from 0 to 1 or 1 to 0, there is a 180° phase shift. The PM signal is still the carrier frequency, but the phase has been changed with respect to the original carrier with a binary 0 input.

The process of phase modulating a carrier with binary data is called *phase-shift keying (PSK)* or *binary phase-shift keying (BPSK)*. The PSK signal shown in Fig. 5-4 uses a 180° phase shift from a reference, but other phase-shift values can be used, for

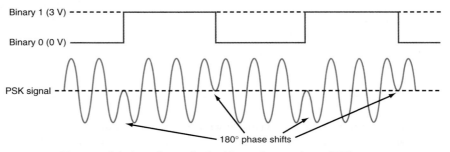

Fig. 5-4 Phase modulation of a carrier by binary data produces PSK.

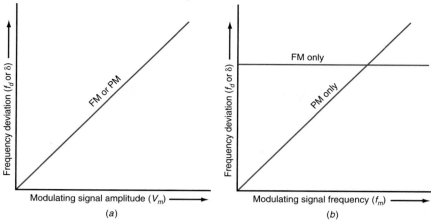

FIG. 5-5 Frequency deviation as a function of (a) modulating-signal amplitude and (b) modulating-signal frequency.

Quadrature amplitude modulation   *QAM*

example, 45°, 90°, 135°, or 225°. The important thing to remember is that no frequency variation occurs. The PSK signal has a constant frequency, but the phase of the signal from some reference changes as the binary modulating signal occurs.

RELATIONSHIP BETWEEN THE
MODULATING SIGNAL AND CARRIER DEVIATION

In FM, the frequency deviation is directly proportional to the amplitude of the modulating signal. The maximum deviation occurs at the peak positive and negative amplitudes of the modulating signal. In PM, the frequency deviation is also directly proportional to the amplitude of the modulating signal. The maximum amount of leading or lagging phase shift occurs at the peak amplitudes of the modulating signal. This effect, for both FM and PM, is illustrated in Fig. 5-5(a).

Now look at Fig. 5-5(b), which shows that the frequency deviation of an FM signal is constant for any value of modulating frequency. Only the amplitude of the modulating signal determines the amount of deviation. But look at how the deviation varies in a PM signal with different modulating-signal frequencies. The higher the modulating-signal frequency, the shorter its period and the faster the voltage changes. Higher modulating voltages result in greater phase shift and this, in turn, produces greater frequency deviation. However, higher modulating frequencies produce a faster rate of change of the modulating voltage and thus greater frequency deviation. In PM, then, the carrier frequency deviation is proportional to both the modulating frequency and amplitude. In FM, frequency deviation is proportional only to the amplitude of the modulating signal, regardless of its frequency.

CONVERTING PM INTO FM

In order to make PM compatible with FM, the deviation produced by frequency variations in the modulating signal must be compensated for. This can be done by passing the intelligence signal through a low-pass *RC* network, as illustrated in Fig. 5-6. This

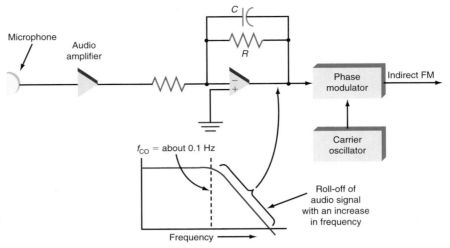

Fig. 5-6 Using a low-pass filter to roll off the audio modulating-signal amplitude with frequency.

low-pass filter, called a *frequency-correcting network, predistorter,* or *1/f filter,* causes the higher modulating frequencies to be attenuated. Although the higher modulating frequencies produce a greater rate of change and thus a greater frequency deviation, this is offset by the lower amplitude of the modulating signal, which produces less phase shift and thus less frequency deviation. The predistorter compensates for the excess frequency deviation caused by higher modulating frequencies. The result is an output that is the same as an FM signal. The FM produced by a phase modulator is called *indirect FM.*

Both FM and PM are widely used in communications systems. A crystal oscillator with high-frequency accuracy and stability is used to generate the carrier. Crystal oscillators cannot, in general, be frequency-modulated over a very wide range. However, their frequency can be "pulled" over a narrow frequency range to produce direct FM. The desired deviation can then be obtained with frequency multipliers. The crystal carrier oscillator can also drive a phase modulator which will produce the desired FM. Practical FM and PM circuits are discussed in Chap. 6.

5-3 MODULATION INDEX AND SIDEBANDS

Any modulation process produces sidebands. When a constant-frequency sine wave modulates a carrier, two side frequencies are produced. The side frequencies are the sum and difference of the carrier and the modulating frequency. In FM and PM, as in AM, sum and difference sideband frequencies are produced. In addition, a large number of pairs of upper and lower sidebands are generated. As a result, the spectrum of an FM or PM signal is usually wider than that of an equivalent AM signal. It is also possible to generate a special narrowband FM signal whose bandwidth is only slightly wider than that of an AM signal.

Figure 5-7 shows the frequency spectrum of a typical FM signal produced by modulating a carrier with a single-frequency sine wave. Note that the sidebands are spaced

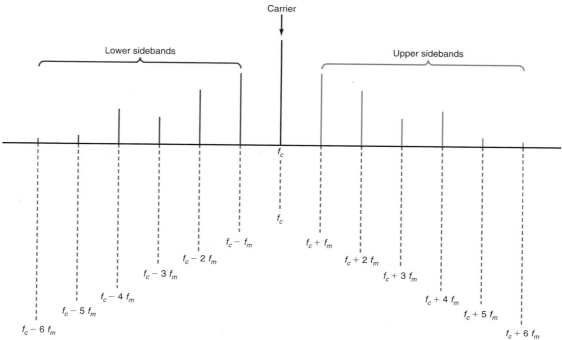

FIG. 5-7 Frequency spectrum of an FM signal. Note that the carrier and sideband amplitudes shown are just examples. The amplitudes depend upon the modulation index m_f.

from the carrier f_c and from one another by a frequency equal to the modulating frequency f_m. If the modulating frequency is 1 kHz, the first pair of sidebands are above and below the carrier by 1000 Hz. The second pair of sidebands are above and below the carrier by 2×1000 Hz = 2000 Hz, or 2 kHz, and so on. Note also that the amplitudes of the sidebands vary. If each sideband is assumed to be a sine wave, with a frequency and amplitude as indicated in Fig. 5-7, and all the sine waves are added together, then the FM signal producing them will be created.

As the amplitude of the modulating signal varies, the frequency deviation changes. The number of sidebands produced, and their amplitude and spacing, depend on the frequency deviation and modulating frequency. Keep in mind that an FM signal has a constant amplitude. Since an FM signal is a summation of the sideband frequencies, the sideband amplitudes must vary with frequency deviation and modulating frequency if their sum is to produce a constant-amplitude but variable-frequency FM signal.

Theoretically, the FM process produces an infinite number of upper and lower sidebands and, therefore, a theoretically infinitely large bandwidth. However, in practice, only those sidebands with the largest amplitudes are significant in carrying the information. Typically any sideband whose amplitude is less than 1 percent of the unmodulated carrier is considered insignificant. Thus FM is readily passed by circuits or communications media with finite bandwidth. Despite this, the bandwidth of an FM signal is usually much wider than an AM signal with the same modulating signal.

DID YOU KNOW?

In FM, only those sidebands with the largest amplitudes are significant in carrying information. Sidebands containing less than 2 percent of the total power have little overall effect on the intelligibility of the signal.

The ratio of the frequency deviation to the modulating frequency is known as the modulation index m_f:

$$m_f = \frac{f_d}{f_m}$$

where f_d is the frequency deviation and f_m is the modulating frequency. Sometimes the lowercase Greek letter delta (δ) is used instead of f_d to represent deviation; then $m_f = \delta/f_m$. For example, if the maximum frequency deviation of the carrier is ± 12 kHz and the maximum modulating frequency is 2.5 kHz, the modulating index is $m_f = 12/2.5 = 4.8$.

In most communications systems using FM, maximum limits are put on both the frequency deviation and the modulating frequency. For example, in standard FM broadcasting, the maximum permitted frequency deviation is 75 kHz and the maximum permitted modulating frequency is 15 kHz. This produces a modulation index of $m_f = 75/15 = 5$.

When the maximum allowable frequency deviation and the maximum modulating frequency are used in computing the modulation index, m_f is known as the *deviation ratio*.

Example 5-2

What is the deviation ratio of TV sound if the maximum deviation is 25 kHz and the maximum modulating frequency is 15 kHz?

$$m_f = \frac{f_d}{f_m} = \frac{25 \text{ kHz}}{15 \text{ kHz}} = 1.667$$

BESSEL FUNCTIONS

Given the modulation index, the number and amplitudes of the significant sidebands can be determined by solving the basic equation of an FM signal. The FM equation, whose derivation is beyond the scope of this book, is $v_{FM} = V_c \sin [2\pi f_c t + m_f \sin (2\pi f_m t)]$, where v_{FM} is the instantaneous value of the FM signal and m_f is the modulation index. The term whose coefficient is m_f is the phase angle of the carrier. Note that this equation expresses the phase angle in terms of the sine wave modulating signal. This equation is solved with a complex mathematical process known as Bessel functions. It is not necessary to show this solution, but the result is as follows:

$$
\begin{aligned}
v_{FM} = V_c \{ & J_0[\sin \omega_c t] + J_1[\sin (\omega_c + \omega_m)t - \sin(\omega_c - \omega_m)t] \\
& + J_2[\sin(\omega_c + 2\omega_m)t + \sin(\omega_c - 2\omega_m)t] \\
& + J_3[\sin(\omega_c + 3\omega_m)t - \sin(\omega_c - 3\omega_m)t] \\
& + J_4[\sin(\omega_c + 4\omega_m)t + \sin(\omega_c - 4\omega_m)t] \\
& + J_5[\sin \ldots] + \ldots \}
\end{aligned}
$$

where $\omega_c = 2\pi f_c$ = carrier frequency
  $\omega_m = 2\pi f_m$ = modulating signal frequency
  V_c = peak value of unmodulated carrier

The FM wave is expressed as a composite of sine waves of different frequencies and amplitudes that when added together give an FM time-domain signal. The first term is the carrier with an amplitude given by a J_n coefficient, in this case J_0. The next term represents a pair of upper and lower side frequencies equal to the sum and difference of the carrier- and modulating-signal frequency. The amplitude of these side frequencies is J_1. The next term is another pair of side frequencies equal to the carrier ± 2 times the modulating-signal frequency. The other terms represent additional side frequencies spaced from one another by an amount equal to the modulating-signal frequency.

The amplitudes of the sidebands are determined by the J_n coefficients, which are, in turn, determined by the value of the modulation index. These amplitude coefficients are computed using the expression

$$J_n(m_f) = \left(\frac{m_f}{2}\right)^n \left[\frac{1}{n} - \frac{(m_f/2)^2}{1!(n+1)!} + \frac{(m_f/2)^4}{2!(n+2)!} - \frac{(m_f/2)^6}{3!(n+1)!} + \cdots\right]$$

where ! = factorial
  n = J number (number of sideband)
  $m_f = \dfrac{f_d}{f_m}$ = frequency deviation

In practice, you do not have to know or calculate these coefficients, since tables giving them are widely available. The Bessel coefficients for a range of modulation indexes are given in Fig. 5-8. The left-hand column gives the modulation index m_f. The remaining columns indicate the relative amplitudes of the carrier and the various pairs of sidebands. Any sideband with a relative carrier amplitude of less than 1 percent (0.01) has been eliminated. Note that some of the carrier and sideband amplitudes have negative signs. This means that the signal represented by that amplitude is simply shifted in phase 180° (phase inversion).

Figure 5-9 shows the curves that are generated by plotting the data in Fig. 5-8. The carrier and sideband amplitudes and polarities are plotted on the vertical axis; the modulation index is plotted on the horizontal axis. As the figures illustrate, the carrier amplitude J_0 varies with the modulation index. In FM, the carrier amplitude, and the amplitudes of the sidebands, changes as the modulating signal frequency and deviation change. In AM, the carrier amplitude remains constant.

Note that at several points in Figs. 5-8 and 5-9, at modulation indexes of about 2.4, 5.5, and 8.7, the carrier amplitude J_0 actually drops to zero. At those points, all of the signal power is completely distributed throughout the sidebands. And as can be seen in Fig. 5-9, the sidebands also go to zero at certain values of the modulation index.

Modulation Index	Carrier	Sidebands (Pairs)															
		1st	2d	3d	4th	5th	6th	7th	8th	9th	10th	11th	12th	13th	14th	15th	16th
0.00	1.00	—	—	—	—	—	—	—	—	—	—	—	—	—	—	—	—
0.25	0.98	0.12	—	—	—	—	—	—	—	—	—	—	—	—	—	—	—
0.5	0.94	0.24	0.03	—	—	—	—	—	—	—	—	—	—	—	—	—	—
1.0	0.77	0.44	0.11	0.02	—	—	—	—	—	—	—	—	—	—	—	—	—
1.5	0.51	0.56	0.23	0.06	0.01	—	—	—	—	—	—	—	—	—	—	—	—
2.0	0.22	0.58	0.35	0.13	0.03	—	—	—	—	—	—	—	—	—	—	—	—
2.5	−0.05	0.50	0.45	0.22	0.07	0.02	—	—	—	—	—	—	—	—	—	—	—
3.0	−0.26	0.34	0.49	0.31	0.13	0.04	0.01	—	—	—	—	—	—	—	—	—	—
4.0	−0.40	−0.07	0.36	0.43	0.28	0.13	0.05	0.02	—	—	—	—	—	—	—	—	—
5.0	−0.18	−0.33	0.05	0.36	0.39	0.26	0.13	0.05	0.02	—	—	—	—	—	—	—	—
6.0	0.15	−0.28	−0.24	0.11	0.36	0.36	0.25	0.13	0.06	0.02	—	—	—	—	—	—	—
7.0	0.30	0.00	−0.30	−0.17	0.16	0.35	0.34	0.23	0.13	0.06	0.02	—	—	—	—	—	—
8.0	0.17	0.23	−0.11	−0.29	−0.10	0.19	0.34	0.32	0.22	0.13	0.06	0.03	—	—	—	—	—
9.0	−0.09	0.24	0.14	−0.18	−0.27	−0.06	0.20	0.33	0.30	0.21	0.12	0.06	0.03	0.01	—	—	—
10.0	−0.25	0.04	0.25	0.06	−0.22	−0.23	−0.01	0.22	0.31	0.29	0.20	0.12	0.06	0.03	0.01	—	—
12.0	−0.05	−0.22	−0.08	0.20	0.18	−0.07	−0.24	−0.17	0.05	0.23	0.30	0.27	0.20	0.12	0.07	0.03	0.01
15.0	−0.01	0.21	0.04	0.19	−0.12	0.13	0.21	0.03	−0.17	−0.22	−0.09	0.10	0.24	0.28	0.25	0.18	0.12

Fig. 5-8 Carrier and sideband amplitudes for different modulation indexes of FM signals based on the Bessel functions.

Example 5-3

What is the maximum modulating frequency that can be used to achieve a modulation index of 2.2 with a deviation of 7.48 kHz?

$$f_m = \frac{f_d}{m_f} = \frac{7480}{2.2} = 3400 \text{ Hz} = 3.4 \text{ kHz}$$

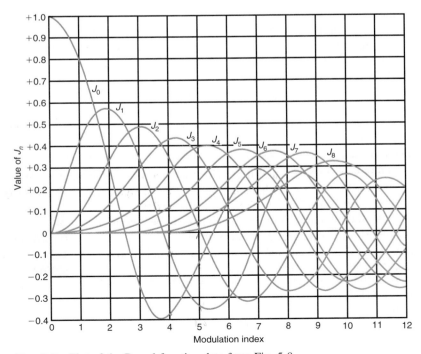

Fig. 5-9 Plot of the Bessel function data from Fig. 5-8.

Figure 5-10 shows several examples of an FM signal spectrum with different modulation indexes. Compare the examples to the entries in Fig. 5-8. The unmodulated carrier in Fig. 5-10(*a*) has a relative amplitude of 1.0. With no modulation, all of the power is in the carrier. With modulation, the carrier amplitude decreases while the amplitudes of the various sidebands increase.

In Fig. 5-10(*d*), the modulation index is 0.25. This is a special case of FM where the modulation process produces only a single pair of significant sidebands like those produced by AM. With a modulation index of 0.25, the FM signal occupies no more spectrum space than an AM signal. This type of FM is called *narrow-band FM,* or *NBFM.* The formal definition of NBFM is any FM system where the modulation index is less than $\pi/2 = 1.57$, or $m_f < \pi/2$. However, for true NBFM with only a single pair of sidebands, m_f must be much less than $\pi/2$. Values of m_f in the 0.2 to 0.25 range will give true NBFM. Common FM mobile radios use a maximum deviation of 5 kHz, with a maximum voice frequency of 3 kHz, giving a modulation index of $m_f = 5$ kHz/3 kHz $= 1.667$. While these systems do not fall within the formal definition of NBFM, they are nonetheless regarded as narrow-band transmissions.

The primary purpose of NBFM is to conserve spectrum space, and NBFM is widely used in radio communications. Note, however, that NBFM conserves spectrum space at the expense of signal-to-noise ratio.

Example 5-4

State the amplitudes of the carrier and first four sidebands of an FM signal with a modulation index of 4. (Use Figs. 5-8 and 5-9.)

$$J_0 = -0.4$$
$$J_1 = -0.07$$
$$J_2 = 0.36$$
$$J_3 = 0.43$$
$$J_4 = 0.28$$

FM SIGNAL BANDWIDTH

As stated previously, the higher the modulation index in FM, the greater the number of significant sidebands and the wider the bandwidth of the signal. When spectrum conservation is necessary, the bandwidth of an FM signal can be deliberately restricted by putting an upper limit on the modulation index.

The total bandwidth of an FM signal can be determined by knowing the modulation index and using Fig. 5-8. For example, assume the highest modulating frequency of a signal is 3 kHz and the maximum deviation is 6 kHz. This gives a modulation index of $m_f = 6$ kHz/3 kHz $= 2$. Referring to Fig. 5-8, you can see that this produces four significant pairs of sidebands. The bandwidth can then be determined with the simple formula

$$BW = 2f_m N$$

where N is the number of significant sidebands in the signal. According to this

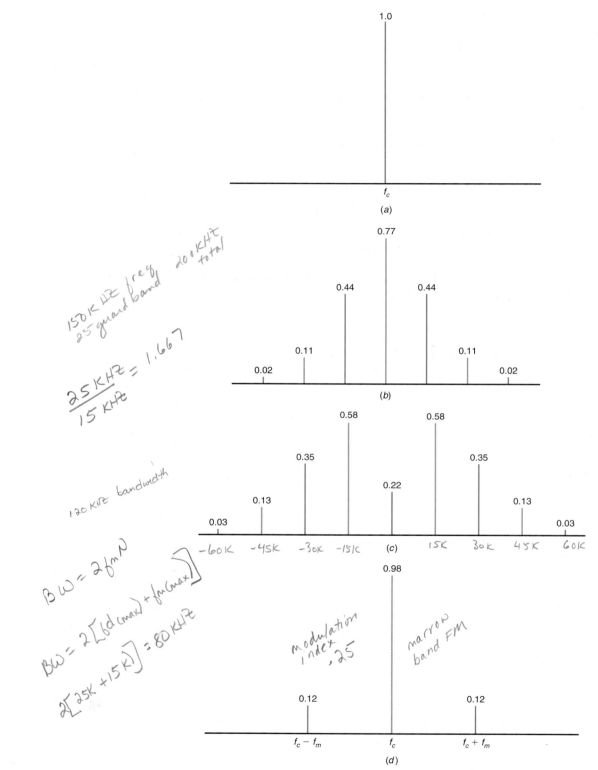

FIG. 5-10 Examples of FM signal spectra. (*a*) Modulation index of 0 (no modulation or sidebands). (*b*) Modulation index of 1. (*c*) Modulation index of 2. (*d*) Modulation index of 0.25 (NBFM).

formula, the bandwidth of our FM signal is BW = 2(3-kHz)(4) = 24 kHz. In general terms, an FM signal with a modulation index of 2 and a highest modulating frequency of 3 kHz will occupy a 24-kHz bandwidth.

Another way to determine the bandwidth of an FM signal is to use what is known as *Carson's rule*. This rule recognizes only the power in the most significant sidebands with amplitudes greater than 2 percent of the carrier (0.02 or higher in Fig. 5-8). This rule is

$$BW = 2[f_{d(\text{max})} + f_{m(\text{max})}]$$

According to Carson's rule, the bandwidth of the FM signal in the previous example would be BW = 2(6 kHz + 3 kHz) = 2(9 kHz) = 18 kHz.

Carson's rule will always give a bandwidth lower than that calculated with the formula BW = $2f_m N$. However, it has been proved that if a circuit or system has the bandwidth calculated by Carson's rule, the sidebands will indeed be passed well enough to ensure full intelligibility of the signal.

So far, all the examples of FM have assumed a single-frequency sine-wave modulating signal. However, as you know, most modulating signals are not pure sine waves, but complex waves made up of many different frequencies. When the modulating signal is a pulse or binary wave train, the carrier is modulated by the equivalent signal, which is a mix of a fundamental sine wave and all of the relevant harmonics as determined by Fourier theory. For example, if the modulating signal is a square wave, the fundamental sine wave and all of the odd harmonics modulate the carrier. Each harmonic produces multiple pairs of sidebands depending on the modulation index. As you can imagine, FM by a square or rectangular wave generates many sidebands and produces a signal with an enormous bandwidth. The circuits or systems that will carry, process, or pass such a signal must have the appropriate bandwidth so as not to distort the signal.

Example 5-5

What is the maximum bandwidth of an FM signal with a deviation of 30 kHz and a maximum modulating signal of 5 kHz as determined by (*a*) Fig. 5-8 and (*b*) Carson's rule.

a. $m_f = \dfrac{f_d}{f_m} = \dfrac{30 \text{ kHz}}{5 \text{ kHz}}$

Figure 5-8 shows nine significant sidebands spaced 5 kHz apart for $f_m = 6$.

$m_f = 6$

$$BW = 2f_m N$$
$$BW = 2(5 \text{ kHz}) \, 9 = 90 \text{ kHz}$$

b. $BW = 2[f_{d(\text{max})} + f_{m(\text{max})}]$
$$= 2(30 \text{ kHz} + 5 \text{ kHz})$$
$$= 2(35 \text{ kHz})$$
$$= 70 \text{ kHz}$$

Percentage of Modulation

In AM, the degree of modulation is usually given as a percentage, which is the ratio of the amplitude of the modulating signal to the amplitude of the carrier. When the two are equal, the ratio is 1 and the percentage of modulation is 100. When the modulating-signal amplitude is greater than the carrier amplitude, overmodulation and distortion result. In contrast, in FM and PM the carrier amplitude remains constant during modulation, and so the percentage-of-modulation indicator used in AM is meaningless. Increasing the amplitude or the frequency of the modulating signal does not cause overmodulation or distortion. Increasing the modulating-signal amplitude simply increases the frequency deviation. This, in turn, increases the modulation index, which simply produces more significant sidebands and a wider bandwidth.

For practical reasons of spectrum conservation and receiver performance, there is usually some limit put on the upper frequency deviation and the upper modulating frequency. The audio in broadcast TV is transmitted by FM. The maximum deviation permitted is 25 kHz and the maximum modulating frequency is 15 kHz. This produces a deviation ratio of $m_f = 25/15 = 1.667$. In standard two-way mobile radio communications using FM, the maximum permitted deviation is usually 5 kHz. The upper modulating frequency is usually limited to 3 kHz, which is high enough for intelligible voice transmission. This produces a deviation ratio of $m_f = 5/3 = 1.667$.

Did You Know?

In contrast to AM, increasing the amplitude or frequency of the modulating signal in FM and PM does not cause overmodulation or distortion. Increasing the modulating signal amplitude simply increases the frequency deviation, which increases the modulation index and results in a wider bandwidth. These can cause adjacent channel interference.

The maximum deviation permitted can be used in a ratio with the actual carrier deviation to produce a percentage of modulation for FM:

$$\text{FM \% modulation} = \frac{\delta_a}{\delta_m} 100$$

where δ_a = actual carrier deviation
δ_m = maximum carrier deviation

In commercial FM broadcasting the maximum allowed deviation is 75 kHz. If the modulating signal is producing only a maximum deviation of 60 kHz, the FM percentage of modulation is (60/75) 100 = 80%.

When maximum deviations are specified, it is important that the percentage of modulation be held to less than 100 percent. The reason for this is that FM stations operate in assigned frequency channels, which are adjacent to channels containing signals from other stations. If the deviation were allowed to exceed the maximum, the number of pairs of sidebands, and the resulting signal bandwidth, would be excessive, causing undesirable adjacent channel interference.

5-4 Noise-Suppression Effects of FM

Noise is interference generated by lightning, motors, automotive ignition systems, and any power-line switching that produces transient signals. Such noise is typically narrow spikes of voltage with very high frequencies. They add to a signal and interfere

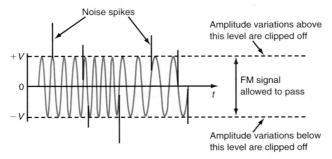

Fig. 5-11 An FM signal with noise.

with it. The potential effect of such noise on an FM signal is shown in Fig. 5-11. If the noise signals were strong enough, they could completely obliterate the information signal.

FM signals, however, have a constant modulated carrier amplitude, and FM receivers contain limiter circuits that deliberately restrict the amplitude of the received signal. Any amplitude variations occurring on the FM signal are effectively clipped off, as shown in Fig. 5-11. This does not affect the information content of the FM signal, since it is contained solely within the frequency variations of the carrier. Because of the clipping action of the limiter circuits, noise is almost completely eliminated. Even if the peaks of the FM signal itself are clipped or flattened and the resulting signal is distorted, no information is lost. In fact, one of the primary benefits of FM over AM is its superior noise immunity. The process of demodulating or recovering an FM signal actually suppresses noise and improves the signal-to-noise ratio.

NOISE AND PHASE SHIFT

The noise amplitude added to an FM signal introduces a small frequency variation, or phase shift, which changes or distorts the signal. Figure 5-12 shows how this works. The carrier signal is represented by a fixed-length (amplitude) phasor S. The noise is usually a short-duration pulse containing many frequencies at many amplitudes and

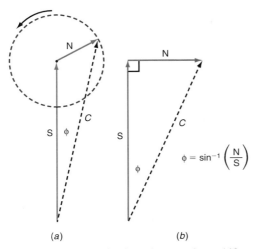

$$\phi = \sin^{-1}\left(\frac{N}{S}\right)$$

(a)   (b)

Fig. 5-12 How noise introduces a phase shift.

phases according to Fourier theory. To simplify the analysis, however, we will assume a single high-frequency noise signal varying in phase. In Fig. 5-12(a), this noise signal is represented as a rotating phasor N. The composite signal of the carrier and the noise, labeled C, is a phasor whose amplitude is the phasor sum of the signal and noise and a phase angle shifted from the carrier by an amount ϕ. If you imagine the noise phasor rotating, you can also imagine the composite signal varying in amplitude and phase angle with respect to the carrier.

The maximum phase shift occurs when the noise and signal phasors are at a right angle to one another, as illustrated in Fig. 5-12(b). This angle can be computed with the arcsine or inverse sine according to the formula

$$\phi = \sin^{-1} \frac{N}{S}$$

It is possible to determine just how much of a frequency shift a particular phase shift produces by using the formula

$$\delta = \phi(f_m)$$

where δ = frequency deviation produced by noise
  ϕ = phase shift in radians
  f_m = frequency of modulating signal

Assume the signal-to-noise ratio (S/N) is 3 to 1 and the modulating signal frequency is 800 Hz. The phase shift is then $\phi = \sin^{-1}$ (N/S) = $\sin^{-1}$ (1/3) = $\sin^{-1}$ (0.3333) = 19.47°. Since there are 57.3° per radian, this angle is ϕ = 19.47/57.3 = 0.34 rad. The frequency deviation produced by this brief phase shift can be calculated as

$$\delta = 0.34(800) = 271.8 \text{ Hz}$$

Just how badly a particular phase shift will distort a signal depends on several factors. Looking at the formula for deviation, you can deduce that the worst-case phase shift and frequency deviation will occur at the highest modulating-signal frequency. The overall effect of the shift depends upon what the maximum allowed frequency shift for the application is. If very high deviations are allowed, meaning a high modulation index, the shift can be small and inconsequential. If the total allowed deviation is small, then the noise-induced deviation can be severe. Remember that the noise interference is of very short duration; thus the phase shift is momentary, and intelligibility is rarely severely impaired. With heavy noise, human speech might be temporarily garbled, but not enough so that it could not be understood.

Assume that the maximum allowed deviation is 5 kHz in the example above. The ratio of the shift produced by the noise to the maximum allowed deviation is

$$\frac{\text{Frequency deviation produced by noise}}{\text{Maximum allowed deviation}} = \frac{271.8}{5000} = 0.0544$$

This is only a bit more than a 5 percent shift. The 5-kHz deviation represents the maximum modulating signal amplitude. The 271.8-Hz shift is the noise amplitude. Therefore this ratio is the noise-to-signal ratio N/S. The reciprocal of this value gives you the FM signal-to-noise ratio:

$$\text{S/N} = \frac{1}{\text{N/S}} = 1/0.0544 = 18.4$$

For FM, a 3 to 1 input S/N translates into an 18.4 to 1 output S/N.

Example 5-6

The input to an FM receiver has an S/N of 2.8. The modulating frequency is 1.5 kHz. The maximum permitted deviation is 4 kHz. What are (a) the frequency deviation caused by the noise and (b) the improved output S/N?

a. $\phi = \sin^{-1}\left(\dfrac{N}{S}\right)\sin^{-1}\left(\dfrac{1}{2.8}\right) = \sin^{-1}(0.3571) = 20.92°$ or 0.3652 rad

$\delta = \phi(f_m) = (0.3652)(1.5 \text{ kHz}) = 547.8 \text{ Hz}$

b. $\dfrac{N}{S} = \dfrac{\text{Frequency deviation produced by noise}}{\text{Maximum allowed deviation}} = \dfrac{547.8}{4000}$

$= 0.13695$

$\dfrac{S}{N} = \dfrac{1}{N/S} = 7.3$

PRE-EMPHASIS

Noise *can* interfere with an FM signal, and particularly with the high-frequency components of the modulating signal. Since noise is primarily sharp spikes of energy, it contains a lot of harmonics and other high-frequency components. These frequencies can be larger in amplitude than the high-frequency content of the modulating signal, causing frequency distortion that can make the signal unintelligible.

Most of the content of a modulating signal, particularly voice, is at low frequencies. In voice communications systems, the bandwidth of the signal is limited to about 3 kHz, which permits acceptable intelligibility. In contrast, musical instruments typically generate signals at low frequencies but contain many high-frequency harmonics which give them their unique sound and must be passed if that sound is to be preserved. Thus a wide bandwidth is needed in high-fidelity systems. Since the high-frequency components are usually at a very low level, noise can obliterate them.

To overcome this problem, most FM systems use a technique known as *pre-emphasis* that helps offset high-frequency noise interference. At the transmitter, the modulating signal is passed through a simple network which amplifies the high-frequency components more than the low-frequency components. The simplest form of such a circuit is a simple high-pass filter of the type shown in Fig. 5-13(a). Specifications dictate a time constant t of 75 μs, where $t = R_1 C$. Any combination of resistor and capacitor (or resistor and inductor) giving this time constant will work. Such a circuit has a cutoff frequency of 2122 Hz; frequencies higher than 2122 Hz will be linearly enhanced. The output amplitude increases with frequency at a rate of 6 dB per octave. The pre-emphasis circuit increases the energy content of the higher-frequency signals so that they become stronger than the high-frequency noise components. This improves the signal-to-noise ratio and increases intelligibility and fidelity.

The pre-emphasis circuit also has an upper break frequency f_u, where the signal enhancement flattens out [see Fig. 5-13(b)], which is computed with the formula

$$f_u = \frac{R_1 + R_2}{2\pi R_1 R_2 C}$$

The value of f_u is usually set well beyond the audio range, and is typically greater than 30 kHz.

To return the frequency response to its normal, "flat" level, a de-emphasis circuit, a simple low-pass filter with a time constant of 75 μs, is used at the receiver (see Fig. 5-13c). Signals above its cutoff frequency of 2122 Hz are attenuated at the rate of 6 dB per octave. The response curve is shown in Fig. 5-13(d). As a result, the pre-emphasis at the transmitter is exactly offset by the de-emphasis circuit in the receiver, providing a flat frequency response. The combined effect of pre-emphasis and de-emphasis is to increase the signal-to-noise ratio for the high-frequency components during transmission so that they will be stronger and not be masked by noise. Figure 5-13e shows the overall effect of pre-emphasis and de-emphasis.

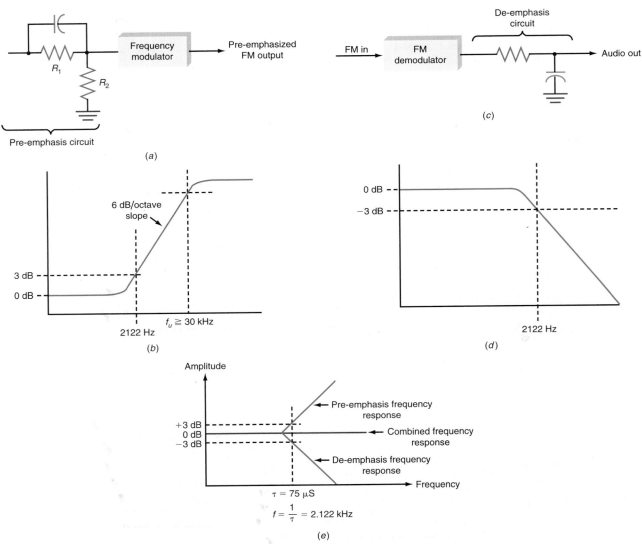

FIG. 5-13   Pre-emphasis and de-emphasis. (a) Pre-emphasis circuit. (b) Pre-emphasis curve. (c) De-emphasis circuit. (d) De-emphasis curve. (e) Combined frequency response.

5-5 FREQUENCY MODULATION VERSUS AMPLITUDE MODULATION

ADVANTAGES OF FM

In general, FM is considered to be superior to AM. Although both AM and FM signals can be used to transmit information from one place to another, FM typically offers some significant benefits over AM.

NOISE IMMUNITY. The main benefit of FM over AM is its superior immunity to noise, made possible by the clipper limiter circuits in the receiver, which effectively strip off all of the noise variations, leaving a constant-amplitude FM signal. Although clipping does not result in total recovery in all cases, FM can nevertheless tolerate a much higher noise level than AM for a given carrier amplitude. This is also true for phase-shift-induced distortion.

CAPTURE EFFECT. Another major benefit of FM is that interfering signals on the same frequency are effectively rejected. Because of the amplitude limiters and the de-modulating methods used by FM receivers, a phenomenon known as the *capture effect* takes place when two or more FM signals occur simultaneously on the same frequency. If one signal is more than twice the amplitude of the other, the stronger signal captures the channel, totally eliminating the weaker signal. With modern receiver circuitry, a difference in signal amplitudes of only 1 dB is usually sufficient to produce the cap-ture effect. In contrast, when two AM signals occupy the same frequency, both signals are generally heard, regardless of their relative signal strengths. When one AM signal is significantly stronger than another, naturally the stronger signal is intelligible; how-ever, the weaker signal is not eliminated, and can still be heard in the background. When the signal strengths of given AM signals are nearly the same, they will interfere with one another, making both of them nearly unintelligible.

Although the capture effect prevents the weaker of two FM signals from being heard, when two stations are broadcasting signals of approximately the same ampli-tude, first one may be captured, and then the other. This can happen, for example, when a driver moving along a highway is listening to a clear broadcast on a particular fre-quency. At some point, the driver may suddenly hear the other broadcast, completely losing the first, then, just as suddenly, hear the original broadcast again. Which one dominates depends on where the car is and on the relative signal strengths of the two signals.

TRANSMITTER EFFICIENCY. A third advantage of FM over AM involves efficiency. Recall that AM can be pro-duced by both low-level and high-level modulation tech-niques. The most efficient is high-level modulation where a class C amplifier is used as the final RF power stage and is modulated by a high-power modulation amplifier. The AM transmitter must produce both very high RF and modulat-ing-signal power. Additionally, at very high power levels, large modulation amplifiers are impractical. Under such

DID YOU KNOW?

Although the ICs in FM are com-plex, they are relatively easy to use and their price is competi-tive with that of AM circuits.

conditions, low-level modulation must be used if the AM information is to be preserved without distortion. The AM signal is generated at a lower level and then amplified with linear amplifiers to produce the final RF signal. Linear amplifiers are either class A or class B, and are far less efficient than class C amplifiers.

FM signals have a constant amplitude, and it is therefore not necessary to use linear amplifiers to increase their power level. In fact, FM signals are always generated at a lower level and then amplified by a series of class C amplifiers to increase their power. The result is greater use of available power because of the high level of efficiency of class C amplifiers.

FM transmitters do not have linear amplifiers

DISADVANTAGES OF FM

EXCESSIVE SPECTRUM USE. Perhaps the greatest disadvantage of FM is that it simply uses too much spectrum space. The bandwidth of an FM signal is, in general, considerably wider than that of an AM signal transmitting similar information. Although it is possible to keep the modulation index low to minimize bandwidth, reducing the modulation index also reduces the noise immunity of an FM signal. In commercial two-way FM radio systems, the maximum allowed deviation is 5 kHz, with a maximum modulating frequency of 3 kHz. This produces a deviation ratio of $5/3 = 1.67$. Deviation ratios as low as 0.25 are possible, although they result in

Portable high-quality "boom" boxes provide music and entertainment for people on the move. These portable entertainment centers now include AM and FM receivers, tape decks, and CD players.

signals that are much less desirable than wide-band FM signals. Both of these deviation ratios are classified as narrowband FM.

Since FM occupies so much bandwidth, it is typically used only in those portions of the spectrum where adequate bandwidth is available, that is, at very high frequencies. In fact, it is rarely used below frequencies of 30 MHz. Most FM communications work is done at the VHF, UHF, and microwave frequencies.

CIRCUIT COMPLEXITY. One major disadvantage of FM in the past involved the complexity of the circuits used for frequency modulation and demodulation in comparison with the simple circuits used for amplitude modulation and demodulation. Today, this disadvantage has almost disappeared due to the use of integrated circuits. While the ICs used in FM transmission are still complex, they require very little effort to use and their price is just as low as comparable AM circuits.

Since the trend in electronic communications is toward higher and higher frequencies and because ICs are so cheap and easy to use, FM has become by far the most widely used modulation method in electronic communications today.

FM AND AM APPLICATIONS

Table 5-1 lists some of the major applications for AM and FM.

TABLE 5-1   COMMON APPLICATIONS FOR AM AND FM

Application	Type of modulation
AM broadcast radio	AM
FM broadcast radio	FM
FM stereo multiplex sound	DSB (AM) and FM
TV sound	FM
TV picture (video)	AM
TV color signals	Quadrature DSB (AM)
Cellular telephone	FM
Cordless telephone	FM
Fax machine	FM, QAM (AM plus PSK)
Aircraft radio	AM
Marine radio	FM and SSB (AM)
Mobile and handheld radio	FM
Citizens' band radio	AM and SSB (AM)
Amateur radio	FM and SSB (AM)
Computer modems	FSK, PSK, QAM (AM plus PSK)
Garage door opener	OOK
TV remote control	OOK
VCR	FM

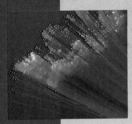

SUMMARY

Frequency deviation in FM is proportional only to the amplitude of the modulating signal regardless of its frequency. In FM, the frequency of the modulating signal determines how many times per second the carrier frequency deviates above and below its nominal center frequency. In frequency shift keying (FSK), as in transmission of serial binary data, the modulating signal is a pulse train or series of rectangular waves. In PM, the amount of phase shift of a constant-frequency carrier is varied in accordance with a modulating signal, and the carrier frequency deviation is proportional to both the modulating frequency and amplitude. Since FM is produced by PM, PM is often referred to as indirect FM.

To make PM compatible with FM, the deviation produced by frequency variations in the modulating signal must be compensated for. The ratio between the maximum permitted frequency deviation and the maximum permitted modulating frequency is the modulation index, or the deviation ratio. Knowing the modulation index, the number of significant sidebands, and the sideband amplitude coefficents as determined by Bessel functions, the basic equation of an FM signal can be solved. The FM wave is expressed as a composite of sine waves of different frequencies and amplitudes that when added together will give an FM time-domain signal.

The bandwidth of an FM signal can be calculated using the modulation index and Bessel functions or by Carson's rule.

One of the primary benefits of FM over AM is its superior immunity to noise. FM receivers contain limiter circuits that restrict the amplitude of the received signal, clipping off any amplitude variations and almost completely eliminating noise. Certain kinds of high-frequency components in the modulating signal can, however, interfere with FM transmission. To overcome this problem, most FM systems use a technique known as pre-emphasis. At the transmitter, the modulating signal is passed through a simple network which amplifies the high-frequency components more than the low-frequency components.

In FM, the stronger of two signals on the same frequency will reject the weaker, through a phenomenon known as the capture effect. In contrast, when one AM signal is significantly stronger than the other, the weaker signal can be heard in the background. A final advantage of FM over AM is efficiency of transmission. FM signals are always generated at a low level and then amplified by a series of highly efficient class C amplifiers.

KEY TERMS

Bessel functions
Binary shift keying (BPSK)
Capture effect
Carson's rule
De-emphasis
Deviation rate

Frequency-correcting
network (1/f filter)
Frequency shift keying
(FSK)
Indirect FM
Modulation index

Narrow band FM
(NBFM)
Percentage of modulation
Phase modulation (PM)
Phase-shift keying (PSK)
Pre-emphasis

REVIEW

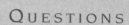

QUESTIONS

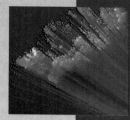

1. What is the general name given to both FM and PM?
2. State the effect on the amplitude of the carrier during FM or PM.
3. What are the name and mathematical expressions for the amount that the carrier varies from its unmodulated center frequency during modulation?
4. State how the frequency of a carrier varies in an FM system when modulating-signal amplitude and frequency change.
5. State how the frequency of a carrier varies in a PM system when modulating-signal amplitude and frequency change.
6. When does maximum frequency deviation occur in an FM signal? A PM signal?
7. State the conditions that must exist for a phase modulator to produce FM.
8. What do you call FM produced by PM techniques?
9. State the nature of the output of a phase modulator during the time that the modulating-signal voltage is constant.
10. What is the name given to the process of frequency modulation of a carrier by binary data?
11. What is the name given to the process of phase modulation of a carrier by binary data?
12. How must the nature of the modulating signal be modified to produce FM by PM techniques?
13. What is the difference between the modulation index and the deviation ratio?
14. Define narrowband FM. What criterion is used to indicate NBFM?
15. What is the name of the mathematical equation used to solve for the number and amplitude of sidebands in an FM signal?
16. What is the meaning of a negative sign on the sideband value in Fig. 5-8?
17. Name two ways that noise affects an FM signal.
18. How is the noise on an FM signal minimized at the receiver?
19. What is the primary advantage of FM over AM?
20. List two additional advantages of FM over AM.
21. What is the nature of the noise that usually accompanies a radio signal?
22. In what ways is an FM transmitter more efficient than a low-level AM transmitter? Explain.
23. What is the main disadvantage of FM over AM? State two ways in which this disadvantage can be overcome.
24. What type of power amplifier is used to amplify FM signals? Low-level AM signals?
25. What is the name of the receiver circuit that eliminates noise?
26. What is the capture effect and what causes it?
27. What is the nature of the modulating signals which are most negatively impacted by noise on an FM signal?
28. Describe the process of pre-emphasis. How does it improve communications performance in the presence of noise? Where is it performed, at the transmitter or receiver?
29. What is the basic circuit used to produce pre-emphasis?
30. Describe the process of de-emphasis. Where is it performed, at the transmitter or receiver?
31. What type of circuit is used to accomplish de-emphasis?
32. What is the cutoff frequency of pre-emphasis and de-emphasis circuits?
33. List four major applications for FM.

PROBLEMS

1. A 162-MHz carrier is deviated by 12 kHz by a 2-kHz modulating signal. What is the modulation index?

2. The maximum deviation of an FM carrier with a 2.5-kHz signal is 4 kHz. What is the deviation ratio?

3. For Problems 1 and 2 above, compute the bandwidth occupied by the signal using the conventional method and Carson's rule. Sketch the spectrum of each signal, showing all significant sidebands and their exact amplitudes.

4. For a single-frequency sine-wave modulating signal of 3 kHz with a carrier frequency of 36 MHz, what is the spacing between sidebands?

5. What are the relative amplitudes of the fourth pair of sidebands for an FM signal with a deviation ratio of 8?

6. At approximately what modulation index does the amplitude of the first pair of sidebands go to zero? Use Fig. 5-8 or 5-9 to find the lowest modulation index that gives this result.

7. The maximum allowed deviation of a TV broadcast sound transmitter is 25 kHz. The actual deviation is 16 kHz. What is the percentage of modulation?

8. An available channel for FM transmission is 30 kHz wide. The maximum allowable modulating signal frequency is 3.5 kHz. What deviation ratio should be used?

9. The signal-to-noise ratio in an FM system is 4 to 1. The maximum modulating frequency is 4 kHz. How much frequency deviation is introduced by the phase shift caused by the noise when the modulating frequency is 650 Hz? What is the real signal-to-noise ratio?

10. A de-emphasis circuit has a capacitor value of 0.02 μF. What value of resistor is needed? Give the closest standard EIA value.

11. Use Carson's rule the determine the bandwidth of an FM channel when the maximum deviation allowed is 5 kHz at frequencies up to 3.333 kHz. Sketch the spectrum, showing carrier and sideband values.

CRITICAL THINKING

1. The AM broadcast band consists of 107 channels for stations 10-kHz wide. The maximum permitted modulating frequency is 5 kHz. Could FM be used on this band? If so, explain what would be necessary to make it happen.

2. A carrier of 49 MHz is frequency-modulated by a 1.5-kHz square wave. The modulation index is 0.25. Sketch the spectrum of the resulting signal. (Assume that only harmonics less than the sixth are passed by the system.)

3. The FM radio broadcast band is allocated the frequency spectrum from 88 to 108 MHz. There are 100 channels spaced 200 kHz apart. The first channel center frequency is 88.1 MHz; the last, or 100th, channel center frequency is 107.9 MHz. Each 200-kHz channel has a 150-kHz modulation bandwidth with 25-kHz "guard bands" on either side of it to minimize the effects of overmodulation (overdeviation). The FM broadcast band permits a maximum deviation of ±75 kHz and a maximum modulating frequency of 15 kHz.

   a. Draw the frequency spectrum of the channel centered on 99.9 MHz, showing all relevant frequencies.

   b. Draw the frequency spectrum of the FM band showing details of the three lowest-frequency channels and the three highest-frequency channels.

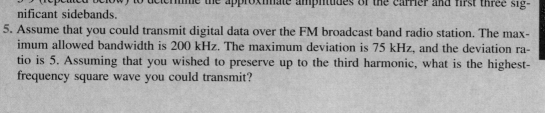

c. Determine the bandwidth of the FM signal using the deviation ratio and the Bessel table.

d. Determine the bandwidth of the FM signal using Carson's rule.

e. Which of the above bandwidth calculations best fits the available channel bandwidth?

4. A 450-MHz radio transmitter uses FM with a maximum allowed deviation of 6 kHz and a maximum modulating frequency of 3.5 kHz. What is the minimum bandwidth required? Use Fig. 5-9 (repeated below) to determine the approximate amplitudes of the carrier and first three significant sidebands.

5. Assume that you could transmit digital data over the FM broadcast band radio station. The maximum allowed bandwidth is 200 kHz. The maximum deviation is 75 kHz, and the deviation ratio is 5. Assuming that you wished to preserve up to the third harmonic, what is the highest-frequency square wave you could transmit?

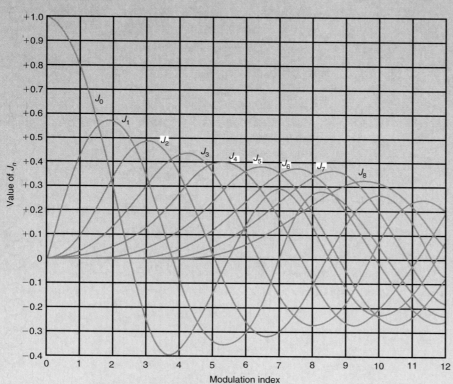

CHAPTER 6

FM CIRCUITS

Objectives

After completing this chapter, you will be able to:

- *Compare and contrast* FM using crystal oscillator circuits with FM using varactors.

- *Explain* the general principles of phase modulator circuits and list the basic techniques for achieving phase shift.

- *Calculate* the total frequency deviation of an FM transmitter knowing the original oscillator frequency and the frequency multiplication factor.

- *Describe* the operation of pulse-averaging discriminators, quadrature detectors, and differential peak detectors.

- *Draw* the block diagram of a phase-locked loop (PLL), *state* what each component does, *explain* the operation of the circuit, and *define* the capture range and the lock range of a PLL.

- *Explain* the operation of a PLL as a frequency demodulator.

Many different circuits have been devised to produce FM and PM signals. There are two different types of frequency modulator circuits, direct circuits and circuits that produce FM indirectly by phase modulation techniques. Direct FM circuits make use of techniques for actually varying the frequency of the carrier oscillator in accordance with the modulating signal. Indirect modulators produce FM via a phase shifter after the carrier oscillator stage. Frequency demodulator or detector circuits convert the FM signal back into the original modulating signal.

Although many of the circuits covered in this chapter are no longer used, they have been included for their historical and theoretical relevance. The circuits of most direct importance for today's practice are varactors, varactor controlled crystal oscillators, and transistor and varactor phase modulators. In the realm of demodulation, the most important circuits are phase-locked loops, quadrature and differential peak detectors, and pulse-averaging discriminators.

6-1 FREQUENCY MODULATORS

A *frequency modulator* is a circuit that varies carrier frequency in accordance with the modulating signal. The carrier is generated by either an *LC* or crystal oscillator circuit, and so a way must be found to change the frequency of oscillation. In an *LC* oscillator, the carrier frequency is fixed by the values of the inductance and capacitance in a tuned circuit, and the carrier frequency can therefore be changed by varying either inductance or capacitance. The idea is to find a circuit or component that converts a modulating voltage into a corresponding change in capacitance or inductance.

When the carrier is generated by a crystal oscillator, the frequency is fixed by the crystal. However, keep in mind that the equivalent circuit of a crystal is an *LCR* circuit with both series and parallel resonant points. Connecting an external capacitor to the crystal allows minor variations in operating frequency to be obtained. Again, the objective is to find a circuit or component whose capacitance will change in response to the modulating signal. The component most frequently used for this purpose is a *varactor*. Also known as a voltage variable capacitor, variable capacitance diode, or varicap, this device is basically a semiconductor junction diode operated in a reverse-bias mode.

VARACTOR OPERATION

A junction diode is created when P- and N-type semiconductors are formed during the manufacturing process. Some electrons in the N-type material drift over into the P-type material and neutralize the holes there [see Fig. 6-1(*a*)], forming a thin area called the *depletion region* where there are no free carriers, holes, or electrons. This region acts like a thin insulator that prevents current from flowing through the device.

If a forward bias is applied to the diode, it will conduct. The external potential forces the holes and electrons toward the junction where they combine and cause a continuous current inside the diode as well as externally. The depletion layer simply disappears [see Fig. 6-1(*b*)]. If an external reverse bias is applied to the diode, as in Fig. 6-1(*c*), no current will flow. The bias actually increases the width of the depletion layer, with the amount of increase depending on the amount of the reverse bias. The

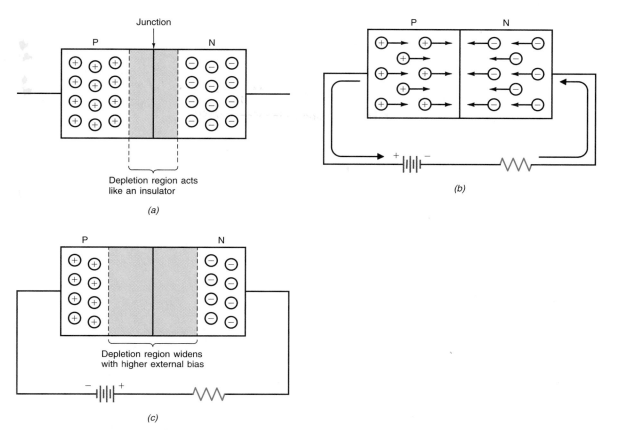

FIG. 6-1 Depletion region in a junction diode.

higher the reverse bias, the wider the depletion layer and the less chance for current flow.

A reverse-biased junction diode acts like a small capacitor. The P- and N-type materials act as the two plates of the capacitor, while the depletion region acts as the dielectric. With all of the active current carriers (electrons and holes) neutralized in the depletion region, it functions just like an insulating material. The width of the depletion layer determines the width of the dielectric and, therefore, the amount of capacitance. If the reverse bias is high, the depletion region will be wide and the dielectric will cause the plates of the capacitor to be widely spaced, producing a low capacitance. Decreasing the amount of reverse bias narrows the depletion region; the plates of the capacitor are effectively closer together, producing a higher capacitance.

All junction diodes exhibit variable capacitance as the reverse bias is changed. However, varactors are designed to optimize this particular characteristic, so that the capacitance variations are as wide and linear as possible. The symbols used to represent varactor diodes are shown in Fig. 6-2.

Varactors are made with a wide range of capacitance values, most units having a nominal capacitance in the 1- to 200-pF range. The capacitance variation range can be as high

(a)   (b)

normally operated in reverse bias mode

FIG. 6-2   Schematic symbols of a varactor diode.

as 12:1. Figure 6-3 shows the curve for a typical diode. A maximum capacitance of 80 pF is obtained at 1 V. With 60 V applied, the capacitance drops to 20 pF, a 4:1 range. The operating range is usually restricted to the linear center portion of the curve.

VARACTOR MODULATORS

Figure 6-4, a carrier oscillator for a transmitter, shows the basic concept of a varactor frequency modulator. L_1 and the capacitance of varactor diode D_1 form the parallel tuned circuit of the oscillator. The value of C_1 is made very large at the operating frequency so that its reactance is very low. As a result, C_1 connects the tuned circuit to the oscillator circuit. C_1 also blocks the DC bias on the base of Q_1 from being shorted to ground through L_1. The values of L_1 and D_1 fix the center carrier frequency.

The capacitance of D_1 is controlled in two ways, through a fixed DC bias and by the modulating signal. In Fig. 6-4, the bias on D_1 is set by the voltage divider potentiometer R_4. Varying R_4 allows the center carrier frequency to be adjusted over a narrow range. The modulating signal is applied through C_5 and the radio frequency choke (RFC). C_5 is a blocking capacitor that keeps the DC varactor bias out of the modulating-signal circuits. The reactance of the RFC is high at the carrier frequency to prevent the carrier signal from getting back into the audio modulating-signal circuits.

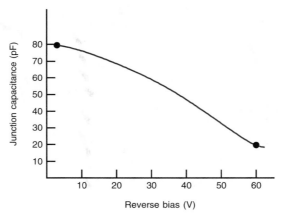

FIG. 6-3   Capacitance versus reverse junction voltage for a typical varactor.

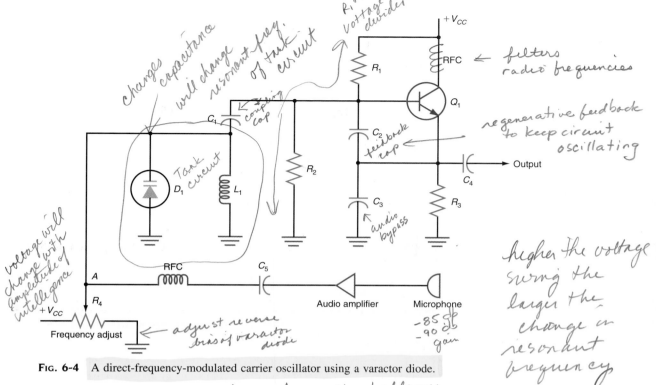

Handwritten annotations:
- changes capacitance
- will change freq.
- resonant freq. of tank circuit
- coupling cap (C₁)
- $R_1 + R_2$ voltage divider
- RFC ← filters radio frequencies
- regenerative feedback to keep circuit oscillating
- C₂ feedback cap ←
- Tank circuit
- audio bypass
- voltage will change with change of amplitude of intelligence
- adjust reverse bias of varactor diode
- higher the voltage swing the larger the change in resonant frequency
- $fL = \dfrac{1}{2\pi\sqrt{LC}}$

Circuit labels: $+V_{CC}$, RFC, R_1, Q_1, Output, C_2, C_4, R_2, C_3, R_3, D_1, L_1, RFC, A, R_4, $+V_{CC}$, Frequency adjust, C_5, Audio amplifier, Microphone, -85 db, -90 db gain

Fig. 6-4 A direct-frequency-modulated carrier oscillator using a varactor diode.

class A emitter follower amplifier (handwritten)

The modulating signal derived from the microphone is amplified and applied to the modulator. As the modulating signal varies, it adds to and subtracts from the fixed-bias voltage. Thus the effective voltage applied to D_1 causes its capacitance to vary. This, in turn, produces the desired deviation of the carrier frequency. A positive-going signal at point A adds to the reverse bias, decreasing the capacitance and increasing the carrier frequency. A negative-going signal at A subtracts from the bias, increasing the capacitance and decreasing the carrier frequency.

Example 6-1

The value of capacitance of a varactor at the center of its linear range is 40 pF. This varactor will be in parallel with a fixed 20-pF capacitor. What value of inductance should be used to resonate this combination to 5.5 MHz in an oscillator?

Total capacitance $C_T = 40 + 20 = 60$ pF

$$f_0 = 5.5 \text{ MHz} = \frac{1}{2\pi\sqrt{LC_T}}$$

$$L = \frac{1}{(2\pi f)^2 C_T} = \frac{1}{(6.28 \times 5.5 \times 10^6)^2 \times 60 \times 10^{-12}}$$
$$= 13.97 \times 10^{-6} \text{ H or } 14 \text{ } \mu\text{H}$$

The main problem with the circuit in Fig. 6-4 is that most *LC* oscillators are simply not stable enough to provide a carrier signal. Even with high-quality components and optimal design, the frequency of *LC* oscillators will vary due to temperature changes, variations in circuit voltage, and other factors. Such instabilities cannot be tolerated in most modern electronic communications systems, where a transmitter must stay on frequency as precisely as possible. *LC* oscillators simply are not stable enough to meet the stringent requirements imposed by the FCC. As a result, crystal oscillators are normally used to set carrier frequency. Not only do crystal oscillators provide a highly accurate carrier frequency, but their frequency stability is superior over a wide temperature range.

FREQUENCY-MODULATING A CRYSTAL OSCILLATOR

It is possible to vary the frequency of a crystal oscillator by changing the value of capacitance in series or in parallel with the crystal. Figure 6-5 shows a typical crystal oscillator. When a small value of capacitance is connected in series with the crystal, the crystal frequency can be "pulled" slightly from its natural resonant frequency. By making the series capacitance a varactor diode, frequency modulation of the crystal oscillator can be achieved. The modulating signal is applied to the varactor diode D_1, which changes the oscillator frequency.

It is important to note that only a very small frequency deviation is possible with frequency-modulated crystal oscillators. Rarely can the frequency of a crystal oscillator be changed more than several hundred hertz from the nominal crystal value. The resulting deviation may be less than the total deviation desired. For example, to achieve a total frequency shift of 75 kHz, which is necessary in commercial FM broadcasting, other techniques must be used. In NBFM communications systems, the narrower deviations are acceptable.

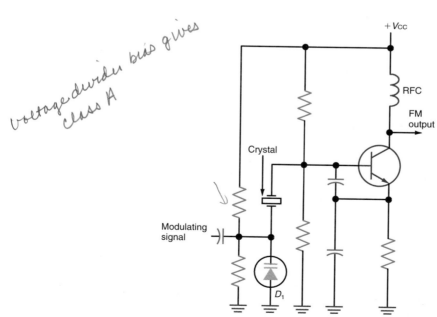

Voltage divider bias gives class A

FIG. 6-5   Frequency modulation of a crystal oscillator with a VVC.

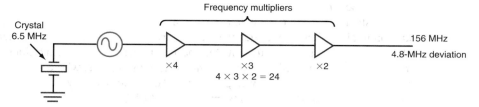

Fig. 6-6 How frequency multipliers increase carrier frequency and deviation.

Although it is possible to achieve a deviation of only several hundred cycles from the crystal oscillator frequency, the total deviation can be increased by using frequency multiplier circuits after the carrier oscillator. A *frequency multiplier circuit* is one whose output frequency is some integer multiple of the input frequency. A frequency multiplier that multiplies a frequency by 2 is called a *doubler*; a frequency multiplier circuit that multiplies an input frequency by 3 is called a *tripler*; and so on. Frequency multipliers can also be cascaded.

When the FM signal is applied to a frequency multiplier, both the carrier frequency of operation and the amount of deviation are increased. Typical frequency multipliers can increase the carrier oscillator frequency by 24 to 32 times. Figure 6-6 shows how frequency multipliers increase carrier frequency and deviation. The desired output frequency from the FM transmitter in the figure is 156 MHz and the desired maximum frequency deviation is 5 kHz. The carrier is generated by a 6.5-MHz crystal oscillator, which is followed by frequency multiplier circuits that increase the frequency by a factor of 24 (6.5 MHz × 24 = 156 MHz). Frequency modulation of the crystal oscillator by the varactor only produces a maximum deviation of 200 Hz. When multiplied by the factor of 24 in the frequency multiplier circuits, this deviation is increased to 200 × 24 = 4800 Hz, or 4.8 kHz, which is close to the desired deviation. Frequency multiplier circuits are discussed in more detail in Chap. 7.

VOLTAGE-CONTROLLED OSCILLATORS

Oscillators whose frequencies are controlled by an external input voltage are generally referred to as *voltage-controlled oscillators (VCOs)*. *Voltage-controlled crystal oscillators* are generally referred to as *VXOs*. Although some VCOs are used primarily in FM, they are also used in other applications where voltage-to-frequency conversion is required.

In high-frequency communications circuits, VCOs are ordinarily implemented with discrete-component transistor and varactor diode circuits. However, many different types of lower-frequency VCOs are in common use, including IC VCOs using *RC* multivibrator-type oscillators whose frequency can be controlled over a wide range by an AC or DC input voltage. These VCOs typically have an operating range of less than 1 Hz to approximately 1 MHz. The output is either a square or triangular wave rather than a sine wave.

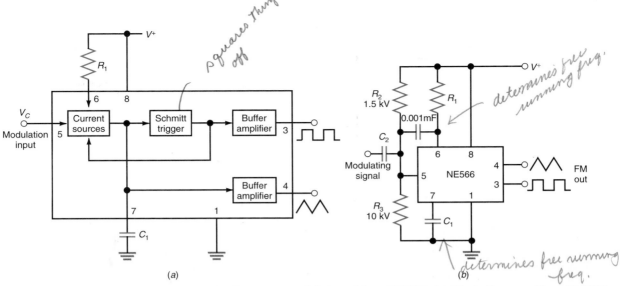

FIG. 6-7 Frequency modulation with an IC VCO. (*a*) Block diagram with an IC VCO. (*b*) Basic frequency modulator using the NE566 VCO.

Figure 6-7(*a*) is a block diagram of one widely used IC VCO, the popular NE566. External resistor R_1 at pin 6 sets the value of current produced by the internal current sources. The current sources linearly charge and discharge external capacitor C_1 at pin 7. An external voltage applied at pin 5, V_C, is used to vary the amount of current produced by the current sources. The Schmitt trigger circuit is a level detector which controls the current source by switching between charging and discharging when the capacitor charges or discharges to a specific voltage level. A linear sawtooth of voltage is developed across the capacitor by the current source. This is buffered by an amplifier and made available at pin 4. The Schmitt trigger output is a square wave at the same frequency available at pin 3. If a sine-wave output is desired, the triangular wave is usually filtered with a tuned circuit resonant to the desired carrier frequency.

A complete frequency modulator circuit using the NE566 is shown in Fig. 6-7(*b*). The current sources are biased with a voltage divider made up of R_2 and R_3. The modulating signal is applied through C_2 to the voltage divider at pin 5. The 0.001-μF capacitor between pins 5 and 6 is used to prevent unwanted oscillations. The center carrier frequency of the circuit is set by the values of R_1 and C_1. Carrier frequencies up to 1 MHz may be used with this IC. If higher frequencies and deviations are necessary, the outputs can be filtered or used to drive other circuits, such as a frequency multiplier. The modulating signal can vary the carrier frequency over nearly a 10:1 range, making very large deviations possible. The deviation is linear with respect to the input amplitude over the entire range.

REACTANCE MODULATORS

Another way to produce direct FM is to use a reactance modulator. This circuit uses a transistor amplifier that acts like either a variable capacitor or inductor. When the circuit is connected across the tuned circuit of an oscillator, the oscillator frequency can be varied by applying the modulating signal to the amplifier.

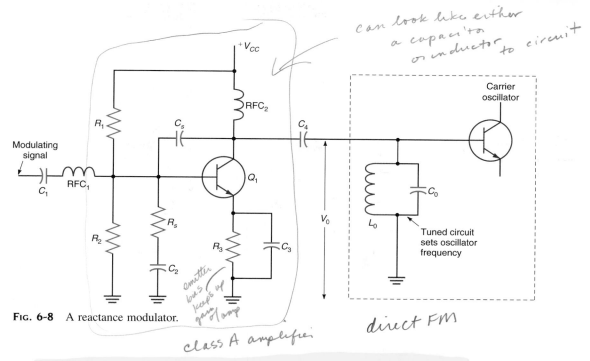

(handwritten annotations on figure:) can look like either a capacitor or inductor to circuit

(handwritten:) class A amplifier

(handwritten:) emitter bias keeps up gain of amp

(handwritten:) direct FM

FIG. 6-8 A reactance modulator.

Figure 6-8 shows a standard reactance modulator, which is basically a common-emitter class A amplifier. Resistors R_1 and R_2 form a voltage divider to bias the transistor into the linear region. R_3 is an emitter bias resistor which is bypassed with capacitor C_3. An RF choke (RFC$_2$), rather than a collector resistor, is used to provide a high-impedance load at the operating frequency. The collector of the transistor is connected to the tuned circuit in the carrier oscillator. Capacitor C_4 has a very low impedance at the oscillator frequency. Its main purpose is to keep the direct current from the collector of Q_1 from being shorted to ground through the oscillator coil L_0. The reactance modulator circuit is connected directly across the parallel tuned circuit that sets the oscillator frequency.

The oscillator signal from the tuned circuit V_0 is connected back to an RC phase-shift circuit made up of C_s and R_s. Capacitor C_2 in series with R_s has a very low impedance at the operating frequency, so it does not affect the phase shift. However, it does prevent R_s from disturbing the DC bias on Q_1. The value of C_s is chosen so that its reactance at the oscillator frequency is about 10 or more times the value of R_s. If the reactance is much greater than the resistance, the circuit acts predominantly capacitively and therefore the current through the capacitor and R_s leads the applied voltage by about 90°. This means that the voltage across R_s that is applied to the base of Q_1 leads the voltage from the oscillator. Since the collector current in Q_1 is in phase with the base current, which, in turn, is in phase with the base voltage, the collector current in Q_1 leads the oscillator voltage V_0 by 90°. Of course, any circuit whose current leads its applied voltage by 90° looks capacitive to the source voltage. This means that the reactance modulator looks like a capacitor to the oscillator tuned circuit.

The modulating signal is applied to the modulator circuit through C_1 and RFC$_1$. The RFC helps keep the RF signal from the oscillator out of the audio circuits from which the modulating signal is derived. The audio modulating signal varies the base voltage and current of Q_1 according to the intelligence to be transmitted, and the collector current varies proportionally. As the collector current amplitude varies, the phase-shift angle changes with respect to the oscillator voltage, which is interpreted by the oscillator as a change in the capacitance. Thus, as the modulating signal changes, the effective capacitance of the circuit varies and the oscillator frequency varies accordingly. An increase in capacitance lowers the frequency, and a decrease in capacitance increases the frequency. The circuit produces direct FM.

If the positions of R_s and C_s in the circuit of Fig. 6-8 are reversed, the current in the phase shifter still leads the oscillator voltage by 90°. However, voltage from across the capacitor is now applied to the base of the transistor, and that voltage lags the oscillator voltage by 90°. With this configuration, the reactance modulator acts like an inductor. The equivalent inductance changes as the modulating signal is applied. Again, the oscillator frequency varies in proportion to the amplitude of the intelligence signal amplitude.

Reactance modulator circuits can produce frequency deviation over a wide range. They are highly linear, and so distortion is minimal. These circuits can also be implemented with field-effect transistors instead of the NPN bipolar component shown in Fig. 6-8. Despite these advantages, reactance modulators are now almost entirely obsolete.

6-2 PHASE MODULATORS

Most modern FM transmitters use some form of phase modulation to produce indirect FM. The reason for using PM instead of direct FM is that the carrier oscillator can be optimized for frequency accuracy and stability. Crystal oscillators or crystal controlled frequency synthesizers can be used to set the carrier frequency accurately and maintain solid stability.

The output of the carrier oscillator is fed to a phase modulator where the phase shift is made to vary in accordance with the modulating signal. Since phase variations produce frequency variations, indirect FM is the result.

Some phase modulators are based upon the phase shift produced by an RC or LC tuned circuit. It should be pointed out that simple phase shifters of this type do not produce linear response over a large range of phase shift. The total allowable phase shift must be restricted to maximize linearity, and multipliers must be used to achieve the desired deviation. The simplest phase shifters are RC networks like those shown in Fig. 6-9(a) and (b). Depending on the values of R and C, the output of the phase shifter can be set to any phase angle between 0 and 90°. In (a), the output leads the input by some angle between 0 and 90°. For example, when X_c equals R, the phase shift is 45°. The phase shift is computed using the formula

$$\phi = \tan^{-1} \frac{X_C}{R}$$

A low-pass RC filter can also be used, as shown in Fig. 6-9(b). Here the output is taken from across the capacitor so it lags the input voltage by some angle between 0 and 90°. The phase angle is computed using the formula

$$\phi = \tan^{-1} \frac{R}{X_C}$$

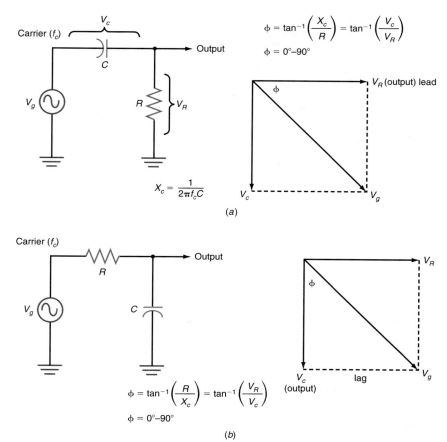

$$\phi = \tan^{-1}\left(\frac{X_c}{R}\right) = \tan^{-1}\left(\frac{V_c}{V_R}\right)$$

$$\phi = 0°\text{–}90°$$

$$X_c = \frac{1}{2\pi f_c C}$$

(a)

$$\phi = \tan^{-1}\left(\frac{R}{X_c}\right) = \tan^{-1}\left(\frac{V_R}{V_c}\right)$$

$$\phi = 0°\text{–}90°$$

(b)

FIG. 6-9   *RC* phase-shifter basics.

VARACTOR PHASE MODULATORS

A simple phase-shift circuit can be used as a phase modulator if the resistance or capacitance can be made to vary with the modulating signal. One way to do this is to replace the capacitor shown in the circuit of Fig. 6-9(*b*) with a varactor. The resulting phase-shift circuit is shown in Fig. 6-10.

In this circuit, the modulating signal causes the capacitance of the varactor to change. If the modulating signal amplitude at the output of amplifier A becomes more positive, it adds to the varactor reverse bias from R_1 and R_2, causing the capacitance to decrease. This causes the reactance to increase; thus the circuit produces less phase shift and less deviation. A more negative modulating signal from A subtracts from the reverse bias on the varactor diode, increasing the capacitance or decreasing the capacitive reactance. This increases the amount of phase shift and the deviation.

With this arrangement, there is an inverse relationship between the modulating signal polarity and the direction of the frequency deviation. This is the opposite of the desired variation. To correct this condition, an inverting amplifier A can be inserted between the modulating signal source and the input to the modulator. Then when the

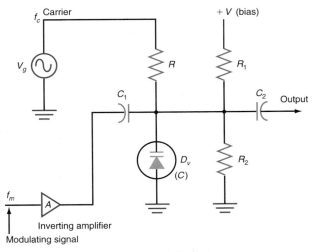

FIG. 6-10 A varactor phase modulator.

modulating signal goes positive, the inverter output and modulator input go negative and the deviation increases. C_1 and C_2 in Fig. 6-10 are DC blocking capacitors and have very low reactance at the carrier frequency. The phase shift produced is lagging, and, as in any phase modulator, the output amplitude and phase vary with a change in the modulating signal amplitude.

TRANSISTOR PHASE MODULATORS

Figure 6-11 shows a transistor used as a variable resistor to create a phase modulator. The circuit is simply a standard common emitter class A amplifier biased into the linear region by resistors R_1 and R_2. With no modulation, the collector current is 1.22 mA. The voltage from the collector to ground is 6.28 V. The transistor from col-

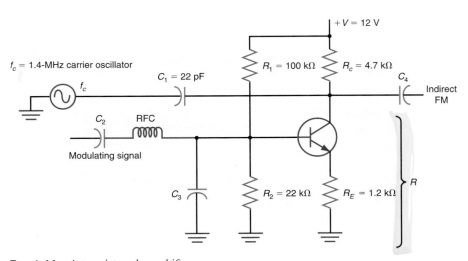

FIG. 6-11 A transistor phase shifter.

lector to ground is acting like a resistor, and has a resistance value of $R = V_C/I_C = 6.28/1.22 \times 10^{-3} = 5147 \, \Omega$. This resistance forms part of the phase shifter, with $C_1 = 22$ pF. With a carrier frequency of 1.4 MHz, the capacitive reactance is $X_C = 1/2\pi f_c C_1 = 1/6.28(1.4 \times 10^6) \, (22 \times 10^{-12}) = 5170 \, \Omega$. The phase shift produced by the circuit is $\phi = \tan^{-1} (5170/5147) = \tan^{-1} (1.004) = 45°$.

Now assume that a modulating signal is applied to the circuit through DC blocking capacitor C_2. The RFC keeps the RF out of the audio modulating circuits. If the signal goes positive, it increases the base current; this increases the collector current which, in turn, increases the voltage across the collector and decreases the voltage between the collector and ground. If the collector current rises by 0.5 mA to 1.72 mA, the voltage at the collector drops to 3.9 V. The transistor now represents a resistance value of $R = 3.9/0.00172 = 2267 \, \Omega$. The new value of phase shift with C_1 is now $\phi = \tan^{-1} (5170/2267) = \tan^{-1} (2.28) = 66.3°$.

If the input signal goes negative, the base current decreases and the collector current decreases proportionately. If the collector current goes down by 0.5 mA, to 0.72 mA, the new output voltage at the collector is 8.6 V. The new value of resistance represented by the transistor is $R = 8.6/0.00072 = 11944 \, \Omega$. The new phase shift is $\phi = \tan^{-1} (5170/11967) = 23.4°$. The total phase shift produced by the circuit is $66.3 - 23.4 = 42.9°$.

A phase shift of 43° can also be represented as $\pm 21.5°$. Expressed in radians, this is a total shift of $43/57.3 = 0.75$ rad or ± 0.375 rad.

Example 6-2

A transmitter must operate at a frequency of 168.96 MHz with a deviation of ± 5 kHz. It uses three frequency multipliers—a doubler, a tripler, and a quadrupler. Phase modulation is used. Calculate (a) the frequency of the carrier crystal oscillator and (b) the phase shift $\Delta \phi$ required to produce the necessary deviation at a 2.8-kHz modulation frequency.

a. The frequency multiplier produces a total multiplication of $2 \times 3 \times 4 = 24$. The crystal oscillator frequency is multiplied by 24 to obtain the final output frequency of 168.96 MHz. Thus the crystal oscillator frequency is

$$f_0 = \frac{168.96}{24} = 7.04 \text{ MHz}$$

b. The frequency multipliers multiply the deviation by the same factor. To achieve a deviation of ± 5 kHz, the phase modulator must produce a deviation of $f_d = 5$ kHz/24 $= \pm 208.33$ Hz. The deviation is computed with $f_d = \Delta \phi \, f_m$; $f_m = 2.8$ kHz

$$\Delta \phi = \frac{f_d}{f_m} = \frac{208.33}{2800} = \pm 0.0744 \text{ rad}$$

Converting to degrees,
$$0.0744 \, (57.3°) = \pm 4.263°$$
The total phase shift is
$$\pm 4.263° = 2 \times 4.263° = 8.526°$$

A simple formula for determining the amount of frequency deviation f_d represented by a specific phase angle is

$$f_d = \Delta\phi\, f_m$$

where $\Delta\phi$ = change in phase angle (rad)
f_m = modulating signal frequency

Assume the lowest modulating frequency for the circuit with the shift of 0.75 rad is 300 Hz. The deviation is $f_d = 0.75(300) = 225$ Hz or ±112.5 Hz. Since this is PM, the actual deviation is also proportional to the frequency of the modulating signal. With the same maximum deviation of 0.75 rad, if the modulating frequency is 3 kHz (3000 Hz), the deviation is $f_d = 0.75(3000) = 2250$ Hz or ±1125 Hz.

To eliminate this effect and to generate real FM, the audio input frequency must be applied to a low-pass filter to roll off the signal amplitude at the higher frequencies. In Fig. 6-11, this is the function of C_3, which, with the output impedance of the driving audio amplifier, creates a low-pass filter.

Example 6-3

For the transmitter in Example 6-2, a phase shifter like that in Fig. 6-9 is used, where C is a varactor and $R = 1$ kΩ. Assume the total phase-shift range is centered on 45°. Calculate the two capacitance values required to achieve the total deviation.

The phase range is centered on 45°, or $45° \pm 4.263° = 40.737°$ and $49.263°$. The total phase range is $49.263 - 40.737 = 8.526°$. If $\phi = \tan^{-1}(R/X_C)$, then $\tan\phi = R/X_c$.

$$X_C = \frac{R}{\tan\phi} = \frac{1000}{\tan 40.737} = 1161\ \Omega$$

$$C = \frac{1}{2\pi f X_C} = \frac{1}{6.28 \times 7.04 \times 10^6 \times 1161} = 19.48\ \text{pF}$$

$$X_C = \frac{R}{\tan\phi} = \frac{1000}{\tan 49.263} = 861\ \Omega$$

$$C = \frac{1}{2\pi f X_C} = \frac{1}{6.28 \times 7.04 \times 10^6 \times 861} = 26.26\ \text{pF}$$

To achieve the desired deviation, the voice signal must bias the varactor to vary over the 19.48- to 26.26-pF range.

FET Phase Modulators

The improved phase modulator shown in Fig. 6-12 uses a phase shifter made up of a capacitor and the variable resistance of a field-effect transistor Q_1. The carrier signal from a crystal oscillator or a phase-locked loop frequency synthesizer is applied directly to the output through C_1 and C_2 and appears across the FET from source to drain. The carrier signal is also applied to the gate of the FET through C_1. The series capac-

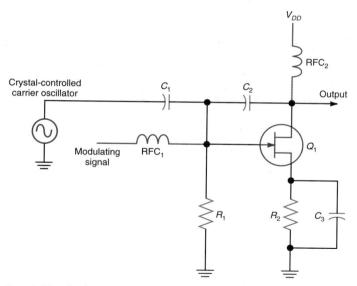

FIG. 6-12 An improved phase modulator.

itance of C_1 and C_2 and the FET source-to-drain resistance produce a leading phase shift of current in the FET and a leading voltage at the output. The carrier signal applied to the gate of the FET also varies the FET current. C_1 and R_1 produce a leading phase shift of less than 90°. The leading voltage across R_1 also controls the drain current in Q_1. When two signals control the FET drain current, the result is a phasor sum of the two currents.

The modulating signal is applied to the gate of the FET. RFC_1 keeps the carrier RF out of the audio circuits. The audio signal now also controls the FET current. This changes the amplitude relationships of the other two controlling inputs, producing a phase shift that is directly proportional to the amplitude of the modulating signal. The carrier output at the FET drain varies in phase and amplitude. The signal is then usually passed on to a class C amplifier or frequency multiplier which removes the amplitude variations but preserves the phase and frequency variations.

ARMSTRONG PHASE MODULATORS

The Armstrong phase modulator is one of the oldest and best types of phase modulator. Originally invented for broadcast transmitters, this circuit is easily implemented today with an IC balanced modulator like the 1496 or a 90° phase shifter, both of which are discussed in Chap. 4. As Fig. 6-13 shows, the balanced modulator produces DSB AM with no carrier, but the carrier is shifted by 90° and added back to the DSB suppressed signal. The result is a phase-modulated signal, not an AM signal.

The best way to visualize what happens in this circuit is to show the phasors of the signals involved. The standard phasors of amplitude modulation are shown in Fig. 6-14(a). If the carrier is suppressed, only the sidebands and the resultant phasor remain. These are shown in Fig. 6-14(b). Now, if the carrier is shifted by 90° and then added back in, the phasors look like those in Fig. 6-14(c). Note that the resultant of the sidebands is at a 90° angle with the reinserted carrier. Adding the DSB signal to

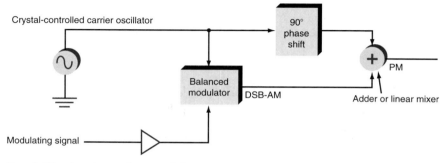

Fig. 6-13 Armstrong phase modulator.

the 90° shifted carrier produces an output signal that is at the carrier frequency but shifted in phase by angle ϕ. As the amplitude of the modulating signal varies, the phase shift varies. The phase shift also varies with the frequency of the modulating signal; however, as explained earlier, this is corrected by a $1/f$ filter. The variation in output amplitude has no practical significance, since all amplitude variations are eliminated as the signal is amplified for transmission. A modulating signal of varying amplitude therefore produces a varying phase shift that also generates FM.

The main problem with the Armstrong modulator is that the total phase shift is very low, meaning that the deviation is very small. Frequency multipliers are required to achieve the desired deviation.

TUNED-CIRCUIT PHASE MODULATORS

Most phase modulators are like the Armstrong modulator in that they are capable of producing only a small amount of phase shift—the total being essentially limited to ±20° due to the narrow range of linearity of the transistor or varactor. The limited phase shift produces, in turn, a limited frequency shift. One technique for solving this problem is to use a parallel tuned circuit to produce the phase shift. At resonance, a parallel resonant circuit acts like a very high-value resistor. Off resonance, the circuit acts inductively or capacitively and, as a result, produces a phase shift between its current and applied voltage.

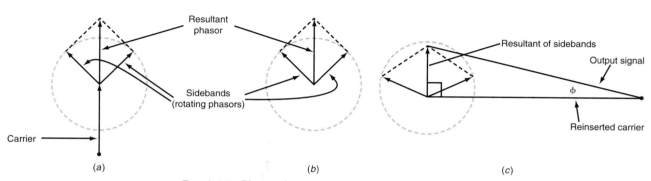

Fig. 6-14 Phasors for Armstrong modulator (a) AM. (b) DSB suppressed carrier. (c) PM output.

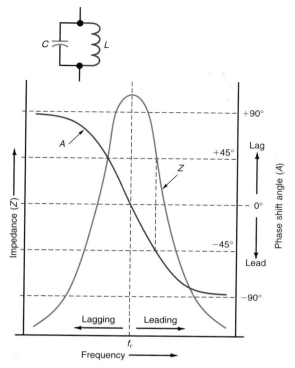

FIG. 6-15 Impedance and phase shift versus frequency of a parallel resonant circuit.

Figure 6-15 shows the basic impedance response curve and the phase variation of a parallel resonant circuit. At the resonant frequency f_r, the inductive and capacitive reactances are equal and their effects cancel one another. The result is an extremely high resistive impedance at f_r. The circuit acts resistively at this point, and the phase angle between the current and the applied voltage is therefore zero.

At frequencies below resonance, X_L decreases and X_C increases. This causes the circuit to act like an inductor, and the current lags the applied voltage. Above resonance, X_L increases and X_C decreases. This causes the circuit to act like a capacitor, and the current leads the applied voltage. If the Q of the resonant circuit is relatively high, the phase shift will be quite pronounced, as shown in Fig. 6-15. The same effect is achieved if the frequency is constant and either L or C is varied. A relatively small change in L or C can produce a significant phase shift. The idea, then, is to cause the inductance or capacitance to vary with the modulating voltage, producing a phase shift.

One of the variety of circuits that have been developed based on this technique is illustrated in Fig. 6-16. In this type of circuit, the parallel tuned circuit is usually part of the output circuit of an RF amplifier driven by the carrier oscillator. A varactor diode D_1 is connected in parallel with the tuned circuit, thereby providing a capacitance change with the modulating signal. The voltage divider, made up of R_1 and R_2, sets the reverse bias on D_1. C_2 acts as a DC blocking capacitor, preventing bias from being applied to the tuned circuit. Its value is very large, so it is, essentially, an

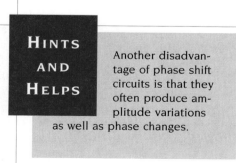

HINTS AND HELPS

Another disadvantage of phase shift circuits is that they often produce amplitude variations as well as phase changes.

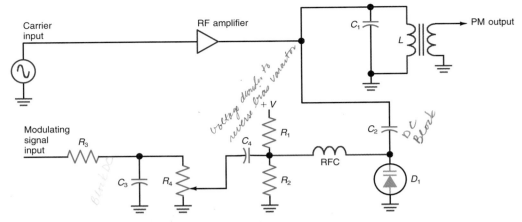

FIG. 6-16   One form of phase modulator.

AC short circuit at the carrier frequency, and the capacitance of D_1 controls the resonant frequency.

The modulating signal is first passed through a low-pass network, made up of R_3 and C_3, which provides the amplitude compensation necessary to produce FM. The modulating signal appears across potentiometer R_4, allowing the desired amount of modulating signal to be tapped off and applied to the phase-shift circuit. Potentiometer R_4 acts as a deviation control. The higher the modulating voltage, the greater the frequency deviation. The modulating signal is applied to the varactor diode through capacitor C_4. With zero modulating voltage, the value of the capacitance of D_1, along with capacitor C_1 and inductor L, sets the resonant frequency of the tuned circuit. C_2 is a DC blocking capacitor with a near-zero impedance at the carrier frequency. The indirect FM output across L is inductively coupled to the output.

When the modulating signal goes negative, it subtracts from the reverse bias of D_1. This increases the capacitance of the circuit and lowers the reactance, making the circuit appear capacitive. Thus a leading phase shift is produced. The parallel LC circuit looks like a capacitor to the output resistance of the carrier amplifier, so the output lags the input. A positive-going modulating voltage decreases the capacitance; the tuned circuit then becomes inductive, producing a lagging phase shift. The LC circuit looks like an inductor to the output resistance of the carrier amplifier, so the output leads the input. The result at the output is a relatively wide phase shift, which, in turn, produces excellent linear frequency deviation.

Phase modulators are relatively easy to implement, but they have two main disadvantages. First, the amount of phase shift they produce, and the resulting frequency deviation, are relatively low. For that reason, the carrier is usually generated at a lower frequency and frequency multipliers are used to increase the carrier frequency and the amount of frequency deviation. Second, all of the phase-shift circuits described above, including the tuned-circuit phase shifter, produce amplitude variations as well as phase changes. When the value of one of the components is changed, the phase shifts but the output amplitude changes as well. Both of these problems are solved by feeding the output of the phase modulator to class C amplifiers used as frequency multipliers. These amplifiers eliminate amplitude variations at the same time they increase the carrier frequency and deviation to the desired final values. (Figure 6-6 illustrated how frequency multipliers increase both carrier frequency and deviation.)

Any circuit that will convert a frequency variation in the carrier back into a proportional voltage variation can be used to demodulate or detect FM signals. Circuits used to recover the original modulating signal from an FM transmission are called demodulators, detectors, or discriminators.

Slope Detectors

The simplest frequency demodulator, the *slope detector,* makes use of a tuned circuit and a diode detector to convert frequency variations into voltage variations. The basic circuit is shown in Fig. 6-17(a). This has the same configuration as the basic AM diode detector described in Chap. 4, although it is tuned differently.

The FM signal is applied to transformer T_1 made up of L_1 and L_2. L_2 and C_1 form a series resonant circuit. Remember that the signal voltage induced into L_2 appears in series with L_2 and C_1 and the output voltage is taken from across C_1. The response curve of this tuned circuit is shown in Fig. 6-17(b). Note that at the resonant frequency f_r the voltage across C_1 peaks. At lower or higher frequencies, the voltage falls off.

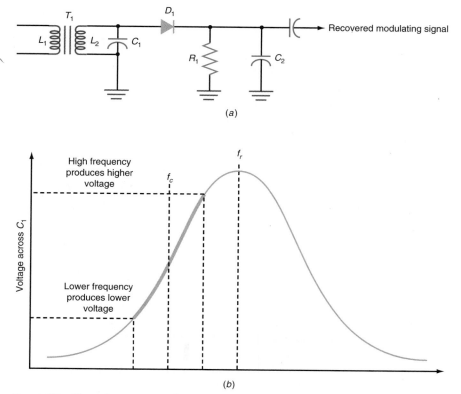

Fig. 6-17   Slope detector operation.

The advent of the Telecommunications Act of 1996 may pave the way for the emergence of telecommunications megaconglomerates. Under this act, a single broadcast company is permitted to own TV stations that broadcast to as many as 35 percent of the nation's households, up from 25 percent under the previous regulations. One company could own an unlimited number of radio stations in the United States, rather than only 20 AM and 20 FM stations. Some restrictions on station ownership within a single local market still exist, however. Perhaps the most obvious change from the consumer's perspective is that broadcasting companies are permitted to provide services other than TV and radio. Also, the act allows telephone companies to compete with cable TV companies.

The Telecommunications Act of 1996 revises or repeals most of the provisions of the Communications Act of 1934. Given the vast number of technological developments in communication technology during the period between 1934 and 1996, it is logical that radical change could benefit the industry and the consumer.

To use the circuit to detect or recover FM, the circuit is tuned so that the center or carrier frequency of the FM signals is approximately centered on the leading edge of the response curve, as shown in Fig. 6-17(b). As the carrier frequency varies above and below its center frequency, the tuned circuit responds as shown in the figure. If the frequency goes lower than the carrier frequency, the output voltage across C_1 decreases. If the frequency goes higher, the output across C_1 goes higher. Thus the AC voltage across C_1 is proportional to the frequency of the FM signal. The voltage across C_1 is rectified into DC current pulses that appear across the load R_1. These are filtered into a varying DC signal that is an exact reproduction of the original modulating signal.

The main difficulty with slope detectors is tuning them so that the FM signal is correctly centered on the leading edge of the tuned circuit. In addition, the tuned circuit does not have a perfectly linear response. It is approximately linear over a narrow range, as Fig. 6-17(b) shows, but for wide deviations, amplitude distortion occurs due to the nonlinearity.

FOSTER-SEELEY DISCRIMINATORS

One of the earliest of the FM demodulators, which is now found only in older equipment, is the *Foster-Seeley discriminator* (see Fig. 6-18). The FM signal is applied to the primary of RF transformer T_1, and the primary and secondary windings are resonated at the carrier frequency with C_1 and C_2. The parallel tuned circuit in the primary of T_1 is connected in the collector of a limiter amplifier Q_1 that removes amplitude variations from the FM signal.

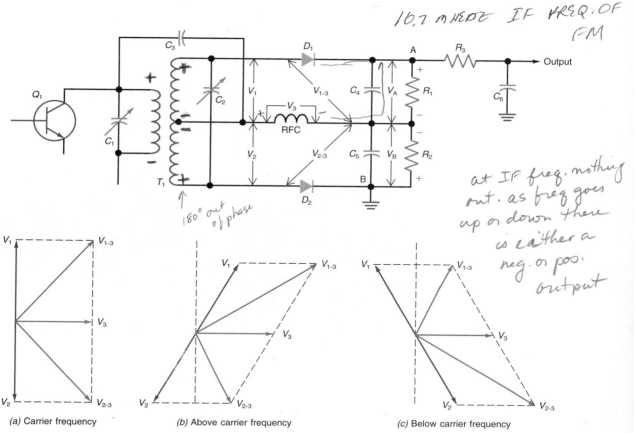

(handwritten annotations:) 10.7 MHEPE IF FREQ. OF FM

(handwritten:) at IF freq. nothing out. as freg goes up or down there is either a neg. or pos. output

(handwritten:) 180° out of phase

(a) Carrier frequency   (b) Above carrier frequency   (c) Below carrier frequency

FIG. 6-18 The Foster-Seeley discriminator. (*a*) At carrier frequency. (*b*) Above carrier frequency. (*c*) Below carrier frequency.

The signal across the primary of T_1 is also passed through capacitor C_3 and appears directly across an RFC. The voltage appearing across the RFC is exactly the same as that across the primary winding simply because C_3 and C_5 are essentially short circuits at the carrier frequency. The voltage across this RFC is designated V_3.

The current flowing in the primary winding of T_1 induces a voltage in the secondary winding. Because the secondary winding is center-tapped, the voltage across the upper portion, V_1, is 180° out of phase with the voltage across the lower portion, V_2. The voltage induced into the secondary winding is 90° out of phase with the voltage across the primary winding. When both the primary and secondary windings of an air-core transformer are tuned resonant circuits, the phase relationship between the voltages across the primary and secondary is 90°. This means that voltages V_1 and V_2 are also 90° out of phase with V_3, the voltage across the RFC. This phase relationship is shown in the vector diagram in Fig. 6-18. In Fig. 6-18(*a*), the input is the unmodulated carrier frequency.

The remainder of the circuit consists of two diode detector circuits similar to those used for AM detection. The voltage applied to D_1, R_1, and C_4 is the sum of voltages V_1 and V_3. The voltage applied to D_2, R_2, and C_5 is the sum of V_2 and V_3. Since these voltages are out of phase with one another, their sums, V_A and V_B, are vector sums; this condition is illustrated in Fig. 6-18(*a*).

On one half cycle of the primary voltage, D_1 conducts and current flows through R_1 and charges C_4. On the next half cycle, D_2 conducts and current flows through R_2

and charges C_5. The voltages across R_1 and R_2 are identical because V_1 and V_2 are the same. Since these two voltages are equal but of the opposite polarity, the voltage between point A and ground is zero. Therefore, at the carrier center frequency with no modulation, the modulator output is zero.

The secondary of T_1 and capacitor C_2 form a series resonant circuit because the voltage induced into the secondary appears in series with that winding. At resonance, the inductive reactance of the secondary winding equals the capacitive reactance of C_2. At that time, the current flowing in the circuit is exactly in phase with the voltage induced into the secondary. The output is derived from a portion of the secondary winding which acts like an inductor. The voltage across the secondary winding is, therefore, 90° out of phase with the current in the circuit.

Should the input frequency change, as would be the case with FM, the secondary winding would no longer be at resonance. For example, if the input frequency increases, the inductive reactance becomes higher than the capacitive reactance, making the circuit inductive. This causes the phase relationship between V_1 and V_2 to change with respect to V_3. If the input is above the resonant frequency, then V_1 leads V_3 by a phase angle less than 90°. Since V_1 and V_2 remain 180° out of phase, V_2 lags V_3 by an angle greater than 90°. This change in phase relationship is shown in Fig. 6-18(b). Computing the new vector sums of $V_1 + V_3$ and $V_2 + V_3$ shows that the voltage applied to D_2 (v_A) is greater than the voltage applied to D_1 (V_B). Therefore, the voltage across R_1 is greater than the voltage across R_1, and the net output voltage is positive with respect to ground.

If the frequency deviation is lower than the center frequency, then voltage V_1 leads by an angle greater than 90° and voltage V_2 lags by an angle less than 90°. The resulting vector additions of $V_1 + V_3$ and $V_2 + V_3$ are shown in Fig. 6-18(c). This time, V_B is greater than V_A. As a result, the voltage across R_2 is greater than the voltage across R_1 and the net output voltage is negative.

As you can see, as the frequency deviates above and below the center frequency, the output at point A increases or decreases and, therefore, the original modulating signal is recovered. This signal is passed through the de-emphasis network consisting of R_3 and C_6, and the output is applied to an amplifier or other circuits as required. Figure 6-19 shows the output voltage occurring across point A and ground with respect to frequency deviation. It is the linear range in the center that produces an accurate reproduction of the original modulating signal.

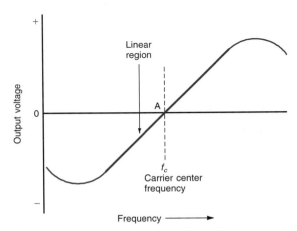

FIG. 6-19   Output voltage of the discriminator.

Discriminator circuits are sensitive to variations in input amplitude, higher inputs producing a greater amount of signal at the output and lower inputs creating less output. Because these variations would be interpreted as frequency changes, resulting in an incorrect reproduction of the modulating signal, all amplitude variations must be removed from an FM signal before it is applied to a discriminator circuit. This is usually handled by limiter circuits.

RATIO DETECTORS

Another once widely used demodulator, the *ratio detector*, is similar in appearance to the discriminator but has several important differences. Refer to Fig. 6-20. The FM signal is applied to the RF transformer T_1 with its center-tapped secondary. As in the discriminator, the FM signal is also passed through capacitor C_3 and applied across the RFC, and the voltages applied to D_1 and D_2 are a composite of $V_1 + V_3$ (V_A) and $V_2 + V_3$ (V_B). The circuit uses two diodes, but the direction of D_2 is reversed from that in the discriminator.

Another major difference in the ratio detector is the use of a very large capacitor C_6 connected across the output. The load resistors R_1 and R_2 are equal in value, and their common connection is at ground. The output is taken from between point C and ground in the circuit. Capacitors C_4 and C_5 and resistors R_1 and R_2 form a bridge circuit. The voltage across capacitors C_4 and C_5 is the bridge input voltage, and the output is taken between points C and D.

With no modulation on the carrier, the voltage applied to D_1 (V_A) is the same as that applied to D_2 (V_B). Therefore, capacitors C_4 and C_5 charge to the same voltage with the polarity shown. Since C_6 is connected across these two capacitors, it charges to the sum of their voltages. Because C_6 is a very large component, usually a tantalum or electrolytic capacitor, it takes several cycles of the input signal for it to charge fully. However, once it charges, it maintains a relatively constant voltage. Since R_1 and R_2 are equal, their voltage drops are equal. The voltage drops across C_4 and C_5 are also equal, and the bridge circuit is therefore balanced. Between points C and D there is zero voltage because their potentials are the same.

Assume, for example, that at the center carrier frequency, the voltage drop across both C_4 and C_5 is 2 V. The charge on C_6 is therefore 4 V, and the voltages across R_1

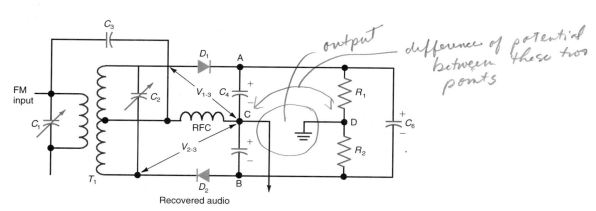

FIG. 6-20 The ratio detector.

and R_2 are 2 V. If the frequency increases, the phase relationship in the circuit changes as described previously for the discriminator circuit (Fig. 6-18(a). This causes the voltage across C_4 to be greater than the voltage across C_5. Assume that the voltage across C_4 is 3 V and the voltage across C_5 is 1 V. The voltages across R_1 and R_2 remain at 2 V because the charge on C_6 does not change. The bridge is now unbalanced, and an output voltage appears between points C and D in the circuit. Using point B as a reference, the voltage at point C is +1 V and the voltage across R_2 is +2 V. The voltage difference at C is therefore -1 V. If the frequency decreases, the phase relationship will be such that the charge on C_5 is greater than the charge on C_4. If the voltage across C_3 is +3 V with respect to B and the voltage across R_2 remains +2 V, then the voltage at point C is +1 V. The bridge is unbalanced, but in the opposite direction, and the output voltage is of the opposite polarity.

The primary advantage of the ratio detector over the discriminator is that, because of the very large capacitor C_6, for all practical purposes it is not sensitive to noise and amplitude variations. Since it takes a long time for the capacitor to charge or discharge, short noise pulses or minor amplitude variations are totally smoothed out. However, the average DC voltage across C_6 is the same as the average signal amplitude. This voltage can, therefore, be used in automatic gain control applications. The ratio detector is not as linear for large frequency deviations as the discriminator.

DIFFERENTIAL PEAK DETECTORS

Today, Foster-Seeley and ratio detector circuits are rarely used in new designs. They have been replaced by IC demodulator circuits that are typically part of a complete FM receiver IC. One popular IC frequency demodulator is the *differential peak detector*, which is used in TV sets and other consumer electronic products. The demodulator is usually only one of many other circuits on the chip. A schematic diagram of this circuit is shown in Fig. 6-21(a).

The differential peak detector is an enhanced differential amplifier. Transistors Q_3 and Q_4 form the differential amplifier; the other transistors are emitter-followers. Transistor Q_7 is a current source, and a tuned circuit is connected between the bases of Q_1 and Q_6. L_1, C_1, and C_2 are discrete components which are external to the chip. The frequency-modulated input is applied to the base of Q_1. The input is a 4.5-MHz FM sound carrier from a TV signal.

Q_1 and Q_6 are emitter-followers that provide high input impedance and power amplification to drive emitter-followers Q_2 and Q_5. Q_2 and Q_5, along with on-chip capacitors C_a and C_b, are peak detectors. C_a and C_b charge and discharge as the voltages at the two inputs vary. For example, assume the input is a carrier sine wave. When the input to Q_1 goes positive, the voltage at the emitter of Q_2 also goes positive and C_a charges to the peak AC value that appears across the tuned circuit. When the peak of the sine wave is past, the voltage declines to zero, then reverses polarity for the negative half cycle of the sine wave. As the voltage drops, capacitor C_a retains the positive peak value as a charge. The input impedance to Q_3 is relatively high, so does not significantly discharge C_a. C_a simply acts as a temporary storage cell for the peak voltage during that cycle.

Capacitor C_b on Q_5 also forms a peak detector circuit. It too charges up to the peak of the input from the opposite end of the tuned circuit. If the 4.5-MHz carrier is unmodulated, the two capacitors charge up to the same value. These equal voltages appear at the inputs to the differential amplifier. The output of a differential amplifier is proportional to the difference of the inputs, and so if the inputs are equal, the difference is zero and so is the output. The output current from Q_4 during this condition is used as the zero reference. A lower or higher current indicates a varying frequency.

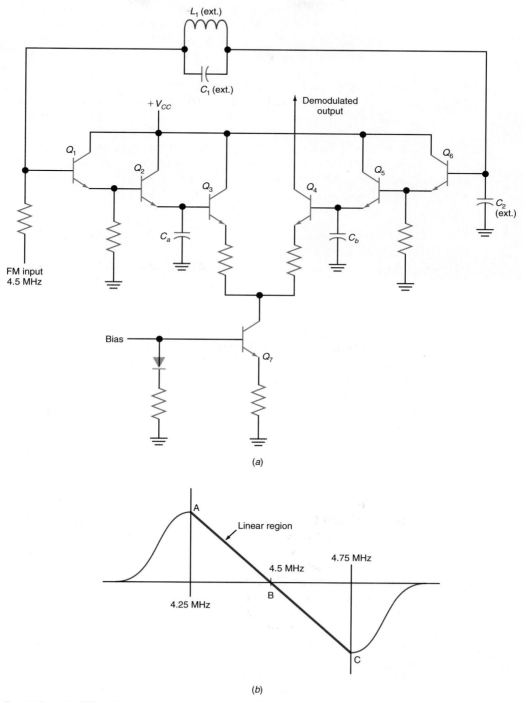

FIG. 6-21 A differential peak detector FM demodulator. (*a*) Circuit. (*b*) Response curve.

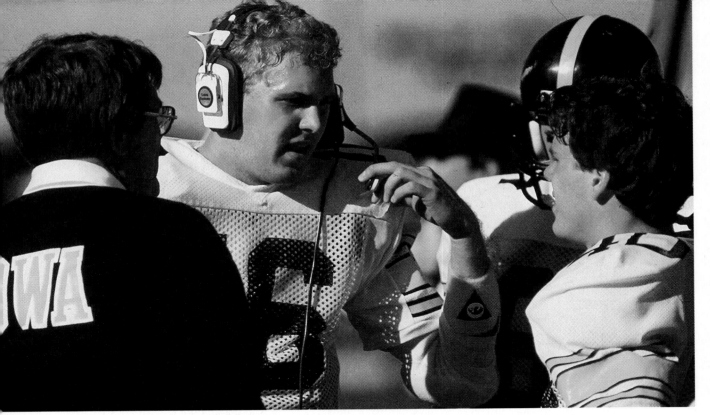

Quarterbacks can receive critical plays from coaches above the playing field. Many teams use in-helmet receivers to "send in" the plays.

The external tuned circuit is designed so that L_1 forms a series resonant circuit with C_2 at a frequency lower than the center frequency, usually 4.5 MHz. A typical series resonant frequency is 4.25 MHz. L_1 forms a parallel resonant circuit with C_1 at a frequency above the center frequency, typically 4.75 MHz.

To see how the circuit works, assume that an input is gradually increasing in frequency from below 4.25 MHz to above 4.75 MHz. At very low deviation frequencies, the effect on the tuned circuit is minimal and the inputs to the differential amplifiers are about equal; therefore the output is near zero. As the input frequency increases, at some point it reaches the series resonant point of L_1 and C_2. When this occurs, the total impedance of the series circuit is resistive and very low. This effectively shorts the input of Q_1 to ground, making it near zero. At the same time, the voltage across C_2 is at its peak value due to resonance, so the input to Q_6 is maximum. The result is a very high output current from Q_4. This condition is illustrated at point A in Fig. 6-21(b).

As the input frequency continues to rise, it eventually reaches the center frequency of 4.5 MHz, at which point the output is zero; this condition is shown at point B in Fig. 6-21(b). The frequency continues to increase, and at some point, L_1 becomes parallel-resonant with C_1. This produces a very high impedance at the base of Q_1, so the voltage is maximum at this time. The lower reactance of C_1 at this higher frequency makes the input to Q_6 very low. As a result, the difference between the two inputs is very high and the output current at Q_4 becomes very high in the opposite (negative) direction. This is point C in Fig. 6-21(b). As the figure shows, as the frequency increases, the variation in output current from point A to point C is very linear. This change in output amplitude with a change in frequency results in excellent demodulation of FM signals.

PULSE-AVERAGING DISCRIMINATORS

Whereas Foster-Seeley discriminators, ratio detectors, and differential peak detectors rely upon the detection and measurement of phase changes inherent in FM, *pulse-averaging discriminators* convert FM signals into constant-amplitude pulses and smooth them into the original signal.

A simplified block diagram of a pulse-averaging discriminator is illustrated in Fig. 6-22. The FM signal is applied to a zero crossing detector or a clipper-limiter which generates a binary voltage-level change each time the FM signal varies from minus to plus or from plus to minus. The result is a rectangular wave containing all of the frequency variations of the original signal but without amplitude variations. The FM square wave is then applied to a one-shot (monostable) multivibrator which generates a fixed-amplitude fixed-width DC pulse on the leading edge of each FM cycle. The duration of the one shot is set so it is less than one-half the period of the highest frequency expected during maximum deviation. The one-shot output pulses are then fed to a simple *RC* low-pass filter which averages the DC pulses to recover the original modulating signal.

The waveforms for the pulse-averaging discriminator are illustrated in Fig. 6-23. At low frequencies, the one-shot pulses are widely spaced; at higher frequencies, they occur very close together. When these pulses are applied to the averaging filter, a DC output voltage is developed the amplitude of which is directly proportional to the frequency deviation.

When a one-shot pulse occurs, the capacitor in the filter charges to the amplitude of the pulse. When the pulse turns OFF, the capacitor discharges into the load. If the *RC* time constant is high, the charge on the capacitor does not decrease much. When the time interval between pulses is long, however, the capacitor loses some of its charge into the load so the average DC output is low. When the pulses occur rapidly, the capacitor has little time between pulses to discharge; the average voltage across it therefore remains higher. As the figure shows, the filter output voltage varies in amplitude with the frequency deviation. The original modulating signal is developed across the filter output. The filter components are carefully selected to minimize the ripple caused by the charging and discharging of the capacitor while at the same time providing the necessary high-frequency response for the original modulating signal.

Some pulse-averaging discriminators generate a pulse every half cycle or at every zero crossing instead of every one cycle of the input. With a greater number of pulses to average, the output signal is easier to filter and contains less ripple.

The pulse-averaging discriminator is a very high-quality frequency demodulator. In the past, its use was limited to expensive telemetry and industrial control applications. Today, with the availability of low-cost ICs, the pulse-averaging discriminator is easily implemented and is used in many electronic products.

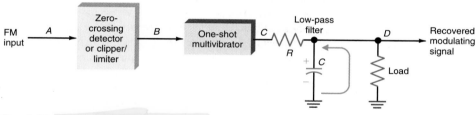

FIG. 6-22 Pulse-averaging discriminator.

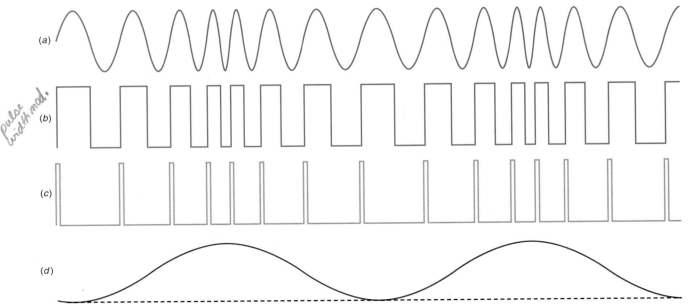

(a)

(b) pulse width mod.

(c)

(d)

FIG. 6-23 (*a*) FM input. (*b*) Output of zero-crossing detector. (*c*) Output of one shot. (*d*) Output of discriminator (original modulating signal).

QUADRATURE DETECTORS

The *quadrature detector* is probably the single most widely used FM demodulator. Its primary application is in TV audio demodulation, although it is also used in some FM radio systems. The quadrature detector uses a phase-shift circuit to produce a phase shift of 90° at the unmodulated carrier frequency. The most commonly used phase-shift arrangement is shown in Fig. 6-24. The frequency-modulated signal is applied through a very small capacitor (C_1) to the parallel tuned circuit, which is adjusted to resonate at the center carrier frequency. At resonance, the tuned circuit appears as a high value of pure resistance. The small capacitor has a very high reactance compared to the tuned circuit impedance. Thus the output across the tuned circuit at the carrier frequency is very close to 90° and leads the input. When frequency modulation occurs, the carrier frequency deviates above and below the resonant frequency of the tuned circuit, resulting in an increasing or decreasing amount of phase shift between the input and the output.

The two quadrature signals are then fed to a phase-detector circuit. The most commonly used phase detector is a balanced modulator using differential amplifiers like those discussed in Chap. 4. The output of the phase detector is a series of pulses whose width varies with the amount of phase shift between the two signals. These signals are averaged in an *RC* low-pass filter to recreate the original modulating signal.

Normally the sinusoidal FM input signals to the phase detector are at a very high level and drive the differential amplifiers in the phase detector into cutoff and saturation. The differential transistors act as switches, so the output is a series of pulses. No limiter is needed if the input signal is large enough. The duration of the output pulse is deter-

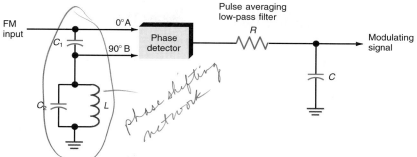

FIG. 6-24 A quadrature FM detector.

mined by the amount of phase shift. The phase detector can be regarded as an AND gate whose output is ON only when the two input pulses are ON and is OFF if either one or both of the inputs are OFF.

Figure 6-25 shows the typical waveforms involved in a quadrature detector. When there is no modulation, the two input signals are exactly 90° out of phase and therefore provide an output pulse with the indicated width. When the FM signal frequency increases, the amount of phase shift decreases, resulting in a wider output pulse. The wider pulses averaged by the *RC* filter produce a higher average output voltage, which corresponds to the higher amplitude required to produce the higher carrier frequency. When the signal frequency decreases, there are more phase shift and narrower output pulses. The narrower pulses, when averaged, produce a lower average output voltage, which corresponds to the original lower-amplitude modulating signal.

Quadrature detectors are usually built into other ICs, such as intermediate-frequency amplifiers and complete receiver ICs, both which are discussed in Chap. 8.

DID YOU KNOW?

The term quadrature refers to a 90° phase shift between two signals.

PHASE-LOCKED LOOPS

A *phase-locked loop* (PLL) is a frequency- or phase-sensitive feedback control circuit used in frequency demodulation, frequency synthesizers, and various filtering and signal-detection applications. All phase-locked loops have the three basic elements shown in Fig. 6-26.

1. A phase detector is used to compare the FM input, sometimes referred to as the reference signal, to the output of a voltage-controlled oscillator (VCO).
2. The VCO frequency is varied by the DC output voltage from a low-pass filter.
3. The low-pass filter smoothes the output of the phase detector into a control voltage that varies the frequency of the VCO.

The primary job of the phase detector is to compare the two input signals and generate an output signal that when filtered will control the VCO. If there is a phase or frequency difference between the FM input and VCO signals, the phase detector output varies in proportion to the difference. The filtered output adjusts the VCO frequency

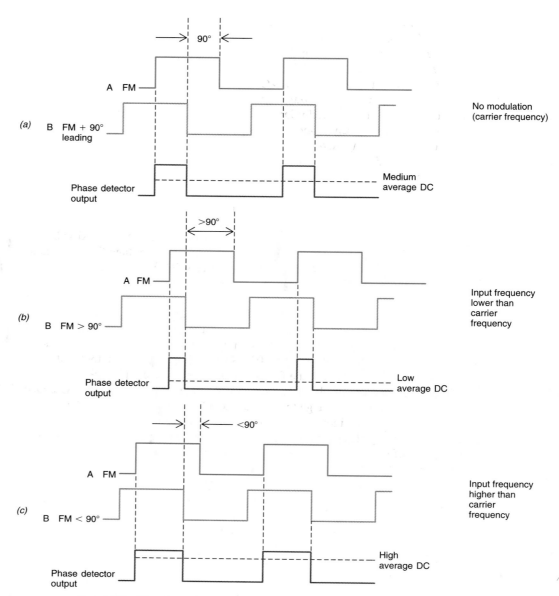

FIG. 6-25   Waveforms in the quadrature detector.

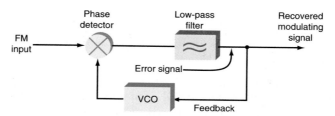

FIG. 6-26   Block diagram of a PLL.

in an attempt to correct for the original frequency or phase difference. This DC control voltage, called the *error signal,* is also the feedback in this circuit.

When no input signal is applied, the phase detector and low-pass filter outputs are zero. The VCO then operates at what is called the *free-running frequency,* its normal operating frequency as determined by internal frequency-determining components. When an input signal close to the frequency of the VCO is applied, the phase detector compares the VCO free-running frequency to the input frequency and produces an output voltage proportional to the frequency difference. Most PLL phase detectors operate just like the one discussed in the section on quadrature detectors. The phase detector output is a series of pulses that vary in width in accordance with the amount of phase shift or frequency difference that exists between the two inputs. The output pulses are then filtered into a DC voltage that is applied to the VCO. This DC voltage is such that it forces the VCO frequency to move in a direction that reduces the DC error voltage. The error voltage forces the VCO frequency to change in the direction that reduces the amount of phase or frequency difference between the VCO and the input. At some point, the error voltage causes the VCO frequency to equal the input frequency; when this happens, the PLL is said to be in a *locked* condition. While the input and VCO frequencies are equal, there is a phase difference between them, usually exactly 90°, which produces the DC output voltage that will cause the VCO to produce the frequency that keeps the circuit locked.

If the input frequency changes, the phase detector and low-pass filter produce a new value of DC control voltage that forces the VCO output frequency to change until it is equal to the new input frequency. Any variation in input frequency is matched by a VCO frequency change, so the circuit remains locked. The VCO in a PLL is, therefore, capable of *tracking* the input frequency over a wide range. The range of frequencies over which a PLL can track an input signal and remain locked is known as the *lock range.* The lock range is usually a band of frequencies above and below the free-running frequency of the VCO. If the input signal frequency is out of the lock range, the PLL will not lock. When this occurs, the VCO output frequency jumps to its free running frequency.

If an input frequency within the lock range is applied to the PLL, the circuit immediately adjusts itself into a locked condition. The phase detector determines the phase difference between the free-running and input frequencies of the VCO and generates the error signal that forces the VCO to equal the input frequency. This action is referred to as *capturing* an input signal. Once the input signal is captured, the PLL remains locked and will track any changes in the input signal as long as the frequency is within the lock range. The range of frequencies over which a PLL will capture an input signal, known as the *capture range,* is much narrower than the lock range, but, like the lock range, is generally centered around the free-running frequency of the VCO (see Fig. 6-27).

This oil tanker captain can use his ship's radio FM for weather updates and navigational information.

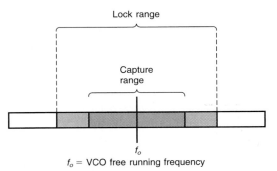

Lock range

Capture
range

f_o

f_o = VCO free running frequency

FIG. 6-27   Capture and lock ranges of a PLL.

The characteristic that causes the PLL to capture signals within a certain frequency range causes it to act like a bandpass filter. Phase-locked loops are often used in signal conditioning applications, where it is desirable to pass signals only in a certain range and reject signals outside of that range. The PLL is highly effective in eliminating the noise and interference on a signal.

The ability of a PLL to respond to input frequency variations makes it useful in FM applications. The PLL's tracking action means that the VCO can operate as a frequency modulator that produces exactly the same FM signal as the input. In order for this to happen, however, the VCO input must be identical to the original modulating signal. The VCO output follows the FM input signal because the error voltage produced by the phase detector and low-pass filter forces the VCO to track it. Thus the VCO output must be identical to the input signal if the PLL is to remain locked. The error signal must be identical to the original modulating signal of the FM input. The low-pass filter cutoff frequency is designed such that it is capable of passing the original modulating signal.

The ability of a PLL to provide frequency selectivity and filtering give it a signal-to-noise ratio superior to that of any other type of FM detector. The linearity of the VCO ensures low distortion and a highly accurate reproduction of the original modulating signal. While PLLs are complex, they are easy to apply because they are readily available in low-cost IC form.

Figure 6-28 is a block diagram of a popular and widely used IC PLL, the 565. The 565 is connected as an FM demodulator. The 565 circuitry is shown inside the dashed lines; all components outside the dashed lines are discrete components. The numbers on the connections are the pin numbers on the 565 IC, which is housed in a standard 14-pin dual in-line package (DIP). The circuit is powered by ± 12-V power supplies.

The low-pass filter is made up of a 3.6-kΩ resistor inside the 565 which terminates at pin 7. A 0.1-μF external capacitor, C_2, completes the filter. Note that the recovered original modulating signal is taken from the filter output. The free-running frequency of the VCO (f_0) is set by external components R_1 and C_1 according to the formula $f_0 = 1.2/4R_1C_1 = 1.2/4(2700)(0.01 \times 10^{-6}) = 11,111$ Hz or 11.11 kHz.

The lock range f_L can be computed with an expression supplied by the manufacturer for this circuit, $f_L = 16f_0/V_S$, where V_S is the total supply voltage. In the circuit of Fig. 6-28, V_S is the sum of the two 12-V supplies, or 24 V, so the total lock range centered around the free-running frequency is $f_L = 16(11.11 \times 10^3)/24 = 7406.7$ Hz, or ± 3703.3 Hz.

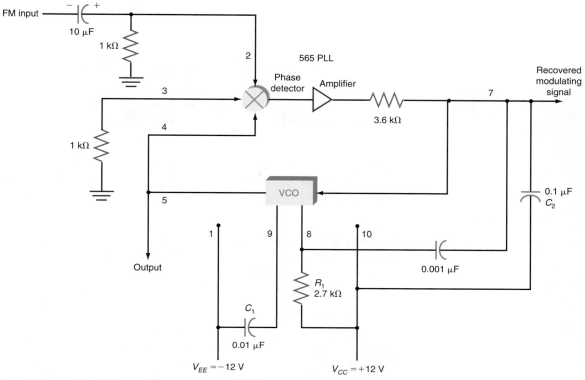

Fig. 6-28 A PLL FM demodulator using the 565 IC.

With this circuit, it is assumed that the unmodulated carrier frequency is the same as the free-running frequency, 11.11 kHz. Of course, it is possible to set this type of circuit to any other desired center frequency simply by changing the values of R_1 and C_1. The upper frequency limitation for the 565 IC is 500 kHz.

Frequency-modulating circuits are designed to convert a modulating voltage into a corresponding change in capacitance or inductance. The varactor is often used in this application. It is basically a semiconductor junction diode operated in a reverse-bias mode.

Because *LC* oscillators are not stable enough to meet the stringent frequency stability requirements imposed by the FCC, crystal oscillators, or oscillators combined with frequency-multiplier circuits, are used to set the carrier frequency.

Simple phase-shift circuits can be used as phase modulators if the resistance or capacitance can be made to vary with the modulating signal. This can be achieved by adding a varactor phase modulator to the circuit.

Any circuit that will convert a frequency variation in the carrier back into a proportional voltage variation can be used for frequency demodulation. The simplest frequency demodulator is the slope detector, but the most widely used detectors today are pulse-averaging discriminators, quadrature detectors, differential peak detectors, and phase-locked loops.

All PLLs have three basic elements: a phase detector, a voltage-controlled oscillator (VCO), and a low-pass filter. The range of frequencies over which the PLL will track an input signal and remain locked is known as the lock range, and the range of frequencies over which the PLL will capture an input signal is known as the capture range.

KEY TERMS

Armstrong phase modulator
Carrier frequency
Depletion region
Differential peak detector
Direct frequency-modulation circuit
Doubler
FET phase modulator
Frequency demodulator
Foster-Seeley discriminator
Frequency modulator
NE566 IC VCO

Phase-modulation circuit
Phase modulator
Phase-locked loop (PLL)
PLL demodulator
Pulse-averaging discriminator
Quadrature detector
Ratio detector
Reactance modulator
Schmitt trigger circuit
Slope detector

Transistor phase modulator
Tripler
Tuned-circuit phase modulator
Varactor
Varactor modulator
Varactor phase modulator
Voltage-controlled crystal oscillator (VXO)
Voltage-controlled oscillator (VCO)

QUESTIONS

1. What parts of the varactor act like the plates of a capacitor?
2. How does capacitance vary with applied voltage?
3. Do varactors operate with forward or reverse bias?
4. What is the main reason why *LC* oscillators are not used in transmitters today?
5. Can the most widely used type of carrier oscillator be frequency-modulated by a varactor?
6. How would the reactance modulator in Fig. 6-8 act if capacitor C_S were replaced by an inductor?

7. What is the main advantage of using a phase modulator rather than a direct frequency modulator?
8. What is the term for frequency modulation produced by PM?
9. What is the advantage of a parallel tuned circuit as a phase shifter over a simple *RC* circuit?
10. What components in Fig. 6-16 compensate for the greater frequency deviation at the higher modulating-signal frequencies?
11. What are the primary applications for quadrature and differential peak detectors?
12. What two IC demodulators use the concept of averaging pulses in a low-pass filter to recover the original modulating signal?
13. Which is probably the best FM demodulator of all of those discussed in this chapter?
14. What is a capture range? What is a lock range?
15. What frequency does the VCO assume when the input is outside the capture range?
16. What type of circuit does a PLL look like over its lock range?

Problems

1. A parallel tuned circuit in an oscillator consists of a 40-μH inductor in parallel with a 330-pF capacitor. A varactor with a capacitance of 50 pF is connected in parallel with the circuit. What is the resonant frequency of the tuned circuit and the oscillator operating frequency?
2. If the varactor capacitance of the circuit in Prob. 1 is decreased to 25 pF, (a) How does the frequency change and (b) What is the new resonant frequency?
3. A phase modulator produces a maximum phase shift of 45°. The modulating frequency range is 300 to 4000 Hz. What is the maximum frequency deviation possible?
4. The FM input to a PLL demodulator has an unmodulated center frequency of 10.7 MHz. (a) To what frequency must the VCO be set? (b) From which circuit is the recovered modulating signal taken?
5. A 565 IC PLL has an external resistor, R_1, of 1.2 kΩ and a capacitor, C_1, of 560 pF. The power supply is 10 V. (a) What is the free-running frequency? (b) The total lock range?
6. A varactor phase modulator like the one in Fig. 6-10 has a resistance value of 3.3 kΩ. The capacitance of the varactor at the center unmodulated frequency is 40 pF and the carrier frequency is 1 MHz. (a) What is the phase shift? (b) If the modulating signal changes the varactor capacitance to 55 pF, what is the new phase shift? (c) If the modulating signal frequency is 400 Hz, what is the approximate frequency deviation represented by this phase shift?

Critical Thinking

1. What circuit must be used ahead of the Foster-Seeley discriminator in order for it to work properly? Does the ratio detector need this circuit too? Explain.
2. Name the three key components of a phase-locked loop and write a brief explanation of how each component works.
3. What happens to an FM signal that has been passed through a tuned circuit that is too narrow, resulting in the higher upper and lower sidebands to be eliminated? What would the output of a demodulator processing this signal look like compared to the original modulating signal?
4. A direct-frequency (DF) modulated crystal oscillator has a frequency of 9.3 MHz. The varactor produces a maximum deviation of 250 Hz. The oscillator is followed by two triplers, a doubler, and a quadrupler. What are the final output frequency and the deviation?
5. Refer to Fig. 6-4. To decrease oscillator frequency, would you adjust potentiometer R_4 closer to $+V_{CC}$ or closer to ground?
6. Refer to Fig. 6-15. If R_1 became open, would the circuit still operate? Explain.

CHAPTER 7

RADIO TRANSMITTERS

Objectives

After completing this chapter, you will be able to:

◆ *Explain* the biasing and operation of, and *compute* the input power generated by, a class C power amplifier.

◆ *Discuss* the operation of frequency synthesizers and *explain* how frequency dividers are used to provide a desired frequency division ratio.

◆ *Calculate* the output frequency of a transmitter knowing the input oscillator frequency and the number and type of multipliers.

◆ *Define* neutralization and *explain* how it can be implemented in a transmitter.

◆ *Compare* the basic operation of vacuum tube amplifiers, linear amplifiers, and class C amplifiers and *list* specific applications in which each is used.

◆ *Diagram* and *explain* the basic design of π, T, and L-type LC impedance-matching networks.

◆ *Describe* the reasons for using speech-processing and speech-compression circuits.

A radio transmitter takes the information to be communicated and converts it into an electronic signal compatible with the communications medium. Typically this process involves carrier generation, modulation, and power amplification. The signal is then fed by wire, coaxial cable, or waveguide to an antenna that launches it into free space. This chapter covers transmitter configurations and the circuits commonly used in radio transmitters, including oscillators, amplifiers, frequency multipliers, impedance matching networks, and speech processing circuits.

7-1 TRANSMITTER FUNDAMENTALS

The transmitter is the electronic unit that accepts the information signal to be transmitted and converts it into an RF signal capable of being transmitted over long distances. Every transmitter has three basic requirements.

1. It must generate a carrier signal of the correct frequency at a desired point in the spectrum.
2. It must provide some form of modulation that causes the information signal to modify the carrier signal.
3. It must provide sufficient power amplification to ensure that the signal level is high enough to carry the desired distance. As part of this process, the transmitter must provide circuits that match the impedance of the power amplifier to that of the antenna for maximum transfer of power.

CONTINUOUS-WAVE OSCILLATORS

The CW oscillator shown in Fig. 7-1 is the simplest type of transmitter. The oscillator generates a carrier signal of the desired frequency. (The frequency here is determined by a crystal.) Information to be transmitted is expressed in a code (the International Morse code) consisting of dots and dashes that represent letters of the alphabet and numbers. Information transmitted in this way is referred to as *continuous-wave (CW)* transmission. A telegraph key, which is simply a convenient hand-operated switch, is connected in series with the emitter to turn the oscillator OFF and ON to produce the dots and dashes. The oscillator produces a short burst of RF energy for a dot and a longer RF burst for a dash. Although this type of transmitter typically has a power of 1 W or less, at the right frequency and with a good antenna it is capable of sending signals around the world.

Although simple transmitters like the one in Fig. 7-1 are not widely used, they are still found in some applications. For example, amateur radio operators (hams) who communicate by radio as a hobby often use low-power transmitters like this as a challenge to see how far they can communicate. Such low-power communications are called *QRP operation.* Using assigned frequencies in the 3.5- to 30-Mhz shortwave range, hams can communicate with other hams across the globe.

Another application for a simple transmitter is a beacon. For example, beacon transmitters are used to track animals in the wild. A tiny battery-powered beacon transmitter is attached by way of a neck collar or some other apparatus to a captured animal, and the animal is then set free. The transmitter provides a continuous signal that

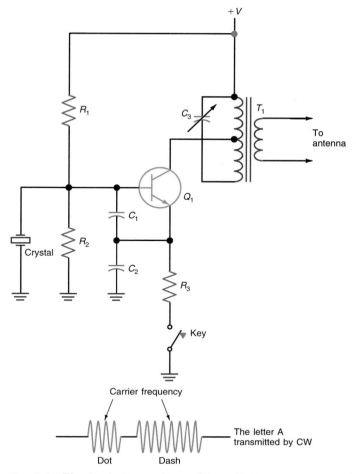

Fig. 7-1 The simplest transmitter, a CW oscillator.

is received and tracked using a directional antenna, allowing the location of the animal to be determined at any time.

Simple single-transistor transmitters that can be modulated are used in specialized applications such as electronic bugging and in the RF modulators used to connect video games to TV sets. Another application for this type of transmitter is in medicine, for example, the small telemetric transmitters that are modulated by heart rate pulses or some other physical characteristic that must be monitored by a doctor to track an illness. The transmitters used in garage door openers have a single transistor but are modulated by binary-coded pulses.

TRANSMITTER CONFIGURATIONS

The CW transmitter can be greatly improved by simply adding a power amplifier to it, as illustrated in Fig. 7-2. The oscillator is still keyed OFF and ON to produce dots and

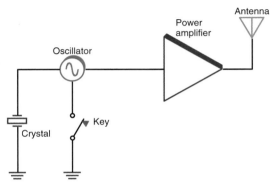

Fɪɢ. 7-2 A more powerful CW transmitter.

dashes, while the amplifier increases the power level of the signal. The result is a stronger signal that carries farther and produces more reliable transmission.

The basic oscillator-amplifier combination shown in Fig. 7-2 is the basis for virtually all radio transmitters. Many other circuits are added depending on the type of modulation used, the power level, and other considerations.

Hɪɢʜ-Lᴇᴠᴇʟ AM Tʀᴀɴsᴍɪᴛᴛᴇʀs. Figure 7-3 shows an AM transmitter using high-level modulation. An oscillator, in most applications a crystal oscillator, generates the final carrier frequency. The carrier signal is then fed to a buffer amplifier whose primary purpose is to isolate the oscillator from the remaining power amplifier stages. The buffer amplifier usually operates at the class A level and provides a modest increase in power output. The main purpose of the buffer amplifier is simply to prevent load changes in the power amplifier stages or in the antenna from causing frequency variations in the oscillator.

The signal from the buffer amplifier is applied to a class C driver amplifier designed to provide an intermediate level of power amplification. The purpose of this circuit is to generate sufficient output power to drive the final power amplifier stage. The final power amplifier, normally just referred to as *the final,* also operates at the class C level at very high power. The actual amount of power depends on the application. For example, in a CB transmitter, the power input is only 5 W. However, AM radio stations operate at much higher powers—say, 250, 500, 1000, 5000, or 50,000 W— and the video transmitter at a TV station operates at even higher power levels.

All of the RF circuits in the transmitter are usually solid-state, meaning they are implemented with either bipolar transistors or field-effect transistors (FETs). Although bipolar transistors are by far the most common type, the use of FETs is increasing because they are now capable of handling high power at high frequencies. Transistors are also typically used in the final as long as the power level does not exceed several hundred watts. Individual RF power transistors can handle up to about 100 W. Many of these can be connected in parallel or in push-pull configurations to increase the power-handling capability to many kilowatts. For higher power levels, vacuum tubes are also still used in some transmitters, but rarely in new designs. Vacuum tubes function into the VHF and UHF ranges, with power levels of 1 kW or more. Most microwave power amplifiers are also some special type of vacuum tube.

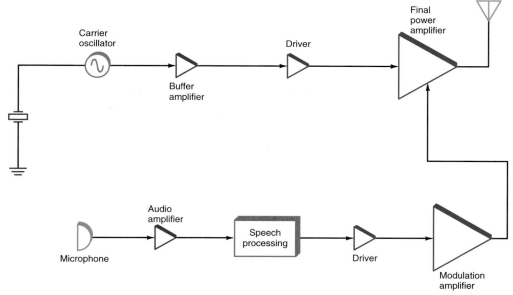

Fig. 7-3 An AM transmitter using high-level collector modulation.

Now, assume that the AM transmitter shown in Fig. 7-3 is a voice transmitter. The input from the microphone is applied to a low-level class A audio amplifier, which boosts the small signal from the microphone to a higher voltage level. (One or more stages of amplification could be used.) The voice signal is then fed to some form of *speech-processing* (filtering and amplitude control) circuit. The filtering ensures that only voice frequencies in a certain range are passed, which helps to minimize the bandwidth occupied by the signal. Most communications transmitters limit the voice frequency to the 300 to 3000 Hz range, which is adequate for intelligible communications. However, AM broadcast stations offer higher fidelity and allow frequencies up to 5 kHz to be used. In practice, many AM stations modulate with frequencies up to 7.5 kHz, and even 10 kHz, since the FCC uses alternate channel assignments within a given region and the outer sidebands are very weak, so no adjacent channel interference occurs.

Speech processors also contain a circuit used to hold the amplitude to some maximum level. High-amplitude signals are compressed and lower-amplitude signals are given more amplification. The result is that overmodulation is prevented, yet the transmitter operates as close to 100 percent modulation as possible. This reduces the possibility of signal distortion and harmonics, which produce wider sidebands that can cause adjacent channel interference, but maintains the highest possible output power in the sidebands.

After the speech processor, a driver amplifier is used to increase the power level of the signal so that it is capable of driving the high-power modulation amplifier. In the AM transmitter of Fig. 7-3, high-level or collector modulation (plate modulation in a tube) is used. As stated previously, the power output of the modulation amplifier must be one-half the input power of the RF amplifier. The high-power modulation amplifier usually operates with a class AB or class B push-pull configuration to achieve these power levels.

LOW-LEVEL FM TRANSMITTERS. In low-level modulation, modulation is performed on the carrier at low power levels, and the signal is then amplified by power amplifiers. This arrangement works for both AM and FM. FM transmitters using this method are far more common than low-level AM transmitters.

Figure 7-4 shows the typical configuration for an FM or PM transmitter. The indirect method of FM generation is used. A stable crystal oscillator is used to generate the carrier signal, and a buffer amplifier is used to isolate it from the remainder of the circuitry. The carrier signal is then applied to a phase modulator like those discussed in Chap. 6. The voice input is amplified and processed to limit the frequency range and prevent overdeviation. The output of the modulator is the desired FM signal.

Most FM transmitters are used in the VHF and UHF range. Because crystals are not available for generating those frequencies directly, the carrier is usually generated at a frequency considerably lower than the final output frequency. To achieve the desired output frequency, one or more frequency multiplier stages are used. A frequency multiplier is a class C amplifier whose output frequency is some integer multiple of the input frequency. Most frequency multipliers increase the frequency by a factor of 2, 3, 4, or 5. Because they are class C amplifiers, most frequency multipliers also provide a modest amount of power amplification.

Not only does the frequency multiplier increase the carrier frequency to the desired output frequency, but it also multiplies the frequency deviation produced by the modulator. Many frequency and phase modulators generate only a small frequency shift, much lower than the desired final deviation. The design of the transmitter must be such that the frequency multipliers will provide the correct amount of multiplication not only for the carrier frequency but also for the modulation deviation. After the frequency multiplier stage, a class C driver amplifier is used to increase the power level sufficiently to operate the final power amplifier, which also operates at the class C level.

Most FM communications transmitters operate at relatively low power levels, typically less than 100 W. All of the circuits, even in the VHF and UHF range, use transistors. For power levels beyond several hundred watts, vacuum tubes must be used. The final amplifier stages in FM broadcast transmitters typically use large

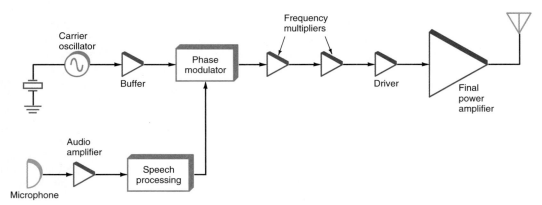

FIG. 7-4   A typical FM transmitter using indirect FM with a phase modulator.

vacuum tube class C amplifiers. In FM transmitters operating in the microwave range, klystrons, magnetrons, and traveling-wave tubes are used to provide the final power amplification.

SSB TRANSMITTERS. A typical SSB transmitter is shown in Fig. 7-5. An oscillator signal generates the carrier, which is then fed to the buffer amplifier. The buffer amplifier supplies the carrier input signal to the balanced modulator. The audio amplifier and speech-processing circuits described previously provide the other input to the balanced modulator. The balanced modulator output—a DSB signal—is then fed to a sideband filter which selects either the upper or lower sideband. Following this, the SSB signal is fed to a mixer circuit, which is used to convert the signal to its final operating frequency. Mixer circuits, which operate like simple amplitude modulators, are used to convert a lower frequency to a higher one or a higher frequency to a lower one. (Mixers are discussed more fully in Chap. 8.)

Typically, the SSB signal is generated at a low RF. This makes the balanced modulator and filter circuits simpler and easier to design. The mixer translates the SSB signal to a higher desired frequency. The other input to the mixer is derived from a local oscillator set at a frequency that, when mixed with the SSB signal, produces the desired operating frequency. The mixer can be set up so that the tuned circuit at its output selects either the sum or difference frequency. The oscillator frequency must be set to provide the desired output frequency. For fixed channel operation, crystals can be used in this local oscillator. However, in some equipment, such as that used by hams, a *variable frequency oscillator (VFO)* is used to provide continuous tuning over a desired range. In most modern communications equipment, a frequency synthesizer is used to set the final output frequency.

The output of the mixer in Fig. 7-5 is the desired final carrier frequency containing the SSB modulation. It is then fed to linear driver and power amplifiers to increase the power level as required. Class C amplifiers distort the signal and therefore cannot be used to transmit SSB or low-level AM of any kind, including DSB. Class A or linear amplifiers must be used to retain the information content in the AM signal.

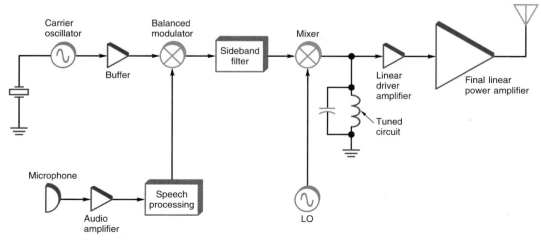

FIG. 7-5 An SSB transmitter.

The starting point for all transmitters is carrier generation. Once generated, the carrier can be modulated, processed in various ways, amplified, and finally transmitted. The source of most carriers in modern transmitters is a crystal oscillator. Conventional *LC* oscillators are simply not accurate or stable enough for real-world operation. A possible exception is in the amateur radio bands, where VFOs may be used to tune continuously to any frequency inside a designated band. However, even here, PLL frequency synthesizers which use a crystal oscillator as the basic stabilizing reference are the equipment of choice.

CRYSTAL OSCILLATORS

Most radio transmitters are licensed by the FCC either directly or indirectly to operate not only within a specific frequency band but also on predefined frequencies or channels. Deviating from the assigned frequency by even a small amount can cause interference with signals on adjacent channels. It can also get the operator a citation from the FCC for violating license conditions. Therefore the transmitter carrier generator must be very precise, operating on the exact frequency assigned, often within very close tolerances. In some radio services, the frequency of operation must be within 0.001 percent of the assigned frequency. In addition, the transmitter must remain on the assigned frequency. It must not drift off or wander from its assigned value despite the many operating conditions, such as wide temperature variations and changes in power supply voltage, that affect frequency.

The only oscillator capable of meeting the precision and stability demanded by the FCC is a crystal oscillator. In fact, the FCC often specifies that a crystal oscillator must be used in a transmitter if it is to be licensed.

A *crystal* is a piece of quartz that has been cut and ground into a thin, flat wafer and mounted between two metal plates. When the crystal is excited by an AC signal across its plates, it vibrates. This action is referred to as the *piezoelectric effect*. The frequency of vibration is determined primarily by the thickness of the crystal. Other factors influencing frequency are the cut of the crystal, that is, the place and angle of cut made in the base quartz rock from which the crystal was derived, and the size of the crystal wafer. Crystals frequencies range from as low as 30 kHz to as high as 100 MHz. As the crystal vibrates or oscillates, it maintains a very constant frequency. Once a crystal has been cut or ground to a particular frequency, it will not change to any great extent even with wide voltage or temperature variations. Even greater stability can be achieved by mounting the crystal in sealed, temperature-controlled chambers known as *crystal ovens*. These devices maintain an absolute constant temperature, ensuring a stable output frequency.

As you saw in Chap. 4, the crystal acts like an *LC* tuned circuit. It can emulate a series or parallel *LC* circuit with a *Q* as high as 30,000. The crystal is simply substituted for the coil and capacitor in a conventional oscillator circuit. The end result is a very precise, stable

DID YOU KNOW?

The only oscillator capable of maintaining the frequency precision and stability demanded by the FCC is a crystal oscillator. In fact, the FCC often requires that a crystal oscillator be used in a transmitter.

oscillator. The precision, or stability, of a crystal is usually expressed in parts per million (ppm). For example, to say that a crystal with a frequency of 1 MHz has a precision of 100 ppm means that the frequency of the crystal can vary from 999,900 to 1,000,100 Hz. Most crystals have tolerance and stability values in the 10- to 1000-ppm range. Expressed as a percentage, the precision is $(100/1,000,000) \times 100 = 0.0001 \times 100 = 0.01$ percent.

You can also use ratio and proportion to figure the frequency variation for a crystal with a given precision. For example, a 24-MHz crystal with a stability of ± 50 ppm has a maximum frequency variation f of $50/1,000,000 = f/24,000,000$. Thus $f = 50(24,000,000)/1,000,000 = 24 \times 50 = 1200$ Hz or ± 1200 Hz.

Example 7-1

What are the maximum and minimum frequencies of a 16-MHz crystal with a stability of 200 ppm?

The frequency can vary as much as 200 Hz for every 1 MHz of frequency or $200 \times 16 = 3200$ Hz.

The possible frequency range is:

$$16,000,000 - 3200 = 15,996,800 \text{ Hz}$$
$$16,000,000 + 3200 = 16,003,200 \text{ Hz}$$

Expressed as a percentage, this stability is $(3200/16,000,000) \times 100 = 0.0002 \times 100 = 0.02$ percent.

In other words, the actual frequency may be different from the designated frequency by as many as 50 Hz for every 1 MHz of designated frequency, or $24 \times 50 = 1200$ Hz.

A precision value given as a percentage can be converted to a ppm value as follows. Assume a 10-MHz crystal has a precision percentage of ± 0.001 percent; 0.001 percent of 10,000,000 is $0.00001 \times 10,000,000 = 100$ Hz. Thus

$$\text{ppm}/1,000,000 = 100/10,000,000$$
$$\text{ppm} = 100(1,000,000)/10,000,000 = 10 \text{ ppm}$$

However, the simplest way to convert from percentage to ppm is to convert the percentage value to its decimal form by dividing by 100 or moving the decimal point two places to the left, then multiplying by 10^6 or moving the decimal point six places to the right. For example, the ppm stability of a 5-MHz crystal with a precision of 0.005 percent is found as follows. First, put 0.005 percent in decimal form: 0.005 percent = 0.00005. Next, multiply by 1 million:

$$0.00005 \times 1,000,000 = 50 \text{ ppm}$$

Example 7-2

A radio transmitter uses a crystal oscillator with a frequency of 14.9 MHz and a frequency multiplier chain with factors of 2, 3, and 3. The crystal has a stability of ± 300 ppm.

a. Calculate the transmitter output frequency.

$$\text{Total frequency multiplication factor} = 2 \times 3 \times 3 = 18$$
$$\text{Transmitter output frequency} = 14.9 \text{ MHz} \times 18$$
$$= 268.2 \text{ MHz}$$

b. Calculate the maximum and minimum frequencies that the transmitter is likely to achieve if the crystal drifts to its maximum extreme.

$$\pm 300 \text{ ppm} = \frac{300}{1,000,000} \times 100 = \pm 0.03\%$$

This variation is multiplied by the frequency multiplier chain, yielding ± 0.03 percent $\times 18 = \pm 0.54$ percent. Now, 268.2 MHz $\times 0.0054 = 1.45$ MHz. Thus the frequency of the transmitter output is 268.2 ± 1.45 MHz. The upper limit is

$$268.2 + 1.45 = 269.65 \text{ MHz}$$

The lower limit is

$$268.2 - 1.45 = 266.75 \text{ MHz}$$

TYPICAL CRYSTAL OSCILLATOR CIRCUITS. The common crystal oscillator shown in Fig. 7-1 is a Colpitts-type oscillator where the feedback is derived from the capacitive voltage divider made up of C_1 and C_2. A popular variation of this circuit is the emitter-follower version shown in Fig. 7-6. Again, the feedback comes from the capacitor voltage divider C_1–C_2. The output is taken from the emitter, which is untuned.

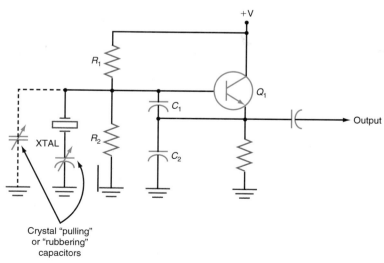

FIG. 7-6 An emitter-follower crystal oscillator.

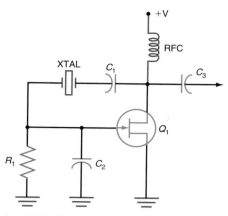

FIG. 7-7 The Pierce crystal oscillator using an FET.

Occasionally you will see a capacitor in series or in parallel with the crystal (not both), as shown in Fig. 7-6. These capacitors can be used to make minor adjustments in the crystal frequency. As discussed previously, it is not possible to effect large frequency changes with series or shunt capacitors, but they can be used to make fine adjustments. The capacitors are called crystal *pulling* capacitors, and the whole process of fine-tuning a crystal is sometimes referred to as *rubbering*.

Field-effect transistors also make good crystal oscillators. Figure 7-7 shows an FET used in the popular Pierce oscillator configuration. Most crystal oscillators are some variation of these basic types. They operate as class A linear amplifiers and generate a clean sine-wave output signal.

OVERTONE OSCILLATORS.   The main problem with crystals is that their upper frequency operation is limited. The higher the frequency, the thinner the crystal must be to oscillate at that frequency. At an upper limit of about 30 MHz, the crystal is so fragile that it becomes impractical to use. However, over the years, operating frequencies have continued to move upward as a result of the quest for more frequency space and greater channel capacity, and the FCC has continued to demand the same stability and precision that are required at the lower frequencies. One way to achieve VHF, UHF, and even microwave frequencies using crystals is by employing frequency multiplier circuits, as described earlier. The carrier oscillator operates on a frequency less than 30 MHz, and multipliers raise that frequency to the desired level. For example, if the desired operating frequency is 163.2 MHz and the frequency multipliers multiply by a factor of 24, the crystal frequency must be 163.2/24 = 6.8 MHz.

Another way to achieve crystal precision and stability at frequencies above 30 MHz is to use what are known as *overtone crystals*. An overtone crystal is cut in a special way so that it optimizes its oscillation at an overtone of the basic crystal frequency. An overtone is like a harmonic as it is usually some multiple of the fundamental vibration frequency. However, the term *harmonic* is usually applied to electrical signals, while the term *overtone* refers to higher mechanical vibration frequencies. Like a harmonic, an overtone is usually some integer multiple of the base vibration frequency. However, most overtones are slightly more or slightly less than the integer value. In a crystal, the second harmonic is the first overtone, the third harmonic the second overtone, and so on. For example, a crystal with a fundamental frequency of 20 MHz would have a second harmonic or first overtone of 40 MHz, and a third harmonic or second overtone of 60 MHz.

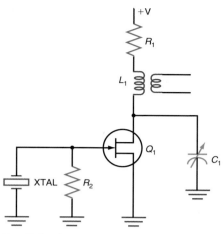

Fig. 7-8 An overtone crystal oscillator.

The term *overtone* is often used as a synonym for harmonic. Most manufacturers refer to their third overtone crystals as *third harmonic crystals*.

The odd overtones are far greater in amplitude than the even overtones. Most overtone crystals oscillate reliably at the third or fifth overtone of the frequency at which the crystal is originally ground. There are also seventh-overtone crystals. Overtone crystals can be obtained with frequencies up to about 100 MHz. A typical overtone crystal oscillator is shown in Fig. 7-8. With this design, a crystal cut for a frequency of, say, 16.8 MHz and optimized for overtone service will have a third overtone oscillation at $3 \times 16.8 = 50.4$ MHz. The tuned output circuit made up of L_1 and C_1 will be resonant at 50.4 MHz.

CRYSTAL SWITCHING. If a transmitter must operate on more than one frequency, as is often the case, but crystal precision and stability are required, multiple crystals can be used and the desired one switched in. The most straightforward way to do this is to use a mechanical rotary switch, as shown in Fig. 7-9. This arrangement works fine at the lower frequencies if the crystals are located close to the switch. The connections between the crystals and the switch and oscillator must be kept short to minimize stray inductance and capacitance, which can affect the feedback and the frequency of operation. At higher frequencies this approach is unacceptable due to excessive distributed stray inductance and capacitance.

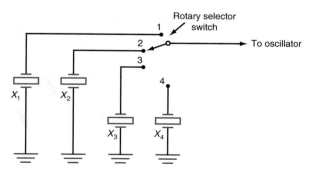

Fig. 7-9 Crystal selection with a rotary switch.

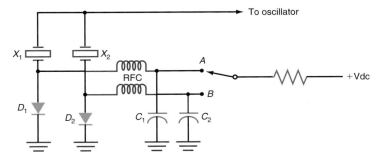

Fig. 7-10 Using diodes to switch crystals.

Another approach to crystal switching, using diode switches, is shown in Fig. 7-10. The mechanical switch is used to apply a DC bias voltage to the diodes to select the desired frequency. Note that a silicon switching diode is connected in series with each crystal. With the switch set to channel A, diode D_1 is forward-biased by the DC voltage applied by the switch. The diode conducts, acting like a very low-value resistor. The diode essentially connects crystal X_1 to ground. The other diode is cut off because no forward bias is applied to it. The RFCs and the capacitors keep the RF signal out of the DC bias circuits.

The diode switching arrangement is fast and reliable and overcomes the problem of long connecting wires between the crystal, switch, and oscillator circuit. The diodes are mounted near the crystals, which in turn are close to the oscillator components, usually on a printed circuit board. The switch can be located any distance away. Since the switch is switching DC and not high-frequency AC at the crystal itself, the length of the wires between the switch and diodes is not a factor.

FREQUENCY SYNTHESIZERS

Frequency synthesizers are variable-frequency generators that provide the frequency stability of crystal oscillators but with the convenience of incremental tuning over a broad frequency range. Frequency synthesizers usually provide an output signal that varies in fixed frequency increments over a wide range. In a transmitter, a frequency synthesizer provides basic carrier generation for channelized operation. Frequency synthesizers are also used in receivers as local oscillators and perform the receiver tuning function.

Using frequency synthesizers overcomes certain cost and size disadvantages associated with crystals. Assume, for example, that a transmitter must operate on 50 channels. Crystal stability is required. The most direct approach is simply to use one crystal per frequency and add a large switch. While such an arrangement works, it has major disadvantages. Crystals are expensive, ranging from $1 to $10 each, and even at the lowest price, 50 crystals may cost more than all of the rest of the parts in the transmitter. The same 50 crystals would also take up a great deal of space, possibly occupying more than ten times the volume of all of the rest of the transmitter parts. With a frequency synthesizer, only one crystal is needed, and the requisite number of channels can be generated using a few tiny ICs.

Over the years, many techniques have been developed for implementing frequency synthesizers with frequency multipliers and mixers. Today, however, most frequency synthesizers use some variation of the *phase-locked loop (PLL)*.

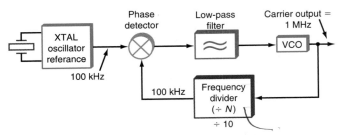

Fig. 7-11 Basic PLL frequency synthesizer.

An elementary frequency synthesizer based on a PLL is shown in Fig. 7-11. Like all phase-locked loops, it consists of a phase detector, a low-pass filter, and a VCO. The input to the phase detector is a reference oscillator. The reference oscillator is normally crystal-controlled to provide high-frequency stability. The frequency of the reference oscillator sets the increments in which the frequency may be changed. Note that the VCO output is not connected directly back to the phase detector, but applied to a frequency divider first. A *frequency divider* is a circuit whose output frequency is some integer submultiple of the input frequency. A divide-by-10 frequency synthesizer produces an output frequency that is one-tenth of the input frequency. Frequency dividers can be easily implemented with digital circuits to provide any integer value of frequency division.

In the PLL in Fig. 7-11, the reference oscillator is set to 100 kHz (0.1 MHz). Assume that the frequency divider is initially set for a division of 10. For a PLL to become locked or synchronized, the second input to the phase detector must be equal in frequency to the reference frequency, for this PLL to be locked, the frequency divider output must be 100 kHz. The VCO output has to be 10 times higher than this, or 1 MHz. One way to look at this circuit is as a frequency *multiplier*: the 100-kHz input is multiplied by 10 to produce the 1 MHz output. In the design of the synthesizer, the VCO frequency is set to 1 MHz so that when it is divided, it will provide the 100-kHz input signal required by the phase detector for the locked condition. The synthesizer output is the output of the VCO. What has been created, then, is a 1-MHz signal source. Because the PLL is locked to the crystal reference source, the VCO output frequency has the same stability as the crystal oscillator. The PLL will track any frequency variations, but the crystal is very stable and the VCO output is as stable as the crystal reference oscillator.

To make the frequency synthesizer more useful, some means must be provided to vary its output frequency. This is done by varying the frequency division ratio. Through various switching techniques, the flip-flops in a frequency divider can be arranged to provide any desired frequency division ratio. The frequency division ratio is normally designed to be manually changed in some way. For example, rotary-switch-controlled logic circuits may provide the correct configuration, or a thumb-wheel switch may be used. Some designs actually incorporate a keyboard where the desired frequency division ratio can be keyed in. In the most sophisticated circuits, a microprocessor generates the correct frequency division ratio and provides a direct frequency readout display.

Varying the frequency division ratio changes the output frequency. For example, in the circuit in Fig. 7-11, if the frequency division ratio is changed from 10 to 11, the VCO output frequency must change to 1.1 MHz. The output of the divider then remains at 100 kHz (1,100,000/11 = 100,000), as necessary to maintain a locked condition. Each incremental change in frequency division ratio produces an output frequency change of 0.1 MHz. This is how the frequency increment is set by the reference oscillator.

A more complex PLL synthesizer, a circuit that generates VHF and UHF frequencies over the 100- to 500-MHz range, is shown in Fig. 7-12. This circuit uses an FET oscillator to generate the carrier frequency directly. No frequency multipliers are needed. The output of the frequency synthesizer can be connected directly to the driver and power amplifiers in the transmitter. This synthesizer has an output frequency in the 390-MHz range, and the frequency can be varied in 30-kHz increments above and below that frequency.

The VCO circuit for the synthesizer in Fig. 17-12 is shown in Fig. 7-13. The frequency of this LC oscillator is set by the values of L_1, C_1, C_2, and the capacitances of the varactor diodes D_1 and D_2, C_a and C_b, respectively. The DC voltage applied to the varactors changes the frequency. Two varactors are connected back to back, and thus the total effective capacitance of the pair is less than either individual capacitance. Specifically, it is equal to the series capacitance C_S, where $C_S = C_a C_b/(C_a + C_b)$. If D_1 and D_2 are identical, $C_S = C_a/2$. A negative voltage with respect to ground is required to reverse-bias the diodes. Increasing the negative voltage increases the reverse bias and decreases the capacitance. This, in turn, increases the oscillator frequency.

Using two varactors allows the oscillator to produce higher RF voltages without the problem of the varactors becoming forward-biased. If a varactor, which is a diode, becomes forward-biased, it is no longer a capacitor. High voltages in the tank circuit of the oscillator can sometimes exceed the bias voltage level and cause forward conduction. When forward conduction occurs, rectification takes place, producing a DC voltage which changes the DC tuning voltage from the phase detector and loop filter. The result is what is called *phase noise*. With two capacitors in series, the voltage required to forward-bias the combination is double that of one varactor. An additional benefit is that two varactors in series produce a more linear variation of capacitance with voltage than one diode. The DC frequency control voltage is, of course, derived by filtering the phase-detector output with the low-pass loop filter.

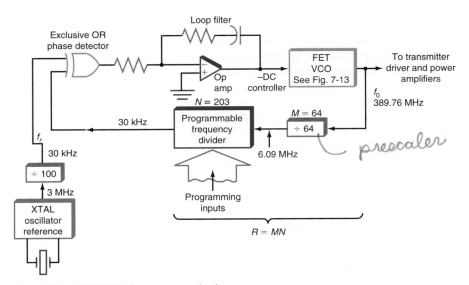

FIG. 7-12 VHF/UHF frequency synthesizer.

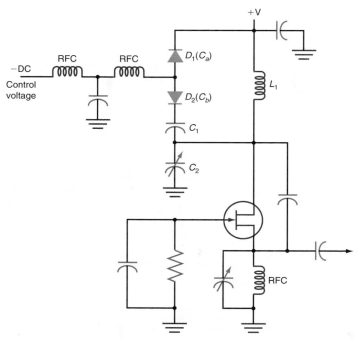

FIG. 7-13 VHF/UHF range VCO.

In most PLLs the phase detector is a digital circuit rather than a linear circuit, since the inputs to the phase detector are usually digital. Remember, one input comes from the output of the feedback frequency divider chain, which is certainly digital, and the other comes from the reference oscillator. In some designs, the reference oscillator frequency is also divided down by a digital frequency divider to achieve the desired frequency step increment. This is the case in Fig. 7-12. Since the synthesizer frequency can be stepped in increments of 30 kHz, the reference input to the phase detector must be 30 kHz. This is derived from a stable 3-MHz crystal oscillator and a frequency divider of 100.

The design shown in Fig. 7-12 uses an exclusive OR gate as a phase detector. Recall that an exclusive OR (XOR) gate generates a binary 1 output only if the two inputs are complementary; otherwise, it produces a binary 0 output.

Figure 7-14 shows how the XOR phase detector works: Remember that the inputs to a phase detector must have the same frequency. This circuit requires that the inputs have a 50 percent duty cycle. The phase relationship between the two signals determines the output of the phase detector. If the two inputs are exactly in phase with one another, the XOR output will be zero, as Fig. 7-14(b) shows. If the two inputs are 180° out of phase with one another, the XOR output will be a constant binary 1 [see Fig. 7-14(c)]. Any other phase relationship will produce output pulses at twice the input frequency. The duty cycle of these pulses indicates the amount of phase shift. A small phase shift produces narrow pulses; a larger phase shift produces wider pulses. Figure 7-14(d) shows a 90° phase shift.

The output pulses are fed to the loop filter (Fig. 7-12), an op amp with a capacitor in the feedback path which makes it into a low-pass filter. This filter averages the phase-detector pulses into a constant DC voltage that biases the VCO varactors. The average DC voltage is proportional to the duty cycle, which is the ratio of the binary 1 pulse time to the period of the signal. Narrow pulses (low duty cycle) produce a

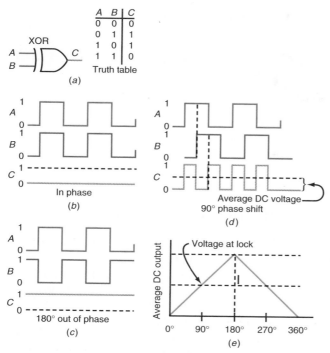

FIG. 7-14 Operation of XOR phase detector.

low average DC voltage, and wide pulses (high duty cycle) produce a high average DC voltage. Figure 7-14(e) shows how the average DC voltage varies with phase shift. Most PLLs lock in at a phase difference of 90°. Then, as the frequency of the VCO changes due to drift or because of changes in the frequency divider ratio, the input to the phase detector from the feedback divider changes, varying the duty cycle. This changes the DC voltage from the loop filter and forces a change of the VCO frequency to compensate for the original change. Note that the XOR produces a positive DC average voltage, but the op amp used in the loop filter inverts this to a negative DC voltage, as required by the VCO.

The output frequency of the synthesizer f_0 and the phase-detector reference frequency f_r are related to the overall divider ratio R as follows:

$$R = \frac{f_0}{f_r} \qquad f_0 = Rf_r \qquad \text{or} \qquad f_r = \frac{f_0}{R}$$

In our example, the reference input to the phase detector, f_r, must be 30 kHz to match the feedback from the VCO output f_0. Assume a VCO output frequency of 389.76 MHz. A frequency divider reduces this amount down to 30 kHz. The overall division ratio is $R = f_0/f_r = 389,760,000/30,000 = 12,992$.

Frequency dividers are usually designed to change the division ratio in integer increments. Programmable or presettable digital counter and divider ICs of the TTL or CMOS variety are available for this purpose. They can be programmed by applying an external binary code from thumbwheel switches, a keypad, a ROM, or a microprocessor. The main problem with divider ICs is that they usually are not made to operate at frequencies over about 50 MHz for the TTL devices and usually much less for the CMOS devices.

To overcome this problem, a special frequency divider called a *prescaler* is normally used between the high-output frequency of the VCO and the programmable part of the divider. The prescaler is usually one or more emitter-coupled logic (ECL) flip-flops or a low-ratio frequency divider that can operate at frequencies up to 1 to 2 GHz. Refer again to Fig. 7-12. The prescaler divides by a ratio of $M = 64$ to reduce the 389.76-MHz output of the VCO to 6.09 MHz, which is well within the range of most programmable frequency dividers. Since we need an overall division ratio of $R = 12,992$ and a factor of $M = 64$ is in the prescaler, the programmable portion of the feedback divider N can be computed. The total division factor is $R = MN = 12,992$. Rearranging, we have $N = R/M = 12,992/64 = 203$.

Now, to see how the synthesizer changes output frequencies when the division ratio is changed, assume the programmable part of the divider is changed by one increment, to $N = 204$. In order for the PLL to remain in a locked state, the phase detector input must remain at 30 kHz. This means that the VCO output frequency must change. The new frequency division ratio is $204 \times 64 = 13,056$. Multiplying this by 30 kHz yields the new VCO output frequency $f_0 = 30,000 \times 13,056 = 391,680,000$ Hz $= 391.68$ MHz. Instead of the desired 30-kHz increment, the VCO output varied by $391,680,000 - 389,760,000 = 1,920,000$ Hz, or a step of 1.92-MHz. This was caused by the prescaler. In order for a 30-kHz step to be achieved, the feedback divider should have changed its ratio from 12,992 to 12,993. Since the prescaler is fixed with a division of 64, the smallest increment step is 64 times the reference frequency, or $64 \times 30,000 = 1,920,000$ Hz. The prescaler solves the problem of having a divider with a high enough frequency capability to handle the VCO output, but forces the use of programmable dividers for only a portion of the total divide ratio. Because of the prescaler, the divider ratio is not stepped in integer increments but in increments of 64. Circuit designers can either live with this or find another solution.

One possible solution is to reduce the reference frequency by a factor of 64. In the example, the reference frequency would become 30 kHz/64 = 468.75 Hz. To achieve this frequency at the other input of the phase detector, an additional division factor of 64 must be included in the programmable divider, making it $N = 203 \times 64 = 12,992$. Assuming the original output frequency of 389.76 MHz, the overall divide ratio is $R = MN = 12,992(64) = 831,488$. This makes the output of the programmable divider equal to the reference frequency, or $f_r = 389,760,000/831,488 = 468.75$ Hz.

This solution is logical, but it has several disadvantages. First, it increases cost and complexity by requiring two more divide-by-64 ICs in the reference and feedback paths. Second, the lower the operating frequency of the phase detector, the more difficult it is to filter the output into DC. Further, the low-frequency response of the filter slows the process of achieving lock. When a change in divider ratio is made, the VCO frequency must change. It takes a finite amount of time for the filter to develop the necessary value of corrective voltage to shift the VCO frequency. The lower the phase-detector frequency, the greater this lock delay time. It has been determined that the lowest acceptable frequency is about 1 kHz, and even this is too low in some applications. At 1 kHz, the change in VCO frequency is very slow as the filter capacitor changes its charge in response to the different duty-cycle pulses of the phase detector. With a 468.75-Hz phase-detector frequency, the loop response becomes even slower. For more rapid frequency changes, a much higher frequency must be used. For spread spectrum and in some satellite applications, the frequency must change in a few microseconds or less, requiring an extremely high reference frequency.

To solve this problem, designers of high-frequency PLL synthesizers created special IC frequency dividers, like the one diagrammed in Fig. 7-15. The VCO output is applied to a special variable-modulus prescaler divider. It is made of emitter-coupled

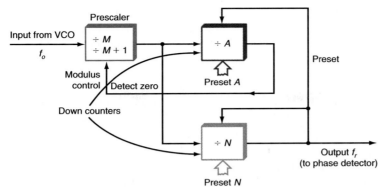

FIG. 7-15 Using a variable-modulus prescaler in a PLL frequency divider.

logic circuits and thus operates at the very high frequencies required. It is designed to have two divide ratios, M and $M + 1$. Some commonly available ratio pairs are 10/11, 64/65, and 128/129. Let's assume the use of a 64/65 counter. The actual divide ratio is determined by the modulus control input. If this input is binary 0, the prescaler divides by M, or 64; if this input is binary 1, the prescaler divides by $M + 1$, or 65. As Fig. 7-15 shows, the modulus control receives its input from an output of counter A. Counters A and N are programmable down-counters used as frequency dividers. The divide ratios are preset into the counters each time a full divide cycle is achieved. These ratios are such that $N > A$. The count input to each counter comes from the output of the variable-modulus prescaler.

A divide cycle begins by presetting the down-counters to A and N and setting the prescaler to $M + 1 = 65$. The input frequency from the VCO is f_0. The input to the down-counters is $f_0/65$. Both counters begin down-counting. Since A is a shorter counter than N, A will decrement to zero first. When it does, its detect-zero output goes high, changing the modulus of the prescaler from 65 to 64. The N counter initially counts down by a factor of A, but continues to down-count with an input of $f_0/64$. When it reaches zero, both down-counters are preset again, the dual modulo prescaler is changed back to a divide ratio of 65, and the cycle starts over.

The total division ratio R of the complete divider in Fig. 7-15 is: $R = MN + A$. If $M = 64$, $N = 203$, and $A = 8$, the total divide ratio is $R = 64(203) + 8 = 12{,}992 + 8 = 13{,}000$. The output frequency is $f_0 = Rf_r = 13{,}000(30{,}000) = 390{,}000{,}000 = 390$ MHz.

Any divide ratio in the desired range can be obtained by selecting the appropriate preset values for A and N. Further, this divider steps the divide ratio one integer at a time so that the step increment in the output frequency is 30 kHz, as desired.

As an example, assume that N is set to 207 and A is set to 51. The total divide ratio is $R = MN + A = 64(207) + 51 = 13{,}248 + 51 = 13{,}299$. The new output frequency is $f_0 = 13{,}299(30{,}000) = 398{,}970{,}000 = 398.97$ MHz.

If the A value is changed by 1, raising it to 52, the new divide ratio is $R = MN + A = 64(207) + 52 = 13{,}248 + 52 = 13{,}300$. The new frequency is $f_0 = 13{,}300(3{,}000) = 399{,}000{,}000 = 399$ MHz. Note that with an increment change in A of 1, R changed by 1 and the final output frequency increased by a 30-kHz (0.03-MHz) increment, from 398.97 to 399 MHz.

The preset values for N and A can be supplied by almost any parallel digital source but are usually supplied by a microprocessor or are stored in a ROM. Although this type of circuit is complex, it achieves the desired results of stepping the output frequency in increments equal to the reference input to the phase detector and allowing the reference frequency to remain high so that the change delay in the output frequency is shorter.

Example 7-3

A frequency synthesizer has a crystal reference oscillator of 10 MHz followed by a divider with a factor of 100. The variable-modulus prescaler has $M = 31/32$. The A and N down-counters have factors of 63 and 285, respectively. What is the synthesizer output frequency?

The reference input signal to the phase detector is
$$\frac{10 \text{ MHz}}{100} = 0.1 \text{ MHz} = 100 \text{ kHz}.$$
The total divide factor R is
$$R = MN + A = 32\,(285) + 63 = 9183$$
The output of this divider must be 100 kHz to match the 100-kHz reference signal to achieve lock. Therefore, the input to the divider, the output of the VCO, is R times 100 kHz, or
$$f_0 = 9183\,(0.1 \text{ MHz}) = 918.3 \text{ MHz}$$

Example 7-4

Demonstrate that the step change in output frequency for the synthesizer in Example 7-3 is equal to the phase-detector reference range, or 0.1 MHz.

Changing the A factor one increment to 64 and recalculating the output yields
$$R = 32\,(285) + 64 = 9184$$
$$f_0 = 9184\,(0.1 \text{ MHz}) = 918.4 \text{ MHz}$$
The increment is $918.4 - 918.3 = 0.1$ MHz.

7-3 POWER AMPLIFIERS

The three basic types of power amplifiers used in transmitters are linear, class C, and switching.

Linear amplifiers provide an output signal that is an identical, enlarged replica of the input. Their output is directly proportional to their input, and they therefore faithfully reproduce an input, but at a higher power level. Most audio amplifiers are linear. Linear RF amplifiers are used to increase the power level of variable-amplitude RF signals such as low-level AM or SSB signals. Linear amplifiers are class A, AB, or B. The class of an amplifier indicates how it is biased.

Class A amplifiers are biased so that they conduct continuously. The bias is set so that the input varies the collector (or drain) current over a linear region of the transistor's characteristics. Thus its output is an amplified linear reproduction of the input. Usually we say that the class A amplifier conducts for 360° of an input sine wave.

Class B amplifiers are biased at cutoff so no collector current flows with zero input. The transistor conducts on only one-half, or 180°, of the sine-wave input. This means that only one-half of the sine wave is amplified. Normally, two class B amplifiers are connected in a push-pull arrangement so that both the positive and negative alternations of the input are amplified.

Class AB linear amplifiers are biased near cutoff with some continuous collector current flow. They conduct for more than 180° but less than 360° of the input. They too are used primarily in push-pull amplifiers and provide better linearity than class B amplifiers, but with less efficiency.

Class A amplifiers are linear but not very efficient. For that reason, they make poor power amplifiers. As a result, they are used primarily as small-signal voltage amplifiers or for low-power amplifications. The buffer amplifiers described previously are class A amplifiers.

Class B amplifiers are more efficient than class A amplifiers, because current flows for only a portion of the input signal, and they make good power amplifiers. However, they distort an input signal because they conduct for only one-half of the cycle. Therefore, special techniques are often used to eliminate or compensate for the distortion. For example, operating class B amplifiers in a push-pull configuration minimizes the distortion.

Class C amplifiers conduct for even less than one-half of the sine-wave input cycle, making them very efficient. The resulting highly distorted current pulse is used to ring a tuned circuit to create a continuous sine-wave output. Class C amplifiers cannot be used to amplify varying-amplitude signals. They will clip off or otherwise distort an AM or SSB signal. However, FM signals do not vary in amplitude and can therefore be amplified with more efficient nonlinear class C amplifiers. This type of amplifier also makes a good frequency multiplier as harmonics are generated in the amplification process.

Switching amplifiers act like ON-OFF or digital switches. They effectively generate a square-wave output. Such a distorted output is undesirable; however, by using high-Q tuned circuits in the output, the harmonics generated as part of the switching process can be easily filtered out. The ON-OFF switching action is highly efficient as current flows only during half of the input cycle and when it does, the voltage drop across the transistor is very low, resulting in low power dissipation. Switching amplifiers are designated class D and class E.

LINEAR AMPLIFIERS

Linear amplifiers are used primarily in AM and SSB transmitters, and both low- and high-power versions are used. Some examples follow.

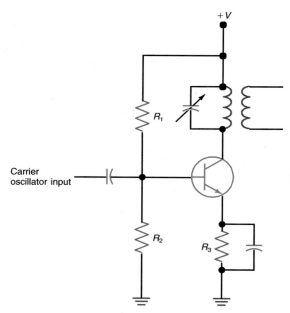

Fig. 7-16 A linear (class A) RF buffer amplifier.

CLASS A BUFFERS. A simple class A buffer amplifier is shown in Fig. 7-16. This type of amplifier is used between the carrier oscillator and the final power amplifier to isolate the oscillator from the power amplifier load, which can change the oscillator frequency. It also provides a modest power increase to provide the driving power required by the final amplifier. Such circuits usually provide milliwatts of power and rarely more than 1 W. The carrier oscillator signal is capacitively coupled to the input. The bias is derived from R_1, R_2, and R_3. The emitter resistor R_3 is bypassed to provide maximum gain. The collector is tuned with a resonant LC circuit at the operating frequency. An inductively coupled secondary loop transfers power to the next stage.

HIGH-POWER LINEAR AMPLIFIERS. A high-power class A linear amplifier is shown in Fig. 7-17. Base bias is supplied by a constant-current circuit that is temperature-compensated. The RF input from a 50-Ω source is connected to the base via an impedance-matching circuit made up of C_1, C_2, and L_1. The output is matched to a 50-Ω load by the impedance-matching network made up of L_2, L_3, C_3, and C_4. When connected to a proper heat sink, the transistor can generate up to 100 W of power up to about 30 MHz. The amplifier is designed for a specific frequency which is set by the input and output tuned circuits. Class A amplifiers have a maximum efficiency of 50 percent. Thus only 50 percent of the DC power is converted to RF, with the remaining 50 percent being dissipated in the transistor. For 100-W RF output, the transistor dissipates 100 W.

Commonly available RF power transistors have an upper power limit of several hundred watts. To produce more power, two or more devices can be connected in parallel, in a push-pull configuration, or in some combination. Power levels of up to several thousand watts are possible with these arrangements.

CLASS B PUSH-PULL AMPLIFIERS. A class B linear power amplifier using push-pull is shown in Fig. 7-18. The RF driving signal is applied to Q_1 and Q_2 through input transformer T_1. It provides impedance-matching and base drive signals to Q_1 and

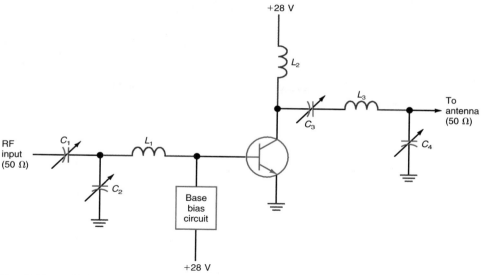

Fig. 7-17 A high-power class A linear RF amplifier.

Q_2 that are 180° out of phase. An output transformer T_2 couples the power to the antenna or load. Bias is provided by R_1 and D_1.

For class B operation, Q_1 and Q_2 must be biased right at the cutoff point. The emitter-base junction of a transistor will not conduct until about 0.6 to 0.8 V forward bias is applied because of the built-in potential barrier. This effect causes the transistors to be

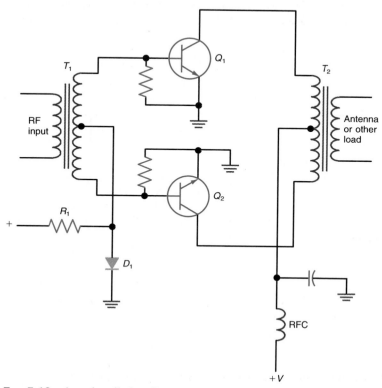

Fig. 7-18 A push-pull class B power amplifier.

naturally biased beyond cutoff, not right at it. A forward-biased silicon diode D_1 has about 0.7 V across it, and this is used to put Q_1 and Q_2 right on the conduction threshold.

On the positive half cycle of the RF input, the base of Q_1 is positive and the base of Q_2 is negative. Q_2 is cut off, but Q_1 conducts, linearly amplifying the positive half cycle. Collector current flows in the upper half of T_2, which induces an output voltage in the secondary. On the negative half cycle of the RF input, the base of Q_1 is negative, so it is cut off. The base of Q_2 is positive, so Q_2 amplifies the negative half cycle. Current flows in Q_2 and the lower half of T_2, completing a full cycle. The power is split between the two transistors.

The circuit in Figure 7-18 is an untuned broadband circuit that can amplify signals over a broad frequency range, typically from 2 to 30 MHz. A low-power AM or SSB signal is generated at the desired frequency, then applied to this power amplifier before being sent to the antenna. With push-pull circuits, power levels of up to 1 kW are possible.

Figure 7-19 shows another push-pull RF power amplifier. It uses two power MOSFETs, can produce an output up to 1 kW over the 10- to 90-MHz range, and has a 12-dB power gain. The RF input driving power must be 63 W to produce the full 1-kW output. Toroidal transformers T_1 and T_2 are used at the input and output for impedance matching. They provide broadband operation over the 10- to 90-MHz range without tuning. The 20-nH chokes and 20-Ω resistors form neutralization circuits that provide out-of-phase feedback from output to input to prevent self-oscillation.

CLASS C AMPLIFIERS

The key circuit in most AM and FM transmitters is the class C amplifier. They are used for power amplification in the form of drivers, frequency multipliers, and final

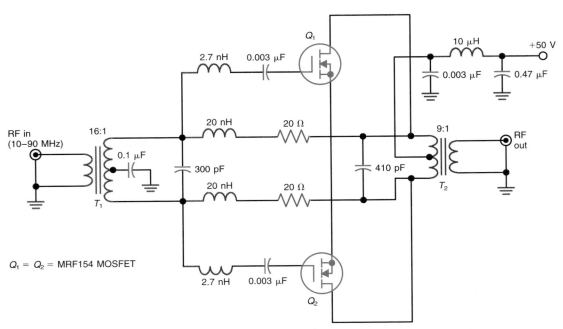

FIG. 7-19 A 1-kW push-pull RF power amplifier using MOSFETs.

amplifiers. Class C amplifiers are biased so they conduct for less than 180° of the input. A class C amplifier typically has a conduction angle of 90 to 150°. Current flows through it in short pulses and a resonant tuned circuit is used for complete signal amplification.

BIASING METHODS. Figure 7-20(*a*) shows one way of biasing a class C amplifier. The base of the transistor is simply connected to ground through a resistor. No external bias voltage is applied. An RF signal to be amplified is applied directly to the base. The transistor conducts on the positive half cycles of the input wave and is cut off on the negative half cycles. Although this sounds like a class B configuration, that is not the case. Recall that the emitter-base junction of a bipolar transistor has a forward voltage threshold of approximately 0.7 V. In other words, the emitter-base junction does

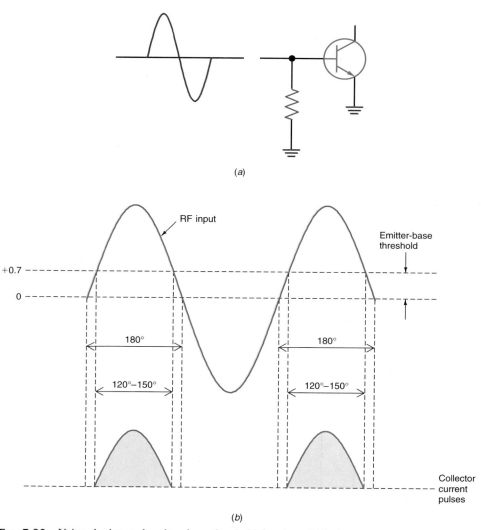

FIG. 7-20 Using the internal emitter-base threshold for class C biasing.

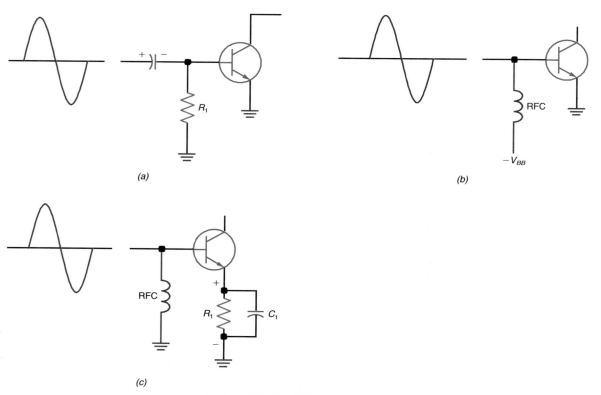
not really conduct until the base is more positive than the emitter by 0.7 V. Because of this, the transistor has an inherent built-in reverse bias. When the input signal is applied, the collector current does not flow until the base is positive by 0.7 V. This is illustrated in Fig. 7-20(b). The result is that collector current flows through the transistor in positive pulses for less than the full 180° of the positive AC alternation.

In many low-power driver and multiplier stages, no special biasing provisions other than the inherent emitter-base junction voltage are required. The resistor between base and ground simply provides a load for the driving circuit. In some cases, a narrower conduction angle than that provided by the circuit in Fig. 7-20(a) must be used. In such cases, some form of bias must be applied. A simple way of supplying bias is with the RC network shown in Fig. 7-21(a). Here the signal to be amplified is applied through capacitor C_1. When the emitter-base junction conducts on the positive half cycle, C_1 charges to the peak of the applied voltage less the forward drop across the emitter-base junction. On the negative half cycle of the input, the emitter-base junction is reverse-biased, so the transistor does not conduct. During this time, however, capacitor C_1 discharges through R_1, producing a negative voltage across R_1 which serves as a reverse bias on the transistor. By properly adjusting the time constant of R_1 and C_1, an average DC reverse-bias voltage can be established. The applied voltage causes the transistor to conduct, but only on the peaks. The higher the average DC bias voltage, the narrower the conduction angle and the shorter the duration of the collector-current pulses. This method is referred to as *signal bias*.

FIG. 7-21 Methods of biasing a class C amplifier. (*a*) Signal bias. (*b*) External bias. (*c*) Self-bias.

Of course, negative bias can also be supplied to a class C amplifier from a fixed DC supply voltage as shown in Fig. 7-21(*b*). After the desired conduction angle is determined, the value of the reverse voltage can be determined and applied to the base through the RFC. The incoming signal is then coupled to the base, causing the transistor to conduct on only the peaks of the positive input alternations. This is called *external bias,* and requires a separate negative DC supply.

Another biasing method is shown in Fig. 7-21(*c*). As in the circuit shown in Fig. 7-21(*a*), the bias is derived from the signal. This arrangement is known as the *self-bias* method. When current flows in the transistor, a voltage is developed across R_1. C_1 is charged and holds the voltage constant. This makes the emitter more positive than the base, which has the same effect as a negative voltage on the base. A strong input signal is required for proper operation.

TUNED OUTPUT CIRCUITS.   All class C amplifiers have some form of tuned circuit connected in the collector as shown in Fig. 7-22. The primary purpose of this tuned circuit is to form the complete AC sine-wave output. A parallel tuned circuit rings, or oscillates, at its resonant frequency whenever it receives a DC pulse. The pulse charges the capacitor, which, in turn, discharges into the inductor. The magnetic field in the inductor increases and then collapses, inducing a voltage which then recharges the capacitor in the opposite direction. This exchange of energy between the inductor and the capacitor, called the *flywheel effect,* produces a damped sine wave at the resonant frequency. If the resonant circuit receives a pulse of current every half cycle, the voltage across the tuned circuit is a constant-amplitude sine wave at the resonant frequency. Even though the current flows through the transistor in short pulses, the class C amplifier output is a continuous sine wave.

Another way to look at the operation of a class C amplifier is to view the transistor as supplying a highly distorted pulse of power to the tuned circuit. According to Fourier theory, this distorted signal contains a fundamental sine wave plus both odd and even harmonics. The tuned circuit acts like a bandpass filter to select the fundamental sine wave contained in the distorted composite signal.

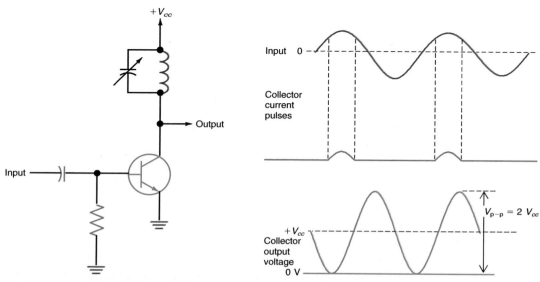

FIG. 7-22 Class C amplifier operation.

The tuned circuit in the collector is also used to filter out unwanted harmonics. The short pulses in a class C amplifier are made up of second, third, fourth, fifth, etc., harmonics. In a high-power transmitter, signals are radiated at these harmonic frequencies as well as at the fundamental resonant frequency. Such harmonic radiation can cause out-of-band interference, and the tuned circuit acts like a selective filter to eliminate these higher-order harmonics. If the Q of the tuned circuit is made high enough, the harmonics will be adequately suppressed.

The Q of the tuned circuit in the class C amplifier should be selected so that it provides adequate attenuation of the harmonics but also has sufficient bandwidth to pass the sidebands produced by the modulation process. Remember that the bandwidth and Q of a tuned circuit are related by the expression

$$\text{BW} = \frac{f_r}{Q} \qquad Q = \frac{f_r}{\text{BW}}$$

If the Q of the tuned circuit is too high, the bandwidth will be very narrow and some of the higher-frequency sidebands will be eliminated. This causes a form of frequency distortion called *sideband clipping* and may make some signals unintelligible or will at least limit the fidelity of reproduction.

One of the main reasons why class C amplifiers are preferred for RF power amplification over class A and class B amplifiers is their high efficiency. Remember, efficiency is the ratio of the output power to the input power. If all of the generated power, the input power, is converted to output power, the efficiency is 100 percent. This doesn't happen in the real world because of losses. But in a class C amplifier more of the total power generated is applied to the load. Because the current flows for less than 180° of the AC input cycle, the average current in the transistor is fairly low, meaning that the power dissipated by the device is low. A class C amplifier functions almost as a transistor switch that is OFF for over 180° of the input cycle. The switch conducts for approximately 90 to 150° of the input cycle. During the time that it conducts, its emitter-to-collector resistance is low. Even though the peak current may be high, the total power dissipation is much less than that in class A and class B circuits. For this reason, more of the DC power is converted to RF energy and passed on to the load, usually an antenna. The efficiency of most class C amplifiers is in the 60 to 85 percent range.

The input power in a class C amplifier is the average power consumed by the circuit, which is simply the product of the supply voltage and the average collector current, or

$$P_{\text{in}} = V_{CC}(I_C)$$

For example, if the supply voltage is 13.5 V and the average DC collector current is 0.7 A, the input power is $P_{\text{in}} = 13.5(0.7) = 9.45$ W.

The output power is the power actually transmitted to the load. The amount of power depends upon the efficiency of the amplifier. The output power can be computed with the familiar power expression

$$P_{\text{out}} = \frac{V^2}{R_L}$$

where V is the RF output voltage at the collector of the amplifier and R_L is the load impedance. When a class C amplifier is set up and operating properly, the peak-to-peak RF output voltage is two times the supply voltage, or $2V_{CC}$ (see Fig. 7-22).

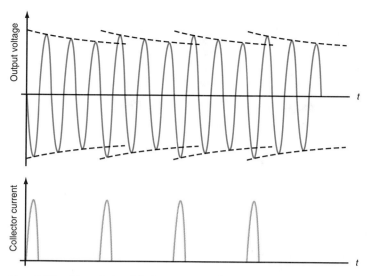

Fig. 7-23 The relationship between transistor current and tuned circuit voltage in a frequency tripler.

FREQUENCY MULTIPLIERS. Any class C amplifier is capable of performing frequency multiplication if the tuned circuit in the collector resonates at some integer multiple of the input frequency. For example, a frequency doubler can be constructed by simply connecting a parallel tuned circuit in the collector of a class C amplifier that resonates at twice the input frequency. When the collector current pulse occurs, it excites or rings the tuned circuit at twice the input frequency. A current pulse flows for every other cycle of the input. A tripler circuit is constructed in exactly the same way, except that the tuned circuit resonates at three times the input frequency, receiving one input pulse for every three cycles of oscillation it produces (see Fig. 7-23).

Multipliers can be constructed to increase the input frequency by any integer factor up to approximately 10. As the multiplication factor gets higher, the power output of the multiplier decreases. For most practical applications, the best result is obtained with multipliers of 2 and 3.

Another way to look at the operation of a class C frequency multiplier is to remember that the nonsinusoidal current pulse is rich in harmonics. Each time the pulse occurs, the second, third, fourth, fifth, and higher harmonics are generated. The purpose of the tuned circuit in the collector is to act as a filter to select the desired harmonic.

In many applications, a multiplication factor greater than that achievable with a single multiplier stage is required. In such cases, two or more multipliers are cascaded. Figure 7-24 shows two multiplier examples. In the first case, multipliers of 2 and 3 are cascaded to produce an overall multiplication of 6. In the second, three multipliers provide an overall multiplication of 30. The total multiplication factor is simply the product of the multiplication factors of the individual stages.

> **HINTS AND HELPS**
>
> Although multipliers can be constructed to increase the input frequency by any integer up to approximately 10, the best results are obtained with multipliers of 2 and 3.

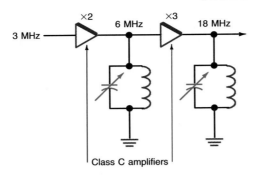

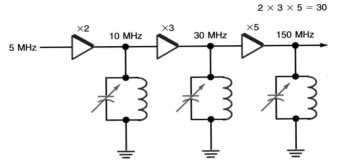

FIG. 7-24 Frequency multiplication with class C amplifiers.

NEUTRALIZATION

One problem that all RF amplifiers have, both linear and class C amplifiers, is *self-oscillation*. When some of the output voltage finds its way back to the input of the amplifier with the correct amplitude and phase, the amplifier oscillates, sometimes at its tuned frequency and in some cases at a much higher frequency. When the circuit oscillates at a higher frequency unrelated to the tuned frequency, the oscillation is referred to as *parasitic oscillation*. In both cases, the oscillation is undesirable, and either prevents amplification from taking place or, in the case of parasitic oscillation, reduces the power amplification and introduces distortion of the signal.

Self-oscillation at the tuned frequency in an amplifier is the result of positive feedback that occurs because of the interelement capacitance of the amplifying device, be it a bipolar transistor, FET, or vacuum tube. In a bipolar transistor this is the collector-to-base capacitance C_{bc} as shown in Fig. 7-25(a). Transistor amplifiers are biased so that the emitter-base junction is forward-biased while the base-collector junction is reversed-biased. As discussed previously, a reversed-biased diode or transistor junction acts like a capacitor. This small capacitance permits output from the collector to be fed back to the base. Depending on the frequency of the signal, the value of the capacitance, and the values of stray inductances and capacitances in the circuit, the signal fed back may be in phase with the input signal and high enough in amplitude to cause oscillation.

This interelement capacitance cannot be eliminated; therefore its effect must be compensated for, or neutralized. In the *neutralization* process, another signal, equal in

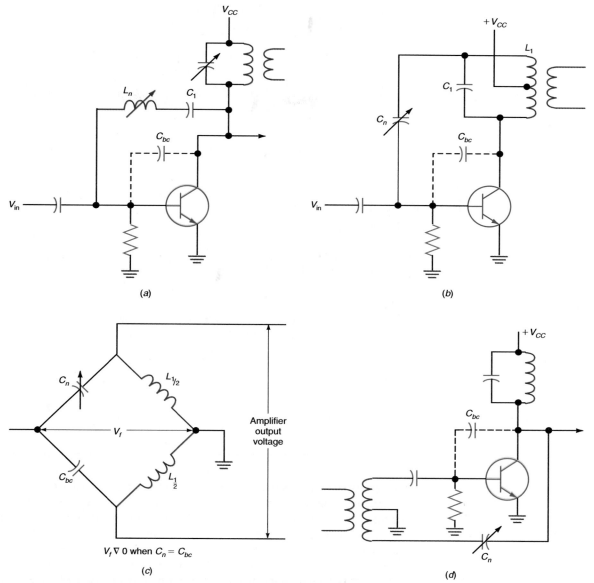

Fig. 7-25 Neutralization circuits. (*a*) Canceling the effect of C_{bc} with an equivalent inductance, L_n. (*b*) Neutralization using a tapped collector coil and a neutralization capacitor, C_n. (*c*) Equivalent circuit of part *b*. (*d*) Neutralization with a tapped input inductor.

amplitude to the signal fed back through C_{bc} and 180° out of phase with it, is fed back. The result is that the two signals cancel.

Several methods of neutralization are shown in Fig. 7-25. In Fig. 7-25(*a*), a signal of equal and opposite phase is provided by the inductor L_n. Capacitor C_1 is a high-value capacitor used strictly for DC blocking to prevent collector voltage from being applied to the base. L_n is made adjustable so that its value can be set to make its reactance equal to the reactance of C_{bc} at the oscillation frequency. As a result, C_{bc} and L_n form a parallel resonant circuit that acts like a very high-value resistor at the resonant frequency. The result is effective cancellation of the positive feedback. The type of neutralization shown in Fig. 7-25(b) uses a tapped collector coil and a neutralization capacitor C_n. The two equal halves of the collector inductance, the junction capacitance C_{bc}, and C_n, form a bridge circuit [Fig. 7-25(c)]. When C_n is adjusted to be equal to C_{bc}, the bridge is balanced and no feedback signal V_f occurs. A variation of this is shown in Fig. 7-25(d), where a center-tapped base input inductor is used.

Parasitic oscillations are usually eliminated by connecting a low value of resistor in the collector or base lead. A value of 10 to 22 Ω is typical. Parasitic oscillations can also be eliminated by putting one or more ferrite beads over the collector or base leads. Another practice is to wind a small inductor on a resistor, creating a parallel R_L circuit which is then placed in the collector or base leads.

SWITCHING POWER AMPLIFIERS

As stated previously, the primary problem with RF power amplifiers is their inefficiency and high power dissipation. In order to generate RF power to transfer to the antenna, the amplifier must dissipate a considerable amount of power itself. For example, a class A power amplifier using a transistor conducts continuously. It is a linear amplifier whose conduction varies as the signal changes. Because of the continuous conduction, the class A amplifier generates a considerable amount of power that is not transferred to the load. No more than 50 percent of the total power consumed by the amplifier can be transferred to the load. Because of the high power dissipation, the output power of a class A amplifier is generally limited. For that reason, class A amplifiers are normally used only in low-power transmitter stages.

To produce greater power output, class B amplifiers are used. Each transistor conducts for 180° of the carrier signal. Two transistors are used in a push-pull arrangement to form a complete carrier sine wave. Since each transistor is conducting for only 180° of any carrier cycle, the amount of power it dissipates is considerably less, and efficiencies of 70 to 75 percent are possible. Class C power amplifiers are even more efficient, since they conduct for less than 180° of the carrier signal, relying on the tuned circuit in the plate or collector to supply power to the load when they are not conducting. With current flowing for less than 180° of the cycle, class C amplifiers dissipate less power and can, therefore, transfer more power to the load. Efficiencies as high as about 85 percent can be achieved, and class C amplifiers are therefore the most widely used type in power amplifiers when the type of modulation permits.

Another way to achieve high efficiencies in power amplifiers is to use a switching amplifier. A *switching amplifier* is a transistor that is simply used as a switch and is either conducting or nonconducting. Both bipolar transistors and enhancement-mode MOSFETs are widely used in switching-amplifier applications. A bipolar transistor as a switch is either cut OFF or saturated. When it is cut OFF, no power is dissipated. When it is saturated, current flow is maximum, but the emitter-collector voltage is extremely low, usually less than 1 V. As a result, power dissipation is extremely low.

When enhancement-mode MOSFETs are used, the transistor is either cut OFF or turned ON. In the cutoff state, no current flows, so no power is dissipated. When the transistor is conducting, its ON resistance between source and drain is usually very low—again, no more than several ohms and typically far less than 1 Ω. As a result, power dissipation is extremely low even with high currents.

The use of switching power amplifiers permits efficiencies of over 90 percent. The current variations in a switching power amplifier are square waves and thus harmonics are generated. However, these are relatively easy to filter out by the use of tuned circuits and filters between the power amplifier and the antenna.

The three basic types of switching power amplifiers, class D, class E, and class S, were originally developed for high-power audio applications, but with the availability of high-power high-frequency switching transistors, they are now widely used in radio transmitter design.

CLASS D AMPLIFIERS. A class D amplifier uses a pair of transistors to produce a square-wave current in a tuned circuit. Figure 7-26 shows the basic configuration of a class D amplifier. Two switches are used to apply both positive and negative DC voltages to a load through the tuned circuit. When switch S_1 is closed, S_2 is open; when S_2 is closed, S_1 is open. When S_1 is closed, a positive DC voltage is applied to the load. When S_2 is closed, a negative DC voltage is applied to the load. Thus the tuned circuit and load receive an AC square wave at the input.

The series resonant circuit has a very high Q. It is resonant at the carrier frequency. Since the input waveform is a square wave, it consists of a fundamental sine wave and odd harmonics. Because of the high Q of the tuned circuit, the odd harmonics are filtered out, leaving a fundamental sine wave across the load. With ideal switches, meaning no leakage current in the OFF state and no ON resistance when conducting, the theoretical efficiency is 100 percent.

Figure 7-27 shows a class D amplifier implemented with enhancement-mode MOSFETs. The carrier is applied to the MOSFET gates 180° out of phase by use of a transformer with a center-tapped secondary. When the input to the gate of Q_1 is positive, the input to the gate of Q_2 is negative. Thus Q_1 conducts and Q_2 is cut off. On the next half cycle of the input, the gate to Q_2 goes positive and the gate of Q_1 goes negative. Q_2 conducts, applying a negative pulse to the tuned circuit. Recall that enhancement-mode MOSFETs are normally nonconducting until a gate voltage higher than a specific threshold value is applied, at which time the MOSFET conducts. The ON resistance is very low. In practice, efficiencies of up to 90 percent can be achieved using a circuit like that in Fig. 7-27.

CLASS E AMPLIFIERS. In class E amplifiers, only a single transistor is used. Both bipolar and MOSFETs can be used, although the MOSFET is preferred because of

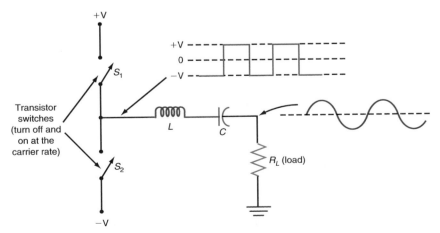

FIG. 7-26 Basic configuration of a class D amplifier.

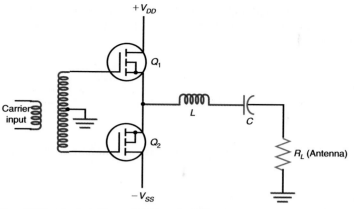

Fig. 7-27 A class D amplifier made with enhancement-mode MOSFETs.

its low drive requirements. Figure 7-28 shows a typical class E RF amplifier. The carrier, which is initially a sine wave, is applied to a shaping circuit which effectively converts it into a square wave. The carrier is usually frequency-modulated. The square-wave carrier signal is then applied to the base of the class E bipolar power amplifier. Q_1 switches OFF and ON at the carrier rate. The signal at the collector is a square wave, which is applied to a low-pass filter and tuned impedance-matching circuit made up of C_1, C_2, and L_1. The odd harmonics are filtered out, leaving a fundamental sine wave which is applied to the antenna. A high level of efficiency is achieved with this arrangement.

CLASS S AMPLIFIERS. Class S amplifiers, which use switching techniques but with a scheme of pulse-width modulation, are found primarily in audio applications but have also been used in low- and medium-frequency RF amplifiers such as those used in AM broadcast transmitters. The low-level audio signal to be amplified is applied to a circuit called a *pulse-width modulator.* A carrier signal at a frequency 5 to 10 times the highest audio frequency to be amplified is also applied to the pulse-width modulator. At the output of the modulator is a series of constant-amplitude pulses whose pulse width or duration varies with the audio signal amplitude. These signals are then applied to a switching amplifier of the class D or class E type. High power and efficiency are achieved because of the switching action. A low-pass filter is connected to

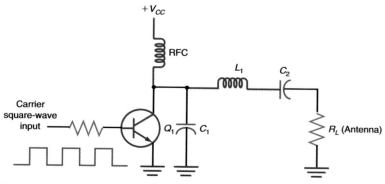

Fig. 7-28 A class E RF amplifier.

the output of the switching amplifier to average and smooth the pulses back into the original audio signal waveform. A capacitor or low-pass filter across the speaker is usually sufficient.

VACUUM TUBE PRINCIPLES

Contrary to what you may think, vacuum tubes are still widely used in electronics, especially in communications. The most common vacuum tube in use today is the cathode ray tube (CRT), which is the heart of all TV sets, computer video monitors, oscilloscopes, spectrum analyzers, and other devices with visual displays. Vacuum tubes are also widely used for RF power amplification. Transistors can provide power amplification up to the 500- to 1000-W range. When power needs exceed several thousand watts, vacuum tubes are the amplifier of choice. Vacuum tube amplifiers are not only simpler and more reliable than transistor amplifiers, but also much less expensive. In many applications, transistors are simply not capable of providing the desired power levels at the very high frequencies in use today. This is particularly true of power amplification in the microwave region. Transistors can provide power amplification up to the 50- to 100-W range in the microwave region. (For power levels beyond that, vacuum tubes such as the magnetron, klystron, and traveling wave tube are used. These are discussed in Chap. 15.)

Like a transistor, a vacuum tube is a current-control device. A small input signal can be made to control a much larger current. Such a device can produce amplification. Figure 7-29(a) shows the schematic diagram for a simple vacuum tube. It consists of two elements, a *filament* and a *plate* or *anode* sealed in a glass or metal tube from which the air has been evacuated. When a low AC or DC voltage is applied to the filament, it becomes hot. This causes the coating on the filament to begin emitting electrons. Remember, as any substance is heated, the atoms become agitated and begin throwing off electrons. Certain substances called *thermionic materials* throw off a considerable number of electrons when heated. The electrons form a cloud around the filament. A positive voltage is applied to the plate or anode with respect to the fila-

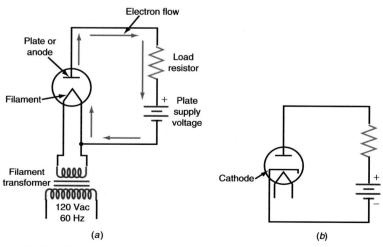

FIG. 7-29 Vacuum tube principles.

ment. The positive charge on the plate begins to attract the electrons, and a current path is established between the filament and the plate.

Figure 7-29(*b*) shows a vacuum tube with a third element, a *cathode,* added. In this type, the cathode is heated by the filament; the filament does not emit the electrons itself. The heated cathode emits the electrons, which are attracted to the positive charge on the plate.

The vacuum tubes shown in Fig. 7-29 are diodes. Like a semiconductor diode, they pass current in only one direction, from filament or cathode to anode. If a negative charge is applied to the plate element, it repels the electrons from the cathode and no current flows. Vacuum tube diodes were once widely used as rectifiers in power supplies and in AM detectors.

TRIODES

To control the current flow in a vacuum tube, a control element called the *grid* is added between the plate and the cathode. The schematic symbol for such a device is shown in Fig. 7-30. Because there are three elements—the cathode, the grid, and the plate—this type of vacuum tube is referred to as a *triode.* The filament heats the cathode, which emits electrons. To establish current flow, a positive voltage is applied to the plate with respect to the cathode. The grid element, which is placed closer to the cathode than to the plate, allows electrons to pass through to the plate. However, if this grid element is made negative with respect to the cathode with a bias voltage, it repels some of the electrons back to the cathode, decreasing the number of electrons reaching the plate. Varying the negative voltage on the grid allows the amount of current flow to be varied. Because the grid is closer to the cathode than the plate, a small control voltage on the grid can produce very large changes in the plate current. Thus the vacuum tube can be made to function as an amplifier.

Figure 7-31 shows a basic triode amplifier circuit. It is similar in many ways to an FET amplifier. A resistor is normally placed in series with the cathode, and current flowing through this resistor develops a voltage with a polarity such that the cathode is positive with respect to the grid. This is equivalent to the grid being negative with respect to the cathode. The grid is, therefore, reverse-biased with respect to the cathode, and no current flows in the grid circuit. However, electrons do pass on through to the plate and to the plate resistor.

When an AC signal is applied to the grid, it varies the reverse bias above and below the level established by the cathode resistor. This causes the plate current to vary in a similar way. The variation in plate current causes an enlarged version of the input signal to appear between the plate and ground. The amplified AC signal rides on the

FIG. 7-30 A triode vacuum tube.

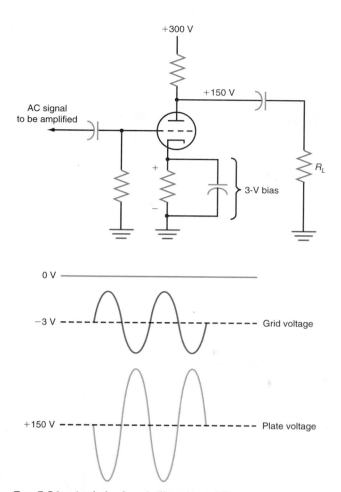

Fig. 7-31 A triode class A (linear) amplifier.

DC plate voltage established by the bias. Note the very high plate voltage supply. Most vacuum tubes require voltages of hundreds or thousands of volts for proper operation.

TETRODES AND PENTODES

One of the main problems with a triode vacuum tube is high interelectrode capacitance. For example, there is considerable grid-plate capacitance in a triode. Because this capacitance has a very low value of reactance at the higher frequencies, the upper-frequency operation of the tube is limited. Grid-plate capacitance also causes self-oscillation.

To eliminate this problem, some tubes have a second grid element installed between the plate and the control grid, which acts like an electrostatic shield that effectively reduces the control grid-plate capacitance to a very low value. Such a tube is called a *tetrode*. Figure 7-32(a) shows a tetrode amplifier circuit. The screen grid is positively biased so that it attracts electrons from the cathode and accelerates them to

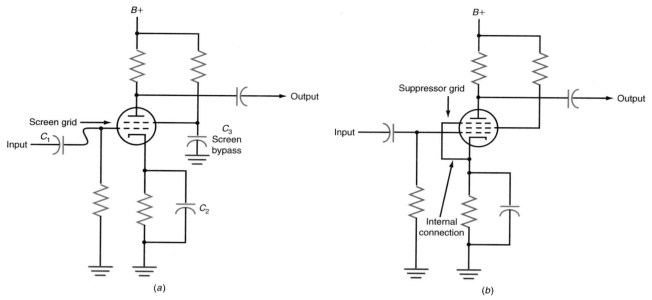

Fig. 7-32 Tetrode and pentode amplifiers. (*a*) Tetrode circuit. (*b*) Pentode circuit.

the plate. The screen grid acts like an intermediate plate, but the electrons flow on through its grid structure. Because the screen grid is positively biased, it does draw a small amount of current. The screen grid is put at ground potential for AC signals with the screen bypass capacitor C_3. With the interelectrode capacitance reduced to a very low level, the tetrode can operate at much higher frequencies and the chance of self-oscillation is reduced.

When the high-velocity electrons from the cathode strike the plate, other electrons in the plate are dislodged. This phenomenon is referred to as *secondary emission*. The electrons attracted back to the positive screen grid cause a kind of reverse current flow between plate and screen grid. This undesirable effect can be eliminated by placing another grid between the plate and screen grid called a *suppressor grid*. This arrangement produces what is called a pentode. The suppressor grid is usually tied internally to the cathode, which is at a more negative potential, so it repels the secondary emission electrons back to the plate [see Fig. 7-32(*b*)].

VACUUM TUBE POWER AMPLIFIERS

At one time, vacuum tubes were widely used for small-signal amplification and other signal-processing operations now taken care of by transistors and ICs. It is rare to find such circuits in use today. Because of their filament current requirements, vacuum tubes use considerably more power than solid-state devices, and they also generally require much higher plate voltages than the collector supply voltage used in transistor circuits. Tubes are also larger, generate more heat, and are more delicate than transistors and other semiconductor devices.

As stated earlier, most vacuum tubes are used as RF power amplifiers. Tubes for such amplifiers are generally large and require very high voltages for operation. A typical vacuum tube amplifier requires plate voltages anywhere from 500 to 3000 V or more for proper operation. Plate currents can be from about 100 mA to several amperes. Obviously, such tubes can generate a considerable amount of output power, and they are often made of ceramic rather than glass or metal and have built-in cooling fans which help dissipate the heat generated on the plate element. Cooling fans or water-cooling arrangements are often required for high-power transmitters.

Vacuum tubes can be biased to operate as linear class A or class AB amplifiers. Most often, however, they are used in push-pull circuits and operate as class B or class C amplifiers for greater efficiency. The grid is usually biased negatively with respect to the cathode so no grid current flows, but in some applications, for example, class C amplifiers, the grid is biased positively to get extra power.

RF power vacuum tubes can easily generate 1 kW or more of RF power. Using multiple tubes in parallel and in push-pull combinations can produce even higher power. Some specially designed vacuum tubes generate tens of thousands and even hundreds of thousands of watts for use in radio broadcast transmitters and powerful short-wave transmitters.

A popular vacuum tube power amplifier configuration is the common grid or grounded grid amplifier using a triode. A typical circuit is shown in Fig. 7-33. The input signal is applied to the cathode or filament rather than to the grid. The grid-to-cathode voltage is still varied by the input signal to control the plate current, but the cathode or filament is above ground. The input signal is coupled through a π-network low-pass filter circuit to the cathode. The input power must be substantial. It is usually 50 to 100 W depending on the tube used and the desired power output. The RF

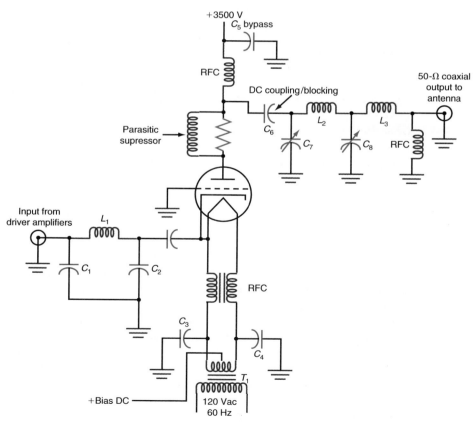

FIG. 7-33 A grounded grid power amplifier.

chokes in the filament leads along with bypass capacitors C_3 and C_4 keep RF out of the filament circuit and transformer. The RFCs also keep the cathode off ground to provide an input impedance for the input signal to be developed. The RFC is a pair of wires bifilar-wound on a ferrite core. In *bifilar winding* the two wires that form the chokes are put alongside one another and wound as one wire around the ferrite core.

The bias voltage is applied at the center tap of the filament transformer secondary. This is positive-bias voltage because the grid must be negative with respect to the cathode, which is the same as the cathode being positive with respect to the grid. The actual amount of bias voltage determines the class of operation—class A for linear applications such as AM and SSB and class C for FM.

The output signal is fed to a π-network output circuit made up of L_2, L_3, C_7, and C_8. It provides impedance matching between the high tube output impedance and the low antenna impedance, usually 50 Ω. C_6 blocks the high-voltage DC from the plate and keeps it off the antenna. Note the R_L network in the plate lead, which is used for parasitic suppression, as described earlier. With 3500 V on the plate and a plate current of 700 mA, the input power is 3500 $\times$ 0.7 = 2450 W (2.45 kW). The output power is about 80 percent of this, or 1960 W.

Grounded grid amplifiers are used primarily at frequencies below 30 MHz, but depending upon the tube they can be used well into the VHF region. They are noncritical and very reliable. Because the grid acts as an electrostatic shield between the plate output and the cathode or filament input, there is little or no undesirable feedback to cause self-oscillation. Thus grounded grid amplifiers do not usually require neutralization and its associated tuning adjustments.

One major disadvantage of the grounded grid amplifier is that it does require high driving power, much higher than when a grounded cathode circuit is used. In addition, if high-level amplitude modulation is used, the circuit cannot be modulated fully to 100 percent. To do this, the driving signal must also be amplitude-modulated.

7-4 IMPEDANCE-MATCHING NETWORKS

Matching networks that connect one stage to another are very important parts of any transmitter. In a typical transmitter, the oscillator generates the basic carrier signal, which is then amplified, usually by multiple stages, before reaching the antenna. Since the idea is to increase the power of the signal, the interstage coupling circuits must permit an efficient transfer of power from one stage to the next. Finally, some means must be provided to connect the final amplifier stage to the antenna, again for the purpose of transferring the maximum possible amount of power. The circuits used to connect one stage to another are known as *impedance-matching* networks. In most cases, they are *LC* circuits, transformers, or some combination. The basic function of a matching network is to provide for an optimum transfer of power through impedance-matching techniques. Matching networks also provide filtering and selectivity. Transmitters are designed to operate on a single frequency or selectable narrow ranges of frequencies. The various amplifier stages in the transmitter must confine the RF generated to these frequencies. In class C, D, and E amplifiers, a considerable number of high-amplitude harmonics are generated. These must be eliminated to prevent spurious radiation from the transmitter. The impedance-matching networks used for interstage coupling accomplish this.

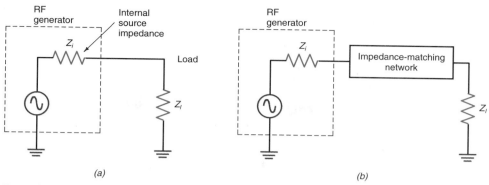

Fig. 7-34 Impedance matching in RF circuits.

(a)

(b)

The basic problem of coupling is illustrated in Fig. 7-34(a). The driving stage appears as a signal source with an internal impedance of Z_i. The stage being driven represents a load to the generator with its internal resistance of Z_l. Ideally, Z_i and Z_l are resistive. Recall that maximum power transfer in DC circuits takes place when Z_i equals Z_l. This basic relationship is essentially true also in RF circuits, but it is a much more complex relationship. In RF circuits, Z_i and Z_l are seldom purely resistive and, in fact, usually include a reactive component of some type. Further, it is not always necessary to transfer maximum power from one stage to the next. The goal is to transfer a sufficient amount of power to the next stage so that it can provide the maximum output of which it is capable.

In most cases, the two impedances involved are considerably different from one another, and therefore a very inefficient transfer of power takes place. To overcome this problem, an impedance-matching network is introduced between the two, as illustrated in Fig. 7-34(b). There are three basic types of LC impedance-matching networks, the L network, the T network, and the π network.

L NETWORKS

L networks consist of an inductor and a capacitor connected in various L-shaped configurations as shown in Fig. 7-35. The circuits in Fig. 7-35(a) and (b) are low-pass filters; those in Fig. 7-35(c) and (d) are high-pass filters. Typically, low-pass networks are preferred so that harmonic frequencies are filtered out.

The L matching network is designed so that the load impedance is matched to the source impedance. For example, the network in Fig. 7-35(a) causes the load resistance to appear larger than it actually is. The load resistance Z_L appears in series with the inductor of the L network. The inductor and the capacitor are chosen to resonate at the transmitter frequency. When the circuit is at resonance, X_L equals X_C. To the generator impedance Z_i, the complete circuit simply appears as a parallel resonant circuit. At resonance, the impedance represented by the circuit is very high. The actual value of the impedance depends upon the L and C values and the Q of the circuit. The higher the Q, the higher the impedance. The Q in this circuit is basically determined by the value of the load impedance. By proper selection of the circuit values, the load impedance can be made to appear as any desired value to the source impedance as long as Z_i is greater than Z_L.

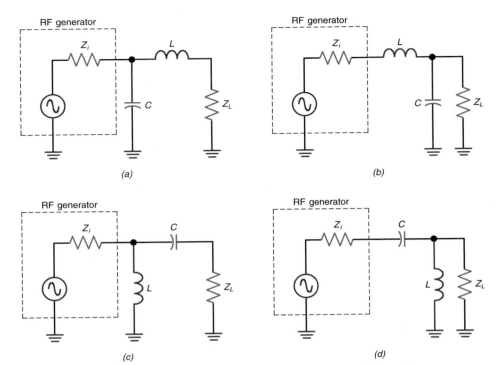

FIG. 7-35 Four L-type impedance-matching networks. (a) $Z_L < Z_i$. (b) $Z_L > Z_i$. (c) $Z_L < Z_i$. (d) $Z_L > Z_i$.

By using the L network shown in Fig. 7-35(*b*), the impedance can be stepped down, or made to appear much smaller than it actually is. With this arrangement, the capacitor is connected in parallel to the load impedance. The parallel combination of C and Z_L has an equivalent series RC combination. C and Z_L appear as equivalent series values C_{eq} and Z_{eq}. The result is that the overall network appears as a series resonant circuit, with C_{eq} and L resonant. Recall that a series resonant circuit has a very low impedance at resonance. The impedance is, in fact, the equivalent load impedance Z_{eq}, which is resistive.

The design equations for L networks are given in Fig. 7-36. Assuming that the internal source and load impedances are resistive, $Z_i = R_i$ and $Z_L = R_L$. The network in Fig. 7-36(*a*) assumes $R_L < R_i$, while the network in Fig. 7-36(*b*) assumes $R_i < R_L$.

Suppose we wish to match a 6-Ω transistor amplifier impedance to a 50-Ω antenna load at 155 MHz. In this case, $R_i < R_L$, so we use the formulas in Fig. 7-35(*b*).

$$X_L = \sqrt{R_i R_L - (R_i)^2} = \sqrt{6(50) - (6)^2} = \sqrt{300 - 36} = \sqrt{264} = 16.25 \ \Omega$$
$$Q = \sqrt{R_L/R_i - 1} = \sqrt{50/6 - 1} = 2.7$$
$$X_C = \frac{R_L R_i}{X_L} = \frac{50(6)}{16.25} = 18.46 \ \Omega$$

To find the values of L and C at 155 MHz, we rearrange the basic reactance formulas as follows:

$$X_L = 2\pi f L$$
$$L = \frac{X_L}{2\pi f} = \frac{16.25}{6.28 \times 155 \times 10^6}$$
$$= \frac{16.25}{6.28 \times 155 \times 10^6} = 16.7 \ \text{nH}$$

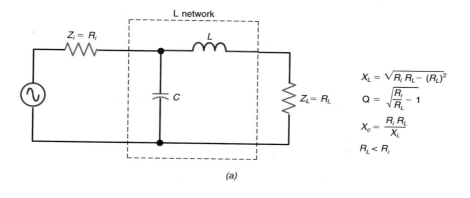

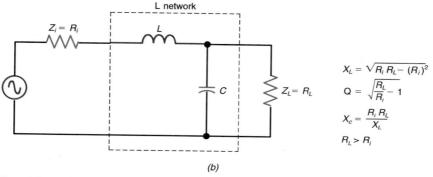

FIG. 7-36   The L-network design equations.

$$X_C = \frac{1}{2\pi f C}$$

$$C = \frac{1}{2\pi f X_C} = \frac{1}{6.28 \times 155 \times 10^6 \times 18.46} = 55.65 \text{ pF}$$

In most cases, internal and stray reactances make the internal impedance and load impedances complex, rather than purely resistive. Figure 7-37 shows an example using the figures given above. Here the internal resistance is 6 Ω, but it includes an internal inductance L_i of 8 nH. There is a stray capacitance C_L of 8.65 pF across the load. The way to deal with these reactances is simply to combine them with the L-network

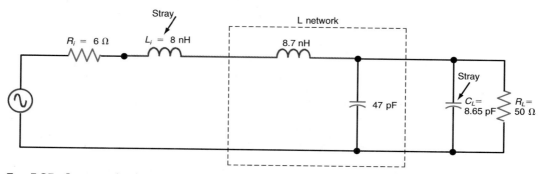

FIG. 7-37   Incorporating internal and stray reactances into a matching network.

values. In the example above, the calculation calls for an inductance of 16.7 nH. Since the stray inductance is in series with the L-network inductance in Fig. 7-37, the values will add. As a result, the L-network inductance must be less than the computed value by an amount equal to the stray inductance of 8 nH, or $L = 16.7 - 8 = 8.7$ nH. If the L-network inductance is made to be 8.7 nH, the total circuit inductance will be correct when it adds to the stray inductance.

A similar thing occurs with capacitance. The circuit calculations above call for a total of 55.65 pF. The L-network capacitance and the stray capacitance add, as they are in parallel. Therefore, the L-network capacitance can be less than the calculated value by the amount of the stray capacitance, or $C = 55.65 - 8.65 = 47$ pF. Making the L-network capacitance 47 pF gives the total correct capacitance when it adds to the stray capacitance.

T AND π NETWORKS

When designing L networks, there is very little control over the Q of the circuit, which is determined by the values of the internal and load impedances and may not always be what is needed to achieve the desired selectivity. To overcome this problem, matching networks using three reactive elements can be used. The three most widely used impedance-matching networks containing three reactive components are illustrated in Fig. 7-38. The network in Fig. 7-38(a) is known as a π network because its configuration resembles the Greek letter π. The circuit in Fig. 7-38(b) is known as a T network because the circuit elements resemble the letter T. The circuit in Fig. 7-38(c) is also a T network, but it uses two capacitors. Note that all are low-pass filters which provide maximum harmonic attenuation. The π and T networks can be designed to either step up or step down the impedance as required by the circuit. The capacitors are usually made variable so the circuit can be tuned to resonance and adjusted for maximum power output.

The most widely used of these circuits is the T network of Fig. 7-38(c). Often called an *LCC* network, it is often used to match the low output impedance of a transistor power amplifier to the higher impedance of another amplifier or an antenna. The design procedure and formulas are given in Fig. 7-39. Suppose once again that a 6-Ω source R_i is to be matched to a 50-Ω load R_L at 155 MHz. Assume a Q of 10. (For class C operation, where many harmonics must be attenuated, it has been determined in practice that a Q of 10 is the absolute minimum needed for satisfactory suppression of the harmonics.) To configure the *LCC* network, the inductance is first calculated:

$$X_L = QR_i$$
$$X_L = 10(6) = 60 \ \Omega$$
$$L = \frac{X_L}{2\pi f} = \frac{50}{6.28 \times 155 \times 10^6} = 51.4 \text{ nH}$$

Next, C_1 is calculated:

$$X_{C_1} = 50\sqrt{\frac{6(101)}{50} - 1} = 50(3.33) = 166.73 \ \Omega$$
$$C_1 = \frac{1}{2\pi f X_C} = \frac{1}{6.28 \times 155 \times 10^6 \times 166.73} = 6.16 \times 10^{-12} = 6.16 \text{ pF}$$

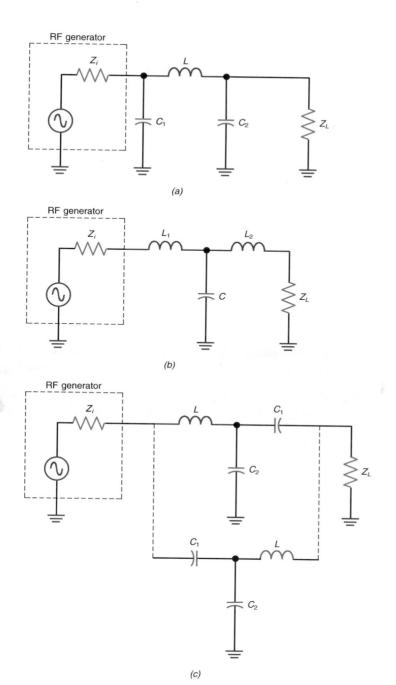

Fig. 7-38 Three-element matching networks. (a) π network. (b) T network.
(c) Two-capacitor T network.

Finally, C_2 is calculated

$$X_{C_2} = \frac{6(10^2 + 1)}{10} \; \frac{1}{1 - [166.73/(10 \times 50)]} = 60.6(1.5) = 91 \; \Omega$$

$$C_2 = \frac{1}{2\pi f X_C} = \frac{1}{6.28 \times 155 \times 10^6 \times 91} = 11.3 \; \text{pF}$$

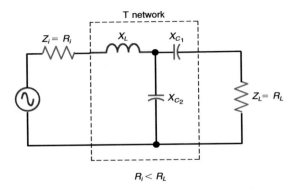

FIG. 7-39 Design equations for an *LCC* T network.

T network

$Z_i = R_i$

X_L X_{C_1}

X_{C_2}

$Z_L = R_L$

$R_i < R_L$

Design Procedure:
1. Select a desired circuit Q
2. Calculate $X_L = QR_i$
3. Calculate X_{C_1}:

$$X_{C_1} = R_i \sqrt{\frac{R_i \, (Q^2 + 1)}{R_L} - 1}$$

4. Calculate X_{C_2}:

$$X_{C_2} = \frac{R_i \, (Q^2 + 1)}{Q} \times \frac{1}{\left(1 - \dfrac{X_{C_1}}{QR_L}\right)}$$

5. Compute final *L* and *C* values:

$$L = \frac{X_L}{2\pi f}$$

$$C = \frac{1}{2\pi f X_C}$$

TRANSFORMERS AND BALUNS

One of the best impedance-matching components is the transformer. Recall that iron-core transformers are widely used at lower frequencies to match one impedance to another. Any load impedance can be made to look like a desired load impedance by simply selecting the correct value of transformer turns ratio. In addition, transformers can be connected in unique combinations called *baluns* to match impedances.

TRANSFORMER IMPEDANCE MATCHING. Refer to Fig. 7-40. The relationship between the turns ratio and the input and output impedances is

$$\frac{Z_i}{Z_L} = \left(\frac{N_P}{N_S}\right)^2 \qquad \frac{N_P}{N_S} = \sqrt{\frac{Z_i}{Z_L}}$$

That is, the ratio of the input impedance Z_i to the load impedance Z_L is equal to the square of the ratio of the number of turns on the primary (N_P) to the number of turns on the secondary (N_S). For example, to match a generator impedance of 6 Ω to a 50-Ω load impedance, the turns ratio should be as shown on the next page.

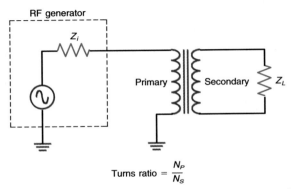

Turns ratio $= \dfrac{N_P}{N_S}$

FIG. 7-40 Impedance matching with an iron-core transformer.

$$\frac{N_P}{N_S} = \sqrt{\frac{Z_i}{Z_L}} = \sqrt{\frac{6}{50}} = \sqrt{0.12} = 0.3464$$

$$\frac{N_S}{N_P} = \sqrt{\frac{1}{N_P/N_S}} = \sqrt{8.33} = 2.887$$

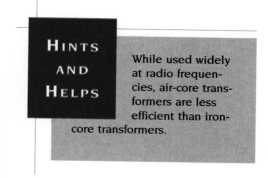

HINTS AND HELPS

While used widely at radio frequencies, air-core transformers are less efficient than iron-core transformers.

This means that there are 2.89 times as many turns on the secondary as on the primary.

The relationship given above holds true only on iron-core transformers. When air-core transformers are used, the coupling between primary and secondary windings is not complete and therefore the impedance ratio is not as indicated. While air-core transformers are widely used at RF frequencies and can be used for impedance matching, they are less efficient than iron-core transformers.

Ferrite (magnetic ceramic) and powdered iron can be used as core materials to provide close coupling at very high frequencies. Both the primary and secondary windings are wound on a core of the chosen material.

The most widely used type of core for RF transformers is the toroid. A *toroid* is a circular, doughnut-shaped core, usually made of a special type of powdered iron. Copper wire is wound on the toroid to create the primary and secondary windings. A typical arrangement is shown in Fig. 7-41. Single-winding tapped coils called *autotransformers* are also used for impedance matching between RF stages. Figure 7-42 shows both impedance step-down and step-up arrangements. Toroids are commonly used in autotransformers.

Unlike air-core transformers, toroid transformers cause the magnetic field produced by the primary to be completely contained within the core itself. This has two important advantages. First, a toroid does not radiate RF energy. Air-core coils radiate because the magnetic field produced around the primary is not contained. Transmitter and receiver circuits using air-core coils are usually contained with magnetic shields to prevent them from interfering with other circuits. The toroid, on the other hand, confines all the magnetic fields and does not require shields. Second, most of the magnetic field produced by the primary cuts the turns of the secondary winding. Thus the basic turns ratio, input-output voltage, and impedance formulas for standard low-frequency transformers apply to high-frequency toroid transformers.

In most new RF designs, toroid transformers are used for RF impedance matching between stages. Further, the primary and secondary windings are sometimes used as inductors in tuned circuits. Alternatively, toroid inductors can be built. Powdered-

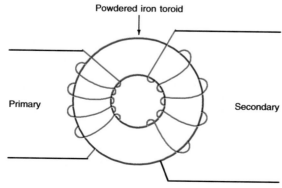

Fig. 7-41 A toroid transformer.

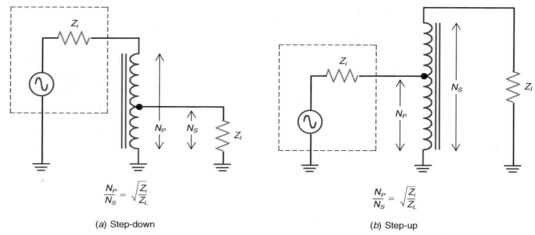

$$\frac{N_P}{N_S} = \sqrt{\frac{Z_i}{Z_L}}$$

(a) Step-down

$$\frac{N_P}{N_S} = \sqrt{\frac{Z_i}{Z_L}}$$

(b) Step-up

FIG. 7-42 Impedance matching in an autotransformer. (*a*) Step down. (*b*) Step up.

iron-core toroid inductors have an advantage over air-core inductors for RF applications because the high permeability of the core causes the inductance to be high. Recall that whenever an iron core is inserted into a coil of wire, the inductance increases dramatically. For RF applications, this means that the desired values of inductance can be created by using fewer turns of wire and thus the inductor itself can be smaller. Further, fewer turns have less resistance, giving the coil a higher Q than that obtainable with air-core coils.

Powdered iron toroids are so effective that they have virtually replaced air-core coils in most modern transmitter designs. They are available in sizes from a fraction of an inch to several inches in diameter. In most applications, a minimum number of turns are required to create the desired inductance.

Figure 7-43 shows a toroid transformer T_1 used for interstage coupling between two class C driver amplifiers. The primary of transformer T_1 is tuned to resonance by capacitor C_1. The capacitor is adjustable, so the exact frequency of operation can be set. The relatively high output impedance of the transistor is coupled with the low input impedance of the next class C stage by a step-down transformer which provides the desired impedance-matching effects. Usually, the secondary winding is only a few turns of wire and is not resonated. The circuit in Fig. 7-43 also shows a similar transformer T_2 used for output coupling to the antenna.

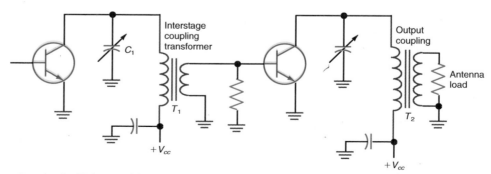

FIG. 7-43 Using toroid transformers for coupling and impedance matching in class C amplifier stages.

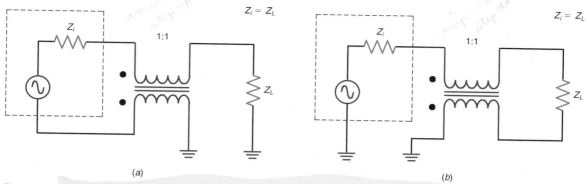

FIG. 7-44 Balun transformers used for connecting balanced and unbalanced loads or genera-tors. (*a*) Balanced to unbalanced. (*b*) Unbalanced to balanced.

BALUNS. Transformers can be connected in unique ways to provide fixed imped-ance-matching characteristics over a wide range of frequencies. One of the most widely used configurations is shown in Fig. 7-44. With this configuration, a transformer is usu-ally wound on a toroid, and the number of primary and secondary turns are equal, giv-ing the transformer a 1:1 turns ratio and a 1:1 impedance-matching ratio. The dots in-dicate the phasing of the windings. Note the unusual way in which the windings are connected. A transformer connected in this way is generally known as a *balun* (from *bal*anced-*un*balanced) because such transformers are normally used to connect a bal-anced source to an unbalanced load or vice versa. In the circuit of Fig. 7-44(*a*), a bal-anced generator is connected to an unbalanced (grounded) load. In Fig. 7-44(*b*), an un-balanced (grounded) generator is connected to a balanced load.

Figure 7-45 shows two ways in which a 1:1 turns ratio balun can be used for im-pedance matching. With the arrangement shown in Fig. 7-44(*a*), an impedance step up is obtained. A load impedance of four times the source impedance Z_i provides a cor-rect match. The balun makes the load of $4Z_i$ look like Z_i. In Fig. 7-45(*b*), an imped-ance step down is obtained. The balun makes the load Z_L look like $Z_i/4$.

Many other balun configurations, offering different impedance ratios, are possi-ble. Several common 1:1 baluns can be interconnected for both 9:1 and 16:1 imped-

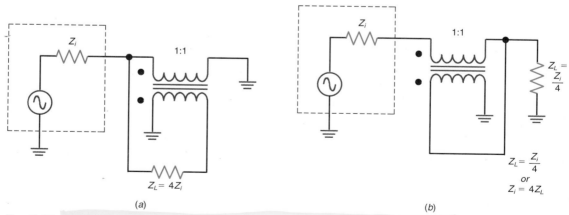

FIG. 7-45 Using a balun for impedance matching. (*a*) Impedance step up. (*b*) Impedance step down.

ance transformation ratios. In addition, baluns can be cascaded so that the output of one appears as the input to the other, and so on. Cascading baluns allow impedances to be stepped up or stepped down by wider ratios.

Note that the windings in a balun are not resonated with capacitors to a particular frequency. The winding inductances are made such that the coil reactances are four or more times that of the highest impedance being matched. This design allows the transformer to provide the designated impedance matching over a tremendous range of frequencies. This broadband characteristic of balun transformers allows designers to construct broadband RF power amplifiers. Such amplifiers provide a specific amount of power amplification over a wide bandwidth and are thus particularly useful in communications equipment that must operate in more than one frequency range. Rather than have a separate transmitter for each desired band, a single transmitter with no tuning circuits can be used.

When using conventional tuned amplifiers, some method of switching the correct tuned circuit into the circuit must be provided. Such switching networks are complex and expensive. Further, they introduce problems, particularly at high frequencies. In order for them to perform effectively, the switches must be located very close to the tuned circuits so that stray inductances and capacitances are not introduced by the switch and the interconnecting leads. One way to overcome the switching problem is to use a broadband amplifier, which does not require switching or tuning. The broadband amplifier provides the necessary amplification as well as impedance matching. However, broadband amplifiers do not provide the filtering necessary to get rid of harmonics. One way to overcome this problem is to generate the desired frequency at a lower power level, allowing tuned circuits to filter out the harmonics, and then provide final power amplification with the broadband circuit. The broadband power amplifier operates as a linear class A or class B push-pull circuit so that the inherent harmonic content of the output is very low.

Figure 7-46 shows a typical broadband linear amplifier. Note that two 4:1 balun transformers are cascaded at the input so that the low base input impedance is made

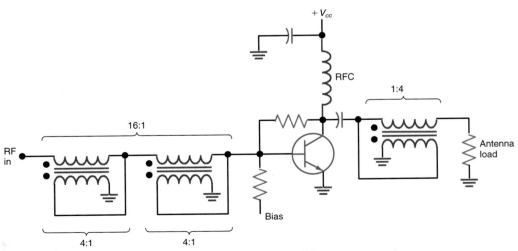

FIG. 7-46 A broadband class A linear power amplifier.

to look like an impedance 16 times higher than it is. The output uses a 1:4 balun that steps up the very low output impedance of the final amplifier to an impedance four times higher to equal the antenna load impedance. In some transmitters, broadband amplifiers are followed by low-pass filters, which are used to eliminate undesirable harmonics in the output.

7-5 SPEECH PROCESSING

Speech processing refers to the ways that the voice signal used in communications is modified before being applied to the modulator. The voice signal from the microphone is usually amplified and applied to circuits that deliberately limit its amplitude and frequency response. The primary purpose of speech-processing circuits is to ensure that overmodulation does not occur and to restrict the bandwidth of the signal. In amplitude modulation, an excessively high voice signal causes overmodulation and severe distortion of the signal. If the amplitude is deliberately limited, overmodulation does not occur.

In an AM transmitter, the distortion caused by overmodulation causes harmonics of the primary frequencies in the voice signal to be generated. These harmonics also modulate the carrier and produce sidebands that extend far beyond the assigned bandwidth of the AM signal, and can interfere with signals on adjacent channels. This interference is generally known as *splatter*. Distortion caused by overmodulation also causes the signal to be generally less intelligible.

Most transmitters are designed to operate in a specific frequency range. This means that the sidebands may not extend beyond certain assigned limits. In AM transmitters, the highest modulating frequency determines the total bandwidth of the AM signal. For example, in AM broadcasting, the channel width is limited to 10 kHz. For this reason, the upper audio frequency limit must be limited to one-half of this, or 5 kHz. Speech-processing circuits are designed to both limit the upper amplitude and frequency allowed.

Speech processing in AM and SSB transmitters is also used to help maintain the average transmitted power at a high level. Recall that when an AM transmitter is modulated, the power from the modulator which appears in the sidebands increases with the percentage of modulation. The more power transmitted in the sidebands, the greater the transmission distance and the more reliable the communications; that is, the strength of the RF signal is determined by the strength of the modulating signal. The same is true of SSB signals: the strength of the modulating signal determines the strength of the transmitted sideband.

Voice signals vary over a wide amplitude range, with the highest peaks of intensity, and the most potential unintelligibility, coming from vowel sounds. The consonants, such as B, K, L, S, T, and V, provide the best detail for good intelligibility. Yet consonant sounds are typically spoken at a much lower amplitude than vowel sounds. Deemphasizing vowel sounds and at the same time emphasizing consonant sounds, through increased amplification, both improves intelligibility and increases the average intensity and power of a voice signal. Special speech-processing circuits have been developed to provide for increased amplification of low-level audio content while at the same time limiting the peaks. This process is generally known as *dynamic compression*.

In FM, it is also necessary to deliberately limit the bandwidth and amplitude of the modulating signal. Recall that the frequency deviation of an FM carrier is directly proportional to the amplitude of the modulating signal. Most FM transmitters are allowed to operate up to a certain maximum deviation limit. In mobile communications transmitters, this frequency deviation is ±5 kHz. In order to ensure that the deviation limit is not exceeded, some means must be provided to prevent the modulating signal amplitude from causing overdeviation. This is comparable to overmodulation in AM transmitters. The frequency content of the modulating signal must also be limited. Recall that an FM signal produces multiple sidebands above and below the carrier. These are spaced from the carrier by the frequency of the modulating signal. The higher the modulating signal frequency, the wider the spacing of the sidebands and the greater the spectrum space occupied by the FM signal. As in most applications, the bandwidth allowed an FM signal is specified. For this reason, the upper frequency limit of the modulating signal is usually limited to prevent the bandwidth from becoming too wide.

CLIPPING AND FILTERING

The basic speech-processing circuit shown in Fig. 7-47 is an amplifier operating in conjunction with a clipper circuit that limits the amplitude swings of the modulating signal and a filter that cuts off modulating frequencies above a certain point. The low-level signal from the microphone is amplified by two stages of audio amplification to raise the signal to a high enough amplitude to make it compatible with the modulator. The output of the amplifier is applied to a diode clipper. Diodes D_1 and D_2 are low-level silicon diodes that conduct when approximately +0.7 V is applied across them. Notice the diodes are connected in an inverse parallel arrangement; that is, one diode is connected to conduct on positive-going AC signals and the other is connected to conduct on negative-going AC signals.

As long as the peak-to-peak amplitude of the voice signal is below approximately 1.4 V, no clipping occurs. If voice peaks cause the signal to be greater, diode D_1 or D_2 conducts, clipping off the peak and holding the amplitude to ±0.7 V. Thus no matter what the amplifier output is, the signal across the diodes will never be greater than 1.4 V peak-to-peak. This deliberate amplitude limitation means that an AM transmitter will not be overmodulated and an FM transmitter will not be overdeviated.

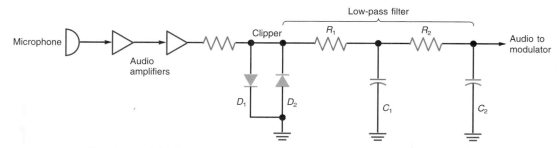

FIG. 7-47 A basic speech-processing circuit incorporating clipping and filtering.

Clipping, however, does cause signal distortion. The deliberate flattening of the signal peaks introduces harmonics. One way to get rid of these harmonics is to pass the clipped signal through a low-pass filter, as shown in Fig. 7-47, where the filter is made up of resistors R_1 and R_2 and capacitors C_1 and C_2. This filter gets rid of the harmonics and thus compensates for the clipping distortion. The other purpose of this type of low-pass filter is to restrict the bandwidth of the modulating signal. Typically, the cutoff frequency of the filter is chosen so that it is somewhere in the 2.5- to 3-kHz range, which is the upper limit for most two-way communications systems.

SPEECH COMPRESSION

Speech-processing circuits are almost always incorporated in FM transmitters and in many AM transmitters as well. However, in AM and SSB transmitters, speech compression is generally preferred over clipping. Speech compression permits the average modulating signal to be higher, thus increasing the average power output. Figure 7-48 shows a typical speech-compression circuit. It consists of two stages of amplification whose gain is controlled automatically by the amplitude of the audio signal. Some of the audio signal is tapped off and applied to a diode rectifier D_1. The audio signal is rectified and filtered by C_2 to produce a DC output, and is amplified by a DC amplifier. The average DC output is proportional to the amplitude of the modulating signal. This DC voltage is fed back to an earlier amplifier stage to control its gain. The gain of a transistor amplifier stage can be controlled by adjusting collector current. This is done by modifying the base bias current. The higher the collector current, the higher the gain. If the amplitude of the modulating signal is high, the DC output from the rectifier will be high and it, in turn, will reduce the gain of the amplifier, thereby limiting the peaks and preventing overmodulation and splatter.

If the amplitude of the modulating signal is very low, the rectifier will produce a very low average DC output. This causes the amplifier gain to be very high, providing extra amplification for low-level signals. This compression circuit is said to have *automatic gain control (AGC)* simply because the gain of the circuit automatically adjusts itself to a level appropriate to the amplitude of the incoming signal. High-amplitude signals are deliberately limited, and low-amplitude signals are provided with extra amplification.

Clipping and compression can be performed in the RF stages of a transmitter rather than in the modulating-signal stages. One common technique using RF compression

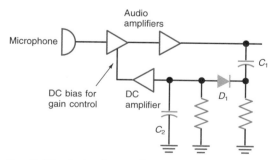

FIG. **7-48**   A simple speech-compression circuit.

for this purpose, which is usually done in SSB transmitters, is referred to as *automatic level control (ALC)*. The technique for RF compression is similar to that used in audio compression. A rectifier circuit is used to tap off a sample of the RF signal at the final power amplifier stage. The RF signal is rectified and converted into a DC voltage that is fed back to control the gain of an earlier stage used to amplify the SSB signal. Again, if the output level exceeds a certain limit, the fed-back DC voltage reduces the gain of the stage and thus automatically controls the amplitude. Low-level signals receive full amplification.

7-6 TYPICAL TRANSMITTER CIRCUIT

Many transmitters used in recent equipment designs are a combination of ICs and discrete component circuits. The transmitter shown in Fig. 7-49 incorporates the most up-to-date techniques. This low-power FM transmitter, which is designed to operate in the 30-MHz range, has an input power of about 3 W and a frequency deviation of 5 kHz for narrowband operation. The circuit is made up of a Motorola MC2833 single-chip FM transmitter IC, a digital shaping circuit, and a pair of power MOSFETs connected in parallel as a class E amplifier. An IC regulator provides a constant DC supply voltage from a battery pack.

The heart of the circuit is the transmitter chip. A more detailed look at this chip is given in Fig. 7-50. It is housed in a standard 16-pin DIP, and contains a microphone amplifier with clipping diodes; an RF oscillator, which is usually crystal-controlled with an external crystal; and a buffer amplifier. Frequency modulation is produced by a variable reactance circuit connected to the oscillator. Also on the chip are two free transistors that can be connected with external components as buffer amplifiers or as multipliers and low-level power amplifiers. This chip is useful up to about 60 to 70 MHz, and at higher ranges if external multipliers are used. The chip is widely used in cordless telephones, which operate in the 46- to 49-MHz band with FM.

As shown in Fig. 7-49, the signal starts with a microphone whose signal is sent to the audio amplifier in the IC at pin 5 through C39. The gain of the amplifier is set with resistor R_{11} on pins 4 and 5. The output of the amplifier is connected to the reactance modulator via C_{38} on pin 3. This circuit connects to the oscillator whose frequency is set by the external crystal between pins 1 and 16. Assume a 10-MHz crystal. The reactance modulator pulls the crystal frequency by a small amount during modulation to produce a frequency variation.

The output of the oscillator is buffered and amplified and appears at pin 14 on the IC. The buffer amplifier has a resonant circuit (L_1 and C_8) in its output tuned to the third harmonic of the crystal or 30 MHz. In addition to multiplying the carrier frequency, the multiplier also multiplies the frequency deviation by a factor of 3 to achieve the desired

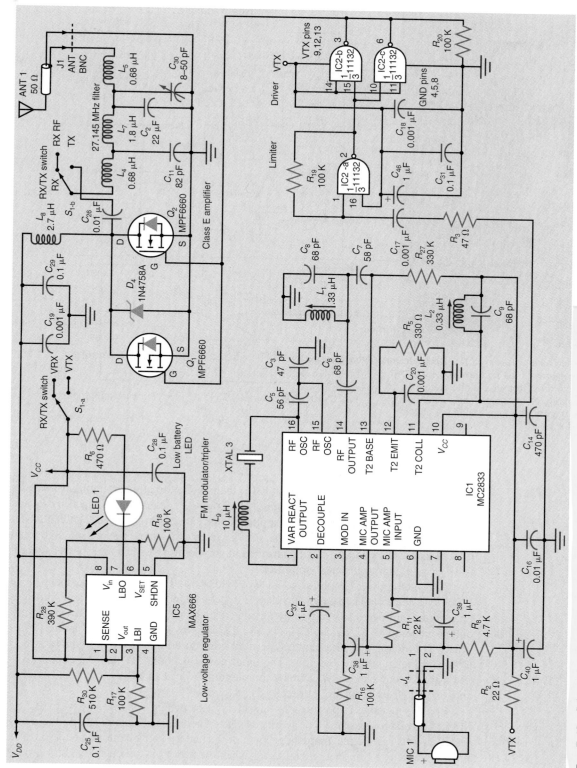

Fig. 7-49 Schematic showing the modulator and tripler, class E amplifier, limiter, driver, and low-voltage regulator sections of the E-Comm transceiver. The key device is IC₁, the FM transmitter chip. (*Courtesy Electronics Now, October 1992.*)

President Franklin Delano Roosevelt (FDR) reached out to the country through his famous "fireside chats." These radio talks from the White House's Diplomatic Reception Room helped ease a nation's concerns during the difficult Depression era and, later, about World War II.

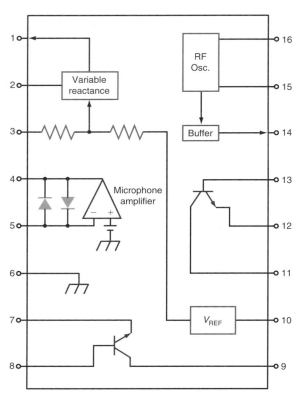

FIG. 7-50 Motorola MC 2833 IC FM VHF transmitter chip.

5-kHz deviation. The resulting FM signal is then applied to a linear amplifier made up of one of the transistors in the IC (pins 11, 12, and 13). Its output is tuned by L_2 and C_9. Next, the FM sine-wave signal is applied to one gate of a high-speed CMOS NAND-gate Schmitt trigger. IC_{2-a} shapes the sine wave into a square wave. Two additional CMOS gates, IC_{2-b} and IC_{2-c} are connected in parallel to provide a high-power square-wave drive signal to the final amplifier.

The final power amplifier uses two MPF6660 RF power enhancement-mode MOSFETs connected in parallel as a class E switching amplifier. Zener diode D_4 provides protection from overloads caused by mismatched antenna impedances. The π-network output made up of L_4, L_7, L_5, C_{11}, C_2, and C_{30} couples the signal to the antenna, providing impedance matching and low-pass filtering to eliminate the harmonics associated with the square-wave output. The input power is about 3W and the circuit gives about 90 percent efficiency, meaning the output power is about $0.9 \times 3 = 2.7$ W. The antenna is a "rubber ducky" vertical.

Finally, the unit is powered by a battery pack which supplies 9.6 V to IC5, a MAX666 voltage regulator. The regulator gives a constant voltage of 8 V to the transmitter circuits despite the gradual dropping of the battery voltage during operation. The IC also contains a component that senses when the battery is too discharged for proper operation and turns on the LED.

This transmitter is part of a handheld unit with a matching receiver. Receiver circuits are covered in Chap 8.

SUMMARY

The starting point for all transmitters is carrier generation, which is almost always accomplished by crystal oscillators. To achieve the frequencies necessary in the VHF, UHF, and even microwave ranges, transmitters use frequency multiplier circuits.

Frequency synthesizers are variable-frequency generators using a phase-locked loop that provide the frequency stability of a crystal oscillator and the convenience of incremental tuning over a broad frequency range. The output of a frequency synthesizer is varied by using switching techniques.

The three basic types of power amplifiers used in transmitters are linear (class A, AB, or B), class C, and switching (class D, class E, and class S). The class of an amplifier is determined by how it is biased. Class A amplifiers are biased so that they conduct continuously, class B amplifiers are biased at cutoff so no collector current flows with zero input, and class C amplifiers conduct for less than one half of the sine-wave input cycle.

Both linear and class C amplifiers face the problem of self-oscillation, which is the result of positive feedback that occurs because of the interelement capacitance of the amplifying device. The process of compensating for this effect is called neutralization.

The circuits used to connect one transmitter stage to another, called impedance-matching networks, provide for an optimum transfer of power as well as filtering and selectivity functions. The basic types of LC impedance-matching networks are the L network, the T network, and the π network.

Speech-processing circuits are used to prevent overmodulation, restrict signal bandwidth, and maximize average transmitted power. The basic circuit is an amplifier operating in conjunction with a clipper circuit that limits the amplitude swings of the modulating signal and a filter that cuts off modulating frequencies above a certain point. Clipping causes signal distortion, but this distortion can be eliminated by passing the clipped signal through a low-pass filter.

REVIEW

KEY TERMS

Automatic gain control	Class C amplifier	Field-effect transistor
Automatic level control	Class D amplifier	Filtering
Autotransformer	Class E amplifier	Final power amplifier
Balun	Class S amplifier	Frequency divider
Cathode	Clipping	Frequency multiplier
Class A amplifier	Colpitts oscillator	Frequency synthesizer
Class AB amplifier	Crystal	Grid
Class B amplifier	Crystal oscillator	Harmonic

Impedance matching	π network	Switching amplifier
L network	Pierce oscillator	T network
Linear amplifier	Piezoelectric effect	Tetrode
Impedance matching	Plate	Toroid
L network	Prescaler	Transformer
Linear amplifier	Push-pull amplifier	Transmitter
MOSFET	QRP operation	Triode
Neutralization	Self-bias method	Tuned output circuit
Oscillator	Self-oscillation	Vacuum tube
Overtone crystal	Single-sideband transmitter	Variable frequency
Pentode	Speech compression	oscillator
Phase-locked loop	Speech processing	

QUESTIONS

1. What circuits are typically part of every radio transmitter?
2. Which type of transmitter does not use class C amplifiers?
3. For how many degrees of an input sine wave does a class B amplifier conduct?
4. What is the name given to the bias for a class C amplifier produced by an input RC network?
5. Why are crystal oscillators used instead of LC oscillators to set transmitter frequency?
6. Name two ways to select crystals with switches. Which is preferred at the higher frequencies?
7. How is the output frequency of a frequency synthesizer changed?
8. What are prescalers and why are they used in VHF and UHF synthesizers?
9. What is the purpose of the loop filter in a PLL?
10. What is the most efficient class of RF power amplifier?
11. What is the approximate maximum power of typical transistor RF power amplifiers?
12. What are parasitics and how are they eliminated in a power amplifier?
13. What is the main reason that switching amplifiers are used?
14. What is the difference between a class D and a class E amplifier?
15. What is a major disadvantage of a switching power amplifier?
16. Why are vacuum tubes still used in RF power amplifiers?
17. What are the advantages and disadvantages of a grounded grid amplifier?
18. Maximum power transfer occurs when what relationship exists between generator impedance Z_i and load impedance Z_L?
19. What is a toroid and how is it used? What components are made from it?
20. What are the advantages of a toroid RF inductor?
21. In addition to impedance matching, what other important function do LC networks perform?
22. What is the name given to a single winding transformer?
23. What is the name given to an RF transformer with a 1:1 turns ratio connected so that it provides a 1:4 or 4:1 impedance matching? Give a common application.
24. Why are untuned RF transformers used in power amplifiers?
25. How is impedance matching handled in a broadband linear RF amplifier?
26. Name the two main techniques used in speech processing.
27. What are the two primary purposes of speech processing?
28. What circuit is frequently used to restrict voice signal amplitude in a speech processor?
29. What are the two main purposes of a low-pass filter in a speech-processing circuit?
30. What is the term used to refer to overmodulation in an AM transmitter that causes distortion and out-of-band operation?

31. What characteristic of an audio amplifier must be controlled to achieve audio compression?
32. How does audio compression improve an AM or SSB signal?
33. What is the name given to the process of adjusting the amplification of a signal based on its amplitude?
34. True or false. Speech processing can be done on the RF signal or the audio signal.

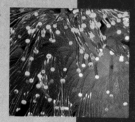

PROBLEMS

1. An FM transmitter has an 8.6-MHz crystal carrier oscillator and frequency multipliers of 2, 3, and 4. What is the output frequency?
2. A crystal has a tolerance of 0.003 percent. What is the tolerance in ppm?
3. A 25-MHz crystal has a tolerance of $\pm$ 200 ppm. If the frequency drifts upward to the maximum tolerance, what is the frequency of the crystal?
4. A frequency synthesizer has a reference frequency of 25 kHz. The frequency divider is set to a factor of 345. What is the output frequency?
5. A frequency synthesizer has an output frequency of 162.7 MHz. The reference is a 1-MHz crystal oscillator followed by a divider of 10. What is the main frequency divider ratio?
6. A frequency synthesizer has an output frequency of 470 MHz. A divide-by-10 prescaler is used. The reference frequency is 10 kHz. What is the frequency step increment?
7. A frequency synthesizer has a variable-modulus prescaler of $M = 10/11$ and divide ratios of the A and N counters of 40 and 260. The reference frequency is 50 kHz. What are the VCO output frequency and the minimum frequency step increment?
8. A class C vacuum tube final amplifier has a plate voltage of 2750 V and a plate current of 660 mA. What is the power input?
9. A class C amplifier has a supply voltage of 36 V and a collector current of 2.5 A. Its efficiency is 80 percent. What is the RF output power?
10. Calculate the L and C values of an L network that is to match a 9-Ω transistor power amplifier to a 75-Ω antenna at 122 MHz.
11. Calculate what L-network components will match a 4-Ω internal resistance in series with an internal inductance of 9 nH to a 72-Ω load impedance in parallel with a stray capacitance of 24 pF at a frequency of 46 MHz.
12. Design an LCC T network that will match 5-Ω internal resistance to a 52-Ω load at 54 MHz. Assume a Q of 12.
13. A transformer has 6 turns on the primary and 18 turns on the secondary. If the generator (source) impedance is 50 Ω, what should the load impedance be?
14. A transformer must match a 2500-Ω generator to a 50-Ω load. What must the turns ratio be?

CRITICAL THINKING

1. Name the five main parts of a frequency synthesizer. Draw the circuit from memory. From which circuit is the output taken?
2. Design an L-network like that in Fig. 7-37 to match a 5.5 Ω transistor amplifier with an internal inductance of 7nH to an antenna with an impedance of 50 Ω and a shunt capacitance of 20 pF. Assume a frequency of 112 MHz.
3. To match a 6-Ω amplifier impedance to a 72-Ω antenna load, what turns ratio (N_P/N_S) must a transformer have?

COMMUNICATIONS RECEIVERS

Objectives

After completing this chapter, you will be able to:

◆ *List* the benefits of a superheterodyne over a TRF receiver and *identify* the function of each component of a superheterodyne, including all selectivity functions.

◆ *Express* the relationship between the IF, local oscillator, and signal frequencies mathematically and calculate any one of them given the other two.

◆ *Explain* how the design of dual-conversion receivers allows them to enhance selectivity and eliminate image problems.

◆ *Describe* the operation of the most common types of mixer circuits.

◆ *List* the major types of external and internal noise and *explain* how each interferes with signals both before and after they reach the receiver.

◆ *Calculate* the noise factor, noise figure, and noise temperature of a receiver.

◆ *Describe* the operation and purpose of the AGC and AVC circuits in a receiver.

◆ *Explain* the operation of squelch circuits.

In radio communications systems, the transmitted signal is very weak when it reaches the receiver, particularly when it has traveled over a long distance. The signal, which has shared the free-space transmission media with thousands of other radio signals, has also picked up noise of various kinds. Radio receivers must provide the sensitivity and selectivity that permit full recovery of the original intelligence signal. The radio receiver best suited to this task is known as the *superheterodyne receiver.* Invented in the early 1900s, the superheterodyne is used today in most electronic communications systems. This chapter reviews some basic principles of signal reception and discusses various superheterodyne circuits in detail.

8-1 BASIC PRINCIPLES OF SIGNAL REPRODUCTION

As discussed in earlier chapters a communications receiver must be able to identify and select a desired signal from the thousands of others present in the frequency spectrum (selectivity) and to provide sufficient amplification to recover the modulating signal (sensitivity). A receiver with good selectivity will isolate the desired signal in the radio frequency spectrum and eliminate or at least greatly attenuate all other signals. A receiver with good sensitivity involves high circuit gain.

SELECTIVITY

Selectivity in a receiver is obtained by using tuned circuits and/or filters. *LC* tuned circuits provide initial selectivity; filters, which are used later in the process, provide additional selectivity.

***Q* AND BANDWIDTH.** *LC* circuits are tuned to resonate at the signal frequency. The ratio of inductive reactance to resistance, *Q*, determines the selectivity of a circuit. The higher the *Q*, the narrower the bandwidth and the better the selectivity. High-*Q* tuned circuits are used to keep the bandwidth narrow to ensure that only the desired signal is passed.

The bandwidth of a tuned circuit is the difference between the upper (f_2) and lower (f_1) cutoff frequencies, which are located at the 3-dB down or 0.707 points on the selectivity curve, as shown in Fig. 8-1. This bandwidth is determined by the resonant frequency f_r and Q according to the relationship BW = f_r/Q. Assume, for example, that a 6.5-μH coil with a resistance of 18 Ω is connected in parallel with a 47-pF capacitor. The circuit resonates at

$$f_r = \frac{1}{2\pi\sqrt{LC}} = \frac{1}{6.28\sqrt{(6.5 \times 10^{-6} \times 47 \times 10^{-12})}} = 9.11 \text{ MHz}$$

The response curve is as shown in Fig. 8-1, and the peak occurs at 9.11 MHz.

The *Q* of the coil and circuit is: $Q = X_L/R$. The inductive reactance is $X_L = 2\pi fL$. $X_L = 6.28 \times 9.11 \times 10^6 \times 6.5 \times 10^{-6} = 372 \ \Omega$. Since the coil resistance is 18 Ω, $Q = X_L/R = 372/18 = 20.67$. The bandwidth is

$$\text{BW} = \frac{f_r}{Q} \ \frac{9.11 \times 10^6}{20.67} = 440{,}735 \text{ Hz or } 440.735 \text{ kHz}$$

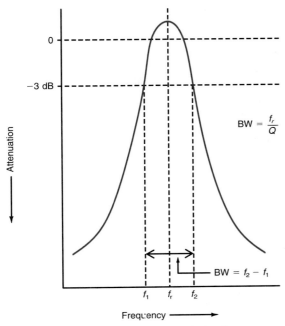

Fig. 8-1 Selectivity curve of a tuned circuit.

This bandwidth is geometrically centered around 9.11 MHz. The upper and lower cutoff frequencies are 9.11 MHz plus or minus approximately half of the bandwidth. Rounding off 440.735 kHz to 441 kHz, we have 441 kHz/2 = 220.5 kHz or 0.22 MHz (rounded).

The upper cutoff frequency is approximately f_2 = 9.11 + 0.22 = 9.33 MHz. The lower cutoff frequency is approximately f_1 = 9.11 − 0.22 = 8.89 MHz. The bandwidth can be verified by computing it with the cutoff frequencies:

$$\text{BW} = (f_2 - f_1) = 9.33 - 8.89 = 0.44 \text{ MHz or } 440 \text{ kHz}$$

Using the approximate cutoff frequencies calculated above, the center frequency f_c is computed as

$$f_c = \sqrt{f_1 f_2} = \sqrt{(9.33)(8.89)} = 9.1073 \text{ MHz} = 9.11 \text{ MHz}$$

The selectivity of a circuit can be improved by narrowing the bandwidth. For example, the required Q for a bandwidth of 8 kHz is $Q = f_r/\text{BW} = 9.11 \times 10^6/8 \times 10^3 = 1139$. This is a very high Q. To increase Q, the coil resistance must be lowered. One way to do this is to use much larger gauge wire. Alternatively, a higher inductance value can be used, but then the capacitor value must also be changed.

Very high Qs are difficult to obtain with a single LC tuned circuit. Larger inductances with lower Qs are a possibility, but there is a limit, since loading an electronic circuit also lowers the Q. Another problem is that the selectivity curve of a tuned circuit has a gradual slope above and below the center frequency. The selectivity is not sharp, that is, the attenuation rate is not steep, and undesired signals above and below the center frequency are passed through,

> ### DID YOU KNOW?
>
> Very high Qs are difficult to obtain using a single LC-tuned circuit.

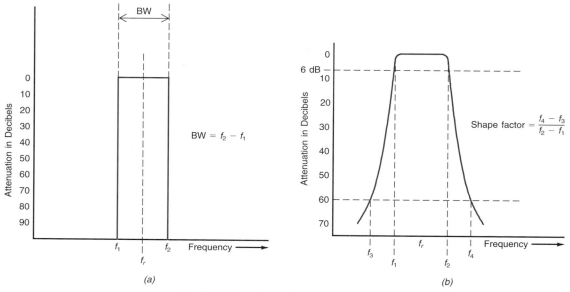

FIG. 8-2 Receiver selectivity response curves. (*a*) Ideal response curve. (*b*) Practical response curve showing shape factor.

although attenuated. The ideal receiver selectivity curve would have perfectly vertical sides, as in Fig. 8-2(*a*). Such a curve cannot be obtained with tuned circuits or any other electronic circuit. Improved selectivity is achieved by cascading tuned circuits or by using crystal, ceramic, or SAW filters. At lower frequencies, digital signal processing (DSP) can provide near ideal response curves. All of these methods are used in communications receivers.

SHAPE FACTOR.   The sides of a tuned circuit response curve are known as *skirts*. The steepness of the skirts, or the skirt selectivity, of a receiver, is expressed as the *shape factor,* the ratio of the 60-dB down bandwidth to the 6-dB down bandwidth. This is illustrated in Fig. 8-2(*b*). The bandwidth at the 60-dB down points is $f_4 - f_3$; the bandwidth of the 6-dB down points is $f_2 - f_1$. Thus the shape factor is $(f_4 - f_3)/(f_2 - f_1)$. Assume, for example, that the 60-dB bandwidth is 8 kHz and the 6-dB bandwidth is 3 kHz. The shape factor is $8/3 = 2.67$, or 2.67:1.

The lower the shape factor, the steeper the skirts and the better the selectivity. The ideal, shown in Fig. 8-2(*a*), is 1; this cannot be obtained in practice.

SENSITIVITY

A communications receiver's sensitivity, or ability to pick up weak signals, is mainly a function of overall gain, the factor by which an input signal is multiplied to produce the output signal. In general, the higher the gain of a receiver, the better its sensitivity. The more gain that a receiver has, the smaller the input signal necessary to produce a desired level of output. High gain in communications receivers is obtained by using multiple amplification stages.

The sensitivity of a communications receiver is usually expressed as the minimum amount of signal voltage input that will produce an output signal that is 10 dB higher than the receiver background noise. Some specifications state a 20-dB S/N ratio. A typical sensitivity figure might be one microvolt input. The lower this figure, the better the sensitivity. Good communications receivers typically have a sensitivity of 0.2 to 1 μV. Consumer AM and FM receivers designed for receiving strong local stations have much lower sensitivity. Typical FM receivers have sensitivities of 5 to 10 μV; AM receivers can have sensitivities of 100 μV or higher.

THE BASIC RECEIVER CONFIGURATION

Figure 8-3(a) shows the simplest radio receiver: a crystal set consisting of a tuned circuit, a diode (crystal) detector, and earphones. The antenna picks up the signal and causes current to flow in the primary winding of coupling transformer T_1. This induces a voltage into the secondary winding which is the inductance in a series resonant circuit. When signal current flows in the primary, a varying magnetic field cuts the turns of the secondary winding, inducing a voltage in them. Each turn of the secondary, therefore, acts like a tiny voltage generator. The total effect is the same as having a signal generator in series with the secondary [see Fig. 8-3(b)]. Generator, secondary winding, and C_1 together form a series resonant circuit. Current flows in the secondary winding, causing a voltage to be developed across the capacitor C_1. When the circuit is tuned to resonance, a high voltage, known as the *resonant rise* or *resonant step-up voltage,* is developed across the capacitor. This voltage, which is significantly higher than the actual voltage induced into the secondary winding, provides gain. The diode or crystal rectifies the signal; the capacitor C_2 filters out the carrier, leaving the original signal, which is heard in the earphones.

The crystal receiver in Fig. 8-3 does not provide the kind of selectivity and sensitivity necessary for modern communications. Only the strongest signals can produce an output, and selectivity is often insufficient to separate incoming signals. The use of earphones is also inconvenient. However, in this configuration, as in more advanced designs, a demodulator is the basic component: it strips the RF carrier from the incoming signal, leaving only the original modulating information. All other circuits in a receiver are designed to improve sensitivity and selectivity, so the demodulator can perform better.

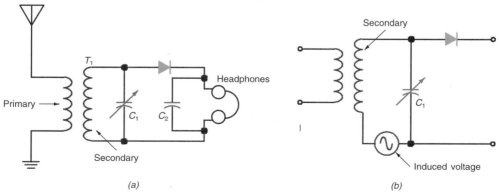

FIG. 8-3 (a) The simplest receiver—a crystal set. (b) The series resonant circuit formed by the secondary.

In the *tuned radio frequency (TRF)* receiver shown in Fig. 8-4 sensitivity has been improved by adding three stages of RF amplification between the antenna and the detector, followed by two stages of audio amplification. The RF amplifier stages increase the gain of the received signal tremendously before it is applied to the detector. The recovered signal is amplified further by the audio amplifiers, which provide sufficient gain to operate a loudspeaker.

Another design improvement is that the RF amplifiers use tuned circuits. Whenever resonant *LC* circuits tuned to the same frequency are cascaded, overall selectivity is improved. The greater the number of tuned stages cascaded, the narrower the bandwidth and the steeper the skirts, as shown in Fig. 8-5. A signal above or below the resonant frequency is attenuated by one tuned circuit, further attenuated by a second tuned circuit, and still further by a third and fourth tuned circuit. The effect is to steepen the skirts.

The main problem with TRF receivers is tracking the tuned circuits. In a receiver, the tuned circuits must be made variable so that they can be set to the frequency of the desired signal. In early receivers, each tuned circuit had a separate capacitor, and multiple dials had to be adjusted to tune in a signal. The solution was to gang the tuning capacitors (Fig. 8-4) so that all would be changed simultaneously when the tuning knob was rotated. This improved performance, but tracking errors still occurred. Differences in capacitors caused slight differences in the resonant frequency of each tuned circuit, increasing the bandwidth. This problem was partially solved by connecting a small trimmer capacitor in parallel with each of the tuning capacitors. The trimmers could then be adjusted to minimize the effects of the frequency differences.

Another problem with TRF receivers is that selectivity varies with frequency. As discussed, the bandwidth of a tuned circuit increases with its resonant frequency, since $BW = f_r/Q$. (Q tends to remain nearly constant because the effective coil resistance increases slightly with frequency as a result of skin effect.) Selectivity was good (narrow) at the low frequencies but poor (broader) at the higher frequencies. Not until the development of the superheterodyne receiver, or superhet, were these basic problems solved.

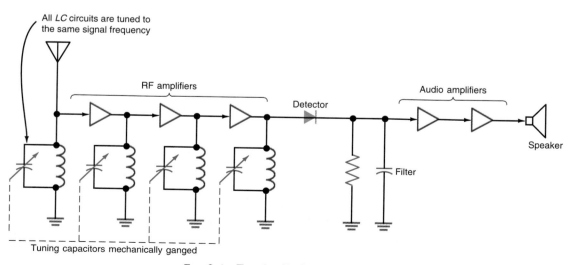

FIG. 8-4 Tuned radio-frequency (TRF) receiver.

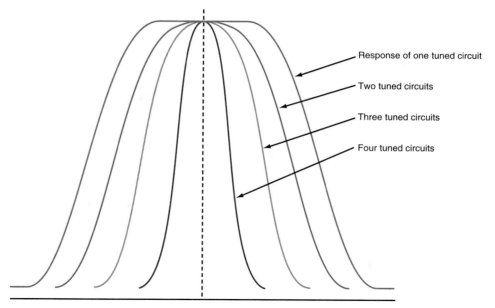

FIG. 8-5 The effect of cascading tuned circuits on selectivity.

Although TRF receivers are no longer used in general communications applications, they are still found in some simple, low-cost single-frequency designs.

Example 8-1

A bandpass filter used in a receiver has upper and lower (3-dB) cutoff frequencies of 4.55 and 4.45 MHz, respectively. Find (a) the center frequency, (b) the circuit bandwidth, and (c) Q.

a. $f_c = \sqrt{f_1 f_2} = \sqrt{(4.55)(4.45)} = 4.5$ MHz

b. $\text{BW} = f_2 - f_1 = 4.55 - 4.45 = 0.1$ MHz or 100 kHz

c. $Q = \dfrac{f_r}{\text{BW}} = \dfrac{4.5}{0.1} = 45$

8-2 SUPERHETERODYNE RECEIVERS

Superheterodyne receivers convert all incoming signals to a lower frequency, known as the *intermediate frequency (IF)*, at which a single set of amplifiers is used to provide a fixed level of sensitivity and selectivity. Most of the gain and selectivity in a superheterodyne receiver are obtained in the IF amplifiers. The key circuit is the mixer, which acts like a simple amplitude modulator to produce sum and difference frequencies. The incoming signal is mixed with a local oscillator signal to produce this con-

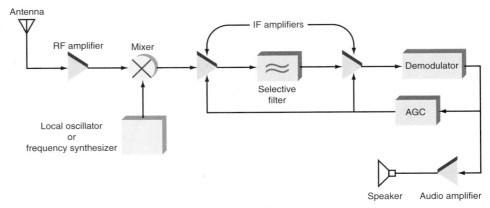

FIG. 8-6 Block diagram of a superheterodyne receiver.

version. Figure 8-6 shows a general block diagram of a superhetrodyne receiver. In the following sections, the basic function of each circuit is examined.

RF AMPLIFIERS

The antenna picks up the weak radio signal and feeds it to the RF amplifier. Because RF amplifiers provide some initial gain and selectivity, they are sometimes referred to as *preselectors.* Tuned circuits help select the desired signal or at least the frequency range in which the signal resides. The tuned circuits in fixed tuned receivers can be given a very high *Q,* so that excellent selectivity can be obtained. However, in receivers that must tune over a broad frequency range, selectivity is somewhat more difficult to obtain. The tuned circuits must resonate over a wide frequency range. Therefore, the *Q,* bandwidth, and selectivity of the amplifier change with frequency.

In communications receivers that do not use an RF amplifier, the antenna is connected directly to a tuned circuit, at the input to the mixer, which provides the desired initial selectivity. This configuration is practical in low-frequency applications where extra gain is simply not needed. (Most of the receiver gain is in the IF amplifier section, and even if relatively strong signals are to be received, additional RF gain is not necessary.) Further, omitting the RF amplifier may reduce the noise contributed by such a circuit. In general, however, it is preferable to use an RF amplifier. RF amplifiers improve sensitivity, because of the extra gain; improve selectivity, because of the added tuned circuits; and improve the S/N ratio. Further, spurious signals are more effectively rejected, minimizing unwanted signal generation in the mixer.

RF amplifiers also minimize oscillator radiation. The local oscillator signal is relatively strong, and some of it can leak through and appear at the input to the mixer. If the mixer input is connected directly to the antenna, some of the local oscillator signal radiates, possibly causing interference to other nearby receivers. The RF amplifier between the mixer and the antenna isolates the two, significantly reducing any local oscillator radiation.

Both bipolar and field-effect transistors can be used as RF amplifiers, but in most modern receivers MOSFETs are used because they generate less noise than bipolar transistors.

Mixers and Local Oscillators

The output of the RF amplifier is applied to the input of the mixer. The mixer also receives an input from a local oscillator or frequency synthesizer. The mixer output is the input signal, the local oscillator signal, and the sum and difference frequencies of these signals. Usually a tuned circuit at the output of the mixer selects the difference frequency, or *intermediate frequency (IF)*. The mixer may be a diode, a balanced modulator, or a transistor. MOSFETs and hot carrier diodes are preferred as mixers because of their low noise characteristics.

The local oscillator is made tunable so that its frequency can be adjusted over a relatively wide range. As the local oscillator frequency is changed, the mixer translates a wide range of input frequencies to the IF. In most receivers, the mixer and local oscillator are separate circuits. In many newer receivers, the local oscillator is a frequency synthesizer. In some simple low-frequency AM radio receivers, the mixing and local oscillator functions are combined within a single circuit called a *converter*, or *autodyne converter*. A bipolar transistor is normally used for this.

IF Amplifiers

The output of the mixer is an IF signal containing the same modulation that appeared on the input RF signal. This signal is amplified by one or more IF amplifier stages, and most of the receiver gain is obtained in these stages. Selective tuned circuits provide fixed selectivity. Since the intermediate frequency is usually much lower than the input signal frequency, IF amplifiers are easier to design and good selectivity is easier to obtain. Crystal, ceramic, or SAW filters are used in most IF sections to obtain good selectivity.

Demodulators

The highly amplified IF signal is finally applied to the demodulator, or detector, which recovers the original modulating information. The demodulator may be a diode detector (for AM), a quadrature detector (for FM), or a product detector (for SSB). The output of the demodulator is then usually fed to an audio amplifier with sufficient voltage and power gain to operate a speaker. For nonvoice signals, the detector output may be sent elsewhere, to a TV picture tube, for example, or a computer.

Automatic Gain Control

The output of a detector is usually the original modulating signal, the amplitude of which is directly proportional to the amplitude of the received signal. The recovered signal, which is usually AC, is rectified and filtered into a DC voltage by a circuit known as the *automatic gain control (AGC)* circuit. This DC voltage is fed back to the IF amplifiers, and sometimes the RF amplifier, to control receiver gain. AGC circuits help maintain a constant output voltage level over a wide range of RF input signal levels; they also help the receiver to function over a wide range, so that strong signals do not produce performance-degrading distortion. Virtually all superheterodyne receivers use some form of AGC.

The amplitude of the RF signal at the antenna of a receiver can range from a fraction of a microvolt to thousands of microvolts; this wide signal range is known as the *dynamic range*. Typically, receivers are designed with very high gain so that weak signals can be reliably received. However, applying a very high-amplitude signal to a receiver causes the circuits to be overdriven, producing distortion and reducing intelligibility.

With AGC, the overall gain of the receiver is automatically adjusted depending on the input signal level. The signal amplitude at the output of the detector is proportional to the amplitude of the input signal; if it is very high, the AGC circuit produces a high DC output voltage, thereby reducing the gain of the IF amplifiers. This reduction in gain eliminates the distortion normally produced by a high-voltage input signal. When the incoming signal is weak, the detector output is low. The output of the AGC is then a smaller DC voltage. This causes the gain of the IF amplifiers to remain high, providing maximum amplification.

Modern communication receivers get their input from signal-receiving dishes that are located in line of sight to one another. They are used primarily for telephones.

8-3 FREQUENCY CONVERSION

As discussed in earlier chapters, frequency conversion is the process of translating a modulated signal to a higher or lower frequency while retaining all the originally transmitted information. In radio receivers, high-frequency radio signals are regularly converted to a lower, intermediate frequency, where improved gain and selectivity can be obtained. This is called *down conversion*. In satellite communications, the original signal is generated at a lower frequency and then converted to a higher frequency for transmission. This is called *up conversion*.

MIXING PRINCIPLES

Frequency conversion is a form of amplitude modulation carried out by a mixer circuit or converter. The function performed by the mixer is called *heterodyning*.

Figure 8-7 is a schematic diagram of a mixer circuit. Mixers accept two inputs. The signal f_s which is to be translated to another frequency is applied to one input, and the sine wave from a local oscillator f_o is applied to the other input. The signal to be translated can be a simple sine wave or any complex modulated signal containing side-

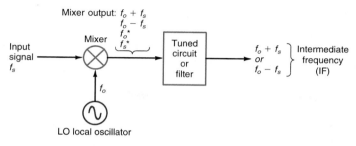

Mixer output: $f_o + f_s$
$f_o - f_s$
f_o^*
f_s^*

Input signal f_s — Mixer — Tuned circuit or filter — $f_o + f_s$ or $f_o - f_s$ Intermediate frequency (IF)

f_o

LO local oscillator

* May or may not be in the output depending upon the type of mixer.

FIG. 8-7 Concept of a mixer.

bands. Like an amplitude modulator, a mixer essentially performs a mathematical multiplication of its two input signals according to the principles discussed in Chaps. 2 and 3. The oscillator is the carrier, and the signal to be translated is the modulating signal. The output contains not only the carrier signal, but also sidebands formed when the local oscillator and input signal are mixed. The output of the mixer, therefore, consists of signals f_s, f_o, $(f_o + f_s)$, and $(f_o - f_s)$ or $(f_s - f_o)$.

The local oscillator signal f_o usually appears in the mixer output, as does the original input signal f_s in some types of mixer circuits. These are not needed in the output, and are therefore filtered out. Either the sum or difference frequency in the output is the desired signal. For example, to translate the input signal to a lower frequency, the lower sideband or difference signal $(f_o - f_s)$ is chosen. The local oscillator frequency will be chosen such that when the information signal is subtracted from it, a signal with the desired lower frequency is obtained. When translating to a higher frequency, the upper sideband or sum signal $(f_o + f_s)$ is chosen. Again, the local oscillator frequency determines what the new higher frequency will be. A tuned circuit or filter is used at the output of the mixer to select the desired signal and reject all the others.

For example, for an FM radio receiver to translate an FM signal at 107.1 MHz to an intermediate frequency of 10.7 MHz for amplification and detection, a local oscillator frequency of 96.4 Mhz is used. The mixer output signals are $f_s = 107.1$ MHz, $f_o = 96.4$ MHz, $(f_o + f_s) = 96.4 + 107.1 = 203.5$ MHz, and $(f_s - f_o) = 107.1 - 96.4 = 10.7$ MHz. Then a tuned circuit selects the 10.7-MHz signal (the IF, or f_{IF}) and rejects the others.

As another example, suppose a local oscillator frequency is needed that will produce an IF of 70 MHz for a signal frequency of 880 MHz. Since the IF is the difference between the input signal and local oscillator frequencies, there are two possibilities:

$$f_o = f_s + f_{IF} = 880 + 70 = 950 \text{ MHz}$$
$$f_o = f_s - f_{IF} = 880 - 70 = 810 \text{ MHz}$$

There are no set rules for deciding which of these to choose. However, at lower frequencies, say those less than about 100 MHz, the local oscillator frequency is traditionally higher than the incoming signals frequency, and at higher frequencies, those above 100 MHz, the local oscillator frequency is lower than the input signal frequency.

Keep in mind that the mixing process takes place on the whole spectrum of the input signal, whether it contains only a single-frequency carrier or multiple carriers and many complex sidebands. In the above example, the 10.7-MHz output signal contains the original frequency modulation. The result is as if the carrier frequency of the input signal is changed as well as all of the sideband frequencies. The frequency-conversion process makes it possible to shift a signal from one part of the spectrum to another, as required by the application.

MIXER AND CONVERTER CIRCUITS

Any diode or transistor can be used to create a mixer circuit, but most modern mixers are sophisticated ICs. This section covers some of the more common and widely used types.

DIODE MIXERS. The primary characteristic of mixer circuits is nonlinearity. Any device or circuit whose output does not vary linearly with the input can be used as a mixer. For example, one of the most widely used types of mixer is the simple but effective diode modulator described in Chap. 3. *Diode mixers* like this are the most common type found in microwave applications.

A diode mixer circuit using a single diode is shown in Fig. 8-8. The input signal, which comes from an RF amplifier or, in some receivers, directly from the antenna is applied to the primary winding of transformer T_1. The signal is coupled to the secondary winding and applied to the diode mixer, and the local oscillator signal is coupled to the diode by way of capacitor C_1. The input and local oscillator signals are linearly added this way and applied to the diode, which performs its nonlinear magic to produce the sum and difference frequencies. The output signals, including both inputs, are developed across the tuned circuit, which acts like a bandpass filter, selecting either the sum or difference frequency and eliminating the others.

SINGLY BALANCED MIXERS. A popular mixer circuit using two diodes is the *singly balanced mixer* illustrated in Fig. 8-9. The input signal is applied to the primary of the transformer, and the local oscillator is applied to the center tap of the secondary. The output signal is developed across an RFC and then applied to a bandpass filter or tuned circuit which selects the sum or difference frequency. In the single- or two-diode mixer, the local oscillator signal is much larger than the input signal, and so the diodes are turned OFF and ON like switches by the local oscillator signal.

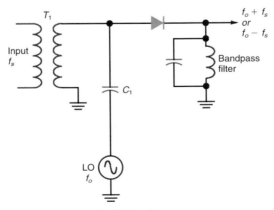

FIG. 8-8 A simple diode mixer.

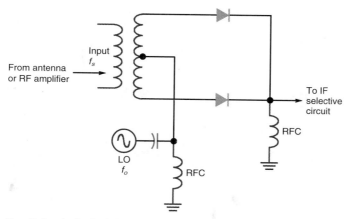

FIG. 8-9 A singly balanced diode mixer.

In small-signal applications, germanium diodes, because of their relatively low turn-on voltage, are used in mixers. Silicon diodes also make excellent RF mixers. The best diode mixers at VHF, UHF, and microwave frequencies are hot carrier or Schottky barrier diodes.

DOUBLY BALANCED MIXER. Balanced modulators are also widely used as mixers. These circuits eliminate the carrier from the output, making the job of filtering much easier. Any of the balanced modulators described previously can be used in mixing applications. Both the diode lattice balanced modulator and the integrated differential amplifier-type balanced modulator are quite effective in mixing applications. A version of the diode balanced modulator shown in Fig. 4-29, known as a *doubly balanced mixer* and illustrated in Fig. 8-10, is probably the single best mixer available, especially for VHF, UHF, and microwave frequencies. The transformers are precision-wound and the diodes matched in characteristics so that a high degree of carrier or local oscillator suppression occurs. In commercial products the local oscillator attenuation is 50 to 60 dB or more.

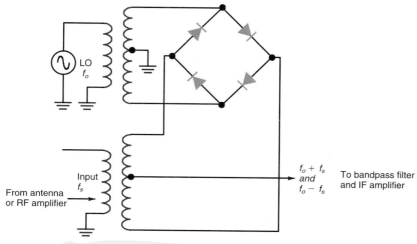

FIG. 8-10 A doubly balanced mixer very popular at high frequencies.

BIPOLAR TRANSISTOR MIXERS. One of the most common applications for a mixer is in radio receivers, where the mixer is used to convert the incoming signal to a lower frequency, making it easier to obtain the high gain and selectivity required. Any of the mixer circuits previously described can be used for this application, as well as special types of mixers that are typically made up of a single transistor biased into the nonlinear range so that it produces analog multiplication. Figure 8-11 shows a mixer of this type, called a *bipolar transistor mixer.*

The primary benefit of transistor mixers over diode mixers is that gain is obtained with a transistor stage. A typical bipolar transistor is biased as if it were a linear amplifier, but the bias is set so that the collector current does not vary linearly with variations in the base current. The class of operation is AB rather than A. This results in analog multiplication, which produces the sum and difference frequencies. In this circuit, both the incoming signal and local oscillator are applied to the base of the transistor. The tuned circuit in the collector usually selects the difference frequency.

FET MIXERS. Improved performance can be obtained by using an FET, as shown in Fig. 8-12. Like the bipolar transistor, the *FET mixer* is biased so that it operates in the nonlinear portion of its range. The input signal is applied to the gate and the local oscillator signal is coupled to the source. Again, the tuned circuit in the drain selects the difference frequency.

Another popular FET mixer, one with a dual-gate MOSFET, is shown in Fig. 8-13. Here the input signal is applied to one gate and the local oscillator is coupled to the other gate. Dual-gate MOSFETs provide superior performance in mixing applications because their drain current I_D is directly proportional to the product of the two gate voltages. Today, bipolar transistors are rarely used in mixing applications except at very low radio frequencies. In receivers built for VHF, UHF, and microwave applications, junction FETs and dual-gate MOSFETs are widely used as mixers because of

square law — non-linear operation

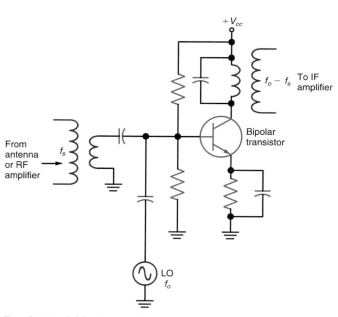

FIG. 8-11 A bipolar transistor mixer.

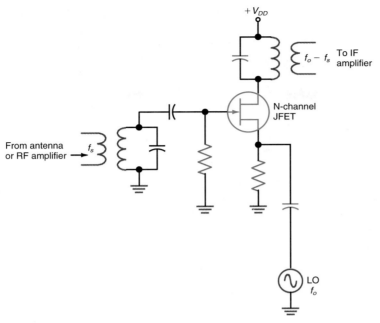

FIG. 8-12 A JFET mixer.

their high gain and low noise. Gallium arsenide FETs are preferred over silicon FETs at the higher frequencies because of their lower noise contribution and higher gain.

One of the best reasons for using an FET mixer is that its characteristic drain current versus gate voltage curve is a perfect square-law function. (Recall that square-law formulas show how upper and lower sidebands and sum and difference frequencies are produced.) With a perfect square-law mixer response, only second-order harmonics are generated in addition to the sum and difference frequencies. Other mixers, such as

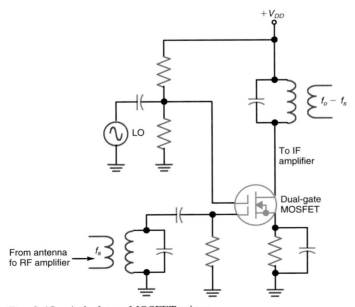

FIG. 8-13 A dual-gate MOSFET mixer.

diodes and bipolar transistors, approximate a square-law function; however, they are nonlinear, so that AM or heterodyning does occur. The nonlinearity is such that higher-order products such as the third, fourth, fifth, and higher harmonics are generated. Most of these can be eliminated by a bandpass filter that selects out the difference or sum frequency for the IF amplifier. However, the presence of higher-order products can cause unwanted low-level signals to appear in the receiver. These signals produce bird-like chirping sounds known as *birdies,* which, despite their low amplitude, can interfere with low-level input signals from the antenna or RF amplifier. FETs do not have this problem, and so FETs are the preferred mixer in most receivers, except for high-frequency applications, where diode mixers are used.

CONVERTERS.   It is possible to combine the functions of the mixer and local oscillator in one circuit, as shown in Fig. 8-14. Known as an *autodyne converter,* the transistor operates as a class C local oscillator and as a mixer. Transformer T_1 and the transistor form a Hartley oscillator. The lower winding of T_1 and C_1 form a tuned circuit that resonates at the desired local oscillator frequency. Tuning is accomplished with C_1; transformer T_2 and the transistor comprise the mixer. T_2 resonates at the difference frequency. The local oscillator signal and input signal combine inside the transistor to produce the four standard mixer outputs. The difference frequency is selected by T_2.

Assume, for example, that the circuit shown in Fig. 8.14 is used in an AM radio. The input is tuned to a radio station on 1230 kHz. The local oscillator is usually set to a frequency higher than the input signal by an amount equal to the IF. To receive the input signal, the local oscillator is tuned to 1685 kHz. The IF is the difference frequency, $f_{\text{IF}} = f_o - f_s = 1685 - 1230 = 455$ kHz, which is a common IF in low-cost receivers.

When both mixing and oscillator functions are performed by a single transistor, the circuit is usually referred to as a *converter.* Converters, which are used only on the lower RF frequencies, are common in AM radios and inexpensive communications receivers operating below 30 MHz.

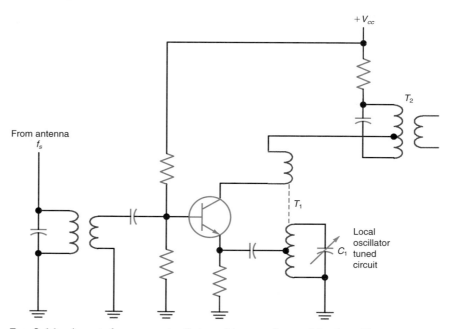

FIG. 8-14   An autodyne converter that combines a mixer and local oscillator.

IC MIXERS. A typical IC mixer, the NE602, is shown in Fig. 8-15(a). The NE602, also known as a *Gilbert transconductance cell,* or *Gilbert cell,* consists of a double balanced mixer circuit made up of two cross-connected differential amplifiers. Although most doubly balanced mixers are passive devices with diodes, as described earlier, the NE602 uses bipolar transistors. Also on the chip is an NPN transistor that can be con-

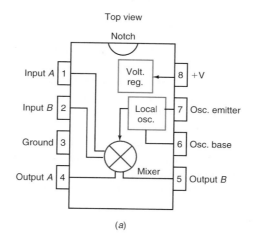

(a)

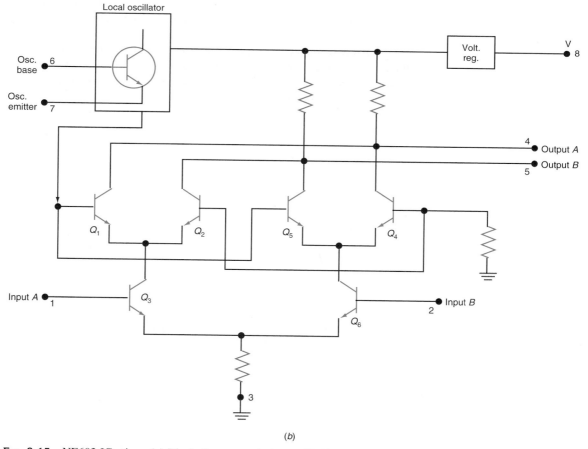

(b)

Fig. 8-15 NE602 IC mixer. (a) Block diagram and pinout. (b) Simplified schematic.

nected as a stable oscillator circuit and a DC voltage regulator. The device is housed in an 8-pin DIP. It operates from a single DC power supply voltage of 4.5 to 8 V. The circuit can be used at frequencies up to 500 MHz, making it useful in HF, VHF, and low-frequency UHF applications. The oscillator, which operates up to about 200 MHz, is internally connected to one input of the mixer. An external LC tuned circuit or a crystal is required to set the operating frequency.

Figure 8-15(b) shows the circuit details of the mixer itself. Bipolar transistors Q_1 and Q_2 form a differential amplifier with current source Q_3, and Q_4 and Q_5 form another differential amplifier with current source Q_6. Note that the inputs are connected in parallel. The collectors are cross-connected; that is, the collector of Q_1 is connected to the collector of Q_4 instead of Q_3, as would be the case for a parallel connection, and the collector of Q_2 is connected to the collector of Q_3. This connection results in a circuit that is like a balanced modulator in that the internal oscillator signal and the input signal are suppressed, leaving only the sum and difference signals in the output. The output may be balanced or single-ended, as required. A filter or tuned circuit must be connected to the output to select the desired sum or difference signal.

A typical 6-V DC circuit using the NE602 IC mixer is shown in Fig. 8-16. R_1 and C_1 are used for decoupling, and a resonant transformer T_1 couples the 72-MHz input signal to the mixer. C_2 resonates with the transformer secondary at the input frequency, and C_3 is an AC bypass connecting pin 2 to ground. External components C_4 and L_1 form a tuned circuit that sets the oscillator to 82 MHz. C_5 and C_6 form a capacitive voltage divider that connects the on-chip NPN transistor as a Colpitts oscillator circuit. C_7 is an AC coupling and blocking capacitor. The output is taken from pin 4 and connected to a ceramic bandpass filter, which provides selectivity. The output—in this case the difference signal, or $82 - 72 = 10$ MHz—appears across R_2. The balanced mixer circuit suppresses the 82-MHz oscillator signal, and the sum signal of 154 MHz is filtered out. The output IF signal plus any modulation that appeared on the input is passed to IF amplifiers for an additional boost in gain prior to demodulation.

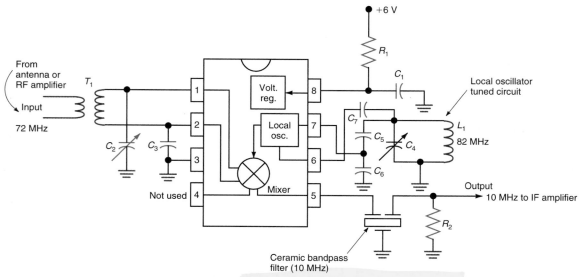

Fig. 8-16   NE602 mixer used for frequency translation.

The local oscillator signal for the mixer comes from either a conventional *LC* tuned oscillator such as a Colpitts or Clapp circuit or from a frequency synthesizer. The simpler continuously tuned receivers use an *LC* oscillator. Channelized receivers use frequency synthesizers.

***LC* OSCILLATORS.** A representative local oscillator for communications receivers is shown in Fig. 8-17. This type of circuit, which is sometimes referred to as a variable-frequency oscillator, or VFO, uses a JFET Q_1 connected as a Colpitts oscillator. Feedback is developed by the voltage divider, which is made up of C_5 and C_6. The frequency is set by the parallel tuned circuit made up of L_1 in parallel with C_1, which is also in parallel with the series combination of C_2 and C_3. The oscillator is set to the center of its desired operating range by a coarse adjustment of trimmer capacitor C_1. Coarse tuning can also be accomplished by making L_1 variable. An adjustable slug-tuned ferrite core moved in and out of L_1 can set the desired frequency range. The main tuning is accomplished with variable capacitor C_3, which is connected mechanically to some kind of dial mechanism that has been calibrated in frequency.

Main tuning can also be accomplished with varactors. For example, C_3 in Fig. 8-17 can be replaced by a varactor, reversed-biased to make it act like a capacitor. Then a potentiometer applies a variable DC bias to change the capacitance and thus the frequency.

The output of the oscillator is taken from across the RFC in the source lead of Q_1 and applied to a direct-coupled emitter follower. The emitter follower buffers the output, isolating the oscillator from load variations which can change its frequency. The emitter follower buffer provides a low-impedance source to connect to the mixer cir-

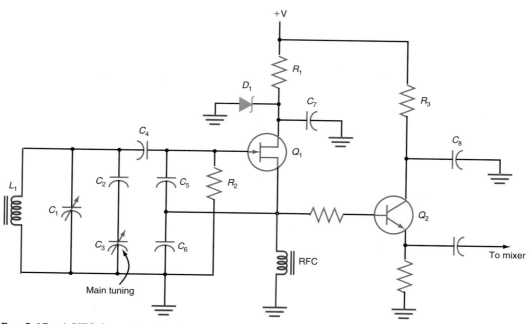

FIG. 8-17 A VFO for receiver local oscillator service.

cuit, which often has a low input impedance. If frequency changes after a desired station is tuned in, which can occur as a result of outside influences such as changes in temperature, voltage, and load, the signal will drift off and no longer be centered in the passband of the IF amplifier. One of the key features of local oscillators is their stability, that is, their ability to resist frequency changes. The emitter follower essentially eliminates the effects of load changes. The zener diode gets its input from the power supply, which is regulated, providing a regulated DC to the circuit and ensuring maximum stability of the supply voltage to Q_1.

Most drift comes from the LC circuit components themselves. Even inductors, which are relatively stable, have slight positive temperature coefficients, and special capacitors that change little with temperature are essential. Usually, negative temperature coefficient (NPO) ceramic capacitors are selected to offset the positive temperature coefficient of the inductor. Capacitors with a mica dielectric are also used.

The oscillator circuit in Fig. 8-17 is most stable at frequencies up to about 10 MHz, with the 5- to 7-MHz range being the most popular. Although it will operate at frequencies up to 30 to 50 MHz, it is far less stable at those frequencies. To achieve the higher frequencies needed to tune some types of receivers, frequency multipliers can be used after the oscillator.

FREQUENCY SYNTHESIZERS. Most new receiver designs incorporate frequency synthesizers for the local oscillator, which provides some important benefits over the simple VFO designs. First, since the synthesizer is usually of the phase-locked loop (PLL) design, the output is locked to a crystal oscillator reference, providing a high degree of stability. Second, tuning is accomplished by changing the frequency division factor in the PLL, resulting in incremental rather than continuous frequency changes. Most communications are channelized; that is, stations operate on assigned frequencies that are a known frequency increment apart, and setting the PLL step frequency to the channel spacing allows every channel in the desired spectrum to be selected simply by changing the frequency division factor.

The former disadvantages of frequency synthesizers—higher cost and greater circuit complexity—have been offset by the availability of low-cost PLL synthesizer ICs which make local oscillator design simple and cheap. Most modern receivers, from AM/FM car radios, stereos, and TV sets to military receivers and commercial transceivers, use frequency synthesis.

The frequency synthesizers used in receivers are largely identical to those described in Chap. 7. However, some additional techniques, such as the use of a mixer in the feedback loop, are employed. The circuit in Fig. 8-18 is a traditional PLL configuration, with the addition of a mixer connected between the VFO output and the frequency divider. A crystal reference oscillator provides one input to a phase detector, which is compared to the output of the frequency divider. Tuning is accomplished by adjusting the frequency division ratio by changing the binary number input to the divider circuit. This binary number can come from a switch, a counter, a ROM, or a microprocessor. The output of the phase detector is filtered by the loop filter into a DC control voltage to vary the frequency of the variable-frequency oscillator which generates the final output that is applied to the mixer in the receiver.

As stated previously, one of the disadvantages of very high frequency PLL synthesizers is that the VFO output frequency is often higher in frequency than the upper operating limit of the variable-modulus frequency divider ICs commonly available. One approach to this problem is to use a prescaler to reduce the VFO frequency before it is applied to the variable-frequency divider. Another is to reduce the VFO

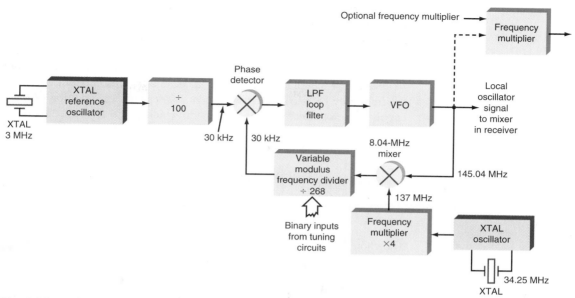

FIG. 8-18 A frequency synthesizer used as a receiver local oscillator.

output frequency to a lower value within the range of the dividers by down-converting it with a mixer, as illustrated in Fig. 8-18. The VFO output is mixed with the signal from another crystal oscillator, and the difference frequency is selected. With some UHF and microwave receivers, it is necessary to generate the local oscillator signal at a lower frequency, then use a frequency multiplier to increase the frequency to the desired higher level. The position of this optional multiplier is shown by the dotted line in Fig. 8-18.

As an example, assume that a receiver must tune to 190.04 MHz and that the IF is 45 MHz. The local oscillator frequency can be either 45 MHz lower or higher than the input signal. Using the lower frequency, we have $190.04 - 45 = 145.04$ MHz. Now when the incoming 190.04-MHz signal is mixed with the 145.04-MHz signal to be generated by the synthesizer in Fig. 8-18, its IF will be the difference frequency of $190.04 - 145.04 = 45$ MHz.

The output of the VFO in Fig. 8-18 is 145.04 MHz. It is mixed with the signal from a crystal oscillator whose frequency is 137 MHz. The crystal oscillator, set to 34.25 MHz, is applied to a frequency multiplier that multiplies by a factor of 4. The 145.04-MHz signal from the VFO is mixed with the 137-MHz signal, and the sum and difference frequencies are generated. The difference frequency is selected, for $145.04 - 137 = 8.04$ MHz. This frequency is well within the range of programmable-modulus IC frequency dividers.

The frequency divider is set to divide by a factor of 268, and so the output of the divider is $8,040,000/268 = 30,000$ Hz or 30 kHz. This signal is applied to the phase detector. It is the same as the other input to the phase detector, as it should be for a locked condition. The reference input to the phase detector is derived from a 3-MHz crystal oscillator divided down to 30 kHz by a divide-by-100 divider. This means that the synthesizer is stepped in 30-kHz increments.

Now, suppose the divider factor is changed from 268 to 269 to tune the receiver. To ensure that the PLL stays locked, the VFO output frequency must change. To achieve 30 kHz at the output of the divider with a ratio of 269, the divider input has to be 269×30 kHz = 8070 kHz = 8.07 MHz. This 8.07-MHz signal comes from the mixer, whose inputs are the VFO and the crystal oscillator. The crystal oscillator input remains at 137 MHz, so the VFO frequency must be 137 MHz higher than the 8.07-MHz output of the mixer, or 137 + 8.07 = 145.07 MHz. This is the output of the VFO and local oscillator of the receiver. With a fixed IF of 45 MHz, the receiver will now be tuned to the IF plus the local oscillator input, or 145.07 + 45 = 190.07 MHz.

Note that the change of divider factor one increment, from 268 to 269, changes the frequency by one 30-kHz increment, as desired. The addition of the mixer to the circuit does not affect the step increment, which is still controlled by the frequency of the reference input frequency.

8-4 INTERMEDIATE FREQUENCY AND IMAGES

The choice of IF is usually a design compromise. The primary objective is to obtain good selectivity. Narrow-band selectivity is best obtained at lower frequencies, particularly when conventional *LC* tuned circuits are used. There are various design benefits of using a low IF. At low frequencies, the circuits are far more stable with high gain. At higher frequencies, circuit layouts must take into account stray inductances and capacitances, as well as the need for shielding, if undesired feedback paths are to be avoided. With very high circuit gain, some of the signal can be fed back in phase and cause oscillation. Oscillation is not as much of a problem at lower frequencies. However, when low IFs are selected, a different sort of problem is faced, particularly if the signal to be received is very high in frequency. This is the problem of images. An *image* is an RF signal that is spaced from the desired incoming signal by a frequency that is two times the intermediate frequency above or below the incoming frequency, or

$$f_i = f_s + 2f_{IF} \quad \text{and} \quad f_i = f_s - 2f_{IF}$$

where f_i = image frequency
  f_s = desired signal frequency
  f_{IF} = intermediate frequency.

This is illustrated graphically in Fig. 8-19. Note that which of the images occurs depends on whether the local oscillator frequency f_o is above or below the signal frequency.

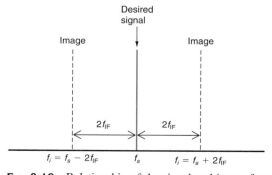

FIG. 8-19 Relationship of the signal and image frequencies.

As stated previously, the mixer in a superhetodyne receiver produces the sum and difference frequencies of the incoming signal and the local oscillator signal. Normally, the difference frequency is selected as the IF. The frequency of the local oscillator is usually chosen to be higher in frequency than the incoming signal by the IF. However, the local oscillator frequency could also be made lower than the incoming signal frequency by an amount equal to the IF. Either choice will produce the desired difference frequency. For the following example, assume that the local oscillator frequency is higher than the incoming signal frequency.

Now, if an image signal appears at the input of the mixer, the mixer will, of course, produce the sum and difference frequencies regardless of the inputs. Therefore, the mixer output will again be the difference frequency at the IF value. Assume, for example, a desired signal frequency of 90 MHz and a local oscillator frequency of 100 MHz. The IF is the difference $100 - 90 = 10$ MHz. The image frequency is $f_i = f_s + 2f_{IF} = 90 + 2(10) = 90 + 20 = 110$ MHz.

If the image appears at the mixer input, the output will be the difference $110 - 100 = 10$ MHz. The IF amplifier will pass it. Now look at Fig. 8-20, which shows the relationships between the signal, local oscillator, and image frequencies. The mixer produces the difference between the local oscillator frequency and the desired signal frequency, or the difference between the local oscillator frequency and the image frequency. In both cases, the IF is 10 MHz. This means that a signal spaced from the desired signal by two times the IF can also be picked up by the receiver and con-

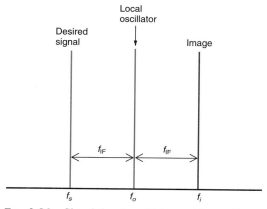

FIG. 8-20 Signal, local oscillator, and image frequencies in a superheterodyne.

verted to the IF. When this occurs, the image signal interferes with the desired signal. In today's crowded RF spectrum, the chances that there will be a signal on the image frequency are high, and image interference can even make the desired signal unintelligible. Superheterodyne design must, therefore, find a way to solve the image problem.

SOLVING THE IMAGE PROBLEM

Image interference occurs only when the image signal is allowed to appear at the mixer input. This is the reason for using high-Q tuned circuits ahead of the mixer, or a selective RF amplifier. If the selectivity of the RF amplifier and tuned circuits is good enough, the image will be rejected. In a fixed tuned receiver designed for a specific frequency, it is possible to optimize the receiver front end for the good selectivity necessary to eliminate images. But many receivers have broadband RF amplifiers that allow many frequencies within a specific band to pass. Other receivers must be made tunable over a wide frequency range. In such cases, selectivity becomes a problem.

Assume, for example, that a receiver is designed to pick up a signal at 25 MHz. The IF is 500 kHz, or 0.5 MHz. The local oscillator is adjusted to a frequency right above the incoming signal by an amount equal to the IF, or $25 + 0.5 = 25.5$ MHz. When the local oscillator and signal frequencies are mixed, the difference is 0.5 MHz, as desired. The image frequency is $f_i = f_s + 2f_{IF} = 25 + 2(0.5) = 26$ MHz. An image frequency of 26 MHz will cause interference to the desired signal at 25 MHz unless it is rejected. The signal, local oscillator, and image frequencies for this situation are shown in Fig. 8-21.

Now, assume that a tuned circuit ahead of the mixer has a Q of 10. Knowing this and the resonant frequency, the bandwidth of the resonant circuit can be calculated as $BW = f_r/Q = 25/10 = 2.5$ MHz. The response curve for this tuned circuit is shown in

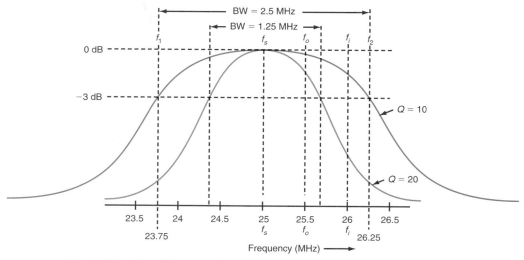

Fig. 8-21 A low IF compared to the signal frequency with low-Q tuned circuits causes images to pass and interfere.

Fig. 8-21. As shown, the bandwidth of the resonant circuit is relatively wide. The bandwidth is centered on the signal frequency of 25 MHz. The upper cutoff frequency is $f_2 = 26.25$ MHz, the lower cutoff frequency is $f_1 = 23.75$ MHz, and the bandwidth is BW $= f_2 - f_1 = 26.25 - 23.75 = 2.5$ MHz. (Remember that the bandwidth is measured at the 3-dB down points on the tuned circuit response curve.)

The fact that the upper cutoff frequency is higher than the image frequency, 26 MHz, means that the image frequency appears in the passband; it would thus be passed relatively unattenuated by the tuned circuit, causing interference.

It is clear how cascading tuned circuits and making them with higher Qs can help solve the problem. For example, assume a Q of 20, instead of the previously given value of 10. The bandwidth at the center frequency of 25 MHz is then $f_s/Q = 25/20 = 1.25$ MHz.

The resulting response curve is shown by the darker line in Fig. 8-21. The image is now outside of the passband and is thus attenuated. Using a Q of 20 would not solve the image problem completely, but still higher Qs would further narrow the bandwidth, attenuating the image even more.

Higher Qs are, however, difficult to achieve and often complicate the design of receivers that must be tuned over a wide range of frequencies. The usual solution to this problem is choosing a higher IF. Assume, for example, that an intermediate frequency of 9 MHz is chosen (the Q is still 10). Now the image frequency is $f_i = 25 + 2(9) = 43$ MHz. A signal at a 43-MHz frequency would interfere with the desired 25-MHz signal if it were allowed to pass into the mixer. But 43 MHz is well out of the tuned circuit bandpass; the relatively low-Q selectivity of 10 is sufficient to adequately reject the image. Of course, choosing the higher intermediate frequency causes some design difficulties, as indicated earlier.

To summarize, the IF is made as high as possible for effective elimination of the image problem, yet low enough to prevent design problems. In most receivers the IF varies in proportion to the frequencies that must be covered. At low frequencies, low values of IF are used. A value of 455 kHz is common for AM broadcast band receivers and for others covering that general frequency range. At frequencies up to about 30 MHz, 3385 kHz and 9 MHz are common IF frequencies. In FM radios that receive 88 to 108 MHz, 10.7 MHz is a standard IF. In TV receivers, an IF in the 40- to 50-MHz range is common. In the microwave region, radar receivers typically use an IF in the 60-MHz range, while satellite communications equipment uses 70- and 140-MHz IFs.

DUAL-CONVERSION RECEIVERS

Another way to obtain selectivity while eliminating the image problem is to use a *dual-conversion superheterodyne receiver.* See Fig. 8-22. The receiver shown in the figure uses two mixers and local oscillators, and so has two IFs. The first mixer converts the incoming signal to a relatively high intermediate frequency for the purpose of eliminating the images; the second mixer converts that IF down to a much lower frequency, where good selectivity is easier to obtain.

Figure 8-22 shows how the different frequencies are obtained. Each mixer produces the difference frequency. The first local oscillator is variable and provides the tuning for the receiver. The second local oscillator is fixed in frequency. Since it need

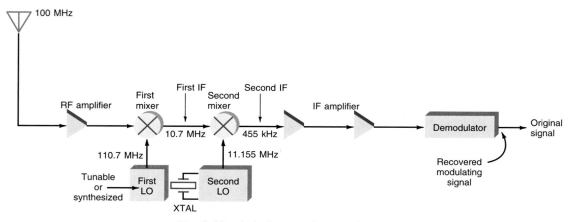

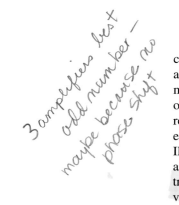

Fig. 8-22 A dual-conversion superheterodyne.

convert only one fixed IF to a lower IF, this local oscillator does not have to be tunable. In most cases, its frequency is set by a quartz crystal. In some receivers, the first mixer is driven by the fixed-frequency local oscillator and tuning is done with the second local oscillator. Dual-conversion receivers are relatively common. Most short-wave receivers and those at VHF, UHF, and microwave frequencies use dual conversion. For example, a CB receiver operating in the 27-MHz range typically uses a 10.7-MHz first IF and a 455-kHz second IF. For some critical applications, triple-conversion receivers are used to further minimize the image problem, although their use is not common. A triple-conversion receiver uses three mixers and three different intermediate frequency values.

Example 8-2

A superheterodyne receiver must cover the range from 220 to 224 MHz. The first IF is 10.7 MHz; the second is 1.5 MHz. Find (*a*) the local oscillator tuning range, (*b*) the frequency of the second local oscillator, and (*c*) the first IF image frequency range. (Assume a local oscillator frequency higher than the input by the IF.)

a. $220 + 10.7 = 230.7$ MHz
$224 + 10.7 = 234.7$ MHz
The tuning range is 230.7 to 234.7 MHz

b. The second local oscillator frequency is 1.5 higher than the first IF.
$$10.7 + 1.5 = 12.2 \text{ MHz}$$

c. The first IF image range is 241.4 to 245.4 MHz.
$$230.7 + 10.7 = 241.4 \text{ MHz}$$
$$234.7 + 10.7 = 245.4 \text{ MHz}$$

Noise consists of an electronic signal that is a mixture of many frequencies at many amplitudes that gets added to a radio or information signal as it is transmitted from one place to another or as it is processed. Noise is *not* the same as interference from other information signals.

When you turn on any AM or FM receiver and tune it to some position between stations, the hiss or static that you hear in the speaker is noise. Noise also shows up on a black-and-white television screen as snow or on a color screen as confetti. If the noise level is high enough and/or the signal is weak enough, the noise can completely obliterate the original signal. Noise that occurs in transmitting digital data causes *bit errors,* and can result in information being garbled or lost.

The noise level in a system is proportional to temperature and bandwidth, and to the amount of current flowing in a component, the gain of the circuit, and the resistance of the circuit. Increasing any of these factors increases noise. Therefore, low noise is best obtained by using low-gain circuits, low direct current, low resistance values, and narrow bandwidths. Keeping the temperature low can also help.

film resistor eliminates noise

Noise is a problem in communications systems whenever the received signals are very low in amplitude. When the transmission is over short distances or high-power transmitters are being used, noise is not usually a problem. But in most communications systems, weak signals are normal, and noise must be taken into account at the design stage. It is in the receiver that noise is the most detrimental because the receiver must amplify the weak signal and recover the information reliably.

Noise can be external to the receiver or originate within the receiver itself. Both types are found in all receivers, and both affect the signal-to-noise ratio.

SIGNAL-TO-NOISE RATIO

The signal-to-noise (S/N) ratio indicates the relative strengths of the signal and the noise in a communications system. The stronger the signal and the weaker the noise, the higher the S/N ratio. If the signal is weak and the noise is strong, the S/N ratio will be low and reception will be unreliable. Communications equipment is designed to produce the highest feasible S/N ratio.

Signals can be expressed in terms of voltage or power. The S/N ratio is computed using either voltage or power values:

$$\text{S/N} = \frac{V_s}{V_n} \quad \text{or} \quad \text{S/N} = \frac{P_s}{P_n}$$

where V_s = signal voltage
V_n = noise voltage
P_s = signal power
P_n = noise power

Assume, for example, that the signal voltage is 1.2 μV and the noise is 0.3 μV. The S/N ratio is 1.2/0.3 = 4. Most S/N ratios are expressed in terms of power rather than voltage. For example, if the signal power is 5 μW and the power is 125 nW, the S/N ratio is $5 \times 10^{-6}/125 \times 10^{-9} = 40$.

The preceding S/N values can be converted to decibels as follows:

For voltage: dB = 20 log (S/N) = 20 log (4) = 20 (0.602) = 12 dB
For power: dB = 10 log (S/N) = 10 log (40) = 10 (1.602) = 16 dB

However it is expressed, if the S/N ratio is less than 1, the dB value will be negative and the noise will be stronger than the signal.

EXTERNAL NOISE

External noise comes from sources over which we have little or no control—industrial, atmospheric, or space. Regardless of its source, noise shows up as a random AC voltage and can actually be seen on an oscilloscope. The amplitude varies over a wide range, as does the frequency. One can say that noise in general contains all frequencies, varying randomly.

All external noise must be dealt with. Atmospheric and space noise are a fact of life and simply cannot be eliminated. Some industrial noise can be controlled at the source, but because there are so many sources of this type of noise, there is no way to eliminate it. The key to reliable communications, then, is simply to generate signals at a high enough power to overcome external noise. In some cases, shielding sensitive circuits in metallic enclosures can aid in noise control.

INDUSTRIAL NOISE. Industrial noise is produced by manufactured equipment, such as automotive ignition systems, electric motors, and generators. Any electrical equipment that causes high voltages or currents to be switched produces transients that create noise. Noise pulses of large amplitude occur whenever a motor or other inductive device is turned on or off. The resulting transients are extremely large in amplitude and rich in random harmonics. Fluorescent and other forms of gas-filled lights are another common source of industrial noise.

ATMOSPHERIC NOISE. The electrical disturbances that occur naturally in the earth's atmosphere are another source of noise. *Atmospheric noise* is often referred to as *static*. Static usually comes from lightning, the electric discharges that occur between clouds or between the earth and clouds. Huge static charges build up on the clouds, and when the potential difference is great enough, an arc is created and electricity literally flows through the air. Lightning is very much like the static charges that we experience during a dry spell in the winter. The voltages involved are, however, enormous, and these transient electrical signals of megavolt power generate harmonic energy that can travel extremely long distances.

Like industrial noise, atmospheric noise shows up primarily as amplitude variations that add to a signal and interfere with it. Atmospheric noise has its greatest impact on signals at frequencies less than 30 MHz.

EXTRATERRESTRIAL NOISE. *Extraterrestrial noise,* solar and cosmic, comes from sources in space. One of the primary sources of extraterrestrial noise is the sun, which radiates a wide range of signals

HINTS AND HELPS

Low noise is best obtained by using low-gain circuits, low direct current, low resistance values, and narrow bandwidths. Keeping temperature low can also help.

in a broad noise spectrum. The noise intensity produced by the sun varies with time. In fact, the sun has a repeatable 11-year noise cycle. During the peak of the cycle, the sun produces an awesome amount of noise that causes tremendous radio signal interference that makes many frequencies unusable for communications. During other years, the noise is at a lower level.

Noise generated by stars outside our solar system is generally known as cosmic noise. Although its level is not as great as that of noise produced by the sun, because of the great distances between those stars and earth, it is nevertheless an important source of noise that must be considered. It shows up primarily in the 10-MHz to 1.5-GHz range, but causes the greatest disruptions in the 15- to 150-MHz range.

INTERNAL NOISE

Electronic components in the receiver such as resistors, diodes, and transistors are major sources of *internal noise*. Internal noise, while low level, is often great enough to interfere with weak signals. The main sources of internal noise in a receiver are thermal noise, semiconductor noise, and intermodulation distortion. Since the sources of internal noise are well known, there is some design control over this type of noise.

THERMAL NOISE. Most internal noise is caused by a phenomenon known as *thermal agitation*, the random motion of free electrons in a conductor caused by heat. Increasing the temperature causes this atomic motion to increase. Since the components are conductors, the movement of electrons constitutes a current flow that causes a small voltage to be produced across that component. Electrons traversing a conductor as current flows experience fleeting impediments in their path as they encounter the thermally agitated atoms. The apparent resistance of the conductor thus fluctuates, causing the thermally produced random voltage we call noise.

You can actually observe this noise by simply connecting a high-value resistor to a very high-gain oscilloscope. The motion of the electrons due to room temperature in the resistor causes a voltage to appear across it. The voltage variation is completely random and at a very low level. The noise developed across a resistor is proportional to the temperature it is exposed to.

Thermal agitation is often referred to as *white noise* or *Johnson noise,* after J. B. Johnson, who discovered it in 1928. Just as white light contains all other light frequencies, white noise contains all frequencies randomly occurring at random amplitudes. A white noise signal therefore occupies, theoretically at least, infinite bandwidth.

In a relatively large resistor at room temperature or higher, the noise voltage across it can be as high as several microvolts. This is the same order of magnitude or higher as many weak RF signals. Weaker-amplitude signals will be totally masked by this noise.

Since noise is a very broad-band signal containing a tremendous range of random frequencies, its level can be reduced by limiting the bandwidth. If a noise signal is fed into a selective tuned circuit, many of the noise frequencies are rejected and the overall noise level goes down. The noise power is proportional to the bandwidth of any circuit to which it is applied. Filtering can reduce the noise level, but does not eliminate it entirely.

The amount of open-circuit noise voltage appearing across a resistor or the input impedance to a receiver can be calculated according to Johnson's formula:

$$v_n = \sqrt{4kTBR}$$

where v_n = rms noise voltage
k = Boltzman's constant (1.38×10^{-23} J/K)
T = temperature, K (°C + 273)
B = bandwidth, Hz
R = resistance, Ω

The resistor is acting like a voltage generator with an internal resistance equal to the resistor value. See Fig. 8-23. Naturally, if a load is connected across the resistor generator, the voltage will decrease as a result of voltage divider action.

Example 8-3

What is the open-circuit noise voltage across a 100-kΩ resistor over the frequency range of direct current to 20 kHz at room temperature (25°C)?

$$v_n = \sqrt{4kTBR}$$
$$= \sqrt{4(1.38 \times 10^{-23})(25 + 273)(20 \times 10^3)(100 \times 10^3)}$$
$$= 5.74 \ \mu V$$

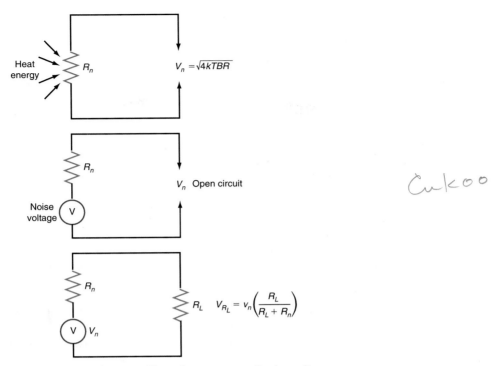

FIG. 8-23 A resistor acts like a tiny generator of noise voltage.

Example 8-4

The bandwidth of a receiver with a 75-Ω input resistance is 6 MHz. The temperature is 29°C. What is the input thermal noise voltage?

$$T = 29 + 273 = 302 \text{ K}$$
$$v_n = \sqrt{4kTBR}$$
$$v_n = \sqrt{4(1.38 \times 10^{-23})(302)(6 \times 10^6)(75)} = 2.74 \ \mu\text{V}$$

Since noise voltage is proportional to resistance value, temperature, and bandwidth, noise voltage can be reduced by reducing resistance, temperature, and bandwidth or any combination to the minimum level acceptable for the given application. In many cases, of course, the values of resistance and bandwidth cannot be changed. One thing, however, that is always controllable to some extent is temperature. Anything that can be done to cool the circuits will greatly reduce the noise. Heat sinks, cooling fans, and good ventilation can help lower noise. Many low-noise receivers for weak microwave signals from spacecraft and in radio telescopes are supercooled; that is, their temperature is reduced to very low (cryogenic) levels with liquid nitrogen or liquid helium.

Thermal noise can also be computed as a power level. Johnson's formula is then

$$P_n = kTB$$

where P_n is the average noise power in watts.

Note that when dealing with power the value of resistance does not enter into the equation.

Example 8-5

What is the average noise power of a device operating at a temperature of 90°F with a bandwidth of 30 kHz?

$$T_C = 5(T_F - 32)/9 = 5(90 - 32)/9 = 5(58)/9 = 290/9 = 32.2°C$$
$$T_K = T_C + 273 = 32.2 + 273 = 305.2 \text{ K}$$
$$P_n = (1.38 \times 10^{-23})(305.2)(30 \times 10^3) = 1.26 \times 10^{-16} \text{ W}$$

SEMICONDUCTOR NOISE. Electronic components such as diodes and transistors are major contributors of noise. In addition to thermal noise, semiconductors produce shot noise, transit-time noise, and flicker noise.

The most common type of semiconductor noise is *shot noise*. Current flow in any device is not direct and linear. The current carriers, electrons or holes, sometimes take random paths from source to destination, whether the destination is an output element, tube plate, or collector or drain in a transistor. It is this random movement that produces the shot effect. Shot noise is also produced by the random movement of electrons or holes across a PN junction. Even though current flow is established by external bias voltages, some random movement of electrons or holes will occur as a result of discontinuities in the device. For example, the interface between the copper lead and the semiconductor material forms a discontinuity that causes random movement of the current carriers.

TEMPERATURE SCALES AND CONVERSIONS

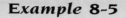

There are three temperature scales in common use: the Fahrenheit scale, expressed in °F; the Celsius scale (formerly centigrade), expressed in °C; and the Kelvin scale, expressed in Kelvins (K). The Kelvin scale, which is used by scientists, is also known as the *absolute scale*. At 0 K (−273.15°C and −459.69°F), or absolute zero, molecular motion ceases.

When calculating noise values, you will frequently need to make conversions from one of these temperature scales to another. The most common conversion formulas are given here.

$$T_C = 5(T_F - 32)/9 \text{ or } T_C = T_K - 273$$
$$T_F = (9T_C/5) + 32$$
$$T_K = T_C + 273$$

Shot noise is also white noise in that it contains all frequencies and amplitudes over a very wide range. The amplitude of the noise voltage is unpredictable, but it does follow a Gaussian distribution curve that is a plot of the probability that specific amplitudes will occur. The amount of shot noise is directly proportional to the amount of DC bias current flowing in a device. The bandwidth of the device or circuit is also important. The rms noise current in a device (I_n) is calculated with the formula

$$I_n = \sqrt{2qIB}$$

where q = charge on an electron, 1.6×10^{-19} C
I = DC current, A
B = bandwidth, Hz

As an example, assume a DC bias current of 0.1 mA and a bandwidth or 12.5 kHz. The noise current is

$$I_n = \sqrt{2(1.6 \times 10^{-19})(0.0001)(12500)} = \sqrt{(4 \times 10^{-19})} = 0.632 \times 10^{-9}$$
$$= 0.632 \text{ nA}$$

Now assume that the current is flowing across the emitter-base junction of a bipolar transistor. The dynamic resistance of this junction (r_e') can be calculated with the expression $r_e' = 0.025/I_e$, where I_e is the emitter current. Assuming an emitter current of 1 mA, we have $r_e' = 0.025/0.001 = 25 \; \Omega$. The noise voltage across the junction is found with Ohm's law:

$$v_n = I_n r_e' = 0.623 \times 10^{-9} \times 25 = 15.8 \times 10^{-9} = 15.8 \text{ nV}$$

This amount of voltage may seem negligible, but keep in mind that the transistor has gain and will therefore amplify this variation, making it larger in the output. Shot noise is normally lowered by keeping the transistor currents low since the noise current is proportional to the actual current. This is not true of MOSFETs, in which shot noise is relatively constant despite the current level.

Another kind of noise that occurs in transistors is called *transit-time noise*. The term *transit time* refers to how long it takes for a current carrier such as a hole or electron to move from the input to the output. The devices themselves are very tiny so the distances involved are minimal, yet the time it takes for the current carriers to move even a short distance is finite. At low frequencies, this time is negligible, but when the frequency of operation is high and the period of the signal being processed is the same order of magnitude as the transit time, problems can occur. Transit-time noise shows up as a kind of random variation of current carriers

While at sea, this lobster fishing boat captain from Prince Edward Island can keep in constant contact with other fishing ships, the coast guard, and his home port by using his ship-to-shore radio.

TYPE OF RESISTOR	NOISE VOLTAGE RANGE, μV
Carbon-composition	0.1–3.0
Carbon film	0.05–0.3
Metal film	0.02–0.2
Wire-wound	0.01–0.2

FIG. 8-24 Flicker noise in resistors.

within a device, occurring near the upper cutoff frequency. Transit-time noise is directly proportional to the frequency of operation. Since most circuits are designed to operate at a frequency much less than the transistor's upper limit, transit-time noise is rarely a problem.

A third type of semiconductor noise, *flicker noise* or excess noise, also occurs in resistors and conductors. This disturbance is the result of minute random variations of resistance in the semiconductor material. It is directly proportional to current and temperature. However, it is inversely proportional to frequency, and for this reason is sometimes referred to as $1/f$ noise. Flicker noise is highest at the lower frequencies, and thus is not pure white noise. Because of the dearth of high-frequency components, $1/f$ noise is also called pink noise.

At some low frequency flicker noise begins to exceed thermal and shot noise. In some transistors this transition frequency is as low as several hundred Hz; in others the noise may begin to rise at a frequency as high as 100 kHz. This information is listed on the transistor data sheet, the best source of noise data.

The amount of flicker noise present in resistors depends on the type of resistor. Figure 8-24 shows the range of noise voltages produced by the various types of popular resistor types. The figures assume a common resistance, temperature, and bandwidth. Because carbon-composition resistors exhibit an enormous amount of flicker noise—an order of magnitude more than the other types—they are avoided in low-noise amplifiers and other circuits. Carbon and metal film resistors are much better, but metal film resistors may be more expensive. Wire-wound resistors have the least flicker noise, but are rarely used because they contribute a large inductance to the circuit, which is unacceptable in RF circuits.

Figure 8-25 shows the total noise voltage variation in a transistor, which is a composite of the various noise sources. At low frequencies, noise voltage is high, due to $1/f$ noise. At very high frequencies, the rise in noise is due to transit-time effects near the upper cutoff frequency of the device. Noise is lowest in the mid-range, where most devices operate. The noise in this range is due to thermal and shot effects, with shot noise sometimes contributing more than thermal noise.

INTERMODULATION DISTORTION. *Intermodulation distortion* results from the generation of new signals and harmonics as the result of circuit nonlinearities. As stated

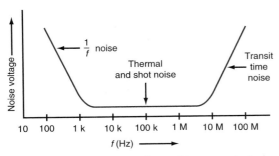

FIG. 8-25 Noise in a transistor with respect to frequency.

previously, circuits can never be perfectly linear, and if bias voltages are incorrect in a given circuit, it is likely to be more nonlinear than intended.

Nonlinearities produce modulation or heterodyne effects. Any frequencies in the circuit mix together, forming sum and difference frequencies. When many frequencies are involved, or with pulses or rectangular waves, the large number of harmonics produces an even larger number of sum and difference frequencies. The resulting products are small in amplitude, but can be large enough to constitute a disturbance that can be regarded as a type of noise. This noise, which is not white or pink, can actually be predicted because the frequencies involved in generating the intermodulation products are known. Because of the predictable correlation between the known frequencies and the noise, intermodulation distortion is also called *correlated noise*. Correlated noise is produced only when signals are present. The types of noise discussed earlier are sometimes referred to as *uncorrelated noise*.

Correlated noise is manifested as the low-level signals called birdies. It can be minimized by good design.

EXPRESSING NOISE LEVELS

The noise quality of a receiver can be expressed as in terms of noise figure, noise factor, noise temperature, and SINAD.

NOISE FACTOR AND NOISE FIGURE. The *noise factor* is the ratio of the S/N power at the input to the S/N power at the output. The device under consideration can be the entire receiver or a single amplifier stage. The noise factor, or noise ratio NR, is computed with the expression

$$NR = \frac{S/N \text{ input}}{S/N \text{ output}}$$

When the noise factor is expressed in decibels, it is called the *noise figure*:

$$NF = 10 \log NR \qquad \text{decibels}$$

Amplifiers and receivers always have more noise at the output than at the input because of the internal noise which is added to the signal. And even as the signal is being

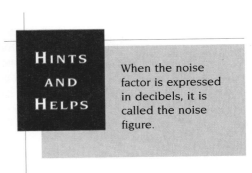

HINTS AND HELPS

When the noise factor is expressed in decibels, it is called the noise figure.

amplified along the way, the noise generated in the process is amplified along with it. The S/N ratio at the output will be less than the S/N ratio of the input, and so the noise figure will always be greater than 1. A receiver that contributed no noise to the signal would have a noise figure of 1, or 0 dB, which is not attainable in practice. A transistor amplifier in a communications receiver usually has a noise figure of several dB. The lower the noise figure, the better the amplifier or receiver. Noise figures of less than about 2 dB are excellent.

Example 8-6

An RF amplifier has S/N of 8 at the input and an S/N of 6 at the output. What are the noise factor and noise figure?

$$NR = \frac{8}{6} = 1.333$$

$$NF = 10 \log 1.3333 = 10(0.125) = 1.25 \text{ dB}$$

NOISE TEMPERATURE.   Most of the noise produced in a device is thermal noise, which is directly proportional to temperature. Therefore, another way to express the noise in an amplifier or receiver is in terms of *noise temperature, T_N*. Noise temperature is expressed in Kelvin. Remember that the Kelvin temperature scale is related to the Celsius scale by the relationship $T_K = T_C + 273$. The relationship between noise temperature and NR is given by

$$T_N = 290(NR - 1)$$

For example, if the noise factor is 1.5, the equivalent noise temperature is $T_N = 290(1.5 - 1) = 290(0.5) = 145$ K. Clearly, if the amplifier or receiver contributes no noise, then NR will be 1, as indicated before. Plugging this value into the expression above gives an equivalent noise temperature of 0 K:

$$T_N = 290(1 - 1) = 290(0) = 0 \text{ K}$$

If the noise ratio is greater than 1, an *equivalent noise temperature* will be produced. The equivalent noise temperature is the temperature to which a resistor equal in value to Z_0 of the device would have to be raised to generate the same V_n as the device generates.

Noise temperature is used only in circuits or equipment that operate at VHF, UHF, or microwave frequencies. The noise factor or noise figure is used at lower frequencies. A good low-noise transistor or amplifier stage typically has a noise temperature of less than 100 K. The lower the noise temperature, the better the device. Often you will see the noise temperature of a transistor given in the data sheet.

SINAD.   Another way of expressing the quality of communications receivers is SINAD—the composite *s*ignal, plus the *n*oise *a*nd *d*istortion contributed by the receiver. In symbolic form,

Example 8-7

A receiver with a 75-Ω input resistance operates at a temperature of 31°C. The received signal is at 89 MHz with a bandwidth of 6 MHz. The received signal voltage of 8.3 μV is applied to an amplifier with a noise figure of 2.8 dB. Find (*a*) the input noise power, (*b*) the input signal power, (*c*) S/N, in dB, (*d*) the noise factor and S/N of the amplifier, and (*e*) the noise temperature of the amplifier.

a. $T_C = 273 + 31 = 304$ K

$v_n = \sqrt{4k\ TBR}$

$= \sqrt{4(1.38 \times 10^{-23})(304)(6 \times 10^6)(75)} = 2.75\ \mu$V

$P_n = \dfrac{(v_n)^2}{R} = \dfrac{(2.75 \times 10^{-6})^2}{75} = 0.1$ pW

b. $P_s = \dfrac{(v_s)^2}{R} = \dfrac{(8.3 \times 10^{-6})^2}{75} = 0.918$ pW

c. $\text{S/N} = \dfrac{P_s}{P_n} = \dfrac{0.918}{0.1} = 9.18$

$\text{dB} = 10 \log \text{S/N} = 10 \log 9.18$

$\text{S/N} = 9.63$ dB

d. $\text{NF} = 10 \log \text{NR}$

$\text{NR} = \text{antilog}\ \dfrac{\text{NF}}{10} = 10^{\text{NF}/10}$

$\text{NF} = 2.8$ dB

$\text{NR} = 10^{2.8/10} = 10^{0.28} = 1.9$

$\text{NR} = \dfrac{\text{S/N input}}{\text{S/N output}}$

$\text{S/N output} = \dfrac{\text{S/N input}}{\text{NR}} = \dfrac{9.18}{1.9} = 4.83$

$\text{S/N (output)} = \text{S/N (amplifier)}$

e. $T_N = 290\ (\text{NR} - 1) = 290(1.9 - 1) = 290\ (0.9) = 261$ K

$$\text{SINAD} = \frac{S + N + D\ (\text{composite signal})}{N + D\ (\text{receiver})}$$

Distortion refers to the harmonics present in a signal caused by nonlinearities.

The SINAD ratio is also used to express the sensitivity of a receiver. Note that the SINAD ratio makes no attempt to discriminate between or separate noise and distortion signals.

To obtain the SINAD ratio, an RF signal modulated by an audio signal (usually of 400 Hz or 1 kHz) is applied to the input of an amplifier or receiver. The composite output is then measured, giving the S + N + D figure. Next, a highly selective notch (band-reject) filter is used to eliminate the modulating audio signal from the output, leaving the noise and distortion, or N + D. The SINAD ratio can then be calculated.

Noise is an important consideration at all communications frequencies, but it is particularly critical in the microwave region because noise increases with bandwidth and impacts high-frequency signals more than low-frequency signals. The limiting factor in most microwave communications systems, such as satellites and radar, is internal noise. In some special microwave receivers, the noise level is reduced by cooling the input stages to the receiver, as mentioned earlier. This technique is called *operating with cryogenic conditions,* the term *cryogenic* referring to very cold conditions approaching absolute zero.

NOISE IN CASCADED STAGES

Noise has its greatest effect at the input to a receiver simply because that is the point at which the signal level is lowest. The noise performance of a receiver is invariably determined in the very first stage of the receiver, usually an RF amplifier or mixer. Design of these circuits must ensure the use of very low-noise components, taking into consideration current, resistance, bandwidth, and gain figures in the circuit. Beyond the first and second stages, noise is basically no longer a problem.

The formula used to calculate the overall noise performance of a receiver or of multiple stages of RF amplification, called *Friis'* formula, is

$$NR = NR_1 + \frac{NR_2 - 1}{A_1} + \frac{NR_3 - 1}{A_1 A_2} + \ldots \frac{NR_n - 1}{A_1 A_2 A_n}$$

where NR = noise ratio

NR_1 = noise ratio of input or first amplifier to receive the signal

NR_2 = noise ratio of second amplifier

NR_3 = noise ratio of third amplifier, and so on

A_1 = power gain of first amplifier

A_2 = power gain of second amplifier

A_3 = power gain of third amplifier, and so on

Note that the noise ratio is used, rather than the noise figure, and so the gains are given in power ratios rather than in decibels.

As an example, consider the circuit shown in Fig. 8-26. The overall noise ratio for the combination is calculated as follows:

$$NR = 1.6 + \frac{4 - 1}{7} + \frac{8.5 - 1}{(7)(12)} = 1.6 + 0.4286 + 0.0893 = 2.12$$

The noise figure is

$$NR = 10 \log NR = 10 \log 2.12 = 10(0.326) = 3.26 \text{ dB}$$

What this calculation means is that *the first stage controls the noise performance for the whole amplifier chain.* This is true even though

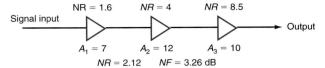

FIG. 8-26 Noise in cascaded stages of amplification.

stage 1 has the lowest NR, because after the first stage, the signal is large enough to overpower the noise. This result is true for almost all receivers and other equipment incorporating multistage amplifiers.

8-6 TYPICAL RECEIVER CIRCUITS

This section focuses on RF and IF amplifiers, AGC and AFC circuits, and other special circuits found in receivers.

RF INPUT AMPLIFIERS

The most critical part of a communications receiver is the front end, which usually consists of the RF amplifier, mixer, and related tuned circuits and is sometimes simply referred to as the tuner. This part of the receiver processes the very weak input signals, increasing their amplitude prior to mixing, and it is essential that low-noise components be used to ensure a sufficiently high S/N ratio. Further, the selectivity should be such that it effectively eliminates images.

In many communications receivers, an RF amplifier is not used, for example, in receivers designed for frequencies lower than about 30 MHz, where the extra gain of an amplifier is not necessary and its only contribution is more noise. In such receivers, the RF amplifier is eliminated and the antenna is connected directly to the mixer input through one or more tuned circuits. The tuned circuits must provide the input selectivity necessary for image rejection. In a receiver of this kind, the mixer must also be of the low-noise variety. Many mixers are MOSFETs, which provide the lowest noise contribution. Low-noise bipolar transistor mixers are used in IC mixers.

Most RF amplifiers are used at frequencies above about 100 MHz. These typically use a single transistor and provide a voltage gain in the 10- to 30-dB range. Bipolar transistors are used at the lower frequencies, while at VHF, UHF, and microwave frequencies FETs are preferred.

An RF amplifier is usually a simple class A circuit. A typical FET circuit is shown in Fig. 8-27. FET circuits are particularly effective because their high input impedance minimizes loading on tuned circuits, permitting the Q of the circuit to be higher and selectivity to be sharper. Most FETs also have a lower noise figure than bipolars.

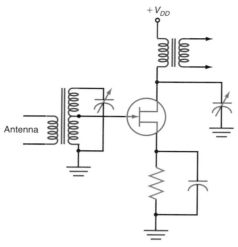

Fig. 8-27 A typical RF amplifier used in receiver front ends.

At microwave frequencies (those above 1 GHz), metal-semiconductor FETs or MESFETs, are used. Also known as GASFETs, these devices are junction field-effect transistors made with gallium arsenide (GaAs). A cross-section of a typical MESFET is shown in Fig. 8-28. The gate junction is a metal-to-semiconductor interface because it is in a Schottky or hot carrier diode. As in other junction FET circuits, the gate-to-source is reverse-biased, and the signal voltage between the source and the gate controls the conduction of current between the source and drain. The actual transit time of electrons through gallium arsenide is far shorter than through silicon, allowing the MESFET to provide high gain at very high frequencies. MESFETs also have an extremely low noise figure, typically 2 dB or less. Most MESFETs have a noise temperature of less than 200 K. The performance of GASFETs is so superior to that of other types of transistors that most designers have virtually stopped using bipolars for this application.

IF AMPLIFIERS

As stated previously, most of the gain and selectivity in a superheterodyne receiver are obtained in the IF amplifier, and choosing the right IF is critical to good design. (For a discussion of typical IF values, see the end of the section "Solving the Image Problem," above.)

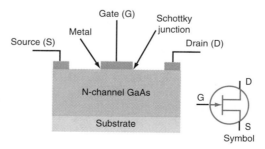

Fig. 8-28 The MESFET configuration and symbol.

TRADITIONAL IF AMPLIFIER CIRCUITS. IF amplifiers, like RF amplifiers, are tuned class A amplifiers capable of providing a gain in the 10- to 30-dB range. Usually two or more IF amplifiers are used to provide adequate overall receiver gain. Figure 8-29 shows a two-stage bipolar IF amplifier. Ferrite-core transformers T_1 and T_2 are used for coupling between stages. Because they are resonant circuits, these transformers also provide the desired selectivity. The dashed lines around the transformers represent the metal can or enclosure that surrounds the transformer components to protect against radiation and undesired feedback. In this circuit the transformers are tuned with trimmer capacitors, although many transformers are tuned with fixed capacitors and variable inductors. The ferrite cores are threaded, allowing their position to be adjusted within the coil thus varying its inductance.

The selectivity in the IF amplifier is provided by the tuned circuits. As indicated earlier, cascading tuned circuits causes the overall circuit bandwidth to be considerably narrowed. High-Q tuned circuits are used, but with multiple tuned circuits, the bandwidth is even narrower. IF amplifiers should be designed so that the selectivity is not too sharp. A too-narrow IF bandwidth will cause sideband cutting, greatly reducing the amplitude of the higher modulating frequencies and thus distorting the received signal. The exact nature of the kinds of signals to be received must be well known so that the bandwidth of the IF amplifier can be appropriately set.

COUPLED CIRCUIT SELECTIVITY. When receiving very broadband signals, it is sometimes necessary to widen the bandwidth of an IF amplifier. There are several ways of doing this. One technique is to connect resistors across the parallel tuned circuits, thereby lowering their Q to a value that will produce the appropriate bandwidth.

Another technique is to use overcoupled tuned circuits. Figure 8-29 shows a two-stage IF amplifier where the coupling between stages is accomplished with double-tuned ferrite-core transformers. Both the primary and secondary windings are resonated with capacitors. The output voltage versus frequency curve for such a double-tuned circuit is strictly dependent on the amount of coupling or mutual

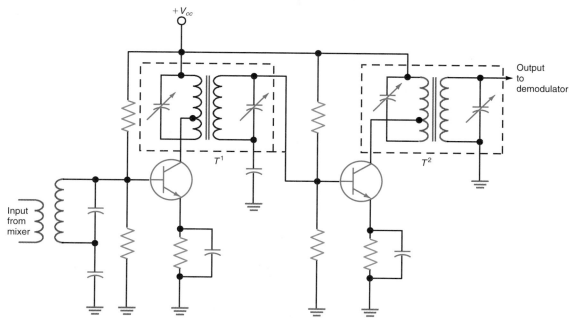

FIG. 8-29 A two-stage IF amplifier using double-tuned transformer coupling for selectivity.

inductance between the primary and secondary windings. That is, the spacing between the windings determines how much of the magnetic field produced by the primary will cut the turns of the secondary. This affects not only the amplitude of the output voltage but also the bandwidth.

Changing the amount of coupling between the primary and secondary windings in IF coupling transformers allows the desired amount of bandwidth to be obtained. Figure 8-30 shows the response curves of a double-tuned transformer for different settings. When the windings are spaced far apart, the coils are said to be *undercoupled*. With this configuration, the amplitude is low and the bandwidth relatively narrow. At some particular degree of coupling, known as *critical coupling,* the output reaches a peak value. In most IF designs, critical coupling provides the best gain if the bandwidth provided is adequate. Moving the coils closer together and increasing the coupling widen the bandwidth further. The output signal amplitude is maximum and will not increase beyond that obtained at critical coupling. This point is usually known as *optimum coupling*. Increasing the amount of coupling still further produces an effect known as *overcoupling*. The result is a double peak output response curve with a considerably wider bandwidth.

CRYSTAL, CERAMIC, AND SAW FILTERS. As discussed previously, in communications receivers in which superior selectivity is required, very sharp crystal filters, usually of the lattice variety, are used to obtain the desired value. Ceramic and mechanical filters can also be used. Such filters are usually packaged as a unit and connected directly at the output of the mixer, before the first IF stage. Ceramic and SAW filters are also widely used to obtain the desired selectivity in IF amplifiers. In fact, few modern receivers use *LC* tuned circuits in the IF section. Instead, they use either crystal, ceramic, or SAW filters. They are typically smaller than *LC* tuned circuits, provide higher selectivity, and require no tuning or adjustment.

LIMITERS. In FM receivers, one or more of the IF amplifier stages is used as a limiter, to remove any amplitude variations on the FM signal before the signal is applied to the demodulator. Typically, limiters are simply conventional class A IF amplifiers. In fact, any amplifier will act as a limiter if the input signal is high enough. When a

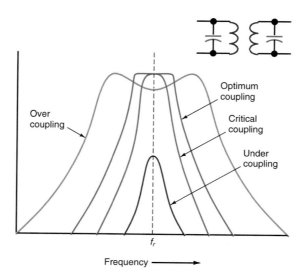

FIG. 8-30 Response curves of a double-tuned air-core transformer for various degrees of coupling.

very large input signal is applied to a single transistor stage, the transistor is alternately driven between saturation and cutoff. For example, in an NPN bipolar class A amplifier, applying a very large positive input signal to the amplifier causes the base bias to increase, thereby increasing the collector current. When a sufficient amount of input voltage is supplied, the transistor reaches maximum conduction, where both the emitter-base and base-collector junctions become forward-biased. At this point, the transistor is saturated and the voltage between the emitter and collector drops to some very small value, typically less than one-tenth of a volt. At this time, the amplifier output is approximately equal to the DC voltage drop across any emitter resistor that may be used in the circuit.

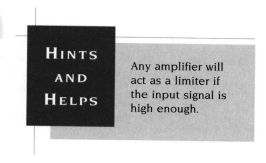

When a very large negative-going signal is applied to the base, the transistor can be driven into cutoff. The collector current drops to zero, and the voltage seen at the collector is simply the supply voltage. Figure 8-31 shows the collector current and voltage for both extremes.

Driving the transistor between saturation and cutoff effectively flattens or clips off the positive and negative peaks of the input signal, removing any amplitude variations. The output signal at the collector is, therefore, a square wave. The most critical part of the limiter design is to set the initial base bias level to that point at which *symmetrical clipping*—that is, equal amounts of clipping on the positive and negative peaks—will occur. The square wave at the collector, which is made up of many undesirable harmonics, is effectively filtered back into a sine wave by the tuned circuit in the collector or the output filter.

AUTOMATIC GAIN CONTROL CIRCUITS

The overall gain of a communications receiver is usually selected on the basis of the weakest signal to be received. In most modern communications receivers, the voltage gain between the antenna and the demodulator is in excess of 100 dB. The RF

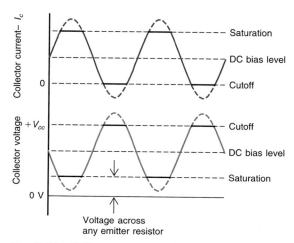

FIG. 8-31 Collector current and voltage in a bipolar limiter IF amplifier circuit.

amplifier usually has a gain in the 5- to 15-dB range. The mixer gain is in the 2- to 10-dB range, although diode mixers, if used, introduce a loss of several dB. IF amplifiers have individual stage gains of 20 to 30 dB. Detectors of the passive diode type may introduce a loss, typically from -2 to -5 dB. The gain of the audio amplifier stage is in the 20- to 40-dB range. Assume, for example, a circuit with the following gains:

RF amplifier	10 dB
Mixer	-2 dB
IF amplifiers (three stages)	27 dB ($3 \times 27 = 81$ total)
Demodulator	-3 dB
Audio amplifier	32 dB

The total gain is simply the algebraic sum of the individual stage gains, or $10 - 2 + 27 + 27 + 27 - 3 + 32 = 118$ dB.

In many cases, gain is far greater than that required for adequate reception. Excessive gain usually causes the received signal to be distorted and the transmitted information to be less intelligible. One solution to this problem is to provide gain controls in the receiver. For example, a potentiometer can be connected at some point in an RF or IF amplifier stage to control the RF gain manually. In addition, all receivers include a volume control in the audio circuit.

The gain controls cited above are used, in part, so that the overall receiver gain does not interfere with the receiver's ability to handle large signals. A more effective way of dealing with large signals, however, is to include AGC circuits. As discussed earlier, the use of AGC gives the receiver a very wide dynamic range, which is the ratio of the largest signal that can be handled to the lowest expressed in decibels. The dynamic range of a typical communications receiver with AGC is usually in the 60- to 100-dB range.

CONTROLLING CIRCUIT GAIN. The gain of a bipolar transistor amplifier is proportional to the amount of collector current flowing. Increasing the collector current from some very low level causes the gain to increase proportionately. At some point, the gain flattens over a narrow collector current range and then begins to decrease as the current increases further. Figure 8-32 shows an approximation of the relationship

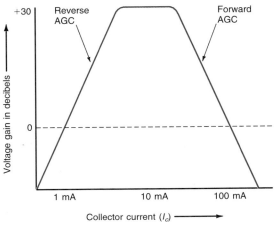

FIG. 8-32 Approximate voltage gain of a bipolar transistor amplifier versus collector current.

between the gain variation and the collector current of a typical bipolar transistor. The gain peaks at 30 dB over the 6- to 15-mA range.

The amount of collector current in the transistor is, of course, a function of the base bias applied. A small amount of base current produces a small amount of collector current, and vice versa. In IF amplifiers the bias level is not usually fixed by a voltage divider, but controlled by the AGC circuit. In some circuits, a combination of fixed voltage divider bias plus a DC input from the AGC circuit controls overall gain. Figure 8-33 shows two common methods of applying AGC to an IF amplifier.

1. The gain can be decreased by decreasing the collector current. An AGC circuit that decreases the current flowing in the amplifier to decrease the gain is called *reverse AGC*.
2. The gain can be reduced by increasing the collector current. As the signal gets stronger, the AGC voltage increases; this increases the base current and, in turn, increases the collector current, reducing the gain. This method of gain control is known as *forward AGC*.

In general, reverse AGC is more common in communications receivers. Forward AGC, which is widely used in TV sets, typically requires special transistors for optimum operation.

In Fig. 8-33(a), the common emitter IF amplifier bias is derived from the voltage divider made up of R_1 and R_2 and the emitter resistor R_3. Resistor R_4 connected to the base accepts a negative DC voltage from the AGC circuit. As the level of the signal

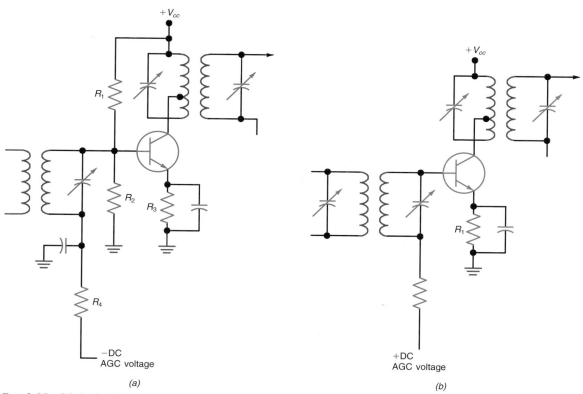

FIG. 8-33 Methods of applying AGC to an IF amplifier.

amplitude increases, the negative DC voltage increases, decreasing the base current. This in turn decreases the collector current and lowers the circuit gain.

The circuit in Fig. 8-33(b) is similar, but the bias for the stage is derived from the emitter resistor R_1 and the AGC circuit itself. In this case, the AGC DC voltage is positive, which sets the bias level. A strong signal increases the positive voltage and, therefore, the base current and collector current. This reduces the circuit gain.

Integrated circuit differential amplifiers are widely used as IF amplifiers. The gain of a differential amplifier is directly proportional to the amount of emitter current flowing. Because of this, the AGC voltage can be conveniently applied to the constant-current source transistor in a differential amplifier. A typical circuit is shown in Fig. 8-34. The bias on constant current source Q_3 is adjusted by R_1, R_2, and R_3 to provide a fixed level of emitter current (I_E) to differential transistors Q_1 and Q_2. Normally, the emitter current value in a constant-gain stage is fixed, and the current divides between Q_1 and Q_2. The gain is easy to control by varying the bias on Q_3. In the circuit shown, increasing the positive AGC voltage increases the emitter current and increases the gain. Decreasing the AGC voltage decreases the gain.

DERIVING THE CONTROL VOLTAGE. The DC voltage used to control the gain is usually derived by rectifying either the IF signal or the recovered information signal after the demodulator. One of the simplest and most widely used methods of AGC voltage generation in an AM receiver is simply to use the output from the diode detector, as shown in Fig. 8-35. The diode detector recovers the original AM information. The voltage developed across R_1 is a negative DC voltage. Capacitor C_1 filters out the IF signal, leaving the original modulating signal. The time constant of R_1 and C_1 is

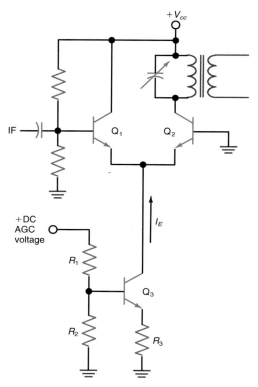

FIG. 8-34 An IF differential amplifier with AGC.

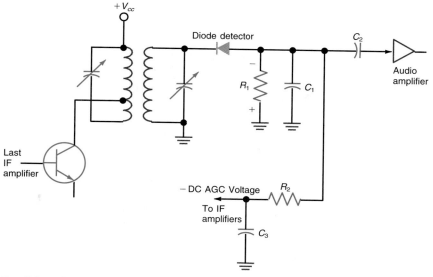

FIG. 8-35 Deriving the AGC voltage from the diode detector in an AM receiver.

adjusted to eliminate the IF ripple yet retain the highest-frequency modulating signal. The recovered signal is passed through C_2 to remove the DC. The resulting AC signal is further amplified and applied to a loudspeaker. The DC voltage across R_1 and C_1 must be further filtered to provide a pure DC voltage. This is done with R_2 and C_3. The time constant of these components is chosen to be very large so that the voltage at the output is pure DC. The DC level varies, of course, with the amplitude of the received signal. The resulting negative voltage is then applied to one or more IF amplifier stages.

In an FM receiver, the DC voltage can usually be derived directly from the demodulator. Both Foster-Seeley discriminator and ratio detector circuits provide convenient starting points for obtaining a DC voltage proportional to the signal amplitude. With additional RC filtering, a DC level proportional to the signal amplitude is derived for use in controlling the IF amplifier gain. As mentioned previously, some FM receivers do not even use AGC because the limiters provide a crude form of gain control by clipping off signal levels higher than a specific amplitude.

In many receivers, a special rectifier circuit devoted strictly to deriving the AGC voltage is used. Figure 8-36 shows a typical circuit of this type. The input, which can be either the recovered modulating signal or the IF signal, is applied to an AGC amplifier. A voltage doubler rectifier circuit made up of D_1, D_2, and C_1 is used to increase the voltage level high enough for control purposes. The RC filter R_1–C_2 removes any

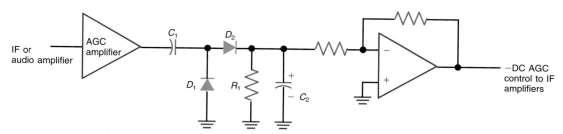

FIG. 8-36 An AGC rectifier and amplifier.

signal variations and produces a pure DC voltage. In some circuits, further amplification of the DC control voltage is necessary; a simple IC op amp like that shown in Figure 8-36 can be used for this purpose. The connection of the rectifier and any phase inversion in the op amp will determine the polarity of the AGC voltage, which can be either positive or negative depending upon the types of transistors used in the IF and their bias connections.

AUTOMATIC FREQUENCY CONTROL

A feedback control circuit similar to AGC that is used in high-frequency receivers, called *automatic frequency control (AFC)*, keeps the local oscillator on frequency. In receivers operating at frequencies above 100 MHz, oscillator stability is a problem. The local oscillators in superheterodyne receivers must be tunable so that stations of any frequency can be selected. This means creating an *LC* variable-frequency oscillator with high stability. Recall that an oscillator's frequency stability is the measure of its ability to remain on the frequency to which it is set despite environmental changes. If an oscillator drifts too far off its set frequency because of temperature changes or other circuit variables, the mixer will not convert the incoming signal to the proper IF value. When that happens, the desired signal will not be picked up or the receiver will be mistuned so that only a portion of the signal passes through the receiver. This, of course, causes lower signal amplitude and distortion. In receivers in which fixed-frequency channel operation is typical, the problem is overcome by using crystal oscillators or a PLL frequency synthesizer for the local oscillator. No AFC is required. In tunable receivers, some form of AFC control is generally used.

In AFC, some of the signal from the output of the demodulator is filtered into a DC voltage and used to control a varactor that, in turn, controls the local oscillator frequency. A typical arrangement is shown in Fig. 8-37. The output from the FM demodulator is a signal whose amplitude varies with frequency deviation. A frequency increase produces a positive output, and a frequency decrease produces a negative output. Using a low-pass filter made up of R_1 and C_1 allows a DC voltage level to be ob-

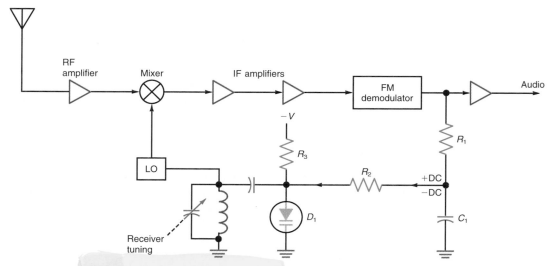

FIG. 8-37 Automatic frequency control.

tained, which is applied to a voltage-variable capacitor D_1 that is connected across the tuned circuit of the local oscillator. It is reverse-biased with a negative voltage $(-V)$ through R_3. Now a drift in the local oscillator frequency will be detected by the demodulator as a frequency variation in the signal. An increase in frequency will cause the local oscillator frequency and the IF to increase. This produces a positive direct current, offsetting the bias on D_1 and causing its capacitance to increase and the oscillator frequency to decrease. The original increase is thus corrected for. Regardless of whether the oscillator drifts higher or lower in frequency, the AFC feedback voltage corrects for the change.

Some FM radios and TV sets have a type of AFC circuit. For example, older FM sets usually include a switch that can be used to turn the AFC off and on. For best results with such sets, initial tuning should be done with the AFC off. Tuning with the AFC on causes the receiver to adjust the oscillator so the signal is properly received even though the oscillator has not been manually set to the correct frequency. The AFC circuit corrects for any tuning error. However, when this occurs, the AFC circuit is not operating at the center of its range and is not able to correct for wide frequency variations. Tuning first to get the signal on channel and then turning on the AFC permits full-range frequency control. In place of AFC, newer FM radios, TV sets, and communications receivers use frequency synthesis for tuning (see Chap. 7).

SQUELCH CIRCUITS

Another circuit found in most communications receivers is a *squelch circuit.* Also called a *muting circuit,* the squelch is used to keep the receiver audio turned off until an RF signal appears at the receiver input. Most two-way communications are short conversations that do not take place continuously. In most cases, the receiver is left on so that should a call be received, it can be heard. When there is no RF signal at the receiver input, the audio output is simply background noise. With no input signal, the AGC sets the receiver to maximum gain, amplifying the noise to a high level. In AM systems such as CB radios, the noise level is relatively high and can be very annoying. The noise level in FM systems can also be high; in some cases listeners may turn down the audio volume to avoid listening to the noise and possibly miss a desired signal. Squelch circuits provide a means of keeping the audio amplifier turned off during the time that noise is received in the background and enabling it when an RF signal appears at the input.

The circuit in Fig. 8-38 shows the basic concept of squelch. The presence of a signal at the input is detected by monitoring the AGC voltage line. The AGC voltage is amplified by a DC amplifier and applied to the base of transistor Q_1, which is simply a switching inverter. This circuit, in turn, drives output switch Q_2, which is connected to the collector of one of the input audio amplifier stages. With no input signal, the AGC voltage is at a very low level, near zero. The DC amplifier output is, therefore, low and Q_1 does not conduct. As a result, Q_2 is turned on by the base current through R_1. Q_2 acts like a short circuit, shunting the audio signal at the collector of Q_3 to ground through diode D_1. As a result, the audio signal from the detector does not get through to the power amplifier stages, and the speaker is quiet.

When an RF signal is received, the output of the DC amplifier is a high positive voltage, turning on Q_1. This shunts base current away from Q_2, so Q_2 cuts off. This allows the audio amplifier Q_3 to operate normally and pass the signal through to the speaker. Most squelch circuits have a built-in level control that allows the circuit thresh-

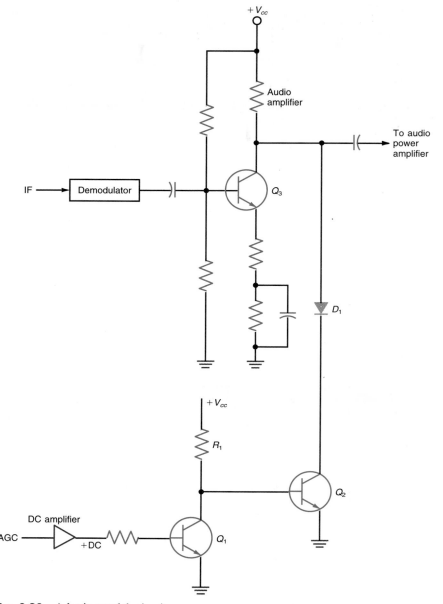

FIG. 8-38 A basic squelch circuit.

old to be adjusted so that it blanks very weak signals; only strong signals will turn the audio on.

NOISE-DERIVED SQUELCH. *Noise-derived* squelch circuits, typically used in FM receivers, amplify the high-frequency background noise when no signal is present and use it to keep the audio turned off. When a signal is received, the noise circuit is overridden and the audio amplifier is turned on.

Figure 8-39 shows a noise-derived squelch circuit used in many communications receivers. The background noise with no signal is taken from the demodulator output and passed through C_1 and potentiometer R_1, which form a high-pass filter. Only frequencies above 6 kHz are passed (most noise is of the high-frequency variety). R_1 also

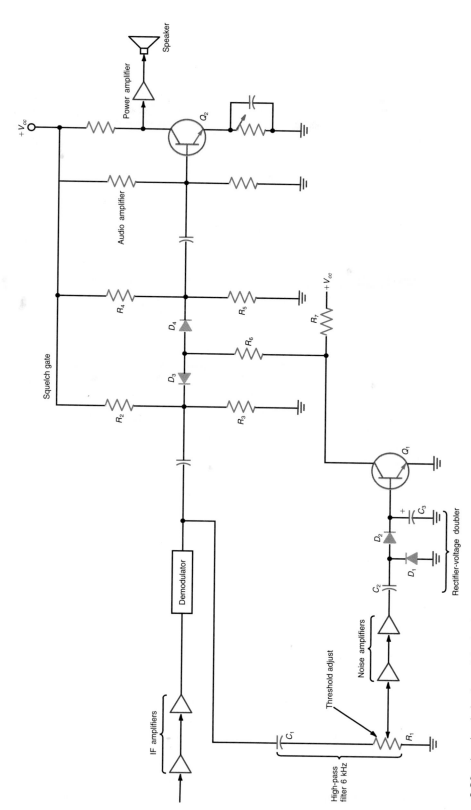

FIG. 8-39 A noise-derived squelch circuit.

serves as a squelch level or muting threshold control. The noise is further amplified by two transistor stages, then rectified into a DC control voltage by a voltage-doubling rectifier circuit made up of C_2, C_3, D_1, and D_2. The rectifier output causes squelch gate Q_1 to saturate when no signal is present and the receiver is picking up noise only.

Q_1 operates the squelch gate, which is made up of D_3 and D_4 and related components. Voltage dividers R_2–R_3 and R_4–R_5 provide a reverse bias to the diodes. With no signal, the noise is amplified and Q_1 saturates as described above. This places the anodes of the diodes at a voltage level below the bias voltage; both diodes are cut off, and so no signal from the demodulator reaches the audio amplifier. When a signal occurs, the audio is not passed by the 6-kHz filter, so Q_1 cuts off. The voltage at the anodes of diodes D_3 and D_4 rises to a level more positive than the bias from the voltage dividers, so the diodes conduct, providing a low-resistance path from the demodulator to the audio amplifier.

CONTINUOUS TONE-CONTROL SQUELCH SYSTEM.

A more sophisticated form of squelch used in some systems is known as the continuous tone-control squelch system (CTCSS). Also known as the continuous tone coded squelch system, it has proprietary versions, such as Motorola's Private Line and GE's Channel Guard. This system is activated by a low-frequency tone transmitted along with the audio. The purpose of CTCSS is to provide some communications privacy on a particular channel. Other types of squelch circuits keep the speaker quiet when no input signal is received; however, in communications systems in which a particular frequency channel is extremely busy, it may be desirable to activate the squelch only when the desired signal is received. This is done by having the desired station transmit a very low-frequency sine wave, usually in the 60- to 254-Hz range, that is linearly mixed with the audio before being applied to the modulator. The low-frequency tone appears at the output of the demodulator in the receiver. It is not usually heard in the speaker, since the audio response of most communications systems rolls off beginning at about 300 Hz, but can be used to activate the squelch circuit.

Figure 8-40 shows a general block diagram of the transmitter and receiver used in a CTCSS system. Most modern transmitters using this system have a choice of multiple tone frequencies, so that different remote receivers can be addressed or keyed up independently, providing a nearly private communications channel. The 52 most commonly used tone frequencies (given in hertz) are listed below:

60.0	100.0	151.4	192.8
67.0	103.5	156.7	196.6
69.3	107.2	159.8	199.5
71.9	110.9	162.2	203.5
74.4	114.8	165.5	206.5
77.0	118.8	167.9	210.7
79.7	120.0	171.3	218.1
82.5	123.0	173.8	225.7
85.4	127.3	177.3	229.1
88.5	131.8	179.9	133.6
91.5	136.5	183.5	241.8
94.8	141.3	186.2	250.3
97.4	146.2	189.9	254.1

At the receiver [see Fig. 8-40(b)], a highly selective bandpass filter tuned to the desired tone selects the tone at the output of the demodulator and applies it to a rectifier and RC filter to generate a DC voltage that operates the squelch circuit.

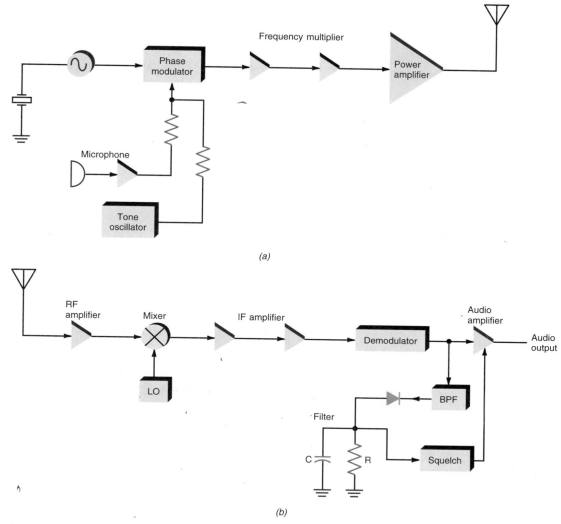

FIG. 8-40 Continuous tone-control squelch.

Signals that do not transmit the desired tone will not trigger the squelch. When the desired signal comes along, the low-frequency tone is received and converted into a DC voltage that operates the squelch and turns on the receiver audio.

Digitally controlled squelch systems, known as *digital coded squelch (DCS)*, are available in some modern receivers. These systems transmit a serial binary code along with the audio. There are 106 different codes used. At the receiver, the code is shifted into a shift register and decoded. If the decode AND gate recognizes the code, the squelch gate is enabled and passes the audio.

SSB AND CONTINUOUS-WAVE RECEPTION

Communications receivers designed for receiving SSB or continuous-wave signals have a built-in oscillator that permits recovery of the transmitted information. This circuit,

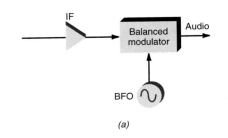

(a)

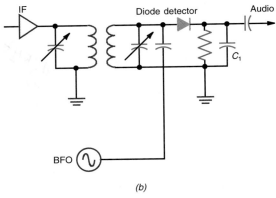

(b)

Fig. 8-41 The use of a BFO.

called the *beat frequency oscillator (BFO),* is usually designed to operate near the IF and is applied to the demodulator along with the IF signal containing the modulation.

Recall that the basic demodulator is a balanced modulator [see Fig. 8-41(*a*)]. A balanced modulator has two inputs, the incoming SSB signal at the intermediate frequency and a carrier which mixes with the incoming signal to produce the sum and difference frequencies, the difference being the original audio. The BFO supplies the carrier signal to the balanced modulator. The term *beat* refers to the difference frequency output. The BFO is set to a value above or below the SSB signal frequency by an amount equal to the frequency of the modulating signal. It is usually made variable so that its frequency can be adjusted for optimum reception. Varying the BFO over a narrow frequency range allows the pitch of the received audio to change from low to high. It is typically adjusted for most natural voice sounds. BFOs are also used in receiving CW code. When dots and dashes are transmitted, the carrier is simply turned OFF and ON for short and long periods of time. The amplitude of the carrier does not vary nor does its frequency; however, the ON-OFF nature of the carrier is, in essence, a form of amplitude modulation.

Consider for a moment what would happen if a CW signal were applied to a diode detector. The output of the diode detector would simply be pulses of DC voltage representing the dots and dashes. When applied to the audio amplifier, the dots and dashes would blank the noise but would not be discernible. To make the dots and dashes audible, the IF signal is mixed with the signal from a BFO. The BFO signal is usually injected directly into the diode detector, as shown in Fig. 8-41(*b*), where the CW signal at the IF signal is mixed, or heterodyned, with the BFO signal. This results in sum and difference frequencies at the output. The sum frequency, of course, is very high, nearly double the IF value, and it is filtered out by capacitor C_1 in the detector. The difference frequency is a low audio frequency. In fact, if the BFO is variable, the difference frequency can be adjusted to any desired audio tone, usually in the 400- to 900-Hz range. Now when the dots and dashes appear at the input to the diode detector, the output will be an audio tone that is amplified and heard in a speaker or earphones. Of course, the BFO is turned off for standard AM signal reception.

INTEGRATED CIRCUITS IN RECEIVERS

In new designs, virtually all receiver circuits are ICs. In fact, most of the circuits in modern receivers are contained on a single IC, and a complete receiver usually consists of three or four ICs at most, plus those discrete components that cannot be easily integrated on a chip. These include coils, transformers, high-capitance and variable capacitors, and crystal and ceramic filters.

IC receivers are typically broken down into three major sections: (1) the tuner, with RF amplifier, mixer, and local oscillator; (2) the IF section, with amplifiers, demodulator, and AGC and muting circuits; and (3) audio power amplifier. The second and third sections are entirely implemented with ICs. The tuner may or may not be. For low-frequency receivers, say those below about 200 MHz, the tuner can be in IC form also.

This CNN reporter at the presidential inauguration can be quickly paged on his "beeper" receiver whenever news breaks.

Higher-frequency receivers require special mixer and local oscillator circuits that are not easily implemented. This is especially true of microwave receivers.

Following is a discussion of two widely used receiver ICs, the 3089 IF system and the DS8911 tuner.

3089 IF System.　Originally developed by RCA and now second-sourced by several other semiconductor manufacturers, the 3089 is one of the oldest and most widely used receiver ICs. The 3089 contains a three-stage IF amplifier, an FM demodulator, and AGC and muting circuits, and is packaged in a standard 16-pin DIP. A block diagram of the system is shown in Fig. 8-42.

The input to this chip comes from the tuner, which consists of the RF amplifier, mixer, and local oscillator. A ceramic filter is normally used to provide the required selectivity. A typical circuit is shown in Fig. 8-43. The mixer output at the IF is applied to a simple common emitter amplifier, which in turn, drives a two-stage ceramic filter. A typical IF is 10.7 MHz. The output of the first filter is applied to another amplifier and then another filter. The ceramic filters provide excellent selectivity but they do have a high insertion loss. The single-stage amplifiers are used to offset this attenuation. No further selectivity is required.

As shown in Fig. 8-42, in the 3089 the output of the second ceramic filter connects to the first IF amplifier. All three IF amplifier stages use differential amplifiers which perform as amplifiers and as symmetrical limiters at the higher signal levels. Note that each IF amplifier has a level detector circuit associated with it. These are AGC circuits that derive a DC control signal from the signal amplitude. The first level detector is used to provide AGC to the RF amplifier in the tuner. It is a delayed AGC,

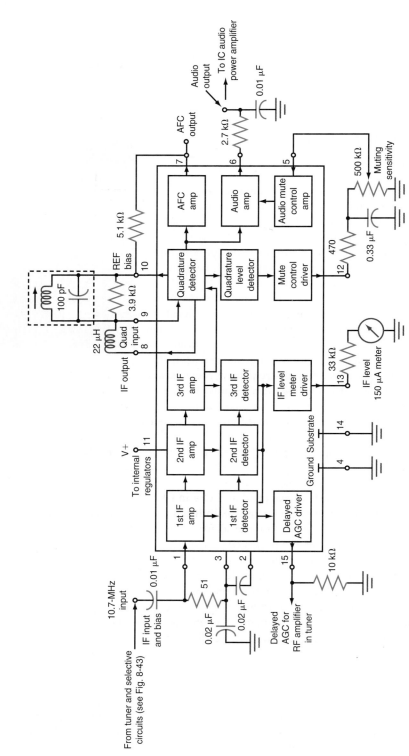

Fig. 8-42 The 3089 IC, an FM receiver IF system.

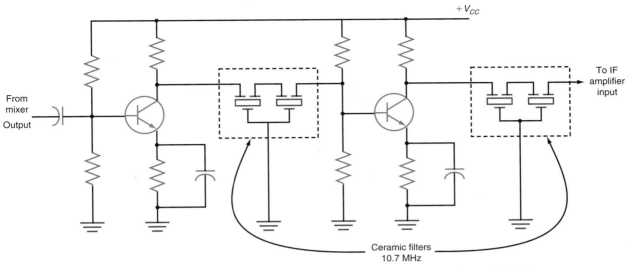

Fig. 8-43 Ceramic bandpass filter provides receiver selectivity between the mixer and the IF amplifier.

which means that it is not fast-acting. The AGC circuit cannot respond instantaneously to a significant signal-level change because of the RC filter usually associated with it. It takes time for the circuit capacitance to charge or discharge to the new level, whether it be higher or lower. This AGC response delay is desirable because it prevents noise and interference from making fast, unexpected changes in receiver gain, which might distort the signal or temporarily desensitize the receiver to a weak signal.

The level detectors all drive an IF-level meter driver circuit, which can be used to operate a DC panel meter. These meters, called *S meters,* provide a way to visually display signal strength. They also function as tuning aids. When you tune a receiver, you are seeking to maximize the signal to the IF amplifier, and when a peak output is tuned for on the meter, the meter indicates that the signal is in the center of the pass-band and is producing maximum signal level.

The demodulator is a standard quadrature detector in which an external 22-μH coil and parallel tuned circuit provide the 90° phase shift required. The quadrature detector also has a level detector, which can be used for squelch or muting. As shown in Fig. 8-42, an external potentiometer is used for muting sensitivity. The output of the potentiometer goes to the audio mute control amplifier, which operates the internal audio amplifier. The 3089 also has an AFC output derived from the demodulator, which is used to control the local oscillator frequency to prevent drift. The audio amplifier receives its signal from the quadrature detector output, and the audio output goes to an IC audio power amplifier, which drives the speaker.

DS8911 TUNER IC. This advanced IC, made by National Semiconductor, contains a mixer and a frequency-synthesized local oscillator. It can be used to create all parts of a tuner circuit except for any RF amplifiers, and is useful for implementing AM, FM, TV sound, and shortwave (SW) receivers. A general block diagram of the DS8911 chip is

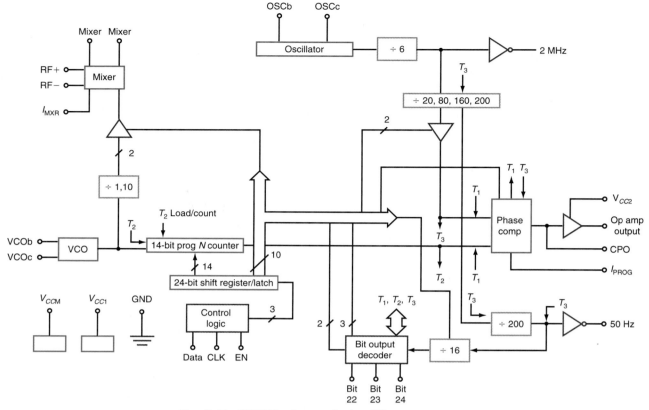

Fig. 8-44 DS8911 mixer-synthesizer IC.

shown in Fig. 8-44. The Gilbert cell mixer shown at the upper left receives its input from the RF amplifier; its output goes to the selective circuits, then on to the IF amplifier and demodulator. The local oscillator, designated VCO in the figure, is applied to the mixer through a divide-by-10 divider. The VCO is part of a built-in PLL frequency synthesizer. The VCO drives a 14-bit counter frequency divider, which then drives the phase comparator. At the output of the phase comparator is an op amp which is used to buffer the phase comparator circuit and which provides a means of connecting the loop filter to the VCO.

The reference oscillator shown at the top of Fig. 8-44 is connected to an external 12-MHz crystal which sets the reference frequency. The output of this oscillator is divided by 6, producing a 2-MHz output which can be used as the clock for an external control microcomputer. The 2-MHz signal is further divided by 20, 80, 160, or 200 to produce reference frequencies of 100, 25, 12.5, or 10 kHz, respectively. This permits the frequency step of the synthesizer to be selected by a binary signal input from an external controller. The 10-kHz signal is further divided by 200 to obtain a 50-Hz signal that can be used to operate an external time-of-day clock.

The frequency of the synthesizer is varied by changing the frequency divider ratio. The 14-bit divider, which is fully programmable from an external controller, has an integral 31/32-modulus prescaler. Normally, this IC is used with a microcontroller that sends it a serial 24-bit control word that is entered into a 24-bit shift register, as shown. Fourteen of those bits go to set the divider ratio and thus the VCO frequency;

the other 10 are control bits that are used for selecting the reference frequency and tuning step increment, for controlling loop gain, and for other functions.

A complete AM/FM receiver using this IC is shown in Fig. 8-45. The antenna drives both RF amplifiers for the AM and FM bands. The outputs of these amplifiers are applied to a transformer T_1; the transformer connects the input signal to the mixer in the DS8911, which converts the signal to an 11.5-MHz IF. The mixer output drives an external transformer T_2 and then an 11.5-MHz ceramic filter that provides the IF selectivity. The IF is split into two sections, one for AM and the other FM. Additional ceramic filters are used, along with a separate IF amplifier section that is usually a chip like the 3089.

An interesting feature of the DS8911 is that it uses a common IF for both AM and FM. Typically, the received signal is converted *down* to an IF. The AM band extends from 535 to 1605 kHz, and a common IF is 455 kHz. The FM band extends from 88 to 108 Mhz, and a common IF is 10.7 MHz. In the DS8911, the FM signal is down-converted to 11.5 MHz and the AM signal is *up-converted* to 11.5 MHz. Up-converting the AM signal eliminates the image problem and even the need for tuned circuits in the AM front end. This greatly simplifies front-end RF amplifier tuning and reduces cost as well as the need for complex manual alignment procedures. As Fig. 8-45 shows, the AM IF is further down-converted in a second mixer, to 450 kHz. The AM receiver is a double-conversion type. The 12-MHz reference oscillator output serves as the second local oscillator.

All operations in the receiver are controlled by a single-chip microprocessor with a control program stored in an on-board ROM. One set of I/O circuits provides the serial control signals for the DS8911, and other I/O circuits connect to external front

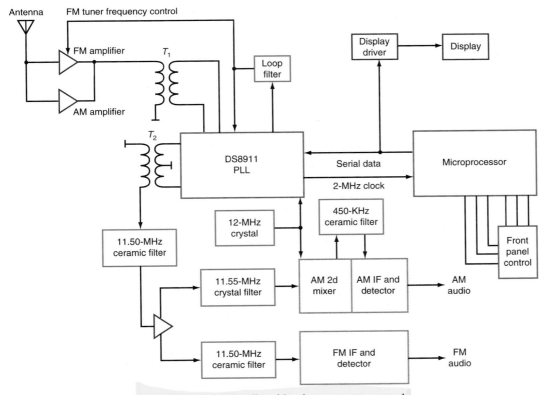

FIG. 8-45 A complete AM/FM radio with microprocessor control.

panel controls such as frequency tuning. Serial data is also supplied to an external chip that converts it into a decimal frequency display on an LCD or fluorescent readout.

A detailed circuit diagram is shown in Fig. 8-46. Refer back to Fig. 8-44 to see the internal connections. The input to the mixer comes from T_1, the input to which is from the RF amplifiers. The mixer output drives T_2, which drives two selective circuits using ceramic filters, one for the AM IF (11.55 MHz) and the other for the FM IF (11.5 MHz). Diodes D_4 and D_5 isolate the two IF sections so they do not interfere with each other. The external 12-MHz crystal for the reference oscillator is shown at the bottom of the circuit in Fig. 8-46.

The VCO frequency is set by external inductor L_3 and capacitors C_8 and C_9, shown at the top of the chip, and varied by changing the capacitance of the back-to-back varactors (D_1). The DC control voltage is, of course, derived from the phase comparator and loop filter. The internal op amp drives the loop filter, made up of R_{11}, C_{11}, and C_{12}, and the DC output from the op amp drives D_1 through R_1.

8-7 RECEIVERS AND TRANSCEIVERS

VHF AIRCRAFT COMMUNICATIONS CIRCUIT

The typical VHF receiver circuit shown in Fig. 8-47 is designed to receive two-way aircraft communications between planes and airport controllers, which take place in the VHF range of 118 to 135 MHz. Amplitude modulation is used. Like most modern receivers, the circuit is a combination of discrete components and ICs.

The signal is picked up by an antenna and fed through a transmission line to input jack J_1. The signal is coupled through C_1 to a tuned filter consisting of the series and parallel tuned circuits made up of L_1–L_5 and C_2–C_6. This broad bandpass filter passes the entire 118- to 135-MHz range.

The output of the filter is connected to an RF amplifier through C_7, which is made up of transistor Q_1 and its bias resistor R_4 and collector load R_5. The signal is then applied to the NE602 IC, U_1–C_8, which contains a balanced mixer and a local oscillator. The local oscillator frequency is set by the circuit made up of inductor L_6 and the related components. C_{14} and D_1 in parallel form the capacitor that resonates with L_6 to set the frequency of the local oscillator. Tuning of the oscillator is accomplished by varying the DC bias on a varactor D_1. Potentiometer R_1 sets the reverse bias, which in turn varies the capacitance to tune the oscillator.

A superheterodyne receiver is tuned by varying the local oscillator frequency, which is set to a frequency above the incoming signal by the amount of the IF. In this receiver, the IF is 10.7 MHz, a standard value for many VHF receivers. To tune the 118- to 135-MHz range, the local oscillator is varied from 128.7 to 145.7 MHz.

The output of the mixer, which is the difference between the incoming signal frequency and the local oscillator frequency, appears at pin 4 of the NE602 and is fed to a ceramic bandpass filter set to the IF of 10.7 MHz. This filter provides most of the receiver's selectivity. The insertion loss of the filter is made up by an amplifier made of Q_2, its bias resistor R_{10}, and collector load R_{11}. The output of this amplifier drives an MC1350 IC through C_{16}. An integrated IF amplifier, U_2 provides extra gain and

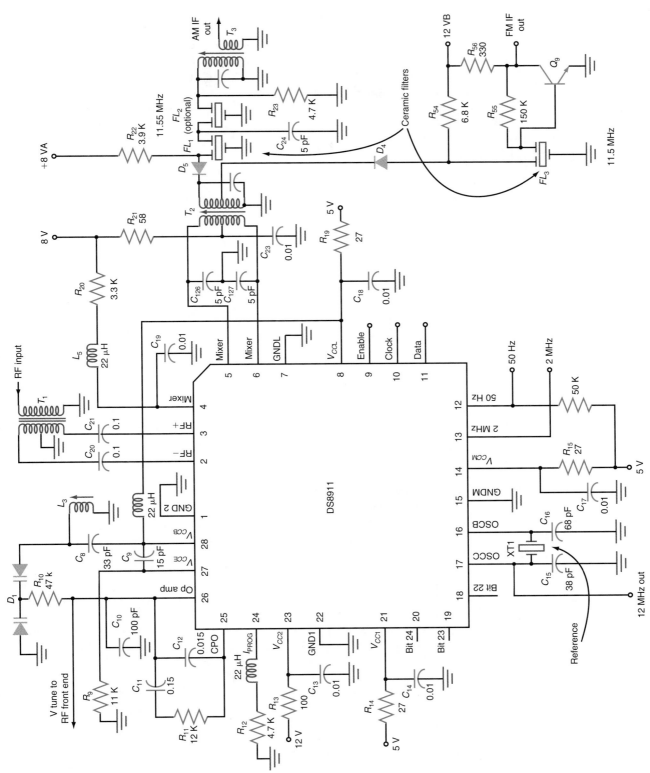

FIG. 8-46 Circuit details of AM/FM receiver.

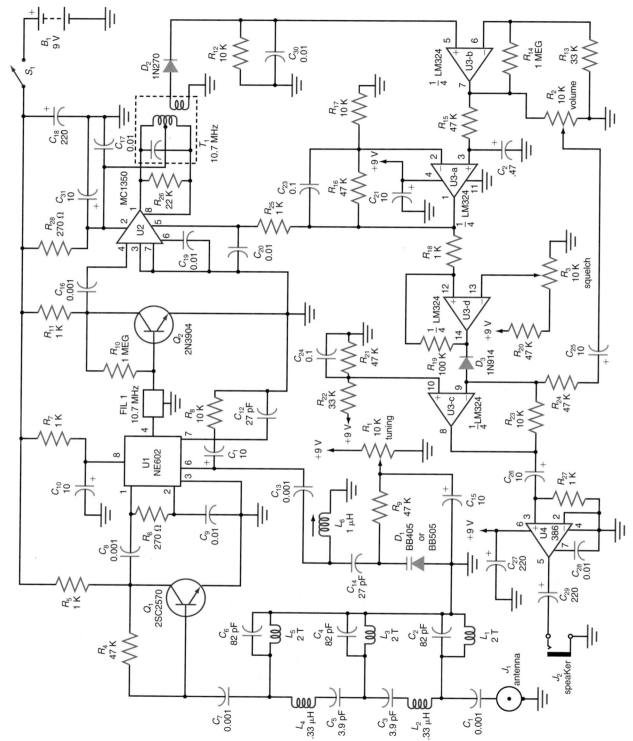

Fig. 8-47 The aviation receiver—a superheterodyne unit built around four ICs—is designed to receive AM signals in the 118- to 135-MHz frequency range. (*Popular Electronics*, January 1991, Gernsback Publications, Inc.)

selectivity. The selectivity comes from the tuned circuit made up of IF transformer T_1. The MC1350 also contains all of the AGC circuitry.

The signal at the secondary of T_1 is then fed to a simple AM diode detector consisting of D_2, R_{12}, and C_{30}. The demodulated audio signal appears across R_{12} and is then fed to op amp U_{3b}, a noninverting circuit biased by R_{13} and R_{14} that provides extra amplification for the demodulated audio and the average DC at the detector output. This amplifier feeds the volume control, potentiometer R_2. The audio signal goes from there through C_{25} and R_{24} to another op amp, U_{3c}. Here the signal is further amplified and fed to the 386 IC power amplifier U_4. This circuit drives the speaker, which is connected via jack J_2.

The audio signal from the diode detector contains the DC level resulting from detection (rectification). Both the audio and the DC are amplified by U_{3b} and further filtered into a nearly pure DC by the low-pass filter made of R_{15} and C_{22}. This DC signal is applied to op amp U_{3a}, where it is amplified into a DC control voltage. This DC at the output pin 1 of U_3 is fed back to pin 5 on the MC 1350 IC to provide AGC control, ensuring a constant comfortable listening level despite wide variations in signal strength.

The AGC voltage from U_{3a} is also fed to an op amp comparator circuit made from amplifier U_{3d}. The other input to this comparator is a DC voltage from potentiometer R_3, which is used as a squelch control. Since the AGC voltage from U_{3a} is directly proportional to the signal strength, it is used as the basis for setting the squelch to a level that will blank the receiver until a signal of a predetermined strength comes along.

If the signal strength is very low or no signal is tuned in, the AGC voltage will be very low or nonexistent. This causes D_3 to conduct, effectively disabling amplifier U_{3c} and preventing the audio signal from the volume control from passing through to the power amplifier. If a strong signal exists, D_3 will be reverse-biased and thus will not interfere with amplifier U_{3c}. As a result, the signal from the volume control passes to the power amplifier and is heard in the speaker.

SINGLE-IC FM RECEIVER

The FM receiver IC chip shown in Fig. 8-48, the popular Motorola MC3363, contains all receiver circuits except for the audio power amplifier, which is a separate chip. Designed to operate at frequencies up to about 200 MHz, this chip is widely used in cordless telephones, paging receivers, and other portable applications such as remote-controlled toys and monitors and short-distance walkie-talkies. The chip is housed in a 28-pin DIP, as shown in Fig. 8-49. This dual-conversion receiver contains two mixers, two local oscillators, a limiter, a quadrature detector, and squelch circuits. The first local oscillator has a built-in varactor that allows it to be controlled by an external frequency synthesizer.

A complete receiver using the MC3363 is shown in Fig. 8-49. The receiver operates in the 30-MHz range and is the companion to the transmitter unit described in Chap. 7 and shown in Fig. 7-49. It uses the transmitters tuned output filters for input selectivity. The output of the tuned circuit in Fig. 7-49 appears at the input to Fig. 8-49, where it feeds into an impedance-matching section made up of L_3 and C_{10}. Diodes D_1 and D_2 provide overload protection for the receiver front end. The signal goes to the transistor internal to the MC3363 on pins 2, 3, and 4, which is the RF amplifier.

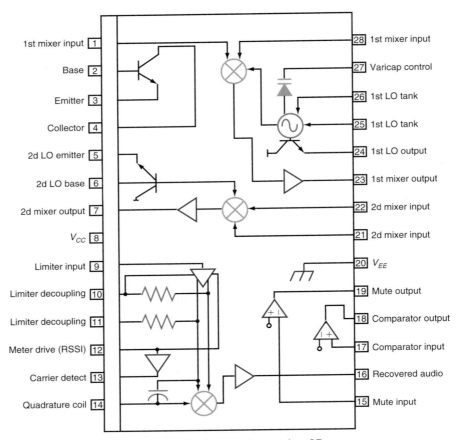

FIG. 8-48 The Motorola MC3363 dual-conversion receiver IC.

The output of the RF amplifier is coupled to the first mixer through R_7 and C_{23}. The receiver is tuned to a single channel, and this frequency is fixed by an external third-overtone-tyne crystal (XTAL1) set to a frequency 10.7 MHz greater than the incoming signal. (For example, for an input signal of 27.125 Mhz, the crystal would have a frequency of $10.7 + 27.125 = 37.845$ MHz.) The crystal is connected to the receiver's first local oscillator on pins 25 and 26.

The output of the first mixer is a 10.7-MHz IF signal at pin 23. It is connected to a 10.7-MHz ceramic filter designated F_2 in Fig. 8-49. The filter output feeds into pin 21, the input to the second mixer. The second mixer is fed by a local oscillator made up of an internal transistor and related components on pins 5 and 6. It too is crystal controlled. A 10.245-MHz crystal (XTAL2) sets the frequency. The first IF and this oscillator produce the second IF, which is the difference between 10.7 and 10.245: 0.455 MHz or 455 kHz. The second mixer output at pin 7 feeds a 455-kHz ceramic filter, providing additional selectivity. The filter output goes to the limiter input at pin 9, and the limiter output drives the quadrature detector. The quadrature tank coil is L10 in the diagram. The quadrature detector output (the recovered audio), is first filtered by an active low-pass filter made up of the internal op amp on pins 15 and 19 and the related resistors and capacitors. This filter cuts off frequencies above 3 kHz.

Finally, the audio signal goes to the audio power amplifier IC4, the MC34119. The squelch circuit in the MC3363 generates a carrier detect signal at pin 13, which is used to mute the IC power amplifier. The carrier detect input is pin 1 on the MC34119, and R_{26} is the volume control.

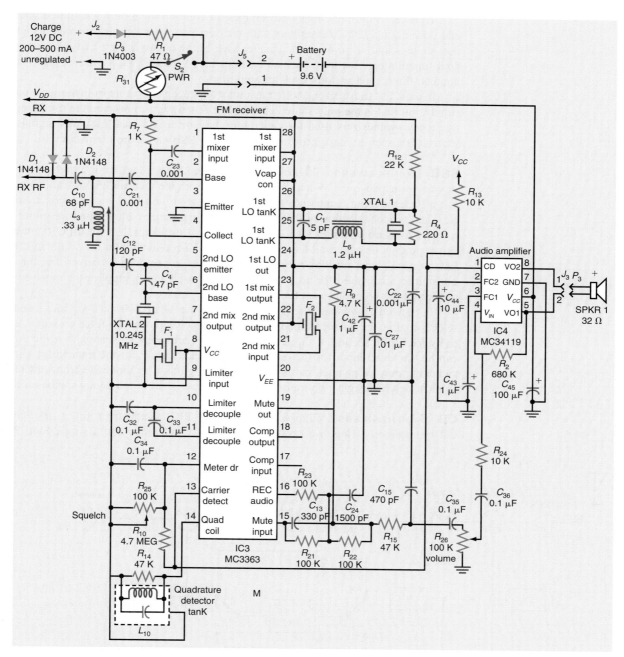

Fig. 8-49 A dual-conversion IC receiver. (*Electronics Now,* October 1992.)

TRANSCEIVERS

In the past, communications equipment was individually packaged in units based on function, and so transmitters and receivers were almost always separate units. Today, most two-way radio communications equipment is packaged such that both transmitter and receiver are in a unit known as a *transceiver.* Transceivers range from large, high-power desk-top units to small, pocket-sized, hand-held walkie-talkies.

Transceivers provide many advantages. In addition to sharing a common housing and power supply, the transmitters and receivers can share circuits, thereby achieving cost savings and, in some cases, smaller size. Some of the circuits that can perform a dual function are antennas, oscillators, frequency synthesizers, power supplies, tuned circuits, filters, and various kinds of amplifiers.

In most FM transceivers, both transmitter and receiver operate from the same power supply, but that is the only shared circuit. In a few designs, however, the oscillators or frequency synthesizers used for generating the carrier and local oscillator signals are shared. Transceivers designed for AM, CW, and SSB can share many circuits.

SSB Transceivers. Figure 8-50 is a general block diagram of a high-frequency transceiver capable of CW and SSB operation. Both the receiver and transmitter make use of heterodyning techniques for generating the IF and final transmission frequencies, and proper selection of these intermediate frequencies allows the transmitter and receiver to share common local oscillators. Local oscillator 1 is the BFO for the receiver product detector and the carrier for the balanced modulator for producing DSB. Later, crystal local oscillator 2 drives the second mixer in the receiver and the first transmitter mixer used for up conversion. Local oscillator 3 supplies the receiver first mixer and the second transmitter mixer.

In transmission mode, the crystal filter (another shared circuit) provides sideband selection after the balanced modulator. In the receive mode, the filter provides selectivity for the IF section of the receiver. Tuned circuits may be shared. A tuned circuit can be the tuned input for the receiver or the tuned output for the transmitter. Circuit switching may be manual, but is often done using relays or diode electronic switches. In most newer designs, the transmitter and receiver share a frequency synthesizer.

CB Synthesizers. Figure 8-51 shows a PLL synthesizer for a 40-channel CB transceiver. Using two crystal oscillators for reference and a single-loop PLL,

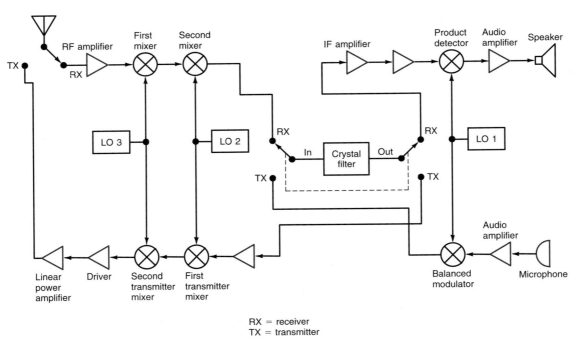

RX = receiver
TX = transmitter

Fig. 8-50 An SSB transceiver showing circuit sharing.

it synthesizes the transmitter frequency and the two local oscillator frequencies for a dual-conversion receiver for all 40-CB channels. The reference crystal oscillator, which operates at 10.24 MHz, is divided by 2 with a flip-flop, then a binary frequency divider that divides by 1024 to produce a 5-kHz frequency (10.24 MHz/2 = 5.12 MHz/1024 = 5 kHz) that is then applied to the phase detector. The channel spacing is, therefore, 5 kHz.

The phase detector drives the low-pass filter and a VCO that generates a signal in the 16.27- to 16.71-MHz range. This is the local oscillator frequency for the first receiver mixer. Assume, for example, that it is desired to receive on CB channel 1, or 26.965 MHz. The programmable divider is set to the correct ratio to produce a 5-kHz output when the VCO is 16.27. The first receiver mixer produces the difference frequency of 26.965 − 16.27 = 10.695 MHz. This is the first intermediate frequency (IF). The 16.27-MHz VCO signal is also applied to mixer A; the other input to this mixer is 15.36 MHz, which is derived from the 10.24-MHz reference oscillator and a frequency tripler (×3). The output of mixer A drives the programmable divider, which feeds the phase detector. The 10.24-MHz reference output is also used as the local oscillator signal for the second receiver mixer. With a 10.695 first IF, the second IF is, then, the difference, or 10.695 − 10.24 MHz = 0.455 MHz or 455 kHz.

The VCO output is also applied to mixer B along with a 10.695-MHz signal from a second crystal oscillator. The sum output is selected, producing the transmit frequency 10.695 + 16.27 = 26.965 MHz. This signal drives the class C drivers and power amplifiers.

Channel selection is achieved by changing the frequency-division ratio on the programmable divider, usually with a rotary switch or a digital keypad that controls a microprocessor. The circuitry in the synthesizer is usually contained on a single IC chip.

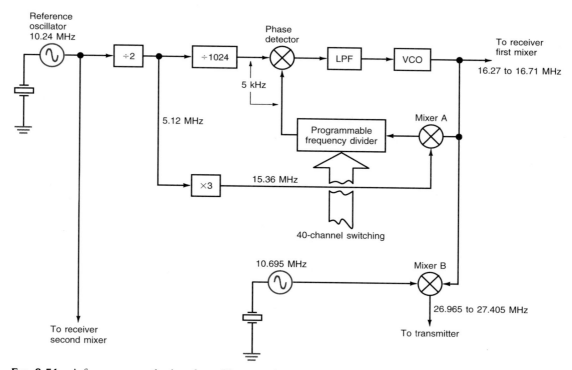

Fig. 8-51 A frequency synthesizer for a CB transceiver.

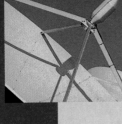

SUMMARY

The two primary requirements for any communications receiver are selectivity—the ability to pick out the desired signal from all others in the frequency spectrum—and sensitivity—the ability to provide sufficient amplification to recover the modulating signal. Superheterodyne receivers convert all incoming signals to a lower frequency known as the intermediate frequency (IF), at which amplifiers provide optimal sensitivity and selectivity.

IC receivers have three major sections: (1) the tuner, with RF amplifier, mixer, and local oscillator; (2) the IF section, with amplifiers, demodulator, and AGC and muting circuits; and (3) the audio power amplifier. In modern receivers all but the tuner are generally implemented with ICs.

Mixers, like amplitude modulators, perform a mathematical multiplication of the two input signals. The output of the mixer is a signal at the IF containing the same modulation that appeared on the input RF signal; this signal is amplified by one or more IF amplifier stages. Popular types of mixer circuits are diode mixers, singly balanced mixers, doubly balanced mixers, bipolar transistor mixers, and FET mixers. Converter circuits perform mixing and oscillator functions with a single transistor.

Two important circuits in superheterodyne receivers are frequency synthesizers for the local oscillator and automatic gain control (AGC) circuits. AGC is a feedback system that automatically adjusts the gain of the receiver based on the amplitude of the received signal. Automatic frequency control (AFC) keeps the local oscillator on frequency.

Noise is random energy that interferes with the desired signal. An important indicator of noise is the S/N ratio, which indicates the relative strengths of the signal and the noise. The stronger the signal and the weaker the noise, the higher the S/N ratio. The two major sources of noise are external (industrial, atmospheric, and extraterrestrial) and internal (thermal and semiconductor). Three figures used for a receiver's noise rating are the noise factor, the ratio of the S/N power at the input to the S/N power at the output; the noise temperature; and SINAD, signal plus noise and distortion.

Squelch, or muting, circuits keep the audio amplifier turned off during the time that noise is received in the background. When an RF signal appears at the input, the audio amplifier is enabled. Two other noise-control circuits are continuous tone-control squelch systems and digital coded squelch systems.

KEY TERMS

3089 IF system
Atmospheric noise
Autodyne converter
Automatic frequency control (AFC)
Automatic gain control (AGC)
Automatic volume control (AVC)
Bandwidth

Beat frequency oscillator (BFO)
Bipolar transistor mixer
Birdies
CB synthesizer
Ceramic filter
Continuous tone-control squelch system (CTCSS)
Critical coupling
Crystal filter

Demodulator (detector)
Diode mixer
Doubly balanced mixer
Down conversion
DS8911 tuner IC
Dual-conversion receiver
External noise
Extraterrestrial noise
FET mixer
Flicker noise

Forward AGC
Frequency conversion
GaAsFET
Gilbert cell
Heterodyning
IC receiver
IF amplifier
Image
Industrial noise
Intermediate frequency (IF)
Intermodulation distortion
LC oscillator
Limiter
Local oscillator
MESFET
Mixer (converter)

NE602 IC mixer
Noise
Noise factor
Noise figure
Noise temperature
Reverse AGC
RF amplifier
SAW filter
Selectivity
Semiconductor noise
Sensitivity
Shape factor
Shot noise
Signal-to-noise ratio (S/N)
SINAD (signal plus noise and distortion)

Singly balanced mixer
Skirt selectivity
Squelch
SSB transceiver
Superheterodyne
Thermal noise
Transceiver
Transit-time noise
Tuned radio frequency (TRF) receiver
Up conversion
VFO (Variable frequency oscillator)
VHF receiver
White (Johnson) noise

QUESTIONS

1. How does decreasing the Q of a resonant circuit affect its bandwidth?
2. How does cascading tuned circuits affect selectivity?
3. What can happen to a modulated signal if the selectivity of a tuned circuit is too sharp?
4. How must the coil resistance be changed to narrow the bandwidth of a tuned circuit?
5. A choice is to be made between two 10.7-MHz IF filters. One has a shape factor of 2.3, the other 1.8. Which has the best selectivity?
6. What type of receiver uses only amplifiers and a detector?
7. What type of receiver uses a mixer to convert the received signal to a lower frequency?
8. What two circuits are used to generate the IF?
9. In what stage are most of the gain and selectivity in a superheterodyne receiver obtained?
10. What circuit in a receiver compensates for a wide range of input signal levels?
11. The mixer output is usually the difference between what two input frequencies?
12. The AGC voltage controls the gain of what two stages of a receiver?
13. What do you call an interfering signal that is spaced from the desired signal by twice the IF?
14. What is the primary cause of images appearing at the mixer input?
15. What advantage does a dual-conversion superhet offer over a single-conversion superhet?
16. How can the image problem best be solved during the design of a receiver?
17. Give the expressions for the outputs of a mixer whose inputs are f_1 and f_2.
18. Name the best type of passive mixer.
19. Name the best type of transistor mixer.
20. The process of mixing is similar to what kind of modulation?
21. What is the primary specification of a VFO used for local oscillator duty?
22. What do you call a single stage that acts as both mixer and local oscillator? Where is it used?
23. What type of local oscillator is used in most modern receivers?
24. Why are mixers sometimes used in frequency synthesizers, as in Fig. 8-18?
25. Name the three primary sources of external noise.

R E V I E W

26. Name the five main types of internal noise that occur in a receiver.
27. What is the primary source of atmospheric noise?
28. List four common sources of industrial noise.
29. What is the main source of internal noise in a receiver?
30. In what units is the signal-to-noise (S/N) ratio usually expressed?
31. How does increasing the temperature of a component affect its noise power?
32. How does narrowing the bandwidth of a circuit affect the noise level?
33. Name the three types of semiconductor noise.
34. True or false. The noise at the output of a receiver is less than the noise at the input.
35. What are the three components that make up SINAD?
36. What stages of a receiver contribute the most noise?
37. What are the advantages and disadvantages of using an RF amplifier at the front end of a receiver?
38. What is the name of the low-noise transistor preferred in RF amplifiers at microwave frequencies?
39. What type of mixers have a loss?
40. How is the selectivity usually obtained in an IF amplifier?
41. In a double-tuned circuit, maximum bandwidth is obtained with what type of coupling?
42. What is the name given to an IF amplifier that clips the positive and negative peaks of a signal?
43. Why is clipping allowed to occur in an IF stage?
44. The gain of a bipolar class A amplifier can be varied by changing what parameter?
45. What is the overall RF–IF gain range of a receiver?
46. What is the process of using the amplitude of an incoming signal to control the gain of a receiver?
47. What is the difference between forward AGC and reverse AGC?
48. How is the gain of a differential amplifier varied to produce AGC?
49. What are two names for the circuit that blocks the audio until a signal is received?
50. Name the two types of signals used to operate the circuit described in Question 49.
51. Describe the purpose and operation of a CTCSS system in a receiver.
52. A BFO is required to receive what two types of signals?
53. What is the source of the signal required at the input to the 3089 receiver IC?
54. Name the three main sources of selectivity for receivers implemented with ICs.
55. What parts of a superheterodyne receiver are implemented by the DS8911 IC?
56. What controls the frequency selection in a receiver using the DS8911?
57. In an FM transceiver, what is the only circuitry commonly shared by the receiver and the transmitter?
58. What are the circuits shared by the transmitter and receiver in an SSB transceiver?
59. The phase-locked loop is often combined with what circuit to produce multiple frequencies in a transceiver?
60. The synthesizer in a transceiver usually generates what three frequencies?

The following questions refer to the receiver in Fig. 8-47.

61. What components or circuits determine the bandwidth of the receiver?
62. As R_1 is varied so that the voltage on the arm of the potentiometer increases toward +9 V, how does the frequency of the local oscillator vary?
63. What component provides most of the gain in this receiver?
64. Is the squelch signal- or noise-derived?
65. Does this receiver contain a BFO?
66. Could this circuit receive CW or SSB signals?
67. If the DC AGC voltage on pin 5 of U_2 is decreased, what happens to the gain of U_2?
68. Where would an audio signal be injected to test the complete audio section of this receiver?
69. What frequency signal would be used to test the IF section of this receiver and where would it be connected?
70. What component would be inoperable if C_{31} became shorted?

PROBLEMS

1. A tuned circuit has a Q of 80 at its resonant frequency of 480 kHz. What is its bandwidth? ◄

2. A parallel LC tuned circuit has a coil of 4 μH and a capacitance of 68 pF. The coil resistance is 9 Ω. What is the circuit bandwidth?

3. A tuned circuit has a resonant frequency of 18 MHz and a bandwidth of 120 kHz. What are the upper and lower cutoff frequencies? ◄

4. What value of Q is needed to achieve a bandwidth of 4 kHz at 3.6 MHz?

5. A filter has a 6-dB bandwidth of 3500 Hz and a 60-dB bandwidth of 8400 Hz. What is the shape factor? ◄

6. A superhet has an input signal of 14.5 MHz. The local oscillator is tuned to 19 MHz. What is the IF?

7. A desired signal at 29 MHz is mixed with a local oscillator of 37.5 MHz. What is the image frequency? ◄

8. A dual-conversion superhet has an input frequency of 62 MHz and local oscillators of 71 and 8.6 MHz. What are the two IFs?

9. What are the outputs of a mixer with inputs of 162 and 189 MHz? ◄

10. What is the most likely IF for a mixer with inputs of 162 and 189 MHz?

11. A frequency synthesizer like the one in Fig. 8-18 has a reference frequency of 100 kHz. The crystal oscillator and the multiplier supply a signal of 240 MHz to the mixer. The frequency divider is set to 1500. What is the VFO output frequency?

12. A frequency synthesizer has a phase detector input reference of 12.5 kHz. The divide ratio is 295. What are the output frequency and the frequency change increment?

13. The signal input power to a receiver is 6.2 nW. The noise power is 1.8 nW. What is the S/N ratio? What is the S/N ratio in decibels?

14. What is the noise voltage produced across a 50-Ω input resistance at a temperature of 25°C with a bandwidth of 2.5 MHz?

15. At what frequencies is noise temperature used to express the noise in a system?

16. The noise ratio of an amplifier is 1.8. What is the noise temperature in Kelvin?

CRITICAL THINKING

1. Why is a noise temperature of 155 K a better rating than a noise temperature of 210 K?

2. An FM transceiver operates on a frequency of 470.6 MHz. The first IF is 45 MHz and the second IF is 500 kHz. The transmitter has a frequency multiplier chain of 2 × 2 × 3. What three signals must the frequency synthesizer generate for the transmitter and two mixers in the receiver?

3. Explain how a digital counter could be connected to the receiver in Fig. 8-47 so that it would read the frequency of the signal to which it was tuned.

4. What is the effect on receiver selectivity if a resistor is connected in parallel with the tuned transformer in Fig. 8-29?

5. The circuits in a superheterodyne receiver have the following gains: RF amplifier, 8 dB; mixer, −2.5 dB; IF amplifier, 80 dB; demodulator, −0.8 dB; audio amplifier, 23 dB. What is the total gain?

6. A superheterodyne receiver receives a signal on 10.8 MHz. It is amplitude-modulated by a 700-Hz tone. The local oscillator is set to the signal frequency. What is the IF? What is the output of a diode detector demodulator?

R
E
V
I
E
W

CHAPTER 9

DIGITAL COMMUNICATIONS TECHNIQUES

Objectives

After completing this chapter, you will be able to:

◆ *Give* a step-by-step account of the transmission of analog signals using digital techniques.

◆ *Explain* how quantizing error occurs, *describe* the techniques used to minimize it, and *calculate* the minimum sampling rate given the upper frequency limit of the analog signal to be converted.

◆ *List* the advantages and disadvantages of the three most common types of analog-to-digital converters.

◆ *Explain* why pulse-code modulation has superseded pulse-amplitude modulation (PAM), pulse-width modulation (PWM), and pulse-position modulation (PPM).

◆ *Draw* and *fully label* a block diagram of a digital signal-processing (DSP) circuit.

Over the last two decades, digital methods of transmitting data have slowly but surely replaced the older, more conventional analog ones. One of the last areas of electronics to adopt digital methods was communications, especially radio communications. Radio communications has remained primarily analog in nature mainly because the type of information to be conveyed (e.g., voice and video) is analog and because of the very high frequencies involved. Until recently, digital circuits were not fast enough to handle the processing of radio signals and the information they convey. Now, however, the availability of fast, low-cost analog-to-digital (A/D) converters and digital-to-analog (D/A) converters and high-speed microprocessors makes the digital processing of analog and radio signals not only possible but, more important, practical.

This chapter begins with the reasons for using digital transmission. Then, the concepts and operation of A/D and D/A converters are summarized. Pulse modulation techniques are then described, and the chapter concludes with an introduction to digital signal processing (DSP), techniques that are truly changing how radio communications is conducted.

9-1 DIGITAL TRANSMISSION OF DATA

Data refers to information to be communicated. Data is in digital form if it comes from a computer. If the data is in the form of voice, video, or some other analog signal, it can be converted into digital form before it is transmitted.

Digital communications were originally limited to the transmission of data between computers. Numerous large and small networks have been formed to support communications between computers, for example, local area networks (LANs), which permit PCs to communicate (see Chap. 12). The use of modems to allow PCs and larger computers to communicate via the telephone system is another example. Now, because analog signals can be readily and inexpensively converted to digital and vice versa, data communications techniques can be used to transmit voice, video, and other analog signals in digital form.

There are three primary reasons for the growth of digital communications systems. First, the increased use of computers has made it necessary to find a way for computers to communicate and exchange data. Second, digital transmission methods offer some major benefits over analog communications techniques. Third, the telephone system, the largest and most widely used communications system, has been converting from analog methods to digital over the years. These reasons are discussed in the following sections.

PROLIFERATION OF COMPUTERS

Since PCs were introduced during the 1970s, their numbers have increased by several orders of magnitude so that tens of millions of PCs are now in use worldwide. Many white-collar workers have PCs on their desks. PCs have speeded up and simplified our work and increased our productivity.

At the same time, the need has arisen for users to share or exchange data and programs. At a simple level, data can be transferred by handing a diskette to another person or by mail. But it is far easier if computers can communicate directly. When computers are connected via some communications medium and programmed appropriately, they can exchange data or share programs and peripherals as well as other resources. The result is increased convenience and usefulness of computers.

Some common examples of computer data communications are as follows:

1. *File transfer.* The transfer of files, records, or whole databases such as company accounting records, customer orders from a sales office to the main office, or bank data transfers.
2. *Electronic mail (E-mail).* Communication between and among individuals by means of a computer. Users send messages to one another as they would letters or memos in printed form.
3. *Bulletin board systems (BBS).* Communication or sharing programs and data by means of a central computer that is the repository of data and programs that users tap into.
4. *Computer-peripheral links.* The use of data communications techniques to send data between a main computer and peripherals. Data is keyed into the computer, sent to and from floppy and hard disk drives for data storage or retrieval, and then sent to a printer.
5. *Information access.* Tapping into remote databases of information by way of the telephone system. On-line services and the Internet are examples.
6. *Local area networks (LANs).* Groups of PCs in an office or company that are connected in order to share data and other resources. This is the fastest-growing segment of data communications.

NONCOMPUTER USES OF DIGITAL COMMUNICATIONS

Among the noncomputer applications of digital techniques is remote control, for example:

1. *TV remote control.* Binary signals generated by pushbuttons modulate an infrared light beam for the purpose of changing channels or volume.
2. *Garage door opener.* Depressing a button on a remote control unit generates a unique binary code that modulates a VHF or UHF radio transmitter for the purpose of opening or closing a garage door.
3. *Carrier current controls.* These systems generate binary codes that modulate a carrier signal which is then superimposed on a 60 Hz-power signal. The AC power lines in a house or building become the transmission medium for control signals from one room or area to another. An example is the X-10 system widely used in homes for the remote control of lights and appliances.
4. *Radio control of models.* Hobbyists build model airplanes, boats, and cars and control them remotely by radio transmission of binary control codes.

BENEFITS OF DIGITAL COMMUNICATIONS

Transmitting information by digital means offers several important advantages over analog methods, as discussed in the next section and in Sec. 9-5.

NOISE IMMUNITY. When a signal is sent over a medium or channel, noise is invariably added to the signal. The S/N ratio decreases, and the signal becomes harder to recover. Noise, which is a voltage of randomly varying amplitude and frequency, easily corrupts analog signals. Signals of insufficient amplitude can be completely obliterated by noise. Some improvement can be achieved with pre-emphasis circuits at the transmitter and de-emphasis circuits at the receiver, and other similar techniques. If the signal is analog FM, the noise can be clipped off at the receiver so that the signal can be more readily recovered, but phase modulation of the signal by the noise will still degrade quality.

Digital signals, which are usually binary, are more immune to noise than analog signals are because the noise amplitude must be much higher than the signal amplitude to make a binary 1 look like a binary 0 or vice versa. If the binary amplitudes for binary 0 and 1 are sufficiently large, the receiving circuitry can easily distinguish between the 0 and 1 levels even with a significant amount of noise (see Fig. 9-1).

At the receiver, circuits can be set up so that the noise is clipped off. A threshold circuit made with a line receiver circuit, an op amp comparator, or a Schmitt trigger will trigger above or below the thresholds to which it is set. If the thresholds are set carefully, only the logic levels will trigger the circuit. Thus a clean output pulse will be generated by the circuit. This process is called *signal regeneration.*

Digital signals, like analog signals, experience distortion and attenuation when transmitted over a cable or by radio. The cable acts like a low-pass filter and thus filters out the higher harmonics in a pulse signal, causing the signal to be rounded and distorted. When a signal is transmitted by radio, its amplitude is seriously reduced. However, digital signals can be transmitted over long distances if the signal is regenerated along the way to restore the amplitude lost in the medium and to overcome the noise added in the process. When the signal reaches its destination, it has almost exactly the same shape as the original. Consequently, with digital transmission, the error rate is minimal.

ERROR DETECTION AND CORRECTION. With digital communications, transmission errors can usually be detected and even corrected. If an error occurs because of a very high noise level, it can be detected by special circuitry. The receiver recognizes that an error is contained in the transmission, and data can be retransmitted. A variety of techniques have been developed to find errors in binary transmissions; some of them are discussed in Chap. 11. In addition, elaborate error-detection schemes have been developed so that the type of error and its location can be identified. This kind of information makes it possible to correct errors before the data is used at the receiver.

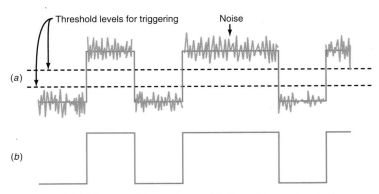

FIG. 9-1 (*a*) Noise on a binary signal. (*b*) Clean binary signal after regeneration.

COMPATIBILITY WITH TIME DIVISION MULTIPLEXING. Digital data communications is adaptable to time division multiplexing schemes. *Multiplexing* is the process of transmitting two or more signals simultaneously on a single communications channel or medium. There are two types of multiplexing: *Frequency division multiplexing,* an analog technique using modulation methods, and *time division multiplexing,* a digital technique. These techniques are discussed further in Chap. 10.

time division — one bit at a time serially

DIGITAL SIGNAL PROCESSING (DSP). DSP is the processing of analog signals by digital methods. This involves converting an analog signal into digital and then processing with a fast digital computer. *Processing* means filtering, equalization, phase shifting, and other traditionally analog methods. Processing also includes data compression techniques that enhance the speed of data transmission and reduce the digital data storage capacity required for some applications. The processing is accomplished by executing unique mathematical algorithms on the computer. The digital signal is then converted back into analog form. DSP permits significant improvements in processing over equivalent analog techniques. But best of all, it permits types of processing that were never available in analog form.

Finally, processing also involves storage of data. Analog data is difficult to store. But digital data is routinely stored in computers using a variety of well-proven digital storage methods and equipment such as RAM, ROM, floppy and hard disk drives, and tape units.

THE DISADVANTAGES OF DIGITAL COMMUNICATIONS

There are some disadvantages to digital communications. The most important is the bandwidth size required by a digital signal. With binary techniques, the bandwidth of a signal can be 10 or more times greater than it would be with analog methods. Also, digital communications circuits are usually more complex than analog circuits. However, although more circuitry is needed to do the same job, the circuits are usually in IC form, are inexpensive, and do not require much expertise or attention on the part of the user.

DID YOU KNOW?

With binary techniques, the bandwidth of a signal may be 10 or more times greater than it would be with analog methods.

9-2 DATA CONVERSION

The key to digital communications is to convert data in analog form into digital form. Special circuits are available to do this. Once it is in digital form, the data can be processed or stored. Data must usually be reconverted to analog form for final consumption by the user; for example, voice and video must be in analog form. Data conversion is the subject of the next section.

Translating an analog signal into a digital signal is called *analog-to-digital (A/D) conversion, digitizing a signal,* or *encoding.* The device used to perform this translation is known as an *analog-to-digital (A/D) converter* or *ADC.* A modern A/D converter is usually a single-chip IC that takes an analog signal and generates a parallel or serial binary output (see Fig. 9-2).

The opposite process is called *digital-to-analog (D/A) conversion.* The circuit used to perform this is called a *digital-to-analog (D/A) converter* (or *DAC*) or a *decoder.* The input to a D/A converter is usually a parallel binary number, and the output is a proportional analog voltage level. Like the A/D converter, a D/A converter is usually a single-chip IC (see Fig. 9-3).

A/D CONVERSION. An analog signal is a smooth or continuous voltage or current variation (see Fig. 9-4). It could be a voice signal, a video waveform, or a voltage representing a variation of some other physical characteristic such as temperature. Through A/D conversion these continuously variable signals are changed into a series of binary numbers.

A/D conversion is a process of sampling or measuring the analog signal at regular time intervals. At the times indicated by the vertical dashed lines in Fig. 9-4, the instantaneous value of the analog signal is measured and a proportional binary number is generated to represent that sample. As a result, the continuous analog signal is translated into a series of discrete binary numbers representing samples.

A key factor in the sampling process is the frequency of sampling *f,* which is the reciprocal of the sampling interval *t* shown in Fig. 9-4. In order to retain the high-frequency information in the analog signal, a sufficient number of samples must be taken so that the waveform is adequately represented. It has been found that the minimum sampling frequency is twice the highest analog frequency content of the signal. For example, if the analog signal contains a maximum frequency variation of 3000 Hz, the analog wave must be sampled at a rate of at least twice this, or 6000 Hz. This minimum sampling frequency is known as the *Nyquist frequency.*

Although theoretically the highest frequency component can be adequately represented by a sampling rate of twice the highest frequency, in practice the sampling rate is much higher than the Nyquist minimum, typically 2.5 to 3 times more. The actual sampling rate depends on the application as well as factors such as cost, complexity, channel bandwidth, and availability of practical circuits.

Assume, for example, that the output of an FM radio is to be digitized. The maximum frequency of the audio in an FM broadcast is 15 kHz. To ensure that the highest frequency is represented, the sampling rate must be twice the highest frequency: $f = 2 \times 15$ kHz $= 30$ kHz. But in practice, the sampling rate is made higher, that is,

[handwritten margin note:] Telephone
300 – 3000 Hz
3–10x faster for sampling
3 KHz – Sample between 6–9 KHz

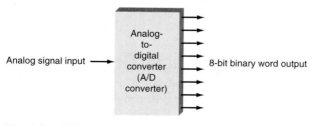

FIG. 9-2   A/D converter.

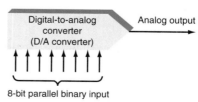

FIG. 9-3   D/A converter.

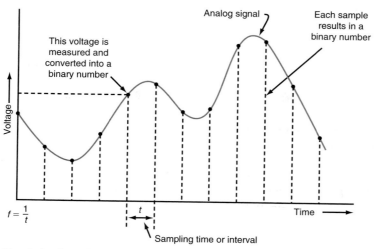

FIG. 9-4 Sampling an analog signal.

3 to 10 times higher, or 3 × 15 kHz = 45 kHz to 10 × 15 kHz = 150 kHz. The sampling rate for compact disk players that store music signals with frequencies up to about 20 kHz is 44.1 kHz or 48 kHz.

Another important factor in the conversion process is that because the analog signal is smooth and continuous, it represents an infinite number of actual voltage values. In a practical A/D converter, it is not possible to convert all analog samples into a precise proportional binary number. Instead, the A/D converter is capable of representing only a finite number of voltage values over a specific range. The samples are converted into a binary number whose value is close to the actual sample value. For example, an 8-bit binary number can represent only 256 states, which may be the converted values from an analog waveform having an infinite number of positive and negative values between +1 V and −1 V.

The physical nature of an A/D converter is such that it divides a voltage range into discrete increments, each of which is then represented by a binary number. The analog voltage measured during the sampling process is assigned to the increment of voltage closest to it. For example, assume that an A/D converter produces 4 output bits. With 4 bits, 2^4 or 16 voltage levels can be represented. For simplicity, assume an analog voltage range of 0 to 15 V. The A/D converter divides the voltage range as shown in Fig. 9-5. The binary number represented by each increment is indicated. Note that although there are 16 levels, there are only 15 increments. The number of levels is 2^N and the number of increments is $2^N - 1$, where N is the number of bits.

Now assume that the A/D converter samples the analog input and measures a voltage of 0 V. The A/D converter will produce a binary number as close as possible to this value, in this case 0000. If the analog input is 8 V, the A/D converter generates the binary number 1000. But what happens if the analog input is 11.7 V, as shown in Fig. 9-5? The A/D converter produces the binary number 1100, whose decimal equivalent is 12. In fact, any value of analog voltage between 11 and 12 V will produce this binary value.

HINTS AND HELPS

The primary benefit of an S/H amplifier is that it stores the analog voltage during the sampling interval, eliminating aperture error caused by input changes during the sampling interval.

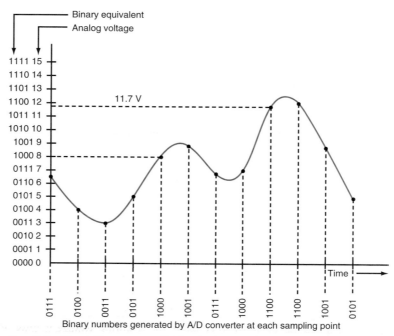

FIG. 9-5 The A/D converter divides the input voltage range into discrete voltage increments.

Sample and hold
Samples and holds
data to alow conversion
from analog to data

As you can see, there is some error associated with the conversion process. This is referred to as *quantizing error.* It can be reduced, of course, by simply dividing the analog voltage range into a larger number of smaller voltage increments. To represent more voltage increments, a greater number of bits must be used. For example, using 8 bits allows the analog voltage range to produce 2^8 or 256 voltage increments. This more finely divides the analog voltage range and thus permits the A/D converter to output a proportional binary number closer to the actual analog value. The greater the number of bits, the greater the number of increments over the analog range and the smaller the quantizing error.

The maximum amount of error can be computed by dividing the voltage range over which the A/D converter operates by the number of increments. Assume a 10-bit A/D converter, with 10 bits, $2^{10} = 1024$ voltage levels, or $1024 - 1 = 1023$ increments. Assume that the input voltage range is from 0 to 6 V. The minimum voltage step increment then is $6/1023 = 5.86 \times 10^{-3} = 5.865$ mV.

As you can see, each increment has a range of less than 6 mV. This is the maximum error that can occur; the average error is one-half that value.

D/A Conversion. In order to retain an analog signal converted to digital, some form of binary memory must be used. The multiple binary numbers representing each of the samples can be stored in random access memory (RAM), on disk, or on magnetic tape. Once they are in this form, the samples can be processed and can be used as data by a microcomputer which can perform mathematical and logical manipulations. This is called *digital signal processing (DSP)* and is discussed in Sec. 9-5.

At some point it is usually desirable to translate the multiple binary numbers back into the equivalent analog voltage. This is the job of the D/A converter, which receives the binary numbers sequentially and produces a proportional analog voltage at the output. Because the input binary numbers represent specific voltage levels, the output of

Example 9-1

An information signal to be transmitted digitally is a rectangular wave with a period of 71.4 μs. It has been determined that the wave will be adequately passed if the bandwidth includes the fourth harmonic. Calculate (*a*) the signal frequency, (*b*) the fourth harmonic, (*c*) the minimum sampling frequency (Nyquist rate).

a. $f = \dfrac{1}{t} = \dfrac{1}{71.4 \times 10^{-6}} = 14{,}006 \text{ Hz} \cong 14 \text{ kHz}$

b. $f_{\text{4th harmonic}} = 4 \times 14 \text{ kHz} = 56 \text{ kHz}$

c. Minimum sampling rate $= 2 \times 56 \text{ kHz} = 112 \text{ kHz}$

the D/A converter has a stairstep characteristic. Figure 9-6 shows the process of converting the 4-bit binary numbers obtained in the conversion of the waveform in Fig. 9-5. If these binary numbers are fed to a D/A converter, the output is a stairstep voltage as shown. Since the steps are very large, the resulting voltage is only an approximation to the actual analog signal. However, the stairsteps can be filtered out by passing the D/A converter output through a low-pass filter with an appropriate cutoff frequency.

If the binary words contain a larger number of bits, the analog voltage range is be divided into smaller increments and the output step increments will be smaller. This leads to a closer approximation to the original analog signal.

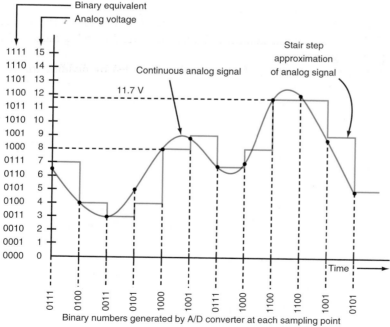

FIG. 9-6 A D/A converter produces a stepped approximation of the original signal.

A D/A converter consists of four major sections, as shown in Fig. 9-7 and described in the next sections.

REFERENCE REGULATORS. The precise reference voltage regulator, a zener diode, receives the DC supply voltage as an input and translates it into a highly precise reference voltage. This voltage is passed through a resistor which establishes the maximum input current to the resistor network and sets the precision of the circuit. The current is called the *full-scale current,* or I_{FS}:

$$I_{FS} = \frac{V_R}{R_R}$$

where V_R = reference voltage
R_R = reference resistor

RESISTOR NETWORKS. The precision resistor network is connected in a unique configuration. The voltage from the reference is applied to this resistor network, which converts the reference voltage into a current proportional to the binary input. The output of the resistor network is a current that is directly proportional to the binary input value and the full-scale reference current. Its maximum value is computed as follows:

$$I_{out} = \frac{I_{FS}(2^N - 1)}{2^N}$$

For an 8-bit D/A converter, N = 8.

Some modern D/A converters use a capacitor network instead of the resistor network to perform the conversion from a binary number to a proportional current.

OUTPUT AMPLIFIERS. The proportional current is then converted by an op amp into a proportional voltage. The output of the resistive network is connected to the summing junction of the op amp. The output voltage of the op amp is equal to the output current of the resistor network multiplied by the feedback resistor value. If the appropriate value of feedback resistance is selected, the output voltage can be scaled to any desired value. The op amp inverts the polarity of the signal:

$$V_{out} = -I_0 R_f$$

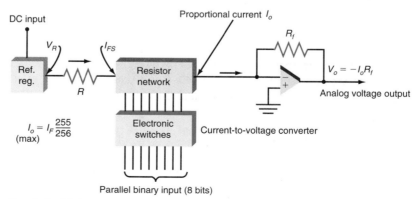

FIG. 9-7 Major components of a D/A converter.

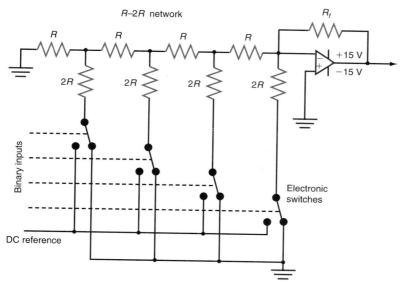

Fig. 9-8 D/A converter with R–$2R$ ladder network.

ELECTRONIC SWITCHES. The resistor network is modified by a set of electronic switches that can be either current or voltage switches and are usually implemented with diodes or transistors. These switches are controlled by the parallel binary input bits from a counter, a register, or a microcomputer output port. The switches turn ON or OFF to configure the resistor network.

All the components shown in Fig. 9-7 are usually integrated onto a single IC chip. The only exception may be the amp, which is often an external circuit.

D/A converters of this type are available in a variety of configurations and can convert 8-, 10-, 12-, 14-, and 16-bit binary words.

The implementation of D/A converter circuitry varies widely. One of the most popular configurations is shown in detail in Fig. 9-8. Only 4 bits are shown, to simplify the drawing. Of particular interest is the resistor network, which uses only two values of resistance and thus is known as an R–$2R$ ladder network. More complex networks have been devised but use a wider range of resistor values which are difficult to make with precise values in IC form. In Fig. 9-8 the switches are shown as mechanical devices, whereas in reality they are transistor switches controlled by the binary input. Many newer D/A converters and A/D converters use a capacitive network instead of the R–$2R$ network.

D/A CONVERTER SPECIFICATIONS. Three important specifications are associated with D/A converters: resolution, error, and settling time.

Resolution refers to the total number of increments the DAC produces over its output voltage range. Resolution, which is directly related to the number of bits, is the smallest voltage increment change possible. It is computed by dividing the reference voltage V_R by the number of output steps, $2^N - 1$. There are one fewer increments than the number of binary states.

For a 10-V reference and an 8-bit D/A converter the resolution is 10 ($2^8 - 1$) = 10/255 = 0.039 V = 39 mV.

For high-precision applications, D/A converters with larger input words should be used. D/A converters with 8 and 12 bits are the most common, but D/A converters with 10, 14, 16, 20, and 24 bits are available.

Error is expressed as a percentage of the maximum, or full-scale, output voltage, which is the reference voltage value. Typical error figures are less than ±0.1 percent. This error should be less than one-half the minimum increment. The smallest increment of an 8-bit D/A converter with a 10-V reference is 0.039 V or 39 mV. Expressed as a percentage, this is $0.039 \div 10 = 0.0039 \times 100 = 0.39$ percent. One-half of this is 0.195 percent. With a 10-V reference, this represents a voltage of $0.00195 \times 10 = 0.0195$ V, or 19.5 mV. A stated error of 0.1 percent of full scale is $0.001 \times 10 = 0.01$ V, or 10 mV.

Settling time is the amount of time it takes for the output voltage of a D/A converter to stabilize to within a specific voltage range after a change in binary input. When a binary input change occurs, a finite amount of time is needed for the electronic switches to turn ON and OFF and for any circuit capacitance to charge or discharge. During the change, the output rings and overshoots, and contains transients from the switching action. The output is thus not an accurate representation of the binary input; it is not usable until it settles down.

Settling time is the time it takes for the D/A converter's output to settle to within ±1/2 least significant bit (LSB) change. In the case of the 8-bit D/A converter described earlier, when the output voltage settles to less than one-half the minimum voltage change of 39 mV or 19.5 mV, the output can be considered stable. Typical settling times are in the 100-ns range. This specification is important because it determines the maximum speed of operation of the circuit. A 100-ns settling time translates into a frequency of $1/100 \times 10^{-9} = 10$ MHz. Operations faster than this result in output errors.

A/D CONVERTERS

A/D conversion begins with the process of sampling, which is usually carried out by a sample and hold (S/H) circuit. The S/H circuit takes a precise measurement of the analog voltage at specified intervals. The A/D converter then converts this instantaneous value of voltage and translates it to a binary number.

S/H CIRCUITS. A *sample and hold (S/H) circuit,* also called a *track/store circuit,* accepts the analog input signal and passes it through, unchanged, during its sampling mode. In the hold mode, the amplifier remembers or memorizes a particular voltage level at the instant of sampling. The output of the S/H amplifier is a fixed DC level whose amplitude is the value at the sampling time.

Figure 9-9 is a simplified drawing of an S/H amplifier. The main element is a high-gain DC differential op amp. The amplifier is connected as a follower with 100 percent feedback. Any signal applied to the noninverting (+) input is passed through unaffected. The amplifier has unity gain and no inversion.

A storage capacitor is connected across the very high input impedance of the amplifier. The input signal is applied to the storage capacitor and the amplifier input through a MOSFET gate. An enhancement-mode MOSFET that acts like an ON-OFF switch is normally used. As long as the control signal to the gate of the MOSFET is held high, the input signal will be connected to the op amp input and capacitor. When the gate is high, the transistor turns on and acts like very low-value resistor, connecting the input signal to the amplifier. The charge on the capacitor follows the input sig-

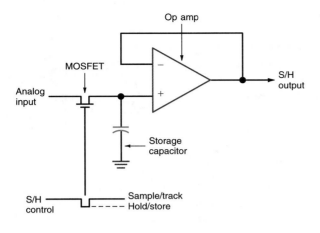
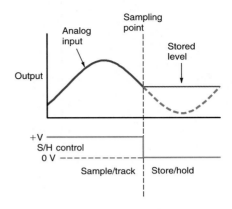

Fig. 9-9 An S/H amplifier.

nal. This is the sample or track mode for the amplifier. The op amp output is equal to the input.

When the S/H control signal goes low, the transistor is cut off, but the charge on the capacitor remains. The very high-input impedance of the amplifier allows the capacitor to retain the charge for a relatively long period of time. The output of the S/H amplifier, then, is the voltage value of the input signal at the instant of sampling, that is, the point at which the S/H control pulse switches from high (sample) to low (hold). The op amp output voltage is applied to the A/D converter for conversion into a proportional binary number.

The primary benefit of an S/H amplifier is that it stores the analog voltage during the sampling interval. In some high-frequency signals, the analog voltage may increase or decrease during the sampling interval; this is undesirable because it confuses the A/D converter and introduces what is referred to as *aperture error*. The S/H amplifier, however, stores the voltage on the capacitor; with the voltage constant during the sampling interval, quantizing is accurate.

There are many ways to translate an analog voltage into a binary number. The next sections describe the most common ones.

COUNTER CONVERTERS. One of the simplest A/D converters is the counter converter shown in Fig. 9-10. It consists of a binary counter, a D/A converter, and an analog voltage comparator. A 4-bit counter and D/A converter are shown for simplicity. The comparator is a very high-gain differential amplifier op amp optimized for switching. The analog signal to be converted is applied to one input of the comparator. This input would normally come from the S/H amplifier output.

The other input to the comparator comes from the D/A converter. If the voltage output of the D/A converter is less than that of the analog input voltage, the comparator output will be a binary 1. This binary 1 controls an AND gate, which, in turn, controls the application of clock pulses to the binary counter. If the output of the D/A converter is equal to or greater than the analog input voltage, the comparator output will be a binary 0.

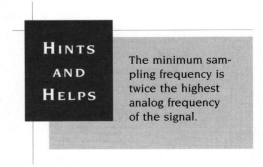

HINTS AND HELPS

The minimum sampling frequency is twice the highest analog frequency of the signal.

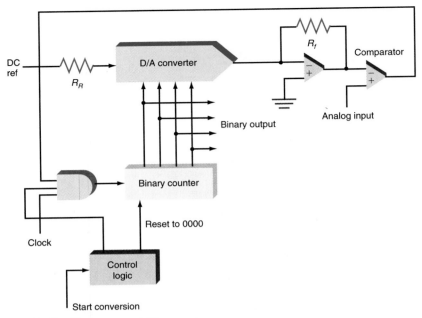

FIG. 9-10 Counter-type A/D converter.

The conversion process begins when a start-conversion pulse is applied to the control logic. This causes the counter to be reset to zero (0000), producing a zero output voltage from the D/A converter. Note, in Fig. 9-10, the symbol used to represent a D/A converter in schematic and logic diagrams. The analog input voltage is greater than the D/A converter output at this time, so the output of the comparator is binary 1. This enables the AND gate, which allows clock pulses to increment the counter. As the counter is incremented, the output from the D/A converter begins to increase step by step as shown in Fig. 9-11. With a 4-bit binary counter and D/A converter, a total of 16 steps or increments can be generated. The counter continues to increment, and the D/A converter output voltage continues to rise. When the D/A converter output voltage is incremented to the point at which it is greater than the analog input voltage value, the comparator switches quickly, producing a binary 0 output and inhibiting the AND gate. No further clock pulses reach the counter. At this time, the binary number retained in the counter is proportional to the analog input value. The counter output is the A/D converter output.

For this example, assume a 5-V reference. The resolution is $5/15 = 0.3333$ V. The smallest step is 0.3333 V, and the error is no more than ±0.16667 V.

Now assume a reference resistor R_R of 5 kΩ on the D/A converter. This means that the reference current $I_{FS} = V_R/R_R = 5/5000 = 0.001 = 1$ mA.

The maximum output current $I_0 = I_{FS}(2^N - 1)/2^N = 1(15/16) = 0.9375$ mA. This current is applied to the op amp summing junction. If a feedback resistor of 5 kΩ is used, the maximum output voltage $V_0 = I_0R_f = 0.9375 \times 10^{-3}(5000) = 4.6375$ V. The minimum output voltage with one increment is:

$$I_0 = \frac{I_{FS}}{(2^N - 1)} = 1(1/15) = 0.667 \text{ mA}$$
$$V_0 = I_0R_f = 0.667 \times 10^{-3}(5000) = 0.3333 \text{ V}$$

Now assume an input voltage to the A/D converter of 3.5 V. The D/A converter will generate a stairstep with 0.3333 increments until 3.5 V is just exceeded. This will take $3.5/0.3333 = 10.5$, or 11 steps. The counter will contain the number 11.

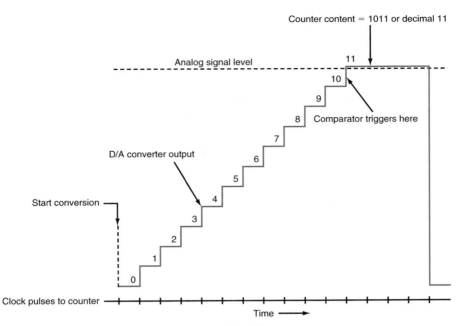

Fig. 9-11 D/A converter output in counter-type A/D converter.

The op amp output voltage is

$$I_0 = I_{FS} = 1(11/15) = 0.7333 \text{ mA.}$$
$$V_0 = I_0 R_f = 0.7333 \times 10^{-3}(5000) = 3.6667 \text{ V}$$

With the op amp output more than the input, the comparator will switch and stop the counter. The binary number in the counter and the A/D converter output is 1011, the number representing 3.6667 V, which is the closest to the actual input of 3.5 V.

The counter A/D converter is extremely simple and effective; however, it suffers from some disadvantages that keep it from being widely used. First, it is slow. If the input voltage is high, the counter may have to count up to the maximum number of possible counts. With an 8-bit D/A there are 255 steps. If the clock is 1 MHz, the count period is 1 μs. This means that a conversion could take up to $255 \times 1 \ \mu s = 255 \ \mu s$. The sampling period can be no less than this amount. As a result, the upper frequency limit of the analog signal is restricted. A sampling period of 255 μs translates to a sampling frequency of $1/255 \times 10^{-6} = 3922$ Hz. Since the sampling frequency has to be at least twice the highest frequency in the input analog signal, the upper frequency limit is $3922/2 = 1961$ Hz. This makes the counter-type A/D converter good only at very low frequencies. Second, the time for a conversion varies with the amplitude of the analog signal, and unequal conversion times can lead to a nonuniform sampling rate. This is corrected by sampling at the speed set by the maximum conversion time. A variation of this technique, called the *successive approximations method,* overcomes these problems.

SUCCESSIVE APPROXIMATIONS CONVERTERS. This converter is an improvement over the counter A/D converter (see Fig. 9-12). This is an 8-bit A/D converter, for it uses an 8-bit D/A converter and an 8-bit *successive approximations register (SAR).* The operation of this ADC is similar to that of the counter A/D converter except for the special SAR, which replaces the counter. Special logic in the register causes each bit

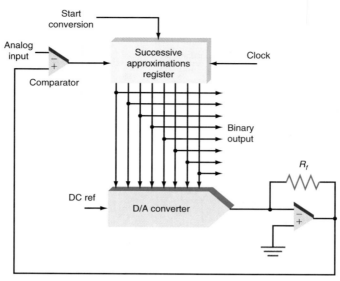

FIG. 9-12 Successive approximations A/D converter.

to be turned on one at a time from MSB to LSB until the closest binary value is stored in the register. The clock input signal sets the rate of turning the bits off and on.

Assume that the SAR is initially reset to zero. When the conversion is started, the MSB is turned on, producing 10 000 000 at the output and causing the D/A converter output to go to half-scale. The D/A converter output is applied to the op amp, which applies it to the comparator along with the analog input. If the D/A converter output is greater than the input, the comparator signals the SAR to turn off the MSB. The next MSB is turned on. The D/A converter output goes to the proportional analog value, which is again compared to the input. If the D/A converter output is still greater than the input, the bit will be turned off; if the D/A converter output is less than the input, the bit will be left at binary 1.

The next MSB is then turned on and another comparison made. The process continues until all 8 bits have been turned ON or OFF and eight comparisons have been made. The output is a proportional 8-bit binary number. With a clock frequency of 200 kHz, the clock period is $1/200 \times 10^3 = 5$ μs. Each bit decision is made during the clock period. For eight comparisons at 5 μs each, the total conversion time is $8 \times 5 = 40$ μs.

Successive approximations converters are fast and consistent. They are available with conversion times from about 2 to 200 μs, and 8-, 10-, 12-, and 16-bit versions are available. Most IC A/D converters in use are the successive approximations type.

FLASH CONVERTERS. A *flash converter* takes an entirely different approach to the A/D conversion process. It uses a large resistive voltage divider and multiple analog comparators. The number of comparators required is equal to $2^N - 1$, where N is the number of desired output bits. A 3-bit A/D converter requires $2^3 - 1 = 8 - 1 = 7$ comparators (see Fig. 9-13).

The resistive voltage divider divides the DC reference voltage range into a number of equal increments. Each tap on the voltage divider is connected to a separate analog comparator. All the other comparator inputs are connected together and driven by the analog input voltage. Some comparators will be ON and others will be OFF, depending on the actual value of input voltage. The comparators operate in such a way

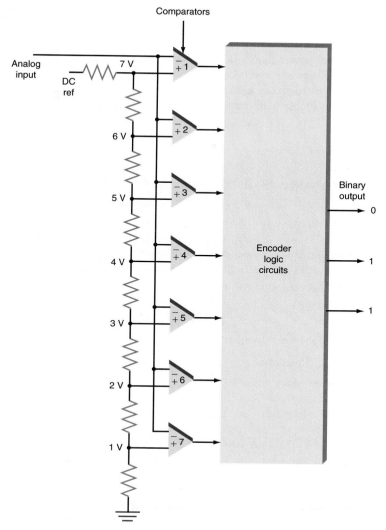

Fig. 9-13 A flash converter.

that, if the analog input is greater than the reference voltage at the divider tap, the comparator output will be binary 1. For example, if the analog input voltage in Fig. 9-13 is 4.5 V, the outputs of comparators 4, 5, 6, and 7 will be binary 1. The other comparator outputs will be binary 0. The encoder logic, which is a special combinational logic circuit, converts the 7-bit input from the comparators into a 3-bit binary output.

The counter and successive approximations converters generate output voltage after the circuits go through their decision-making process. The flash converter, on the other hand, produces a binary output almost instantaneously. Counters do not have to be incremented and a sequence of bits in a register do not have to be turned on and off. Instead, the flash converter produces an output as fast as the comparators can switch and the signals can be translated into binary levels by the logic circuits. Comparator switching and logic propagation delays are extremely short. Flash converters, therefore, are the fastest type of A/D converter. Conversion speeds of less than 100 ns are typical, and speeds of less than 50 ns are possible. Flash A/D converters are complicated and expensive because of the large number of analog comparators required for large binary numbers. The total number of comparators required is based upon the power of 2. An 8-bit flash converter has $2^8 - 1 = 255$ comparator circuits. Obviously, ICs requiring this many components are large and difficult to make. They also consume much more power than a digital circuit because the comparators are linear circuits. Yet for high-speed conversions, they are the best choice. With the high speed they can achieve, high-frequency signals like video signals can be easily digitized. Flash converters are available with output word lengths of 6, 8, 10, and 12 bits.

Example 9-2

The voltage range of an A/D converter that uses 14-bit numbers is -6 to $+6$ V. Find (*a*) the number of discrete levels (binary codes) that are represented, (*b*) the number of voltage increments used to divide the total voltage range, and (*c*) the resolution of digitization expressed as the smallest voltage increment.

a. $2^N = 16,384$

b. $2^N - 1 = 2^{14} - 1 = 16,384 - 1 = 16,383$

c. The total voltage range is -6 to $+6$ V, or 12 V; thus,

$$\text{Resolution} = \frac{12}{16,383} = 0.7325 \text{ mV or } 732.5 \ \mu V$$

modem–serial transmitting device

9-3 PARALLEL AND SERIAL TRANSMISSION

There are two ways to move binary bits from one place to another: transmit all bits of a word simultaneously or send only 1 bit at a time. These methods are referred to respectively as parallel transfer and serial transfer.

PARALLEL TRANSFER

In *parallel* data transfers, all the bits of a code word are transferred simultaneously (see Fig. 9-14). The binary word to be transmitted is usually loaded into a register containing one flip-flop for each bit. Each flip-flop output is connected to a wire to carry that bit to the receiving circuit, which is usually also a storage register. As can be seen in the figure, in parallel data transmission, there is one wire for each bit of information to be transmitted. This means that a multiwire cable must be used. Multiple parallel lines that carry binary data are usually referred to as a *data bus*. All eight lines are referenced to a common ground wire.

Parallel data transmission is extremely fast because all the bits of the data word are transferred simultaneously. The speed of parallel transfer depends on the propagation delay in the sending and receiv-

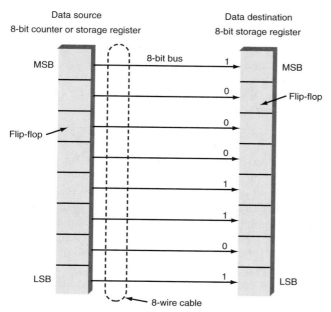

Fig. 9-14 Parallel data transmission.

ing of logic circuits and any time delay introduced by the cable. Such data transfers can occur in only a few nanoseconds in many applications.

Parallel data transmission is not practical for long-distance communications. To transfer an 8-bit data word from one place to another, eight separate communications channels are needed, one for each bit. Although multiwire cables can be used over limited distances (usually no more than 20 ft or so), for long-distance data communications they are impractical because of cost and signal attenuation. And, of course, parallel data transmission by radio would be even more complex and expensive, because one transmitter and receiver would be required for each bit.

SERIAL TRANSFER

Data transfers in communications systems are made serially; each bit of a word is transmitted one after another (see Fig. 9-15). This figure shows the code 10011101 being transmitted 1 bit at a time. The LSB is transmitted first, and the MSB last. The MSB is on the right, indicating that it was transmitted later in time than the LSB. Each bit is transmitted for a fixed interval of time t. The voltage levels representing each bit ap-

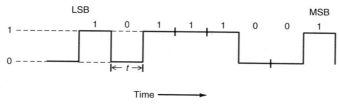

Fig. 9-15 Serial data transmission.

pear on a single data line (with respect to ground) one after another until the entire word has been transmitted. For example, the bit interval may be 10 μs, which means that the voltage level for each bit in the word appears for 10 μs. It would therefore take 80 μs to transmit an 8-bit word.

SERIAL-PARALLEL CONVERSION

Because both parallel and serial transmission occur in computers and other equipment, there must be techniques for converting between parallel and serial and vice versa. Such data conversions are usually taken care of by shift registers (see Fig. 9-16).

A *shift register* is a sequential logic circuit made up of a number of flip-flops connected in cascade. The flip-flops are capable of storing a multibit binary word, which is usually loaded in parallel into the transmitting register. When a clock pulse (CP) is applied to the flip-flops, the bits of the word are shifted from one flip-flop to another in sequence. The last (right-hand) flip-flop in the transmitting register ultimately stores each bit in sequence as it is shifted out.

UART–
USART–
parallel to serial
data transfer

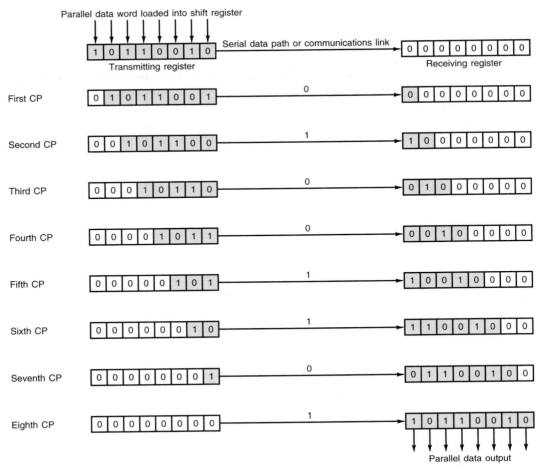

FIG. 9-16   Parallel-to-serial and serial-to-parallel data transfers with shift registers.

The serial data word is then transmitted over the communications link and is received by another shift register. The bits of the word are shifted into the flip-flops one at a time until the entire word is contained within the register. The flip-flop outputs can then be observed and the data stored in them transferred in parallel to other circuits. These serial-parallel data transfers take place inside circuits referred to as *interfaces* in the computer and peripheral equipment.

9-4 PULSE-CODE MODULATION

The most widely used technique for digitizing information signals for electronic data transmission is pulse-code modulation (PCM). PCM signals are serial digital data. There are two ways to generate them. The more common is to use an S/H circuit and traditional A/D converter to sample and convert the analog signal into a sequence of binary words, convert the parallel binary words into serial form, and transmit the data serially, one bit at a time. The second way is to use a special method of A/D conversion that generates a serial data signal directly.

TRADITIONAL PCM

In traditional PCM, the analog signal is sampled and converted into a sequence of parallel binary words. A successive approximations A/D converter is the most common way of doing this. The parallel binary output word is converted into a serial signal by a shift register (see Fig. 9-17). Each time a sample is taken, an 8-bit word is generated by the A/D converter. This word must be transmitted serially before another sample is taken and another binary word is generated. The clock and start conversion signals are synchronized so that the resulting output signal is a continuous train of binary words.

Figure 9-18 shows the timing signals. The start conversion signal triggers the S/H to hold the sampled value and starts the A/D converter. Once the conversion is complete, the parallel word from the A/D converter is transferred to the shift register. The clock pulses start shifting the data out a bit at a time. When one 8-bit word has been transmitted, another conversion is initiated and the next word is transmitted. In Fig. 9-18, the first word sent is 01010101; the second word is 00110011.

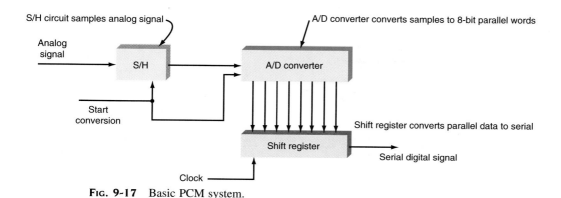

FIG. 9-17 Basic PCM system.

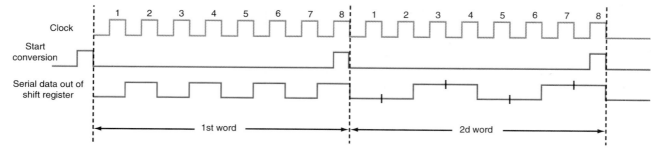

FIG. 9-18   Timing signals for PCM.

At the receiving end of the system, the serial data is shifted into a shift register (see Fig. 9-19). The clock signal is derived from the data to ensure exact synchronization with the transmitted data. The process of clock recovery will be discussed in Chap. 11. Once one 8-bit word is in the register, the D/A converter converts it into a proportional analog output. Thus the analog signal is reconstructed one sample at a time as each binary word representing a sample is converted into the corresponding analog value. The D/A converter output is a stepped approximation of the original signal. This signal may be passed through a low pass filter to smooth out the steps.

DELTA MODULATION

Delta modulation is a special form of A/D conversion that results in a continuous serial data signal being transmitted. The delta modulator looks at a sample of the analog input signal, compares it to a previous sample, and then transmits a 0 or a 1 if the sample is less than or more than the previous sample.

In Fig. 9-20, the analog signal is sampled by an S/H circuit, as in any other form of A/D converter. The sample is also applied to a comparator as in other A/D converter circuits. The other input to the comparator comes from a D/A converter driven by an up-down counter. The counter counts up (increments) or down (decrements) depending on the output state of the comparator. The comparator output is also the serial data signal representing the analog value.

Assume that the counter is initially at zero. This means that the D/A converter output will be zero. The analog input and S/H output are at some nonzero value, causing the comparator output to be binary 1. A binary 1 output sets the counter to count up.

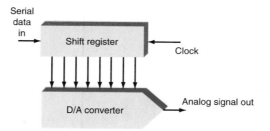

FIG. 9-19   PCM to analog translation at the receiver.

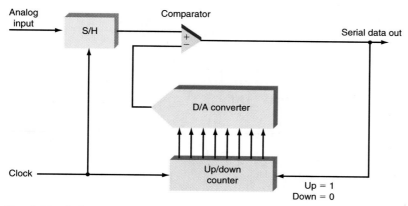

Fɪɢ. 9-20 Delta modulator.

The clock increments the counter, causing the D/A converter output to rise a step at a time. As long as the D/A converter output is less than the analog input value, the comparator output will be binary 1 and the counter will continue to count up and the D/A converter output to rise a step at a time. When the D/A converter output exceeds the analog input by one increment, the comparator output switches to binary 0. Figure 9-21 shows the various signals in the circuit.

If the analog input decreases, the comparator output will be binary 0. The comparator compares the current analog sample to the previous sample that appears at the D/A converter output. The D/A converter output is always one clock period behind. If the analog signal continues to decrease, the comparator output remains binary 0, as Fig. 9-21 shows. If the analog signal is constant, it will not change with each sample; therefore, the comparator just switches between 0 and 1.

Basically, a delta modulator is a 1-bit A/D converter. It does not transmit the absolute value of a sample. Instead, it transmits a 0 or 1, indicating whether the new sample is higher or lower than the previous sample. The resolution of the D/A converter establishes the minimum value of the step.

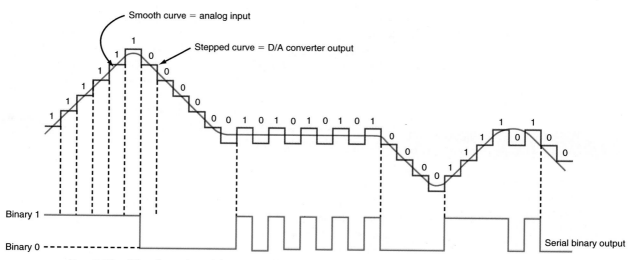

Fɪɢ. 9-21 Waveforms in a delta modulator A/D converter.

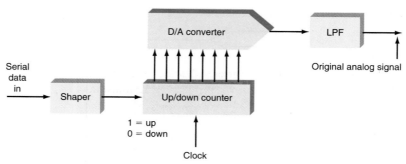

FIG. 9-22 A delta demodulator.

A delta demodulator is shown in Fig. 9-22. It is really a type of D/A converter. The serial data signal controls an up-down counter. A clock steps the counter, which drives a D/A converter. The D/A converter reproduces the stepped approximation shown in Fig. 9-21. A low-pass filter (LPF) on the output of the D/A converter removes the steps and smoothes the wave into its original form.

COMPANDING

Companding is a process of signal compression and expansion that is used to overcome problems of distortion and noise in the transmission of audio signals.

The range of voice amplitude levels in the telephone system is approximately 1000 to 1. In other words, the largest-amplitude voice peak is approximately 1000 times the smallest voice signal or 1000:1, representing a 60-dB range. If a quantizer with 1000 increments were used, very high quality analog signal representation would be achieved. For example, an A/D converter with a 10-bit word can represent 1024 individual levels. A 10-bit A/D converter would provide excellent signal representation. If the maximum peak audio voltage were 1 V, the smallest voltage increment would be 1/1023 of this, or 0.9775 mV.

As it turns out, it is not necessary to use that many quantizing levels for voice, and in most practical PCM systems, a 7- or 8-bit A/D converter is used for quantizing. One popular format is to use an 8-bit code, where 7 bits represent 128 amplitude levels and the eighth bit designates polarity (0 = +, 1 = −). Overall, this provides 255 levels; approximately half are positive, and the remainder negative.

Although the analog voltage range of the typical voice signal is approximately 1000:1, lower-level signals predominate. Most conversations take place at a low level, and the human ear is most sensitive in the low-amplitude range. Thus the upper end of the quantizing scale is not often used.

Since most signals are low level, quantizing error is relatively large. That is, small increments of quantization become a large percentage of the lower-level signal. This is a small amount of the peak amplitude value, of course, but this fact is irrelevant when the signals are low in amplitude. The increased quantizing error can produce garbled or distorted sound.

In addition to their potential for increasing quantizing error, low-level signals are also susceptible to noise. Noise represents random

DID YOU KNOW?

Companding is the most common means of overcoming the problems of quantizing error and noise.

spikes or voltage impulses added to the signal. The result is static that interferes with the low-level signals and makes intelligibility difficult.

Companding is the most common means of overcoming the problems of quantizing error and noise. At the transmitting end of the system, the voice signal to be transmitted is compressed; that is, its dynamic range is decreased. The lower-level signals are emphasized, and the higher-level signals are de-emphasized. Compression can take place prior to quantizing. But in some systems, companding is accomplished digitally in the A/D converter by means of unequal quantizing steps, small ones at low levels and larger ones at higher levels.

At the receiving end, the recovered signal is fed to an expander circuit that does the opposite, de-emphasizing the lower-level signals and emphasizing the higher-level signals, thereby returning the transmitted signal to its original condition. Companding greatly improves the quality of the transmitted signal.

Originally, companding circuits were analog, and the concept is most easily understood when described in analog terms. One type of compression circuit is a nonlinear amplifier that amplifies lower-level signals more than it does upper-level signals. Figure 9-23 illustrates the companding process. The curve shows the relationship between the input and output of the compander. At the lower input voltages, the gain of the amplifier is high and produces high output voltages. As the input voltage increases, the curve begins to flatten, producing proportionately lower gain. The nonlinear curve compresses the upper-level signals while bringing the lower-level signals to a higher amplitude. Such compression greatly reduces the dynamic range of the audio signal. Compression reduces the customary ratio of 1000:1 to approximately 60:1. The degree of compression can be controlled by careful design of the gain characteristics of the compression amplifier, in which case the 60-dB voice range can be reduced to more like 36 dB.

In addition to minimizing quantizing error and the effects of noise, compression lowers the dynamic range so that fewer binary bits are required to digitize the audio

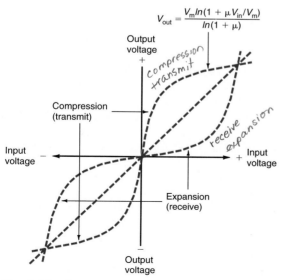

$$V_{out} = \frac{V_m \ln(1 + \mu V_{in}/V_m)}{\ln(1 + \mu)}$$

FIG. 9-23 Compression and expansion curves.

signal. A 64:1 voltage ratio could be easily implemented with a 6-bit A/D converter, but in practice, a 7-bit A/D converter is used.

Two basic types of companding are used in telephone systems: the *μ-law* (pronounced "mu law") compander and the A-law compander. The two companders differ slightly in their compression and expansion curves. The $μ$-law compander is used in telephone systems in the United States and Japan, and the A-law compander is used in European telephone networks. The two are incompatible, but conversion circuits have been developed to convert $μ$-law to A-law and vice versa. According to international telecommunications regulations, users of $μ$-law companders are responsible for the conversions. The voltage formulas for both are as follows:

$μ$-law:
$$V_{out} = \frac{V_m \ln (1 + μV_{in}/V_m)}{\ln (1 + μ)}$$

A-law:
$$V_{out} = \frac{1 + \ln AV_{in}/V_m}{(1 + \ln A)}$$

where V_{out} = output voltage
V_m = maximum possible input voltage
V_{in} = instantaneous value of input voltage

The value of $μ$ is usually 255; A is usually 87.6.

Example 9-3

The input voltage of a compander with a maximum voltage range of 1 V and a $μ$ of 255 is 0.25. What are the output voltage and gain?

$$V_{out} = \frac{V_m \ln (1 + μV_{in}/V_m)}{\ln (1 + μ)}$$
$$= \frac{1 \ln [1 + (255 \, (0.25)/1)]}{\ln (1 + 255)} = \frac{\ln 64.75}{\ln 256} = \frac{4.17}{5.55} = 0.75 \text{ V}$$
$$\text{Gain} = \frac{V_{out}}{V_{in}} = \frac{0.75}{0.25} = 3$$

Example 9-4

The input to the compander of Example 9-3 is of 0.8 V. What are the output voltage and gain?

$$V_{out} = \frac{V_m \ln (1 + μV_{in}/V_m)]}{\ln (1 + μ)}$$
$$= \frac{[1 \ln [1 + 255(0.8)]/1]}{\ln (1 + 255)} = \frac{\ln 205}{\ln 256} = \frac{5.32}{5.55} = 1.02 \text{ V}$$
$$\text{Gain} = \frac{V_{out}}{V_{in}} = \frac{0.96}{0.8} = 1.2$$

As the examples show, the gain of a compander is higher at the lower input voltages than at the higher input voltages.

Older companding circuits used analog methods such as the nonlinear amplifiers described earlier. Today, most companding is digital. One method is simply to use a nonlinear A/D converter. These converters provide a greater number of quantizing steps at the lower levels than at the higher levels, providing compression. On the receiving end, a matching nonlinear D/A converter is used to provide the opposite compensating expansion effect.

CODECS

coder/decoder

Both ends of the communications link in telephone systems have transmitting and receiving capability. All A/D and D/A conversion and related functions such as serial-to-parallel and parallel-to-serial conversion as well as companding are usually taken care of by a single large-scale IC chip known as a *codec*. One codec is used at each end of the communications channel. Codecs are usually combined with digital multiplexers and demultiplexers; clock and synchronizing circuits complete the system. These elements are discussed in Chap. 10.

Figure 9-24 is a simplified block diagram of a codec. The analog input is sampled by the S/H amplifier at an 8-kHz rate. The samples are quantized by the successive approximations type of A/D converter. Compression is done digitally in the A/D converter. The parallel A/D converter output is sent to a shift register to create the serial data output which usually goes to one input of a digital multiplexer.

The serial digital input is generally derived from a digital demultiplexer. The clock shifts the binary words representing the voice into a shift register for serial-to-parallel conversion. The 8-bit parallel word is sent to the D/A converter, which has built-in digital expansion. The analog output is then buffered, and it may be filtered externally. Most codecs are made with a complementary metal oxide semiconductor (CMOS) and come in a 16-pin dual inline (DIP) IC package.

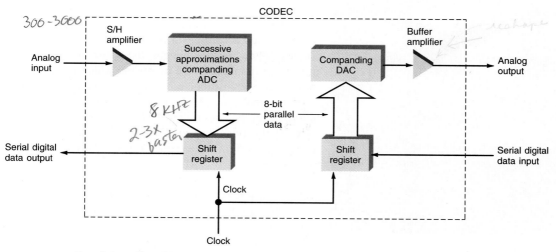

FIG. 9-24 Simplified block diagram of an IC codec.

Pulse modulation is the process of changing a binary pulse signal to represent the information to be transmitted.

Probably the primary benefits of transmitting information by binary techniques arise from the great noise tolerance and the ability to regenerate the degraded signal. Any noise that gets added to the binary signal along the way is usually clipped off. Further, any distortion of the signal can be eliminated by reshaping the signal with a Schmitt trigger, comparator, or similar circuit. If information can be transmitted on a carrier consisting of binary pulses, these aspects of binary techniques can be used to improve the quality of communications. Pulse-modulation techniques were developed to take advantage of these qualities. The information signal, usually analog, is used to modify a binary (ON–OFF) or pulsed carrier in some way.

With pulse modulation the carrier is not transmitted continuously but in short bursts whose duration and amplitude correspond to the modulation. The duty cycle of the carrier is usually made short so that the carrier is off for a longer time than the bursts. This arrangement allows the *average* carrier power to remain low, even when high peak powers are involved. For a given average power, the peak power pulses can travel a longer distance and more effectively overcome any noise in the system.

There are three basic forms of pulse modulation: *pulse-amplitude modulation (PAM), pulse-width modulation (PWM),* and *pulse-position modulation (PPM).*

Figure 9-25 shows an analog modulating signal and the various waveforms produced by PAM, PWM, and PPM modulators. In all three cases, the analog signal is sampled by an S/H circuit as it would be in A/D conversion. The sampling points are shown on the analog waveform. The sampling time t is constant and subject to the Nyquist conditions described earlier. The sampling rate of the analog signal must be at least 2 times the highest frequency component of the analog wave.

The PAM signal in Fig. 9-25 is a series of constant-width pulses whose amplitudes vary in accordance with the analog signal. The pulses are usually very narrow compared to the period of sampling; this means that the duty cycle is low. The PWM signal is binary in amplitude (has only two levels). The width or duration of the pulses varies according to the amplitude of the analog signal: At low analog voltages, the pulses are narrow; at the higher amplitudes, the pulses get wider. In PPM, the pulses change position according to the amplitude of the analog signal. The pulses are very narrow. These pulse signals may be transmitted in a baseband form, but in most applications they modulate a high-frequency radio carrier. They turn the carrier on and off in accordance with their shape.

Of the three types of pulse modulation, PAM is the simplest and least expensive to implement. On the other hand, because the pulses vary in amplitude, they are far more susceptible to noise, and clipping techniques to eliminate noise cannot be used because they would also remove the modulation. PWM and PPM are binary and offer the full benefits of pulse modulation.

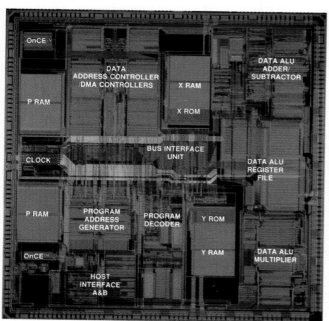

Integrated circuits, such as this Motorola 96002 Media Engine™, Floating Point Digital Signal Processor, are rapidly replacing analog circuits in communications equipment.

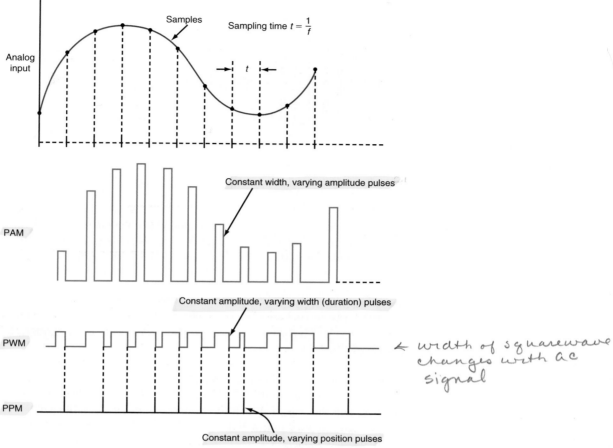

Fig. 9-25 Types of pulse modulation.

Although the techniques of pulse modulation have been known for decades, their development surged in the late 1950s and 1960s as a result of military missile development and the space program. Pulse-modulation techniques were widely used in telemetry systems. Telemetry, a system of monitoring and measuring at a distance, allows scientists and engineers to monitor physical characteristics such as temperature, speed, acceleration, and pressure in a remote missile or spacecraft. Pulse-modulation techniques are also used for remote-control purposes, for example, in model airplanes, boats, and cars.

Today pulse-modulation techniques have been largely superseded by more advanced digital techniques such as pulse-code modulation (PCM), in which actual binary numbers representing the digital data are transmitted. The three pulse modulation techniques are discussed in more detail in the following sections.

PAM

Sampling is the process of "looking at" an analog signal for a brief time. During this very short sampling interval, the amplitude of the analog signal is allowed to be passed

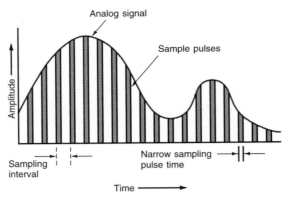

FIG. 9-26 Sampling and analog signal to produce pulse-amplitude modulation.

or stored. If multiple samples of the analog signal are taken at a periodic rate, most of the information contained in the analog signal is passed. The resulting signal is a series of samples or pulses that vary in amplitude according to the variation of the analog signal.

PAM MODULATOR. Figure 9-26 shows an analog signal which is being sampled. The resulting output is a series of pulses whose amplitudes are the same as those of the analog signal during the sample period. This process is known as *pulse-amplitude modulation (PAM)*.

The basic circuit for generating PAM is illustrated in Fig. 9-27. An astable clock oscillator drives a one-shot multivibrator which generates a narrow, fixed-width pulse. This pulse is applied to a gate circuit, a switch that opens and closes in accordance with the one-shot signal. When the one-shot signal is OFF, the gate is closed and the analog signal applied to it will not pass. When the clock triggers the one shot once per cycle, the gate opens for a short time, allowing the analog signal to pass through. The gate circuit can be constructed with diodes or can be an arrangement of bipolar or field-effect transistors (FETs).

Remember, in order for the recovered signal to be an accurate representation of the original, the sampling rate must be high enough to ensure that rapid fluctuations are sampled a sufficient number of times. The sampling rate must be at least 2 times the highest-frequency component of the original signal in order for the signal to be adequately represented. If the sampled signal is a simple sine wave, the minimum sampling frequency can be twice the sine-wave frequency. A 2-kHz sine wave would have to be sampled a minimum of 2×2 kHz, or 4 kHz. A more complex signal containing harmonics up to 650 kHz would have to be sampled at a 2×650 kH = 1300 kHz =

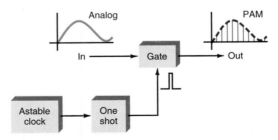

FIG. 9-27 A pulse-amplitude modulator.

1.3 MHz rate or higher. The higher the sampling rate, the better the representation. To ensure good fidelity, most systems sample at a rate higher than the minimum 2 times. The actual value depends upon the application, but typically the sampling rate is 3 to 10 times the highest-frequency component in the analog signal. A sampling rate of 10 times the maximum analog bandwidth is ideal, for it provides excellent representation of the signal.

In telephone communications, the upper frequency value of the voice content is assumed to be 3 kHz. This dictates a 2×3 kHz, or 6 kHz, sampling rate. In practice, the sampling rate for audio in telephone systems is 8 kHz, because of the waveform complexity. The higher sampling rate provides more faithful reproduction.

DEMODULATING PAM. To recover the original information, the transmitted pulses are passed through a low-pass filter. The upper cutoff frequency of the low-pass filter is selected to pass the highest-frequency components contained within the analog signal. All higher frequencies are eliminated. Since the pulses themselves represent a composite of many high-frequency harmonics, they are effectively filtered out; that is, they are smoothed into a continuous analog signal that is virtually identical in information content to the original transmitted signal. The filtering process is similar to that used in a simple AM diode detector.

PWM

Pulse-width modulation (PWM), also known as *pulse-duration modulation (PDM)*, is perhaps the most widely used of the three pulse-modulation techniques. The basic PWM process is illustrated in Fig. 9-28. The constant-frequency clock oscillator drives the PWM modulator. The other input to the modulator is the analog information signal to be transmitted. The modulator modifies the width or duration of the clock pulses in accordance with the modulating signal. The output is a varying pulse-width signal. A typical modulator is a monostable or one-shot multivibrator whose pulse duration can be modified by the analog input signal.

MODULATING PWM SIGNALS. Figure 9-29 shows one simple but effective technique for developing PWM signals. The modulator is a 555 timer IC, connected as a one-shot multivibrator, consisting of a control flip-flop that is set or reset by a pair of comparators. The comparators are biased with direct current from the supply voltage V_{CC} and an internal voltage divider of three 5-kΩ resistors. The upper comparator has

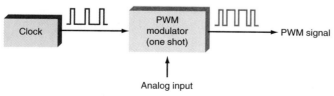

FIG. 9-28 Basic PWM modulator.

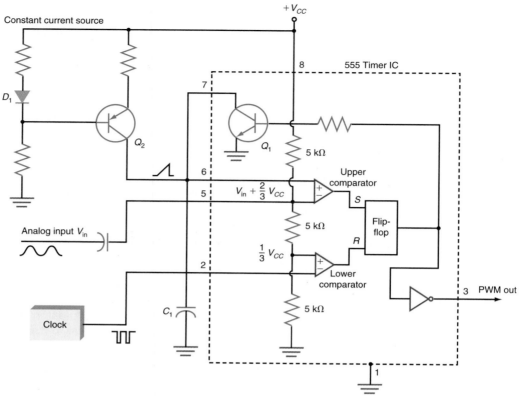

FIG. 9-29 PWM modulator.

a bias reference of two-thirds of V_{CC}, and the lower comparator has a DC reference of one-third of V_{CC}. The analog input voltage is applied to the upper comparator at pin 5, where this analog voltage adds to and subtracts from the two-thirds of V_{CC} bias.

The flip-flop is initially set so that its output is high and the output of the inverter at pin 3 is near zero. The flip-flop turns on Q_1, which shorts the external capacitor C_1, preventing it from charging. When a trigger pulse from the clock occurs, the lower comparator changes state and resets the flip-flop. The PWM output at pin 3 goes high and Q_1 turns OFF. At this time, the capacitor C_1 can charge. Q_2 and related components form a constant-current source that charges the capacitor. With a constant charging current, the voltage across the capacitor is a linear ramp. The capacitor continues to charge until the threshold of the upper comparator is exceeded; at this time, the flip-flop is set. The output pulse goes low and Q_1 turns ON, shorting and discharging the capacitor.

The duration of the output pulse is determined by how long the capacitor charges. This, in turn, is determined by the amplitude of the input signal. The greater the input signal, the higher the voltage threshold on the upper comparator and the longer the capacitor must charge to equal it. Therefore, the output pulse duration is directly proportional to the amplitude of the analog input. The relevant waveforms are shown in Fig. 9-30.

DEMODULATING PWM SIGNALS. The demodulation of a PDM signal is extremely simple. The demodulator circuit is simply a low-pass filter that averages the pulses.

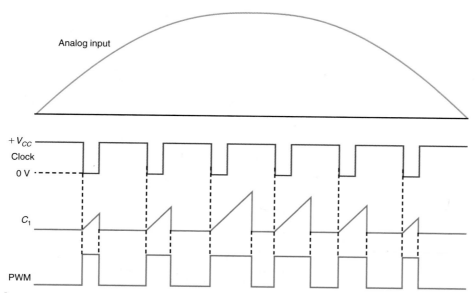

Fig. 9-30 Waveforms for a PWM modulator.

During periods of narrow pulses, the average voltage at the output of the filter is low; during wide pulses, the average output is high. The filter smooths the varying-width pulses back into the original analog modulated signal.

PPM

A PPM signal is easily derived from a PWM signal by the addition of a couple of simple circuits [see Fig. 9-31(a)]. The PWM signal is passed through an *RC* differentiator which creates very narrow positive and negative pulses [see Fig. 9-31(b)]. The half-wave diode rectifier causes the positive pulses to be clipped off, and the negative pulses are used to trigger a one-shot multivibrator. The one shot produces constant-amplitude, constant-width pulses whose position varies depending upon the modulating signal. The one-shot output is the PPM signal.

PULSE MODULATION IN PERSPECTIVE

As mentioned, these three pulse-modulation techniques have been mostly superseded by PCM methods. They are still used in electronics but more for control purposes than for communications. For example, PWM is widely used in power supply switching regulators and in motor speed control circuits. PPM is almost never used in modern electronics.

PAM still plays a role, though not alone. PAM sampling techniques measure points along a continuous analog curve. The samples are then converted into a proportional

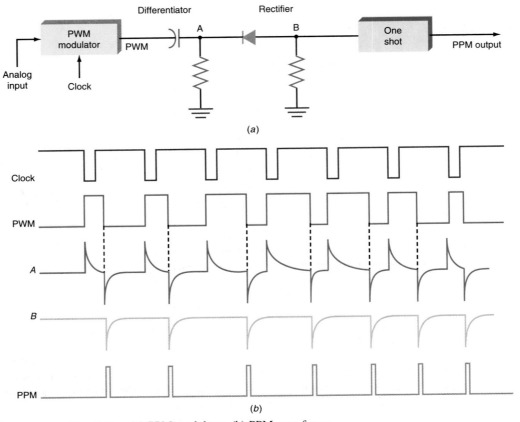

Fig. 9-31 (*a*) PPM modulator. (b) PPM waveforms.

binary number that becomes PCM. PAM is still used because it lends itself to time multiplexing methods, which are covered in Chap. 10.

9-6 Digital Signal Processing

As previous chapters have emphasized, communications involves a great deal of signal processing. To carry out communications, analog signals must be processed in some way; for example, they may be amplified or attenuated. Often they must be filtered to remove undesirable frequency components. They must be shifted in phase and modulated or demodulated. Or they may have to be mixed, compared, or analyzed to determine their frequency components. Thousands of circuits have been devised to process analog signals, and many of them have been described in this book.

Although analog signals are still widely processed by analog circuits, increasingly they are being converted to digital for transmission or for final application. As described earlier in this chapter, there are several important advantages to transmitting and using data in digital form. One advantage is that signals can now be manipulated by *digital signal processing (DSP)*.

DSP is the use of a fast digital computer to perform processing on digital signals. Any digital computer with sufficient speed and memory can be used for DSP. The superfast 32-bit reduced instruction set computing (RISC) processors are especially adept at DSP. However, DSP is best implemented with processors developed specifically for this application because they differ in organization and operation from traditional microprocessors.

The basic DSP technique is shown in Fig. 9-32. An analog signal to be processed is fed to an A/D converter, where it is converted into a series of binary numbers which are stored in a read-write random-access memory (RAM). (See Fig. 9-33.) A program, usually stored in a read-only memory (ROM), performs mathematical and other manipulations on the data. Most digital processing involves complex mathematical algorithms that are executed in *real time;* that is, the output is produced simultaneously with the occurrence of the input. With real-time processing, the processor must be extremely fast so that it can perform all the computations on the samples before the next sample comes along.

> **HINTS AND HELPS**
>
> DSP is best implemented using processors designed specifically for this application, although any digital computer with enough speed and memory can be used.

The processing results in another set of data words which are also stored in RAM. They can then be used or transmitted in digital form, or they may be fed to a D/A converter where they are converted back into an analog signal. The output analog signal then looks as though it has undergone processing by an analog circuit.

Almost any processing operation that can be done with analog circuits can also be done with DSP. The most common is filtering, but equalization, companding, phase shifting, and modulation can also be programmed on a DSP computer.

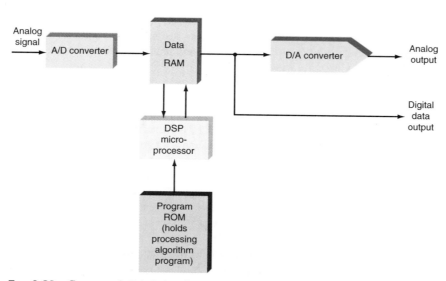

FIG. 9-32 Concept of digital signal processing (DSP).

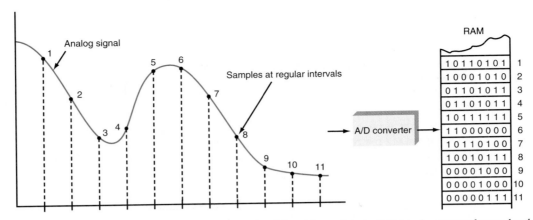

Fig. 9-33 Converting the analog signal into binary data in RAM to be operated upon by the DSP processor.

DSP Processors

When DSP was first developed, during the 1960s, only the largest and fastest mainframe computers were able to handle it, and even then, in some applications real-time processing could not be achieved. As computers got faster, more sophisticated processing could be performed, and in real time. However, only the most demanding of applications could afford a fast mainframe or minicomputer. For example, NASA used DSP to process and enhance the digital video data from remote exploratory spacecraft like the *Voyager*, which passed by Mars and Jupiter. Oil companies used DSP in the 1960s and 1970s to process geological data to determine whether oil deposits were present in structures beneath the earth.

With the appearance of 16- and 32-bit microprocessors the use of DSP became practical for many applications, and, finally, in the 1980s special microprocessors optimized for DSP were developed.

Most computers and microprocessors use an organization known as the *Von Neumann architecture*. Physicist John Von Neumann is generally credited with creating the so-called *stored program concept* that is the basis of operation of all digital computers. Binary words representing computer instructions are stored sequentially in a memory to form a program. The instructions are fetched and executed one at a time at high speed. The program usually processes data in the form of binary numbers that are stored in the same memory. The key feature of the Von Neumann arrangement is that both instructions and data are stored in a common memory space. That memory space may be read-write RAM or ROM or some combination thereof. But the important point is that there is only one path between the memory and the CPU and therefore only one data or instruction word can be accessed at a time. This has the effect of greatly limiting execution speed. This shortcoming is generally referred to as the *Von Neumann bottleneck.*

DSP microprocessors work in a similar way, but they use a variation called the *Harvard architecture*. In a Harvard architecture microprocessor, there are two memories, a program or instruction memory, usually a ROM, and a data memory, which is a RAM. Also, there are two data paths into and out of the CPU between the memories. Because both instructions and data can be accessed simultaneously, very high-speed operation is possible.

DSP microprocessors are designed to perform the math operations common to DSP. Most DSP is a combination of multiplication and addition operations on the data words developed by the A/D converter and stored in RAM. DSP processors carry out addition and multiplication faster than any other type of CPU, and most of them combine these operations in a single instruction for even greater speed.

Finally, DSP microprocessors are designed to operate at the highest speeds possible. Clock speeds up to 100 MHz are not uncommon. Some DSP processors are available as just the CPU chip, but others combine the CPU with data RAM and a program ROM on chip. Some even include the A/D and D/A converter circuits. If the desired processing program is written and stored in ROM, a complete single-chip DSP circuit can be created for customized analog signal processing by digital techniques.

DSP APPLICATIONS

FILTERING. The most common DSP application is filtering. A DSP processor can be programmed to perform bandpass, low-pass, high-pass, and band-reject filter operations. With DSP, the filters can have characteristics far superior to those of equivalent analog filters: Selectivity can be better, and the passband or reject band can be customized to the application. Further, the phase response of the filter can be controlled more easily than with analog filters.

COMPRESSION. Data compression is a process that reduces the number of binary words needed to represent a given analog signal. It is often necessary to convert a video analog signal into digital for storage and processing. Digitizing a video signal with an A/D converter produces an immense amount of binary data. If the video signal contains frequencies up to 4 MHz, the A/D converter must sample at 8 MHz or faster. Assuming a sampling rate of 8 MHz with an 8-bit A/D converter 8 million bytes per second will be produced. Digitizing 1 minute of video is equal to 60 s $\times$ 8 Mbytes, or 480 million bytes of data. This amount of data exceeds the capacity of most computer RAM, although a larger hard disk could store this data. In terms of data communications, it would take a great deal of time to transmit this amount of data serially.

To solve this problem, the data is compressed. Numerous algorithms have been developed to compress data. The data is examined for redundancy and other characteristics, and a new group of data, based upon various mathematical operations, is created. Data can be compressed by a factor of up to 100; in other words, the compressed data is 1/100 its original size. With compression, 480 million bytes of data becomes 4.8 million bytes. This is still a lot, but it is now within the capabilities of RAM and disk storage components.

A DSP chip does the compression on the data received from the A/D converter. The compressed version of the data is then stored or transmitted. In the case of data communications, compression greatly reduces the time needed to transmit data.

When the data is needed, it must be decompressed. A reverse-calculation algorithm is used to reconstruct the original data. Again, a special DSP chip is used for this purpose.

SPECTRUM ANALYSIS. *Spectrum analysis* is the process of examining a signal to determine its frequency content. Recall that all nonsinusoidal signals are a combination of a fundamental sine wave to which have been added harmonic sine waves of different frequency, amplitude, and phase. An algorithm known as the *discrete Fourier transform (DFT)* can be used in a DSP processor to analyze the frequency content of an input signal. The analog input signal is converted to a block of digital data, which is then processed by the DFT program. The result is a frequency domain output that indicates the content of the signal in terms of sine-wave frequencies, amplitudes, and phases.

The DFT is a complex program that is long and time-consuming to run. In general, computers are not fast enough to perform DFT in real time as the signal occurs. Therefore, a special version of the algorithm has been developed to speed up the calculation. Known as the *fast Fourier transform (FFT),* it permits real-time signal spectrum analysis.

OTHER APPLICATIONS. As mentioned, DSP can do almost everything analog circuits can do, for example, phase shifting, equalization, and signal averaging. *Signal averaging* is the process of sampling a recurring analog signal which is transmitted in the presence of noise. If the signal is repeatedly converted to digital and the mathematical average of the samples is taken, the signal-to-noise ratio is greatly improved. Since the noise is random, an average of it tends to be zero. The signal, which is constant and unchanging, averages into a noise-free version of itself.

DSP can also be used for signal synthesis. Waveforms of any shape or characteristics can be stored as digital bit patterns in a memory. Then when it is necessary to generate a signal with a specific shape, the bit pattern is called up and transmitted to the DAC, which generates the analog version. This type of technique is used in voice and music synthesis.

DSP is widely used in fax machines, CD players, modems, and a variety of other common electronic products. Its use in communications is increasing as DSP processors become even faster. Some fast DSP processors have been used to perform all the normal communications receiver functions from the IF stages through signal recovery. All-digital or "software" receivers are a reality now and are the receivers of the near future.

HOW DSP WORKS

The advanced mathematical techniques used in DSP are beyond the scope of this book and certainly beyond the knowledge required of electronics technicians and technologists in their jobs. For the most part, it is sufficient to know that the techniques exist. However, without getting bogged down in the math, it is possible to give some insight into the workings of a DSP circuit. For example, it is relatively easy to visualize the digitizing of an analog signal into a block of sequential binary words representing the amplitudes of the samples, and then to imagine that the binary words representing the analog signal are stored in a RAM (see Fig. 9-33). Once the signal is in digital form, it can be processed in many different ways. Two common applications are filtering and spectrum analysis.

FILTER APPLICATIONS. One of the most popular DSP filters is called a *finite impulse response (FIR)* filter. It is also called a *nonrecursive filter.* (A nonrecursive filter

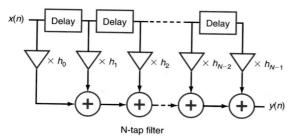

Fig. 9-34 A block diagram showing the processing algorithm of a nonrecursive FIR filter.

is one whose output is a function of the sum of products of the current as well as the past input samples.) A program can be written to create a low-pass, high-pass, band-pass, or band-reject filter of the FIR type. The algorithm of such a filter has the mathematical form $Y = \Sigma a_i b_i$. In this expression, Y is the binary output, which is the summation Σ of the products of a and b. The terms a and b represent the binary samples, and i is the number of the sample. Usually these samples are multiplied by coefficients appropriate to the type of filter and the results summed.

Figure 9-34 is a graphical representation of what goes on inside the filter. The term $X(n)$, where n is the number of the sample, represents the input data samples from

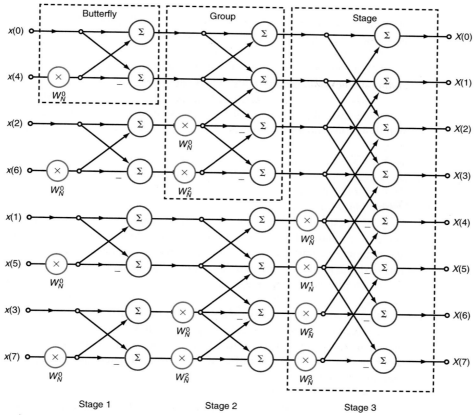

Fig. 9-35 The fast Fourier transform decimation in time.

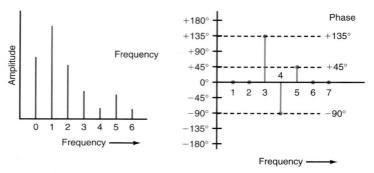

Fig. 9-36 Output plot of an FFT spectrum analysis.

RAM. The boxes labeled DELAY represent delay lines. (A *delay line* is a circuit that delays a signal or sample by some constant time interval.) In reality, nothing is being delayed. Rather, the circuit generates samples that occur one after another at a fixed time interval equal to the sampling time, which is a function of the A/D converter clock frequency. In effect, the output of the delay boxes in Fig. 9-34 is the sequential samples which occur one after another at the sampling rate which is equivalent to a series of delays.

Note that the samples are multiplied by some constant represented by the term h_n. These constants, or coefficients, are determined by the algorithm and the type of filter desired. After the samples have been multiplied by the appropriate coefficient, they are summed. The first two samples are added, this sum is added to the next-multiplied sample, that sum is added to the next sample, and so on. The result is the output Y, which is a value made up of the sum of products of the other samples. These new data samples are also stored in RAM. This block of new data is sent to the D/A converter at whose output the filtered analog signal appears.

Another type of DSP filter is the *infinite impulse response (IIR)* filter, a recursive filter that uses feedback: each new output sample is calculated using both the current output and past samples (inputs).

DIT/FFT. As indicated earlier, a DSP processor can perform spectrum analysis by using the discrete or fast Fourier transform (FFT). Figure 9-35 illustrates the processing that takes place with FFT. It is called a *decimation in time (DIT)*. The $x(n)$ values at the input are the samples, which are processed in three stages. In the first stage, a so-called butterfly operation is performed on pairs of samples. Some of the samples are multiplied by a constant and then added. At the second stage, some of the outputs are multiplied by constants, and new pairs of sums, called *groups,* are formed. Then a similar process is performed to create the final outputs, called *stages.* These outputs are converted into new values that can be plotted in the frequency domain.

In the graph in Fig. 9-36, the horizontal axis in the upper plot is frequency and the vertical axis is the amplitude of the DC and AC sine-wave components that make up the sampled wave. A 0 frequency component is represented by a vertical line indicating the DC component of a signal. The 1 indicates the amplitude of the fundamental sine wave making up the signal. The other values, at 2, 3, 4, and so on, are the amplitudes of the harmonics. In the lower plot, the phase angle of the sine waves is given for each harmonic. A negative value indicates phase inversion of the sine wave (180°).

SUMMARY

Transmitting data using digital techniques offers a number of advantages over analog processing: high immunity to noise, excellent error detection and correction capabilities, compatibility with time division multiplexing techniques, and the use of digital signal processing (DSP) circuits.

Before analog signals can be transmitted digitally, they must be converted into digital signals by analog-to-digital (A/D) conversion, in which the signal is translated into a series of discrete binary numbers representing samples. A sufficient number of samples must be taken to retain the high-frequency information in the analog signal.

Modern A/D converters are usually single-chip ICs that take an analog signal and generate a parallel binary output. Since A/D converters can represent only a finite number of voltage values over a specific range, the samples are converted into binary numbers whose values are close to the actual sample values. The error associated with the conversion process, quantizing error, can be reduced by dividing the analog voltage range into a larger number of smaller voltage increments. A process of signal compression and expansion known as companding is used to overcome the problems of quantizing error and noise.

D/A converters receive the binary signals sequentially and produce a proportional analog voltage at the output. D/A converters have four major sections: a regulator, a resistance network, an output amplifier, and electronic switches. The three most important specifications associated with D/A converters are resolution, error, and settling time.

The most common A/D conversion circuits are the counter converter, the successive approximations converter, and the flash converter. For high-speed conversions, flash converters are the circuits of choice, offering conversion speeds of less than 50 ns.

In parallel data transfers, all of the bits of a code word are transferred simultaneously. In serial data transfers, each bit of the word is transmitted in sequence. The conversion between parallel and serial and serial and parallel is accomplished using shift registers.

In pulse modulation the information signal, usually analog, is used to modify a binary or pulsed carrier in some way. There are three basic forms of pulse modulation: pulse-amplitude modulation (PAM), pulse-width modulation (PWM), and pulse-position modulation (PPM). Today, these techniques have been almost entirely superseded by the more sophisticated and effective pulse-code modulation.

In digital signal processing (DSP), very fast, specially designed computers control the conversion process. The analog signal to be processed is fed to an A/D converter where it is converted into a series of binary numbers, stored in RAM, and executed in real time. Programs for filtering, equalization, companding, phase shifting, modulation, and so on are written for DSP computers.

KEY TERMS

A-law compander	Codec	Data bus
Analog-to-digital (A/D) conversion	Companding	Data communications
	Compression	Decimation in time (DIT)
Bulletin board systems (BBS)	Counter converter	Delay line
	Data	Delta modulation

Digital signal processing
  (DSP)
Digital-to-analog (D/A)
  conversion
Discrete Fourier transform
  (DFT)
Electronic mail (E-mail)
Error
Expansion amplifier
Fast Fourier transform (FFT)
Files transfer
Filtering
Finite impulse response
  (FIR) filter (nonrecur-
  sive filter)
Flash converter
Harvard architecture
Infinite impulse response
  (IIR) filter (recursive filter)

Local area networks
  (LANs)
μ-law compander
Noise immunity
Nyquist frequency
Output amplifier
Parallel transfer
Pulse-amplitude modula-
  tion (PAM)
Pulse-code modulation
  (PCM)
Pulse-duration modulation
  (PDM)
Pulse-position modulation
  (PPM)
Pulse-width modulation
  (PWM)
Quantizing error
Reference regulator

Resistor network
Resolution
Sample-and-hold circuit
  (track/store circuit)
Sampling
Serial transfer
Serial-parallel conversion
Settling time
Shift register
Signal regeneration
Spectrum analysis
Successive approximations
  converter
Successive approximations
  register (SAR)
Von Neumann architecture
Von Neumann bottleneck

QUESTIONS

1. Name the four primary benefits of using digital techniques in communications. Which of these is probably the most important?
2. What is data conversion? Name two basic types.
3. What is the name given to the process of measuring the value of an analog signal at some point in time?
4. What is the name given to the process of assigning a specific binary number to an instantaneous value on an analog signal?
5. What is another name commonly used for A/D conversion?
6. Describe the nature of the signals and information obtained when an analog signal is converted into digital form.
7. Describe the nature of the output waveform obtained from a D/A converter.
8. Name the four major components in a D/A converter.
9. What is the name of the most commonly used resistor network in a D/A converter?
10. What type of circuit is commonly used to translate the current output from a D/A converter into a voltage output?
11. Name three types of A/D converters and state which is the most widely used.
12. What A/D converter circuit sequentially turns the bits of the output on one at a time in sequence from MSB to LSB in seeking a voltage level equal to the input voltage level?
13. What is the fastest type of A/D converter? Briefly describe the method of conversion used.
14. What type of A/D converter generates a serial output data signal directly from the conversion process?
15. What circuit is normally used to perform serial-to-parallel and parallel-to-serial data conversion?
16. What circuit performs the sampling operation prior to A/D conversion and why is it so important?

17. What process converts an analog signal into sequential binary numbers and transmits them serially?
18. What is the name given to the process of compressing the dynamic range of an analog signal at the transmitter and expanding it later at the receiver?
19. What is the general mathematical shape of a companding curve?
20. Name the three basic types of pulse modulation. Which type is not binary?
21. What basic circuit is used to create PAM?
22. What type of circuit is used to demodulate PAM?
23. What is the primary disadvantage of PAM compared to other types of pulse modulation?
24. What is another name for PWM?
25. What kind of circuit is used to demodulate PWM?
26. For the basic PWM circuit described in the text, answer the following questions:
    a. What type of IC circuit is used?
    b. What charges the capacitor in the circuit?
    c. To what circuit is the analog modulating input signal applied in the IC?
27. Name the two basic circuits used to convert PWM to PPM.
28. Where was pulse modulation first used?
29. Name two common noncommunications applications for PWM.
30. Describe briefly the techniques known as digital signal processing (DSP).
31. What type of circuit performs DSP?
32. Briefly describe the basic mathematical process used in the implementation of DSP.
33. Give the names for the basic architecture of non-DSP microprocessors and for the architecture normally used in DSP microprocessors and briefly describe the difference between the two.
34. Name five common processing operations that take place with DSP. What is probably the most commonly implemented DSP application?
35. Briefly describe the nature of the output of a DSP processor that performs the discrete Fourier transform or the fast Fourier transform.
36. Name the two types of filters implemented with DSP and explain how they differ.

PROBLEMS

1. A video signal contains light variations that change at a frequency as high as 3.5 MHz. What is the minimum sampling frequency for A/D conversion? ◆
2. A D/A converter has a 12-bit binary input. The output analog voltage range is 0 to 5 V. How many discrete output voltage increments are there and what is the smallest voltage increment?

CRITICAL THINKING

1. List three major types of communications services that are not yet digital but could eventually be and explain how digital techniques could be applied to those applications.
2. Explain how an all-digital receiver would process the signal of an analog AM radio broadcast signal.
3. What type of A/D converter would work best for video signals with a frequency content up to 5 MHz? Why?

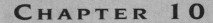

MULTIPLEXING AND DEMULTIPLEXING

Objectives

After completing this chapter, you will be able to:

◆ *Explain* why multiplexing techniques are necessary in telemetry, telephone systems, and today's radio and TV broadcasting.

◆ *Compare* frequency division multiplexing with time division multiplexing.

◆ *Trace* the steps in the transmission and reception of multiplexed signals.

◆ *List* the major subtypes of time division multiplexing.

◆ *Define* pulse-code modulation, *draw* the diagram of a typical PCM multiplexer, and *state* the primary benefit of PCM over other forms of pulse modulation.

A communications channel, or link, between two points is established whenever a cable is connected or a radio transmitter and receiver are set up between the two points. When there is only one link, only one function—whether it involves signal transmission or control operations—can be performed at a time. For two-way communications, a half-duplex process is set up: both ends of the communications link can send and receive but not at the same time.

Transmitting two or more signals simultaneously can be accomplished by running multiple cables or setting up one transmitter-receiver pair for each channel, but this is an expensive approach. In fact, a single cable or radio link can handle multiple signals simultaneously using a technique known as multiplexing, which permits hundreds or even thousands of signals to be combined and transmitted over a single medium. Multiplexing has made simultaneous communications more practical and economically feasible, helped conserve spectrum space, and allowed new, sophisticated applications to be implemented.

10-1 MULTIPLEXING PRINCIPLES

Multiplexing is the process of simultaneously transmitting two or more individual signals over a single communications channel. In effect, it increases the number of communications channels so that more information can be transmitted. There are many instances in communications where it is necessary or desirable to transmit more than one voice or data signal simultaneously. An application may require multiple signals, or cost savings can be gained by using a single communication channel to send multiple information signals. Three applications that would be prohibitively expensive or impossible without multiplexing are telemetry; telephone systems, including satellite functions; and modern radio and TV broadcasting.

In *telemetry*, the physical characteristics of a given application are monitored by sensitive transducers which generate electrical signals that vary in response to changes in the status of the various physical characteristics. The sensor-generated information can be sent to a central location for monitoring, or can be used as feedback in a closed-loop control system. Most spacecraft and many chemical plants, for example, use telemetry systems to monitor characteristics such as temperature, pressure, speed, light level, flow rate, liquid level, and others.

The use of a single communications channel for each characteristic being measured in a telemetry system would not be practical, both because of the multiple possibilities for signal degradation and because of the high cost. Consider, for example, monitoring a space shuttle flight. Wire cables are obviously out of the question, and multiple transmitters impractical. If a deep-space probe were used, it would be necessary to use multiple transducers, and many transmitters would be required to send the signals back to earth. Because of cost, complexity, and equipment size and weight, this approach would not be feasible. Clearly, telemetry is an ideal application for multiplexing, where the various information signals can all be sent over a single channel.

A second well-known multiplexing application is the global satellite system that links telephones across the United States and all over the world. If each link from one telephone to another required its own

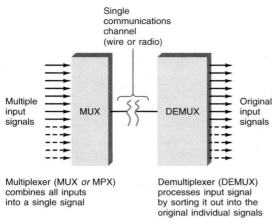

Single
communications
channel
(wire or radio)

Multiple
input
signals

MUX

DEMUX

Original
input
signals

Multiplexer (MUX *or* MPX)
combines all inputs
into a single signal

Demultiplexer (DEMUX)
processes input signal
by sorting it out into the
original individual signals

Fɪɢ. 10-1 Concept of multiplexing.

two-way channel, the number of such channels—and the subsequent cost—would be astronomical. Finally, modern FM stereo broadcasting requires multiplexing techniques, as does the transmission of stereo sound and color in TV.

Multiplexing is accomplished by an electronic circuit known as a multiplexer. A simple multiplexer is illustrated in Fig. 10-1.

Multiple input signals are combined by the multiplexer into a single composite signal that is transmitted over the communications medium. Alternatively, multiplexed signals can modulate a carrier before transmission. At the other end of the communications link, a demultiplexer is used to process the composite signal to recover the individual signals.

The two basic types of multiplexing are frequency division multiplexing (FDM) and time division multiplexing (TDM). Two variations of these basic methods are frequency division multiple access and time division multiple access. In general, FDM systems are used for analog information and TDM systems are used for digital information. Of course, TDM techniques are also found in many analog applications because the processes of A/D and D/A conversion are so common. The primary difference between these techniques is that in FDM, individual signals to be transmitted are assigned a different frequency within a common bandwidth. In TDM, the multiple signals are transmitted in different time slots.

10-2 Frequency Division Multiplexing

In *frequency division multiplexing (FDM)* multiple signals share the bandwidth of a common communications channel. Remember that all channels have specific bandwidths, and some are relatively wide. A coaxial cable, for example, has a bandwidth of 200 to 500 MHz. The bandwidths of radio channels vary, and are usually determined by FCC regulations and the type of radio service involved. Regardless of the type of channel, a wide bandwidth can be shared for the purpose of transmitting many signals at the same time.

Figure 10-2 shows a general block diagram of an FDM system. Each signal to be transmitted feeds a modulator circuit. The carrier for each modulator (f_c) is on a different frequency. The carrier frequencies are usually equally spaced from one another over a specific frequency range. These carriers are referred to as *subcarriers*. Each input signal is given a portion of the bandwidth. The resulting spectrum is illustrated in Fig. 10-3. Any of the standard kinds of modulation can be used, including AM, SSB, FM, or PM. The FDM process divides up the bandwidth of the single channel into smaller, equally spaced channels, each capable of carrying information in sidebands.

The modulator outputs containing the sideband information are added together algebraically in a linear mixer; no modulation or generation of sidebands takes place. The resulting output signal is a composite of all the modulated subcarriers. This signal can be used to modulate a radio transmitter, or can itself be transmitted over the single communications channel. Alternatively, the composite signal can become one input to another multiplexed system.

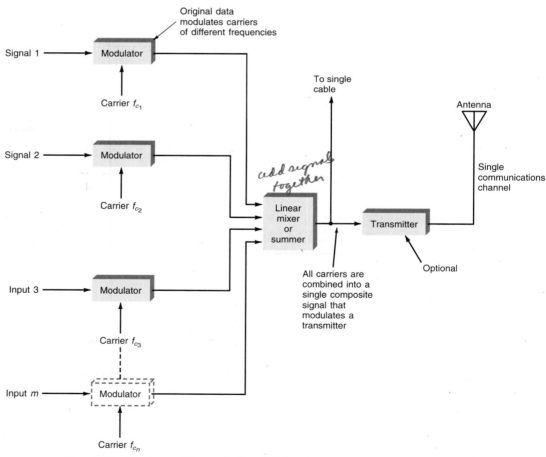

FIG. 10-2   The transmitting end of an FDM system.

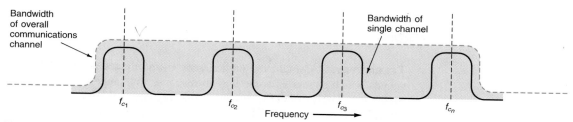

FIG. 10-3 Spectrum of an FDM signal. The bandwidth of a single channel is divided into smaller channels.

RECEIVER-DEMULTIPLEXERS

The receiving portion of an FDM system is shown in Fig. 10-4. A receiver picks up the signal and demodulates it, recovering the composite signal. This is sent to a group of bandpass filters, each centered on one of the carrier frequencies. Each filter passes only its channel and rejects all others. A channel demodulator then recovers each original input signal.

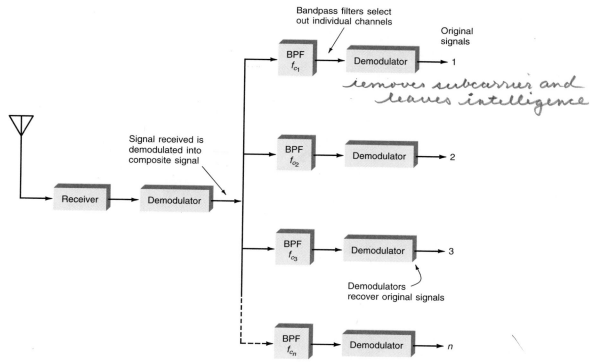

removes subcarrier and leaves intelligence

FIG. 10-4 The receiving end of an FDM system.

TELEMETRY. As indicated earlier, sensors in telemetry systems generate electrical signals which change in some way in response to changes in physical characteristics. An example of a sensor is a *thermistor,* a device used to measure temperature. A thermistor's resistance varies inversely with temperature: as the temperature increases, the resistance decreases. The thermistor is usually connected into some kind of a resistive network, such as a voltage divider or bridge, and also to a DC voltage source. The result is a DC output voltage, which varies in accordance with temperature and which is transmitted to a remote receiver for measurement, readout, and recording. The thermistor becomes one channel of an FDM system.

Other sensors have different kinds of outputs. Many simply have varying DC outputs, while others are AC in nature. Each of these signals is typically amplified, filtered, and otherwise conditioned before being used to modulate a carrier. All of the carriers are then added together to form a single multiplexed channel.

The conditioned transducer outputs are normally used to frequency-modulate a subcarrier. The varying direct or alternating current changes the frequency of an oscillator operating at the carrier frequency. Such a circuit is generally referred to as a *voltage-controlled oscillator (VCO)* or a *subcarrier oscillator (SCO)*. To produce FDM, each VCO operates at a different center or carrier frequency. The outputs of the subcarrier oscillators are added together. A diagram of such a system is shown in Fig. 10-5.

Figure 10-6(*a*) shows a block diagram of a typical VCO circuit. The VCOs are available as single IC chips. The popular 566 VCO IC consists of a dual-polarity cur-

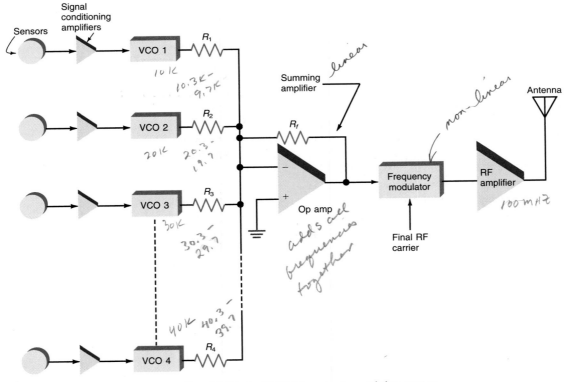

FIG. 10-5 An FDM telemetry transmitting system.

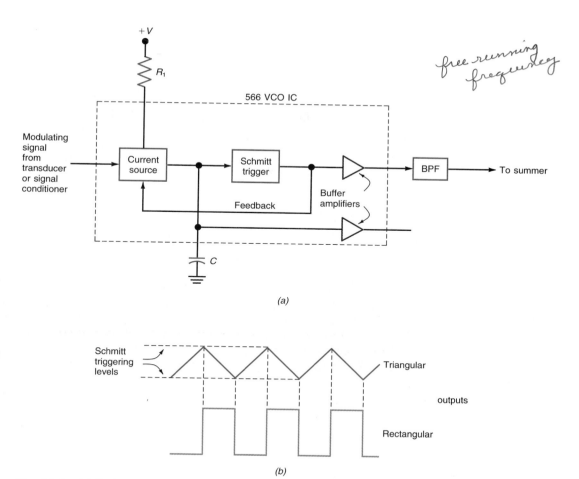

free running frequency

(a)

(b)

FIG. 10-6 (a) Typical IC VCO circuit. (b) Waveforms.

rent source that linearly charges and discharges an external capacitor C. The current value is set by an external resistor R_1. Together R_1 and C set the operating or center carrier frequency, which can be any value up to about 1 MHz.

The current source can be varied by an external signal, either DC or AC, which is the modulating signal from a transducer or other source. The input signal varies the charging and discharging current, thereby varying the carrier frequency. The result is direct FM.

The current source output is a linear triangular waveform which is buffered by an amplifier for external use and fed to an internal Schmitt trigger. The Schmitt trigger generates a rectangular pulse at the operating frequency that is fed to a buffer amplifier for external use.

The Schmitt trigger output is also fed back to the current source, where it controls whether the capacitor is charged or discharged. For example, the VCO may begin by charging the capacitor. When the Schmitt trigger senses a specific level on the triangular wave, it switches the current source. Discharging then occurs. The waveforms in Fig. 10-6(b) show this feedback action, which is what creates a free-running, astable oscillator.

Most VCOs are astable multivibrators whose frequency is controlled by the input from the signal conditioning circuits. The frequency of the VCO changes linearly in proportion to the input voltage. Increasing the input voltage causes the VCO frequency to increase. The rectangular or triangular output of the VCO is usually filtered into a sine wave by a bandpass filter centered on the unmodulated VCO center frequency. This can be either a conventional *LC* filter or an active filter made with an op amp and *RC* input and feedback networks. The resulting sinusoidal output is applied to the linear mixer.

The linear mixing process in an FDM system can be accomplished with a simple resistor network, as shown in Fig. 10-7. However, such networks greatly attenuate the signal, and some voltage amplification is usually required for practical systems. A way to achieve the mixing and amplification at the same time is to use an op amp summer, such as that shown in Fig. 10-5. Recall that the gain of each input is a function of the ratio of the feedback resistor, R_f, to the input resistor value (R_1, R_2, etc.). The output is given by the expression $V_{out} = -[V_1(R_f/R_1) + V_2(R_f/R_2) + V_3(R_f \cdots V_n(R_f/R_n)]$.

In most cases, the VCO FM output levels are the same, and all input resistors on the summer amplifier are therefore equal. If variations do exist, amplitude corrections can be accomplished by making the summer input resistors adjustable. The output of the summer amplifier does invert the signal; however, this has no effect on the content.

The composite output signal is then typically used to modulate a radio transmitter. Again, most telemetry systems use FM, although it is possible to use other kinds of modulation schemes. A system that uses FM of the VCO subcarriers as well as FM of the final carrier is usually called an *FM/FM system*.

Most FM/FM telemetry systems conform to standards established many years ago by an organization known as the Inter-Range Instrumentation Group (IRIG). Figure 10-8 lists IRIG FM subcarrier bonds, giving the center frequency, lower and upper limits, maximum frequency deviations, and frequency response for each. Note that a frequency deviation of ±7.5 percent is used on most of the channels, and this increases to ±15 percent on the upper frequency channels. The frequency response column gives the upper frequency range that the modulating signal can have on each channel. On channel 1, for example, with a 400-Hz center frequency, the maximum signal frequency that can be used is 6 Hz. Most of the lower-frequency channels are used for DC or very low-frequency AC signals.

This set of standards is known as the *proportional bandwidth FM/FM system*. A fixed percentage of frequency deviation is specified, so that the bandwidth is proportional to the carrier frequency. The higher the carrier frequency, the wider the bandwidth over which the modulating signal can occur.

Constant-bandwidth FM telemetry channels are also used. Carrier frequencies in the same approximate range as those given in Fig. 10-8 are used. However, a fixed deviation of ±2 kHz is typically specified, creating multiple channels with a 4-kHz band-

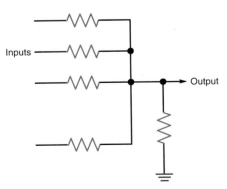

FIG. 10-7 Resistive summing network.

Band Number	Center Frequency (Hz)	Lower Limit (Hz)	Upper Limit (Hz)	Maximum Deviation (%)	Frequency Response (cps)
1	400	370	430	%7.5	6.0
2	560	518	602	—	8.4
3	730	675	785	—	11
4	960	888	1,032	—	14
5	1,300	1,202	1,399	—	20
6	1,700	1,572	1,828	—	25
7	2,300	2,127	2,473	—	35
8	3,000	2,775	3,225	—	45
9	3,900	3,607	4,193	—	59
10	5,400	4,995	5,805	—	81
11	7,350	6,799	7,901		110
12	10,500	9,712	11,288		160
13	14,500	13,412	15,588	—	220
14	22,000	20,350	23,650	—	330
15	30,000	27,750	32,250	—	450
16	40,000	37,000	43,000	—	600
17	52,500	48,562	56,438	—	790
18	70,000	64,750	75,250	—	1,050
A.	22,000	18,700	25,300	%15	660
B.	30,000	25,500	34,500	—	900
C.	40,000	34,000	46,000	—	1,200
D.	52,500	44,625	60,375	—	1,600
E.	70,000	59,500	80,500	—	2,100

FIG. 10-8   The IRIG FM subcarrier bands and specifications.

telemetric — uses FM — as noise free as possible

width. These are spaced throughout the frequency spectrum, with some guard space between channels to minimize interference.

The receiving end of a telemetry system is shown in Fig. 10-9. A standard superheterodyne receiver tuned to the RF carrier frequency is used to pick up the signal. An FM demodulator reproduces the original composite multiplexed signal, which is then fed to a demultiplexer. The demultiplexer divides the signals and reproduces the original inputs.

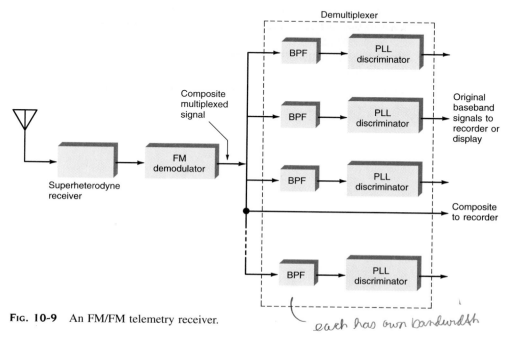

FIG. 10-9   An FM/FM telemetry receiver.

each has own bandwidth

The output of the first FM demodulator is fed simultaneously to multiple band-pass filters, each of which is tuned to the center frequency of one of the specified sub-channels. Each filter passes only its subcarrier and related sidebands and rejects all the others. The demultiplexing process is, then, essentially one of using filters to sort the composite multiplex signal back into its original components. The output of each filter is the subcarrier oscillator frequency with its modulation.

These signals are then applied to FM demodulators. Also known as *discriminators,* these circuits take the FM signal and recreate the original DC or AC signal produced by the transducer. The original signals are then measured or processed to provide the desired information from the remote transmitting source. In most systems, the multiplexed signal is sent to a data recorder where it is stored for possible future use. The original telemetry output signals can be graphically displayed on a strip chart recorder or otherwise converted into usable outputs.

The demodulator circuits used in typical FM demultiplexers are either of the phase-locked loop (PLL) or pulse-averaging type. PLL circuits have superior noise performance over the simpler pulse-averaging types. A PLL discriminator is also used to demodulate the receiver output.

FDM telemetry systems, which are inexpensive and highly reliable, are still widely used in aircraft and missile instrumentation, and for monitoring of medical devices, such as pacemakers.

TELEPHONE SYSTEMS. For decades, telephone companies have used FDM to send multiple telephone conversations over a minimum number of cables. In this application, the original voice signal, in the 300- to 3000-Hz range, is used to modulate a subcarrier. Each subcarrier is on a different frequency, and these subcarriers are then added together to form a single channel. This multiplexing process is repeated at several levels, so an enormous number of telephone conversations can be carried over a single communications channel assuming a wide enough bandwidth.

The frequency plan for a typical telephone multiplex system is shown in Fig. 10-10. Here the voice signal amplitude-modulates one of 12 channels in the 60- to 108-kHz range. The carrier frequencies begin at 64 kHz, with a spacing of 4 kHz.

The type of modulation used is single-sideband (SSB) suppressed carrier AM. Only the lower sideband (LSB) is transmitted. The LSBs are represented by the shaded triangles in Fig. 10-10. For example, channel 12 has a carrier frequency of 64 kHz. With AM, a modulating signal up to 4 kHz would produce sidebands down to 60 kHz and up to 68 kHz. The upper sideband is eliminated. In some parts of the telephone multiplex system, the upper sideband is selected.

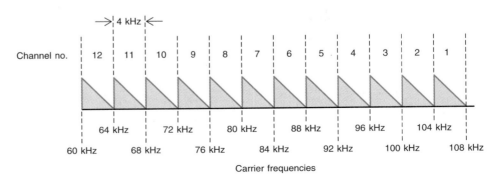

FIG. 10-10 Basic group frequency plan for FDM telephone system.

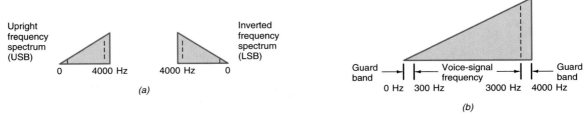

Fig. 10-11 (*a*) Upper and lower sidebands. (*b*) Sideband showing guard bands.

Figure 10-11(*a*) shows the details of the sidebands. The LSB sideband is said to be inverted because the highest modulating frequencies in each channel are at the lower-frequency end of the channel bandwidth. The USB is said to be upright where the highest modulating frequencies appear at the higher-frequency part of the channel. Only the 300- to 3000-Hz voice frequencies are passed, so guard bands are provided, as Fig. 10-11(*b*) shows.

Figure 10-12 shows the multiplexing circuits. The voice signal is applied to a balanced modulator along with a carrier. The output of the balanced modulator consists of the upper and lower sideband frequencies. The carrier is suppressed by the balanced modulator, and a highly selective filter is used to pass the lower sideband. All 12 SSB signals are then summed in a linear mixer to produce a single-frequency multiplexed signal. A set of 12 modulated carriers, as shown here, is generally referred to as a *basic group*.

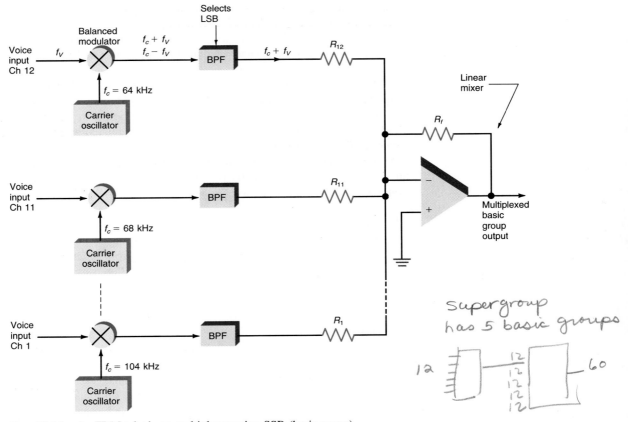

Fig. 10-12 An FDM telephone multiplexer using SSB (basic group).

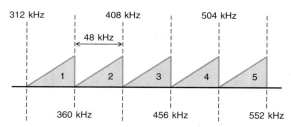

Fig. 10-13 Five basic groups are used to form a super group.

If more than 12 voice channels are needed, multiple basic groups are used, and their outputs are further multiplexed onto higher-frequency subcarriers. In the telephone system, as many as five 12-channel basic groups can be combined. Carrier frequencies in the 312- to 504-kHz range, spaced 48 kHz apart, are used. SSB is used, and the upper sidebands are selected. This frequency plan is shown in Fig. 10-13. The multiplexing process is similar, but in this case the output of each basic group modulates the higher-frequency carriers which are again summed to create an even more complex single-channel signal. This composite signal, called a *super group,* consists of 60 voice signals (5 basic groups, each carrying 12 channels). Pilot carriers are transmitted with each group to make demodulation simpler.

Up to 10 super groups can be used to modulate subcarriers in the 600- to 2540-kHz range, allowing a total of $5 \times 12 \times 10 = 600$ voice channels to be carried. The output of these 10 multiplexers is referred to as a *master group.* The process can continue, with six master groups being further combined into one jumbo group for a total of 3600 channels. These three jumbo groups can then be multiplexed again into one final output for a total of 10,800 voice channels.

Of course, bandwidth requirements increase with each level of multiplexing, and a bandwidth of many megahertz is required to deal with this 10,800-channel composite signal.

The receiving end of a system with a super-group input is shown in Fig. 10-14. Bandpass filters select out the various channels, and balanced modulators are used to reinject the carrier frequency and produce the original voice input. The output is 12 basic groups, which are demultiplexed and demodulated into the original 60 voice signals.

This analog multiplex method is no longer widely used in the telephone system because it has been essentially replaced by more reliable digital PCM multiplex systems.

FM STEREO BROADCASTING.   In recording original stereo, two microphones are used to generate two separate audio signals. The two microphones pick up sound from a common source, such as a voice or orchestra, but from different directions. The separation of the two microphones provides sufficient differences in the two audio signals to provide more realistic reproduction of the original sound. When reproducing stereo, the two signals can come from a cassette tape, a CD, or some other source. These two independent signals must somehow be transmitted by a single transmitter. This is done through FDM techniques.

Figure 10-15 is a general block diagram of a stereo FM multiplex modulator. The two audio signals, generally called the left (L) and right (R) signals, originate at the two microphones shown in the figure. These two signals are fed to a combining circuit, where they are used to form sum ($L + R$) and difference ($L - R$) signals. The $L + R$ signal is a linear algebraic combination of the left and right channels. The composite signal it produces is the same as if a single microphone were used to pick up

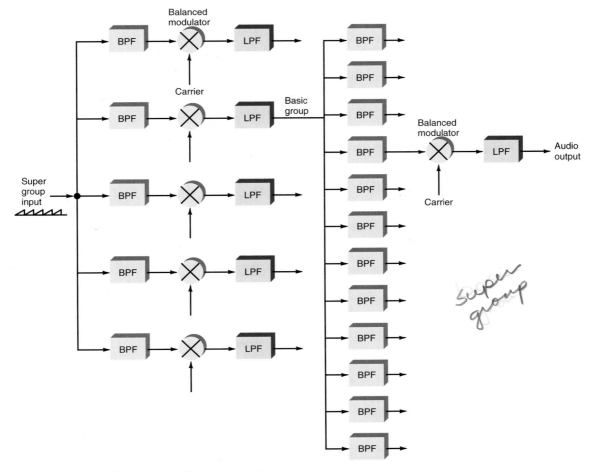

Fɪɢ. 10-14 Demultiplexing the telephone signals.

the sound. It is this signal that a monaural receiver will hear. The frequency response is 50 Hz to 15 kHz.

The combining circuit inverts the right channel signal, thereby subtracting it from the left channel to produce the $L - R$ signal. These two signals, the $L + R$ and $L - R$, are transmitted independently and recombined later in the receiver to produce the individual right- and left-hand channels.

The $L - R$ signal is used to amplitude-modulate a 38-kHz carrier in a balanced modulator. The balanced modulator suppresses the carrier, but generates upper and lower sidebands. The resulting spectrum of the composite modulating signal is shown in Fig. 10-16. As shown, the frequency range of the $L + R$ signal is from 50 Hz to 15 kHz. Since the frequency response of an FM signal is 50 to 15 kHz, the sidebands of the $L - R$ signal are in the frequency range of 38 kHz ± 15 kHz or 23 kHz to 53 kHz. This DSB suppressed carrier signal is algebraically added to, and transmitted along with, the standard $L + R$ audio signal.

Also transmitted with the $L + R$ and $L - R$ signals is a 19-kHz pilot carrier, which is generated by an oscillator whose output also modulates the main transmitter. Note that the 19-kHz oscillator drives a frequency doubler to generate the 38-kHz carrier for the balanced modulator.

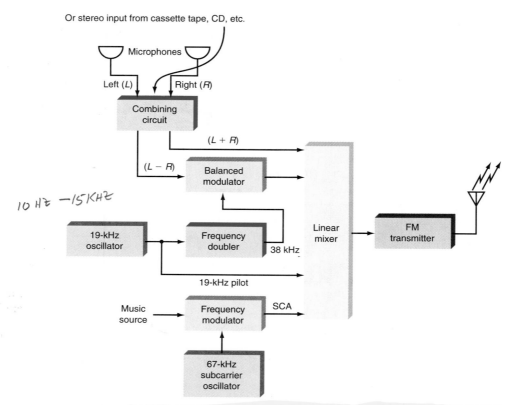

Fig. 10-15 General block diagram of an FM stereo multiplex modulator, multiplexer, and transmitter.

Some FM stations also broadcast one or more additional signals, referred to as subsidiary communications authorization (SCA) signals. The basic SCA signal is a separate subcarrier of 67 kHz which is frequency-modulated by audio signals, usually music. SCA signals are also used to transmit weather, sports, and financial information. Special SCA FM receivers can pick up these signals. The SCA portion of the system is generally used for broadcasting background music for elevators, stores, offices, and restaurants. If an SCA system is being used, the 67-kHz subcarrier with its music

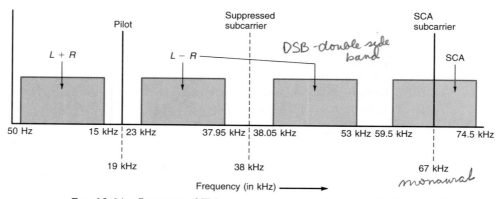

Fig. 10-16 Spectrum of FM stereo multiplex broadcast signal. This signal frequency-modulates the RF carrier.

modulation will also be added to the $L - R$ and $L + R$ signals to modulate the FM transmitter. Not all stations transmit SCA, but some transmit several channels, using additional, higher-frequency subcarriers.

As in other FDM systems, all the subcarriers are added with a linear mixer to form a single signal (Fig. 10-16). This signal is used to frequency-modulate the carrier of the broadcast transmitter. Again, note that FDM simply takes a portion of the frequency spectrum. There is sufficient spacing between adjacent FM stations so that the additional information can be accommodated. Keep in mind that each additional subcarrier reduces the amount by which the main $L + R$ signal can modulate the carrier, since the maximum total modulation voltage is determined by the legal channel width.

At the receiving end, the demodulation is accomplished with a circuit similar to that illustrated in Fig. 10-17. The FM superheterodyne receiver picks up the signal, amplifies it, and translates it to an intermediate frequency, usually 10.7 MHz. It is then demodulated. The output of the demodulator is the original multiplexed signal. The additional circuits now sort out the various signals and reproduce them in their original form.

The original audio $L + R$ signal is extracted simply by passing the multiplex signal through a low-pass filter. Only the 50- to 15-kHz original audio is passed. This signal is fully compatible with monaural FM receivers without stereo capability. In a stereo system, the $L + R$ audio signal is fed to a linear matrix or combiner where it is mixed with the $L - R$ signal to create the two separate L and R channels.

The multiplexed signal is also applied to a bandpass filter that passes the 38-kHz suppressed subcarrier with its sidebands. This is the $L - R$ signal that modulates the 38-kHz carrier. This signal is fed to a balanced modulator for demodulation.

The 19-kHz pilot carrier is extracted by passing the multiplexed signal through a narrow bandpass filter. This 19-kHz subcarrier is then fed to an amplifier and frequency doubler circuit which produces a 38-kHz carrier signal that is fed to the balanced modulator. The output of the balanced modulator, of course, is the $L - R$ audio signal. This is fed to the linear resistive combiner along with the $L + R$ signal.

If an SCA signal is being used, a separate bandpass filter centered on the 67-kHz subcarrier will extract the signal and feed it to an FM demodulator. The demodulator output is then sent to a separate audio amplifier and speaker.

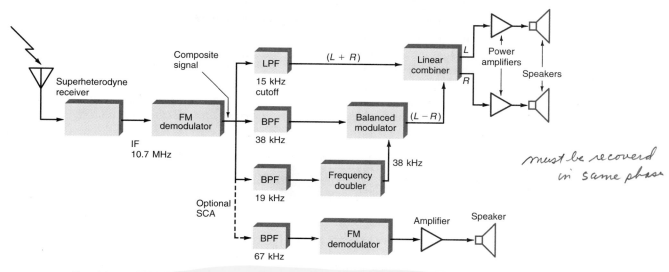

FIG. 10-17 Demultiplexing and recovering the FM stereo and SCA signals.

The $L + R$ and $L - R$ audio signals are recovered and fed to the linear matrix or combiner that both adds and subtracts these two signals. Addition produces the left-hand channel: $(L + R) + (L - R) = 2L$. Subtraction produces the right-hand channel: $(L + R) - (L - R) = 2R$. The left- and right-hand audio signals are then sent to separate audio amplifiers and ultimately the speakers.

All the circuitry used in the demultiplexing process is usually contained in a single IC. In fact, most FM receivers contain a single chip that includes the IF, demodulator, and demultiplexer. Note that the multiplexing and demultiplexing of FM stereo in a TV set are exactly as described above, but with a different IF.

Cable TV is an example of FDM. TV signals include video and audio to modulate carriers using analog methods. Each channel uses a separate set of carrier frequencies, which can be added together to produce FDM. The cable box acts as a tunable filter to select the desired channel.

Example 10-1

A cable TV service uses a single coaxial cable with a bandwidth of 350 MHz to transmit multiple TV signals to subscribers. Each TV signal is 6 MHz wide. How many channels can be carried?

Total channels = 350/6 = 58.33 or 58

10-3 TIME DIVISION MULTIPLEXING

In FDM, multiple signals are transmitted over a single channel, each signal being allocated a portion of the spectrum within that bandwidth. In *time division multiplexing (TDM),* each signal can occupy the entire bandwidth of the channel. However, each signal is transmitted for only a brief period of time. In other words, the multiple signals take turns transmitting over the single channel, as diagrammed in Fig. 10-18. Here, each of the four signals being transmitted over a single channel is allowed to use the channel for a fixed period of time, one after another. Once all four have been transmitted, the cycle repeats.

TDM can be used with both digital and analog signals. For example, if the data consists of sequential bytes, one byte of data from each source can be transmitted during the time interval assigned to a particular channel. Each of the time slots shown in

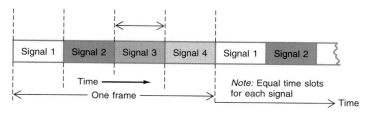

FIG. 10-18   The basic TDM concept.

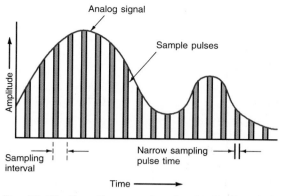

Fɪɢ. 10-19 Sampling an analog signal to produce pulse-amplitude modulation.

Fig. 10-18 might contain a byte from each of four sources. One channel would transmit 8 bits, then halt, while the next channel transmitted 8 bits. The third channel would then transmit its data word, and so on. The cycle would repeat itself at a high rate of speed. Using this technique, the data bytes of individual channels can be interleaved. ← *put together* The resulting single-channel signal is a digital bit stream that is deciphered and reassembled at the receiving end.

The transmission of digital data by TDM is straightforward in that the incremental digital data is already broken up into chunks which can easily be assigned to different time slots. TDM can also be used to transmit continuous analog signals, be they voice, video, or telemetry-derived. This is accomplished by sampling the analog signal repeatedly at a high rate, then converting the samples to proportional binary numbers and transmitting them serially.

As discussed in Chap. 9, sampling an analog signal creates pulse-amplitude modulation (PAM). As shown in Fig. 10-19, the analog signal is converted into a series of constant-width pulses whose amplitude follows the shape of the analog signal. The original analog signal is recovered by passing it through a low-pass filter. In TDM using PAM, a circuit called a multiplexer (MUX or MPX) samples multiple analog signal sources; the resulting pulses are interleaved, then transmitted over a single channel.

PAM Multiplexers

The simplest time multiplexer operates like a single-pole multiple-position mechanical or electronic switch that sequentially samples the multiple analog inputs at a high rate of speed. A basic mechanical rotary switch is shown in Fig. 10-20. The switch arm dwells momentarily on each contact, allowing the input signal to be passed through to the output. It then switches quickly to the next channel, allowing that channel to pass for a fixed duration. The remaining channels are sampled in the same way. After each signal has been sampled, the cycle repeats. The result is that four analog signals are sampled, creating pulse-amplitude-modulated signals that are interleaved with one another. The speed of sampling is directly related to the speed of rotation, and the dwell time of the switch arm on each contact depends on the speed of rotation and the duration of contact.

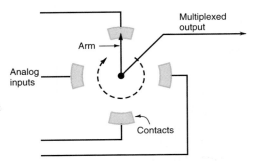

FIG. 10-20 Simple rotary-switch multiplexer.

Figure 10-21 illustrates how four different analog signals are sampled by this technique. Signals A and C are continuously varying analog signals, signal B is a positive-going linear ramp, and signal D is a constant DC voltage.

COMMUTATOR SWITCHES. Multiplexers in early TDM/PAM telemetry systems used a form of rotary switch known as a *commutator*. Multiple switch segments were attached to the various incoming signals while a high-speed brush rotated by a DC motor rapidly sampled the signals as it passed over the contacts. (Commutators have now been totally replaced by electronic circuits, which are discussed in the next section.)

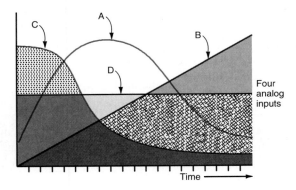

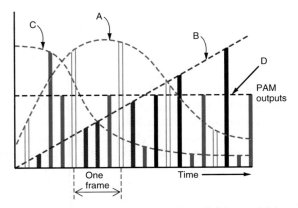

FIG. 10-21 Four-channel PAM time division multiplexer.

In practice, the duration of the sample pulses is shorter than the time which is allocated to each channel. For example, assume it takes the commutator or multiplexed switch 1 ms to move from one contact to another. The contacts can be set up so that each sample is 1 ms long. Typically, the duration of the sample is set to be about half the channel period value, in this example, 0.5 ms.

One complete revolution of the commutator switch is referred to as a *frame*. In other words, during one frame, each input channel is sampled one time. The number of contacts on the multiplexed switch or commutator sets the number of samples per frame. The number of frames completed in one second is called the *frame rate*. Multiplying the number of samples per frame by the frame rate yields the commutation rate or multiplex rate, which is the basic frequency of the composite signal, the final multiplexed signal that is transmitted over the communications channel.

In Fig. 10-21, the number of samples per frame is 4. Assume that the frame rate is 100 frames per second. The period for one frame, therefore, is 1/100 = 0.01 s = 10 ms. During that 10-ms frame period, each of the four channels is sampled once. Assuming equal sample durations, each channel is thus allotted 10/4 = 2.5 ms. (As indicated earlier, the full 2.5-ms period would not be used. The sample duration during that interval might be, for example, only 1 ms). Since there are four samples taken per frame, the commutation rate is 4 × 100 or 400 pulses per second.

ELECTRONIC MULTIPLEXERS. In practical TDM/PAM systems, electronic circuits are used instead of mechanical switches or commutators. The multiplexer itself is usually implemented with FETs, which are nearly ideal ON–OFF switches and can turn off and on at very high speeds. A complete four-channel TDM/PAM circuit is illustrated in Fig. 10-22.

The multiplexer is an op amp summer circuit with MOSFETs on each input resistor. When the MOSFET is conducting, it has a very low ON resistance and therefore acts as a closed switch. When the transistor is OFF, no current flows through it, and it, therefore, acts as an open switch. A digital pulse applied to the gate of the MOSFET turns the transistor ON. The absence of a pulse means that the transistor is OFF. The control pulses to the MOSFET switches are such that only one MOSFET is turned on at a time. These MOSFETs are turned on in sequence by the digital circuitry illustrated.

All of the MOSFET switches are connected in series with resistors (R_1–R_4); this, in combination with the feedback resistor (R_f) on the op amp circuit, determines the gain. For the purposes of this example, assume that the input and feedback resistor are all equal in value, meaning that the op amp circuit has a gain of 1. Since this

"Communications rooms" like this rely upon multiplexing techniques to link cities, states, and foreign countries.

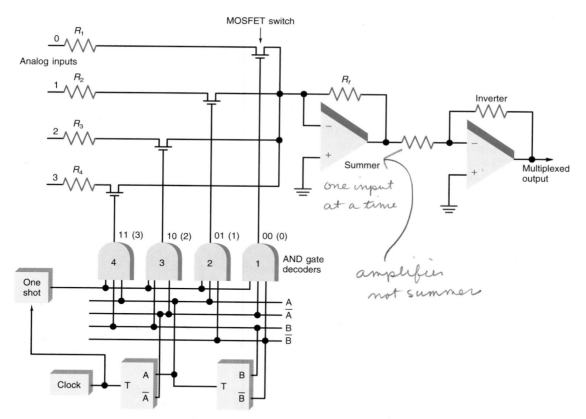

FIG. 10-22 A time division multiplexer used to produce pulse-amplitude modulation.

op amp summing circuit inverts the polarity of the analog signals, it is followed by another op amp inverter which again inverts, restoring the proper polarity.

All the circuitry shown in Fig. 10-22 is usually contained on a single IC chip. MOSFET multiplexers are available with 4, 8, and 16 inputs, and these may be grouped to handle an even larger number of analog inputs.

The digital control pulses are developed by the counter and decoder circuit shown in Fig. 10-22. Since there are four channels, four counter states are needed. Such a counter can be implemented with two flip-flops, representing four discrete states—00, 01, 10, and 11—which are the binary equivalents of the decimal numbers 0, 1, 2, and 3. The four channels can, therefore be labeled 0, 1, 2, and 3.

A clock oscillator circuit triggers the two flip-flop counters. The clock and flip-flop waveforms are illustrated in Fig. 10-23. The flip-flop outputs are applied to decoder AND gates that are connected to recognize the four binary combinations 00, 01, 10, and 11. The output of each decoder gate is applied to one of the multiplexed FET gates.

The one-shot multivibrator diagrammed in Fig. 10-22 is used to trigger all the decoder AND gates at the clock frequency. It produces an output pulse whose duration has been set to the desired sampling interval, in this case 1 ms.

Each time the clock pulse occurs, the one shot generates its pulse, which is applied simultaneously to all four AND decoder gates. At any given time, only one of the gates is enabled. The output of the enabled gate is a pulse whose duration is the same as that of the one shot.

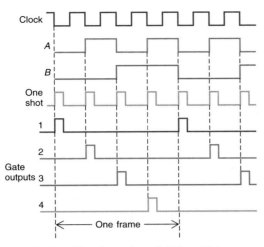

FIG. 10-23 Waveforms for a PAM multiplexer.

When the pulse occurs, it turns on the associated MOSFET and allows the analog signal to be sampled and passed through the op amps to the output. The output of the final op amp is a multiplexed PAM signal like that in Fig. 10-21.

PAM/FM AND PAM/FM/PM SYSTEMS. The multiplexed PAM signal is not usually transmitted as a baseband signal over the single channel. Instead, these varying-amplitude pulses are used to modulate a carrier which is then transmitted over the communications medium. AM can be used, but in most systems, the PAM signal is used to frequency-modulate either an RF carrier for radio transmission or a subcarrier which, in turn, modulates a final RF carrier. Two such arrangements are shown in Fig. 10-24. In the first arrangement, referred to as a *PAM/FM* system, the PAM signals frequency-modulate a carrier. In the second arrangement, referred to as *PAM/FM/PM*, the PAM signals frequency-modulate a subcarrier. In the figure, two sets of PAM signals are shown, each frequency-modulating a different subcarrier. These subcarriers are then linearly mixed and used to phase-modulate the RF carrier, which is the final signal transmitted. PAM/FM/PM systems use a combination of TDM and FDM schemes to create the final composite signal.

PAM DEMULTIPLEXERS

Once the composite signal is received, it must be demodulated and demultiplexed. Figure 10-25 shows the receiver and its demodulation and demultiplexing arrangements for PAM/FM and PAM/FM/PM. In the PAM/FM system [Fig. 10-25(*a*)], the signal is picked up by the receiver, which ultimately sends the signal to an FM demodulator that recovers the original PAM data. The PAM signal is then demultiplexed into the original analog signals. In the PAM/FM/PM system [Fig. 10-25(*b*)], two levels of demodulation are required before the PAM signal is available. A PM demodulator recovers the PAM/FM signals, which are passed through bandpass filters centered on the subcarrier frequencies. The output of each bandpass filter is fed to a FM demodulator, where each group of PAM signals is recovered then demultiplexed.

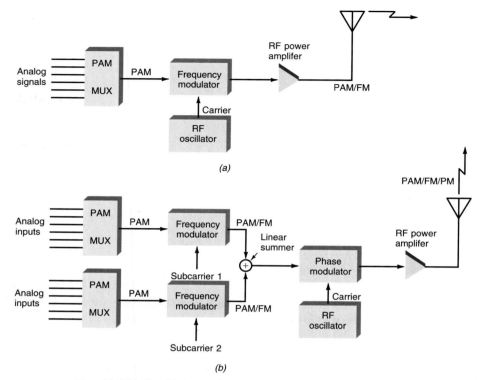

FIG. 10-24 Combining analog (frequency) and digital (PAM) multiplexing. (*a*) PAM/FM. (*b*) PAM/FM/PM.

send multiple channels of analog signal in digital format

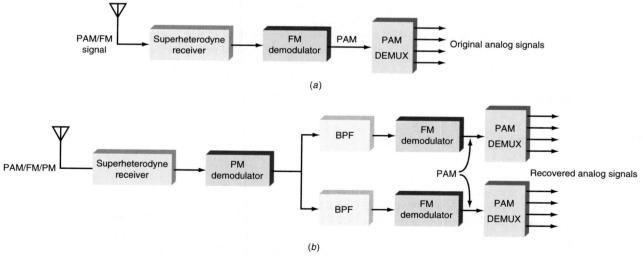

FIG. 10-25 Demultiplexing PAM signals.

Once the composite PAM signal is recovered, it is applied to a demultiplexer (DEMUX). The demultiplexer is, of course, the reverse of a multiplexer. It has a single input and multiple outputs, one for each original input signal. Typical DEMUX circuitry is shown in Fig. 10-26. A four-channel demultiplexer has a single input and four outputs. Most demultiplexers use FETs driven by a counter-decoder. The individual PAM signals are sent to op amps, where they are buffered and possibly amplified. They are then sent to low-pass filters, where they are smoothed into the original analog signals.

The main problem encountered in demultiplexing is synchronization. That is, for the PAM signal to be accurately demultiplexed into the original sampled signals, the clock frequency used at the receiver demultiplexer must be identical to that used at the transmitting multiplexer. In addition, the sequence of the demultiplexer must be identical to that of the multiplexer so that when channel 1 is being sampled at the transmitter, channel 1 is turned on in the receiver demultiplexer at the same time. Such synchronization is usually carried out by a special synchronizing pulse included as a part of each frame. Some of the circuits used for clock frequency and frame synchronization are discussed in the following sections.

CLOCK RECOVERY. Instead of using a free-running clock oscillator set to the identical frequency of the transmitter system clock, the clock for the demultiplexer is derived from the received PAM signal itself. The circuits shown in Fig. 10-27, called *clock recovery circuits,* are typical of those used to generate the demultiplexer clock pulses.

In Fig. 10-27(*a*), the PAM signal has been applied to an amplifier-limiter circuit, which first amplifies all received pulses to a high level and then clips them off at a fixed level. The output of the limiter is thus a constant-amplitude rectangular wave whose output frequency is equal to the commutation rate. This is the frequency at which the PAM pulses occur and is determined by the transmitting multiplexer clock.

The rectangular pulses at the output of the limiter are applied to a bandpass filter, which eliminates the upper harmonics, creating a sine-wave signal at the transmitting clock frequency. This signal is applied to the phase detector circuit in a PLL along with the input from a voltage controlled oscillator (VCO). The VCO is set to operate at the frequency of the PAM pulses. However, the VCO frequency is controlled by a DC error voltage applied to its input. This input is derived from the phase detector output, which is filtered by a low-pass filter into a DC voltage.

The phase detector compares the phase of the incoming PAM sine wave to the VCO sine wave. If a phase error exists, the phase detector produces an output voltage which is filtered to provide a varying direct current. The system is stabilized or locked when the VCO output frequency is identical to that of the sine-wave frequency derived from the PAM input. The difference is that the two are shifted in phase by 90°.

Should the PAM signal frequency change for some reason, the phase detector picks up the variation and generates an error signal which is used to change the frequency of the VCO to match. Because of the closed-loop feature of the

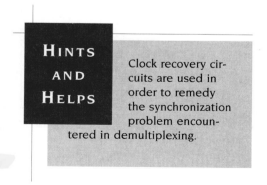

HINTS
AND
HELPS

Clock recovery circuits are used in order to remedy the synchronization problem encountered in demultiplexing.

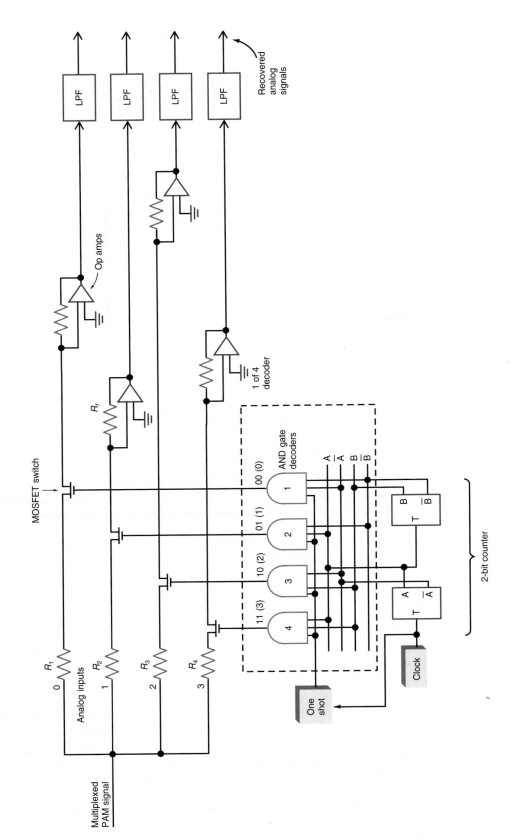

Fig. 10-26 A PAM demultiplexer.

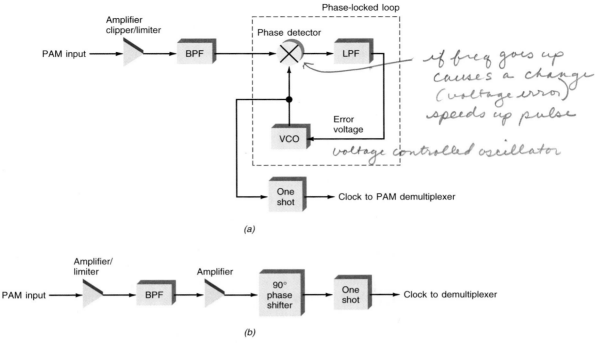

(a)

(b)

Fig. 10-27 Two PAM clock recovery circuits. (*a*) Closed loop. (*b*) Open loop.

Handwritten annotations:
- *if freq goes up causes a change (voltage error) speeds up pulse*
- *voltage controlled oscillator*

system, the VCO automatically tracks even minute frequency changes in the PAM signal, ensuring that the clock frequency used in the demultiplexer will always perfectly match that of the original PAM signal.

The output signal of the VCO is applied to a one-shot pulse generator which creates rectangular pulses at the proper frequency. These are used to step the counter in the demultiplexer; the counter generates the gating pulses for the FET demultiplexer switches.

A simpler, open-loop clock pulse circuit is shown in Fig. 10-27(*b*). Again, the PAM signal is applied to an amplifier-limiter and then a bandpass filter. The sine-wave output of the bandpass filter is amplified and applied to a phase-shift circuit which produces a 90° phase shift at the frequency of operation. This phase-shifted sine wave is then applied to a pulse generator, which, in turn, creates the clock pulses for the demultiplexer. One disadvantage of this technique is that the phase-shift circuit is fixed to create a 90°shift at only one frequency, and so minor shifts in input frequency produce clock pulses whose timing is not perfectly accurate. However, in most systems in which frequency variations are not great, the circuit operates reliably.

FRAME SYNCHRONIZATION. After clock pulses of the proper frequency have been obtained, it is necessary to synchronize the multiplexed channels. This is usually done with a special synchronizing (sync) pulse applied to one of the input channels at the transmitter. In the four-channel system discussed previously, only three actual signals are transmitted. The fourth channel is used to transmit a special pulse whose characteristics are unique in some way so that it can be easily recognized. The amplitude of the pulse can be higher than the highest-amplitude data pulse, or the width of the pulse can be wider than those pulses derived by sampling the input signals. Special circuits are then used to detect the sync pulse.

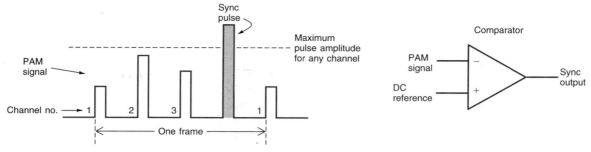

Fig. 10-28 Frame sync pulse and comparator detector.

Figure 10-28 shows an example of a sync pulse that is higher in amplitude than the maximum pulse value of any data signal. The sync pulse is also the last to occur in the frame. At the receiver, a comparator circuit is used to detect the sync pulse. One input to the comparator is set to a DC reference voltage that is slightly higher than the maximum amplitude possible for the data pulses. When a pulse greater than the reference amplitude occurs—that is, the sync pulse—the comparator immediately generates an output pulse, which can then be used for synchronization. Alternatively, it is possible not to transmit a pulse during one channel interval, leaving a blank space in each frame that can then be detected for purposes of synchronization.

When the sync pulse is detected at the receiver, it acts as a reset pulse for the counter in the demultiplexer circuit. At the end of each frame, the counter is reset to zero (channel 0 is selected). When the next PAM pulse occurs, the demultiplexer will be set to the proper channel. Clock pulses then step the counter in the proper sequence for demultiplexing.

Finally, at the output of the demultiplexer, separate low-pass filters are applied to each channel to recover the original analog signals. Figure 10-29 shows the complete PAM demultiplexer.

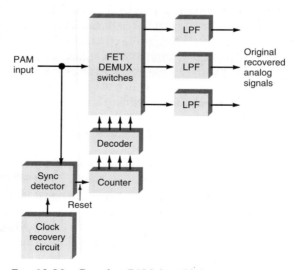

Fig. 10-29 Complete PAM demultiplexer.

The most popular form of TDM uses pulse-code modulation (PCM) (see Sec. 9-3), in which multiple channels of digital data are transmitted in serial form. Each channel is assigned a time slot in which to transmit one binary word of data. The data streams from the various channels are interleaved and transmitted sequentially.

PCM MULTIPLEXERS

When PCM is used to transmit analog signals, the signals are sampled with a multiplexer as described previously for PAM, then converted by an A/D converter into a series of binary numbers, where each number is proportional to the amplitude of the analog signal at the various sampling points. These binary words are converted from parallel to serial format, then transmitted.

At the receiving end, the various channels are demultiplexed and the original sequential binary numbers recovered, stored in a digital memory, and then transferred to a D/A converter which reconstructs the analog signal. (When the original data is strictly digital, D/A conversion is not required, of course.)

Any binary data, multiplexed or not, can be transmitted by PCM. Most long-distance space probes such as the *Mariner* and *Voyager* have on-board video cameras whose output signals are digitized and transmitted back to earth in binary format. Such PCM video systems make possible the transmission of graphic images over incredible distances. In computer multimedia presentations, video data is often digitized and transmitted by PCM techniques to a remote source.

Figure 10-30 shows a general block diagram of the major components in a PCM system, where analog voice signals are the initial inputs. The voice signals are applied to A/D converters, which generate an 8-bit parallel binary word (byte) each time a sample is taken. Since the digital data must be transmitted serially, the A/D converter output is fed to a shift register, which produces a serial data output from the parallel input. In telephone systems, a codec takes care of the A/D parallel-to-serial conversion. The clock oscillator circuit driving the shift register operates at the desired bit rate.

The multiplexing is done with a simple digital MUX. Since all of the signals to be transmitted are binary in nature, a multiplexer constructed of standard logic gates can be used. A binary counter drives a decoder that selects the desired input channel.

The multiplexed output is a serial data waveform of the interleaved binary words. This baseband digital signal can then be encoded and transmitted directly over a twisted pair of cables, a coaxial cable, or a fiber-optic cable. Alternatively, the PCM binary signal can be used to modulate a carrier. If AM is used, the output is PCM/AM; if FM is used, the output is PCM/FM. The output of the modulator is then fed to a transmitter for radio communications, or subjected to further modulation.

Figure 10-31 shows the details of a four-input PCM multiplexer. Typically, the inputs to such a multiplexer come from an A/D converter. The binary inputs are applied to a

DID YOU KNOW?

PCM video systems make the transmission of graphic images over great distances possible.

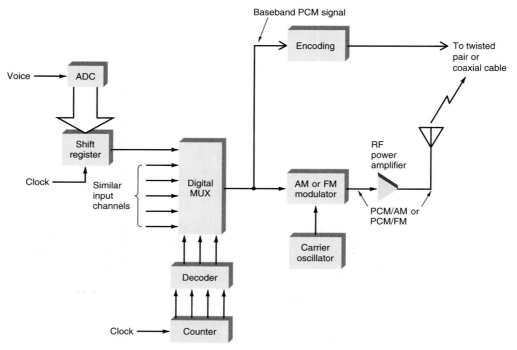

Fɪɢ. 10-30 A PCM system.

shift register, which can be loaded from a parallel source like the A/D converter or another serial source. In most PCM systems, the shift registers are part of a codec.

The multiplexer itself is the familiar digital circuit known as a data selector. It is made up of gates 1 through 5. The serial data is applied to gates 1 to 4; only one gate at a time is enabled, by a 1-of-4 decoder. The serial data from the enabled gate is passed through to OR gate 5 and appears in the output.

Now, assume that all of shift registers are loaded with the bytes to be transmitted. The 2-bit counter AB is reset, sending the 00 code to the decoder. This turns on the 00 output, enabling gate 1. Note that the decoder output also enables gate 6. The clock pulses begin to trigger shift register 1, and the data is shifted out a bit at a time. The serial bits pass through gates 1 and 5 to the output. At the same time, the bit counter, which is a divide-by-8 circuit, keeps track of the number of bits shifted out. When 8 pulses have occurred, all 8 bits in shift register 1 have been transmitted. After counting 8 clock pulses, the bit counter recycles and triggers the 2-bit counter. Its code is now 01. This enables gate 2 and gate 7. The clock pulses continue, and now the contents of shift register 2 are shifted out a bit at a time. The serial data passes through gates 2 and 5 to the output. Again, the bit counter counts 8 bits, then recycles after all bits in shift register 2 have been sent. This again triggers the 2-bit counter, whose code is now 10. This enables gates 3 and 8. The contents of shift register 3 are now shifted out. The process repeats for the contents shift register 4.

When the contents of all four shift registers have been transmitted once, one PCM frame has been formed (see Fig. 10-31). The data in the shift registers is then updated (the next sample of the analog signal converted), and the cycle is repeated.

If the clock in Fig. 10-31 is 64 kHz, the bit rate is 64 kilobits per second and the bit interval is $1/64{,}000 = 15.625\ \mu s$. With 8 bits per word, it takes $8 \times 15.625 = 125\ \mu s$ to transmit one word. This means that the word rate is $1/125 \times 10^{-6} = 8$ kilobytes per second. If the shift registers get their data from an A/D converter, the sampling rate is

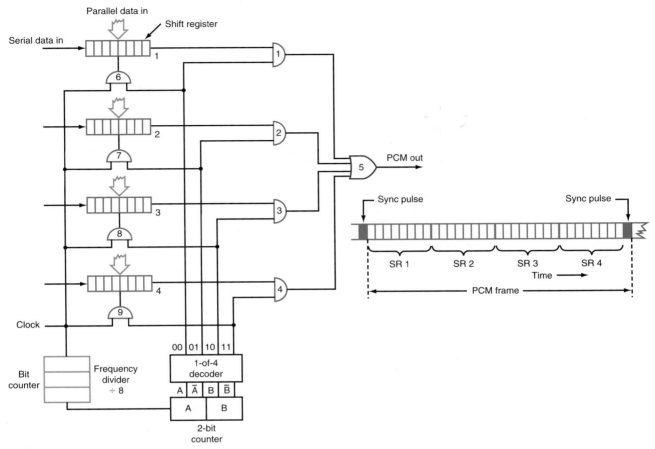

Fig. 10-31 PCM multiplexer.

125 μs or 8 kHz. This is the rate used in telephone systems for sampling voice signals. Assuming a maximum voice frequency of 3 kHz, the minimum sampling rate is twice that, or 6 kHz, and so a sampling rate of 8 kHz is more than adequate to accurately represent and reproduce the analog voice signal. (Serial data rates are further explained in Chap. 11.)

As shown in Fig. 10-31, a sync pulse is added to the end of the frame. This signals the receiver that one frame of four signals has been transmitted and that another is about to begin. The receiver uses the sync pulse to keep all of its circuits in step so that each original signal can be accurately recovered.

PCM DEMULTIPLEXERS

At the receiving end of the communications link, the PCM signal is demultiplexed and converted back into the original data (see Fig. 10-32). The PCM baseband signal may come in over a cable, in which case the signal is regenerated and reshaped prior to being applied to the demultiplexer. Alternatively, if the PCM signal has modulated a carrier and is being transmitted by radio, the RF signal will be picked up by a receiver and then demodulated. The original serial PCM binary waveform is recovered and fed

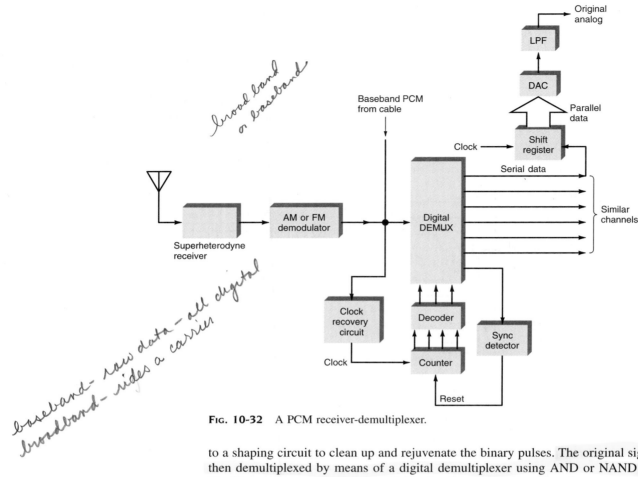

broad band or baseband

baseband — raw data — all digital

broadband — rides a carrier

Fig. 10-32 A PCM receiver-demultiplexer.

to a shaping circuit to clean up and rejuvenate the binary pulses. The original signal is then demultiplexed by means of a digital demultiplexer using AND or NAND gates. The binary counter and decoder driving the demultiplexer are kept in step with the receiver through a combination of clock recovery and sync pulse detector circuits similar to those used in PAM systems. The sync pulse is usually generated and sent at the end of each frame. The demultiplexed serial output signals are fed to a shift register for conversion to parallel data, sent to a D/A converter, and then a low-pass filter. The shift register and D/A converter are usually part of a codec. The result is a highly accurate reproduction of the original voice signal.

T-1 Systems

The most commonly used PCM system is the T-1 system developed by Bell Telephone for transmitting telephone conversations by high-speed digital links. The T-1 system multiplexes 24 voice channels onto a single line using TDM techniques. Each serial digital word (8-bit words, 7 bits of magnitude and one bit representing polarity) from the 24 channels is then transmitted sequentially. Each channel is sampled at an 8-kHz rate, producing a 125-μs sampling interval. During the 125-μs interval between analog samples on each channel, 24 8-bit words, each representing one sample from each of the inputs, are transmitted. The channel sampling interval is 125 μs/24 = 5.2 μs, which corresponds to a rate of 192 kHz. This represents a total of 24 × 8 = 192 bits.

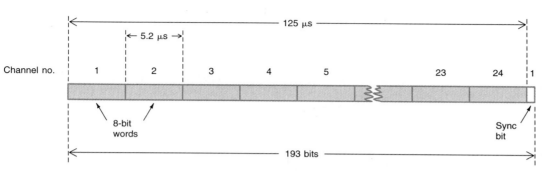

FIG. 10-33 The T-1 frame format, serial data.

An additional bit—a frame sync pulse—is added to this stream to keep the transmitting and receiving signals in synchronization with one another. The 24 8-bit words and the synchronizing bit form one frame of 193 bits. This sequence is carried out repeatedly. The total bit rate for the multiplexed signal is 193×8 kHz $= 1544$ kHz $= 1.544$ MHz. Figure 10-33 shows one frame of a T-1 signal.

The T-1 signal can be transmitted via cable, coaxial cable, twisted pair cable, or fiber-optic cable, or can be used to modulate a carrier for radio transmission. For example, for long-distance telephone calls, T-1 signals are sent to microwave relay stations, where they frequency-modulate a carrier for transmission over long distances. T-1 signals are also transmitted via satellite.

Example 10-2

A special PCM system uses 16 channels of data, one of which is used for identification (ID) and synchronization. The sampling rate is 3.5 kHz. The word length is 6 bits. Find (a) the number of available data channels, (b) the number of bits per frame, and (c) the serial data rate.

a. 16 (total no. of channels) $-$ 1 (channel used for ID) $= 15$ (for data)
b. Bits per frame $= 6 \times 16 = 96$
c. Serial data rate $=$ sampling rate $\times$ no. bits/frame $= 3.5$ kHz $\times 96 = 336$ kHz

BENEFITS OF PCM

PCM is reliable, inexpensive, and highly resistant to noise. In PCM, the transmitted binary pulses all have the same amplitude and, like FM signals, can be clipped to reduce noise. Further, even when signals have been degraded due to noise, attenuation, or distortion, all the receiver has to do is determine if a pulse was transmitted or not. Amplitude, width, frequency, phase shape, etc., do not affect reception. Thus PCM signals are easily recovered and rejuvenated, no matter what the circumstances. PCM is so superior to other forms of pulse modulation and multiplexing for transmission of data that it has replaced them all in communications applications.

SUMMARY

Multiplexing is the process of transmitting multiple signals simultaneously over a single communications channel. Three applications that would be prohibitively expensive or impossible without multiplexing are telemetry; telephone systems, including satellite functions; and modern radio and TV broadcasting.

The two basic types of multiplexing are frequency division multiplexing (FDM) and time division multiplexing (TDM). In FDM, multiple signals are assigned a different frequency within a common bandwidth. In TDM, multiple signals are transmitted in different time slots.

FDM requires both mixing and amplification, which is accomplished by using a circuit known as an operational amplifier summer. The composite output signal is then typically used to modulate a radio transmitter. Most telemetry systems use FM modulation. FM/FM systems use frequency modulation of the VCO subcarriers as well as frequency modulation of the final carrier. At the receiving end, the signal is demultiplexed by demodulator circuits, typically either of the PLL or pulse-averaging type.

TDM can be used with both digital and analog signals. In TDM transmission of analog signals, the analog signal is repeatedly sampled at a high rate, creating pulse-amplitude modulation (PAM), then converted to proportional binary numbers and transmitted serially. PAM can be used to time-multiplex several analog signals. In practical TDM/PAM systems, electronic circuits are used instead of mechanical switches or commutators. In PAM/FM systems, the signals frequency-modulate a carrier; in PAM/FM/PM systems, the signals frequency-modulate subcarriers, which are linearly mixed and used to phase-modulate an RF carrier. Once the composite PAM signal is obtained, it is applied to a demultiplexer, which recovers the original signals.

Today virtually all TDM systems use the digital technique known as pulse code modulation (PCM), a popular example of which is the T-1 system developed by Bell Telephone for transmitting telephone conversations by high-speed digital links. PCM is highly immune to noise, reliable, and, thanks to low-cost ICs, relatively inexpensive to implement.

KEY TERMS

Demultiplexer
FM/FM system
Frame synchronization
Frequency division multiplexing (FDM)
Inter-Range Instrumentation Group (IRIG)
Multiplexer
Multiplexing
One-shot multivibrator

PAM demultiplexer
PCM demultiplexer
Proportional bandwidth
Pulse-code modulation (PCM)
Subcarrier oscillator (SCO)
Subsidiary Communications Authorization (SCA) signal
Sync pulse

T-1 system
Telemetry
Time division multiplexing (TDM)
Voltage controlled oscillator (VCO)

QUESTIONS

1. Is multiplexing the process of transmitting multiple signals over multiple channels?
2. State the primary benefit of multiplexing.
3. What is the name of the circuit used at the receiver to recover multiplexed signals?
4. What is the basic principle of FDM?
5. What circuit combines the multiple signals in an FDM system?
6. What is the name of the circuit to which the signals to be frequency-multiplexed are applied?
7. What kind of modulation do most telemetry multiplex systems use?
8. Name three places where telemetry is used.
9. What is the best demodulator for FDM telemetry work?
10. What is the mathematical designation of the monaural signal in FM stereo multiplexing?
11. Name the four signals transmitted in FM stereo multiplex.
12. What type of modulation is used on the $L - R$ channel?
13. What type of modulation is used by the SCA signals in stereo systems?
14. What is the basic circuit used for demultiplexing in FDM systems?
15. In TDM, how do multiple signals share a channel?
16. What type of modulation is produced when an analog signal is sampled at high speed?
17. What type of circuit is used to recreate the clock pulses at the receiver in a PAM system?
18. In a PAM system, how are the multiplexer and demultiplexer kept in step with one another?
19. What is the time period called in a PAM or PCM system when all channels are sampled once?
20. What circuit is used to demodulate a PAM signal?
21. What type of switch is used in an electronic multiplexer?
22. How is the desired input to a time division multiplexer selected?
23. Explain the operation of a PAM/FM transmitter.
24. What type of clock recovery circuit tracks PAM frequency variations?
25. How are analog signals transmitted in a PCM system?
26. What is the name of the IC that converts A/D and D/A signals in a PCM telephone system?
27. What is the standard audio sampling rate in a PCM telephone system?
28. What is the standard word size in a PCM telephone system?
29. What is the main benefit of PCM over PAM, FM, and other modulation techniques?
30. What is the total number of bits in one T-1 frame?
31. Does a T-1 system use baseband or broadband techniques? Explain.

PROBLEMS

1. How many 6-MHz-wide TV channels can be multiplexed on a 400-MHz coaxial cable? ◆
2. What is the minimum sampling rate for a signal with a 14-kHz bandwidth?
3. State the bit rate and maximum number of channels in a T-1 telephone TDM system. ◆

CRITICAL THINKING

1. How long does it take to transmit one PCM frame with 16 bytes, no synchronization, and a clock rate of 46 MHz?
2. A special PCM system uses 12-bit words and 32 channels. Data is transmitted serially with no sync pulse. The time duration for one bit is 488.28 ns. How many bits per frame are transmitted and at what rate?
3. Can FDM be used with binary information signals? Explain.

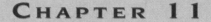

THE TRANSMISSION OF BINARY DATA IN COMMUNICATIONS SYSTEMS

Objectives

After completing this chapter, you will be able to:

◆ *Explain* the difference between asynchronous and synchronous data transmission.

◆ *State* the relationship between communications channel bandwidth and data rate in bits per second.

◆ *Name* the four basic types of encoding used in the serial transmission of data.

◆ *Describe* the function of FSK, PSK, and QAM.

◆ *Explain* the need for and types of communication protocols.

◆ *Compare* and *contrast* redundancy, parity, block-check sequences, cyclical redundancy checks, and forward error correction.

◆ *Explain* the operation and benefits of spread spectrum systems.

Over the last two decades, the proliferation of applications that send digital data over communications channels has resulted in the need for efficient methods of transmission, conversion, and reception of digital data. As with analog data, the amount of digital information that can be transmitted is proportional to the bandwidth of the communications channel and the time of transmission. This chapter further develops some of the basic concepts and equipment used in the transmission of digital data, including some mathematical principles that apply to transmission efficiency, and discusses the development of digital standards, error detection and correction methods, and spread spectrum techniques.

11-1 DIGITAL CODES

Data processed and stored by computers can be numerical (e.g., accounting records, spreadsheets, and stock quotations) or text (e.g., letters, memos, reports, and books). As previously discussed, the signals used to represent computerized data are digital, rather than analog, in nature. Even before the advent of computers, digital codes were used to represent data.

AN EARLY DIGITAL CODE: THE MORSE CODE

The earliest forms of electronic communications were digital in nature. The first electronic communications system was the telegraph, which was invented in 1832 and first demonstrated successfully in 1844 by Samuel F. B. Morse. A telegraph hand key acting as a switch connects to a remote receiver or sounder. The sounder is nothing more than a magnetic coil which attracts an armature that makes a clicking sound. Whenever the telegraph key is depressed, current flows through the remote coil, producing a magnetic field that attracts the armature and makes a click. When the key is opened, the armature makes another click. For two-way communications, both stations have a key and a sounder so that either station can transmit and receive. The operation is not strictly binary even though there are two voltage leads, since there are three states: ON (dash), ON (dot), and OFF (space).

The inventor of the telegraph, Samuel Morse, created a special code using this ON–OFF capability. Known as the Morse Code, it is a series of dots and dashes that represent letters of the alphabet, numbers, and punctuation marks. These are summarized in Fig. 11-1. When the armature is closed for a short duration, a dot is produced. When the armature is attracted for a longer time, a dash is produced. With special training, humans can easily send and receive messages at speeds ranging from 15 to 20 words per minute to 70 to 80 words per minute.

The earliest radio communications was also carried out using the Morse Code of dots and dashes to send messages. A key turned the carrier of a transmitter OFF and ON to produce the dots and dashes. There were detected at the receiver and converted by an operator back into letters and numbers making up the message. This type of radio communications is known as continuous-wave (CW) transmission.

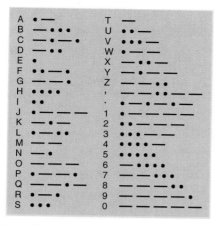

A	• —	T	—
B	— • • •	U	• • —
C	— • — •	V	• • • —
D	— • •	W	• — —
E	•	X	— • • —
F	• • — •	Y	— • — —
G	— — •	Z	— — • •
H	• • • •	,	— — • • — —
I	• •	.	• — • — • —
J	• — — —	1	• — — — —
K	— • —	2	• • — — —
L	• — • •	3	• • • — —
M	— —	4	• • • • —
N	— •	5	• • • • •
O	— — —	6	— • • • •
P	• — — •	7	— — • • •
Q	— — • —	8	— — — • •
R	• — •	9	— — — — •
S	• • •	0	— — — — —

FIG. 11-1   The Morse code. A dot (.) is a short click; a dash (—) is a long click.

AN EARLY BINARY CODE: THE BAUDOT CODE

Binary codes consist of patterns of 0s and 1s. Each pattern represents a number, a letter of the alphabet, or some special symbol such as punctuation or a mathematical operation. Binary codes can also represent control operations such as line feed on a printer or the ringing of a bell or buzzer.

One of the first binary data codes was the Baudot (pronounced *baw dough*) code used in the early teletype machines, a device for sending and receiving coded signals over a communications link. With teletype machines, it was no longer necessary for operators to learn the Morse Code. Whenever a key on the typewriter keyboard is pressed, a unique code is generated and transmitted to the receiving machine, which recognizes and then prints the corresponding letter, number, or symbol.

The 5-bit Baudot code is shown in Fig. 11-2. With 5 bits, $2^5 = 2 \times 2 \times 2 \times 2 \times 2 = 32$ different symbols can be represented. The different 5-bit combinations, together with two shift codes, can generate the 26 letters of the alphabet, 10 numbers, and various punctuation marks. If the message is preceded by the letter shift code (11011), all the following codes are interpreted as letters of the alphabet. Sending the figure shift code (11111) causes all of the following characters to be interpreted as numbers or punctuation marks. The Baudot code is rarely used today, having been supplanted by codes that can represent more characters and symbols.

MODERN BINARY CODES

For modern data communications, information is transmitted using a system in which the numbers and letters to be represented are coded, usually by way of a keyboard, and the binary word representing each character is stored in a computer memory. The message can also be stored on magnetic tape or disk. The following sections describe some widely used codes for transmission of digitized data.

Character Shift		Binary Code				
Letter	Figure	Bit: 4	3	2	1	0
A	–	1	1	0	0	0
B	?	1	0	0	1	1
C	:	0	1	1	1	0
D	$	1	0	0	1	0
E	3	1	0	0	0	0
F	!	1	0	1	1	0
G	&	0	1	0	1	1
H	#	0	0	1	0	1
I	8	0	1	1	0	0
J	'	1	1	0	1	0
K	(	1	1	1	1	0
L	)	0	1	0	0	1
M	.	0	0	1	1	1
N	,	0	0	1	1	0
O	9	0	0	0	1	1
P	0	0	1	1	0	1
Q	1	1	1	1	0	1
R	4	0	1	0	1	0
S	bel	1	0	1	0	0
T	5	0	0	0	0	1
U	7	1	1	1	0	0
V	;	0	1	1	1	1
W	2	1	1	0	0	1
X	/	1	0	1	1	1
Y	6	1	0	1	0	1
Z	"	1	0	0	0	1
Figure shift		1	1	1	1	1
Letter shift		1	1	0	1	1
Space		0	0	1	0	0
Line feed (LF)		0	1	0	0	0
Blank (null)		0	0	0	0	0

FIG. 11-2 The Baudot code.

AMERICAN STANDARD CODE FOR INFORMATION INTERCHANGE. The most widely used data communications code is the 7-bit binary code known as the *American Standard Code for Information Interchange* (abbreviated ASCII and pronounced *ass-key*), which can represent 128 numbers, letters, punctuation marks, and other symbols (see Fig. 11-3). With ASCII, a sufficient number of code combinations is available to represent both upper- and lowercase letters of the alphabet.

The first ASCII codes listed in Fig. 11-3 have two- and three-letter designations. These codes initiate operations or provide responses for inquiries. For example, BEL or 0000111 will ring a bell or a buzzer; CR is carriage return; SP is a space, such as that between words in a sentence; ACK means "acknowledge that a transmission was received"; STX and ETX are start and end of text, respectively; and SYN is a synchronization word that provides a way to get transmission and reception in step with one another. The meanings of all the letter codes are given at the bottom of the table.

HEXADECIMAL VALUES. Binary codes are often expressed using their *hexadecimal,* rather than decimal, values. To convert a binary code to its hexadecimal equivalent, first divide the code into 4-bit groups, starting at the least significant bit on the right and working to the left. (Assume a leading zero on each of the codes.) The hex equivalents for the binary codes for the decimal numbers 0 through 9 and the letters A through F are given in Fig. 11-4.

Two examples of ASCII-to-hex conversion are as follows:

1. The ASCII code for the number 4 is 0110100. Add a leading zero to make 8 bits, then divide into 4-bit groups: 00110100 = 0011 0100 = hex 34.

Bit:	6	5	4	3	2	1	0		Bit:	6	5	4	3	2	1	0		Bit:	6	5	4	3	2	1	0
NUL	0	0	0	0	0	0	0		+	0	1	0	1	0	1	1		V	1	0	1	0	1	1	0
SOH	0	0	0	0	0	0	1		,	0	1	0	1	1	0	0		W	1	0	1	0	1	1	1
STX	0	0	0	0	0	1	0		-	0	1	0	1	1	0	1		X	1	0	1	1	0	0	0
ETX	0	0	0	0	0	1	1		.	0	1	0	1	1	1	0		Y	1	0	1	1	0	0	1
EOT	0	0	0	0	1	0	0		/	0	1	0	1	1	1	1		Z	1	0	1	1	0	1	0
ENQ	0	0	0	0	1	0	1		0	0	1	1	0	0	0	0		[	1	0	1	1	0	1	1
ACK	0	0	0	0	1	1	0		1	0	1	1	0	0	0	1		E	1	0	1	1	1	0	0
BEL	0	0	0	0	1	1	1		2	0	1	1	0	0	1	0		]	1	0	1	1	1	0	1
BS	0	0	0	1	0	0	0		3	0	1	1	0	0	1	1		∧	1	0	1	1	1	1	0
HT	0	0	0	1	0	0	1		4	0	1	1	0	1	0	0		—	1	0	1	1	1	1	1
NL	0	0	0	1	0	1	0		5	0	1	1	0	1	0	1		¡	1	1	0	0	0	0	0
VT	0	0	0	1	0	1	1		6	0	1	1	0	1	1	0		a	1	1	0	0	0	0	1
FF	0	0	0	1	1	0	0		7	0	1	1	0	1	1	1		b	1	1	0	0	0	1	0
CR	0	0	0	1	1	0	1		8	0	1	1	1	0	0	0		c	1	1	0	0	0	1	1
SO	0	0	0	1	1	1	0		9	0	1	1	1	0	0	1		d	1	1	0	0	1	0	0
SI	0	0	0	1	1	1	1		:	0	1	1	1	0	1	0		e	1	1	0	0	1	0	1
DLE	0	0	1	0	0	0	0		;	0	1	1	1	0	1	1		f	1	1	0	0	1	1	0
DC1	0	0	1	0	0	0	1		<	0	1	1	1	1	0	0		g	1	1	0	0	1	1	1
DC2	0	0	1	0	0	1	0		=	0	1	1	1	1	0	1		h	1	1	0	1	0	0	0
DC3	0	0	1	0	0	1	1		>	0	1	1	1	1	1	0		i	1	1	0	1	0	0	1
DC4	0	0	1	0	1	0	0		?	0	1	1	1	1	1	1		j	1	1	0	1	0	1	0
NAK	0	0	1	0	1	0	1		@	1	0	0	0	0	0	0		k	1	1	0	1	0	1	1
SYN	0	0	1	0	1	1	0		A	1	0	0	0	0	0	1		l	1	1	0	1	1	0	0
ETB	0	0	1	0	1	1	1		B	1	0	0	0	0	1	0		m	1	1	0	1	1	0	1
CAN	0	0	1	1	0	0	0		C	1	0	0	0	0	1	1		n	1	1	0	1	1	1	0
EM	0	0	1	1	0	0	1		D	1	0	0	0	1	0	0		o	1	1	0	1	1	1	1
SUB	0	0	1	1	0	1	0		E	1	0	0	0	1	0	1		p	1	1	1	0	0	0	0
ESC	0	0	1	1	0	1	1		F	1	0	0	0	1	1	0		q	1	1	1	0	0	0	1
FS	0	0	1	1	1	0	0		G	1	0	0	0	1	1	1		r	1	1	1	0	0	1	0
GS	0	0	1	1	1	0	1		H	1	0	0	1	0	0	0		s	1	1	1	0	0	1	1
RS	0	0	1	1	1	1	0		I	1	0	0	1	0	0	1		t	1	1	1	0	1	0	0
US	0	0	1	1	1	1	1		J	1	0	0	1	0	1	0		u	1	1	1	0	1	0	1
SP	0	1	0	0	0	0	0		K	1	0	0	1	0	1	1		v	1	1	1	0	1	1	0
!	0	1	0	0	0	0	1		L	1	0	0	1	1	0	0		w	1	1	1	0	1	1	1
"	0	1	0	0	0	1	0		M	1	0	0	1	1	0	1		x	1	1	1	1	0	0	0
#	0	1	0	0	0	1	1		N	1	0	0	1	1	1	0		y	1	1	1	1	0	0	1
$	0	1	0	0	1	0	0		O	1	0	0	1	1	1	1		z	1	1	1	1	0	1	0
%	0	1	0	0	1	0	1		P	1	0	1	0	0	0	0		{	1	1	1	1	0	1	1
&	0	1	0	0	1	1	0		Q	1	0	1	0	0	0	1		¦	1	1	1	1	1	0	0
'	0	1	0	0	1	1	1		R	1	0	1	0	0	1	0		}	1	1	1	1	1	0	1
(	0	1	0	1	0	0	0		S	1	0	1	0	0	1	1			1	1	1	1	1	1	0
)	0	1	0	1	0	0	1		T	1	0	1	0	1	0	0		DEL	1	1	1	1	1	1	1
*	0	1	0	1	0	1	0		U	1	0	1	0	1	0	1									

NUL = null
SOH = start of heading
STX = start of text
ETX = end of text
EOT = end of transmission
ENQ = enquiry
ACK = acknowledge
BEL = bell
BS = back space
HT = horizontal tab
NL = new line

VT = vertical tab
FF = form feed
CR = carriage return
SO = shift-out
SI = shift-in
DLE = data link escape
DC1 = device control 1
DC2 = device control 2
DC3 = device control 3
DC4 = device control 4
NAK = negative acknowledge

SYN = synchronous
ETB = end of transmission block
CAN = cancel
SUB = substitute
ESC = escape
FS = field separator
GS = group separator
RS = record separator
US = unit separator
SP = space
DEL = delete

FIG. 11-3 The popular ASCII code.

2. The letter w in ASCII is 1110111. Add a leading zero to get 01110111; 01110111 = 0111 0111 = hex 77.

EXTENDED BINARY CODED DECIMAL INTERCHANGE CODE. The *Extended Binary Coded Decimal Interchange Code* (EBCDIC, pronounced *ebb-see-dick*), developed by IBM, is an 8-bit code allowing a maximum of 256 characters to be represented. Its primary use is in IBM and IBM-compatible computing systems and equipment. It is not as widely used as ASCII.

DID YOU KNOW?

The transmitters used in garage door openers are modulated by binary-coded pulses.

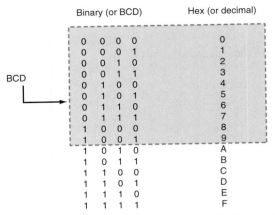

Binary (or BCD)   Hex (or decimal)

BCD

0 0 0 0	0
0 0 0 1	1
0 0 1 0	2
0 0 1 1	3
0 1 0 0	4
0 1 0 1	5
0 1 1 0	6
0 1 1 1	7
1 0 0 0	8
1 0 0 1	9
1 0 1 0	A
1 0 1 1	B
1 1 0 0	C
1 1 0 1	D
1 1 1 0	E
1 1 1 1	F

FIG. 11-4   The hexadecimal (hex) and BCD designations.

PURE BINARY CODE VS. BCD.   When the information to be transmitted is other than numbers and text, the pure binary code is normally used. The data to be transmitted can be the digital representation of voice or video as developed by an A/D converter. The standard binary code is normally used. The 2s complement of negative numbers is used when the data is bipolar; that is, it can be negative or positive in value. An 8-bit word or byte length is the most common, but other lengths are used depending upon the required precision for a specific application.

   An alternative to straight binary code is *binary coded decimal* (BCD). Some test instruments produce output data in BCD, as do some industrial monitors and controllers. Numbers are transmitted as 4-bit codes, which are the same as the number codes for hex in Fig. 11-4. Each transmitted byte usually contains two BCD digits. For example, the number 95 is transmitted as 1001 0101, but without the space.

11-2 PRINCIPLES OF DIGITAL TRANSMISSION

As indicated in Chap. 9, data can be transmitted in two ways: parallel and serial.

SERIAL TRANSMISSION

Parallel data transmission is not practical for long-distance communications. Data transfers in long-distance communications systems are made serially; each bit of a word is transmitted one after another (see Fig. 11-5). The figure shows the ASCII code for the letter M (1001101) being transmitted 1 bit at a time. The LSB is transmitted first, and the MSB last. The MSB is on the right, indicating that it was transmitted later in time. Each bit is transmitted for a fixed interval of time t. The voltage levels representing each bit appear on a single data line one after another until the entire word has been transmitted. For example, the bit interval may be 10 μs, which means that the voltage level for each bit in the word appears for 10 μs. It would therefore take 70 μs to transmit a 7-bit ASCII word.

EXPRESSING THE SERIAL DATA RATE.   The speed of data transfer is usually indicated as number of bits per second (bps or b/s). Some data rates take place at relatively slow speeds, usually a few hundred or several thousand bits per second. Personal computers, for example, communicate at 9600, 14,400, 19,200, or 28,800 bps over the telephone lines. However, in some data communications systems like local area networks, bit rates as high as tens or hundreds of million bits per second are used.

The speed of serial transmission is, of course, related to the bit time of the serial data. The speed in bits per second is the reciprocal of the bit time t, or bps $= 1/t$. For example, assume a bit time of 104.17 μs. The speed is bps $= 1/104.17 \times 10^{-6} = 9600$ bps.

If the speed in bits per second is known, the bit time can be found by rearranging the formula: $t = 1/$bps. For example, the bit time at 230.4 kbps (230,400 bps) is $t = 1/230,400 = 4.34 \times 10^{-6} = 4.34$ μs.

Another term used to express the data speed in digital communications systems is *baud rate*. Baud rate is the number of signaling elements or symbols that occur in a given unit of time, such as 1 s. A *signaling element* is simply some change in the binary signal transmitted. In many cases it is simply a binary logic voltage level change, either 0 or 1. In that case, the baud rate is equal to the data rate in bits per second. In summary, the baud rate is the reciprocal of the smallest signaling interval.

The signaling element can also be one of several discrete signal amplitudes or phase shifts, each of which represents 2 or more data bits. Several unique modulation schemes have been developed so that each symbol or baud can represent multiple bits. The number of symbol changes per unit of time is no higher than the straight binary bit rate, but more bits per unit time are transmitted. As a result, higher bit rates can be transmitted over telephone lines or other severely bandwidth-limited communications channels which would not ordinarily be able to handle them. Several of these modulation methods are discussed in the section on modems.

Assume, for example, a system that represents two bits of data as different voltage levels. With 2 bits there are $2^2 = 4$ possible levels, and a discrete voltage is assigned to each.

00	0 V
01	1 V
10	2 V
11	3 V

In this system each of the four symbols is one of four different voltage levels. Each level is a different symbol representing 2 bits. Assume, for example, that it is desired to transmit the decimal number 201, which is binary 11001001. The number can be transmitted serially as a sequence of equally spaced pulses that are either ON (1) or

$$T = \frac{1}{b}$$
$$b = \frac{1}{t}$$

RS-232 —
protocol
set of rules

FIG. 11-5   Serial transmission of the ASCII letter M.

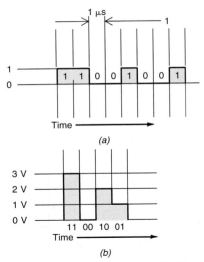

FIG. 11-6 The bit rate can be higher than the baud rate when each symbol represents two or more bits. (*a*) Bit rate = baud rate = 1 bit/μs = 1,00,000 bps = 1 Mb/s. (*b*) Baud rate = 1 symbol/μs = 1,000,000 baud; bit rate = 2 bits/baud = 2,000,000 bps.

OFF (0) [see Fig. 11-6(*a*)]. If each bit interval is 1 μs, the bit rate is $1/1 \times 10^{-6}$ = 1,000,000 bps (1 Mb/s). The baud rate is also 1 million bps.

Using the four-level system, we could also divide the word into 2-bit groups and transmit the appropriate voltage level representing each. The number 11001001 would be divided into these groups: 11 00 10 01. Thus the transmitted signal would be voltage levels of 3, 0, 2 and 1 V, each occurring for a fixed interval of, say, 1 μs [see in Fig. 11-6(*b*)]. The baud rate is still 1 million because there is only one symbol or level per time interval (1 μs). However, the bit rate is 2 million bps—double the baud rate—because each symbol represents two bits. The total transmission time is also shorter. It would take 8 μs to transmit the 8-bit binary word, but only 4 μs to transmit the four-level signal. The bit rate is greater than the baud rate because multiple levels are used.

Because of the sequential nature of serial data transmission, naturally it takes longer to send data this way than it does to transmit it by parallel means. However, with a high-speed logic circuit, even serial data transfers can take place at very high speeds. Currently that data rate is as high as 100 billion bits per second on copper wire cable and even higher on fiber-optic cable. So while serial data transfers are slower than parallel transfers, they are fast enough for most communications applications.

ASYNCHRONOUS TRANSMISSION – *one character at a time*

In asynchronous transmission each data word is accompanied by start and stop bits that indicate the beginning and ending of the word. Asynchronous transmission of an ASCII character is illustrated in Fig. 11-7. When no information is being transmitted, the communications line is usually high, or binary 1. In data communications terminology, this is referred to as a *mark*. To signal the beginning of a word, a start bit, a binary 0 or

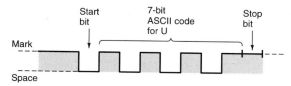

Fig. 11-7 Asynchronous transmission with start and stop bits.

space, as shown in the figure, is transmitted. The start bit has the same duration as all other bits in the data word. The transmission from mark to space indicates the beginning of the word and allows the receiving circuits to prepare themselves for the reception of the remainder of the bits.

After the start bit, the individual bits of the word are transmitted. In this case, the 7-bit ASCII code for the letter U, 1010101, is transmitted. Once the last code bit has been transmitted, a stop bit is included. The stop bit may be the same duration as all other bits and again is a binary 1 or mark. In some systems, two stop bits are transmitted, one after the other, to signal the end of the word.

Most low-speed digital transmission (the 1200- to 38,400-baud range) is asynchronous. This technique is extremely reliable, and the start and stop bits ensure that the sending and receiving circuits remain in step with one another. The minimum separation between character words is 1 stop plus 1 start bit, as Fig. 11-8 shows. There can also be time gaps between characters or groups of characters, as the illustration shows, and thus the stop "bit" may be of some indefinite length.

The primary disadvantage of asynchronous communications is that the extra start and stop bits effectively slow down the data transmission. This is not a problem in low-speed applications such as those involving certain printers and plotters. But when huge volumes of information must be transmitted, the start and stop bits represent a significant percentage of the bits transmitted. A 7-bit ASCII character plus start and stop bits is 9 bits. Two of the 9 bits are not data. This represents 2/9 = 222, or 22.2 percent inefficiency. Removing the start and stop bits and stringing the ASCII characters end to end allow many more data words to be transmitted per second.

SYNCHRONOUS TRANSMISSION

The technique of transmitting each data word one after another without start and stop bits, usually in multiword blocks, is referred to as *synchronous* data communications. To maintain synchronization between transmitter and receiver, a group of synchronization bits is placed at the beginning of the block and the end of the block. Figure 11-9 shows one arrangement. Each block of data can represent hundreds or even thou-

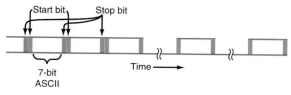

Fig. 11-8 Sequential words transmitted asynchronously.

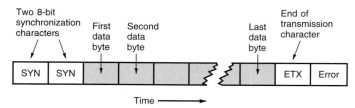

Two 8-bit synchronization characters

First data byte

Second data byte

Last data byte

End of transmission character

| SYN | SYN | | | | ⌇⌇ | | ETX | Error |

Time ⟶

FIG. 11-9 Synchronous data transmission.

sands of 1-byte characters. At the beginning of each block is a unique series of bits that identifies the beginning of the block. In Fig. 11-9, two 8-bit synchronous (SYN) codes signal the start of a transmission. Once the receiving equipment finds these characters, it begins to receive the continuous data, the block of sequential 8-bit words or bytes. At the end of the block, another special ASCII code character, ETX, signals the end of transmission. The receiving equipment looks for the ETX code; detection of this code is how the receiving circuit recognizes the end of the transmission. An error detection code usually appears at the very end of the transmission.

The special synchronization codes at the beginning and end of a block represent a very small percentage of the total number of bits being transmitted, especially in relation to the number of start and stop bits used in asynchronous transmission. Synchronous transmission is therefore much faster than asynchronous transmission.

An important consideration in synchronous transmission is how the receiving station keeps track of the individual bits and bytes, especially when the signal is noisy, since there is no clear separation between them. This is done by transmitting the data at a fixed, known, precise clock rate. Then the number of bits can be counted to keep track of the number of bytes or characters transmitted. For every 8 bits counted, 1 byte is received. The number of received bytes is also counted.

Synchronous transmission assumes that the receiver knows or has a clock frequency identical to the transmitter clock. Usually, the clock at the receiver is derived from the received signal, so that it is precisely the same frequency as, and in synchronization with, the transmitter clock.

Example 11-1

A block of 256 sequential 12-bit data words is transmitted serially in 0.16 s. Calculate (a) the time duration of one word, (b) the time duration of one bit, and (c) the speed of transmission in bits/s.

a. $t_{\text{word}} = \dfrac{0.016}{256} = 0.000625 = 625 \ \mu s$

b. $t_{\text{bit}} = \dfrac{625 \ \mu s}{12 \text{ bits}} = 52.0833 \ \mu s$

c. $\text{bps} = \dfrac{1}{t} = 1/52.0833 \times 10^{-6} = 19200 \text{ bps or } 19.2 \text{ kbps}$

Whether digital signals are being transmitted by baseband methods or broadband methods (see Sec. 1-4), before the data is put on the medium, it is usually encoded in some way to make it compatible with the medium or to facilitate some desired operation connected with the transmission. The primary encoding methods used in data communications are summarized below.

Nonreturn to Zero. In the nonreturn to zero (NRZ) method of encoding the signal remains at the binary level assigned to it for the entire bit time. Figure 11-10(a), which shows unipolar NRZ, is a slightly different version of Fig. 11-5. The logic levels are 0 and +5 V. When a binary 1 is to be transmitted, the signal stays at +5 V for the entire bit interval. When a binary 0 is to be sent, the signal stays at 0 V for the total bit time. In other words, the voltage does not return to zero *during* the binary 1 interval.

In *unipolar* NRZ, the signal has only a positive polarity. A *bipolar* NRZ signal has two polarities, positive and negative, as shown in Fig. 11-10(b). The voltage levels are +12 and −12 V. The popular RS-232 serial computer interface uses bipolar NRZ, where a binary 1 is a negative voltage between −3 and −25 V and a binary 0 is a voltage between +3 and +25 V.

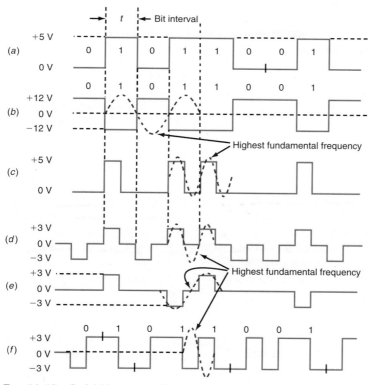

FIG. 11-10 Serial binary encoding methods. (*a*) Unipolar NRZ. (*b*) Bipolar NRZ. (*c*) Unipolar RZ. (*d*) Bipolar RZ. (*e*) Bipolar RZ-AMI. (*f*) Manchester or biphase.

The NRZ method is normally generated inside computers, at low speeds, when asynchronous transmission is being used. It is not popular for synchronous transmission because there is no voltage or level change when there are long strings of sequential binary 1s or 0s. If there is no signal change it is difficult for the receiver to determine just where one bit ends and the next one begins. If the clock is to be recovered from the transmitted data in a synchronous system, there must be more frequent changes, preferably one per bit. NRZ is usually converted into another format, such as RZ or Manchester, for synchronous transmissions.

RETURN TO ZERO. In *return to zero (RZ)* encoding [see Fig. 11-10(c) and (d)] the voltage level assigned to a binary 1 level returns to zero during the bit period. *Unipolar RZ* is illustrated in Fig. 11-10(c). The binary 1 level occurs for 50 percent of the bit interval, and the remaining bit interval is zero. Only one polarity level is used. Pulses occur only when a binary 1 is transmitted; no pulse is transmitted for a binary 0.

Bipolar RZ is illustrated in Fig. 11-10(d). A 50 percent bit interval +3-V pulse is transmitted during a binary 1, and a −3-V pulse is transmitted for a binary 0. Because there is one clearly discernible pulse per bit, it is extremely easy to derive the clock from the transmitted data. For that reason, bipolar RZ is preferred over unipolar RZ.

A popular variation of the bipolar RZ format is called *alternative mark inversion (AMI)* [see Fig. 11-10(e)]. During the bit interval, binary 0s are transmitted as no pulse. Binary 1s, also called marks, are transmitted as alternating positive and negative pulses. One binary 1 is sent as a positive pulse, the next binary 1 as a negative pulse, the next as a positive pulse, and so on.

MANCHESTER. Manchester encoding, also referred to as *biphase encoding,* can be unipolar or bipolar. It is widely used in LANs. In the Manchester system a binary 1 is transmitted first as a positive pulse, for one half of the bit interval, and then as a negative pulse, for the remaining part of the bit interval. A binary 0 is transmitted as a negative pulse for the first half of the bit interval and a positive pulse for the second half of the bit interval [see Fig. 11-10(f)]. The fact that there is a transition at the center of each 0 or 1 bit makes clock recovery very easy. However, because of the transition in the center of each bit, the frequency of a Manchester-encoded signal is two times an NRZ signal, doubling the bandwidth requirement.

CHOOSING A CODING METHOD. The choice of an encoding method depends on the application. For synchronous transmission, RZ and Manchester are preferred because the clock is easier to recover. Another consideration is average DC voltage buildup on the transmission line. When unipolar modes are used, a potentially undesirable average DC voltage builds up on the line because of the charging of the line capacitance. To eliminate this problem, bipolar methods are used, where the positive pulses cancel the negative pulses and the DC voltage is averaged to zero. Bipolar RZ or Manchester is preferred if DC buildup is a problem.

DC buildup is not always a problem. In some applications the average DC value is used for signaling purposes. An example is an Ethernet LAN, which uses the direct current to detect when two or more stations are trying to transmit at the same time.

Other encoding methods exist but are not as widely used as those described here. The encoding schemes used to encode the serial data recorded on magnetic floppy and hard disks are an example.

Craig McCaw, one of the foremost innovators of the cellular communications industry, is credited with increasing the accessibility and efficiency of cellular communications. After the death of his father in 1969, McCaw took over his family's cable TV company. Soon after assuming control, Craig McCaw knew that his small cable company would have trouble in a market controlled by much larger and growing companies. McCaw saw cellular communications as the technology whereby his business could grow and thrive. McCaw's first exposure to paging had occurred years earlier when his father was president of the World's Fair in Seattle. McCaw borrowed billions to buy up struggling cellular systems across the country. Before selling his company to AT&T in 1994, McCaw Cellular owned U.S. franchises that covered a population of 100 million people. McCaw then began to build a network of low-flying communications satellites that would link telephone and Internet networks around the world. This communications network would bring such services to underdeveloped areas where other communications lines have never been extended. (*Fortune*, May 27, 1996, pp. 62–72)

11-3 TRANSMISSION EFFICIENCY

Transmission efficiency—that is, the accuracy and speed with which information, be it voice or video, analog or digital, is sent and received over communications media—is the basic subject matter of a field known as *information theory*. Information theorists seek to determine, mathematically, the likelihood that a given amount of data being transmitted under a given set of circumstances (e.g., medium, bandwidth, speed of transmission, noise, and distortion) will be transmitted accurately.

HARTLEY'S LAW

The amount of information that can be sent in a given transmission is dependent on the bandwidth of the communications channel and the duration of transmission. A bandwidth of only about 3 kHz is required to transmit voice so that it is intelligible and recognizable. However, because of the high frequencies and harmonics produced by musical instruments, a bandwidth of 15 kHz to 20 kHz is required to transmit music with full fidelity. Music inherently contains more information than voice, and so requires greater bandwidth. A picture signal contains more information than a voice or music signal. Therefore, greater bandwidth is required to transmit it. A typical TV signal contains both voice and picture; therefore, it is allocated 6 MHz of spectrum space.

4.5 M – video
1.5 M – audio

The greater the bandwidth of a channel, the greater the amount of information that can be transmitted in a given time. It is possible to transmit the same amount of information over a narrower channel, but it must be done over a longer period of time. This general concept is known as *Hartley's law,* and the principles of Hartley's law also apply to the transmission of binary data. The greater the number of bits transmitted in a given time, the greater the amount of information that is conveyed. But the higher the bit rate, the wider the bandwidth needed to pass the signal with minimum distortion. Narrowing the bandwidth of a channel causes the harmonics in the binary pulses to be filtered out, degrading the quality of the transmitted signal and making error-free transmission more difficult.

TRANSMISSION MEDIA; BANDWIDTH

The two most common types of media used in data communications are wire cable and radio. Two types of wire cable are used, coaxial and twisted-pair (see Fig. 11-11). The coaxial cable shown in Fig. 11-11(*a*) has a center conductor surrounded by an insulator over which is a braided shield. The entire cable is covered with a plastic insulation.

A twisted-pair cable is two insulated wires twisted together. The one shown in Fig. 11-13(*b*) is an *unshielded twisted-pair (UTP)* cable, but a shielded version is also available. Coaxial cable and shielded twisted-pair cables are usually preferred, as they provide some protection from noise and crosstalk. Crosstalk is the undesired transfer of signals from one unshielded cable to another adjacent one caused by inductive or capacitive coupling.

The bandwidth of any cables is determined by their physical characteristics. All wire cables act as low-pass filters because they are made up of the wire that has inductance, capacitance, and resistance. The upper cutoff frequency of a cable depends upon the cable type, its inductance and capacitance per foot, its length, the sizes of the conductor, and the type of insulation.

Coaxial cables have a wide usable bandwidth, ranging from 200 to 300 MHz for smaller cables to 500 MHz to 1 GHz for larger cables. The bandwidth decreases dras-

[handwritten margin note:] at higher freq. capacitance become a factor

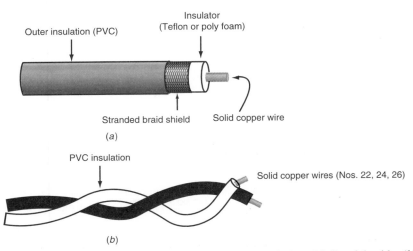

FIG. 11-11   Types of cable used for digital data transmission. (*a*) Coaxial cable. (*b*) Twisted-pair cable, unshielded (UTP).

tically with length. Twisted-pair cable has a narrower band-width, from a few kHz to over 100 MHz. Again, the actual bandwidth depends on the length and other physical characteristics.

The bandwidth of a radio channel is determined by how much spectrum space is allotted to the application by the FCC. At the lower frequencies, limited bandwidth is available, usually several kilohertz. At higher frequencies, wider bandwidths are available, from hundreds of kilohertz to many megahertz.

As discussed in Chap. 2, binary pulses are rectangular waves made up of a fundamental sine wave plus many harmonics. The channel bandwidth must be wide enough to pass all of the harmonics and preserve the waveshape. Most communications channels or media act as low-pass filters. Voice-grade telephone lines, for example, act as a low-pass filter with an upper cutoff frequency of about 3000 Hz. Harmonics higher in frequency than the cutoff are filtered out, resulting in signal distortion. Eliminating the harmonics rounds the signal off (see Fig. 11-12).

If the filtering is particularly severe, the binary signal is, essentially, converted into its fundamental sine wave. If the cutoff frequency of the cable or channel is equal to or less than the fundamental sine-wave frequency of the binary signal, the signal at the receiving end of the cable or radio channel will be a greatly attenuated sine wave at the signal fundamental frequency. However, the data is not lost, assuming the S/N ratio is high enough. The information is still transmitted reliably, but in the minimum possible bandwidth. The sine-wave signal shape can easily be restored to a rectangular wave at the receiver by amplifying it to offset the attenuation of the transmission medium, then squaring it with a Schmitt-trigger comparator, or other waveshaping circuit.

The upper cutoff frequency of any communications medium is approximately equal to the channel bandwidth. It is this bandwidth that determines the information capacity of the channel. The channel capacity C, expressed in bits per second, is twice the channel bandwidth B, in hertz:

$$C = 2B$$

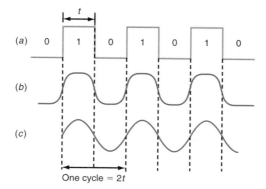

FIG. 11-12 Bandwidth limitations filter out the higher harmonics in a binary signal, distorting it. (*a*) Original binary signal. (*b*) Distortion or rounding due to filtering (bandwidth limiting). (*c*) Severe distortion.

For direct communications, this deaf worker is able to use a keyboard and specialized TDD telephone. *TDD* and *TTY* are equivalent terms. They are telephone devices called *text telephones* that are used by people who are deaf or cannot speak. The individual can keyboard a message and read the "conversation." Such devices could aid an estimated 25 million people in the United States.

The bandwidth B is usually the same as the upper cutoff (3-dB down) frequency of the channel. This is the maximum theoretical limit, and it assumes that no noise is present.

For example, the maximum theoretical bit capacity for a 10-kHz-bandwidth channel is $C = 2B = 2(10,000) = 20,000$ bps.

You can see why this is by considering the bit time in comparison to the period of the fundamental sine wave. A 20,000-bps (20 kbps) binary signal has a bit period of $t = 1/20,000 = 50 \times 10^{-6} = 50\ \mu s$.

It takes two bit intervals to represent a full sine wave with alternating binary 0s and 1s, one for the positive alternation and one for the negative alternation (see Fig. 11-12). The two bit intervals make a period of $50 + 50 = 100\ \mu s$. This sine-wave period translates into a sine-wave frequency of $f = 1/t = 1/100\ \mu s = 1/100 \times 10^{-6} = 10,000$ Hz (10 kHz), which is exactly the cutoff frequency or bandwidth of the channel.

Ideally, the shape of the binary data should be preserved as much as possible. Although the data can usually be recovered if it is degraded to a sine wave, recovery is far more reliable if the rectangular waveshape is maintained. This means that the channel must be able to pass at least some of the lower harmonics contained in the signal. As a general rule of thumb, if the bandwidth is roughly 5 to 10 times the data rate, the binary signal is passed with little distortion. For example, to transmit a 230.4-kbps serial data signal, the bandwidth should be at least 5×230.4 KHz (1.152 MHz) to 10×230.4 kHz (2.304 MHz).

The encoding method used also affects the required bandwidth for a given signal. For NRZ encoding, the bandwidth required is as described above. However, the bandwidth requirement for RZ is twice that for NRZ. This is because the fundamental frequency contained in a rectangular waveform is the reciprocal of the duration of one cycle of the highest-frequency pulse, regardless of the duty cycle. The dashed lines in Fig. 11-10 show the highest fundamental frequencies for NRZ, RZ, RZ-AMI, and Manchester. AMI has a lower fundamental frequency than RZ. The rate for Manchester encoding is twice that for NRZ and AMI.

As an example, assume an NRZ bit interval of 100 ns, which results in a bps data rate of $1/t = 1/100$ ns $= 1/100 \times 10^{-9} = 10$ Mbps.

Alternating binary 1s and 0s produces a fundamental sine-wave period of twice the bit time, or 200 ns, for a bandwidth of $1/t = 1/200 \times 10^{-9} = 5$ MHz. This is the same as that computed with the formula

$$B = \frac{C}{2} = \frac{10\ \text{Mbps}}{2} = 5\ \text{MHz}$$

Looking at Fig. 11-10, you can see that the RZ and Manchester pulses occur at a faster rate, the cycle time being 100 ns. The RZ and Manchester pulses are one-half the bit time of 100 ns, or 50 ns. The bit rate or channel capacity associated with this

reciprocal

100 μs — 10 K bps
bandwidth 5 K will
allow transmit
10,000 BPS

time is $1/50 \text{ ns} = 1/50 \times 10^{-9} = 20 \text{ Mbps}$. The bandwidth for this bit rate is $C/2 = 20 \text{ Mbps}/2 = 10 \text{ Mhz}$.

Thus the RZ and Manchester encoding schemes require twice the bandwidth. This tradeoff of bandwidth for some benefit, such as ease of clock recovery, may be desirable in certain applications.

MULTIPLE CODING LEVELS

Channel capacity can be modified by using multiple-level encoding schemes that permit more bits per symbol to be transmitted. Remember that it is possible to transmit data using more than just two binary voltage levels or symbols. Multiple voltage levels can be used, as illustrated earlier in Fig. 11-6(b). Other schemes, such as using different phase shifts for each symbol, are used. Consider the equation

$$C = 2B \log_2 N$$

where N is the number of different encoding levels per time interval. The implication is that for a given bandwidth, the channel capacity, in bits per second, will be greater if more than two encoding levels are used per time interval.

Refer back to Fig. 11-6, where two levels or symbols (0 or 1) were used in transmitting the binary signal. The bit or symbol time is 1 μs. The bandwidth needed to transmit this 1,000,000-bps signal can be computed from $C = 2B$, or $B = C/2$. Thus a minimum bandwidth of $1,000,000 \text{ bps}/2 = 500,000 \text{ Hz}$ (500 kHz) is needed.

The same result is obtained with the new expression. If $C = 2B \log_2 N$, then $B = C/2 \log_2 N$.

The logarithm of a number to the base 2 can be computed with the expression

$$\log_2 N = \frac{\log_{10} N}{\log_{10} 2}$$
$$= \frac{\log_{10} N}{0.301}$$
$$= 3.32 \log_{10} N$$

where N is the number whose logarithm is to be calculated. The base 10 or common logarithm can be computed on any scientific calculator. With two coding levels (binary 0 and 1 voltage levels), the bandwidth is

$$B = \frac{C}{2 \log_2 N} = \frac{1,000,000 \text{ bps}}{2(1)} = 500,000 \text{ Hz}$$

Note that $\log_2 2$ for a binary signal is simply 1.

Now we continue, using $C = 2B \log_2 N$. Since $\log_2 N = \log_2 2 = 1$,

$$C = 2B(1) = 2B$$

Now let's see what a multilevel coding scheme does. Again, we start with $B = C/2 \log_2 N$. The channel capacity is 2 million bps, as shown in Fig. 11-6(b), because each symbol (level) interval is 1 μs long. But here N, the number of levels, = 4.

Therefore two bits are transmitted per symbol. The bandwidth is then 2,000,000 bps/2 $\log_2$ 4. Since $\log_2$ 4 = 3.32 $\log_{10}$ 4 = 3.32 (0.602) = 2, then

$$B = \frac{2,000,000}{2(2)} = \frac{2,000,000}{4} = 500,000 \text{ Hz} = 500 \text{ kHz}$$

By using a multilevel (4) coding scheme, we can transmit at twice the speed in the same bandwidth. The data rate is 2 Mbps with four levels of coding in a 500-kHz bandwidth compared to 1 Mbps with only two symbols (binary).

To transmit even higher rates within a given bandwidth, more voltages levels can be used, where each level represents three, four, or even more bits per symbol. The multilevel approach need not be limited to voltage changes; frequency changes and phase changes can also be used. Even greater increases in speed can be achieved if changes in voltage levels are combined with changes in phase.

IMPACT OF NOISE IN THE CHANNEL

Another important aspect of information theory is the impact of noise on a signal. As discussed in earlier chapters, increasing bandwidth increases the rate of transmission, but also allows more noise to pass, and so the choice of a bandwidth is a tradeoff.

The relationship between channel capacity, bandwidth, and noise is summarized in what is known as the *Shannon-Hartley theorem:*

$$C = B \log_2 (1 + \text{S/N})$$

where C = channel capacity, bps
B = bandwidth, Hz
S/N = signal-to-noise ratio

Assume, for example, that the maximum channel capacity of a voice-grade telephone line with a bandwidth of 3100 Hz and an S/N of 30 dB is to be calculated.

First, 30 dB is converted to a power ratio. If dB = 10 log P, where P is the power ratio, then P = antilog (dB/10). Antilogs are easily computed on a scientific calculator. A 30-dB S/N ratio translates to a power ratio of

$$P = \text{antilog} \frac{30}{10} = \text{antilog } 3 = 1000$$

The channel capacity is then

$$C = B \log_2 (1 + \text{S/N}) = 3100 \log_2 (1 + 1000) = 3100 \log_2 (1001)$$

The base 2 logarithm of 1001 is

$$\log_2 1001 = 3.32 \log_{10} 1001 = 3.32(3) = 9.97 \text{ or about } 10$$

Therefore, the channel capacity is

$$C = 3100(10) = 31,000 \text{ bps}$$

A bit rate of 31,000 bps is surprisingly high for such a narrow bandwidth. In fact, it appears to conflict with what we learned earlier, that is, that maximum channel capacity is twice the channel bandwidth. If the bandwidth of the voice-grade line is 3100 Hz, then the channel capacity is $C = 2B = 2(3100) = 6200$ bps. That rate is for a binary (two-level) system only, and it assumes no noise. How, then, can the Shannon-Hartley theorem predict a channel capacity of 31,000 bps when noise is present?

The Shannon-Hartley expression says that it is *theoretically* possible to achieve a 31,000-bps channel capacity on a 3100-Hz bandwidth line. What it doesn't say is that multilevel encoding is needed to do so. Going back to the basic channel capacity expression, $C = 2B \log_2 N$, we have a C of 31,000 bps and a B of 3100 Hz. The number of coding or symbol levels has not been specified. Rearranging the formula,

$$\log_2 N = \frac{C}{2B} = \frac{31,000}{2(3100)} = \frac{31,000}{6200} = 5$$

Therefore,

$$N = \text{antilog}_2 5$$

The antilog of a number is simply the value of the base raised to the number, in this case, 2^5, or 32.

Thus a channel capacity of 31,000 can be achieved by using a multilevel encoding scheme, one that uses 32 different levels or symbols per interval, instead of a two-level (binary) system. The baud rate of the channel is still $C = 2B = 2(3100) = 6200$ baud. But because a 32-level encoding scheme has been used, the bit rate is 31,000 bps. As it turns out, the maximum channel capacity is very difficult to achieve in practice. Typical systems limit the channel capacity to one-third to one-half the maximum to ensure more reliable transmission in the presence of noise.

Example 11-2

The bandwidth of a communications channel is 12.5 kHz. The S/N ratio is 25 dB. Calculate (*a*) the maximum theoretical data rate in bps, (*b*) the maximum theoretical channel capacity, and (*c*) the number of coding levels N needed to achieve the maximum speed. [For part (*c*), use the $\boxed{y^x}$ key on a scientific calculator.]

a. $C = 2B = 2(12.5 \text{ kHz}) = 25 \text{ kbps}$

b. $C = B \log_2 (1 + S/N) = B\, 3.32 \log_{10} (1 + S/N)$

$25 \text{ dB} = 10 \log P \qquad$ where $P = $ S/N power ratio

$\log P = \dfrac{25}{10} = 2.5$

$P = \text{antilog } 2.5 = \log^{-1} 2.5 = 316.2$

$C = 12,500 \,(3.32) \log_{10} (316.2 + 1)$

$\quad = 41,500 \log_{10} (317.2)$

$\quad = 41,500 \,(2.5)$

$\quad = 103,805.3 \text{ bps or } 103.8 \text{ kbps}$

c. $C = 2B \log_2 N$

$\log_2 N = C/2B$

$N = \text{antilog}_2 \, C/2B$

$N = \text{antilog}_2 \,(103,805.3)/2(12,500) = \text{antilog}_2 \,(4.152)$

$\quad = 2^{4.152} = 17.78 \text{ or } 17 \text{ levels or symbols}$

Telephone networks, originally designed to carry voice signals, are now widely used to carry digital information as well, to linking computers and computer networks across the globe.

There are two primary problems in transmitting digital data over the telephone network. The first is that binary signals are usually switched DC pulses (i.e., the binary 1s and 0s are represented by pulses of either single or dual polarity), whereas polarity telephone lines are designed to carry only AC analog signals. Voice and other analog signals are processed by numerous components, circuits, and equipment along the way. Usually only AC signals of a specific frequency range are allowed, voice frequencies in the 300- to 3000-Hz range being the most common. If a binary signal were applied directly to the telephone network, it simply would not pass. The transformers, capacitors, and other AC circuitry virtually ensure that no DC signals get through in a recognizable form. Second, binary data is usually transmitted at high speeds. This high-speed data would be filtered out by the limited-bandwidth telephone media.

The question is, then, How does digital data get transmitted over the telephone network? The answer is by using broadband communications techniques involving modulation, which are implemented by a modem, a device containing both a *mo*dulator and a *dem*odulator. Modems convert binary signals into analog signals capable of being transmitted over the telephone lines and demodulate such analog signals, reconstructing the equivalent binary output. Figure 11-13 shows two ways that modems are commonly used in digital data transmission. In Fig. 11-13(*a*), two mainframe computers exchange data by speaking through modems. While one modem is transmitting, the other is receiving. Full duplex operation is also possible. In Fig. 11-13(*b*), a remote video terminal or personal computer is using a modem to communicate with a large mainframe computer. Modems are also the interface between the millions of computer terminals that are links in the so-called information superhighway, the Internet, or on a bulletin board system.

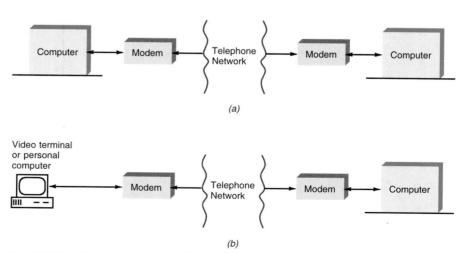

FIG. 11-13   How modems permit digital data transmission on the telephone network.

Figure 11-14 is a complete block diagram of a typical modem. It consists of both transmitter and receiver sections. Most modern modems are implemented using digital signal processing (DSP) techniques, and thus consist of only a few chips.

Physically, modems are packaged in two ways. In *external modems* the circuitry is packaged in a small housing, separate from the computer, containing its own power supply and usually front-panel LEDs to indicate various operational modes and conditions. The modem connects to the computer by way of an RS-232 serial interface and to the telephone line by way of modular telephone connector, such as the RJ-11. *Internal modems* are packaged on a single small printed circuit board and are designed to plug into the PC bus. The modem takes its power from the PC power supply, and a modular connector attaches the modem to the telephone line. No additional interfacing is needed.

Most modems today are of the internal type. In fact, many laptop and notebook computers have long come with the modem built right into the motherboard of the computer. Some modems plug into portable computers by way of a special interface called the PCMCIA bus.

The data to be transmitted is stored in the computer's RAM. It is formatted there by the communications software installed with the computer. It is then sent a byte at a time to the modem. Most data transfers and operations inside a computer are

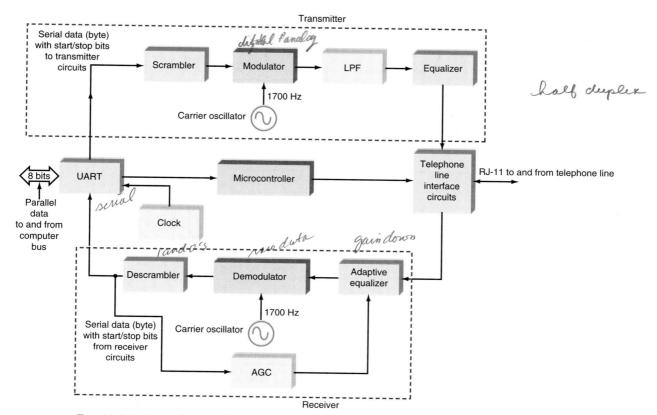

FIG. 11-14 Block diagram of a modem.

parallel, but since long-distance data communication is done using serial binary data, the modem's first job is to convert parallel data into serial data. This is done with shift registers, as described earlier. It is usually carried out by a special large-scale IC called a *UART* (*Universal Asynchronous Receiver/Transmitter*), a digital IC that performs parallel-to-serial conversion for transmission and serial-to-parallel conversion for reception.

Figure 11-15 shows a general block diagram of a UART. Parallel data from the computer data bus is fed into and out of a bidirectional data buffer, which is usually a storage register with appropriate level-shifting circuits. The data, usually parallel 8 bits, is then put on an internal data bus. Before transmission, this data is stored first in a buffer storage register and then sent to a shift register. The internal circuitry adds start and stop bits, which signal the beginning and end of the word, making modem operation strictly asynchronous. The parity bit is used for error detection. A clock signal shifts the data out serially, one bit at a time.

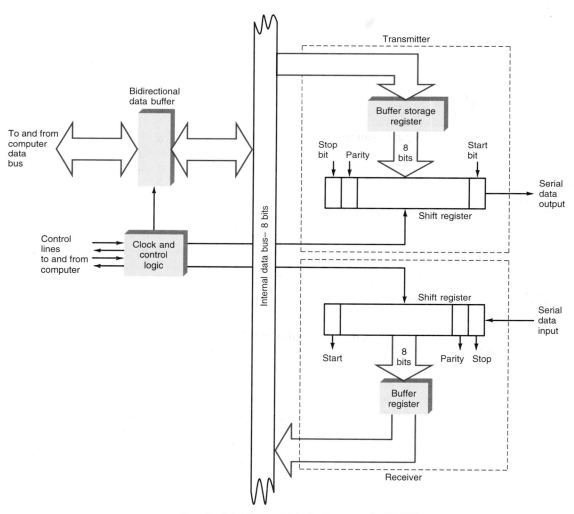

FIG. 11-15 General block diagram of a UART.

The receive section of the UART is at the bottom of Fig. 11-15. Serial data is shifted into a shift register, where the start, stop, and parity bits are stripped off. The remaining data is transferred to a buffer register, to the internal data bus, and through the bidirectional data buffer to the computer in parallel form. The clock and control logic circuits in the UART control all internal shifting and data transfer operations under the direction of control signals from the computer.

Referring again to Fig. 11-14, you can see that the serial data from the UART is passed through a scrambler circuit. This is not to encrypt the data for security, but simply to ensure that the data is random, with plenty of 0 to 1 and 1 to 0 transitions. This ensures that sufficient transitions occur so that clocking operations are reliable when long strings of binary 0s or 1s occur. Some circuits, especially adaptive equalizers and echo suppressors, rely on the randomness of the binary data for proper operation. The scrambler is a shift register with feedback that takes the serial data and scrambles it into a random binary pulse train.

The random serial data is sent to the modulator. The serial data, usually in NRZ form, modulates a low-frequency sine-wave carrier inside the telephone bandwidth that extends from about 300 to 3000 Hz. Typical carrier frequencies are in the 1700-Hz range.

The output of the modulator is filtered to band-limit it, then fed to an *equalizer* circuit, which precompensates for the attenuation and distortion that the signal will receive as it is passed through the telephone system. Since the characteristics of the telephone are predictable (within limits), the equalizer processes the modulated signal in such a way to overcome the attenuation and distortion so that it will appear normal at the receiving end. (The equalizer used in modems is what is called a *compromise equalizer;* it only partially corrects the problems because the exact extent of the distortion is not known.) The equalizer output is sent to the interface circuits that connect the signal to the telephone line.

During receive operations, the signal is picked off the telephone line, passed through the interface circuits, and fed to the receiver section. It first encounters an adaptive equalizer. The adaptive equalizer adjusts itself automatically to compensate for the amplitude attenuation and distortion of the signal. AGC circuits are normally used in the receiver to keep the gain constant over what can be a wide range of received signal amplitudes.

The signal is then demodulated, resulting in an NRZ serial digital signal. This is passed through a descrambler, which produces the opposite effect of the transmit scrambler. The descrambler output is the original serial data signal. This is sent to the UART, where it is translated into a parallel byte that the computer can store and use.

Although not shown in Fig. 11-14, data compression and decompression circuits are now being used in some modems. A digital signal-processing (DSP) circuit takes the binary data and compresses it prior to modulation so that the transmitted signal is many bits less than the actual message length. Because fewer bits are transmitted in a given period of time, overall transmission speed is higher. At the receiver, a DSP decompressor restores the shorter signal to its original length after it is demodulated.

Another feature not depicted in Fig. 11-14 is error detection and correction. Many of the newer modem types incorporate circuitry that can detect bit transmission errors and correct them as they occur. Such schemes, while complex and expensive, greatly increase the reliability of transmission, especially high-speed transmission in a noisy environment.

A modem is typically controlled by its own internal microprocessor-based controller. This miniature single-chip computer executes commands given through the communications software in the computer, which arrive over the computer bus. In external modems, the commands arrive serially, via the UART. The microcontroller lets the user tell the modem exactly what to do. It also implements automatic functions such as auto-dial and auto-answer operations which most modems are capable of.

MODULATION SCHEMES

As stated earlier, modems convert binary signals to a form that can be transmitted over the analog telephone network. And even though the bandwidth of the telephone network is essentially restricted to 3 kHz, a variety of tricky modulation techniques permit serial data transmission rates as high as about 40 kbps. These techniques fall into three major categories: frequency-shift keying (FSK), phase-shift keying (PSK), and quadrature amplitude modulation (QAM).

FSK is used primarily in low-speed modems capable of transmitting data up to 300 baud or bps, a relatively slow data rate that is rarely seen today. Most communications data rates via the telephone system are 9600 baud or higher, and rates of 14,400, 19,200, and 28,800 bps are common. To achieve these higher speeds, PSK, QAM, and variations of these are used.

FSK. The oldest and simplest form of modulation used in modems is FSK. In FSK, two sine-wave frequencies are used to represent binary 0s and 1s. For example, a binary 0, usually called a space in data communications jargon, has a frequency of 1070 Hz. A binary 1, referred to as a mark, is 1270 Hz. These two frequencies are alternately transmitted to create the serial binary data. The resulting signal looks something like that shown in Fig. 11-16. Both of the frequencies are well within the 300- to 3000-kHz bandwidth normally associated with the telephone system, as illustrated in Fig. 11-17.

The simultaneous transmit and receive operations that are carried out by a modem, known as full-duplex operation, require that another set of frequencies be defined. These are also indicated in Fig. 11-17. A binary 0 or space is 2025 Hz; a binary 1 or mark is 2225 Hz. These tones are also within the telephone bandwidth but are spaced far enough from the other frequencies so that selective filters can be used to distinguish between the two. The 1070- and 1270-Hz tones are used for transmitting (originate), and the 2025- and 2225-Hz tones are used for receiving (answer).

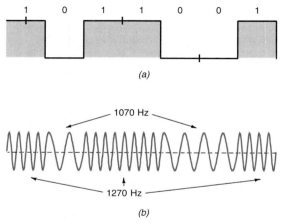

FIG. 11-16 Frequency-shift keying. (*a*) Binary signal. (*b*) FSK signal.

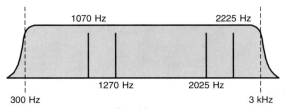

FIG. 11-17 The FSK signals within the telephone audio bandpass.

Figure 11-18 is a block diagram of the modulator and demodulator sections of an FSK modem. Each modem contains an FSK modulator and an FSK demodulator so that both send and receive operations can be achieved. Bandpass filters at the inputs to each modem separate the two tones. For example, in the upper modem, a bandpass filter allows frequencies between 1950 and 2300 Hz to pass. This means that 2025- and 2225-Hz tones will be passed, but the 1070- and 1270-Hz tones generated by the internal modulator will be rejected. The lower modem has a bandpass filter that accepts the lower-frequency tones while rejecting the upper-frequency tones generated internally.

A wide variety of modulator and demodulator circuits are used to produce and recover FSK. Virtually all of the circuits described in Chap. 6 have been or could be used. A typical FSK modulator is simply an oscillator whose frequency can be switched between two frequencies, usually by switching in different capacitor values. Both *RC* and *LC* oscillators are used. At the lower audio frequencies, *RC* oscillators are preferred because of their simplicity. A variety of demodulators, including PLLs, pulse averaging discriminators, and others, are used.

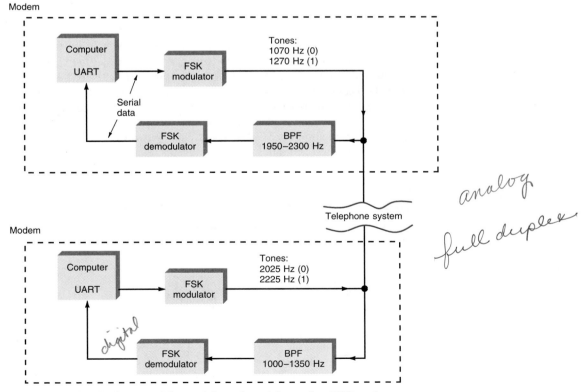

FIG. 11-18 Block diagram of the modulator-demodulator of an FSK modem.

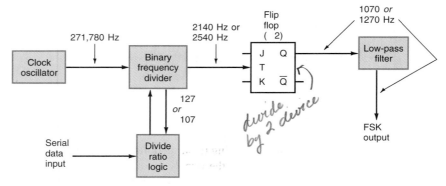

Fig. 11-19 A digital FSK modulator.

Most modems now use digital techniques because they are simpler and more adaptable to IC implementation. One type of digital FSK modulator is shown in Fig. 11-19. A clock oscillator generates a clock at a frequency of 271,780 Hz. It is applied to a binary frequency divider, which is usually some form of binary counter with various feedback logic gates for setting the divide ratio. This frequency divider is set up so that it divides by two different integer values: one divide ratio produces the mark frequency and the other produces the space frequency.

To transmit a space at 1070 Hz, the divide ratio logic in Fig. 11-19 produces frequency division by 127. That is, when the serial binary input is zero, the frequency divider output is 1/127th of its input, so the output frequency is 271,780/127, or 2140 Hz. This is fed to a single flip-flop which divides the frequency by 2, producing the desired 1070-Hz output. The flip-flop produces a 50 percent duty cycle square wave; that is, its ON and OFF times are equal in length. This is done because the duty cycle of the frequency divider is other than a 50 percent duty cycle, which when converted to sine wave will produce distortion. The flip-flop output is passed through a low-pass filter which removes the higher odd harmonics, producing a 1070-Hz sine-wave tone.

When a binary 1 is applied to the divide ratio logic, the frequency divider divides by 107, producing an output frequency of 2540 Hz, which, when divided by 2 in the flip-flop, produces the desired 1270-Hz output. The low-pass filter removes the higher-frequency harmonics, producing a sine-wave output.

A digital method of FSK demodulation is shown in Fig. 11-20. The sine-wave FSK signal is applied to a limiter which removes any amplitude variations and shapes the

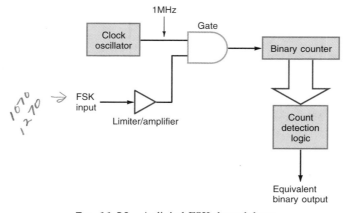

Fig. 11-20 A digital FSK demodulator.

signal into a square wave. The square wave is then applied to a gate circuit, which is used to turn a 1-MHz clock signal off and on. A binary counter counts or accumulates the 1-MHz clock pulses.

When a low-frequency space signal is applied, the period of the signal will be long, allowing the gate to open and the binary counter to accumulate the clock pulses. The detection logic, which is a set of binary gates, determines whether the number in the binary counter is above or below some predetermined value. For the low-frequency tone input, the number in the counter will be higher than the counter value when the high-frequency or mark signal is applied to the input. The detection logic produces a binary 0 or binary 1 output, depending upon whether the number in the counter is above or below a specific value between the two limits. Such FSK modems including modulator and demodulator come packaged in a single IC.

PSK.　In PSK, the binary signal to be transmitted changes the phase shift of a sine-wave character depending upon whether a binary 0 or binary 1 is to be transmitted. (Recall that phase shift is a time difference between two sine waves of the same frequency.) Figure 11-21 illustrates several examples of phase shift. A phase shift of 180°, the maximum phase difference that can occur, is known as a phase reversal, or phase inversion.

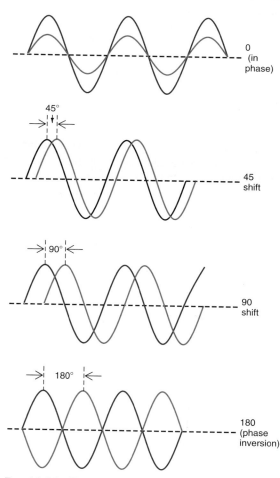

FIG. 11-21　Examples of phase shift.

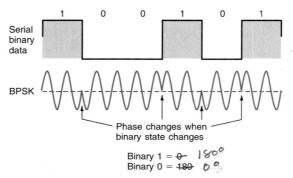

Phase changes when
binary state changes

Binary 1 = ~~0~~ $180°$
Binary 0 = ~~180~~ $0°$

FIG. 11-22   Binary phase shift keying (BPSK).

Figure 11-22 illustrates the simplest form of PSK, binary phase-shift keying (BPSK). During the time that a binary 0 occurs, the carrier signal is transmitted with one phase; when a binary 1 occurs, the carrier is transmitted with a 180° phase shift.

Figure 11-23 shows one kind of circuit used for generating BPSK, a standard lattice ring modulator or balanced modulator used for generating DSB signals. The carrier sine wave, usually 1600 or 1700 Hz, is applied to the input transformer T_1 while the binary signal is applied to the transformer center taps. The binary signal provides a switching signal for the diodes. When a binary 0 appears at the input, A is + and B is −, so diodes D_1 and D_4 conduct. They act as closed switches, connecting the secondary of T_1 to the primary of T_2. The windings are phased so that the BPSK output is in phase with the carrier input.

When a binary 1 appears at the input, A is − and B is +, so diodes D_1 and D_4 are cut off while diodes D_2 and D_3 conduct. This causes the secondary of T_1 to be connected to the primary of T_2 but with the interconnections reversed. This introduces a 180° phase-shift carrier at the output.

Demodulation of a BPSK signal is also done with a balanced modulator. A version of the diode ring or lattice modulator can be used as shown in Fig. 11-24. This is actually the same circuit as that in Fig. 11-23, but the output is taken from the center taps. The BPSK and carrier signals are applied to the transformers. IC balanced modulators can also be used at the lower frequencies. The modulator and demodulator circuits are identical to the doubly balanced modulators used for mixers. They are available as fully wired and tested components for frequencies up to about 1 GHz.

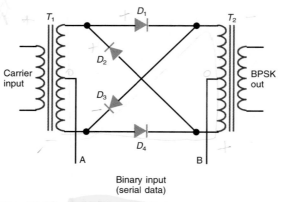

FIG. 11-23   A BPSK modulator.

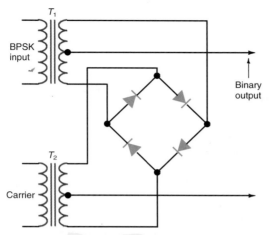

Fig. 11-24 A BPSK demodulator.

The key to demodulating BPSK is that a carrier with the correct frequency and phase relationship must be applied to the balanced modulator along with the BPSK signal. Typically the carrier is derived from the BPSK signal itself, using a carrier recovery circuit like that shown in Fig. 11-25. A bandpass filter ensures that only the desired BPSK signal is passed. The signal is then squared or multiplied by itself by a balanced modulator or analog multiplier by applying the same signal to both inputs. Squaring removes all the 180° phase shifts, resulting in an output that is twice the input signal frequency ($2f$). A bandpass filter set at twice the carrier frequency passes this signal only. The resulting signal is applied to the phase detector of a PLL. Note that a ×2 frequency multiplier is used between the VCO and phase detector, ensuring that the VCO frequency is at the carrier frequency. Use of the PLL means that the VCO will track any carrier frequency shifts. The result is a signal with the correct frequency and phase relationship for proper demodulation. The carrier is applied to the balanced modulator-demodulator along with the BPSK signal. The output is the recovered binary data stream.

DPSK. To simplify the demodulation process, a version of binary PSK called *differential phase-shift keying (DPSK)* can be used. In DPSK, there is no absolute carrier phase reference. Instead, the transmitted signal itself becomes the phase reference. In demodulating DPSK, the phase of the received bit is compared to the phase of the previously received bit.

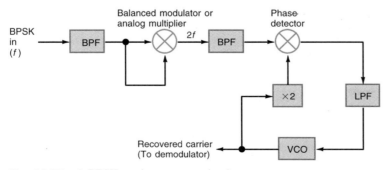

Fig. 11-25 A BPSK carrier recovery circuit.

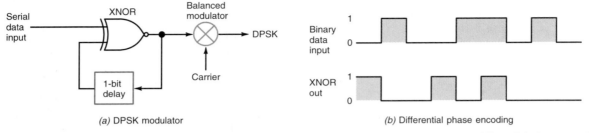

(a) DPSK modulator

(b) Differential phase encoding

FIG. 11-26 The DPSK process. (a) DPSK modulator. (b) Differential phase encoding.

In order for DPSK to work, the original binary bit stream must undergo a process known as *differential phase coding,* in which the serial bit stream passes through an inverted exclusive NOR circuit (XNOR), as shown in Fig. 11-26(a). Note that the XNOR output is applied to a 1-bit delay circuit before being applied back to the input. The delay can simply be a clocked flip-flop or a delay line. The resulting bit pattern permits the signal to be recovered because the present bit phase can be compared with the previously received bit phase.

In Fig. 11-26(b), the input binary word to be transmitted is shown along with the output of the XNOR. An XNOR circuit is simply a 1-bit comparator that produces a binary 1 output when both inputs are alike and a binary 0 output when the two bits are different. The output of the circuit is delayed a 1-bit interval by storing it in a flip-flop. Therefore, the XNOR inputs are the current bit plus the previous bit. The XNOR signal is then applied to the balanced modulator along with the carrier to produce a BPSK signal.

Demodulation is accomplished with the circuit shown in Fig. 11-27. The DPSK signal is applied to one input of the balanced modulator and a 1-bit delay circuit, either a flip-flop or a delay line. The output of the delay circuit is used as the carrier. The resulting output is filtered by a low-pass filter to recover the binary data. Typically the low-pass filter output is shaped with a Schmitt trigger or comparator to produce clean, high-speed binary levels.

QPSK. The main problem with BPSK and DPSK is that the speed of data transmission in a given bandwidth is limited. One way to increase the binary data rate while not increasing the bandwidth required for the signal transmission is to encode more than one bit per phase change. There is a symbol change for each bit change with BPSK and DPSK, so the baud (symbol) rate is the same as the bit rate. In BPSK and DPSK, each binary bit produces a specific phase change. An alternative approach is to use combinations of two or more bits to specify a particular phase shift, so that a symbol change (phase shift) represents multiple bits. Because more bits per baud are encoded, the bit rate of data transfer can be higher than the baud rate, yet the signal will not take up additional bandwidth.

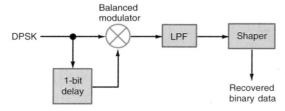

FIG. 11-27 A DPSK demodulator.

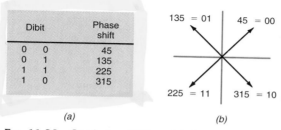

Dibit		Phase shift
0	0	45
0	1	135
1	1	225
1	0	315

135 = 01   45 = 00

225 = 11   315 = 10

(a)   *(b)*

FIG. 11-28 Quadrature PSK modulation. (*a*) Phase angle of carrier for different pairs of bits. (*b*) Phasor representation of carrier sine wave.

One commonly used system for doing this is known as *quadrature, quarternary,* or *quadra phase PSK (QPSK* or *4-PSK*). In QPSK, each pair of successive digital bits in the transmitted word is assigned a particular phase, as indicated in Fig. 11-28(*a*). Each pair of serial bits, called a *dibit,* is represented by a specific phase. A 90° phase shift exists between each pair of bits. Other phase angles can also be used as long as they have a 90° separation. For example, it is common to use phase shifts of 45°, 135°, 225°, and 315° as shown in Fig. 11-28(*b*).

A circuit for producing QPSK is shown in Fig. 11-29. It consists of a 2-bit shift register implemented with flip-flops, commonly known as a *bit splitter.* The serial binary data train is shifted through this register, and the bits from the two flip-flops are applied to balanced modulators. The carrier oscillator is applied to balanced modulator 1 and through a 90° phase shifter to balanced modulator 2. The outputs of the balanced modulators are linearly mixed to produce the QPSK signal.

The output from each balanced modulator is a BPSK signal. With a binary 0 input, the balanced modulator produces one phase of the carrier. With a binary 1 input, the carrier phase is shifted 180°. The output of balanced modulator 2 also has two phase states, 180° out of phase with one another. The 90° carrier phase shift at the input causes

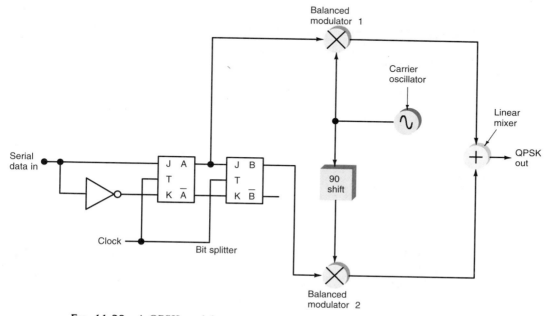

FIG. 11-29 A QPSK modulator.

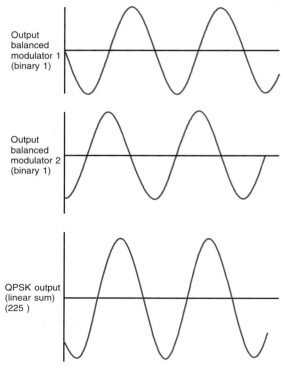

Output
balanced
modulator 1
(binary 1)

Output
balanced
modulator 2
(binary 1)

QPSK output
(linear sum)
(225)

Fig. 11-30 How the modulator produces the correct phase by adding two signals.

the outputs from balanced modulator 2 to be shifted 90° from those of balanced modulator 1. The result is four different carrier phases, which are combined two at a time in the linear mixer. The result is four unique output phase states.

Figure 11-30 shows the outputs of one possible set of phase shifts. Note that the carrier outputs from the two balanced modulators are shifted 90°. When the two carriers are algebraically summed in the mixer, the result is an output sine wave that has a phase shift of 225°, which is halfway between the phase shifts of the two balanced modulator signals.

A demodulator for QPSK is illustrated in Fig. 11-31. The carrier recovery circuit is similar to the one described previously. The carrier is applied to balanced modulator 1 and is shifted 90° before being applied to balanced modulator 2. The outputs of the two balanced modulators are filtered and shaped into bits. The two bits are combined in a shift register and shifted out to produce the originally transmitted binary signal.

Encoding still more bits per phase change produces higher data rates. In 8-PSK, for example, 3 serial bits are used to produce a total of eight different phase changes. In 16-PSK, 4 serial input bits produce 16 different phase changes, for an even higher data rate.

QAM. One of the most popular modulation techniques used in modems for increasing the number of bits per baud is *quadrature amplitude modulation (QAM).* QAM uses both amplitude and phase modulation of a carrier; not only are different phase shifts produced, but the amplitude of the carrier is also varied.

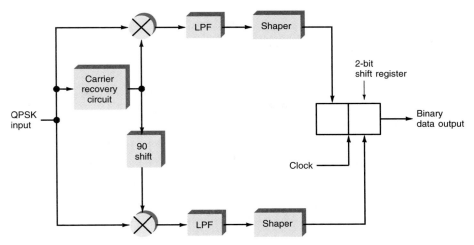

FIG. 11-31 A QPSK demodulator.

In 8-QAM, there are four possible phase shifts, as in QPSK, and two different carrier amplitudes, so that eight different states can be transmitted. With eight states, 3 bits can be encoded for each baud or symbol transmitted. Each 3-bit binary word transmitted uses a different phase-amplitude combination.

Figure 11-32 is a *constellation diagram* of an 8-QAM signal showing all possible phase and amplitude combinations. The points in the diagram indicate the eight possible phase-amplitude combinations. Note that there are two amplitude levels for each phase position. Point A shows a low carrier amplitude with a phase shift of 225°. It represents 100. Point B shows a higher amplitude and a phase shift of 315°. This sine wave represents 011.

A block diagram of an 8-QAM modulator is shown in Fig. 11-33. The binary data to be transmitted is shifted serially into the 3-bit shift register. These bits are applied in pairs to two 2-to-4-level converters. A 2-to-4-level converter circuit, basically a sim-

8-QAM
Bandwidth
tripled

1 signal
3 pulse 3X
as much
data

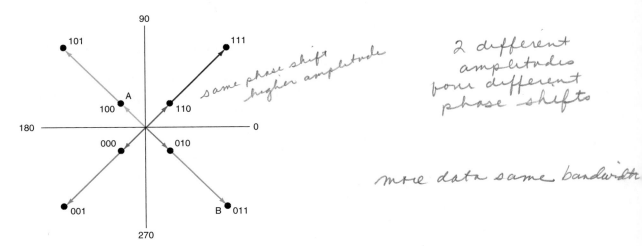

2 different
amplitudes
four different
phase shifts

more data same bandwidth

Note: Each vector has a specific amplitude and phase shift and represents one 3-bit word.

FIG. 11-32 A constellation diagram of a QAM signal.

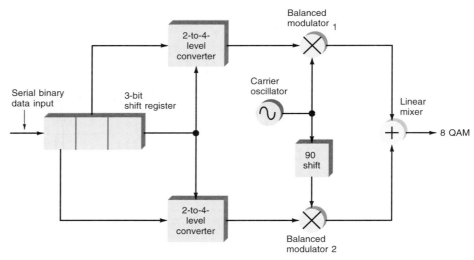

FIG. 11-33 An 8-QAM modulator.

ple D/A converter, translates a pair of binary inputs into one of four possible DC output voltage levels. The idea is to produce four voltage levels corresponding to the different combinations of 2 input bits, that is, four equally spaced voltage levels. These are applied to the two balanced modulators fed by the carrier oscillator and a 90° phase shifter, as in a QPSK modulator. Each balanced modulator produces four different output phase-amplitude combinations. When these are combined in the linear mixer, eight different phase-amplitude combinations are produced. The most critical part of the circuit is the 2-to-4-level converters; these must have very precise output amplitudes so that when they are combined in the linear summer, the correct output and phase combinations are produced.

A 16-QAM signal can also be generated by encoding four input bits at a time. The result is 8 different phase shifts and 2 amplitude levels, producing a total of 16 different phase-amplitude combinations.

TRELLIS CODE MODULATION. *Trellis code modulation (TCM)* is a special form of QAM that facilitates error detection and correction. It is also known as *TCQAM*. At speeds above 9600 bps, bit errors become more common due to high-speed modulation methods and noise. TCM is also known as *convolutional coding,* and trellis codes are also known as *Vitberi codes,* after the individual generally credited with discovering them. In TCM, the data to be transmitted is passed through an encoder which produces a unique code, the trellis code, that is then sent to the modulator and transmitted. TCM modulators usually encode 6, 7, or 8 bits per baud or symbol change. These codes contain many invalid bit combinations.

At the receiver, the decoding process recovers the original data. During decoding, if an invalid code is picked up, it means that an error occurred. With appropriate circuitry in the receiver, the error can be corrected in real time. With TCM, data rates of 19,200 to 38,400 bps through the 3-kHz telephone bandwidth have been obtained.

BPSK, QPSK, QAM, TCM, and other techniques are also widely used to transmit digital data in microwave and satellite radio communications. With very high-frequency carriers, bit rates of millions of bits per second can easily be achieved with minimum transmission error in a noisy environment. The circuits used for modulation and demodulation in such systems are referred to as *radio* or *rf modems.*

A special form of the rf modem is the cable TV modem, which is designed to transmit high-speed digital data over two-way cable TV systems. Because of the wide bandwidth provided by the cable, data rates of 4 to 25 Mbps are possible. Such modems are expected to replace slower telephone system modems in coming years.

MODEM CLASSIFICATION AND STANDARDS

In data communications involving modems, the sending and receiving modems must be fully compatible if data is to be exchanged. Over the years, many different standards to facilitate data exchange have evolved, and new standards continue to emerge as designers push the technology to achieve ever faster and more reliable data transmission over the telephone network.

The earliest standards were developed by the Bell Telephone Company, which manufactured two widely used modems—the Bell 103, an FSK modem, and the Bell 202 and 212 modems, both PSK modems—which are rarely seen today.

Currently, the International Telecommunications Union (ITU), formerly called the CCITT, sponsors, negotiates, and maintains modem and many other communications standards. Modem standards are designated by a special V.xx symbol. The more common ones are described in Fig. 11-34, and a comparison chart is given in Fig. 11-35. The standards that apply to the modems used in fax machines are referred to as *group modems*.

When you purchase a modem, be sure that its designation conforms to the application you have in mind. Modems are usually capable of operating in several different modes. The modem will automatically adjust itself to the highest speed possible but will drop back to a lower speed or different mode if the receiving modem cannot handle it.

> **DID YOU KNOW?**
>
> Modems are usually capable of operating in several different modes. The modem automatically adjusts itself to the highest speed possible but will drop back to a lower speed if the receiving modem cannot handle the higher speed.

V.17—Describes an application-specific modulation scheme for Group 3 fax which provides two-wire, half-duplex, trellis-coded transmission at 7200, 9600, 12000, and 14400 bps. Despite its low number, this is a recently introduced standard.

V.19—Describes early DTMF modems for low-speed parallel transmission. This standard is essentially obsolete.

V.21—Provides the specifications for 300 bps FSK serial modems (based on BELL103).

V.22—Provides the specifications for 1200 bps (600 baud) PSK modems (based upon BELL212A).

FIG. 11-34 ITU modem standards. (*Electronics Now, September 1994.*) *Continues on next page.*

V.22bis—Describes 2400 bps modems operating at 600 baud with QAM.

V.23—Describes the operation of an unusual FM modem operating at 1200/75 bps. The host transmits at 1200 bps and receives at 75 bps, while the remote modem transmits at 75 bps and receives at 1200 bps. In Europe, V.23 supports some videotext applications.

V.24—This is known as EIA RS-232C in the U.S. V.24 defines only the functions of the serial port circuits, EIA-232E (the current version) also defines electrical characteristics and connectors.

V.25—Defines automatic answering equipment and parallel automatic dialing. It also defines the answer tone that modems send.

V.25bis—Defines serial automatic calling and answering, which is the ITU (CCITT) equivalent of AT commands. This is the current ITU standard for modem control by computers via serial interface. The Hayes AT command set is used primarily in the US.

V.26—Defines a 2400-bps, PSK, full-duplex modem operating at 1200 baud.

V.26bis—Defines a 2400-bps, PSK, half-duplex modem operating at 1200 baud.

V.26terbo—Defines a switchable 2400/1200-bps, PSK, full-duplex modem operating at 1200 baud.

V.27—Defines a 4800-bps, PSK modem operating at 1600 baud.

V.27bis—Defines a more advanced 4800/2400-bps, PSK modem operating at 1600/1200 baud.

V.27terbo—Defines a 4800/2400-bps, PSK modem commonly used in half-duplex mode at 1600/1200 baud to handle Group 3 fax rather than computer modems.

V.28—Defines the electrical characteristics and connections for V.24 (RS-232). Where the RS-232 specification defines all necessary parameters, the ITU breaks the specifications down into two separate documents.

V.29—Defines a 9600/7200/4800-bps, PSK/QAM modem operating at 2400 baud. This type of modem often implements Group 3 fax rather than computer communications.

V.32—Defines the first of the truly modern modems as a 9600/4800-bps, QAM, full-duplex modem operating at 2400 baud. This standard also incorporates trellis coding and echo cancellation to produce a stable, reliable, high-speed modem.

V.32bis—A fairly new standard extending the V.32 specification to define a 4800/7200/ 9600/12000/14400 bps TC-QAM full-duplex modem operating at 2400 baud. Trellis coding, automatic transfer

FIG. 11-34   (continued).

rate negotiation, and echo cancellation make this type of modem one of the most popular and least expensive for general PC communication.

V.32terbo—Continues to extend the V.32 specification by adding advanced techniques to implement a 14400/16800/19200-bps, TCQAM, full-duplex modem operating at 2400 baud. Unlike V.32bis, V.32terbo is not widely popular because of the high cost of compatible components.

V.32fast—The informal name for a standard that the ITU has not yet completed. When finished, a V.32fast modem will probably replace V.32bis with speeds up to 28,800 bps. It is anticipated that this will be the last analog protocol, eventually giving way to all-digital protocols as local telephone systems become entirely digital. It is expected

that V.32fast will be renamed V.34 on completion and acceptance.

V.33—Defines a specialized 14400-bps, TCQAM, full-duplex modem operating at 2400 baud.

V.34—The ratified version of V.32fast. It provides for data speeds up to 128 kbps with transmission rates as high as 3429 baud.

V.36—Defines a specialized 48000-bps "group" modem that is rarely used commercially. This type of modem requires several conventional telephone lines.

V.42—Defines a two-stage process of detection and negotiation for LAPM error control.

V.42bis—Extends the V.42 standard to include data compression.

Fig. 11-34 (continued).

11-5 Error Detection and Correction

When transmitting high-speed binary data over a communications link, be it a cable or radio, errors will occur. These errors are changes in the bit pattern caused by interference, noise, or equipment malfunctions. Such errors will cause incorrect data to be received. To ensure reliable communications, schemes have been developed to detect and correct bit errors.

The number of bit errors that occur for a given number of bits transmitted is referred to as the *bit error rate (BER)*. The bit error rate is similar to a probability in that it is the ratio of the number of bit errors to the total number of bits transmitted. If there is one error for 100,000 bits transmitted, the BER is $1:100,000 = 10^{-5}$. The bit error rate depends on the equipment, the environment, and other considerations. The BER

Standard	Data Rate (bps)	Modulation Technique	Baud Rate (baud)
Bell 103	300	FSK	300
Bell 212A	1200	PSK	600
V-17	7200–14400	QAM	7400
V.21	300	FSK	300 (Bell 103)
V.22bis	2400	QAM	600 (Bell 212A)
V.23	1200/75	FM	2100/1300 & 450/390
V.26	2400	PSK (full-duplex)	1200
V.26bis	2400	PSK (half-duplex)	1200
V.26terbo	2400/120	PSK (full-duplex)	1200
V.27	4800	PSK	1600
V.27bis	PSK	1600/1200	
V.27terbo	4800/2400	PSK	1600/120 (Group 3 fax)
V.29	9600/7200/4800	PSK/QAM	2400 (Group 3 fax)
V.32	9600/4800	QAM	2400
V.32bis	4800–14400	TCQAM	2400
V.32terbo	14400–19200	TCQAM	2400
V.32fast	to 28800	TCQAM	to 3429
V.33	14400	TCQAM	2400
V.36	48000	Group-line modem	
V.37	72000	Group-line modem	

Fig. 11-35   Modem standards comparison chart. (*Electronics Now, September, 1994.*)

is an average over a very large number of bits. The BER for a given transmission depends on specific conditions. When high speeds of data transmission are used in a noisy environment, bit errors are inevitable. However, if the S/N ratio is favorable, the number of errors will be extremely small. The main objective in error detection and correction is to maximize the probability of 100 percent accuracy.

Example 11-3

Data is transmitted in 512-byte blocks or packets. Eight sequential packets are transmitted. The system BER is $2:10,000$ or 2×10^{-4}. On average, how many errors can be expected in this transmission?

$$8 \text{ packets} \times 512 \text{ bytes} = 4096 \text{ bytes}$$
$$4096 \text{ bytes} \times 8 \text{ bits} = 32,768 \text{ bits}$$
$$32,768/10,000 = 3.2768 \text{ sets of } 10,000 \text{ bits}$$
$$\text{Average number of errors} = 2 \times 3.2768 = 6.5536$$

Many different methods have been used to ensure reliable error detection, including redundancy, special coding and encoding schemes, parity checks, block checks, and cyclical redundancy check.

REDUNDANCY. The simplest way to ensure error-free transmission is to send each character or each message multiple times until it is properly received. This is known as *redundancy*. For example, a system may specify that each character will be transmitted twice in succession. Entire blocks or messages can be treated in the same way.

ENCODING METHODS. Another approach is to use an encoding scheme like the RZ-AMI described earlier, whereby successive binary 1 bits in the bit stream are transmitted with alternating polarity. If an error occurs somewhere in the bit stream, then two or more binary 1 bits with the same polarity are likely to be transmitted successively. If the receiving circuits are set to recognize this characteristic, single bit errors can be detected.

The trellis codes discussed earlier are another example of the use of special coding to detect errors. Since many bit patterns are invalid in trellis modulation, if a bit error occurs, one of the invalid codes will appear, signaling an error that can then be corrected.

SPECIAL CODES. Another approach to reliable transmission is to use special codes which permit automatic checking for accuracy. One such code is the ARQ exact-count code, a 7-bit binary code used for representing letters of the alphabet, numbers, and other symbols (see Fig. 11-36). Each 7-bit word contains exactly three binary 1s. One way to determine if a character has been received properly is to count the number of binary 1s in each character received. If the count is 3, the character has most likely been sent correctly. Should noise occur and cause one of the bits to change, the number of 1s will be a different value, signaling an error. Detection of this incorrect number of 1s could cause the system to repeat the character or the entire block.

PARITY. One of the most widely used systems of error detection is known as *parity*, in which each character transmitted contains one additional bit, known as a *parity bit*. The bit may be a binary 0 or binary 1, depending upon the number of 1s and 0s in the character itself.

Two systems of parity are normally used, odd and even. *Odd parity* means that the total number of binary 1 bits in the character, including the parity bit, is odd. *Even parity* means that the number of binary 1 bits in the character, including the parity bit, is even. Examples of odd and even parity are indicated below. The seven left-hand bits are the ASCII character, and the right-hand bit is the parity bit.

Odd parity: 10110011
00101001
Even parity: 10110010
00101000

The parity of each character to be transmitted is generated by a parity generator circuit. The parity generator is made up of several levels of exclusive OR (XOR)

Binary Code							Character	
Bit: 1	2	3	4	5	6	7	Letter	Figure
0	0	0	1	1	1	0	Letter shift	
0	1	0	0	1	1	0	Figure shift	
0	0	1	1	0	1	0	A	-
0	0	1	1	0	0	1	B	?
1	0	0	1	1	0	0	C	:
0	0	1	1	1	0	0	D	(WRU)
0	1	1	1	0	0	0	E	3
0	0	1	0	0	1	1	F	%
1	1	0	0	0	0	1	G	@
1	0	1	0	0	1	0	H	£
1	1	1	0	0	0	0	I	8
0	1	0	0	0	1	1	J	(bell)
0	0	0	1	0	1	1	K	(
1	1	0	0	0	1	0	L	)
1	0	1	0	0	0	1	M	.
1	0	1	0	1	0	0	N	,
1	0	0	0	1	1	0	O	9
1	0	0	1	0	1	0	P	0
0	0	0	1	1	0	1	Q	1
1	1	0	0	1	0	0	R	4
0	1	0	1	0	1	0	S	'
1	0	0	0	1	0	1	T	5
0	1	1	0	0	1	0	U	7
1	0	0	1	0	0	1	V	=
0	1	0	0	1	0	1	W	2
0	0	1	0	1	1	0	X	/
0	0	1	0	1	0	1	Y	6
0	1	1	0	0	0	1	Z	+
0	0	0	0	1	1	1	(blank)	
1	1	0	1	0	0	0	(space)	
1	0	1	1	0	0	0	(line feed)	
1	0	0	0	0	1	1	(carriage return)	

Fig. 11-36 The ARQ exact-count code.

circuits, as shown in Fig. 11-37. Normally the parity generator circuit monitors the shift register in a UART in the computer or modem. Just before transmitting the data in the register by shifting it out, the parity generator circuit generates the correct parity value, inserting it as the last bit in the character. In an asynchronous system, the start bit comes first, followed by the character bits, the parity bit, and finally one or more stop bits (see Fig. 11-38).

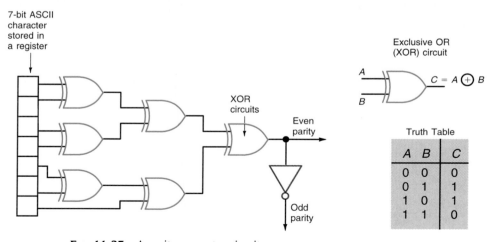

Fig. 11-37 A parity generator circuit.

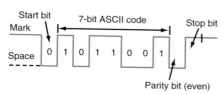

FIG. 11-38 How parity is transmitted.

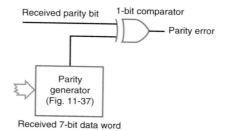

FIG. 11-39 Parity checking at the receiver.

At the receiving modem or computer, the serial data word is transferred into a shift register in a UART. A parity generator in the receiving UART produces the parity on the received character. It is then compared to the received parity bit in an XOR circuit, as shown in Fig. 11-39. If the internally generated bit matches the transmitted and received parity bit, it is assumed that the character was transmitted correctly. The output of the XOR will be 0, indicating no error. If the received bit does not match the parity bit generated from the received data word, the XOR output will be 1, indicating an error. The system signals the detection of a parity error to the computer. The action taken will depend on the result desired: the character may be retransmitted, an entire block of data may be transmitted, or the error may simply be ignored.

The individual-character parity method of error detection is sometimes referred to as the *vertical redundancy check (VRC)*. To display characters transmitted in a data communications systems, the bits are written vertically (see Fig. 11-40). The bit at the bottom is the parity, or VRC, bit for each vertical word.

Parity checking is useful only for detecting single bit errors. If two or more bit errors occur, the parity circuit may not detect it. If an even number of bit changes occurs, the parity circuit will not give a correct indication.

BLOCK-CHECK CHARACTER. The *horizontal* or *longitudinal redundancy check (LRC)* is the process of logically adding, by exclusive ORing, all the characters in a specific block of transmitted data. To add the characters, the top bit of the first vertical word in Fig. 11-40 is exclusive-ORed with the top bit of the second word. The result of this operation is exclusive ORed with the top bit of the third word, and so on until all the bits in a particular horizontal row have been added. There are no carries to the next bit position. The final bit value for each horizontal row then becomes one bit in a character known as the *block-check character (BCC),* or the *block-check*

		D	A	T	A		C	O	M	LRC or BCC
Character										
(LSB)		0	1	0	1	0	1	1	1	1
		0	0	0	0	0	1	1	0	1
ASCII		1	0	1	0	0	0	1	1	0
		0	0	0	⓪	0	0	1	1	0
Code		0	0	1	0	0	0	0	0	1
		0	0	0	0	1	0	0	0	1
(MSB)		1	1	1	1	0	1	1	1	1
Parity → or VRC (odd)		1	1	0	1	0	0	0	1	0

FIG. 11-40 Vertical and horizontal redundancy checks.

sequence (BCS). Each row of bits is done the same way to produce the BCC. All the characters transmitted in the text, as well as any control or other characters, are included as part of the BCC. Exclusive ORing all bits of all characters is the same as binary addition without a carry of the codes.

The BCC is computed by circuits in the computer or modem as the data is transmitted, and its length is usually limited to 8 bits, so that carries from one bit position to the next are ignored. It is appended to the end of a series of bytes that comprise the message to be transmitted. At the receiving end, the computer computes its own version of the BCC on the received data and compares it to the received BCC. Again, the two should be the same.

When both the parity on each character and the BCC are known, the exact location of a faulty bit can be determined. The individual character parity bits and the BCC bits provide a form of coordinate system that allows a particular bit error in one character to be identified. Once it is identified, the bit is simply complemented to correct it. The VRC identifies the character containing the bit error while the LRC identifies the bit that contains the error.

Assume, for example, that a bit error occurs in the fourth vertical character from the left in Fig. 11-40. The fourth bit down from the top should be 0, but because of noise, it is received as a 1. This causes a parity error. With odd parity, and a 1 in the fourth bit, the parity bit should be 0, but it is a 1.

Next, the logical sum of the bits in the fourth horizontal row from the top will be incorrect because of the bit error. Instead of 0 it will be 1. All of the other bits in the BCC will be correct. It is now possible to pinpoint the location of the error because both the vertical column where the parity error occurred and the horizontal row where the BCC error occurred are known. The error can be corrected by simply complementing (inverting) the bit from 1 to 0. This operation can be programmed in software or implemented in hardware.

CYCLICAL REDUNDANCY CHECK. The *cyclic redundancy check (CRC)* is a mathematical technique used in synchronous data transmission that effectively catches 99.9 percent or more of transmission errors. The mathematical process implemented by CRC is essentially a division. The entire string of bits in a block of data is considered to be one giant binary number which is divided by some preselected constant. CRC is expressed by the equation

$$\frac{M(x)}{G(x)} = Q(x) + R(x)$$

where $M(x)$ is the binary block of data, called the *message function,* and $G(x)$ is the generating function. The *generating function* is a special code which is divided into the binary message string. The outcome of the division is a quotient function $Q(x)$ and a remainder function $R(x)$. The quotient resulting from the division is ignored; the remainder is known as the CRC character and is transmitted along with the data.

For convenience of calculation, the message and generating functions are usually expressed as an algebraic polynomial. For example, assume an 8-bit generating function of 10000101. The bits are numbered such that the LSB is 0 and the MSB is 7.

7	6	5	4	3	2	1	0
1	0	0	0	0	1	0	1

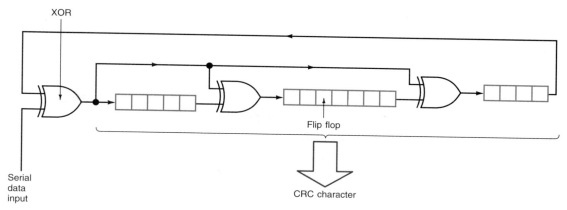

FIG. 11-41 A CRC error-detection circuit made with a 16-bit shift register and XOR gates.

The polynomial is derived by expressing each bit position as a power of x, where the power is the number of the bit position. Only those terms where binary 1s appear in the generating function are included in the polynomial. The polynomial resulting from the above number is

$$G(x) = x^7 + x^2 + x^0 \qquad \text{or} \qquad G(x) = x^7 + x^2 + 1$$

The CRC mathematical process can be programmed using a computer's instruction set. It can also be computed by a special CRC hardware circuit consisting of several shift registers into which XOR gates have been inserted at specific points (see Fig. 11-41). The data to be checked is fed into the registers serially. There is no output, since no output is retained. The data is simply shifted in a bit at a time; when the data has all been transmitted, the contents will be the remainder of the division, or the desired CRC character. Since a total of 16 flip-flops are used in the shift register, the CRC is 16 bits long and can be transmitted as two sequential 8-bit bytes. The CRC is computed as the data is transmitted, and the resulting CRC is appended to the end of the block. Because CRC is used in synchronous data transmission, no start and stop bits are involved.

At the receiving end, the CRC is computed by the receiving computer and compared to the received CRC characters. If the two are alike, the message has been correctly received. Any difference indicates an error, which triggers retransmission or some other form of corrective action. CRC is probably the most widely used error detection scheme in synchronous systems. Both 16- and 32-bit CRCs are used. Parity and BCC methods are used primarily in asynchronous systems.

ERROR CORRECTION

As stated previously, the easiest way to correct transmission errors—to retransmit any character or block of data that has an error in it—is time-consuming and wasteful. A number of efficient error-correction schemes have been devised to complement the parity and BCC methods described above. The process of detecting and correcting errors at the receiver so that retransmission is not necessary is called *forward error correction (FEC)*. The most popular FEC method is the use of Hamming codes. Hamming was a researcher at Bell Labs who discovered that if extra bits were added to a transmitted word, these extra bits could be processed such that bit errors could be identified and corrected. These extra bits, like several types of parity bits, are known as *Hamming bits* and together they form a *Hamming code.* To determine exactly where the error is, a sufficient number of bits must be added. The minimum number of Hamming bits is computed with the expression

$$2^n \geq m + n + 1$$

where m = number of bits in data word
n = number of bits in Hamming code

Assume, for example, an 8-bit character word and some smaller number of Hamming bits (say, 2). Then

$$2^n \geq m + n + 1$$
$$2^2 \geq 8 + 2 + 1$$
$$4 \geq 11$$

Two Hamming bits are insufficient, and so are three. When $n = 4$,

$$2^4 \geq 8 + 4 + 1$$
$$16 \geq 13$$

Thus 4 Hamming bits must be transmitted along with the 8-bit character. Each character requires $8 + 4 = 12$ bits. These Hamming bits can be placed anywhere within the data string. Assume the placement shown below, where the data bits are shown as a 0 or 1 and the Hamming bits are designated with an H. The data word is 01101010. Note that the bits are numbered from right to left.

12	11	10	9	8	7	6	5	4	3	2	1
H	0	1	H	1	0	H	1	0	H	1	0

One way to look at Hamming codes is simply as a more sophisticated parity system, where the Hamming bits are parity bits derived from some but not all of the data bits. Each Hamming bit is derived from different groups of the data bits. (Recall that parity bits are derived from the data by XOR circuits.) One technique used to determine the Hamming bits is discussed below.

At the transmitter, a circuit is used to determine the Hamming bits. This is done by first expressing the bit positions in the data word containing binary 1s as a 4-bit binary number ($n = 4$). For example, the first binary 1 data bit appears at position 2, so its position code is just the binary code for 2, or 0010. The other data bit positions with a binary 1 are 5 = 0101, 8 = 1000, and 10 = 1010.

Next, the transmitter circuitry logically adds (XORs) these codes together.

Position code 2	0010
Position code 5	0101
XOR sum	0111
Position code 8	1000
XOR sum	1111
Position code 10	1010
XOR sum	0101

This final sum is the Hamming code bits from left to right. Position code 12 is 0, position code bit 9 is 1, position code 6 is 0, and position code 3 is 1. These bits are interested in their proper position. The complete 12-bit transmitted word is:

12	11	10	9	8	7	6	5	4	3	2	1
H	0	1	H	1	0	H	1	0	H	1	0

12	11	10	**9**	8	7	**6**	5	4	**3**	2	1
0	0	1	**1**	1	0	**0**	1	0	**1**	1	0

The Hamming bits are shown in boldface type.

Now assume that an error occurs in bit position 10. The binary 1 is received as a binary 0. The received word is

12	11	10	**9**	8	7	**6**	5	4	**3**	2	1
0	0	0	**1**	1	0	**0**	1	0	**1**	1	0

The receiver recognizes the Hamming bits and treats them as a code word, in this case 0101. The circuitry then adds (XORs) this code with the bit number of each position in the word containing a binary 1, positions 2, 5, and 8.

The Hamming code is then added to the binary numbers representing each position with a 1.

Hamming code	0101
Position code 2	0010
XOR sum	0111
Position code 5	0101
XOR sum	0010
Position code 8	1000
XOR sum	1010

This final sum is a code that identifies the bit position of the error, in this case bit 10 (1010). To correct the bit, it is simply complemented from 0 to 1.

Note that the Hamming code method does not work if an error occurs in one of the Hamming bits itself.

For the Hamming code method of error detection and correction to work when 2 or more bit errors occur, more Hamming bits must be added. This increases overall transmission time, storage requirements at the transmitter and receiver, and the complexity of the circuitry. The benefit, of course, is that errors are reliably detected at the sending end. The transmitter never has to resend the data, which may in fact be impossible in some applications. Not all applications require such rigid data-correction practices.

Other methods for forward error correction involve complex computer and digital circuitry, and are beyond the scope of this chapter. Most of these are built into the large-scale ICs that implement today's data communications operations.

11-6 PROTOCOLS

Protocols are rules and procedures used to ensure compatibility between the sender and receiver of serial digital data regardless of the hardware and software being used. They are used to identify the start and end of a message, identify the sender and the receiver, state the number of bytes to be transmitted, state a method of error detection,

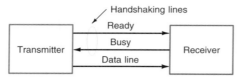

Fig. 11-42 The handshaking process in data communications.

and for other functions. Various protocols, and various levels of protocols, are used in data communications.

The simplest form of protocol is the asynchronous transmission of data using a start bit and a stop bit (refer to Fig. 11-7) framing a single character, with a parity bit between the character bit and the stop bit. The parity bit is part of the protocol, but may or may not be used. In data communications, however, a message is more than one character. As discussed previously it is composed of blocks, groups of letters of the alphabet, numbers, punctuation marks, and other symbols in the desired sequence. In synchronous data communications applications, the block is the basic transmission unit.

To identify a block, one or more special characters are transmitted prior to the block and after the block. These additional characters, which are usually represented by 7- or 8-bit codes, perform a number of functions. Like the start and stop bits on a character, they signal the beginning and end of the transmission. But they are also used to identify a specific block of data, and they provide a means for error checking and detection.

Some of the characters at the beginning and end of each block are used for handshaking purposes. These characters give the transmitter and receiver status information. Figure 11-42 illustrates the basic handshaking process. For example, a transmitter may send a character indicating that it is ready to send data to a receiver. Once the receiver has identified that character, it responds by indicating its status, for example, by sending a character representing "busy" back to the transmitter. The transmitter will continue to send its ready signal until the receiver signals back that it is not busy, or is ready to receive.

At that point, data transmission takes place. Once the transmission is complete, some additional handshaking takes place. The receiver acknowledges that it has received the information. The transmitter then sends a character indicating that the transmission is complete, which is usually acknowledged by the receiver.

A common example of the use of such control characters is the XON and XOFF protocol used between a printer and a computer. XON is usually the ASCII character DC1, and XOFF is the ASCII DC3 character. A printer that is ready and able to receive data will send XON to the computer. If the printer is not able to receive data, it sends XOFF. When the computer detects XOFF, it immediately stops sending data until XON is again received.

ASYNCHRONOUS PROTOCOLS

Three popular protocols used for asynchronous ASCII-coded data transmission between personal computers, via modem, are Xmodem, Kermit, and MPN.

XMODEM. In Xmodem, the data transmission procedure begins with the receiving computer transmitting a negative acknowledge (NAK) character to the transmitter. NAK

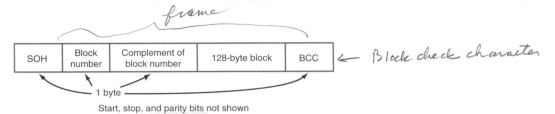

Fig. 11-43   Xmodem protocol frame.

is a 7-bit ASCII character that is transmitted serially back to the transmitter every 10 s until the transmitter recognizes it. Once the transmitter recognizes the NAK character, it begins sending a 128-byte block of data, known as a *frame* (packet) of information (see Fig. 11-43). The frame begins with a start of header (SOH) character, which is another ASCII character meaning that the transmission is beginning. This is followed by a header, which usually consists of two or more characters preceding the actual data block which give auxiliary information. In Xmodem, the header consists of two bytes designating the block number. In most messages, several blocks of data are transmitted and each is numbered sequentially. The first byte is the block number in binary code. The second byte is the complement of the block number; that is, all bits have been inverted. Then the 128-byte block is transmitted. At the end of the block, the transmitting computer sends a check-sum byte, which is the BCC, or binary sum of all the binary information sent in the block. (Keep in mind that each character is sent along with its start and stop bits, since Xmodem is an asynchronous protocol.)

The receiving computer looks at the block data and also computes the check sum. If the check sum of the received block is the same as that transmitted, it is assumed that the block was received correctly. If the block was received correctly, the receiving computer sends an acknowledge (ACK) character—another ASCII code—back to the transmitter. Once ACK is received by the transmitter, the next block of data is sent. When a block has been received incorrectly due to interference or equipment problems, the check sums will not match and the receiving computer will send a NAK code back to the transmitter. A transmitter that has received NAK automatically responds by sending the block again. This process is repeated until each block, and the entire message, has been sent without errors.

When the entire message has been sent, the transmitting computer sends an end-of-transmission (EOT) character. The receiving computer replies with an ACK character, terminating the communications.

KERMIT.   Another popular asynchronous protocol is Kermit (see Fig. 11-44). The transmission begins with an SOH character followed by a length (LEN) character, which tells how long the block of data is. A block can be up to 94 bytes long. Next is a packet sequence number (SEQ). There can be up to 63 blocks, and these are given a sequence number so that both transmitter and receiver can keep track of long messages.

Next in the packet is a data-type (TYPE) designator. This byte may contain an ASCII control code like ACK, NAK, EOT, EOF, or any one of a number of special codes used in the send-receive handshaking process.

The data block, which can be up to 94 bytes long, comes next. At the end of the packet is an error detection code. It can be a 1-byte check sum, a 2-byte check sum, or a 16-bit CRC.

DID YOU KNOW?

Kermit is a very reliable protocol because it requires that every packet which is sent is acknowledged by the receiver as read correctly.

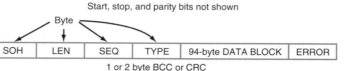

Start, stop, and parity bits not shown

| SOH | LEN | SEQ | TYPE | 94-byte DATA BLOCK | ERROR |

1 or 2 byte BCC or CRC

FIG. 11-44   Kermit asynchronous protocol.

Kermit is a very reliable protocol, since it requires that every packet sent be acknowledged by the receiver as being read correctly.

MNP.   *Microcom Networking Protocols (MNPs)* are a series of protocols developed by the manufacturer Microcom to be used with asynchronous modems. They specify ways to handle error detection and correction and how to specify whether or not data compression is used. There are ten classes of protocols. Not all modems support these protocols, but in recent years, most manufacturers have adopted them. They are reasonably easy to implement because they can be programmed into the control microcomputer used in most modems. The Microcom protocols are subdivided into classes.

The MNP class 3 protocol strips the start and stop bits from the characters in the message, which speeds up the transmission by a factor of about 8. The receiving modem puts the start and stop bits back in.

The MNP class 4 protocol uses data compression to speed up transmission. It also checks for a noisy line, and if the noise is low, sends larger blocks of data. If the noise is high, smaller blocks are sent.

The MNP class 5 protocol uses data compression.

The MNP class 6 protocol offers automatic speed increases if the line can handle it without error.

The MNP class 9 protocol offers an improved error correction method.

SYNCHRONOUS PROTOCOLS

Protocols used for synchronous data communications are more complex than asynchronous protocols. However, like the asynchronous Xmodem and Kermit systems, they use various control characters for signaling purposes at the beginning and ending of the block of data to be transmitted.

BISYNC.   IBM's *Bisync protocol,* which is widely used in computer communications, usually begins with the transmission of two or more ASCII sync (SYN) characters (see Fig. 11-45). These characters signal the beginning of the transmission and are also used to initialize the clock timing circuits in the receiving modem. This ensures proper synchronization of the data transmitted a bit at a time.

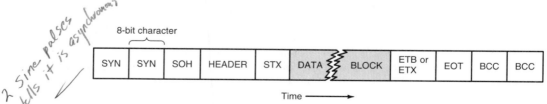

8-bit character

| SYN | SYN | SOH | HEADER | STX | DATA | BLOCK | ETB or ETX | EOT | BCC | BCC |

Time ⟶

FIG. 11-45   Bisync synchronous protocol.

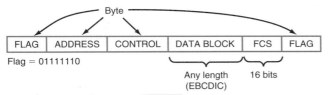

FLAG | ADDRESS | CONTROL | DATA BLOCK | FCS | FLAG

Flag = 01111110

Any length (EBCDIC)   16 bits

FIG. 11-46   The SDLC frame format.

After the SYNC characters, a start-of-header (SOH) character is transmitted. The header is a group of characters that typically identifies the type of message to be sent, the number of characters in a block (usually up to 256), and a priority code or some specific routing destination. The end of the header is signaled by a start-of-text (STX) character. At this point, the desired message is transmitted, one byte at a time. No start and stop bits are sent. The 7- or 8-bit words are simply strung together one after another, and the receiver must sort them out into individual binary words which are handled on a parallel basis farther along on the receiving circuit in the computer.

At the end of a block an end-of-block (ETB) character is transmitted. If the block is the last one in a complete message, an end-of-text (ETX) character is transmitted. An end-of-transmission (EOT) character, which signals the end of the transmission, is followed by an error detection code, usually a 1- or 2-byte BCC.

SDLC.   One of the most flexible and widely used synchronous protocols is the *synchronous data link control (SDLC)* protocol (see Fig. 11-46). SDLC is used in networks that are interconnections of multiple computers. All frames begin and end with a flag byte with the code 01111110 or hex 7E, which is recognized by the receiving computer. A sequence of binary 1s starts the clock synchronous process. Next comes an address byte that specifies a specific receiving station. Each station on the network is assigned an address number. The address hex FF indicates that the message to follow is to be sent to all stations on the network.

A control byte following the address allows the programmer or user to specify how the data will be sent and how it will be dealt with at the receiving end. It allows the user to specify the number of frames, how the data will be received, and so on.

The data block (all codes are EBCDIC, not ASCII) comes next. It can be any length, but 256 bytes is typical. The data is followed by a frame-check sequence (FCS), a 16-bit CRC. A flag ends the frame.

SDLC is widely used in networking mainframe computers and minicomputers. A more sophisticated SDLC-type system, which permits interface between a larger number of different software and hardware configurations, called high-level data link control (HDLC), is also available.

THE OPEN SYSTEMS INTERCONNECTION MODEL

As you have seen, there are many types and variations of protocols. If there is to be widespread compatibility between different systems, then some industrywide standardization is necessary. The ability of one hardware-software configuration to communicate with another, different system is known as *interoperability*. Only if all manufacturers and users adopt the same standards can true interoperability be achieved. One organization that has attempted to standardize data communications procedures is the *International Organization for Standardization (ISO)*. ISO has come up with a framework, or hierarchy, that defines how data can be communicated. This hierarchy, known as the *open systems interconnection (OSI)* model, is designed to establish general interoperability guidelines for developers of communications systems and protocols.

The OSI hierarchy is made up of seven levels, or layers (see Fig. 11-47). Each layer is defined by software (or, in one case, hardware) and is clearly distinct from the other layers. These layers are not really protocols themselves, but they provide a way to define and partition protocols to make data transfers in a standardized way. Each layer is designed to handle messages that it receives from a lower layer or an upper layer. Each layer also sends messages to the layer above or below it according to specific guidelines.

As shown in the figure, the highest level is the application layer, which interfaces with the user's application program. The lowest level is the physical layer, where the electronic hardware, interfaces, and media are defined.

Layer 1: Physical layer. The physical connections and electrical standards for the communication system are defined here. This layer specifies interface characteristics such as binary voltage levels, encoding methods, data transfer rates, and the like. A typical physical standard is the popular RS-232 or RS-422 standard.

Layer 2: Data link. This layer defines the framing information for the block of data. It identifies any error detection and correction methods, as well as any synchronizing and control codes relevant to communications. The data-link layer includes basic protocols, such as HDLC, SDLC, etc.

Layer 3: Network. This layer determines network configuration and the route the transmission can take. In some systems, there may be several paths for the data to traverse. The network layer determines the specific data routing and switching methods that can occur in the system, for example, selection of a dial-up line, a private leased line, or some other dedicated path.

Layer 4: Transport. Included in this layer are multiplexing, if any; error recovery; partitioning of data into smaller units so that they can be handled more efficiently; and addressing and flow-control operations.

Layer 5: Session. This layer handles such things as management and synchronization of the data transmission. It typically includes network log-on and log-off procedures, as well as user authorization, and determines the availability of the network for processing and storing the data to be transmitted.

Layer 6: Presentation. This layer deals with the form and syntax of the message. It defines data formatting, encoding and decoding, encryption and decryption, synchronization, and other characteristics. It defines any code translations required, and sets the parameters for any graphics operations.

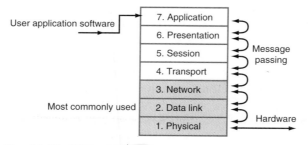

FIG. 11-47 The seven OSI layers.

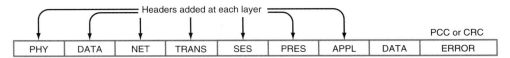

PHY	DATA	NET	TRANS	SES	PRES	APPL	DATA	ERROR

PCC or CRC

FIG. 11-48   The data as transmitted at the physical layer.

Layer 7: Applications.   This layer is the overall general manager of the network or the communications process. Its primary function is to format and transfer files between the communications message and the user's applications software.

The basic process is that information is added or removed as data is transmitted from one layer to another (see Fig. 11-48). Assume, for example, that the applications program you are using contains some data that you wish to send to another computer. That data will be transmitted in the form of some kind of serial packet or frame, or a sequence of packets. The applications layer attaches the packet to some kind of header or preamble before sending it to the next level. At the presentation level, more headers and other information are added. At each of the lower levels, headers and related information are appended until the data message is almost completely encased in a much larger packet. Finally, at the physical level, the data is transmitted to the other system. As Fig. 11-48 shows, the message that is actually sent may contain more header information than actual data.

At the receiving end, the header information gets stripped off at the various levels as the data is transferred to successive levels. The data comes in at the physical level and goes up through the various layers until it gets to the applications layer, where it is finally used.

Note that most modern data communications applications need only the first two or three layers, and the seventh layer, to fully define a given data exchange.

The primary benefit of the OSI standards is that if they are incorporated into data communications equipment and software, compatibility between systems and equipment is ensured. As more and more computers and networks are interconnected, for example, on the Internet, true interoperability becomes more and more important.

11-7 SPREAD SPECTRUM

Spread spectrum (SS) is a modulation and multiplexing technique that distributes a signal and its sidebands over a very wide bandwidth. Traditionally, the efficiency of a modulation or multiplexing technique is determined by how little bandwidth it uses. The continued growth of all types of radio communications, the resulting crowding, and the finite bounds of usable spectrum space have made everyone in the world of data communications sensitive to how much bandwidth a given signal occupies. Designers of communications systems and equipment typically do all in their power to minimize the amount of bandwidth a signal takes. How, then, can a scheme that spreads a signal over a very wide piece of the spectrum be of value? The answer to this question is the subject of this section.

After World War II, spread spectrum was developed primarily by the military because it is a secure communications technique that is essentially immune to jamming. In the mid-1980s, the FCC authorized use of spread spectrum in civilian applications. Currently, unlicensed operation is permitted in the 902- to 928-MHz, 2.4- to 2.483-GHz, and 5.725- to 5.85-GHz ranges, with 1 W of power. Spread spectrum on these frequencies is now being widely incorporated into a variety of commercial communications systems. One of the most important of these new applications is wireless data communications. Numerous LANs and portable personal computer modems use SS techniques, as does a new class of cordless telephones in the 900-MHz range. Future-design cellular telephones will probably use SS.

There are two basic types of spread spectrum, *frequency hopping (FH)* and *direct sequence (DS)*. In frequency-hopping SS, the frequency of the carrier of the transmitter is changed according to a predetermined sequence, called pseudorandom, at a rate higher than the serial binary data modulating the carrier. In direct-sequence SS, the serial binary data is mixed with a higher-frequency pseudorandom binary code at a faster rate and the result is used to phase-modulate a carrier. Let's examine each technique in more detail.

FREQUENCY-HOPPING SS

Figure 11-49 shows a block diagram of a frequency-hopping SS transmitter. The serial binary data to be transmitted is applied to a conventional two-tone FSK modulator, and the modulator output is applied to a mixer. Also driving the mixer is a frequency synthesizer. The output signal from the bandpass filter after the mixer is the difference between one of the two FSK sine waves and the frequency of the frequency synthesizer. As the figure shows, the synthesizer is driven by a pseudorandom code generator, which is either a special digital circuit or the output of a microprocessor.

The pseudorandom code is a serial pattern of binary 0s and 1s that changes in a random fashion. The randomness of the 1s and 0s makes the serial output of this circuit appear like digital noise. Sometimes the output of this generator is called *pseudorandom noise (PSN)*. The binary sequence is actually predictable, since it does repeat after many bit changes (hence "pseudo"). The randomness is sufficient to minimize the possibility of someone accidentally duplicating the code, but the predictability allows the code to be duplicated at the receiver.

PSN sequences are usually generated by a shift register circuit similar to that shown in Fig. 11-50. In the figure, eight flip-flops in the shift register are clocked by an external clock oscillator. The input to the shift register is derived by XORing two or more

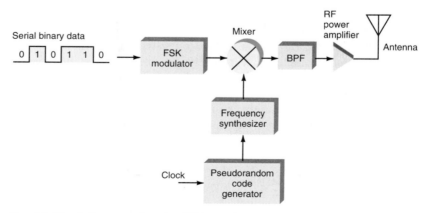

FIG. 11-49 A frequency-hopping SS transmitter.

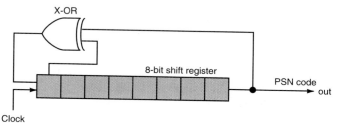

Fig. 11-50 A typical PSN code generator.

of the flip-flop outputs. It is this connection that produces the pseudorandom sequence. The output is taken from the last flip-flop in the register. Changing the number of flip-flops in the register and/or which outputs are XORed and fed back changes the code sequence. Alternatively, a microprocessor can be programmed to generate pseudorandom sequences.

In a frequency-hopping SS system, the rate of synthesizer frequency change is higher than the data rate. This means that while the data bit and the FSK tone it produces remain constant for one data interval, the frequency synthesizer switches frequencies many times during this period. See Fig. 11-51, where the frequency synthesizer changes frequencies four times for each bit time of the serial binary data. The time that the synthesizer remains on a single frequency is called the *dwell time*. The frequency synthesizer puts out a random sine-wave frequency to the mixer, and the mixer creates a new carrier frequency for each dwell interval. The resulting signal, whose frequency rapidly jumps around, effectively scatters pieces of the signal all over the band. Specifically, the carrier randomly switches between dozens or even hundreds of frequencies over a given bandwidth. The actual dwell time on any frequency varies with the application and data rate, but it can be as short as 10 ms. Currently, FCC regulations specify that there be a minimum of 50 hopping frequencies and that the dwell time not exceed 400 ms.

Figure 11-52 shows a random frequency hop sequence. The horizontal axis is divided into dwell-time increments. The vertical axis is the transmitter output frequency, divided into step increments of the PLL frequency synthesizer. As shown, the signal is spread out over a very wide bandwidth. Thus a signal that occupies only a few kilohertz of spectrum can be spread out over a range that is 10 to 10,000 times that wide. Because an SS signal does not remain on any one

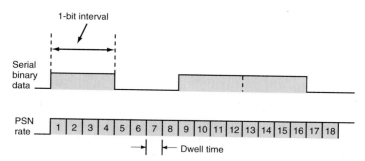

Fig. 11-51 Serial data and the PSN code rate.

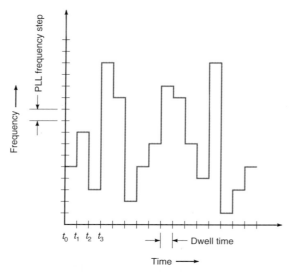

FIG. 11-52 A pseudorandom frequency hop sequence.

frequency for a long period of time, but jumps around randomly, it does not interfere with a traditional signal on any of the hopping frequencies. An SS signal actually appears to be more like noise to a conventional narrow-bandwidth receiver. A conventional receiver picking up such a signal will not even respond to a signal of tens of milliseconds duration. In addition, a conventional receiver cannot receive an SS signal because it does not have a wide enough bandwidth and it cannot follow or track its random-frequency changes. Therefore, the SS signal is as secure as if it were scrambled.

Two or more SS transmitters operating over the same bandwidth but with different pseudorandom codes hop to different frequencies at different times, and do not typically occupy a given frequency simultaneously. Thus SS is also a kind of multiplexing, as it permits two or more signals to use a given bandwidth concurrently without interference. In effect, SS permits more signals to be packed in a given band than any other type of modulation or multiplexing.

A frequency-hopping receiver is shown in Fig. 11-53. The very-wideband signal picked up by the antenna is applied to a broadband RF amplifier, then to a conventional mixer. The mixer, like that in any superheterodyne receiver is driven by a local oscillator. In this case the local oscillator is a frequency synthesizer like the one used at the transmitter. The local oscillator at the receiving end must have the same pseudorandom code sequence as that generated by the transmitter so that it can receive the signal on the correct frequency. The signal is thus reconstructed as an IF signal which contains the original FSK data. The signal is then applied to an FSK demodulator, which reproduces the original binary data train.

One of the most important parts of the receiver is the circuit that is used to acquire and synchronize the transmitted signal with the internally generated pseudorandom code. The problem of getting the two codes into step with one another is solved by a preamble signal and code at the beginning of the transmission. Once synchronization has been established, the code sequences occur in step. This characteristic makes the SS technique extremely secure and reliable. Any receiver not having the correct code cannot receive the signal.

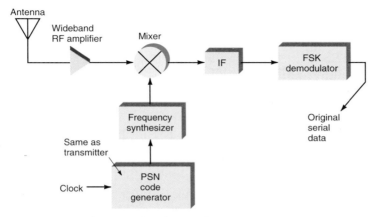

FIG. 11-53 A frequency-hopping receiver.

Many stations can share a common band if, instead of assigning each station a single frequency to operate on, each is given a different pseudorandom code within the same band. This permits a transmitter to selectively transmit to a single receiver without other receivers in the band being able to pick up the signal.

DIRECT-SEQUENCE SS

A block diagram of a direct-sequence SS transmitter is shown in Fig. 11-54. The serial binary data is applied to an XOR gate along with a serial pseudorandom code that occurs faster than the binary data. Figure 11-55 shows typical waveforms. One bit time for the pseudorandom code is called a *chip,* and the rate of the code is called the *chipping rate.* The chipping rate is faster than the data rate.

The signal developed at the output of the XOR gate is then applied to a PSK modulator, typically a BPSK device. The carrier phase is switched between zero and 180° by the 1s and 0s of the XOR output. QPSK and other forms of PSK can also be used. The PSK modulator is generally some form of balanced modulator. The signal phase modulating the carrier, being much higher in frequency than the data signal, causes the modulator to produce multiple, widely spaced sidebands whose strength is such that the complete signal takes up a great deal of the spectrum. Thus the resulting signal is

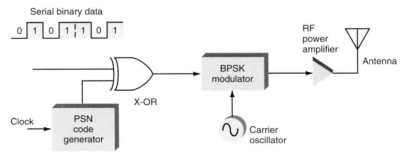

FIG. 11-54 A direct-sequence SS transmitter.

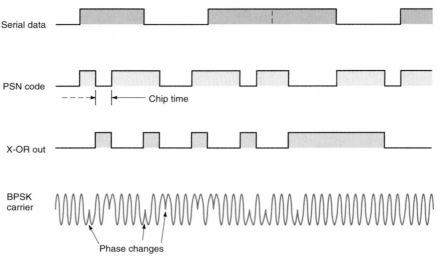

Fig. 11-55 Data signals in direct-sequence SS.

spread. Because of its randomness, the signal appears to be nothing more than wide-band noise to a conventional narrowband receiver.

One type of direct-sequence receiver is shown in Fig. 11-56. The broadband SS signal is amplified, mixed with a local oscillator, and then translated down to a lower IF frequency in mixer 1. For example, the SS signal at an original carrier of 902 MHz might be translated down to an IF of 70 MHz. The IF signal is then compared to an IF signal that is produced in mixer 3 using a PSN sequence that is similar to that trans-mitted. This comparison process, called *correlation,* takes place in mixer 2. If the two signals are identical, the correlation is 100 percent. If the two signals are not alike in any way, the correlation is 0. The correlation process in the mixer produces a signal that is averaged in the low-pass filter at the output of mixer 2. The output signal will be a high average value if the transmitted and received PSN codes are alike.

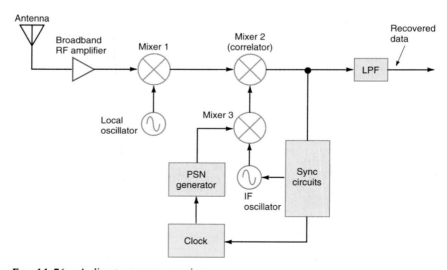

Fig. 11-56 A direct-sequence receiver.

The signal out of mixer 2 is fed to a synchronization circuit, which must recreate the exact frequency and phase of the carrier so that demodulation can take place. The synchronization circuit varies the clock frequency so that the PSN code output frequency varies, seeking the same chip rate as the incoming signal. The clock drives a PSN code generator containing the exact code used at the transmitter. The PSN code in the receiver is the same as that of the received signal, but the two are out of sync with one another. Adjusting the clock by speeding it up or slowing it down eventually causes the two to come into synchronization.

The PSN code produced in the receiver is used to phase-modulate a carrier at the IF in mixer 3. Like all of the other mixers, this one is usually of the double balanced diode ring type. The output of mixer 2 is a BPSK signal similar to that being received. It is compared to the received signal in mixer 3, which acts as a correlator. The output of mixer 3 is then filtered to recover the original serial binary data. The received signal is said to be *despread.*

Direct-sequence SS is also called *code division multiple access (CDMA),* or SS multiple access. The term *multiple access* applies to any technique that is used for multiplexing many signals on a single communications channel. CDMA is used in

This woman is working at a computer phone center for deaf persons. She types in information received through a normal telephone connection from an individual who wishes to communicate with a deaf person. The message is transmitted over telephone lines. The deaf individual can then read the message and use a TDD telephone to carry on a conversation.

satellite systems so that many signals can use the same transponder. CDMA is also being considered as a method for future cellular telephone service, as it permits more users to occupy a given band than other methods.

OTHER SS METHODS

Although frequency-hopping and direct-sequence SS are the two most widely used types, there are other methods. In one, called *chirp,* the signal modulates a carrier that is linearly swept over the band being used. Chirp is used primarily in radar systems and in other distance-ranging equipment. In *time-hopping* SS the carrier is keyed ON and OFF by the pseudorandom code sequence. The data signal modulates the carrier in the usual way.

Hybrid systems can also be created by combining two types of SS. The combination will depend on the application, but the most common hybrid combines frequency-hopping and direct-sequence SS.

BENEFITS OF SS

Spread spectrum is being used in more and more applications in data communications as its benefits are discovered and as new components and equipment become available to implement it. Some major benefits of SS are as follows:

- *Security.* SS prevents unauthorized listening. Unless a receiver has a very wide bandwidth and the exact pseudorandom code and type of modulation, it cannot intercept an SS signal.

- *Resistance to jamming and interference.* Jamming signals are typically restricted to a single frequency, and jamming one frequency does not interfere with an SS signal. Similarly, unintentional interference from a signal occupying the same band is greatly minimized and in most cases virtually eliminated.

- *Band sharing.* Many users can share a single band with little or no interference. (As more and more signals use a band, the background noise produced by the switching of many signals increases, but not enough to prevent highly reliable communications.)

- *Resistance to fading.* Frequency-selective fading occurs during signal propagation because signals of different frequencies arrive at a receiver at slightly different times. SS virtually eliminates wide variations of signal strength due to reflections and other phenomena during propagation.

- *Precise timing.* Use of the pseudorandom code in SS provides a way to precisely determine the start and end of a transmission. Thus SS is a superior method for radar and other applications that rely on accurate knowledge of transmission time to determine distance.

SUMMARY

Data stored and processed by computers is encoded in binary digital form according to some standardized system, such as the American Standard Code for Information Interchange (ASCII). The signaling elements (bits, baud, or symbols) that comprise digitally stored information can be transmitted synchronously or asynchronously. In asynchronous communications, each data word is accompanied by stop and start bits that identify the beginning and ending of the word. In synchronous transmission, start and stop bits are not used; data is transferred in multiword blocks in step with a master clock oscillator.

The speed of data transfer ranges from less than 10,000 bits per second (for the slower personal computers) to tens or hundreds of million bits per second. Data speed in digital communications systems can also be expressed in terms of baud rate. Several unique modulation schemes have been developed so that multiple bits can be included in a given symbol.

As with analog data, the amount of digital information that can be transmitted is proportional to the bandwidth of the communications channel and the time of transmission. Information theorists seek to determine, mathematically, the likelihood that a given amount of data being transmitted under a given set of circumstances (e.g., medium, bandwidth, speed of transmission, noise and distortion) will be transmitted accurately.

Before digital serial data is sent, it is usually encoded according to the system that will facilitate the desired result. Some common encoding techniques are nonreturn to zero, return to zero, and Manchester.

Digital data is transmitted over the analog telephone network using a modem, which both converts binary signals into analog signals and demodulates analog signals to reconstruct the equivalent binary output. The three most common modulation and demodulation techniques used in today's modems are frequency shift keying (FSK); phase shift keying (PSK), including binary phase shift keying, differential phase shift keying, and quadrature phase shift keying; and quadrature amplitude modulation (QAM), including 8-QAM, 16-QAM, and trellis code modulation.

To reach a goal of 100 percent accuracy in transmission, both error detection and error correction methods must be used. Common detection techniques are redundancy, various special encoding schemes, parity, block-check sequences, and cyclical redundancy checks. Forward error correction is the process of detecting and correcting errors at the receiver without retransmission; the most popular FEC method is the use of Hamming codes.

Protocols are designed to ensure hardware and software compatibility between the sender and receiver of serial digital data. All asynchronous protocols use start and stop bits, and some include procedures such as hardware handshaking and the transmission of ASCII control characters. Three popular asynchronous protocols are Xmodem, Kermit, and MPN. Synchronous protocols, for example, IBM's Bisync and synchronous data link control (SDLC), do not use start and stop bits. The open systems interconnection (OSI) model has been developed to provide a way to define and partition protocols so that data transfer can be truly standardized.

One way to deal with bandwidth limitations is spread spectrum (SS), a technique that distributes a signal and its sidebands over a very wide bandwidth and offers the benefits of improved security, resistance to jamming, efficient band sharing, and resistance to fading. The two basic SS types are frequency hopping and direct sequence.

REVIEW

KEY TERMS

Acknowledge (ACK) character

American Standard Code for Information Interchange (ASCII)

Asynchronous data transmission

Asynchronous protocol

Baseband transmission

Baudot code

Binary check code (BCC)

Binary code

Binary phase-shift keying (BPSK)

Bit error rate (BER)

Block

Block-check character (BCC)

Block-check sequence (BCS)

Broadband transmission

Cyclical redundancy check (CRC)

Differential phase-shift keying (DPSK)

Direct-sequence (DS) SS

8-QAM

End-of-block (ETB) character

End-of-transmission (EOT) character

Extended Binary Coded Decimal Interchange Code (EBCDIC)

Forward error correction (FEC)

Frame (packet) of information

Frequency hopping (FH) SS

Frequency shift keying (FSK)

Hamming code

Handshaking

Hartley's law

High level data link control (HDLC) protocol

International Organization for Standardization (ISO)

Kermit protocol

Longitudinal redundancy check (LRC)

Manchester code

Microcom Networking Protocol (MNP)

Modem

Morse code

Nonreturn to zero (NRZ) encoding

Open Systems Interconnection (OSI) protocol

Parallel data transfer

Parity

PCMCIA bus

Phase-shift keying (PSK)

Protocol

Quadrature (quarternary or quadra) phase PSK (QPSK or 4-PSK)

Quadrature amplitude modulation (QAM)

Redundancy

Return to zero (RZ) encoding

Serial data transfer

Shannon-Hartley theorem

Spread spectrum (SS)

Start of header (SOH) character

Start of text (STX) character

Start-stop bits

Synchronous data link control (SDLC) protocol

Synchronous data transmission

Time-hopping SS

Trellis-code modulation

Universal asynchronous receiver transmitter (UART)

Xmodem protocol

XNOR circuit

QUESTIONS

1. Name the earliest form of binary data communications.
2. What do you call dot-dash code transmission by radio?
3. How do you distinguish between uppercase and lowercase letters in the Morse code?
4. What is the Morse code for the characters C, 7, and ??

5. What 5-bit code is used in teletype systems?
6. What system of data communications uses typewriterlike devices to send and receive coded messages?
7. If the Baudot character 11011 is sent, then 00110, what character is received?
8. What is the most widely used binary data code that uses 7 bits to represent characters?
9. What ASCII character is transmitted to ring a bell?
10. Name the two ways that bits are transmitted from one place to another.
11. What is used to signal the beginning and end of the transmission of a character in asynchronous transmissions?
12. What is the name given to the number of symbols occurring per second in a data transmission?
13. How can there be more than one bit per baud in data transmissions?
14. Which is faster, asynchronous transmission or synchronous transmission? Explain.
15. How is a message sent using synchronous data transmission?
16. In serial data transmission, what are the special names given to a binary 0 and a binary 1?
17. What encoding method is used in most standard digital logic signals?
18. What occurrence in serial data transmission makes it difficult to detect the clock rate when NRZ is used?
19. What two encoding methods are best for clock recovery?
20. What is a benefit of the RZ-AMI method of encoding?
21. Give two names for the encoding system that makes use of a transition at the center of each bit.
22. What is information theory?
23. What two factors does the amount of information that can be sent in a given transmission depend on?
24. True or false. Multilevel or multisymbol binary encoding schemes permit more data to be transmitted in less time assuming a constant symbol interval.
25. For a given bandwidth system, what is the advantage of using a multisymbol encoding scheme?
26. Name the major components and circuits of a modem.
27. What is the function of a modem?
28. What kind of modulation is used in low-speed modems?
29. What is the bandwidth over which modems must transmit?
30. What are the typical originate and answer frequencies used in a full-duplex FSK modem?
31. Why is a scrambler needed in a modem?
32. What is the name of the IC used to perform serial-to-parallel and parallel-to-serial conversions and other operations in a modem?
33. What do you call the modulation method that represents bits as different phase shifts of a carrier?
34. In BPSK, what phase shifts are used to represent binary 0 and binary 1?
35. What basic circuit is used to produce BPSK?
36. What circuit is used to demodulate BPSK?
37. What circuit is used to generate the carrier to be used in demodulating a BPSK signal?
38. What type of PSK does not require a carrier recovery circuit for demodulation?
39. Name the key circuit used in a DPSK modulator.
40. How many different phase shifts are used in QPSK?
41. In QPSK, how many bits are represented by each phase shift?
42. How many bits are represented by each phase shift in 16-PSK?
43. Is carrier recovery required in a QPSK demodulator?
44. When QPSK is used, is the bit rate faster than the baud (symbol) rate?
45. What circuit is used to create a dibit in a QPSK modulator?
46. What do you call a circuit that converts a 2-bit binary code into one of four DC voltage levels?
47. QAM is a combination of what two types of modulation?
48. With QAM, can a 28,800-bps signal be transmitted within a 3000-Hz bandwidth?

49. What is trellis code modulation? Why is it used?
50. What organization establishes and maintains modem standards?
51. What are the two key specifications given by modem standards?
52. State the maximum bit and baud rate and modulation method used by a modem with a V.32bis standard.
53. List five methods of error detection.
54. Name one simple but time-consuming way to ensure an error-free transmission.
55. Name a special self-checking error code used to ensure error-free transmission.
56. What is the most common cause of error in data transmission?
57. What is the name given to the ratio of the number of bit errors to the total number of bits transmitted?
58. What serial bit encoding method makes it possible to detect single bit errors?
59. What is the bit added to a transmitted character to help indicate an error called?
60. What is the name of a number added to the end of a data block to assist in detecting errors and how is it derived?
61. What is another name for parity?
62. What logic circuit is the basic building block of a parity generator circuit?
63. What error-detection system uses a BCC at the end of a data block?
64. Describe the process of generating a CRC.
65. What basic circuit generates a CRC?
66. True or false. If a bit error can be identified, it can be corrected.
67. Describe the procedure for checking the accuracy of a block transmission at the receiver using a CRC.
68. What is the name of one popular type of error correcting code?
69. What will be the outcome of XORing Hamming bits if no error occurs in the data word during transmission?
70. What is the name given to the rules and procedures that describe how data will be transmitted and received?
71. What is the process of exchanging signals between transmitter and receiver to indicate status or availability?
72. What is a popular protocol used with PCs and printers? What control characters are used?
73. Name two popular asynchronous protocols used in PC communications via modem.
74. What do MNP protocols define and where are they used?
75. What is the string of characters making up a message or part of a message to be transmitted called?
76. What do synchronous protocols usually begin with? Why?
77. Name two common synchronous protocols.
78. What appears at the very end of most protocol packets?
79. What does interoperability mean?
80. Name one way to achieve interoperability.
81. Name the seven OSI levels in order from highest to lowest.
82. What are the three or four most used OSI levels?
83. What is spread spectrum?
84. Name the two main types of spread spectrum.
85. What circuit generates the transmitter frequency in a frequency-hopping SS system?
86. In a frequency-hopping SS system, what circuit selects the frequency produced by the synthesizer?
87. True or false. The hop rate is slower than the bit rate of the digital data.
88. How does an SS signal appear to a narrowband receiver?
89. How are two or more stations using spread spectrum and sharing a common band identified and distinguished from one another?

90. What do you call the length of time that a frequency-hopping SS transmitter stays on one frequency?
91. What circuit is used to generate a PSN signal?
92. What is the purpose of a PSN signal?
93. In a direct-sequence SS transmitter, the data signal is mixed with a PSN signal in what kind of circuit?
94. True or false. In a direct-sequence SS system, the chip rate is faster than the data rate.
95. What type of modulation is used with direct-sequence SS?
96. What is the most difficult part of spread spectrum communications?
97. What do you call the process of comparing one signal to another in an effort to obtain a match?
98. Name the two main benefits of spread spectrum.
99. What are the frequencies allocated to spread spectrum by the FCC for civilian applications?
100. Name several important applications for spread spectrum.
101. How are voice signals transmitted via spread spectrum?
102. What is another name for direct-sequence SS?

PROBLEMS

1. What is the name of the 8-bit character code used in IBM systems? ◆
2. A serial pulse train has a 70-μs bit time. What is the data rate in bps?
3. The data rate of a serial bit stream is 14,400 bps. What is the bit interval? ◆
4. The speed of a serial data transmission is 2.5 Mbps. What is the actual number of bits that occur in 1 s?
5. What is the channel capacity, in bits per second, of a 30-kHz-bandwidth binary system, assuming no noise? ◆
6. If an 8-level encoding scheme is used in a 30-kHz-bandwidth system, what is the channel capacity in bits per second?
7. What is the channel capacity, in bits per second, of a 15-MHz channel with an S/N ratio of 28 dB? (Assume a noiseless channel.) ◆
8. What is the minimum allowable bandwidth that can transmit a binary signal with a bit rate of 350 bps?
9. What is the BER if 4 errors occur in the transmission of 500,000 bits? ◆
10. Write the correct parity bits for each number:
    a. odd 1011000 __
    b. even 1011101 __
    c. odd 0111101 __
    d. even 1001110 __
11. Determine the four Hamming bits for the 8-bit number 11010110. Use the format illustrated in the text.

CRITICAL THINKING

1. Describe in detail a simple data communications system that will monitor the temperature of a remote inaccessible location and display temperature on a personal computer.
2. Suggest a future application for spread spectrum techniques and explain why SS would be appropriate for that application.
3. Name three or more common data communications applications that you may use.

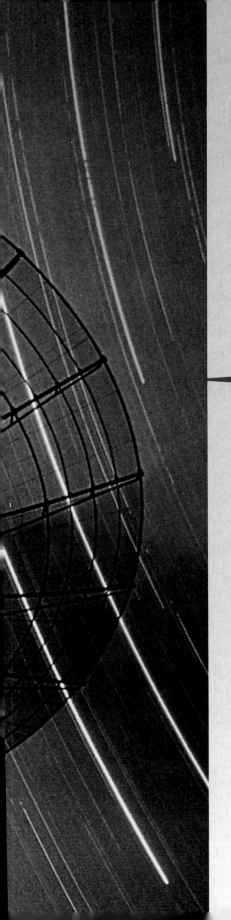

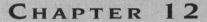

LOCAL AREA NETWORKS

Objectives

After completing this chapter, you will be able to:

- *Define* LAN, *explain* the basic purposes of LANs, and *describe* some specific LAN applications.

- *Draw* diagrams of LANs arranged according to the three basic topologies: star, ring, and bus.

- *Describe* how repeaters, transceivers, hubs, bridges, routers, and gateways are used in LANs.

- *Explain* the basic difference between client-server and peer-to-peer LANs and give the advantages and disadvantages of each.

- *Calculate* transmission speed given the length of the data to be transmitted and the characteristics of the LAN over which the data is to be sent.

- *Compare* Ethernet and Token-Ring transmission media, topologies, configurations, signal-coding methods, and protocols governing transmission, reception, access, and frame format.

Personal computers were originally designed to free individual users from the need to share software and hardware. Each PC has its own CPU, software, and storage capacity, and PC owners purchase applications for their own private use. In many instances, however, it is desirable for one PC user to communicate with another user, or to have access to special equipment, data, or other resources not available to the individual PC. To make this possible, PCs are equipped with a special I/O interface. Each PC becomes, in effect, a transceiver capable of both transmitting and receiving data. PCs (or larger computers) linked for purposes of communication and/or sharing of data and peripherals form a network. This chapter provides an introduction to networks in general and local area networks (LANs) in particular.

12-1 NETWORK FUNDAMENTALS

A *network* is a communications system with two or more stations that can communicate with one another. Chapter 11 discussed how computers communicate with another over the telephone network via modems. Figure 12-1 shows an example of a simple network of this type. Usually the host computer in such a network is a large minicomputer or mainframe with many I/O ports and many modems connected to several telephone trunk lines. The host computer, referred to as a multiuser or timeshared computer, can handle many transmissions between remote PCs via the telephone network. Each of the remote PCs can tap the large database of files and information available on the host.

Computers can also be connected to one another for the exchange of data without a mini or mainframe. For example, two PCs can communicate through the common RS-232 serial interface that all PCs are equipped with. Such an interface and some simple communications software permit two-way communications.

When it is desired to have each computer communicate to two or more additional computers, the interconnections can become complex. As Fig. 12-2 indicates, if four computers are to be interconnected, there must be three links to each PC.

The number of links L required between N PCs (nodes) is determined using the formula

$$L = \frac{N(N-1)}{2}$$

Assume, for example, there are six PCs. The number of links is $L = 6(6-1)/2 = 6(5)/2 - 30/2 = 15$.

The number of links or cables increases in proportion to the number of nodes involved. The type of arrangement shown in Fig. 12-2 is obviously expensive and impractical. Some special type of network wiring must be used, a combination of hard-

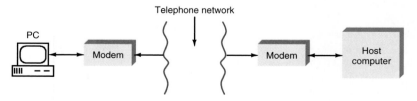

FIG. 12-1 Communicating by way of the telephone network.

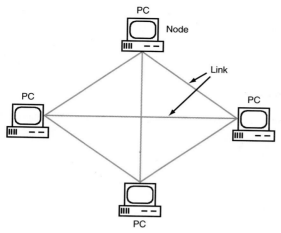

Fig. 12-2 A network of four PCs.

ware and software that permits multiple computers to be connected together inexpensively and simply with the minimum number of links necessary for communication.

Example 12-1

An office with 20 PCs is to be wired so that any computer can communicate with any other. How many interconnecting wires (or links, L) are needed?

$$L = \frac{N(N-1)}{2} = \frac{20(20-1)}{2} = \frac{20(19)}{2} = 190$$

TYPES OF NETWORKS

Each computer or user in a network is referred to as a *node*. The interconnection between the nodes is referred to as the *communications link*. In most computer networks, each node is a personal computer, but in some cases a peripheral device such as a laser printer can be a node. There are three basic types of electronic networks in common use: LANs, wide-area networks (WANs), and metropolitan-area networks (MANs). Let's take a brief look at each.

WIDE-AREA NETWORKS (WANS). A WAN covers a significant geographical area. Local telephone systems are WANs, as are the many long-distance telephone systems linked together across the country and to WANs in other countries. Each telephone set is, in effect, a node in a network that links local offices and central offices. Any node can contact any other node in the system. Telephone systems use twisted-pair wire and cable, as well as microwave relay networks, satellites, and fiber-optic cabling.

There are also WANs that are not part of the public telephone networks, for example, corporate LANs set up to permit independent intercompany communications regardless of where the various subsidiaries and company divisions, sales offices, and manufacturing plants may be. The special communications, command, and control networks set up by the military are also WANs.

METROPOLITAN-AREA NETWORKS (MANs). MANs are smaller networks that generally cover a city, town, or village. Cable TV systems are MANs. The cable TV company receives signals from multiple sources, including local TV stations, as well as special programming from satellites, and assembles all these signals into a single composite signal that is placed on a coaxial cable. The cable is then channeled to each subscriber home. The cable TV channel selector boxes are all nodes in the system. Most existing cable systems are simplex or one-way transmission systems; however, many cable companies are beginning to add two-way communications capability.

A MAN can also be a local telephone company, or a special network setup by a government organization for communicating around a city or county.

LOCAL-AREA NETWORKS (LANs). A LAN is the smallest type of network in general use. It consists primarily of personal computers interconnected within an office or building. LANs can have as few as 3 to 5 users, although most systems connect 10 to 1000 users.

Small LANs can be used by a company to interconnect several offices in the same building; in such cases wiring can be run between different floors of the building to make the connection. Larger LANs can interconnect several buildings within a complex, for example, large companies with multiple buildings, military installations, and college campuses.

Some LANs link mainframes or minicomputers to terminals consisting of a keyboard and video monitor that allow them to communicate with the computer. The terminals themselves are not standalone PCs, but simply input-output devices. Most large minicomputers or mainframes can handle multiple users with many terminals. In addition, PCs can be set up to emulate a terminal. Built-in interface cards or emulator boards plus specialized software allow PCs to be used as standard LAN terminals in addition to performing as standalone computers.

Other LANs consist of multiple PCs that are linked both to each other and to a minicomputer or mainframe. This allows each user on the LAN to have access to the big computer as well as continue to operate independently.

LAN TOPOLOGIES

The *topology* of a network describes the basic communications paths between, and methods used to connect, the nodes on a network. The three most common topologies used in LANs are star, ring, and bus. Of these, the bus and ring configurations are the most common. Ethernet network uses a bus, and the popular Token Ring network uses the ring topology.

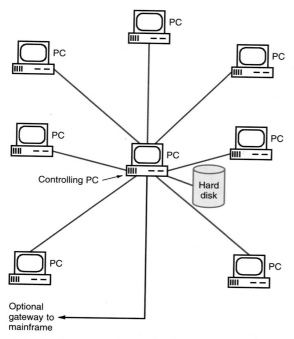

Controlling PC

Hard disk

Optional gateway to mainframe

FIG. 12-3 A star LAN configuration.

STAR TOPOLOGY. A basic *star* configuration consists of a central controller node *server* and multiple individual stations connected to it (Fig. 12-3). The resulting system resembles a multipointed star. The central or controlling PC, often referred to as the *server,* is typically larger and faster than the other PCs, and contains a large hard drive where shared data and programs are stored. Any communications between two PCs passes through the central controller. The server and its software manages the link-up of individual computers and the transfer of data between them.

A star-type LAN is extremely simple and straightforward. New nodes can be quickly and easily added to the system. Further, the failure of one node does not disable the entire system. Of course, if the server node goes down, the network is disabled, although the individual PCs will continue to operate independently. Star networks generally require more cable than other network topologies, and the fact that all communications must pass through the node does place some speed restrictions on the transfer of data.

RING TOPOLOGY. In a ring configuration, one PC in the LAN acts as a server or main control computer, and all the computers are simply linked together in a single closed loop (Fig. 12-4). Usually, data is transferred around the ring in only one direction, passing through each node. Therefore, there is some amplification and regeneration of the data at each node, permitting long transmission distances between nodes.

The ring topology is easily implemented and is low in cost. Expansion is generally simple, since a new node can be inserted in the ring at almost any point. The computer that has a message to send identifies the message with a code or address identifying the destination node, places it on the ring, and sends it to the next computer in the loop. Essentially, each PC in the ring receives and retransmits the message until it reaches the target PC. At that point, the receiving node accepts the message.

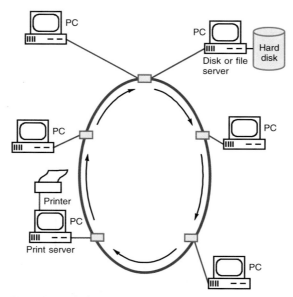

Fig. 12-4   A ring LAN configuration.

The downside of a ring network is that a failure in a single node generally causes the entire network to go down. It is also somewhat difficult to diagnose problems on a ring. Despite these limitations, the ring configuration is widely used.

BUS TOPOLOGY.   A *bus* is simply a common cable to which all of the nodes are attached. Figure 12-5 shows the bus configuration. The bus is bidirectional in that signals can be transmitted in either direction between any two of the nodes; however, only one of the nodes can transmit at a given time. A signal to be transmitted can be destined for a single node, or transmitted or broadcast to all nodes simultaneously.

The primary advantage of the bus is that it is faster than any of the other topologies. The wiring is simple, and the bus can be easily expanded.

OTHER TOPOLOGIES.   There are many variations of the basic topologies discussed above, for example, the so-called *daisy chain* topology, which is simply a ring that has been broken. Another variation is called *tree* topology. In some cases, this topology is simply a bus design in which each node has multiple interconnections to other nodes through a star interconnection. In others, the network consists of branches from one node to two or more other nodes.

LAN APPLICATIONS

The common denominator of all LANs is the communication of information. Nodes are linked together to transmit and receive information, software, databases, personal messages, or anything else that can be put in binary form.

The earliest LANs were developed primarily as a way to control costs. For example, in an office with multiple PCs, the cost of permitting each computer to have its own printer might be prohibitive. Networking is an inexpensive way to link all the computers in the office together to a single computer to which the printer is attached. Users

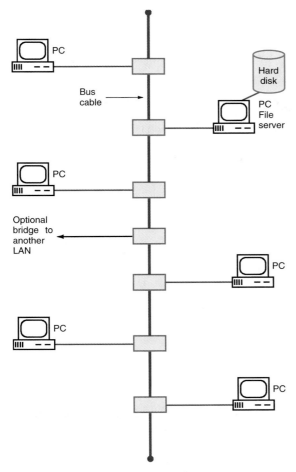

FIG. 12-5 A bus LAN configuration.

share the printer by transmitting information to be printed to the main computer, which controls print operations for all the office computers.

Another example is the sharing of large files and databases. Networks can be used to connect all the PCs in an office to a central computer with a large hard drive containing the data to be shared, so that any computer can access it. Today, networks are used for many applications other than centralizing and sharing expensive peripherals such as printers and for permitting many users access to a large database.

E-MAIL. One of the most popular networks applications is e-mail, the sending and receiving of electronic messages and memos. Users simply key memos on their computers and send them to other users over the network. In many organizations e-mail has all but replaced interoffice paper and telephone messages. In others, e-mail supplements the normal interoffice communications procedures.

Most network software provides a means of simulating mailboxes, which essentially consist of allocated storage locations on the hard disk of the server. All messages are forwarded to the server, which holds them. The server then notifies the recipient that mail is present in the mail box. When the receiving user turns on his or her computer, the e-mail message is given on the screen, at which time that user can access the mail.

When 28-year-old Dennis C. Hayes formed D. C. Hayes Associates, Inc., in 1978, modems were simply translators that had to be opened prior to each communication in order to rather laboriously set a series of switches. In the early days of his venture, Hayes hand-assembled and soldered products on a borrowed dining room table at his home. Then, in 1981 the release of the Hayes™ Smartmodem™ started a communications revolution for personal computer users by providing a system element that was easier to integrate into the computer environment. This device utilized the Hayes Standard AT Command Set, a software command language interface developed to control modem features such as Auto Dial and Auto Answer. Today this command language is used by major modem manufacturers and communications developers. Hayes's company, now known as Hayes Microcomputer Products, Inc., continues to be a leading supplier of computer communications hardware and software products.

GROUPWARE. *Groupware* is a set of programs that facilitate the efforts of two or more individuals working on a single project. Groupware provides ways to share common databases, exchange messages related to a particular group task, manage work flow (e.g., for project management), and maintain a common calendar. The most popular and widely used groupware program is Lotus Notes.

CLIENT-SERVER AND PEER-TO-PEER LANs

Most LANs conform to one of two general configurations: client-server or peer-to-peer. In the *client-server* type, one of the computers in the network, the server, essentially runs the LAN and determines how the system operates. The server, which is the largest and fastest computer in the network, manages printing operations of a central printer and controls access to a very large hard drive or bank of hard drives containing databases, files, and other information which the clients, the other computers on the network, can access.

The server controls clients' access to the printer, the hard drive, the files and databases, other nodes, and various other services. In many client-server systems, the server is a minicomputer such as a DEC VAX or an IBM AS/400. The server can also be a large mainframe. In one client-server variation, a large, fast server PC acts as the link between the client PCs and the minicomputer or mainframe. In client-server networks, the client PCs also have standalone capabilities.

In a *peer-to-peer* system, any PC can serve as either client or server. Any PC on a peer-to-peer network can have access to any other PC's files and connected peripherals. For example, a laser printer or drum plotter connected to any one of the PCs can be used by any of the others on the network.

Peer-to-peer LANs are smaller and less expensive than the client-server variety, and provide a simple way to provide network communications in an office. The total number of interconnected PCs is generally small, usually less than 20 machines, and the system is relatively easy to set up. Peer-to-peer networks also have some disadvantages. They usually have lower performance, meaning lower speed transmission capability. There are also manageability and security problems related to the ability of any user to access any other user's files. Protecting individual users' software and private files from access is more difficult with peer-to-peer networks than it is with client-server systems.

12-2 LAN HARDWARE

All LANs are a combination of hardware and software. The primary hardware devices are the computers themselves and the cables and connectors that link them together. Additional pieces of hardware unique to networks include network interface cards, repeaters, hubs and concentrators, bridges, routers, gateways, and many other special interfacing devices. This section is an overview of the specific types of hardware involved in networking.

CABLES

Most LANs use some type of copper wire cable to carry data from one computer to another via baseband transmission. The digital data stored in the computer is converted to serial binary data, and voltages representing binary 1s and 0s are transmitted directly over the cable from one computer to another. The three basic cable types used in LANs are coaxial cable, twisted pair, and fiber-optic cable. Most local area networks started out using coax cable, but the use of twisted pair has increased over the years. Today the usage is probably equally split between coaxial cable and twisted-pair installations. Fiber-optic cable is used in higher-speed, secure networks which are not as widespread. Less than 5 percent of all LANs use fiber-optic cable.

COAXIAL CABLE AND TWISTED PAIR. *Coaxial cable* is far superior to twisted pair as a communications medium. Its extremely wide bandwidth permits very high-speed bit rates. While loss is generally high, attenuation is usually offset by using repeaters that boost the signal level and regenerate the signal waveshape. The major benefit of coaxial cable is that it is completely shielded, so that external noise has little or no effect on it.

Coaxial cable is shown in Fig. 12-6. It consists of a thin center conductor surrounded by an insulating material that is, in turn, completely encircled by a shield. The shield can be criss-crossed wire braid or solid metal foil. Surrounding the shield is an outer sheath, usually made of PVC.

DID YOU KNOW?

Networks have become a factor in the construction of new office buildings because channels or chambers called *plenums*, through which cables are run between floors or across ceilings, are often included.

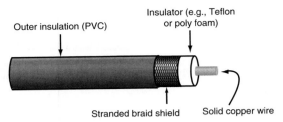

FIG. 12-6   Coaxial cable.

Coaxial cable comes in a wide variety of sizes and shapes, which are usually designated by a letter-number combination such as RG-8/U. Figure 12-7 shows the most common types of coaxial cable used in LANs. As shown in the figure, the different types have specific characteristic impedances, attenuation factors, and amounts of capacitance per foot. Since a coaxial cable is two conductors separated by an insulator, the cable itself acts as a capacitor. The longer the cable, the higher its capacitance. This distributed capacitance can distort and attenuate the signal.

Twisted pair, as the name implies, is simply two insulated copper wires twisted together loosely to form a cable [see Fig. 12-8(*a*)]. Telephone companies use twisted pair to connect individual telephones to the central office. The wire itself is solid copper, 22, 24, or 26 gauge. The insulation is usually PVC. Twisted pair by itself has a characteristic impedance of about 100 Ω, but the actual impedance depends on how tightly or how loosely the cable is twisted and can be anywhere from about 70 to 150 Ω.

There are two basic types of twisted-pair cables in use in LANs: unshielded (*UTP*) [Fig. 12-8(*a*)] and shielded (*STP*) [Fig. 12-8(*b*)]. UTP cables are highly susceptible to noise, particularly over long cable runs. Most twisted pairs are contained within a common cable sheath along with several other twisted pairs, and crosstalk, the coupling between adjacent cables, can also be a problem; again, this is especially true when transmission distance is great.

STP cables, which are more expensive than UTP cables, have a metal foil or braid shield around them forming a third conductor. The shield is usually connected to ground

Designation	Impedance (Ω)	OD (in)	Capacitance/ft (pF)	Attenuation
RG-8/U	52	0.405	29.5	0.68 dB/ft @ 10 MHz
RG-58/U	53.5	0.195	28.5	1.4 dB/ft @ 10 MHz
RG-59/U	75	0.242	21.0	3.8 dB/ft @ 100 MHz
RG-62/U	93	0.242	13.5	0.46 dB/ft @ 2.5 MHz

FIG. 12-7   Common types of coaxial cable used in LANs.

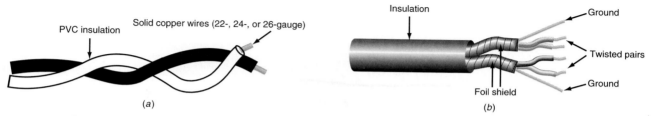

FIG. 12-8 Types of twisted pair. (*a*) Twisted pair, unshielded (UTP). (*b*) Multiple shielded twisted pair (STP).

and, therefore, provides protection from external noise and crosstalk. STP cables are, therefore, routinely used for long cable runs.

STP cables are also typically grouped together, and two, four, or more pairs can be contained within a single cable housing [Fig. 12-8(*b*)]. When multiple lines must be run between computers and connecting devices, multiple cables are the linking medium of choice.

Many newer office buildings are constructed with special vertical channels or chambers, called *plenums,* through which cables are run between floors or across ceilings. Cable used this way, called *plenum cable,* must be made of fireproof material that will not emit toxic fumes if it catches fire. Plenum cable can be either coaxial or twisted pair.

FIBER-OPTIC CABLE. Fiber-optic cable is a nonconducting cable consisting of a glass or plastic center cable surrounded by a plastic cladding encased in a plastic outer sheath (Fig. 12-9). Most fiber-optic cables are extremely thin glass, and they are usually bundled together. Special fiber-optic connectors are required to attach them to the network equipment. Fiber-optic cables are covered in more detail in Chap. 18.

CONNECTORS

All cables used in networks have special terminating connectors that provide a fast and easy way to connect and disconnect the equipment from the cabling and maintain the characteristics of the cable through the connection.

COAXIAL CABLE CONNECTORS. Coaxial cables in networks use two types of connectors, N connectors and BNC connectors, as Fig. 12-10 shows. The N connectors are widely used in RF applications, and BNC connectors are commonly used for attaching test leads to measuring instruments such as oscilloscopes. N connectors are

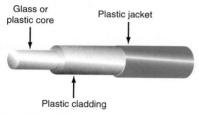

FIG. 12-9 Fiber-optic cable.

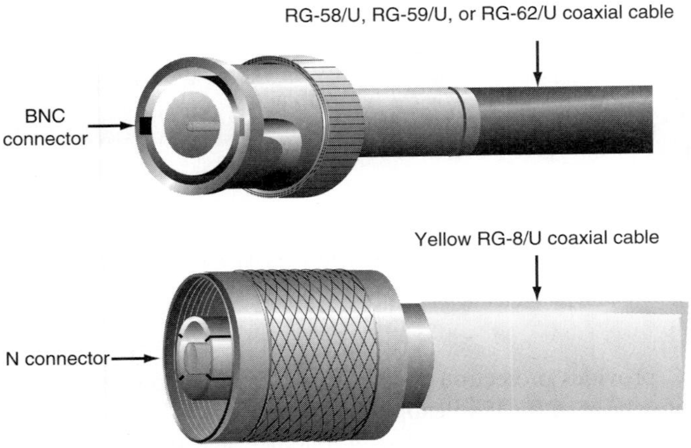

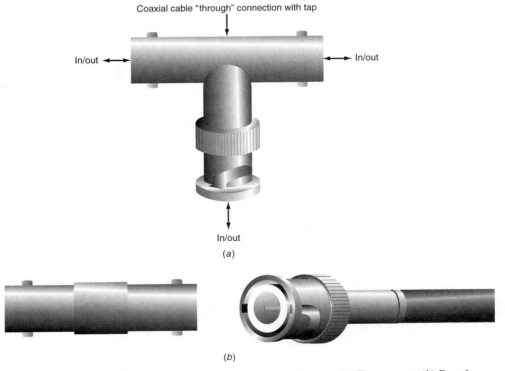

RG-58/U, RG-59/U, or RG-62/U coaxial cable

BNC connector

Yellow RG-8/U coaxial cable

N connector

FIG. 12-10 Common coaxial connectors.

used on larger cables, such as RG-8/U. BNC connectors are used on smaller cables, such as RG-58/U and RG-59/U. The BNC plug attaches to a mating jack with a simple twist of the outer housing; the N connector has continuous threads.

A variety of connector accessories are also available to facilitate connections. One of the most popular is the BNC *T connector*, which is used to interconnect two cables to the network hardware [Fig. 12-11(*a*)]. T connectors provide a convenient way to break into an existing coaxial cable for attaching an additional node.

Coaxial cable "through" connection with tap

In/out

In/out

In/out

(*a*)

(*b*)

FIG. 12-11 BNC connector accessories and adapters. (*a*) T connector. (*b*) Barrel connector.

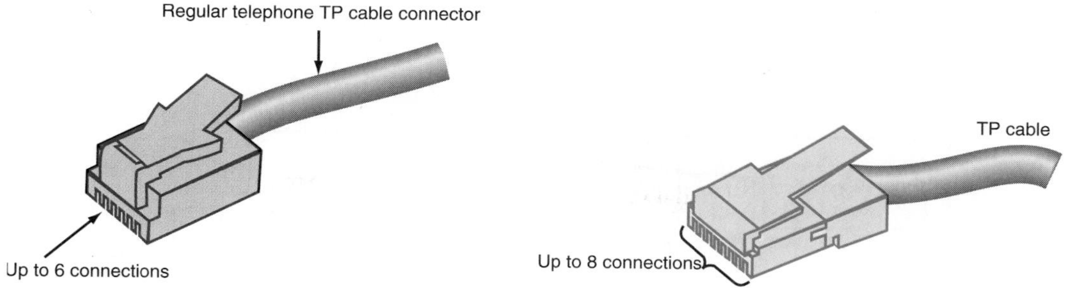

Regular telephone TP cable connector

Up to 6 connections

(a)

TP cable

Up to 8 connections

(b)

Fig. 12-12 Modular (telephone) connectors used with twisted pair. (a) RJ-11. (b) RJ-45.

The *barrel connector* shown in Fig. 12-11(b) also provides a convenient way to connect two coaxial cables. This is usually done to extend a cable's length or to interconnect two existing cables end-to-end.

A *terminator* is a special connector containing a resistor whose value is equal to the characteristic impedance of the coaxial cable. In LANs, the characteristic impedance of the coaxial cable is about 50 Ω. Thus the terminator is a 50-Ω resistor. Providing the correct termination value prevents reflection on the cable.

TWISTED-PAIR CONNECTORS. The connectors used on twisted pairs can be standard telephone jacks or a special larger version. Most telephones attach to an outlet by way of an RJ-11 modular plug [Fig. 12-12(a)]. Standard variations are the RJ-12, RJ-13, and RJ-14. These contain up to six terminals; three twisted pairs can be terminated on a six-terminal plug.

A larger modular connector known as the RJ-45 is also widely used in terminating twisted pairs [see Fig. 12-12(b)]. The RJ-45 contains eight connectors, so it can be used to terminate four twisted pairs. In most cases, one jack is used to terminate only one or two twisted-pair cables. Matching jacks on the equipment or wall outlets are used with these connectors.

FIBER-OPTIC CONNECTORS. A wide range of connectors are available to terminate fiber-optic cables. Like electrical connectors, these are designed to provide a fast and easy way to attach or remove cables.

NETWORK INTERFACE CARDS

Network interface cards (NIC), provide the I/O interface between each node on a network and the network wiring. These cards usually plug into the PC bus and provide connectors at the rear of the computer for attaching the cable connectors (see Fig. 12-13). NICs perform a variety of tasks. For example, when the PC wishes to transmit information over the network, it takes data stored in RAM to be transmitted and converts it into a serial data format. This serial information is usually stored within RAM on the NIC. Logic circuitry in the NIC then groups the information into frames or packets, the format of which is defined by the communications protocol used by the LAN. Once the packet or frame has been formed, the binary data is encoded, normally using the Manchester code, and then sent to a logic-level converter which

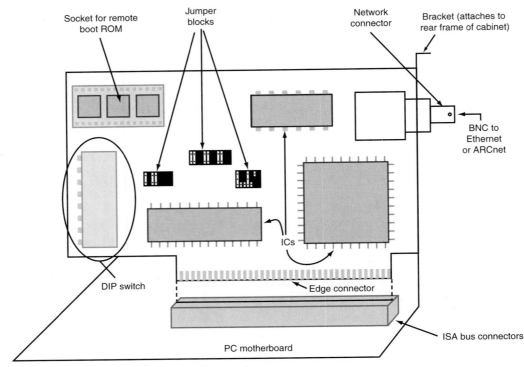

Socket for remote boot ROM

Jumper blocks

Network connector

Bracket (attaches to rear frame of cabinet)

BNC to Ethernet or ARCnet

ICs

DIP switch

Edge connector

ISA bus connectors

PC motherboard

FIG. 12-13 A network interface card.

generates the proper binary 0 and binary 1 voltage levels that are sent over the coaxial cables or twisted pairs.

Upon reception, the destination NIC recognizes when it is being addressed, that is, when data is being sent to it. The NIC performs logic-level conversion and decoding, recovering the serial frame or packet of information; performs housekeeping functions such as error detection and correction; and places the recovered data in a buffer storage memory. The data is then converted from serial to parallel, where it is transferred to the computer RAM and used by the software.

The NIC is the key hardware component in any LAN. It completely defines the protocols and performance characteristics of the LAN. Despite their sophistication, NICs are relatively low in price and available from many manufacturers. Today, because so many PCs are networked, the trend is to build NIC chips into the PC motherboards.

REPEATERS

When signals from an NIC must travel a long distance over coaxial cables or twisted pairs, the binary signal is greatly attenuated by the resistance of the wires and distorted by the capacitance of the cable. In addition, it can pick up noise along the way. As a result, the signal can be too distorted to be received reliably.

A common solution to this problem is to use one or more repeaters along the way (see Fig. 12-14). A *repeater* is an electronic circuit that takes a partially degraded

FIG. 12-14 Concept of a repeater.

signal, boosts its level, shapes it up, and sends it on its way. Over long transmission distances, several repeaters may be required.

Repeaters are small, inexpensive devices that can be inserted into a line with appropriate connectors or built into other LAN equipment. Most repeaters are really transceivers, bidirectional circuits that can both send and receive data. Transceiver repeaters can receive signals from either direction and transmit them in the opposite direction.

HUBS

A *hub* is a LAN accessory that facilitates the interconnections of the cables to the nodes. Whether the network topology used by a LAN is bus, ring, or star, the actual wiring usually resembles a star. This is because the cabling for most networks today is permanently installed in walls, ceilings, and plenums. Bus and ring topologies, where cables logically run between individual PCs, are not convenient for plenum wiring, as they do not provide an easy way to modify the network to add or remove nodes in different parts of the office or building.

The device that facilitates such wiring is the hub, a central connecting box designed to receive the cable inputs from the various PC nodes and connect them to the server (see Fig. 12-15). In most cases, hub wiring physically resembles a star because all the cabling comes into a central point, or hub. However, the hub wiring is such that it can logically implement either the bus or the ring configuration. That is, inside the hub the wiring connects the nodes into a miniature ring or bus.

Hubs can be passive devices, containing no electronic circuits involved in transmission or reception. Such hubs, which contain connectors and possibly switches, can be regarded as network versions of common AC power-line bus strips that allow many AC plugs to be conveniently connected. Some hubs are active devices, containing repeaters.

Many devices used as hubs have special names. For example, the basic hub in certain Token Ring networks is referred to as a *multistation access unit (MAU)*.

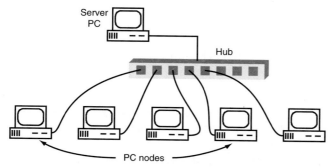

FIG. 12-15 A hub facilitates interconnections to the server.

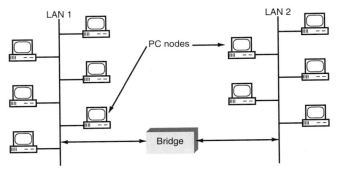

FIG. 12-16 A bridge connects two LANs.

BRIDGES

A *bridge* is a box of electronic equipment that is connected as a node on the network and performs bidirectional communications between two LANs (see Fig. 12-16).

Most companies get into networking by building a small LAN in a department or office. The success of that LAN leads to the establishment of others within the organization. In most cases the company will find it necessary to establish a communications link between the separate LANs, for sharing of information and other resources. This can be accomplished with a bridge.

Another case is when one LAN becomes too big. Most LANs are designed for a maximum upper limit of nodes. The reason for this is that the greater the number of nodes, the longer and more complex the wiring. Further, when many individuals attempt to use a LAN simultaneously, performance deteriorates greatly, leading to network delays. One way to deal with this problem is to break a large LAN into two or more smaller LANs. It is first determined which nodes communicate with other nodes the most, and a logical breakdown into individual LANs is then made. Communication between all users is maintained by interconnecting the separate LANs with bridges. The result of this is improved overall performance.

A bridge is generally designed to interconnect two LANs with the same protocol, for example, two Ethernet networks or two Token-Ring networks. However, there are bridges that are able to accomplish protocol conversion so that two LANs with different protocols can converse.

Remote bridges are special bridges used to connect two LANs that are separated by a long distance. A bridge can use the telephone network to connect LANs in two different parts of the country, or can connect two LANs on a large campus or the grounds of a big military base through a fiber-optic cable.

ROUTERS

Like bridges, routers are designed to connect two networks, generally of the same type and with similar protocols. The main difference between bridges and routers is that routers are intelligent devices with decision-making and switching capabilities.

The basic function of a router is to expedite traffic flow on both networks and maintain maximum performance. When many users are accessing a network at the same time, conflicts occur and speed performance is degraded. Routers are designed to recognize traffic buildup and provide automatic switching to reroute transmissions in a different direction if possible. If transmission is blocked in one direction, the router can switch transmission through other nodes or other paths in the network.

Devices called *brouters* are a combination of a bridge and a router. There are many different types of routers and brouters for the wide variety of networks in use. They can switch, perform protocol conversion, and serve as communications managers between two LANs.

GATEWAYS

A gateway is another internetwork device that acts as an interface between two LANs or between a LAN and a larger computer system. The primary benefit of a gateway is that it can connect networks with incompatible protocols and configurations. The gateway acts as a two-way translator that allows systems of different types to communicate.

Figure 12-17 shows a typical gateway system, one designed to interconnect one or more PC-based LANs to a minicomputer or mainframe. There are many different types of gateways available depending upon the equipment and protocols involved. Most gateways are computers and are sometimes referred to as gateway servers.

PRIVATE BRANCH EXCHANGES (PBXs)

A *PBX* is a telephone switching system dedicated to a single customer (see Fig. 12-18). It provides telephone switching and links inside an organization for local communications but also connects to the standard public telephone network. You will hear private branch exchanges referred to as PABXs (private automatic branch exchanges) and DPBXs (digital private branch exchanges). All switching operations are computer-controlled. PABXs have replaced PBXs controlled by human operators manipulating pushbuttons or patch cords. In the newer DPBXs, all switching is by IC transistor switches that replace mechanical relays in older PBXs and PABXs.

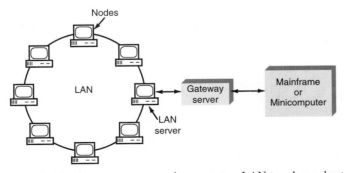

FIG. 12-17 A gateway commonly connects a LAN to a larger host computer.

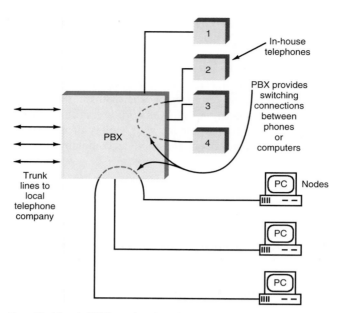

Fig. 12-18   A PBX used to implement a slow star LAN.

PBXs provide all the dialing and switching functions required to interconnect any internal telephone to any other internal telephone within a company, even when hundreds or thousands of phones are involved. Thus full telephone communications is provided in-house without using the external telephone company. The PBX also connects to one or more outside trunk lines that go to the regular telephone system. The number of external trunk lines depends upon how many internal phones are present and the amount of outside calling that is done.

The newer PABXs are extremely small, fast, and efficient. And many have been designed so that they can carry digital signals as well as the analog voice signals of the telephone. Because of this capability, digital automatic private branch exchanges can be used for networking operations. Computers in the network can communicate with one another through the PBX and its standard telephone wiring, so that separate wires and switching systems are not necessary. With appropriate NICs, such systems work efficiently in many applications. The PBX-network configuration is a star topology, with the PBX serving as the central node and switching system. The primary limitation of these systems is that the data transfer rate is limited to less than several hundred thousand bits per second.

MODEMS

As discussed in Chap. 11, modems are interfaces between PCs and standard telephone systems. They convert the binary signals of the computer into audio-frequency analog signals compatible with the telephone system and at the other end, convert the analog signals back into digital signals.

Modems are widely used in networking. The most common application is one in which remote PCs use modems to link up to a central minicomputer or mainframe that

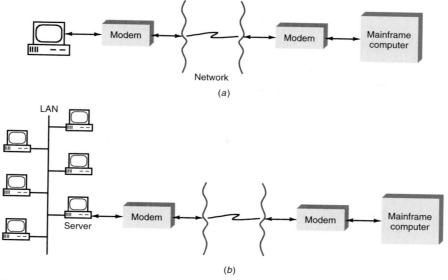

FIG. 12-19 Modems. (*a*) Used with a single PC. (*b*) Used with a LAN.

maintains large databases and other services (see Fig. 12-19). Although modems are not generally used in LANs, they can be. For example, the server or one of the PCs on a network can be connected via modem to a remote computer, providing a link to that system for anyone on the network. Individual modems for each PC node are not needed. When the server in a LAN controls remote modem calls, it is sometimes referred to as a *communications server.*

WIRELESS LANs

One of the most complex and expensive parts of any LAN is the cabling. The cables themselves, as well as their installation and maintenance, are expensive, especially when the LAN is being installed in an existing building. In addition, in large, growing organizations, LAN needs change regularly. New users must be added, and the network must be reconfigured during expansion, reorganizations, and moves. One way to avoid the expense and headache of running and maintaining LAN cabling is to use wireless LANs, which communicate through radio and infrared techniques.

Each PC in a wireless LAN must contain a wireless modem or transceiver. This device can be like an NIC that plugs in the computer, or it can be an external device that communicates through the standard I/O ports on a computer. In either case, the modem transceiver converts the serial binary data from the computer into infrared or radio signals for transmission, and converts the received infrared or radio signals back into binary data. Wireless LANs operate like cable-connected LANs in that any node can communicate with any other node. They do have significant speed limitations: presently, most wireless LANs have a top speed of about 10 Mbps. However, this is fast enough for most applications, including e-mail, sharing printing facilities, and exchanging files and databases.

INFRARED. The simplest and least expensive forms of wireless LANs are created by using infrared transceivers. The infrared portion of the electromagnetic spectrum lies right below that of visible light. Therefore, infrared transmissions are treated more like light transmission than anything else.

The remote control on your TV set is, in effect, a digital infrared communications system. The remote control contains a digital circuit that produces serial binary codes when any one of the buttons on the remote is pressed. The serial digital codes then turn an infrared light beam off and on. The infrared light beam is generated by one or more LEDs. These look like any other standard LED except they have clear housings and the light they emit is infrared and thus not visible to the human eye. An infrared-sensitive diode or transistor in the TV set picks up the infrared signals and converts them into binary voltages that are then recovered. The received codes are translated by the microcontroller in the TV set which, in turn, produces the desired action, either channel changing or volume adjustment.

Infrared LAN transceivers work in the same way. The binary data from the computer to be transmitted is sent to an infrared transmitter, which uses multiple LEDs to transmit the information to all other computers in the network. Each computer in the network has a transceiver which can both send and receive such infrared signals.

The primary limitation of infrared LANs is transmission distance. Transmission distances of over several hundred feet between units are not possible, and, in general, transmission must be line of sight. This means that the transmitting LED must be able to "see" the receiving diode or transistor; there can be no physical barrier between the sending and receiving units. Only through careful positioning of transceivers in an infrared wireless LAN can reliable communications be established and maintained. Wireless infrared transceivers are thus used primarily to interconnect the computers within a single office.

amplitude or phase modulation

RADIO. Radio wireless LANs are far more versatile than infrared wireless LANs. Because radio waves can penetrate the walls and floors of buildings, there is greater flexibility in placing and interconnecting the various computers. Wireless LANs use microwave radio transmitters and receivers to communicate. Because high frequencies are used, the antennas are small and practically unnoticeable. Most of the newer wireless LANs operate in special microwave-frequency bands set aside by the FCC specifically for LANs, and no special licenses for operations are required. However, the output power of the transceivers is strictly limited to prevent long-distance communications. Although wireless communication can be maintained within large buildings, distances greater than several thousand feet are not possible. The microwave signals undergo significant attenuation when attempting to penetrate walls and floors, particularly in large structures using steel-reinforced concrete.

A radio LAN transceiver contains a modem that converts the binary data from the computer into the signals that modulate the microwave carrier. Spread spectrum techniques are used in most wireless transceivers. As discussed earlier, spread spectrum systems translate digital data into signals that modulate a carrier in such a way that the transmitted signal occupies an entire band of frequencies rather than a single-frequency channel. Such methods make the transmission secure and permit dozens or even hundreds of users to transmit over the same band at the same time.

Networks, like any computer application, require both hardware and software to work. The most important network software package is the *network operating system (NOS)*. In addition, there are special software packages available for networks, such as e-mail and groupware.

Network Operating Systems

An operating system is a program that resides on a computer and manages all of its resources, including the programs and files stored on the hard disks or on diskettes. Most operating systems, the most common being Microsoft MS-DOS, also manage   N T other input-output operations, such as those which control the keyboard, the video display, and the I/O ports.

Most networks are of the client-server variety, in which one of the computers is designated the host or server. The network operating system is installed on that server, and establishes and controls communications between the server and all other nodes. In addition, the network operating system manages access to the data, files, and software stored on the file server's hard disks. A request from a PC node for data is accepted by and handled by the network operating system.

In addition to managing the files, the network operating system also provides security in that it can permit or deny access to selected files or to the database for each operating node. The network operating system also handles printer sharing, file backup, and a variety of administrative tasks, such as network configuration.

One of the most widely used network operating systems is Novell NetWare. Once this operating system has been installed on a LAN computer, that computer becomes the server, and cannot function as a standalone PC. The server is a full-time network manager. NetWare is used with many popular networking configurations, including Ethernet, Token Ring, and others.

Another popular network operating systems is Banyan Vines, which is used for computers running the Unix operating system. Microsoft's LAN Manager is another widely used network operating system. Like NetWare, it converts one PC into a server. IBM also has numerous network operating systems, including one which is built into its popular OS/2 operating system.

No special software is generally included in the PCs on a client-server network. The server communicates to the nodes via the NICs, which take advantage of the built-in operating system in each workstation node. The individual PC operating system on each node, however, must be configured so that it recognizes the existence of the network.

In peer-to-peer LANs, special operating system software is installed on each of the PCs in the nodes. Some of the more popular peer-to-peer operating system software are Microsoft Windows for Workgroups (WFW), Novell Netware Lite and Novell Personal Netware, Artisoft LANtastic, and several others. The Apple Macintosh computer operating system, currently System 7, contains built-in software for operating a peer-to-peer network using Apple's LocalTalk hardware. It also works with Ethernet, Token Ring, and other network configurations.

12-4 ETHERNET LANs

One of the oldest and still one of the most widely used of all LAN types is Ethernet. Ethernet, which was developed by Xerox Corporation at Palo Alto Research Center in the 1970s, was based on the Aloha wide-area satellite network implemented at the University of Hawaii in the late 1960s.

In 1980, Xerox joined with Digital Equipment Corporation and Intel to sponsor a joint standard for Ethernet. The collaboration resulted in a definition that became the basis for the IEEE 802.3 standard. [The Institute of Electrical and Electronics Engineers (IEEE) establishes and maintains a wide range of electrical, electronic, and computing standards. The 802.X series relates to LANs.]

TOPOLOGY

Ethernet uses the bus topology. Network nodes simply tap into the bus cable (see Fig. 12-20). The bus may be a large coaxial cable, like RG-8/U, a small coaxial cable, like RG-58/U, or a twisted pair. Information to be transmitted from one user to the other can move in either direction on the bus, but only one node can transmit at any given time. The bus coaxial cable has a special terminating connector at each end containing a resistor whose value is equal to its characteristic impedance. This prevents signal reflections, which cause signal loss and significant data errors. For the RG-8 and RG-58 cables used in Ethernet, this value is 53 Ω.

ENCODING

Ethernet uses baseband data-transmission methods. This means that the serial data to be transmitted is placed directly on the bus media. Before transmission, however, the

Straight raw data — not modulated — all marks & spaces

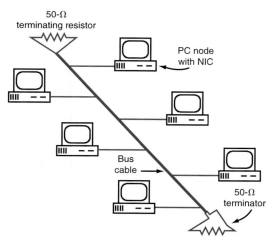

FIG. 12-20   The Ethernet bus.

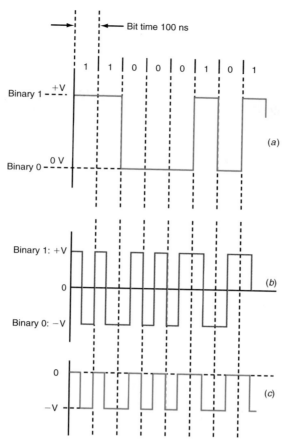

FIG. 12-21 Manchester encoding eliminates DC shift on baseband lines and provides a format from which a clock signal can be extracted. (*a*) Standard DC binary coding (unipolar NRZ). (*b*) Manchester AC encoding. (*c*) DC Manchester.

binary data is encoded into a unique variation of the binary code known as the Manchester code (see Fig. 12-21).

Figure 12-21(*a*) shows normal serial binary data made up of standard DC binary pulses. This serial binary data, which is developed internally by the computer, is known as unipolar NRZ. This data is encoded using the Manchester format by circuits on the NIC. Manchester coding can be AC or DC. In AC Manchester, shown in Fig. 12-21(*b*), the signal voltage is switched between a minus level and a plus level. Each bit, whether it is a binary 0 or binary 1, is transmitted as a combination of a positive pulse followed by a negative pulse or vice versa. For example, a binary 1 is a positive pulse followed by a negative pulse. A binary 0 is a negative pulse followed by a positive pulse. During each bit interval a positive-to-negative or negative-to-positive voltage transition takes place.

This method of encoding prevents the DC voltage level on the transmission cable from building up to an unacceptable level. With standard binary pulses, an average DC voltage will appear on the cable, the value of which depends on the nature of the binary data on the cable, especially the number of binary 1s or 0s occurring in sequence. Large numbers of binary 1 bits will cause the average DC to be very high. Long strings of binary 0s will cause the average voltage to be low. The resulting binary signal rides up and down on the average DC level, which can cause transmission errors and signal

← prevents DC levels from occurring at baseline

interpretation problems. With Manchester encoding, the positive and negative switching for each bit interval causes the direct current to be averaged out.

Ethernet uses DC Manchester [Fig. 12-21(*c*)]. The average DC level is about -1 V, and the maximum swing is about zero to -2 V. The average DC is used to detect the presence of two or more signals transmitting simultaneously.

Another benefit of Manchester encoding is that the transition in the middle of each bit time provides a means of detecting and recovering the clock signal from the transmitted data.

SPEED

The standard transmission speed for Ethernet LANs is 10 Mbps. As discussed in Chap. 9, the time for each bit interval is the reciprocal of the speed, or $1/f$, where f is the transmission speed or frequency. With a 10-Mbps speed, the bit time is $1/10 \times 10^6 = 0.1 \times 10^{-6} = 0.1$ μs (100 ns) [see Fig. 12-21(*a*)].

TRANSMISSION MEDIUM

The standard transmission medium for Ethernet is coaxial cable. However, both coaxial-cable and twisted-pair versions of Ethernet are in operation.

COAXIAL CABLE. The two main types of coaxial cable used in Ethernet networks are RG-8/U and RG-58/U. *RG-8/U* cable has a characteristic impedance of 53 Ω and is approximately 0.4 in. in diameter. This large coaxial cable, referred to as *thick cable,* was originally designed for antenna transmission lines in RF systems but is now widely used in Ethernet LANs. It is usually bright yellow in color. Large type-N coaxial connectors are used to make the interconnections.

Figure 12-22 shows an Ethernet LAN using thick coaxial cable. Transceivers connect the nodes to the bus cable by way of NICs. A repeater extends the bus cable. Note the terminators at the end of the cable.

When thick network cable is used, it does not attach directly to the NICs. Instead, the inputs and outputs from the NIC terminate in an *attachment user interface (AUI).* This 15-pin connector, known as a *DB-15* or *DIX connector,* plugs into a matching connector on the NIC. The other end of the cable attaches to a transceiver unit, which is connected to the bus. The transceiver takes care of amplification and signal shaping in both transmit and receive operations.

Ethernet systems using thick coaxial cable are generally referred to as *10Base-5* systems, where *10* means a 10-Mbps speed, *Base* means baseband operation, and the *5* designates a 500-m maximum distance between nodes, transceivers, or repeaters. Ethernet LANs using thick cable are also referred to as *Thicknet.*

In 10Base-5 Ethernet LANs, the cable is generally one long continuous bus that is routed from node to node. The availability of a wide variety of coaxial connectors permits shorter cables to be interconnected from computer to computer. The cable between the PC node and the transceiver can be up to 164 ft long, making node placement and wiring easy to implement. Depending on the distance between nodes, repeaters may have to be inserted to boost and rejuvenate the signal along the way. The

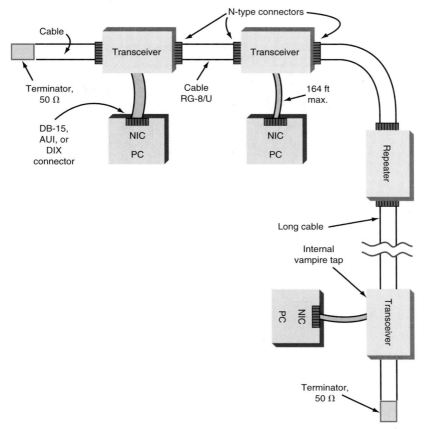

FIG. 12-22 The Ethernet (10Base-5) bus.

bus is relatively easy to reconfigure; adding new nodes is not even difficult. The bus cable can be cut and new connectors installed, allowing the bus to pass through while at the same time connecting to the new NICs. Special connectors referred to as "vampire" taps can pierce the coaxial cable without cutting it.

Ethernet systems implemented with thinner coaxial cable are known as *10Base-2,* or *Thinnet* systems; here the *2* indicates the maximum 200-m (actually, 185 m) run between nodes or repeaters. The most widely used thin cable is RG-58/U. RG-58/U, used for antenna transmission lines for CB radios and cellular telephones as well as for LANs, is slightly less than 0.25 in. in diameter. It is much more flexible and easier to work with than RG-8/U cable.

Figure 12-23 shows a 10Base-2 LAN with RG-58/U. Transceivers are not used. The cable is attached directly to the NIC in each node by way of a BNC T connector, which lets the bus pass through while making the connection to the NIC. Repeaters can be used over long distances. The cable is terminated at each end with resistor terminators.

TWISTED PAIR.　More recent versions of Ethernet use twisted pair. The twisted-pair version of Ethernet is referred to as a *10Base-T network,* where the T stands for twisted pair. The twisted pair used in 10Base-T systems is standard 22-, 24-, or 26-gauge solid copper wire with RJ-45 modular connectors. The PC nodes connect to a hub, as shown

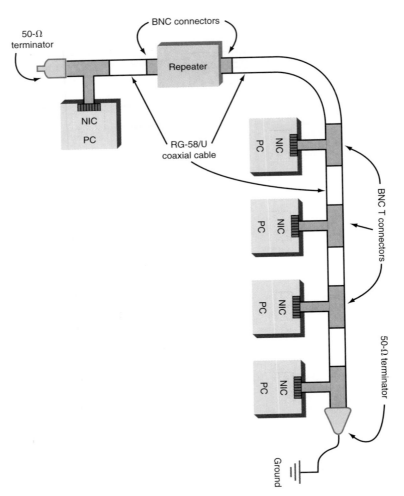

Fig. 12-23 10Base-2 coaxial Ethernet bus.

in Fig. 12-24, which provides a convenient way to connect all nodes. Physically, a 10-Base-T LAN looks like a star, but the bus is implemented inside the hub itself. It is usually easier and cheaper to install 10Base-T LANs than it is to install coaxial Ethernet systems, but the transmission distances are usually more limited.

COMPARING ETHERNET VERSIONS. The choice of medium for an Ethernet LAN depends upon the size of the LAN (i.e., the number of nodes), the distance between nodes, and the total length of the bus. In general, the longer the bus and the greater the distance between nodes, the thicker the cable should be. Thick coaxial cable has less attenuation than thinner coaxial cable, and thin coaxial cable generally has less attenuation and distortion than twisted pair.

When thick coaxial cable is used, the maximum total length of an Ethernet bus is 2286 m (2500 yards) and the longest distance that can be run between nodes or repeaters is 500 m. When thin cable is used, the maximum distance between nodes or repeaters is 185 m. The total length of a thin coaxial Ethernet bus is 925 m. With twisted pairs, the maximum distance between nodes or repeaters is 100 m; the total permissible length with repeaters is 2500 m. For all Ethernet systems, a maximum of 1024 nodes is suggested.

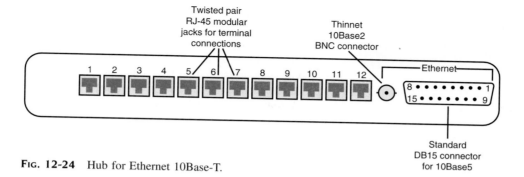

FIG. 12-24 Hub for Ethernet 10Base-T.

Because noise and crosstalk can be problems when twisted pair is used, twisted-pair Ethernet is usually restricted to smaller networks in less noisy environments.

100-MBPS ETHERNET

The most recent development in Ethernet technology is a modified version that permits a data rate of 100 Mbps, 10 times the normal Ethernet rate. Two versions have been developed, 100Base-T and 100VG-AnyLAN. Both use twisted pair and have the same access method (see below) and packet size. The 100VG-AnyLAN version has the IEEE standard number 802.12. The trend for all new designs is to adopt 100-Mbps Ethernet. Many NICs support both 10- and 100-Mbps rates.

FIBER-OPTIC INTERREPEATER LINKS (FOIRLs)

FOIRLs are fiber-optic communication channels with repeaters at each end that are designed to interconnect two Ethernet networks. If two networks that must work together are far apart, or if some of the nodes in a network exceed distance limits imposed by the network configuration and hardware, FOIRLs may be the answer.

FOIRL repeaters are attached to an Ethernet bus, coaxial cable, or twisted pair, just as a regular node would be attached, a fiber-optic cable links the FOIRL to a remote repeater in the other network on another node. FOIRLs can be 500 to 1000 m apart depending upon the exact configuration of the network. A FOIRL cable is actually two cables, one for transmitting and the other for receiving. Normally, the repeaters at each end of the cable attach to an Ethernet hub where all the other nodes terminate.

ACCESS METHOD

Access method refers to the protocol used for transmitting and receiving information on a bus. Ethernet uses an access method known as *carrier sense multiple access with collision detection* (CSMA/CD), and the IEEE 802.3 standard is primarily devoted to a description of CSMA/CD.

Whenever one of the nodes on an Ethernet system has data to transmit to another node, the software sends the data to the NIC in the PC. The NIC builds a *packet,* a unit of data formed with the information to be transmitted. The completed packet is stored in RAM on the NIC while the sending node monitors the bus. Since all the nodes or PCs in a network monitor activity on the bus, whenever a PC is transmitting information on the bus, all the PCs detect (sense) what is known as the carrier. The carrier in this case is simply the Ethernet data being transmitted at a 10-Mbps (or 100-Mbps) rate. If a carrier is present, none of the nodes will attempt to transmit.

When the bus is free, the sending station initiates transmission. The transmitting node sends one complete packet of information. In a sense, the transmitting node "broadcasts" the data on the bus so that any of the nodes can receive it. However, the packet contains a specific binary address defining the destination or receiving node. When the packet is received, it is decoded by the NIC and the recovered data is stored in the computer's RAM.

Although a node will not transmit if a carrier is sensed, at times two or more nodes may attempt to transmit at the same time or almost the same time. When this happens, a *collision* occurs. If the stations that are attempting to transmit sense the presence of more than one carrier on the bus, both will terminate transmission. The CSMA/CD algorithm calls for the sending stations to wait a brief interval of time, then attempt to transmit again. The waiting interval is determined randomly, making it statistically unlikely that both will attempt retransmission at the same time. Once a transmitting node gains control of the bus, no other station will attempt to transmit until the transmission is complete. This method is called a *contention system* because the nodes compete, or *contend,* for use of the bus.

In Ethernet LANs with few nodes and little activity, gaining access to the bus is not a problem. However, the greater the number of nodes and the heavier the traffic, the greater the number of message transmissions and potential collisions. The contention process takes time; when many nodes attempt to use a bus simultaneously, delays in transmission are inevitable. Although the initial delay might be only tens or hundreds of microseconds, delay times increase when there are many active users. Delay would become an unsurmountable problem in busy networks were it not for the packet system, which allows users to transmit in short bursts. The contention process is worked out completely by the logic in the NICs.

PACKET PROTOCOLS

Figure 12-25(*a*) shows the packet (frame) protocol for the original Ethernet system, and Figure 12-25(*b*) shows the packet protocol defined by IEEE standard 802.3. The 802.3 protocol is described here.

The packet is made up of two basic parts: (1) the frame, which contains the data plus addressing and error detection codes, and (2) an additional 8 bytes (64 bits) at the beginning, which contain the preamble and the *start frame delimiter (SFD).* The preamble consists of 7 bytes of data that help to establish clock synchronization within the NIC, and the SFD announces the beginning of the packet itself.

The *destination address* is a 6-byte, 48-bit code that designates the receiving node. This is followed by a 6-byte source address that identifies the sending node. Next is a 2-byte field that specifies how many bytes will be sent in the data field. Finally, the

Ethernet (original)

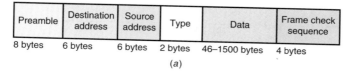

Preamble	Destination address	Source address	Type	Data	Frame check sequence
8 bytes	6 bytes	6 bytes	2 bytes	46–1500 bytes	4 bytes

(a)

IEEE 802.3

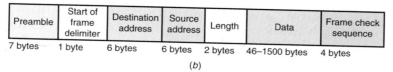

Preamble	Start of frame delimiter	Destination address	Source address	Length	Data	Frame check sequence
7 bytes	1 byte	6 bytes	6 bytes	2 bytes	46–1500 bytes	4 bytes

(b)

FIG. 12-25 Ethernet frame formats.

data itself is transmitted. Any integer number of bytes in the range from 46 to 1500 bytes can be sent in one packet. Longer messages are divided up into as many separate packets as required to send the data.

Finally, the packet and frame end in a 4-byte frame check sequence generated by putting the entire transmitted data block through a cyclical redundancy check (CRC). The resulting 32-bit word is a unique number designating the exact combination of bits used in the data block. At the receiving end, the NIC again generates the CRC from the data block. If the received CRC is the same as the transmitted CRC, no transmission error has occurred. Should a transmission error occur, the software at the receiving end will be notified.

It is important to point out that data transmission in Ethernet systems is synchronous: the bytes of data are transmitted end to end without start and stop bits. This speeds up data transmission, but puts the burden of sorting the data out on the receiving equipment. The clocking signals to be used by the digital circuits at the receiving end are derived from the transmitted data itself to ensure proper synchronization and counting of bits, bytes, fields, and frames.

Example 12-2

How fast can a 1500-byte block of data be transmitted on (a) a 10-Mbps Ethernet (IEEE 803.2) packet and (b) a 16-Mbps Token-Ring packet? [Note: For the Ethernet IEEE 803.2 format, use Fig. 12-25(b); for the Token-Ring format, use Fig. 12-29(a).]

a. Time for 1 bit $= 1/10 \times 10^6 = 100$ ns
   Time for 1 byte $= 8 \times 100 = 800$ ns
   Time for 1526 bytes $= 1526 \times 800$ ns $= 1220.800$ ns
   (1.2208 ms)

b. Time for 1 bit $= 1/16 \times 10^6 = 62.5$ ns
   Time for 1 byte $= 8 \times 62.5 = 500$ ns
   Time for 1521 bytes $= 1521 \times 500 = 760,500$ ns (0.7605 ms)

12-5 TOKEN-RING LAN

Another widely used LAN configuration is the Token Ring, originally developed by IBM. IBM offered several low-performance LANs in the early 1980s, right after the introduction of the original PC in August 1981, none of which ever became popular. In 1982, IBM presented a new LAN architecture and in the subsequent years, developed it fully into a commercial product line. IBM's Token-Ring LAN was announced in 1985; since then, the Token-Ring configuration has been established as a standard. Its IEEE standard number is 802.5.

TOPOLOGY

In the Token-Ring configuration, all of the nodes or PCs in the network are connected end to end in a continuous circle or loop (see Fig. 12-26). The data in the network travels only in one direction on the ring. The transmitted information passes through the NICs of each PC in the loop. Like Ethernet and most other standard networks, Token Ring uses baseband transmission; the binary data is placed directly on the cable. When the ring is not transmitting data, a 3-byte token code, the access key, is passed continuously around the ring.

ENCODING

A modified version of the Manchester coding scheme is used.

SPEED

The original Token Ring introduced by IBM ran at a speed of 4 Mbps, which is still fast enough to provide excellent communications in small to medium-sized LANs. In 1989 IBM announced and began shipping a 16-Mbps Token-Ring system. It accommodates more nodes at higher speeds and thus provides better performance for large, active networks. Today both the 4- and 16-Mbps versions are in use. Many NICs contain circuitry that can accommodate either speed.

TRANSMISSION MEDIUM

Token-Ring LANs use twisted pair. In general, the smaller, lower-speed systems are unshielded twisted pair and the larger, faster systems use shielded twisted pair. However, either type of cable can be used with either system. Connections are usually made using RJ-45 modular connectors.

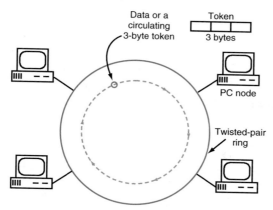

Fig. 12-26 A Token-Ring LAN.

Because the data transmitted around the ring must pass through the NIC of all PCs in the loop, two twisted-pair cables are required for the connection at each node (see Fig. 12-27). One twisted pair, the *ring in (RI)*, comes into the card; another twisted pair, the *ring out (RO)*, carries the data out to the next node. Both twisted pairs are contained within a single cable and attach to the NIC by way of a DB-25 or DB-9 connector. The cable on this connector is usually terminated with the RJ-45 modular connector, which plugs into a wall jack that attaches to the ring cabling.

It is extremely difficult to run twisted pair around an office area or building in a single continuous loop as required by the ring topology. To make wiring simpler, two twisted pairs in a single cable, as described earlier, are used. These cables terminate at a wiring hub referred to as a *multistation access unit (MAU)*. The input and output of each NIC of each node appears at the MAU. The wiring inside the MAU connects the PCs in a logical ring, as shown in Fig. 12-28. The cable from each PC contains two twisted pairs, one for the RI connections and the other for the RO, which terminate at the MAU. The RI and RO connections are made inside the MAU, creating a daisy-chain ring.

Each MAU hub is designed to accommodate eight nodes. In LANs with more nodes, multiple MAUs can be interconnected to form a larger ring. When MAUs are

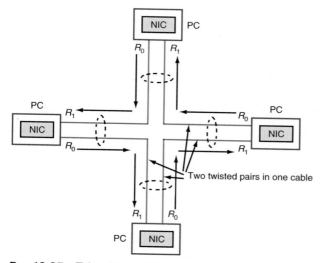

Fig. 12-27 Token-Ring wiring.

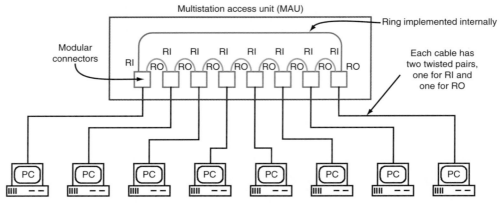

FIG. 12-28 Wiring of the Token Ring through the MAU.

used as wiring hubs, the Token-Ring LAN takes on the physical configuration of a star, despite the fact that it is wired as a ring.

The MAU also contains circuitry that detects when a PC node is ON. When a PC is ON, the NIC sends a signal to the MAU. Inside the MAU, a relay is activated that connects the PC into the ring. If the PC node is OFF, the MAU does not get a signal. Thus the relay in the MAU shorts the connections to the PC and closes the loop for all of the PCs that are turned on.

Because each node is, in effect, a repeater, there are no specific distance limitations on wiring between nodes or on total ring length. The signals coming in are read and repeated at each node; thus no dedicated repeaters are required. Despite this, there are some general guidelines about wiring distances. IBM, for example, recommends a cable length between nodes and the MAU less than 50 m (150 ft). Other companies have used cable lengths between node and MAU of 100 to 150 m (350 to 500 ft). The maximum practical distance between hubs and MAUs seems to be 150 m (500 ft).

IBM also recommends that the number of nodes for the 4-Mbps Token Ring be limited to 72, and that the number of nodes for the 16-Mbps system not exceed 250.

ACCESS METHOD

The access method used by Token-Ring systems is *token passing*. A *token* is a unique binary word passed from one node to another around the ring. When no message is being sent, the token simply rotates around the ring from node to node. Whenever a node desires to transmit information, it captures the token. The NIC builds a packet or frame of information and transmits the data on the ring. This information passes through the NIC of each node on the ring. The receiving station designated by an address in the transmitted information recognizes its address and captures the transmitted data. After the packet of information has been received, it continues on around the ring until it comes back to the transmitting node, which takes the data off the ring. (Messages are not permitted to make the loop more than once.) The sending station then releases the token, which begins to circulate again.

The token-passing method of access is completely different from CSMA/CD. There is no contention. The station desiring to transmit captures the token, then transmits. Once it has sent one packet of data, the token passes to the next station in sequence.

If the next station does not have data to transmit, the token continues to pass on around the ring until a station wanting to transmit captures the token and sends its message.

PACKET PROTOCOL

Figure 12-29 shows the packet protocol for a Token-Ring LAN. The token frame format [Fig. 12-29(a)] consists of 3 bytes—a starting delimiter (SD), an access control (AC) byte, and an ending delimiter (ED). Each byte contains unique patterns of bits that the NIC recognizes. The actual token is one bit in the AC byte.

A typical complete data frame is shown in Fig. 12-29(b). It begins with the SD and AC bytes, followed by a frame control (FC) byte, a 6-byte destination address, and then a 6-byte source address. Some frames permit a variable-length block containing routing information if the LAN calls for it.

The data (message information) is transmitted next in a synchronous format. The block of data is limited to the number of bytes that can be transmitted in the maximum length of time allotted to the node that has the token, or 10 ms, a relatively long time within the context of data transmission rates of 4 or 16 Mbps.

The data is followed by a 4-byte (32-bit) CRC frame check sequence that is generated to catch transmission errors. The packet ends with the ending delimiter and frame status bytes. If the message is too long to be accommodated in one packet, the NIC assembles additional frames to be transmitted sequentially until all of the data is sent. Remember, if another node captures the token between packets, there will be a wait before the next packet can be sent.

12-6 OTHER NETWORK TYPES

While Ethernet and Token-Ring configurations dominate the world of LANs today, many other special networks and technologies are used. The following section discusses some of the more popular and widely used of these, including ARCnet, LocalTalk, MAP, FDDI, and broadband.

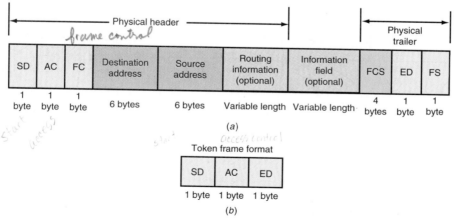

FIG. 12-29 Token-Ring packet format. (a) Token. (b) Data frame.

The oldest commercial networking product line is ARCnet, which was developed by the Datapoint Corporation in the 1970s for use with its own proprietary line of computers and released commercially in 1977. Although Ethernet and Token Ring are now more popular than ARCnet, many organizations prefer the ARCnet technology because it is inexpensive and extremely simple to install and use. For certain applications, it is the best LAN choice.

The basic configuration of an ARCnet LAN is a token-passing bus. This unique format is fully documented in IEEE standard 802.4. The individual PCs can simply be connected together along the bus, but a more typical arrangement is shown in Fig. 12-30. Two or more PCs are connected to each hub, and the hubs themselves are connected along the single bus. This might be more accurately described as a bus interconnection of stars. Each hub is the center of a star and each hub attaches to the bus. Most hubs are active, meaning that they contain repeaters. However, passive hubs like the one in Fig. 12-30 can be used to connect three to four nodes over very short distances. The hub wiring ensures that all nodes appear on the logical bus.

The cabling used in ARCnet LANs is RG-62A/U. This cable has a characteristic impedance of 93 Ω, offers very low loss, and is small, flexible, and inexpensive. BNC connectors are used for connections. The cable run between two active hubs cannot exceed 2000 ft (610 m), and the total length of the bus must not exceed 20,000 ft (6000 m). ARCnet systems are limited to 256 nodes. The longest permissible cable run between a passive hub and one of the PC nodes is 100 ft (30 m).

Until recently the speed of data transmission on the ARCnet system was 2.5 Mbps. However, the newest version of ARCnet transmits at 20 Mbps.

The token passing in ARCnet is similar to that in Token-Ring LANs. All the stations or nodes on the bus have an address of 0 to 255. The token is circulated from PC to PC continuously in address numerical order until a station wishes to transmit. Numerical order, rather than physical location, determines sequence.

When a PC wishes to transmit information, it takes the token and makes the data transfer. Data transfer is by way of a 512-byte packet made up of a destination address, source address, and up to 508 bytes of data. The packet is placed on the bus and transferred from PC to PC with each NIC unit repeating the transmission, as in a

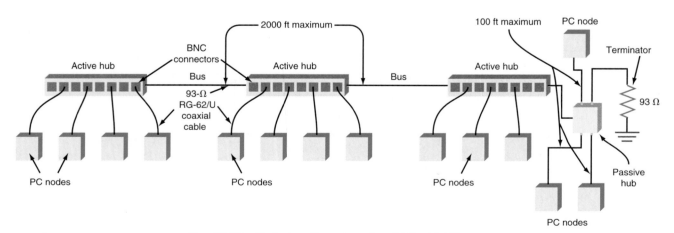

FIG. 12-30 Typical configuration of an ARCnet LAN.

Token Ring. When the receiving computer recognizes its address, it extracts the information; the information continues moving along the sequence until it comes back to the transmitting computer, which removes it and again places the token on the bus to continue to circulate. ARCnet components can be used with almost any of the available network operating systems.

MACINTOSH LANs

LocalTalk is a peer-to-peer LAN implemented by the software in Apple's popular operating system. This simple, inexpensive LAN capability is built into all Macintosh computers and thus requires no special NIC or software. The current version is System 7. Using LocalTalk, any Macintosh computer can communicate with any other Macintosh computer on the LocalTalk network and share printers and other shared peripherals.

The basic topology of LocalTalk is a bus; it is implemented with shielded twisted pair. The signals conform to EIA Standard RS-422, which specifies a balanced line to minimize noise pickup. The maximum data rate is 230.4 kbps, the maximum length of the bus is 1000 ft (300 m), and the maximum number of nodes is 32. The access method is CSMA/CD.

Special LocalTalk adapter units are used to connect to the local bus. These are simply coupling transformers that plug into a serial output port on the back of the Macintosh computer, as shown in Fig. 12-31. The transformers provide low-cost isolation for each node and at the same time permit the 230.4-kbps data rate. The LocalTalk adapters are simply a small box containing the transfer and two connectors which allow daisy chaining the units to form the bus.

The LocalTalk bus is slow and limited to a few stations, but it is widely used in the Apple Macintosh community. Many small LANs are implemented in offices using this bus. All the hardware and software, except for the LocalTalk adapters, is built in to every Macintosh computer. Only the wiring and the adapter modules are needed to interconnect computers and initiate LAN operation.

> ### DID YOU KNOW?
>
> A simple, peer-to-peer LAN called *LocalTalk* is built into all Macintosh computers.

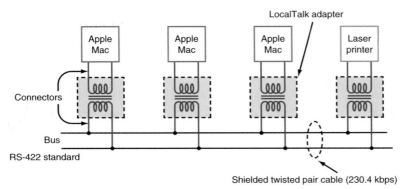

FIG. 12-31 The Apple LocalTalk LAN.

Fiber Digital Data Interface (FDDI)

FDDI is a high-speed fiber-optic cable network offering a data transmission rate of 100 Mbps. While not yet a formal standard, FDDI is well defined and its use is growing.

In FDDI systems, the digital data is transmitted into high-speed light pulses by laser diodes. The wiring consists of two fiber-optic cables that are bundled together [see Fig. 12-32(a)]. The basic FDDI topology is a ring, and the access method is token passing. NICs installed in each computer contain fiber-optic transmitters and receivers that repeat the data transmission. Only one of the fiber-optic cables is used at any given time, and data circulates from node to node around the ring.

The other fiber-optic cable is primarily a reserve path that is used if the main ring fails. The second ring has its own set of fiber-optic transmitters and receivers, but the data travels in the opposite direction from the primary ring. Built-in automatic circuitry senses when one ring fails and causes immediate switching to the other. As Fig. 12-32 shows, if a break occurs, a loopback is implemented inside the node to provide a continuous path for the data using both cables. The availability of two rings provides redundancy and extremely high reliability.

FDDI offers many advantages. The primary benefit of FDDI is speed. Its high-speed transmission rate allows more users to access the network and transmit high volumes of data with little or no loss of network performance. A second major benefit of FDDI is security; it is not possible to tap into or monitor information on an FDDI ring. Fiber-optic cables are also completely immune to electrical noise. When a network must be implemented over long distances in noisy environments, FDDI is an excellent choice. FDDI rings can be extremely large; use of repeaters allows the fiber-optic cables to carry data reliably over very long distances. Finally, FDDI components, including NICs, repeaters, cables, and connectors, are readily available from a number of vendors, and FDDI LANs can be used with a variety of network operating systems.

The downside of FDDI is its high cost. The cost of a fiber-optic network is considerably higher than that of networks using coaxial cable or twisted pair. However,

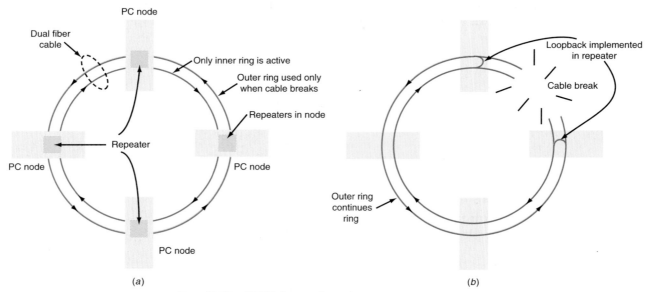

(a)   (b)

FIG. 12-32 FDDI ring configuration.

despite the fact that the newer Ethernet and 100VG-Any-LAN systems offer 100-Mbps data transmission rates and are cheaper than FDDI, FDDI may be the system of choice when security and long-distance reliability are primary considerations.

One special application for FDDI is its use as a backbone LAN for supporting or interconnecting two or more other LANs. A backbone network is a central network whose nodes are other LANs (see Fig. 12-33). The nodes in the individual LANs communicate with one another, but also have the capability of communicating with nodes in other LANs. The FDDI backbone LAN, rather than bridges or routers, serves as the intermediate connection link. Data from one LAN is placed on the backbone and circulated until it is picked off by another LAN containing a node that is the message destination.

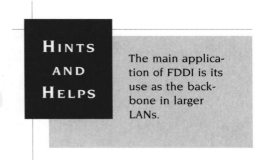

HINTS AND HELPS

The main application of FDDI is its use as the backbone in larger LANs.

Backbone LANs are becoming more prevalent as the use of LANs continues to grow. It provides a convenient way to interconnect a variety of different LANs for organizationwide communications.

Some LAN vendors also implement a standard wire cable version of the FDDI system known as the copper digital data interface (CDDI). CDDI has all of the basic features of FDDI, including a 100-Mbps data rate, except that twisted pair is used instead of fiber-optic cable.

CDDI systems are considerably cheaper than FDDI, but do not offer comparable security or noise immunity. CDDI is an option for short-distance applications in which cost is a major consideration.

BROADBAND LANs

All the LANs discussed in the previous sections use baseband transmission. There are also broadband LANs, in which the binary data to be transmitted is used to modulate a carrier, which is then placed on the transmission medium. Using many different carrier frequencies permits many different digital signals to be transmitted simultaneously.

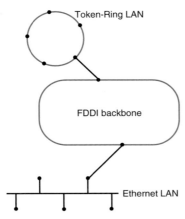

FIG. 12-33 An FDDI backbone.

In other words, broadband modulation techniques permit the multiplexing of many digital data sources on a single cable. Frequency division multiplexing is used.

The hundreds of cable TV systems around the country are broadband networks. In most cases, these cable systems are MANs. The cable TV station collects video signals from a variety of sources. For example, it uses TV receivers connected to antennas to pick up signals from local TV stations. The audio and video signals from those stations are recovered by demodulation in the receiver and used to modulate new carriers. The carrier frequencies used can be the same as those used by the local TV stations or different. The cable station also picks up and demodulates a variety of signals from satellites containing premium movie channels and many cable networks. These signals are used to modulate new carriers on different frequencies. All the carriers are then simply added together and placed on a common coaxial cable. Because the frequencies are different, these signals do not interfere with one another. The jumbled composite signal on the cable is easily unscrambled using filters and demodulation techniques.

The basic concept described above is, basically, the same as that used in broadband LANs. However, instead of transmitting video and audio signals, digital data from computers is used. High-speed digital data is used to modulate the carriers with FSK or PSK techniques. Most broadband network channels are 6 MHz wide, allowing very high-speed digital data to be reliably transmitted. Even higher speeds can be used by adopting many of the high-speed principles used in conventional analog modems.

Figure 12-34 is a simplified block diagram of a complete broadband LAN. The computer nodes contain NICs that convert the computer data into packets for transmission and read and interpret the received packets. The NICs contain or are connected to special broadband modems. The digital data modulates a carrier which is then placed on the cable. Simultaneously, a receiver with a filter and demodulator allows the NIC to receive data on another channel.

The coaxial cable in broadband networks usually has a 75-Ω characteristic impedance. Depending on the size of the network and the transmission distances involved, this cable can be large (RG-6/U or RG-11/U) or small (the familiar RG-59/U used in cable TV installations). The greater the number of stations and the greater the length of the cable, the larger the cable used. The smaller RG-59/U cable is more common, and either BNC or TV-type F connectors are used for interconnections.

Most broadband LANs make use of low-cost, widely available standard TV cable. However, special NICs must be used. The very wide bandwidth of coaxial cable, up to 500 MHz, means that the cable can carry as many as 80 6-MHz-wide broadband channels. Figure 12-35 shows the spectrum of a broadband LAN; the carrier frequencies,

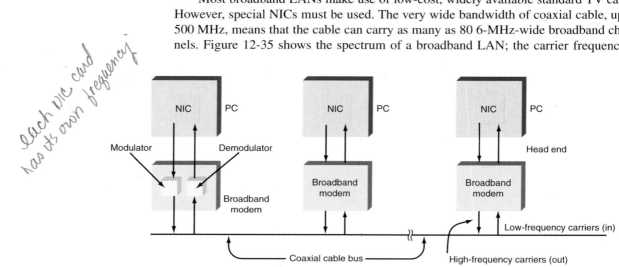

each nic card has its own frequency

FIG. 12-34 A broadband LAN.

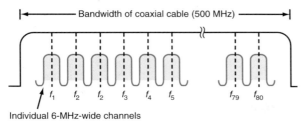

Bandwidth of coaxial cable (500 MHz)

f_1 f_2 f_2 f_3 f_4 f_5 f_{79} f_{80}

Individual 6-MHz-wide channels

FIG. 12-35 The spectrum of a broadband LAN.

labeled f_1 to f_{80}, are centered on 6-MHz-wide parts of the bandwidth. Each PC and NIC modulates a different carrier frequency. A filter at the receiving node selects the frequency of the sending node, rejecting all other signals. The demodulator in the modem recovers the data for the NIC. All data transmission take place through a single PC set up to serve as what is called a *head end*. The head end performs all frequency translations between the sending and receiving nodes. For example, the sending PC may transmit on 55 MHz. The signal goes to the head end, where it is demultiplexed and demodulated. The recovered serial baseband signal is used to modulate a new carrier at, say, 220 MHz, and this signal is sent to the receiving PC.

The primary benefit of a broadband LAN is that it can carry many digital data signals simultaneously. It is not necessary to use the access methods used by other LANs. True full-duplex communications can be carried out by many users at the same time. Another benefit of broadband LANs is that they can be set up to carry both audio and video signals as well as digital data. The tricky part of implementing broadband LAN systems is selecting receiver filters that will receive data from a specific transmitter or multiple transmitters.

MANUFACTURING AUTOMATION PROTOCOL. *Manufacturing automation protocol (MAP)* defines a special LAN designed for manufacturing companies, factories, and plants. In fully automated factories computer-controlled machines perform the manufacturing jobs. The controlling computers are sometimes linked in a LAN to form a completely integrated manufacturing system. In turn, the integrated factory system may be part of an even larger network that includes all design, accounting, inventory, and other company functions. A mainframe or minicomputer is often the primary host computer for such a system, which is then connected by a backbone LAN to several other LANs. MAP is often used in the factory part of the larger network.

Factories are electrically noisy environments in which most standard LANs would not perform well. Shielded twisted pair and coaxial cable provide some noise protection, but MAP goes one step further by using a broadband modulation method that is virtually insensitive to noise.

Office LAN systems permit office staffs to use a variety of software, to access inventory and database systems, to directly access telephone lines for distant communications, and to share peripheral equipment such as laser printers.

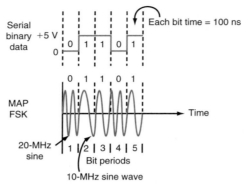

Fig. 12-36 MAP LAN signal.

A MAP LAN uses the bus topology and a token-passing access method. Coaxial cable is the medium of choice. The serial binary data frequency-modulates a sine-wave carrier. The modulation is a type of FSK referred to as *phase-coherent frequency modulation* (see Fig. 12-36). If the serial data is a binary 1, the carrier is a single cycle consisting of a 10-MHz sine wave. If the data is a binary 0, then two cycles, both consisting of a 20-MHz sine wave, are transmitted. The modulation is said to be *phase-coherent*, since the sine waves are continuous, the 10- and 20-MHz sine waves starting and stopping at the zero baseline without discontinuities. With this arrangement, the data rate is 10 Mbps.

A variation of the MAP system uses a 5-Mbps data rate. Here a binary 1 is a 5-MHz sine-wave cycle and a binary 0 is two 10-MHz sine-wave cycles.

MAP systems provide effective noise immunity. As discussed in earlier chapters, noise is caused by amplitude variations, usually high-voltage spikes that occur randomly and add to the binary signal being transmitted. In a baseband system, the noise spikes can be misinterpreted as binary 0s or 1s. With an FM system, the signal is passed through a clipper limiter circuit that strips off all noise spikes. The only variable of interest is the frequency variation, which tells us if a 0 or 1 has been transmitted.

IBM PC NETWORK. Another example of a broadband LAN is one of IBM's original LAN systems, PC Network (Fig. 12-37). PC Network features a 2-Mbps data rate, CSMA/CD access method, and tree topology. The cable is 75-Ω coaxial. The PC nodes are all connected to a translator unit through various couplers and splitters. The translator performs all the modulation, demodulation, and multiplexing operations.

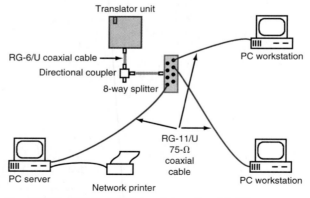

Fig. 12-37 IBM's PC Network broadband LAN.

SUMMARY

Networks are formed by computers linked by a communications medium for purposes of information exchange and sharing of data and peripherals. The three basic types of electronic network in common use are wide area networks, metropolitan area networks, and local area networks (LANs). LANs are digital data communications systems designed for computers located within a restricted geographical area.

There are three basic LAN topologies, or ways in which the links in the network are interconnected: the star, the ring, and the bus. Most LANs have either a client-server or a peer-to-peer configuration. In client-server systems, the largest and fastest computer in the network, the server, runs the LAN; client access to shared data is controlled by the server. In a peer-to-peer system, any PC can serve as either client or server, and files on any PC in the network can be accessed by any of the other PCs.

LANs are usually connected by cables: coaxial cable, twisted pair, and fiber-optic cable are the usual media. Specialized LAN hardware includes network interface cards (the special I/O interface between the computer and the network wiring), repeaters, hubs, concentrators, bridges, routers, gateways, and PBXs. Some LANs are wireless, for example, those using infrared transceivers and those using microwave radio transmitters and receivers to communicate.

Different LAN systems use different network operating systems, transmission media, topologies, configurations, signal-coding methods, and protocols governing transmission, reception, access, and frame format. Speed of transmission and such parameters as maximum number of nodes in a network and distance between nodes vary widely.

The two most widely used baseband systems are Ethernet and token ring. Ethernet uses the CSMA/CD access method; token ring uses the token-passing access method. Other popular baseband systems are ARCnet, LocalTalk, fiber digital data interface (FDDI), and manufacturing automation protocol. FDDI can be used as a backbone LAN for supporting or interconnecting two or more other LANs. Cable TV systems are broadband LANs.

KEY TERMS

10Base-2 Ethernet LAN
10Base-5 Ethernet LAN
10Base-T Ethernet LAN
Access method
ARCnet
Automatic user interface (AUI)
Backbone LAN
Barrel connector
Baseband LAN
BNC connector
Bridge
Broadband LAN
Brouter
Bus topology

Carrier sense multiple access with collision detection (CSMA/CD)
Client-server configuration
Coaxial cable
Coaxial connector
Copper digital data interface (CDDI)
Cyclical redundancy check (CRC)
Daisy chain topology
Digital private branch exchange (DPBX)
E-mail
Encoding

Ethernet LAN
Fiber digital data interface (FDDI)
Fiber-optic cable
Fiber-optic connector
Fiber-optic interrepeater link (FOIRL)
Frame
Gateway
Groupware
Hub
Infrared LAN
Institute of Electrical and Electronics Engineers (IEEE)

REVIEW

LAN Manager
Local area network (LAN)
LocalTalk
Manchester code
Manufacturing automation
 protocol (MAP)
Metropolitan area network
 (MAN)
Modem
Multistation access unit
 (MAU)
N connector
Network interface card
 (NIC)
Network operating system
 (NOS)
Novell NetWare

Packet
PC Network
Peer-to-peer configuration
Plenum cable
Preamble
Private automatic branch
 exchange (PABX)
Private branch exchange
 (PBX)
Radio LAN
Remote bridge
Repeater
RG-8/U coaxial cable
RG-58/U coaxial cable
Ring topology
RJ-11 connector
RJ-45 connector

Router
Shielded twisted pair
 (STP)
Star topology
Start frame delimiter (SFD)
T connector
Terminator
Token passing
Token Ring
Topology
Tree topology
Twisted pair
Unipolar NRZ
Unshielded twisted pair
 (UTP)
Wide area network (WAN)
Wireless LAN

QUESTIONS

1. What is the main purpose of a LAN?
2. Name an example of a WAN other than the telephone system.
3. What is a common example of a MAN?
4. What is a typical number of PCs in a LAN used by a corporation?
5. What is the upper limit on the number of users on a LAN?
6. What is the name given to each PC in a network?
7. Name the three common LAN topologies.
8. What two topologies are the most popular?
9. What is the name given to the main controlling PC in a LAN?
10. Briefly describe how a client-server LAN functions.
11. What is a peer-to-peer LAN?
12. What is probably the most common application on a LAN?
13. What is groupware?
14. What is the main advantage of coaxial cable over twisted pair?
15. Name the two types of twisted pair.
16. What is one of the primary specifications of coaxial cable?
17. What type of connector is used with (a) large coaxial cable and (b) small coaxial cable?
18. Name the two types of connectors used with twisted pair.
19. What is the controller board used to connect a PC to a network?
20. What accessory is added to an Ethernet network if the distance between nodes is long or if the overall cable length is long and the signal is overly attenuated or distorted?

21. What piece of equipment is used to connect two LANs using the same formats or protocols?
22. Define PBX and explain how it can be used to create a LAN.
23. How is a modem used in a LAN?
24. What are the two technologies used to implement wireless LANs? Which is the most flexible and useful?
25. What is the most important piece of software used in a LAN and where does it reside?
26. Name the basic topology of Ethernet.
27. What is the name of the line encoding method used with Ethernet and why is it used?
28. Name the three types of cables used with Ethernet and tell what their designations are.
29. What is the access method used by Ethernet called? (Give the full name and the abbreviation.)
30. Explain briefly how a station gains access to the LAN when Ethernet is used.
31. What is the maximum length of data that can be transmitted in one Ethernet packet?
32. Describe the function of a FOIRL.
33. What topology does a Token-Ring LAN physically resemble?
34. What is the speed of the Token Ring?
35. What is the bit time for the data on a Token Ring operating at its fastest speed?
36. What is the maximum size of the data block in a Token-Ring frame?
37. What method of error detection is used in Token Ring?
38. What types of connectors are used in Token Ring?
39. Briefly describe the configuration and basic specifications of an ARCnet LAN.
40. Give the configuration access method and speed of the Apple Macintosh network.
41. What does FDDI stand for?
42. Describe the FDDI cable.
43. Name one primary application for FDDI.
44. Define broadband operation.
45. What is the transmission medium of a broadband network and what is its bandwidth?
46. What is the primary advantage of a broadband LAN?
47. Where is the MAP network used?
48. Describe the technical characteristics of a MAP.
49. List the basic features of one PC LAN using broadband techniques.
50. What is the main control PC (server) in a broadband LAN called?

PROBLEMS

1. What are the basic data transmission rate of Ethernet and its bit rate interval?
2. Currently, what is the fastest Ethernet speed?

CRITICAL THINKING

1. Networks are usually thought of in terms of general-purpose PCs in a LAN. However, other types of devices and computers are networked. Give one example.
2. Other than speed of transmission, what three key factors influence how fast two nodes in a LAN can communicate?
3. How many bytes can be transmitted in the 10 ms allowed by a LAN at the fastest Token-Ring speed?

TRANSMISSION LINES

Objectives

After completing this chapter, you will be able to:

◆ *Name* the different types of parallel-wire and coaxial-cable transmission lines and *list* some specific applications in which each is used.

◆ *Explain* the circumstances under which transmission lines can be used as tuned circuits and reactive components.

◆ *Define* characteristic impedance and *calculate* the characteristic impedance of a transmission line using several different methods.

◆ *Compute* the length of a transmission-line segment using wavelengths and fractions of wavelengths.

◆ *Define* standing wave ratio (SWR), *explain* its significance for transmission-line design, and *calculate* SWR using impedance values or the reflection coefficient.

◆ *State* the criterion for a perfectly matched line and describe conditions that produce an improperly matched line.

◆ *Use* the Smith chart to make transmission-line calculations.

Transmission lines in communications carry telephone signals, computer data in LANs, TV signals in cable TV systems, and signals from a transmitter to an antenna or from an antenna to a receiver. Transmission lines are critical links in any communications system. They are more than pieces of wire or cable. Their electrical characteristics are critical, and must be matched to the equipment for successful communications. Transmission lines are also circuits. At very high frequencies where wavelengths are short, transmission lines act like resonant circuits and even reactive components. At VHF, UHF, and microwave frequencies, most tuned circuits and filters are implemented with transmission lines. This chapter gives basic transmission-line principles—theory, behavior, and applications.

13-1 TRANSMISSION-LINE BASICS

The two primary requirements of a transmission line are (1) the line should introduce minimum attenuation and distortion to the signal, and (2) the line should not radiate any of the signal as radio energy. All transmission lines and connectors are designed with these requirements in mind.

TYPES OF TRANSMISSION LINES

PARALLEL-WIRE LINES. Parallel-wire line is made of two parallel conductors separated by a space of one-half to several inches. Figure 13-1(a) shows a two-wire balanced line in which insulating spacers have been used to keep the wires separated. Such lines are rarely used today. A variation of parallel line is the 300-Ω twin-lead type shown in Fig. 13-1(b), where the spacing between the wires is maintained by a continuous plastic insulator. This type of line is still used in some TV and FM radio antenna installations.

COAXIAL CABLE. The most widely used type of transmission line is coaxial cable, which consists of a solid center conductor surrounded by a dielectric material, usually a plastic insulator such as Teflon [see Fig. 13-1(c). An air or gas dielectric, where the center conductor is held in place by periodic insulating spacers, can also be used. Over the insulator is a second conductor, a tubular braid or shield made of fine wires. An outer plastic sheath protects and insulates the braid. Coaxial cable comes in a variety of sizes, from approximately one-quarter inch to several inches in diameter.

BALANCED VERSUS UNBALANCED LINES

Transmission lines can be balanced or unbalanced. A *balanced line* is one where neither wire is connected to ground. Instead, the signal on each wire is referenced to ground. The same current flows in each wire with respect to ground, although the

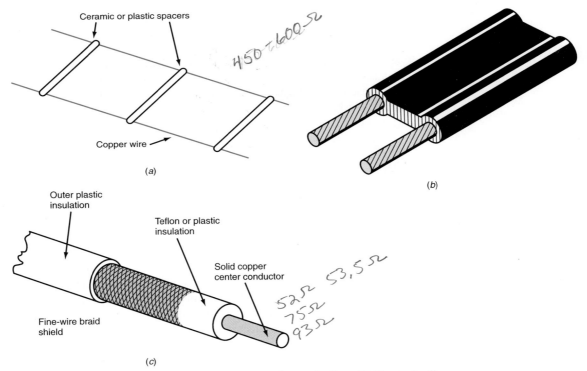

Ceramic or plastic spacers

450 ~ 600 Ω

Copper wire

(a)

(b)

Outer plastic
insulation

Teflon or plastic
insulation

Solid copper
center conductor

52 Ω 53.5 Ω
75 Ω
93 Ω

Fine-wire braid
shield

(c)

FIG. 13-1 Common types of transmission lines. (a) Open-wire line. (b) Open-wire line called 300-Ω twin lead. (c) Coaxial cable.

direction of current in one wire is 180° out of phase with the current in the other wire. In an *unbalanced line,* one conductor is connected to ground.

Most open-wire line has a balanced configuration. A typical feed arrangement is shown in Fig. 13-2(a). The driving generator and the receiving circuit are center-tapped transformers in which the center taps are grounded. Balanced-line wires offer significant protection from noise pickup and crosstalk. Because of the identical polarities of the signals on balanced lines, any external signal inducted into the cable appears on

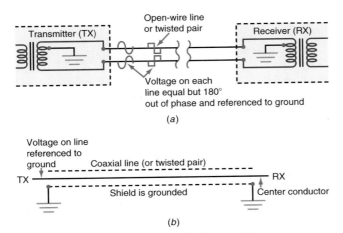

Open-wire line
or twisted pair

Transmitter (TX)   Receiver (RX)

Voltage on each
line equal but 180°
out of phase and referenced to ground

(a)

Voltage on line
referenced to
ground   Coaxial line (or twisted pair)

TX   RX

Shield is grounded   Center conductor

(b)

FIG. 13-2 (a) Balanced line. (b) Unbalanced line.

both wires simultaneously but cancels at the receiver. This is called common mode rejection, and noise reduction can be as great as 60 to 70 dB.

Figure 13-2(*b*) shows an unbalanced line. Coaxial cables are unbalanced lines; the current in the center conductor is referenced to the braid, which is connected to ground. Coaxial cable and shielded twisted pair provide significant but not complete protection from noise pickup or crosstalk from inductive or capacitive coupling due to external signals. Unshielded lines may pick up signals and crosstalk and can even radiate energy, resulting in an undesirable loss of signal.

It is sometimes necessary or desirable to convert from balanced to unbalanced operation or vice versa. This is done with a device called a *balun*, from "*bal*anced-*un*balanced."

WAVELENGTH OF CABLES

The two-wire cables that carry 60-Hz power-line signals into homes are transmission lines as are the wires connecting the audio output of stereo receivers to stereo speakers. At these low frequencies, the transmission line simply acts as a carrier of the AC voltage. For these applications, the only characteristic of the cable of interest is resistive loss. The size and electrical characteristics of low-frequency lines can vary widely without affecting performance. An exception is conductor size, which determines current-carrying capability and voltage drop over long distances. Their electrical length is typically short compared to 1 wavelength of the frequency they carry. A pair of current-carrying conductors is not considered to be a transmission line unless it is at least 0.1 λ long at the signal frequency.

Cables used to carry RF energy are not simply resistive conductors but are complex interconnections of inductors and capacitors as well as resistors. Furthermore, whenever the length of a transmission line is the same order of magnitude or greater than the wavelength of the transmitted signal, the line takes on special characteristics which requires a more complex analysis.

As discussed earlier, wavelength is the length or distance of one cycle of an AC wave or the distance that an AC wave travels in the time required for one cycle of that signal. Mathematically, wavelength is the ratio of the speed of light to the frequency of the signal: $\lambda = 300,000,000/f$, where 300 million is the speed of light, in meters per second, in free space or air (300,000,000 m/s $\approx$ 186,400 mi/s) and f is in hertz. This is also the speed of a radio signal.

The wavelength of a 60-Hz power-line signal is, then,

$$\lambda = 300,000,000/60 = 5 \times 10^6, \text{ m}$$

That's an incredibly long distance—several thousand miles. Practical transmission-line distances at such frequencies are, of course, far smaller. At radio frequencies, however, say 3 MHz or more, the wavelength becomes considerably shorter. The wavelength at 3 MHz is $\lambda = 300,000,000/3,000,000 = 100$ m, a distance of a little less than 300 ft, the length of a football field. That is a very practical distance. As frequency gets higher, wavelength gets shorter. At higher frequencies, the wavelength formula is simplified to $\lambda = 300/f$, where frequency is in megahertz. A 50-MHz signal has a wavelength of 6 m. Using feet instead of meters, the wavelength formula becomes $\lambda = 984/f$, where f is in megahertz (λ is now expressed in feet).

If the wavelength is known, frequency can be computed as follows:

$$f = 300/\lambda \quad \text{or} \quad f = 984/\lambda$$

meters [handwritten, pointing to $f=300/\lambda$]   *feet* [handwritten, pointing to $f=984/\lambda$]   *in megahertz* [handwritten]

The distance represented by a wavelength in a given cable depends on the type of cable. The speed in a cable can be anywhere from 0.5 to 0.95 of the speed of light waves (radio waves) in space, and the signal wavelength in a cable will be proportionally less than the wavelength of that signal in space. Thus the calculated length of cables is shorter than wavelengths in free space. This is discussed later.

Example 13-1

For an operating frequency of 450 MHz, what length of a pair of conductors is considered to be a transmission line? (A pair of conductors does not act like a transmission line unless it is at least 0.1 λ long.)

$$\lambda = \frac{984}{450} = 2.19 \text{ ft}$$
$$0.1\,\lambda = 2.19(0.1) = 0.219 \text{ ft (2.628 in)}$$

Example 13-2

Calculate the physical length of the transmission line in Example 13-1 at 3/8 λ long.

$$\frac{3}{8}\,\lambda = \frac{2.19(3)}{8} = 0.82 \text{ ft (9.84 in)}$$

CONNECTORS

In microwave range — a resistor has capacitance [handwritten note in margin]

Most transmission lines terminate in some kind of *connector,* a device that connects the cable to a piece of equipment or to another cable. An ordinary AC power plug and outlet are basic types of connectors. Special connectors are used with parallel lines and coaxial cable. Connectors, ubiquitous in communications equipment, are often taken for granted. This is unfortunate, because they are a common failure point in many applications.

COAXIAL-CABLE CONNECTORS. Coax cable requires special connectors that will maintain the characteristics of the cable. Although the inner conductor and shield braid could theoretically be secured with screws as parallel lines, the result would be a drastic change in electrical attributes, resulting in signal attenuation, distortion and other problems. Thus coax connectors are designed not only to provide a convenient way to attach and disconnect equipment and cables but also to maintain the physical integrity and electrical properties of the cable.

The choice of a coaxial connector depends on the type and size of cable, the frequency of operation, and the application. The most common types are the PL-259 or UHF, BNC, F, SMA, and N-type connectors.

The PL-259 connector is shown in Fig. 13-3(*a*); the internal construction and connection principles for the PL-259 are shown in Fig. 13-3(*b*). The body of the connector is designed to fit around the end of a coaxial cable and to provide convenient ways to attach the shield braid and the inner conductor. The inner conductor is soldered to a male pin that is insulated from the body of the connector which is soldered or crimped to the braid. A coupler fits over the body; it has inner threads that permit the connector to attach to matching screw threads on a connector called the SO-239, which is mounted on a female chassis.

The PL-259, which is also referred to as a *UHF connector,* can be used up to low UHF frequencies (less than 500 MHz), although it is more widely used at HF and VHF frequencies. It can accommodate both large (up to 0.5 in) and small (0.25 in) coaxial cable.

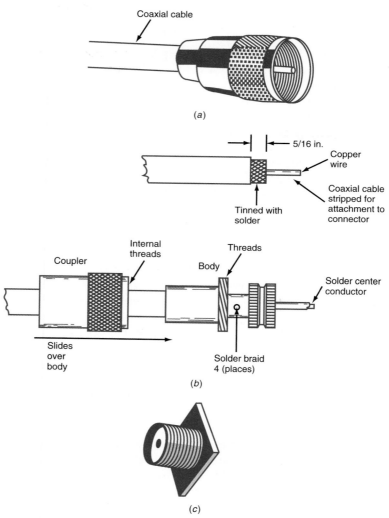

FIG. 13-3 UHF connectors. (*a*) PL-259 male connector. (*b*) Internal construction and connections for the PL-259. (*c*) SO-239 female chassis connector.

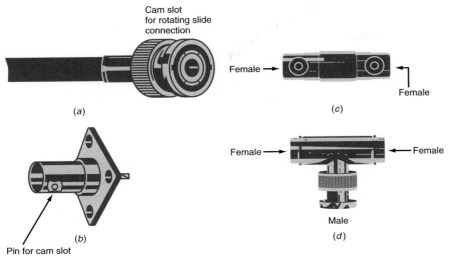

Cam slot
for rotating slide
connection

Female → ← Female

(a)

(c)

Female → ← Female

Male

(b)

(d)

Pin for cam slot

FIG. 13-4 BNC connectors. (*a*) Male. (*b*) Female. (*c*) Barrel connector. (*d*) T connector.

Another very popular connector is the BNC connector (Fig. 13-4). BNC connectors are widely used on 0.25-in coaxial cables for attaching test instruments, such as oscilloscopes, frequency counters, spectrum analyzers, and many others, to the equipment being tested. BNC connectors are also widely used on 0.25-in coaxial cables in LANs and some UHF radios.

In BNC connectors the center conductor of the cable is soldered or crimped to a male pin and the shield braid is attached to the body of the connector. An outer shell or coupler rotates and physically attaches the connector to a mating female connector by way of a pin and cam channel on the rotating coupling [see Fig. 13-4(*b*)].

One of the many variations of BNC connectors is the *barrel connector,* which allows two cables to be attached to one another end to end, and the T coupler, which permits taps on cables [see Fig. 13-4(*a*) and (*d*)]. Another variation is the SMA connector, which uses screw threads instead of the cam slot and pin (Fig. 13-5). The SMA connector is characterized by the hexagonal shape of the body of the male connector. Like the BNC connector, it is used with smaller coaxial cable.

The least expensive coaxial cable connector is the F-type connector, which is widely used for TV sets, VCRs, and cable TV. The cable plug and its matching chassis jack are shown in Fig. 13-6. The shield of the coaxial cable is crimped to the connector, and the solid wire center conductor of the cable, rather than a separate pin, is used as the

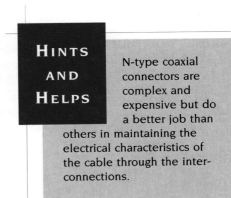

HINTS AND HELPS

N-type coaxial connectors are complex and expensive but do a better job than others in maintaining the electrical characteristics of the cable through the interconnections.

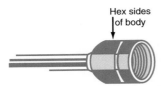

Hex sides of body

FIG. 13-5 SMA connector.

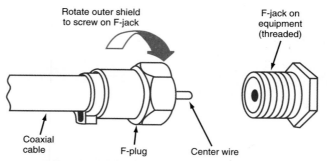

FIG. 13-6 The F connector used on TV sets, VCRs, and cable TV boxes.

connection. A hex-shaped outer ring is threaded to attach the plug to the mating jack.

Another inexpensive coaxial connector is the well-known RCA phonograph connector (Fig. 13-7), which is used primarily in audio equipment. Originally designed over 60 years ago to connect phonograph pick-up arms from turntables to amplifiers, these versatile and low-cost devices can be used at radio frequencies and have been used for TV set connections in the low VHF range.

The best-performing coaxial connector is the N-type connector (Fig. 13-8), which is used mainly on large coaxial cable at the higher frequencies, both UHF and microwave. N-type connectors are complex and expensive, but do a better job than other connectors in maintaining the electrical characteristics of the cable through the interconnections.

CHARACTERISTIC IMPEDANCE

When the length of a transmission line is longer than several wavelengths at the signal frequency, the two parallel conductors of the transmission line appear as a complex impedance. The wires exhibit considerable series inductance whose reactance is significant at high frequencies. In series with this inductance is the resistance of the wire or braid making up the conductors, which includes inherent ohmic resistance plus any resistance due to skin effect. Further, the parallel conductors form a distributed capacitance with the insulation, which acts as the dielectric. In addition, there is a shunt

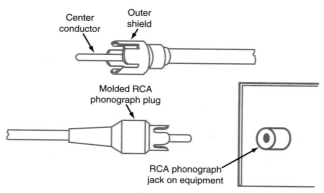

FIG. 13-7 RCA phonograph connectors are sometimes used for RF connectors up to VHF.

FIG. 13-8 N-type coaxial connector.

or leakage resistance or conductance (G) across the cable as the result of imperfections in the insulation between the conductors. The result is that to a high-frequency signal, the transmission line appears as a distributed low-pass filter consisting of series inductors and resistors and shunt capacitors and resistors [Fig. 13-9(a)]. This is called a *lumped* model of a distributed line.

In the simplified equivalent circuit in Fig. 13-9(b) the inductance, resistance, and capacitance have been combined into larger equivalent lumps. The shunt leakage resistance is very high and has negligible effect, so it is ignored. In short segments of the line, the series resistance of the conductors can sometimes be ignored because it is so low as to be insignificant. Over longer lengths, however, this resistance is responsible for considerable signal attenuation. The effects of the inductance and capacitance are considerable, and in fact determine the characteristics of the line.

An RF generator connected to such a transmission line sees an impedance that is a function of the inductance, resistance, and capacitance in the circuit—the *characteristic* or *surge impedance* Z_0. If we assume the length of the line is infinite, this impedance is resistive. The characteristic impedance is also purely resistive for a finite length of line if a resistive load equal to the characteristic impedance is connected to the end of the line.

DETERMINING Z_0 FROM INDUCTANCE AND CAPACITANCE. For an infinitely long transmission line, the characteristic impedance Z_0 is given by the formula $Z_0 = \sqrt{L/C}$, where Z_0 is in ohms, L is the inductance of the transmission line for a given length, and C is the capacitance for that same length. The formula is valid even for finite lengths if the transmission line is terminated with a resistor equal to

[handwritten margin notes: higher freq decreases capacitance increases inductance]

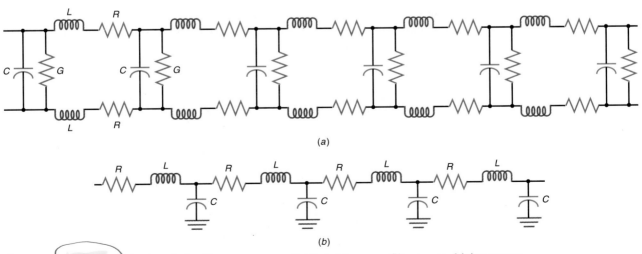

(a)

(b)

FIG. 13-9 A transmission line appears as a distributed low-pass filter to any driving generator. (a) A distributed line with lumped components. (b) Simplified equivalent circuit.

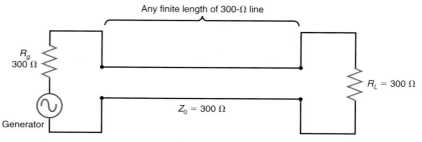

FIG. 13-10 A transmission line whose load is resistive and equal to the surge impedance appears as an equal resistance to the generator.

the characteristic impedance (see Fig. 13-10). This is the normal connection for a transmission line in any application. In equation form,

$$R_L = Z_0$$

If the line, load, and generator impedances are made equal, as is the case with matched generator and load resistances, the criterion for maximum power transfer is met.

An impedance meter or bridge can be used to measure the inductance and capacitance of a section of parallel line or coaxial cable to obtain the values needed to calculate the impedance. Assume, for example, that a capacitance of 0.0022 μF (2200 pF) is measured for 100 ft. The inductance of each conductor is measured separately and then added, for a total of 5.5 μH. (Resistance is ignored because it does not enter into the calculation of characteristic impedance; however it will cause signal attenuation over long distances.) The surge impedance is then:

$$Z_0 = \sqrt{(L/C)} = \sqrt{(5.5 \times 10^{-6})/(2200 \times 10^{-12})} = \sqrt{2500} = 50 \ \Omega$$

In practice, it is unnecessary to make these calculations because cable manufacturers always specify impedance.

The characteristic impedance of a cable is independent of length. We calculated it using value of L and C for 100 ft, but 50 Ω is the correct value for 1 ft or 1000 ft. Note that the actual impedance approaches the calculated impedance only if the cable is several wavelengths or more in length as terminated in its characteristic impedance. For line lengths less than 0.1 λ, characteristic impedance does not matter.

DETERMINING Z_0 FROM PHYSICAL DIMENSIONS. Another way to calculate the characteristic impedance of a transmission line is to first calculate the inductance and capacitance of the line from the physical dimensions of the conductors. An approximate expression for the characteristic impedance of a parallel two-wire transmission line is:

$$Z_0 \approx 276 \log_{10} \frac{2S}{D} \qquad \text{or} \qquad Z_0 \approx \frac{276}{\sqrt{\epsilon}} \log_{10} \frac{2S}{D}$$

where S = spacing between centers of parallel conductors
D = diameter of conductors
ϵ = dielectric constant of insulation between conductors

These equations are good for the case where $S \gg D$ (refer to Fig. 13-11).

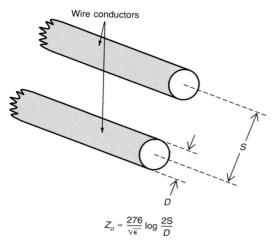

Wire conductors

$$Z_o \approx \frac{276}{\sqrt{\epsilon}} \log \frac{2S}{D}$$

FIG. 13-11 The characteristic impedance Z_0 of a two-wire transmission line depends on its physical dimensions.

The dielectric constants of some common cable insulators are given in the accompanying table. The most common insulator is polyethylene (PE). There is also a foam version of this plastic that greatly reduces capacitance per foot and attenuation. Its dielectric constant, 1.23, is roughly half of that of regular solid polyethylene. Teflon [polytetrafluoroethylene (PTFE)] is also widely used. PVC is the material most often used for the outer covering of coaxial cable and insulation on most other wires.

Dielectric Insulator	ϵ
Air	1.0
Polyethylene (PE)	2.27
Polystyrene	2.5
Polyvinyl chloride (PVC)	3.3
Teflon (PTFE)	2.1

For an example of how to compute characteristic impedance from physical dimensions, assume two 0.023-in.-diameter wires spaced 0.5 in. apart, with a continuous polyethylene insulation between them. The characteristic impedance is:

$$Z_0 = \frac{276}{\sqrt{2.27}} \log \frac{2 \times 0.5}{0.023}$$
$$= \frac{276}{1.5} \log 43.5$$
$$= 184(1.64)$$
$$= 301 \ \Omega$$

The characteristic impedance of a coaxial cable (see Fig. 13-12) is given by the expression

$$Z_0 = \frac{138}{\sqrt{\epsilon}} \log \frac{D}{d}$$

where D = inside diameter of outer braid shield
d = diameter of inner conductor

DID YOU KNOW?

Most transmission lines come with standard fixed values of characteristic impedance.

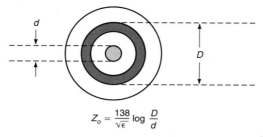

$$Z_o = \frac{138}{\sqrt{\epsilon}} \log \frac{D}{d}$$

FIG. 13-12 The characteristic impedance of a coaxial cable.

Assume that the inner diameter of the shield is 0.2 in. and the center conductor has a diameter of 0.06 in. If the insulation is Teflon, the dielectric constant is 2.1. The impedance is

$$Z_0 = \frac{138}{\sqrt{2.1}} \log \left(\frac{0.2}{0.06} \right)$$

$$= \frac{138}{1.45} \log 3.333$$

$$= 95.23(0.523)$$

$$= 50 \ \Omega$$

Most transmission lines come with standard fixed values of characteristic impedance. For example, the widely used twin-lead balanced line [Fig. 13-1(b)] has a characteristic impedance of 300 Ω. Open-wire line [Fig. 13-1(a)], which is no longer widely used was made with impedances of 450 and 600 Ω. As discussed in Chap. 12, the common characteristic impedances of coaxial cable are 52, 53.5, 75, and 93 Ω.

VELOCITY FACTOR

An important consideration in transmission-line applications is that the speed of the signal in the transmission line is slower than the speed of a signal in free space. The velocity of propagation of a signal in a cable is less than the velocity of propagation of light in free space by a fraction called the velocity factor, the ratio of the velocity in the transmission line V_p to the velocity in free space V_c:

$$\text{VF} = \frac{V_p}{V_c} \qquad \text{or} \qquad \text{VF} = \frac{V_p}{c}$$

where $V_c = c = 300,000,000$ m/s.

Velocity factors in transmission lines vary from approximately 0.5 to 0.9. The velocity factor of a coaxial cable is typically in the 0.6 to 0.8 range. Open-wire line has a VF of about 0.9, and 300-Ω twin-lead line has a velocity factor of about 0.8.

CALCULATING VELOCITY FACTOR. The velocity factor in a line can be computed with the expression $\text{VF} = 1/\sqrt{\epsilon}$, where ϵ is the dielectric constant of the insulating material. For example, if the dielectric in a coaxial cable is Teflon, the dielectric constant is 2.1 and the velocity factor is $1/\sqrt{2.1} = 1/1.45 = 0.69$. That is, the speed of the signal in the coaxial cable is 0.69 times the speed of light, or $0.69 \times 300,000,000 = 207,000,000$ m/s (128,616 mi/s).

If a lossless (zero-resistance) line is assumed, an approximation of the velocity of propagation can be computed with the expression

$$V_p = \frac{l}{\sqrt{LC}}$$

where l is the length or total distance of travel of the signal in feet or some other unit of length and L and C are given in that same unit. Assume, for example, a coaxial cable with a characteristic impedance of 50 Ω and a capacitance of 30 pF/ft. The inductance per foot is 0.075 μH or 75 nH. The velocity of propagation per foot in this cable is

$$V_p = \frac{1}{\sqrt{(75 \times 10^{-9} \times 30 \times 10^{-12})}} = 6.7 \times 10^8 \text{ ft/s}$$

or 126,262 mi/s, or 204×10^6 m/s.

The velocity factor is then

$$\text{VF} = \frac{V_p}{V_c} = \frac{204 \times 10^6}{300 \times 10^6} = 0.68$$

CALCULATING TRANSMISSION-LINE LENGTH. The velocity factor must be taken into consideration when computing the length of a transmission line in wavelengths. It is sometimes necessary to use a one-half or one-quarter wavelength of a specific type of transmission line for a specific purpose, for example, impedance matching, filtering, and tuning.

The formula given earlier for one wavelength of a signal in free space is $\lambda = 984/f$. This expression, however, must be modified by the velocity factor in order to arrive at the true length of a transmission line: The new formula is

$$\lambda = 984 \frac{\text{VF}}{f}$$

95% of ¼ wavelength [handwritten annotation]

Suppose, for example, that we want to find the actual length in feet of a quarter-wavelength segment of coaxial cable with a VF of 0.65 at 30 MHz. Using the formula, $\lambda = 984(\text{VF}/f) = 984(0.65/30) = 21.32$ ft. The length in feet is one quarter of this, or $21.32/4 = 5.33$ ft.

The correct velocity factor for calculating the correct length of a given transmission line can be obtained from manufacturers' literature and various handbooks.

TIME DELAY

Because the velocity of propagation of a transmission line is less than the velocity of propagation in free space, it is logical to assume that any line will slow down or delay any signal applied to it. A signal applied at one end of a line appears some time later at the other end of the line. This is called the *time delay* or *transit time* for the line. A transmission line used specifically for the purpose of achieving delay is called a *delay line*.

Figure 13-13 shows the effect of time delay on a sine-wave signal and a pulse train. The output sine wave appears later in time than the input, so it is shifted in phase. The effect is the same as if a lagging phase shift were introduced by a reactive circuit. In the case of the pulse train, the pulse delay is determined by a factor that depends on the type and length of the delay line.

The amount of delay time is a function of a line's inductance and capacitance. The opposition to changes in current offered by the inductance plus the charge and discharge time of the capacitance leads to a finite delay. This delay time is computed with the expression

$$T_d = \sqrt{LC}$$

where T_d is in seconds and L and C are the inductance and capacitance, respectively, per unit length of line. If L and C are given in terms of feet, the delay time will be per foot. For example, if the capacitance of a particular line is 30 pF/ft and its inductance is 0.075 μH/ft, the delay time is

$$T_d = \sqrt{(0.075 \times 10^{-6} \times 30 \times 10^{-12})} = 1.5 \times 10^{-9} \text{ or } 1.5 \text{ ns/ft}$$

A 50-foot length of this line would introduce $1.5 \times 50 = 75$ ns of delay.

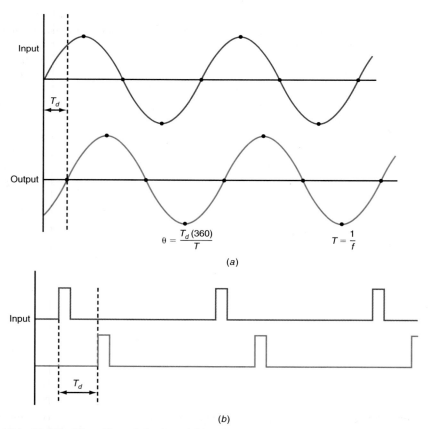

$$\theta = \frac{T_d (360)}{T} \qquad T = \frac{1}{f}$$

(a)

(b)

FIG. 13-13 The effect of the time delay of a transmission line on signals. (*a*) Sine-wave delay causes a lagging phase shift. (*b*) Pulse delay.

Time delay introduced by a coaxial cable can also be calculated using the formula

$$t_d = 1.016\sqrt{\epsilon}$$

where t_d is the time delay in nanoseconds per foot and ϵ is the dielectric constant of the cable.

For example, the total time delay introduced by a 75-ft cable with a dielectric constant of 2.3 is

$$t_d = 1.016\sqrt{\epsilon} = 1.016\sqrt{2.3} = 1.016(1.517) = 1.54 \text{ ns/ft}$$

for a total delay of $1.54(75) = 115.6$ ns.

To determine the phase shift represented by the delay, the frequency and period of the sine wave must be known. The period or time T for one cycle can be determined with the well-known formula $T = 1/f$, where f is the frequency of the sine wave. Assume a frequency of 4 MHz. The period is

$$T = \frac{1}{4 \times 10^6} = 250 \times 10^{-9} = 250 \text{ ns}$$

The phase shift of the previously described 50-ft line with a delay of 75 ns is given by the formula

$$\Theta = \frac{360T_d}{T} = \frac{360(75)}{250} = 108°$$

Transmission-line delay is usually ignored in RF applications, and it is virtually irrelevant in radio communications. However, in high-frequency applications where timing is important, transmission-line delay can be significant. For example, in LANs, the time of transition of the binary pulses on a coaxial cable is often the determining factor in calculating maximum allowed cable length.

Some applications require exact timing and sequencing of signals, especially pulses. A coaxial delay line can be used for this purpose. Obviously, a large roll of coaxial cable is not a convenient component in modern electronic equipment. As a result, artificial delay lines have been developed. These are made up of individual inductors and capacitors connected as a low-pass filter to simulate a distributed transmission line. Alternatively, a more compact distributed delay line can be constructed consisting of a coil of insulated wire wound over a metallic form. The coil of wire provides the distributed inductance and at the same time acts as one plate of a distributed capacitor. The metallic form is the other plate. Such artificial delay lines are widely used in TV sets, oscilloscopes, radar units, and many other pieces of electronic equipment.

TRANSMISSION-LINE SPECIFICATIONS

Figure 13-14 summarizes the specifications of several popular types of coaxial cable. Most coaxial cables are designated by an alphanumeric code beginning with the letters RG or a manufacturer's part number. The primary specifications are characteristic impedance and attenuation. Other important specifications are maximum breakdown voltage rating, capacitance per foot, velocity factor, and outside diameter in inches. The attenuation is the amount of power lost per 100 ft of cable expressed in decibels at

Type of cable	Z_0 (Ω)	VF, %	C, PF/ft	Outside diameter, in	V_{max}, rms	Attenuation, dB/100 ft*
RG-8/U	52	66	29.5	0.405	4000	2.5
RG-8/U foam	50	80	25.4	0.405	1500	1.6
RG-11/U	75	66	20.6	0.405	4000	2.5
RG-11/U foam	75	80	16.9	0.405	1600	1.6
RG-58A/U	53.5	66	28.5	0.195	1900	5.3
RG-59/U	73	66	21.0	0.242	2300	3.4
RG-62A/U	93	86	13.5	0.242	750	2.8
RG-214/U	50	66	30.8	0.425	5000	2.5
9913	50	84	24.0	0.405	—	1.3
Twin lead (open line)	300	82	5.8	—	—	0.55

*At 100 MHz.

Fig. 13-14   Table of common transmission-line characteristics.

100 MHz. Attenuation is directly proportional to cable length and increases with frequency. Detailed charts and graphs of attenuation versus frequency are also available, so that users can predict the losses for their applications. Attenuation versus frequency for four coaxial cable types is plotted in Fig. 13-15. The loss is significant at very high frequencies. However, the larger the cable, the lower the loss. For purposes of

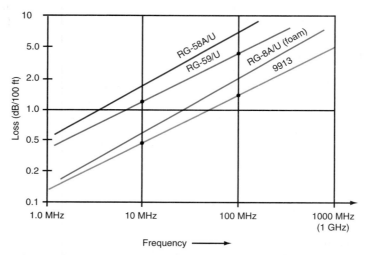

Fig. 13-15   Attenuation versus frequency for common coaxial cables. Note that both scales on the graph are logarithmic.

comparison, look at the characteristics of 300-Ω twin-lead cable listed in Fig. 13-14. Note the low loss compared to coaxial cable.

Example 13-3

A 165-ft section of RG-58A/U at 100 MHz is being used to connect a transmitter to an antenna. Its attenuation for 100 feet at 100 MHz is 5.3 dB. Its input power from a transmitter is 100 W. What is the total attenuation and the output power to the antenna?

Cable attenuation = 5.3 dB/100 ft = 0.053 dB/ft

Total attenuation = 0.053 × 165 = 8.745 dB or −8.745

$$dB = 10 \log\left(\frac{P_{out}}{P_{in}}\right)$$

$$\frac{P_{out}}{P_{in}} = \text{antilog} \frac{dB}{10} \quad \text{and} \quad P_{out} = P_{in} = \text{antilog} \frac{dB}{10}$$

$$P_{out} = 100 \text{ antilog} \frac{-8.745}{10} = 100 \text{ antilog } (0.8745)$$

$$= 100(0.143) = 14.3 \text{ W}$$

The loss in a cable can be significant, especially at the higher frequencies. In Example 13-3, the transmitter put 100 W into the line, but at the end of the line, the output power—the level of the signal that is applied to the antenna—was only 14.3 W. A major loss of 85.7 W dissipated as heat in the transmission line.

Several things can be done to minimize loss. First, every attempt should be made to find a way to shorten the distance between the transmitter and the antenna. If that is not feasible, it may be possible to use a larger cable. For the application in Example 13-3, the RG-58A/U cable was used. This cable has a characteristic impedance of 53.5 Ω, so any value near that would be satisfactory. One possibility would be the RG-8/U, with an impedance of 52 Ω and an attenuation of only 2.5 dB/100 ft. An even better choice would be the 9913 cable, with an impedance of 50 Ω and an attenuation of 1.3 dB/100 ft.

When considering the relationship between cable length and attenuation, remember that a transmission line is a low-pass filter whose cutoff frequency depends both on distributed inductance and capacitance along the line and on length. The longer the line, the lower its cutoff frequency, meaning that higher-frequency signals beyond the cutoff frequency are rolled off at a rapid rate.

This is illustrated in Fig. 13-16, which shows attenuation curves for four lengths of a popular type of coaxial cable. Remember that the cutoff frequency is the 3-dB down point on a frequency-response curve. If we assume that an attenuation of 3 dB is the same as a 3-dB loss, then we can estimate the cutoff frequency of different lengths of cable. The 3-dB down level is marked on the graph. Now, note the cutoff frequency for different lengths of cable. The shorter cable (100 ft) has the highest cutoff frequency, making the bandwidth about 30 MHz. The 200-ft cable has a cutoff of about 8 MHz; the 500-ft cable has a cutoff of approximately 2 MHz, and the 1000-ft cable has a cutoff of approximately 1 MHz. The higher frequencies are severely attenuated by the

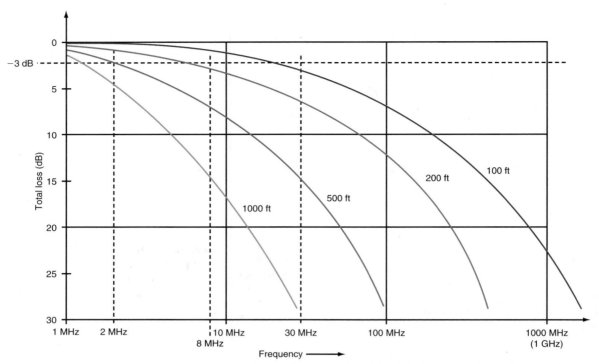

Fig. 13-16 Attenuation versus length for RG-58A/U coaxial cable. Note that both scales on the graph are logarithmic.

cable as it gets longer. It should be clear why it is important to use the larger, lower-loss cables for longer runs despite cost and handling inconvenience.

Finally, a gain antenna can be used to offset cable loss. These antennas are discussed in Chap. 14.

Example 13-4

A 150-ft length of RG-62A/U coaxial cable is used as a transmission line. Find (*a*) the load impedance that must be used to terminate the line to avoid reflections, (*b*) the equivalent inductance per foot, (*c*) the time delay introduced by the cable, (*d*) the phase shift that occurs on a 2.5-MHz sine wave, and (*e*) the total attenuation in decibels. (Refer to Fig. 13-14.)

a. The characteristic impedance is 93 Ω; therefore the load must offer a resistance of 93 Ω to avoid reflections.

b. $Z_0 = \sqrt{\dfrac{L}{C}}$ $Z_0 = 93\ \Omega$ $C = 13.5\ \text{pF/ft}$

$L = CZ_0^2 = 13.5 \times 10^{-12} \times (93)^2 = 116.76\ \text{nH/ft}$

c. $t_d = \sqrt{LC} = \sqrt{116.76 \times 10^{-9} \times 13.5 \times 10^{-12}} = 1.256\ \text{ns/ft}$

$150\ \text{ft} \times 1.256 = 188.3\ \text{ns}$

d. $T = \dfrac{1}{f} = \dfrac{1}{2.5} \times 10^6 = 400$ ns

$\theta = \dfrac{188.3(360)}{400} = 169.47°$

e. Attenuation = 2.8 dB/100 ft = 0.028 dB/ft

150 ft $\times$ 0.028 = 4.2 dB

13-2 STANDING WAVES

When a signal is applied to a transmission line, it appears at the other end of the line some time later because of the propagation delay. If a resistive load equal to the characteristic impedance of a line is connected at the end of the line, the signal is absorbed by the load and power dissipated as heat. If the load is an antenna, the signal is converted into electromagnetic energy and radiated into space.

If the load at the end of a line is an open circuit or a short circuit or has an impedance other than the characteristic impedance of the line, the signal is not fully absorbed by the load. When a line is not terminated properly, some of the energy is reflected from the end of the line and actually moves back up the line, toward the generator. This *reflected* voltage adds to the forward or incident generator voltage and forms a composite voltage that is distributed along the line. This pattern of voltage and its related current constitute what is called a *standing wave*.

Standing waves are not desirable. The reflection indicates that the power produced by the generator is not totally absorbed by the load. In some cases, for example, a shorted or open line, no power gets to the load because it is all reflected back to the generator. The following sections examine in detail how standing waves are generated.

THE RELATIONSHIP BETWEEN REFLECTIONS AND STANDING WAVES

Figure 13-17 will be used to illustrate how reflections are generated and how they contribute to the formation of standing waves. Part (*a*) shows how a dc pulse propagates along a transmission line made up of identical *LC* sections. A battery (generator) is used as the input signal along with a switch to create an ON-OFF DC pulse.

The transmission line is open at the end rather than being terminated in the characteristic impedance of the line. An open transmission line will, of course, produce a reflection and standing waves. Note that the generator has an internal impedance of R_g, which is equal to the characteristic impedance of the transmission line. Assume a transmission-line impedance of 75 Ω and an internal generator resistance of 75 Ω. The 10 V supplied by the generator is, therefore, distributed equally across the impedance of the line and the internal resistance.

Now assume that the switch is closed to connect the generator to the line. As you know, connecting a DC source to reactive components such as inductors and capaci-

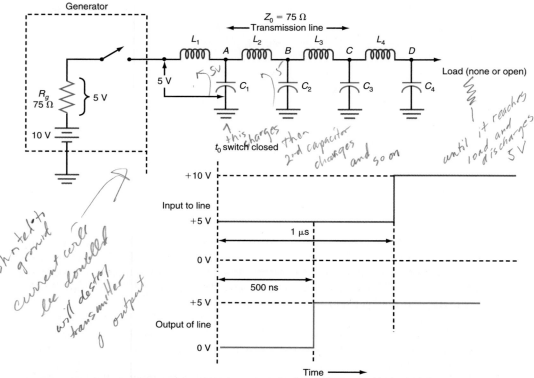

FIG. 13-17 How a pulse propagates along a transmission line.

[Handwritten annotations near the figure: "5V", "this charges then 2nd capacitor charges and so on", "until it reaches load and discharges 5V", "short to ground", "current coil", "be doubled", "will destroy transmitter", "0 output", "open", "capacitive at end — voltage to will discharge 10 volts back to source", "cancel out", "double voltage", "0 output", "reflecting signal back", "open — voltage device", "short — current device"]

tors produces transient signals as the inductors oppose changes in current while the capacitors oppose changes in voltages. Capacitor C_1 initially acts as a short circuit when the switch is closed, but soon begins to charge toward the battery voltage through L_1. As soon as the voltage at point A begins to rise, it applies a voltage to the next section of the transmission line made up of C_2 and L_2. Therefore, C_2 begins to charge through L_2. The process continues on down the line until C_4 charges through L_4, and so on. The signal moves down the line from left to right as the capacitors charge. Lossless (zero-resistance) components are assumed, and so the last capacitor C_4 eventually charges to the supply voltage.

For the purposes of this illustration, assume that the length of the line and its other characteristics are such that the time delay is 500 ns: 500 ns after the switch is closed, an output pulse will occur at the end of the line. At this time, the voltage across the output capacitance C_4 is equal to 5 V or half of the supply voltage.

The instant that the output capacitance charges to its final value of 5 V, all current flow in the line ceases, causing any magnetic field around the inductors to collapse. The energy stored in the magnetic field of L_4 is equal to the energy stored in the output capacitance C_4. Therefore, a voltage of 5 V is induced into the inductor. The polarity of this voltage will be in such a direction that it adds to the charge already on the capacitor. Thus the capacitor will charge to 2 times the applied 5-V voltage, or 10 V.

A similar effect then takes place in L_3. The magnetic field across L_3 collapses, doubling the voltage charge on C_3. Next, the magnetic field around L_2 collapses, charging C_2 to 10 V. The same effect occurs in L_1 and C_1. Once the signal reaches the right end of the line, a reverse charging effect takes place on the capacitors from right to left. The effect is as if a signal were moving from output to input. This moving charge from right to left is the reflection, or *reflected wave*, and the input wave from the generator to the end of the line is the *incident wave*.

It takes another 500 ns for the reflected wave to get back to the generator. At the end of 1 μs, the input to the transmission line goes more positive by 5 V, for a total of 10 V.

Figure 13-17(b) shows the waveforms for the input, output, and reflected voltages with respect to time. Observing the waveforms, follow the previously described action with the closing of the switch at time t_0. Since both the characteristic impedance of the line and the internal generator resistance are 75 Ω, half the battery voltage appears at the input to the line at point A. This voltage propagates down the line, charging the line capacitors as it goes, until it reaches the end of the line and fully charges the output capacitance. At that time the current in the inductance begins to cease with magnetic fields collapsing and inducing voltages that double the output voltage at the end of the line. Thus after 500 ns the output across the open end of the line is 10 V.

The reflection begins and now moves back down the line from right to left; after another 500 ns, it finally reaches the line input, the input to the line to jump to 10 V. Once the reflection stops, the entire line capacitance is fully charged to 10 V, as might be expected.

The preceding description concerned what is known as an *open-circuit load*. Another extreme condition is a *short-circuit load*. For this situation, assume a short across C_4 in Fig. 13-17(a).

When the switch is closed, again 5 V is applied to the input of the line, which is then propagated down the line as the line capacitors charge. Since the end of the line is short-circuited, inductor L_4 is, in effect, the load for this line. The voltage on C_3 is then applied to L_4. At this point, the reflection begins. The current in L_4 collapses, inducing a voltage which is then propagated down the line in the opposite direction. The voltage induced in L_4 is equal and opposite to the voltage propagated down the line. Therefore, this voltage is equal and opposite to the voltage on C_3, which causes C_3 to be discharged. As the reflection works its way back down the line from right to left, the line capacitance is continually discharged until it reaches the generator. It takes 500 ns for the charge to reach the end of the line and another 500 ns for the reflection to move back to the generator. Thus in a total of 1 μs, the input voltage switches from 5 to 0 V. Of course, the voltage across the output short remains zero during the entire time.

Open and shorted transmission lines are sometimes used to create special effects. In practice, however, the load on a transmission line is neither infinite nor is it zero Ω; rather, it is typically some value in between. The load may be resistive or may have a reactive component. Antennas typically do not have a perfect resistance value. Instead they frequently have a small capacitive or inductive reactance. Thus the load impedance is equivalent to a series RC or RL circuit with an impedance of the form $R \pm jX$. If the load is not exactly resistive and is equal to the characteristic impedance of the line, a reflection is produced, the exact voltage levels depending on the complex impedance of the load. Usually some of the power is absorbed by the resistive part of the line; the mismatch still produces a reflection, but the reflection is not equal to the original signal, as in the case of a shorted or open load.

In most communications applications, the signal applied to a transmission line is an AC signal. This situation can be analyzed by assuming the signal to be a sine wave. The effect of the line on a sine wave is like that described above in the discussion based on an analysis of Fig. 13-17. If the line is terminated in a resistive load equal to the characteristic impedance of the line, the sine-wave signal is fully absorbed by the load and no reflection occurs.

MATCHED LINES

Ideally, a transmission line should be terminated in a load that has a resistive impedance equal to the characteristic impedance of the line. This is called a *matched line*. For example, a 50-Ω coaxial cable should be terminated with a 50-Ω resistance, as shown in Fig. 13-18. If the load is an antenna, then that antenna should look like a resistance of 50 Ω. When the load impedance and the characteristic impedance of the line match, the transmission goes smoothly and maximum power transfer—less any resistive losses in the line—takes place. The line can be any length. One of the key objectives in designing antenna and transmission-line systems is to ensure this match.

AC voltage (or current) at any point on a matched line is a constant value (disregarding losses). A correctly terminated transmission line, therefore, is said to be *flat*. For example, if a voltmeter is moved down a matched line from generator to load and the rms voltage values plotted, the resulting wavelength versus voltage line will be flat (see Fig. 13-19). Resistive losses in the line would, of course, produce a small voltage drop along the line, giving it a downward tilt for greater lengths.

If the load impedance is different from the line characteristic impedance, not all of the power transmitted is absorbed by the load. The power not absorbed by the load is reflected back toward the source. The power sent down the line toward the load is called forward or incident power; the power not absorbed by the load is called *reflected power*. The signal actually on a line is simply the algebraic sum of the forward and reflected signals.

Reflected power can represent a significant loss. If a line has a 3-dB loss end-to-end, only a 3-dB attenuation of the reflected wave will occur by the time the signal reaches the generator. What happens there depends upon the relative impedance of the generator and the line. Only part of the reflected energy is dissipated in the line. In cases in which the mismatch between load resistive impedance and line impedance is great, the reflected power can be high enough to actually cause damage to the transmitter or the line itself.

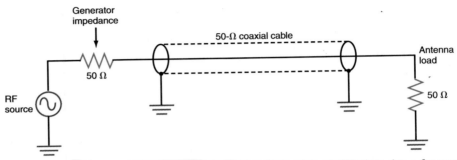

FIG. 13-18 A transmission line must be terminated in its characteristic impedance for proper operation.

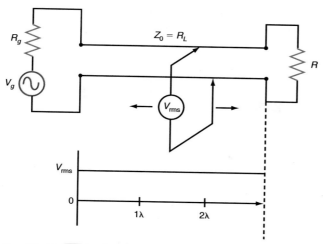

FIG. 13-19   The voltage on a matched line is constant over length.

SHORTED LINES

The condition for a short circuit is illustrated in Fig. 13-20. The graph below the transmission line shows the plot of voltage and current at each point on the line that would be generated by using the values given by a voltmeter and ammeter moved along the line. As would be expected in the case of a short at the end of a line, the voltage is zero when the current is maximum. All of the power is reflected back toward the generator. Looking at the plot, you can see that the voltage and current variations distribute themselves according to the signal wavelength. The fixed pattern, which is the result of a composite of the forward and reflected signals, repeats every half wavelength. The actual voltage and current levels at the generator are dependent on signal wavelength and the line length.

The phase of the reflected voltage at the generator end of the line depends upon the length of the line. If the line is some multiple of a quarter wavelength, the reflected wave will be in phase with the incident wave and the two will add together, producing a signal at the generator that is twice the generator voltage. If the line length is some multiple of a half wavelength, the reflected wave will be exactly 180° out of

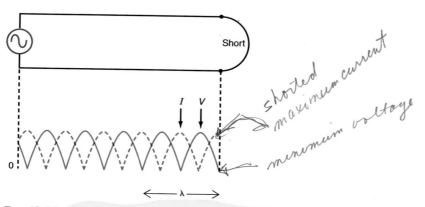
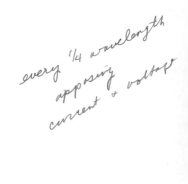

FIG. 13-20   Standing waves on a shorted transmission line.

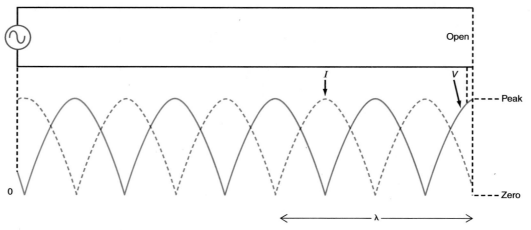

FIG. 13-21 Standing waves on an open-circuit transmission line.

phase with the incident wave and the two will cancel, giving a zero voltage at the generator. In other words, the effects of the reflected wave can simulate an open or short circuit at the generator.

OPEN LINES

Figure 13-21 shows the standing waves on an open-circuited line. With an infinite impedance load, the voltage at the end of the line is maximum when the current is zero. All the energy is reflected, setting up the stationary pattern of voltage and current standing waves shown.

MISMATCHED (RESONANT) LINES

Most often, lines do not terminate in a short or open circuit. Rather, the load impedance does not exactly match the transmission line impedance. Further, the load, usually an antenna, will probably have a reactive component, either inductive or capacitive, in addition to its resistance. Under these conditions, the line is said to be *resonant*. Such a mismatch produces standing waves, but the amplitude of these waves is lower than the standing waves resulting from short or open circuits. The distribution of these standing waves looks like that shown in Fig. 13-22. Note that the voltage or current never goes to zero, as it does with an open or shorted line.

CALCULATING THE STANDING WAVE RATIO

The magnitude of the standing waves on a transmission line is determined by the ratio of the maximum current to the minimum current, or the ratio of the maximum voltage to the minimum voltage, along the line. These ratios are referred to as the *standing wave ratio (SWR)*.

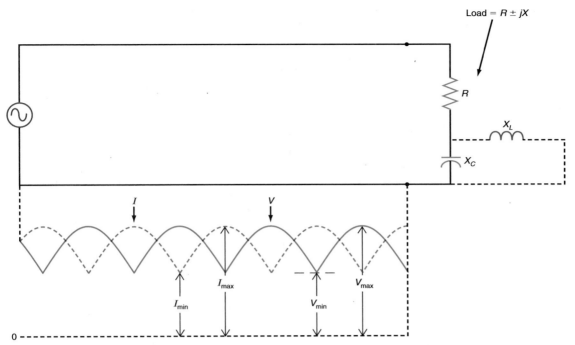

Fig. 13-22 Transmission line with mismatched load and the resulting standing waves.

$$\text{SWR} = \frac{I_{\max}}{I_{\min}} = \frac{V_{\max}}{V_{\min}}$$

Under the shorted and open conditions described earlier, the current or voltage minimums are zero. This produces an SWR of infinity. It means that no power is dissipated in the load; all of it is reflected.

In the ideal case, there are no standing waves. The voltage and current are constant along the line, so there are no maximum or minimums (or the maximum and minimums are the same). Therefore, the SWR is 1.

Measuring the maximum and minimum values of voltage and current on a line is not practical in the real world, so other ways of computing the SWR have been devised. For example, the SWR can be computed if the impedance of the transmission line and the actual impedance of the load are known. The SWR is the ratio of the load impedance (Z_l) to the characteristic impedance (Z_0), or vice-versa.

If $Z_l > Z_0$:

$$\text{SWR} = \frac{Z_l}{Z_0}$$

If $Z_0 > Z_l$:

$$\text{SWR} = \frac{Z_0}{Z_l}$$

For example, if a 75-Ω antenna load is connected to a 50-Ω transmission line, the SWR is 75/50 = 1.5. Since the standing wave is really the composite of the original inci-

> ### DID YOU KNOW?
>
> Some solid-state systems automatically shut down if the SWR is greater than 2.

dent wave added to the reflected wave, the SWR can also be defined in terms of those waves. The ratio of the reflected voltage wave V_r to the incident voltage wave V_i is called the reflection coefficient: $\Gamma = V_r/V_i$. The reflection coefficient provides information on current and voltage along the line.

If a line is terminated in its characteristic impedance, then there is no reflected voltage, so $V_r = 0$ and $\Gamma = 0$. If the line is open or shorted, then total reflection occurs. This means that V_r and V_i are the same, so $\Gamma = 1$. The reflection coefficient really expresses the percentage of reflected voltage to incident voltage. If Γ is 0.5, for example, the reflected voltage is 50 percent of the incident voltage, and the reflected power is 25 percent of the incident power.

If the load is not matched but also is not an open or short, the line will have voltage minimums and maximums, as described previously. These can be used to obtain the reflection coefficient using the formula

$$\Gamma = \frac{V_{\max} - V_{\min}}{V_{\max} + V_{\min}}$$

The SWR is obtained from the reflection coefficient according to the equation

$$SWR = \frac{1 + \Gamma}{1 - \Gamma}$$

Example 13-5

An RG-11/U foam coaxial cable has a maximum voltage standing wave of 52 V and a minimum voltage of 17 V. Find (*a*) the SWR, (*b*) the reflection coefficient, and (*c*) the value of a resistive load.

a. $SWR = \dfrac{V_{\max}}{V_{\min}} = \dfrac{52}{17} = 3.05$

b. $\Gamma = \dfrac{V_{\max} - V_{\min}}{V_{\max} + V_{\min}} = \dfrac{52 - 17}{52 + 17} = \dfrac{35}{69}$

$\Gamma = 0.51$

or

$SWR = \dfrac{1 + \Gamma}{1 - \Gamma}$

$\Gamma = \dfrac{SWR - 1}{SWR + 1} = \dfrac{3.05 - 1}{3.05 + 1} = \dfrac{2.05}{4.05} = 0.51$

c. $SWR = 3.05 \qquad Z_0 = 75\ \Omega \qquad SWR = \dfrac{Z_l}{Z_0} = \dfrac{Z_0}{Z_l}$

$Z_l = Z_0\ SWR = 75(3.05) = 228.75\ \Omega$

or

$Z_l = \dfrac{Z_0}{SWR} = \dfrac{75}{3.05} = 24.59\ \Omega$

If the load matches the line impedance, then $\Gamma = 0$. The formula above gives an SWR of 1, as expected. With an open or shorted load, $\Gamma = 1$. This produces an SWR of infinity.

The reflection coefficient can also be determined from the line and load impedances:

$$\Gamma = \frac{Z_l - Z_0}{Z_l + Z_0}$$

For the example of an antenna load of 75 Ω and a coaxial cable of 50 Ω, the reflection coefficient is $\Gamma = (75 - 50)/(75 + 50) = 25/125 = 0.2$.

The importance of the SWR is that it gives a relative indication of just how much power is lost in the transmission line and the generator. This assumes that none of the reflected power is re-reflected by the generator. In a typical transmitter, some power is reflected and sent to the load again.

The curve in Fig. 13-23 shows the relationship between the percentage of reflected power and the SWR. Naturally when the standing wave ratio is 1, the percentage of reflected power is zero. But as a line and load mismatch grows, reflected power increases. When the SWR is 1.5, the percentage of reflected power is 4 percent. This is still not too bad, as 96 percent of the power gets to the load.

It is possible to compute reflected power P_r given the SWR and the incident power P_l. Since $\Gamma^2 = P_r/P_i$, $P_r = \Gamma^2 P_i$. Knowing the SWR, you can compute Γ, then solve using the equation above.

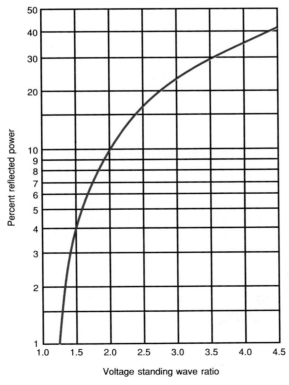

FIG. 13-23 Percentage of reflected power on a transmission line for different SWR values.

For SWR values of 2 or less, the percentage of reflected power is less than 10, which means that 90 percent gets to the load. For most applications this is acceptable. For SWR values higher than 2, the percentage of reflected power increases dramatically, and measures must be taken to reduce the SWR to prevent potential damage. Some solid-state systems shut down automatically if the SWR is greater than 2. The most common approach to reducing SWR is to add or include a π, L, or T *LC* network to offset antenna reactance and other resistive components and produce a match. Antenna length can also be adjusted to improve the impedance match.

Example 13-6

The line input to the cable in Example 13-5 is 30 W. What is the output power? (See Fig. 13-22; disregard attenuation due to length.)

The percentage of reflected power with SWR of 3.05 is about 22.

$$P_r = 30(0.22) = 6.6 \text{ W}$$
$$P_{out} = 30 - 6.6 = 23.4 \text{ W}$$

13-3 TRANSMISSION LINES AS CIRCUIT ELEMENTS

The standing-wave conditions resulting from open- and short-circuited loads must usually be avoided when working with transmission lines. However, with quarter- and half-wavelength transmissions, these open- and short-circuited loads can be used as resonant or reactive circuits.

RESONANT CIRCUITS AND REACTIVE COMPONENTS

Consider the shorted quarter-wavelength ($\lambda/4$) line shown in Fig. 13-24. At the load end, voltage is zero and current is maximum. But one-quarter wavelength back, at the generator, the voltage is maximum and the current is zero. To the generator, the line appears as an open circuit, or at least a very high impedance. The key point here is that this condition exists at only one frequency, the frequency at which the line is exactly one-quarter wavelength. Because of this frequency sensitivity, the line acts like an *LC* tuned or resonant circuit, in this case, a parallel resonant circuit because of its very high impedance at the reference frequency.

With a shorted half-wavelength line, the standing-wave pattern is like that shown in Fig. 13-25. The generator sees the same conditions as at the end of the line, that is, zero voltage and maximum current. This represents a short, or very low impedance. That condition occurs only if the line is exactly one-half wavelength long at the generator frequency. In this case, the line looks like a series resonant circuit to the generator.

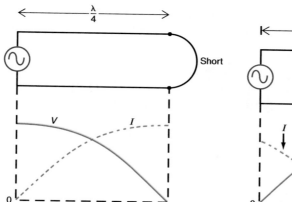

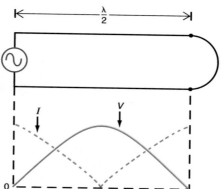

Fig. 13-24 A shorted quarter-wavelength line acts like a parallel resonant circuit.

Fig. 13-25 A shorted half-wavelength line acts like a series resonant circuit.

If the line length is less than one-quarter wavelength at the operating frequency, the shorted line looks like an inductor to the generator. If the shorted line is between one-quarter and one-half wavelength, it looks like a capacitor to the generator. All these conditions repeat with multiple quarter or half wavelengths of shorted line.

Similar results are obtained with an open line, as shown in Fig. 13-26. To the generator, a quarter-wavelength line looks like a series resonant circuit and a half-wavelength line looks like a parallel resonant circuit, just the opposite of a shorted line. If the line is less than a quarter wavelength, the generator sees a capacitance. If the line is between a quarter and a half wavelength, the generator sees an inductance. These characteristics repeat for lines that are some multiple of quarter or half wavelengths.

Figure 13-27 is a summary of the conditions represented by open and shorted lines of lengths up to one wavelength. The horizontal axis is length, in wavelengths, and the vertical axis is the reactance of the line, in ohms, expressed in terms of the line's characteristic impedance. The solid curves are shorted lines and the dashed curves are open-circuit lines.

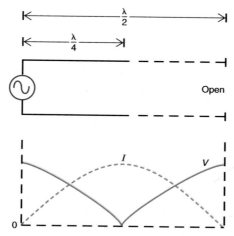

Fig. 13-26 One-quarter and one-half wavelength open lines look like resonant circuits to a generator.

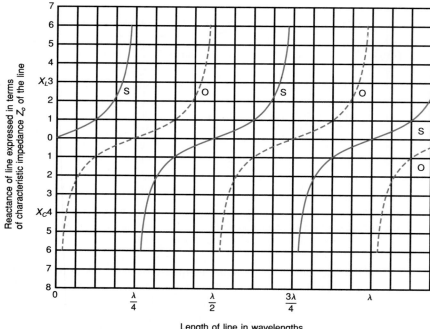

Fig. 13-27 Summary of impedance and reactance variations of shorted and open lines for lengths up to one wavelength.

If the line acts like a series resonant circuit, its impedance is zero. If the line is of such a length that it acts like a parallel resonant circuit, its impedance is near infinity. If the line is some intermediate length, it is reactive. For example, consider a shorted one-eighth-wavelength line. The horizontal divisions represent one-sixteenth wavelength, so two of these represent one-eighth wavelength. Assume that the line has a characteristic impedance of 50 Ω. At the one-eighth-wavelength point on the left-hand solid curve is a reading of 1. This means that the line acts like an inductive reactance of $1 \times Z_0$, or $1 \times 50 = 50\ \Omega$. An open line about three-eighths wavelength would have the same effect, as the left-most dashed curve in Fig. 13-26 indicates.

How could a capacitive reactance of 150 Ω be created with the same 50-Ω line? First, locate the 150-Ω point on the capacitive reactance scale in Fig. 13-27. Since 150/50 = 3, the 150-Ω point is at $X_C = 3$. Next, draw a line from this point to the right until it intersects with two of the curves. Then read the wavelength from the horizontal scale. A capacitive reactance of 150 Ω with a 50-Ω line can be achieved with an open line somewhat longer than 1/32 wavelength or a shorted line a bit longer than 9/32 wavelength.

STRIPLINE AND MICROSTRIP

At low frequencies (below about 300 MHz), the characteristics of open and shorted lines discussed in the previous sections have little significance. At low frequencies the lines are just too long to be used as reactive components or as filters and tuned circuits. However, at UHF (300 to 3000 MHz) and microwave (1 GHZ and greater) frequencies the length of a half wavelength is less than 1 ft; the values of inductance and capacitance become so small that it is difficult to realize them physically with standard coils and capacitors. Special transmission lines constructed with copper patterns on a printed circuit board (PCB), called *microstrip* or *stripline,* can be used as tuned circuits, filters, phase shifters, reactive components, and impedance-matching circuits at these high frequencies.

A PCB is a flat insulating base made of fiberglass or some other insulating base material to which is bonded copper on one or both sides and sometimes in several layers. Teflon is used as the base for some PCBs in microwave applications. In microwave ICs, the base is often alumina or even sapphire. The copper is etched away in patterns to form the interconnections for transistors, ICs, resistors, and other components. Thus pointto-point connections with wire are eliminated. Diodes, transistors, and other components are mounted right on the PCB and connected directly to the formed microstrip or stripline.

MICROSTRIP. Microstrip is a flat conductor separated by an insulating dielectric from a large conducting ground plane [Fig. 13-28(*a*)]. The microstrip is usually a quarter or half wavelength long. The ground plane is the circuit common. This type of microstrip is equivalent to an unbalanced line. Shorted lines are usually preferred over open lines. Microstrip can also be made in a two-line balanced version [Fig. 13-28(*b*)].

The characteristic impedance of microstrip, like any transmission line, is dependent on its physical characteristics. It can be calculated using the formula

$$Z = \frac{87}{\sqrt{\epsilon + 1.41}} \ln\left(\frac{5.98h}{0.8w + t}\right)$$

where Z = characteristic impedance
ϵ = dielectric constant
w = width of copper trace
t = thickness of copper trace
h = distance between trace and ground plane (dielectric thickness) of the dielectric

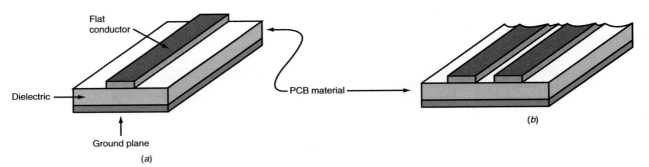

FIG. 13-28 Microstrip. (*a*) Unbalanced. (*b*) Balanced.

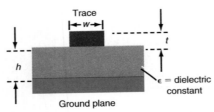

FIG. 13-29 Dimensions for calculating characteristic impedance.

Any units of measurement can be used (e.g., inches or millimeters), as long as all dimensions are in the same units. (See Fig. 13-29.) The dielectric constant of the popular FR-4 fiberglass PC board material is 4.5. The value of ϵ for Teflon is 3.

The characteristic impedance of microstrip with the dimensions $h = 0.0625$ in, $w = 0.1$ in, $t = 0.003$ in, and $\epsilon = 4.5$ is

$$Z = \frac{87}{\sqrt{4.5 + 1.41}} \ln\left[\frac{5.98(.0625)}{0.8(0.1) + 0.003}\right]$$
$$= 35.8 \ln(4.5) = 35.8(1.5)$$
$$= 53.9 \ \Omega$$

STRIPLINE. Stripline is a flat conductor sandwiched between two ground planes (Fig. 13-30). It is more difficult to make than microstrip; however it does not radiate as microstrip does. Radiation produces losses. The length is one-quarter or one-half wavelength at the desired operating frequency, and shorted lines are more commonly used than open lines.

The characteristic impedance of stripline is given by the formula

$$Z = \frac{60}{\epsilon} \ln\left(\frac{4d}{0.67\pi w(0.8 + t/h)}\right)$$

Figure 13-31 shows the dimensions required to make the calculations.

Even tinier microstrip and striplines can be made using monolithic, thin-film, and hybrid IC techniques. When these are combined with diodes, transistors, and other components, microwave integrated circuits (MICs) are formed.

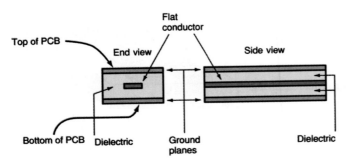

FIG. 13-30 Stripline.

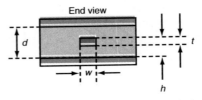

FIG. 13-31 Dimensions for calculating stripline impedance.

Example 13-7

A microstrip transmission line is to be used as a capacitor of 4 pF at 800 MHz. The PCB dielectric is 3.6. The microstrip dimensions are $h = 0.0625$ in, $w = 0.13$ in, and $t = 0.002$ in. What are (a) the characteristic impedance of the line and (b) the reactance of the capacitor?

a. $Z_0 = \dfrac{87}{\sqrt{\epsilon + 1.41}} \ln\left(\dfrac{5.98h}{0.8w + t}\right)$

$= \dfrac{87}{\sqrt{3.6 + 1.41}} \ln\left[\dfrac{5.98(0.0625)}{0.8(0.13) + 0.002}\right]$

$= \dfrac{87}{2.24} \ln\left(\dfrac{0.374}{0.106}\right)$

$= (38.8)(1.26) = 48.9 \; \Omega$

b. $X_C = \dfrac{1}{2\pi fC} = \dfrac{1}{6.28}(800 \times 10^6)(4 \times 10^{-12}) = 49.76 \; \Omega$

Example 13-8

What is the length of the transmission line in Example 13-7?
  Refer to Fig. 13-27. Take the ratio of X_C to Z_0

$$\frac{X_C}{Z_0} = \frac{49.76}{48.9} = 1.02 \approx 1$$

Locate 1 on the X_C vertical region of the graph. Move to the right to encounter the dashed curve for an open line. Read the wavelength on the lower horizontal axis of $\frac{3}{16}$ or $\frac{1}{8}$ λ.

$$\lambda = \frac{984}{800} = 1.23 \; \text{ft} \qquad \frac{\lambda}{8} = \frac{1.23}{8} = 0.15375 \; \text{ft}$$

$$0.15375 \times 12 \; \text{in} = 1.845 \; \text{in}$$

The velocity of propagation is

$$V_p = \frac{1}{\sqrt{\epsilon}} = \frac{1}{\sqrt{3.6}} = 0.527$$

$$\frac{\lambda}{8} = 1.845 \; \text{in} \times V_p = 1.845 \times 0.527 = 0.972 \; \text{in}$$

The mathematics required to design and analyze transmission lines is complex, whether the line is a physical cable connecting a transceiver to an antenna or is being used as a filter or impedance-matching network. This is because the impedances involved are complex ones, involving both resistive and reactive elements. The impedances are in the familiar rectangular form, $R + jX$. Computations with complex numbers such as this are long and time-consuming. Further, many calculations involve trigonometric relationships. Although no individual calculation is difficult, the sheer volume of the calculations can lead to errors.

In the 1930s, one clever engineer decided to do something to reduce the chance of error in transmission-line calculations. The engineer's name was Philip H. Smith, and in January 1939 he published the Smith chart, a sophisticated graph that permits visual solutions to transmission-line calculations.

Today, of course, the mathematics of transmission-line calculations is not a problem, because of the wide availability of electronic computing options. Transmission-line equations can be easily programmed into a scientific and engineering calculator for fast, easy solutions. Personal computers and RISC workstations provide an ideal way to make these calculations, either by using special mathematics software packages or using BASIC, Fortran, or another language to write the specific programs. The math software packages commonly available today for engineering and scientific computation also provide graphical outputs if desired.

Despite the availability of all of the computing options today, the Smith Chart is still used. Its unique format provides a more or less standardized way of viewing and solving transmission-line and related problems. Further, a graphical representation of an equation conveys more information than is gained by a simple inspection of the equation. For these reasons, it is desirable to become familiar with the Smith chart.

Figure 13-32 is a Smith chart. This imposing graph is created by plotting two sets of orthogonal (at right angles) circles on a third circle. Smith chart graph paper is available from the original publisher, Analog Instruments Company, and from the American Radio Relay League. Smith chart graph paper is also available in some college and university bookstores.

The first step in creating a Smith chart is to plot a set of eccentric circles along a horizontal line as shown in Fig. 13-33. The horizontal axis is the pure resistance or zero reactance line. The point at the far left end of the line represents zero resistance and the point at the far right represents infinite resistance. The resistance circles are centered on and pass through this pure resistance line. The circles are all tangent to one another at the infinite resistance point, and the centers of all the circles fall on the resistance line.

Each circle represents all the points of some fixed resistance value. Any point on the outer circle represents a resistance of $0 \, \Omega$. The other circles have other values of resistance. The $R = 1$ circle passes through the exact center of the resistance line and is known as the *prime center*. Values of pure resistance and the characteristic impedance of transmission line are plotted on this line.

The most common transmission-line impedance in use today is $50 \, \Omega$. For that reason, most of the impedances and reactances are in the $50\text{-}\Omega$ range. It is convenient, then, to have the value of $50 \, \Omega$ located at the prime center of the chart. This means that all points on the $R = 1$ circle represent $50 \, \Omega$, all points on the $R = 0.5$ circle represent $25 \, \Omega$, all points on the $R = 2$ circle represent $100 \, \Omega$, and so on.

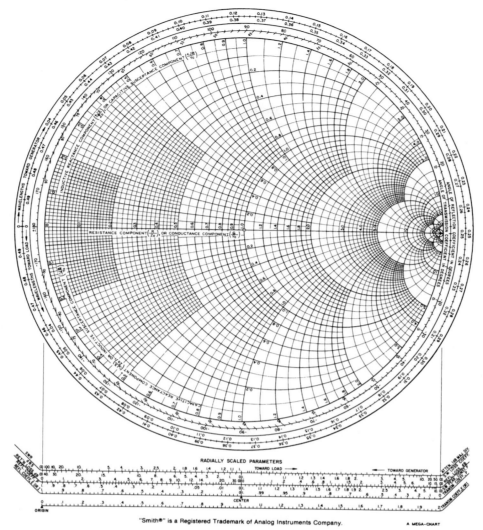

FIG. 13-32 The Smith chart.

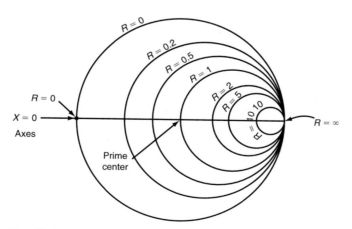

FIG. 13-33 Resistance circles on a Smith chart.

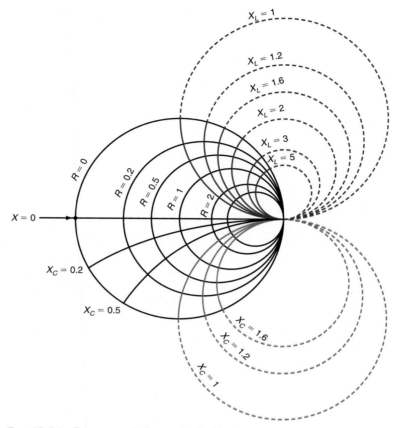

FIG. 13-34 Reactance circles on the Smith chart.

Smith charts are in what is called *normalized* form, with $R = 1$ at the prime center. Users customize the Smith chart for specific applications by assigning a different value to the prime center.

The remainder of the Smith chart is completed by adding reactance circles, as shown in Fig. 13-34. Like the resistive circles, these are eccentric, with all circles meeting at the infinite resistance point. Each circle represents a constant reactance point, with the inductive-reactance circles at the top and the capacitive-reactance circles at the bottom. Note that the reactance circles on the chart are incomplete. Only those segments of the circles within the $R = 0$ line are included on the chart. Like the resistive circles, the reactive circles are presented in normalized form. Compare Figs. 13-33 and 13-34 to the complete Smith chart in Fig. 13-32 before proceeding.

PLOTTING AND READING IMPEDANCE VALUES

The Smith chart in Fig. 13-35 shows several examples of plotted impedance values:

$$Z_1 = 1.5 + j0.5$$
$$Z_2 = 5 - j1.6$$
$$Z_3 = 0.2 + j3$$
$$Z_4 = 0.4 - j0.36$$

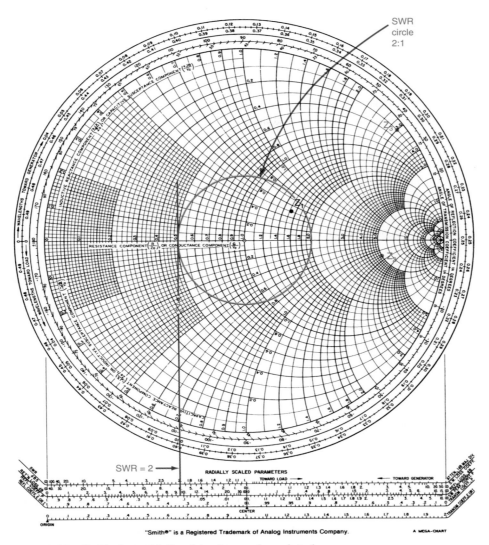

Fig. 13-35 Smith chart with four impedance values plotted.

Locate each of these points on the Smith chart and be sure that you understand how each was obtained.

The impedance values plotted on Fig. 13-35 are normalized values. Scaling the Smith chart to a particular impedance range requires multiplying those values by some common factor. For example, many Smith charts are plotted with 50 Ω at the prime center. The above-listed impedances for a prime-center value of 50 Ω would be as follows:

$$Z_1 = 75 + j25$$
$$Z_2 = 250 - j80$$
$$Z_3 = 25 + j150$$
$$Z_4 = 20 - j18$$

To solve problems with real impedance values, you put them into normalized form by dividing the resistive and reactive numbers by a factor equal to the resistance value at prime center. Then plot the numbers.

When reading values from a normalized Smith chart, convert them to standard impedance form by multiplying the resistance and reactance by a factor equal to the resistance of the prime center.

WAVELENGTH SCALES

The three scales on the outer perimeter of the Smith chart in Fig. 13-32 are wavelength toward the generator, wavelength toward the load, and angle of reflection coefficient in degrees. The scale labeled "wavelength toward generator" begins at the zero resistance and zero reactance line and moves clockwise around to the infinite resistance position. One-half of a circular rotation is 90°, and on the Smith chart scale, it represents a quarter wavelength. One full rotation is a half wavelength. The transmission-line patterns of voltage and current distributed along a line repeat every half wavelength.

The scale labeled "wavelength toward load" also begins at the zero resistance–zero reactance point and goes in a counterclockwise direction for one complete rotation, or one-half wavelength. The infinite resistance point is the quarter-wavelength mark.

The reflection coefficient (the ratio of the reflected voltage to the incident voltage) has a range of 0 to 1, but it can also be expressed as an angle from zero to 360°, from 0 to positive 180°, or from 0 to minus 180°. The zero marker is at the infinite resistance point on the right hand of the resistance line.

SWR CIRCLE

The SWR of a transmission line plotted on the Smith chart is a circle. If the load is resistive and matched to the characteristic impedance of the line, the standing wave ratio is 1. This is plotted as a single point at the prime center of the Smith chart. The impedance of the line is flat at 50 Ω or any other normalized value. However, if the load is not perfectly matched to the prime impedance, standing waves will exist. The SWR in such cases is represented by a circle whose center is the prime center.

To draw an SWR circle, first calculate the SWR using one of the previously given formulas. For this example, assume an SWR of 2. Starting at the prime center, move to the right on the resistance line until the value 2 is encountered. Then, using a drawing compass, place the point at the center and draw a circle through the 2 mark to the right of prime center. The circle should also pass through the 0.5 mark to the left of prime center. The red circle drawn on Fig. 13-35 represents a plot of the impedance variations along an unmatched or resonant transmission line. The variations in the voltage and current standing waves mean there is a continuous variation in the impedance along the line. In other words, the impedance at one point on the unmatched line is different from the impedance at all other points on the line. All the impedance values appear on the SWR circle.

The SWR circle can also be used to determine the maximum and minimum voltage points along the line. For example, the point at which the SWR circle crosses the resistance line to the right of the prime center indicates the point of maximum or peak voltage of the standing wave in wavelengths from the load. This is also the maximum impedance point on the line. The point at which the SWR circle passes through the resistance line to the left of prime center indicates the minimum voltage and impedance points.

The linear scales printed at the bottom of Smith charts are used to find SWR, dB loss, and reflection coefficient. For example, to use the linear SWR scale, simply draw a straight line tangent to the SWR circle and perpendicular to the resistance line on the left side of the Smith chart. Make the line long enough so that it intersects the SWR scale at the bottom of the chart. This has been done on Fig. 13-35. Note that the SWR circle for a value of 2 can be read from the linear SWR scale.

USING THE SMITH CHART: EXAMPLES

As discussed above, when the load does not match the characteristic impedance in a given application, the length of the line becomes a part of the total impedance seen by the generator. The Smith chart provides a way to find this impedance. Once the impedance is known, an impedance-matching circuit can be added to compensate for the conditions and make the line flat and the SWR as close to 1 as possible.

EXAMPLE 1 FOR FIG. 13-36. The operating frequency for a 24-ft piece of RG-58A/U coaxial cable is 140 MHz. The load is resistive, with a resistance of 93 Ω. What is the impedance seen by a transmitter?

The first step is to find the number of wavelengths represented by the 24 ft of cable. A 140-MHz signal has a wavelength of

$$\lambda = 984/f = 984/140 = 7.02 \text{ ft}$$

Remember that coaxial cable has a velocity factor less than 1 due to the slowdown of RF signals in a cable. The velocity factor of RG-58A/U is 0.66. Therefore, one wavelength at 140 MHz is actually as shown in the following expression.

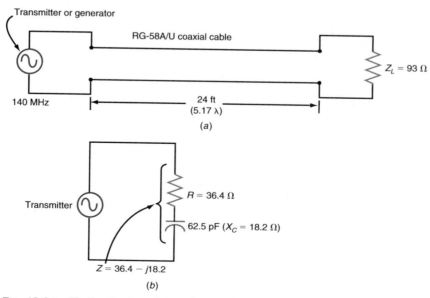

FIG. 13-36 Finding the impedance of a mismatched pair. (*a*) Actual circuit. (*b*) Equivalent circuit.

$$\lambda = 7.02 \times 0.66 = 4.64 \text{ ft}$$

The number of wavelengths represented by the 24-ft cable is $24/4.64 = 5.17$ wavelengths.

As indicated earlier, the impedance variations along a line repeat every half wavelength and therefore every full wavelength; thus for purposes of calculation, we need only the 0.17 part of the above value.

Next, we normalize the Smith chart to the characteristic impedance of the coaxial cable, which is 53.5 Ω. The value of the prime center is 53.5 Ω. The SWR is then computed:

$$\text{SWR} = \frac{Z_l}{Z_0} = \frac{93}{53.5} = 1.74$$

This SWR is now plotted on the Smith chart. See Fig. 13-37, where point X represents the resistive load of 93 Ω.

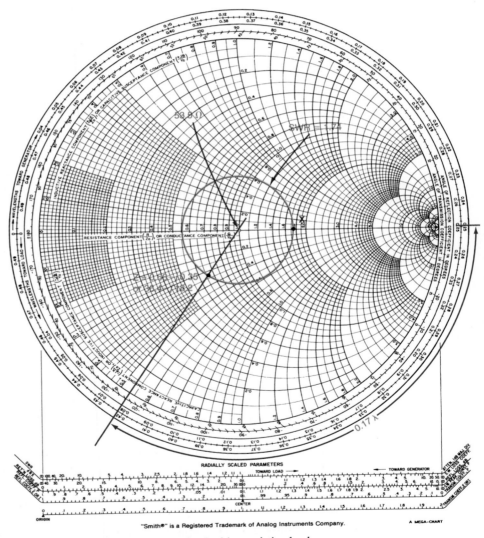

FIG. 13-37 Finding the generator load with a resistive load.

To find the impedance at the transmitter end of the coaxial cable, we move along the line 5.17 wavelengths from the load back to the transmitter or generator. Remember that one full rotation around the Smith chart is one-half wavelength, since the values repeat every half wavelength. Starting at point X, then, move in a clockwise direction (toward the generator) around the SWR circle for 10 complete revolutions, which represents 5 wavelengths. This brings you back to point X.

Continue the clockwise rotation for an additional 0.17 wavelength, stop there, and mark that point on the SWR circle. Draw a line from the prime center to the location that represents 0.17 wavelength. The marking on the right-hand side of point X is the 0.25-wavelength mark. You need to go 0.17 wavelength from that, or $0.25 + 0.17 = 0.42$. Stop at the 0.42 mark on the lower part of the circle. Draw a line from there to the prime center. Refer again to Fig. 13-37.

The point at which the line cuts the SWR circle is the impedance that the generator sees. Reading the values from the chart, you have $R \approx 0.68$ and $X \approx 0.34$. Since the point is in the lower half of the chart, the reactance is capacitive. Thus the impedance at this point is $0.68 - j0.34$.

To find the actual value, convert from the normalized value by multiplying by the impedance of the prime center, or 53.5:

$$Z = 53.5(0.68 - j0.34) = 36.4 - j18.2 \ \Omega$$

The transmitter is seeing what appears to be a 36.4-Ω resistor in series with a capacitor with a reactance of 18.2 Ω of reactance. The equivalent capacitance value is (using $X_C = 1/2\pi fC$ or $C = 1/2\pi fX_C$)

$$C = 1/6.28(140 \times 10^6)(18.2) = 62.5 \ \text{pF}$$

The equivalent circuit is shown in Fig. 13-36(*b*).

EXAMPLE 2 FOR FIG. 13-38.   An antenna is connected to the 24-ft 53.5-Ω RG-58A/U line described in example 1. The load is $40 + j30 \ \Omega$. What impedance does the transmitter see?

The prime center is 53.5 Ω, as before. Before plotting the load impedance on the chart, you must normalize it by dividing by 53.5:

$$Z_l = \frac{40 + j30}{53.5} = 0.75 + j.56 \ \Omega$$

The plot for this value is shown in Fig. 13-38. Remember that all impedances fall on the SWR circle. You can draw the SWR circle for this example simply by placing the compass on the prime center with a radius out to the load impedance point and rotating 360°.

The SWR is obtained from the chart at the bottom of the figure by extending a line perpendicular to resistance axis down to the SWR line. The SWR is about 2:1.

The next step is to draw a line from the prime center through the plotted load impedance so that it intersects the wavelength scales on the perimeter of the curve. Refer again to Fig. 13-38. Since our starting point of reference is the load impedance, we use the "wavelengths toward generator" scale to determine the impedance at the input to the line. At point X is the wavelength value of 0.116.

Now, since the generator is 5.17 wavelengths away from the load and since the readings repeat every half wavelength, you move toward the generator in the clockwise direction 0.17 wavelength: 0.116 and $0.17 = 0.286$. This point is marked Y in Fig. 13-38.

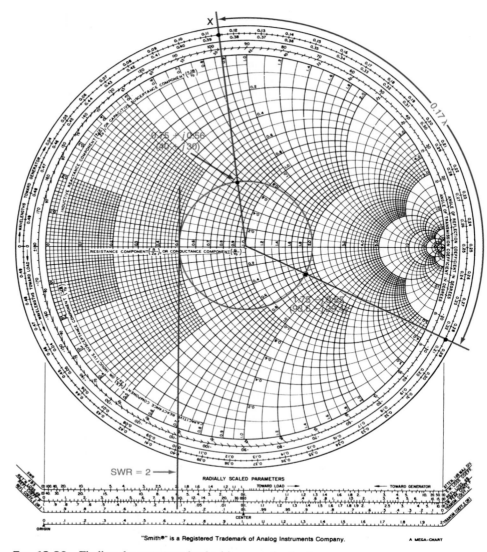

FIG. 13-38 Finding the generator load with a complex load.

Draw a line from point *Y* through the prime center. The point at which it intersects the SWR circle represents the impedance seen by the generator. The normalized value is $1.75 - j0.55$. Correcting this for the 53.5 Ω prime center, we get

$$Z = 53.5(1.75 - j0.55) = 93.6 - j29.4 \ \Omega$$

This is the impedance that the generator sees.

EXAMPLE 3 FOR FIG. 13-39. In many cases the antenna or other load impedance is not known. If it is not matched to the line, the line modifies this impedance so the transmitter sees a different impedance. One way to find the overall impedance as well as the antenna or other load impedance is to measure the combined impedance of the load and the transmission line at the transmitter end, using an impedance bridge. Then the Smith chart can be used to find the individual impedance values.

A 50-ft RG-11/U foam dielectric coaxial cable with a characteristic impedance of 75 Ω and a velocity factor of 0.8 has an operating frequency of 72 MHz. The load is an antenna whose actual impedance is unknown. A measurement at the transmitter end of the cable gives a complex impedance of 82 + j43. What is the impedance of the antenna?

One wavelength at 72 MHz is 984/72 = 13.67 feet. Taking the velocity factor into account yields

$$\lambda = 13.67 \times 0.8 = 10.93 \text{ ft}$$

The length of the 50-ft cable, in wavelengths, is 50/10.93 = 4.57. As before, it is the 0.57-wavelength value that is of practical use.

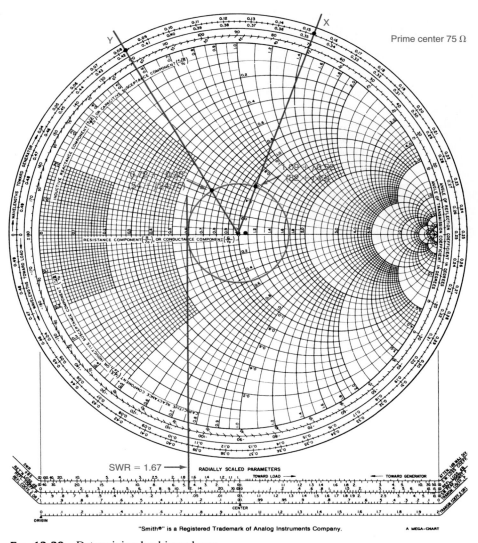

Prime center 75 Ω

SWR = 1.67

"Smith®" is a Registered Trademark of Analog Instruments Company.

Fig. 13-39 Determining load impedance.

For every antenna you see here, there is a transmission line. Coaxial cable is the most common, but waveguides are used for microwaves.

First, the measured impedance is normalized and plotted. The prime center is 75 Ω. $Z = (82 + j43)/75 = 1.09 + j0.52$ is plotted on the Smith chart in Fig. 13-39. The SWR circle is plotted through this point from prime center. A tangent line is then drawn from the circle to the linear SWR chart on the bottom. The SWR is 1.67.

Next, a line is drawn from prime center to the normalized impedance point and through the wavelength plots on the perimeter of the graphs. It intersects the "wavelengths toward load" scale at point X, or 0.346 wavelength. Refer to the figure.

The plotted impedance is what is seen at the generator end, so it is necessary to move around the graph to the load in the counterclockwise direction to find the load impedance. We move a distance equal to the length of the cable, which is 4.57 wavelengths. Nine full rotations from point X represents 4.5 wavelengths, which brings us back to point X. We then rotate 0.07 wavelength more, counterclockwise, to complete the length. This puts us at the $0.346 + .07 = 0.416$ wavelength point, which is designated point Y in Fig. 13-42. We then draw a line from Y to the prime center. The point at which this line intersects with the SWR circle is the actual antenna impedance. The normalized value is $0.72 + j0.33$. Multiplying by 75 gives the actual value of the antenna impedance:

$$Z = 75(0.72 + j0.33) = 54 + j24.75 \ \Omega$$

SUMMARY

Transmission lines are two-wire cables that connect two ends of a communication system, a transmitter to an antenna, or an antenna to a receiver. At high frequencies, they also act like resonant circuits and even reactive components.

The two basic types of wire transmission line are parallel lines, such as twin-lead wire, unshielded twisted pair, and shielded twisted pair, and coaxial cables. Each type is designed to be used with specific types of connectors. Transmission lines can be balanced or unbalanced.

When cables are used to carry RF energy, calculation of wavelength becomes important, and the distance represented by a wavelength in a given transmission depends upon the type of cable used. When the length of a transmission line is no longer than several wavelengths at the signal frequency, an RF generator connected to the transmission line sees a characteristic impedance, which is a function of the inductance, resistance, and capacitance in the circuit. Most transmission lines come with standard fixed values of characteristic impedance and attenuation. Some applications require exact timing and sequencing of signals, especially pulses. A coaxial delay line can be used for this purpose.

A transmission line that is terminated with a load that has a resistive impedance equal in value to the characteristic impedance of the line is called a matched line. The forward and reflected signals on an incorrectly terminated transmission line produce standing waves, which interfere with signal transmission. The standing wave ratio (SWR) is the ratio of the maximum current or voltage along a line to the minimum current or voltage along the line. The ideal-case SWR for transmission is 1.

Standing wave conditions can be useful. For example, quarter- or half wavelength segments of transmission line with an open- or short-circuit load can be used in place of resonant or reactive circuits.

Calculating impedances for transmission-line applications can be complex, as they involve both resistive and reactive elements. The Smith chart offers a graphical method for obtaining the necessary impedance values. Once they have been established, some component, such as an impedance-matching circuit, can be added to compensate for the conditions and bring the SWR as close to 1 as possible.

KEY TERMS

Balanced transmission line
Balun
BNC connector
Characteristic (surge)
    impedance
Coaxial cable
Coaxial connector
Connector
Cutoff frequency
Dielectric constant
F-type connector

Microstrip
Parallel-line connector
Parallel-wire line
PL-259 connector
Polyethylene (PE) dielectric
Polyvinylchloride (PVC)
RCA phonograph connector
Reflection coefficient
Resonant circuit
Shielded twisted pair
    (STP)

SMA connector
Smith chart
Standing-wave ratio
    (SWR)
Stripline
Teflon (PTFE)
Time delay (transit time)
Transmission line
Unbalanced transmission
    line
Velocity factor

SUMMARY

QUESTIONS

1. Name the two basic types of transmission lines. Which is the most widely used?
2. What is the name for a transmission line that has one of its two conductors connected to ground?
3. Name a popular type of balanced line.
4. What is the name given to the distance that a signal travels during one cycle?
5. Name a popular UHF connector used for coaxial cable.
6. What is the coaxial cable connector commonly used on test equipment and in LANs?
7. What is the best coaxial cable connector for UHF and microwave applications?
8. What type of coaxial connector is widely used for cable TV and VCR connections?
9. Describe the equivalent circuit of a transmission line.
10. What determines the characteristic impedance of a transmission line?
11. What is another name for characteristic impedance?
12. What is the importance of the velocity factor in determining cable lengths?
13. How does the cutoff frequency of a coaxial cable vary with its length?
14. Describe the pattern of the current and voltage along a properly matched transmission line.
15. Describe what happens if a transmission line is not terminated in its characteristic impedance.
16. What is the effect on transmitted power when load and transmission-line impedances are mismatched?
17. Under which two conditions will all incident power on a line be reflected?
18. What is the ratio of the reflected voltage to the incident voltage on a transmission line called?
19. What do you call a mismatched transmission line?
20. How does the length of a matched transmission line affect the SWR?
21. Name two ways to implement a series resonant circuit with a transmission line.
22. How do transmission lines that are less than $\lambda/4$ or between $\lambda/4$ and $\lambda/2$ at the operating frequency act?
23. One complete revolution around a Smith chart represents how many wavelengths?

PROBLEMS

1. Compute the wavelength in meters at a frequency of 350 MHz.
2. At what frequency does a line 3.5 in long represent a half wavelength?
3. A coaxial cable has a capacitance of 30 pF/ft and an inductance of 78 nH/ft. What is the characteristic impedance?
4. A coaxial line has a shield braid with an inside diameter of 0.2 in and a center conductor with a diameter of 0.057 in. What is the characteristic impedance? (Dielectric is positive.)
5. What is the velocity factor of a coaxial cable with a dielectric constant of 2.5?
6. A manufacturer's data sheet states that the velocity factor for a specific coaxial cable is 0.7. What is the dielectric constant?
7. Determine the length in feet of a quarter-wavelength coaxial cable with a velocity factor of 0.8 at a frequency of 49 MHz.
8. What is the time delay introduced by a coaxial cable with the specifications given in Problem 7 if the length is 65 ft?
9. What is the time delay of 120 ft of a coaxial cable with a dielectric constant of 0.7?
10. How much phase shift is introduced by a coaxial cable 15 ft long to a 10-MHz sine-wave signal? The dielectric constant is 2.9.
11. What is the attenuation of 350 ft of RG-11 U coaxial cable at 100 MHz?
12. What is the approximate attenuation of 270 ft of type 9913 coaxial cable at 928 MHz?

13. A transmitter has an output power of 3 W that is fed to an antenna through an RG-58A/U cable 20 ft long. How much power reaches the antenna?

14. What should be the transmission-line impedance for optimum transfer of power from a generator to a 52-Ω load?

15. A 52-Ω coaxial cable has a 36-Ω antenna load. What is the SWR?

16. If the load and line impedances of the cable in Prob. 15 are matched, what is the SWR?

17. The maximum voltage along a transmission line is 170 V and the minimum voltage is 80 V. Calculate the SWR and the reflection coefficient.

18. The reflection coefficient of a transmission line is 0.75. What is the SWR?

19. What are the reflection coefficient and SWR of an open or shorted transmission line?

20. A transmission line has an SWR of 1.65. The power applied to the line is 50 W. What is the amount of reflected power?

21. What must be the value of the load impedance on a transmission line if the voltage variations along the line go to zero at regular intervals?

22. An open-wire transmission line 6 in long acts as a parallel resonant circuit at which frequency? (The dielectric is air.)

23. A coaxial cable has a velocity factor of 0.68. How long is a half wavelength of this cable at 133 MHz?

24. Calculate the impedance of a microstrip line whose dimensions are $h = 0.05$ in, $w = 0.125$ in, and $t = 0.002$ in. The dielectric constant is 4.5.

25. What is the length of one quarter wavelength of the microstrip in Prob. 24 at a frequency of 915 MHz?

26. Plot, in normalized form, the following impedances on a Smith chart: $Z_1 = 80 + j25$, $Z_2 = 35 - j98$, assuming a prime central value of 52 Ω.

27. Plot the load impedance $104 - j58$ on a coaxial line on a Smith chart. Assume a characteristic impedance value of 75 Ω. Then find the SWR.

28. If the operating frequency is 230 MHz and the cable length is 30 ft for Prob. 27, what is the impedance at the generator if the velocity factor is 0.66? Use the Smith chart and show all work.

29. What does the SWR circle on a Smith chart look like if a 52-Ω load is connected to a 52-Ω coaxial cable?

CRITICAL THINKING

1. Describe the operation of a transmission line, one-quarter or one-half wavelength at the operating frequency, if the line is shorted or open. How can such a line be used?

2. At what frequencies do transmission lines implemented on a printed circuit board become practical?

3. Compare the characteristics of microstrip and stripline. Which is preferred and why?

4. A 22-ft coaxial cable with an impedance of 50 Ω has a velocity factor of 0.78. The operating frequency is 400 MHz. The impedance of the antenna load combined with the cable impedance as measured at the generator end of the cable is $74 + j66$. What is the antenna impedance as determined on the Smith chart? What is the SWR?

5. Assume that a square wave of 10 MHz is applied to a 200-ft-long RG-58A/U coaxial cable. It is properly terminated. Describe the signal at the load and explain why the signal appears as it does. (See Fig. 13-15.)

6. Explain how a coaxial transmission line (RG-59/U) could be used as a filter to eliminate interference at the input to a receiver. Assume a frequency of 102.3 MHz. Design the filter and calculate size.

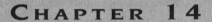

ANTENNAS AND WAVE PROPAGATION

Objectives

After completing this chapter, you will be able to:

◆ *Describe* the characteristics of a radio wave.

◆ *Compute* the length of quarter-wavelength, half-wavelength, and full-wavelength antennas given frequency of operation.

◆ *Name* the basic antenna types and *give* the characteristics of each.

◆ *Explain* how arrays are used to create directivity and gain.

◆ *Describe* ways in which antenna design can be modified to produce an optimal match between the impedances of a transmitter and an antenna.

◆ *Describe* the characteristics of ground waves, sky waves, and space waves.

◆ *Compute* signal strength.

In wireless communications systems, an RF signal generated by a transmitter is sent into free space and eventually picked up by a receiver. The interface between the transmitter and free space and between free space and the receiver is the antenna. At the transmitting end the antenna converts the transmitter RF power into electromagnetic signals that can propagate over long distances; at the receiving end the antenna picks up the electromagnetic signals and converts them into signals for the receiver.

The incredible variety of antennas types used in radio communications are all based on a few key concepts. This chapter introduces all of the most popular and widely used antennas used in HF, VHF, and UHF applications. Microwave antennas are covered in Chap. 15. Also discussed are the characteristics of free space and its ability to propagate signals over long distances. A study of wave propagation—how radio signals are affected by the earth and space in moving from transmitting antenna to receiving antenna—is critical to understanding how to ensure reliable communications over the desired distance at specific frequencies under widely varying conditions.

14-1 ANTENNA FUNDAMENTALS

RADIO WAVES

A radio signal is called an *electromagnetic wave* because it is made up of both electric and magnetic fields. Whenever voltage is applied to the antenna, an electric field is set up. At the same time, this voltage causes current to flow in the antenna, producing a magnetic field. The electric and magnetic fields are at right angles to one another. These electric and magnetic fields are emitted from the antenna and propagate through space over very long distances at the speed of light.

MAGNETIC FIELDS. A *magnetic field* is an invisible force field created by a magnet. An antenna is a type of electromagnet. A magnetic field is generated around a conductor when current flows through it. Figure 14-1 shows the magnetic field, or flux,

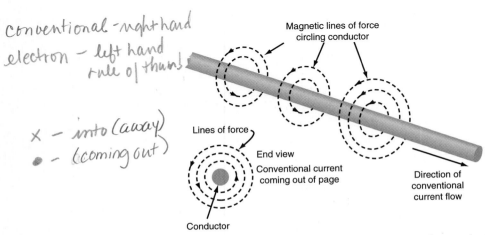

FIG. 14-1 Magnetic field around a current-carrying conductor. Magnetic field strength H in ampere-turns per meter = $H = I/2\pi d$.

around a wire carrying a current. Although the magnetic field is a continuous force field, for calculation and measurement purposes it is represented as individual lines of force. This is how the magnetic field appears in most antennas. The strength and direction of the magnetic field depend upon the magnitude and direction of the current flow.

The strength of a magnetic field (H) produced by a wire antenna is expressed by

$$H = \frac{I}{2\pi d}$$

where I = current (A)
d = distance from wire (m)

The SI unit for magnetic field strength is ampere-turns per meter.

ELECTRIC FIELD. An electric field is also an invisible force field produced by the presence of a potential difference between two conductors. A common example in electronics is the electric field produced between the plates of a charged capacitor (Fig. 14-2). Of course, an electric field exists between any two points across which a potential difference exists.

The strength of an electric field (E) is expressed by

$$E = \frac{q}{4\pi\epsilon d^2}$$

where q = charge between the two points (C)
ϵ = permittivity
d = distance between conductors (m)

H field —
magnetic field

The SI unit for electric field strength is volts per meter.

Permittivity is the dielectric constant of the material between the two conductors. The dielectric is usually air or free space, which has an ϵ value of approximately $8.85 \times 10^{-12}\ \epsilon_r$, where ϵ_r is the dielectric constant of the medium.

MAGNETIC AND ELECTRIC FIELDS IN A TRANSMISSION LINE. Figure 14-3(a) shows the electric and magnetic fields around a two-wire transmission line. Note that at any given time, the wires have opposite polarities. During one-half cycle of the AC input, one wire is positive and the other is negative. During the negative half cycle, the polarity reverses. This means that the direction of the electric field between the

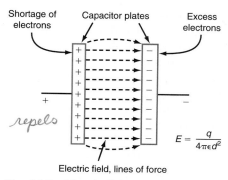

FIG. 14-2 Electric field across the plates of a capacitor.

electric field
from positive to neg
(conventional)

pos. holes pulled towards
electrons

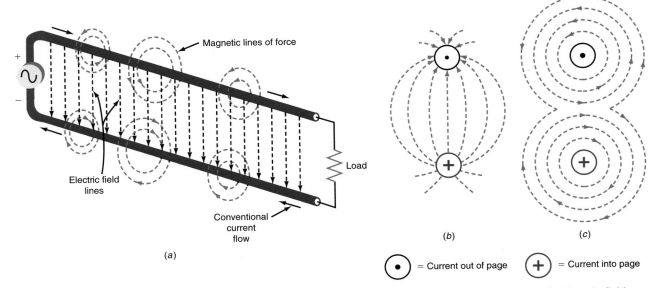

(a)

(b)

(c)

= Current out of page   = Current into page

FIG. 14-3 (a) Magnetic and electric fields around a transmission line. (b) Electric field. (c) Magnetic fields.

(electric)
ε –field
determines
polarity

wires reverses once per cycle. Figure 14-3(b) is a detail of an electric field around conductors.

Note also that the direction of current flow in one wire is always opposite that of the direction in the other wire. Therefore, the magnetic fields combine, as shown in Fig. 14-3(c). The magnetic field lines aid one another directly between the conductors, but as the lines of force spread out, the direction of the magnetic field from one conductor is opposite that of the other conductor, so the fields tend to cancel one another. The cancellation is not complete, but the resulting magnetic field strength is extremely small. While the magnetic and electric fields are shown separately in Fig. 14-3(b) and (c) for clarity, remember that they occur simultaneously and at right angles to one another.

A transmission line, like an antenna, is made up of a conductor or conductors. However, transmission lines unlike antennas, do not radiate radio signals efficiently. The configuration of the conductors in a transmission line is such that the electric and magnetic fields are contained. The closeness of the conductors keeps the electric field primarily concentrated in the transmission-line dielectric. The magnetic fields mostly cancel one another. The electric and magnetic fields do extend outward from the transmission line, but the small amount of radiation that does occur is extremely inefficient.

Figure 14-4(a) shows the electric fields associated with a coaxial cable, and Fig. 14-4(b) shows the magnetic fields associated with a coaxial cable. The electric field lines are fully contained by the outer shield of the cable, so none are radiated. The direction of the electric field lines reverses once per cycle.

The magnetic field around the center conductor passes through the outer shield. However, note that the magnetic field produced by the outer conductor is in the opposite direction of the field produced by the inner conductor. Since the amplitude of the current in both conductors is the same, the magnetic field strengths are equal. The inner and outer magnetic fields cancel one another, and so a coaxial cable does not radiate any electromagnetic energy. That is why coaxial cable is the preferred transmission line for most applications.

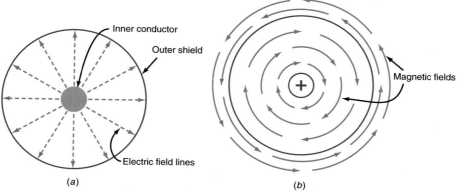

FIG. 14-4 Electric and magnetic fields in a coaxial cable (cross-sectional end view). (*a*) Electric field. (*b*) Magnetic fields.

ANTENNA OPERATION

As stated above, an antenna acts as the interface between a transmitter or receiver and free space. It either radiates or senses an electromagnetic field. But the question is, What exactly is an antenna, and what is the relationship between an antenna and a transmission line? Further, how are the electric and magnetic fields produced?

THE NATURE OF AN ANTENNA. If a parallel-wire transmission line is left open, the electric and magnetic fields escape from the end of the line and radiate into space [Fig. 14-5(*a*)]. This radiation, however, is inefficient and unsuitable for reliable transmission or reception.

The radiation from a transmission line can be greatly improved by bending the transmission-line conductors so they are at a right angle to the transmission line, as shown in Fig. 14-5(*b*). The magnetic fields no longer cancel, and in fact, aid one another. The electric field spreads out from conductor to conductor (Fig. 14-6). The result is an antenna. Optimum radiation occurs if the segment of transmission wire con-

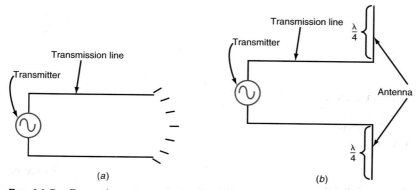

FIG. 14-5 Converting a transmission line into an antenna. (*a*) An open transmission line radiates a little. (*b*) Bending the open transmission line at right angles creates an efficient radiation pattern.

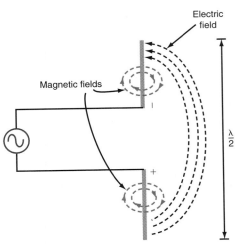

FIG. 14-6 The electric and magnetic fields around the transmission-line conductors when an antenna is formed.

verted into an antenna is a quarter wavelength long at the operating frequency. This makes an antenna that is one-half wavelength long.

An antenna, then, is a conductor or pair of conductors to which is applied the AC voltage at the desired frequency. In Fig. 14-5, the antenna is connected to the transmitter by the transmission line that was used to form the antenna. In most practical applications, the antenna is remote from the transmitter and receiver, and a transmission line is used to transfer the energy between antenna and transmitter or receiver. It is sometimes useful, however, to analyze an antenna as if the conductors were connected directly to the generator or transmitter, as in Fig. 14-7. The voltage creates an electric field and the current creates a magnetic field. Figure 14-7(*a*) shows the magnetic field for one polarity of the generator, and Fig. 14-7(*b*) shows the accompanying electric field. Figure 14-7(*c*) and (*d*) shows the magnetic and electric fields, respectively, for the opposite polarity of the generator.

The magnetic fields vary in accordance with the applied signal from the generator, which is usually a modulated sine-wave carrier. The sinusoidal electric field chang-

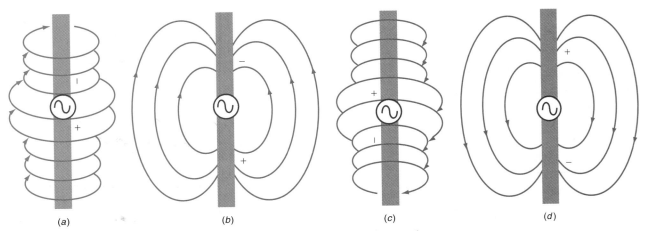

(*a*) (*b*) (*c*) (*d*)

FIG. 14-7 Electric and magnetic fields around an antenna.

ing over time is similar to a current which causes the generation of the sinusoidal magnetic field. A sinusoidally varying magnetic field produces an electric field. Thus the two fields support and sustain one another. The ratio of the electric field strength of a radiated wave to the magnetic field strength is a constant. It is called the impedance of space, or the *wave impedance,* and is 377 Ω. The resulting fields are radiated into space at the speed of light (3×10^8 m/s or 186,400 mi/s).

The antenna that is radiating electromagnetic energy appears to the generator as an ideally resistive electrical load so that the applied power is consumed as radiated energy. In addition to the resistive component, an antenna can have a reactive component. The resistive component is called the *antenna radiation resistance.* This resistance does not dissipate power in the form of heat, as in electronic circuits. Instead, the power is dissipated as radiated electromagnetic energy.

THE ELECTROMAGNETIC FIELD. The electric and magnetic fields produced by the antenna are at right angles to one another and both are perpendicular to the direction of propagation of the wave. This is illustrated in several ways in Fig. 14-8. Figure 14-8(*a*) shows the basic right-angle relationship. Now, assume that you are looking at a small area of the space around an antenna and that the signal is moving either directly out of the page toward you or into the page away from you. Figure 14-8(*b*) is a view of the field lines in Fig. 14-8(*a*), but the perspective is shifted 90° so that you are getting an edge view. Figure 14-8(*c*) shows the variation in strength of the electric and magnetic fields as they move outward from the antenna. Note that the amplitude and direction of the magnetic and electric fields vary in a sinusoidal manner depending upon the frequency of the signal being radiated.

POLARIZATION. *Polarization* refers to the orientation of magnetic and electric fields with respect to the earth. If an electric field is parallel to the earth, the electromagnetic wave is said to be *horizontally polarized;* if the electric field is perpendicular to the earth, the wave is vertically polarized. Antennas that are horizontal to the earth produce horizontal polarization, and antennas that are vertical to the earth produce vertical polarization.

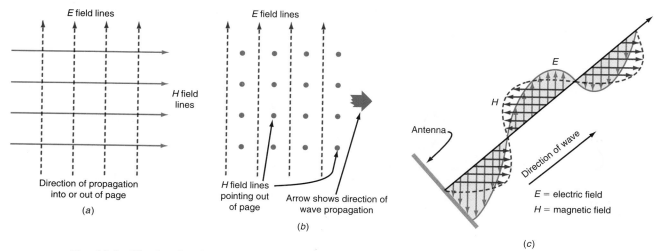

FIG. 14-8 Viewing the electromagnetic wave emitted by an antenna.

A radio wave is an electromagnetic wave, that is, one made up of both an electric field and a magnetic field. In your earlier electronic studies, you learned that a magnetic field is created when electrons flow through a conductor. You also learned that an electric field exists between two oppositely charged bodies. The big question is, how do these two fields get together to form a radio wave, and specifically how are they propagated through space?

In the 1870s, a Scottish physicist, James Clerk Maxwell, wrote a book predicting the existence of electromagnetic waves. Using the basic electrical theories of Faraday, Ohm, Ampère, and other electrical researchers at the time, he postulated that a rapidly changing field of one type would produce the other type of field and vice versa over time. Since one type of field produces the other, the two coexist in a self-sustaining relationship. Maxwell's equations express this relationship mathematically.

Later, in the 1880s, German physicist Heinrich Hertz proved Maxwell's theories by generating radio waves, radiating and manipulating them, and detecting them at a distance. Thus the "wireless" or radio was discovered.

Maxwell defined four basic mathematical relationships between the strength and densities of the electric and magnetic fields as they vary over time. These relationships involve partial differential equations and are, therefore, beyond the scope of this text. However, Maxwell's equations are regularly taught as part of most electrical engineering degree curricula.

Maxwell's equations tell us that an electric field changing with time acts like charges in motion or current flow that, in turn, set up a magnetic field. As the magnetic field changes over time, it sets up an electric field. The electric and magnetic fields interact with one another and sustain one another as they propagate through space at the speed of light. This explains how the electromagnetic wave can exist and move through space after it leaves the antenna or other component or apparatus that initially generates it.

Some antennas produce *circular polarization,* where the electric and magnetic fields rotate as they leave the antenna. There can be right-hand circular polarization (RHCP) and left-hand circular polarization (LHCP); the type depends on the direction of rotation as the signal leaves the antenna. An electric field can be visualized as rotating as if the antenna were connected to a large fan blade. The electric field and the accompanying magnetic field rotate at the frequency of the transmitter, with one full rotation occurring in one cycle of the wave. Looking from the transmitter to the distant receiver, RHCP gives a clockwise rotation to the electric field and LHCP gives a counterclockwise rotation.

For optimal transmission and reception the transmitting and receiving antennas must both be of the same polarization. Theoretically, a vertically polarized wave will produce 0 V in a horizontal antenna and vice versa. But during transmission over long distances, the polarization of waves changes slightly due to the various propagation effects in free space. Thus even when the polarization of the transmitting and receiving antennas is not matched, a signal is usually received.

A vertical or horizontal antenna can receive circular polarized signals, but the signal strength is reduced. When circular polarization is used at both transmitter and receiver, both must use either left- or right-hand polarization if the signal is to be received.

HINTS AND HELPS

When circular polarization is used at both transmitter and receiver, the transmitter and receiver must use either left- or right-hand polarization if the signal is to be received.

ANTENNA RECIPROCITY

The term *antenna reciprocity* means that the characteristics and performance of an antenna are the same whether the antenna is radiating or intercepting an electromagnetic signal. A transmitting antenna takes a voltage from the transmitter and converts it into an electromagnetic signal. A receiving antenna has a voltage induced into it by the electromagnetic signal that passes across it. The voltage is then connected to the receiver. In both cases, the properties of the antenna—gain, directivity, frequency of operation, etc.—are the same. However, an antenna used for transmitting high power, such as in a radio or TV broadcast station, must be constructed of materials that can withstand the high voltages and currents involved. A receiving antenna, no matter what the design, can be made of wire. But a transmitting antenna for high-power applications might, for example, be designed in the same way but be made of larger, heavier material, such as metal tubing.

In most communications systems, the same antenna is used for both transmitting and receiving, and these can occur at different times or be simultaneous. An antenna can transmit and receive at the same time as long as some means is provided for keeping the transmitter energy out of the front end of the receiver. A device called a *diplexer* is used for this purpose.

cannot transmit and receive at the same frequency

DID YOU KNOW?

An antenna can transmit and receive at the same time only if a device such as a diplexer is used to keep the transmitter energy out of the receiver.

THE BASIC ANTENNA

An antenna can be a length of wire, a metal rod, or a piece of tubing. Many different sizes and shapes are used. The length of the conductor is dependent on the frequency of operation. Antennas radiate most effectively when their length is directly related to the wavelength of the transmitted signal. Most antennas have a length that is some fraction of a wavelength. One-half and one-quarter wavelengths are most common.

½ wavelength dipole

An important criterion for radiation is that the length of the conductor be approximately one-half or one-quarter wavelength of the AC signal. A 60-Hz sine-wave signal has a wavelength of $\lambda = 300,000,000/60 = 5,000,000$ m. Since 1 mi $\approx$ to 1609.34 m, one wavelength of a 60-Hz signal is about $5,000,000/1609.34 = 3106.86$ mi. A half wavelength is 1553.43 mi. Very little radiation of an electromagnetic field occurs if antenna wires are less than this length.

The same is true of wires carrying audio signals. A 3-kHz audio signal has a wavelength of $300,000,000/3000 = 100,000$ m, or 62.14 mi. This wavelength is so long compared to the length of the wire normally carrying such signals that little radiation occurs.

However, as frequency is increased, wavelength decreases. At frequencies from about 100 kHz up to 100 GHz, the wavelength is within the range of practical conductors and wires. It is within this range that long-distance radiation occurs. For example, a 300-MHz UHF signal has a wavelength of 1 m, a very practical length.

The other factor that determines how much energy is radiated is how the conductors carrying the signal are arranged. If they are in the form of a cable such as a transmission line with a generator at one end and a load at the other, as shown in Fig. 14-3, very little radiation occurs at any frequency.

As seen in Figs. 14-5 and 14-6, an open transmission line can be made into an antenna simply by bending the conductors out at a right angle with the transmission line. This concept is illustrated again in Fig. 14-9. Such a line has a standing wave such that the voltage is maximum at the end of the line and the current is minimum. One-quarter wave back from the open end, there is a voltage minimum and a current maximum, as shown in the figure. By bending the conductors at a right angle to the transmission line at the quarter-wave point, an antenna is formed. The total length of the antenna is one-half wavelength at the frequency of operation. Note the distribution of the voltage and current standing waves on the antenna. At the center the voltage is minimum and the current is maximum.

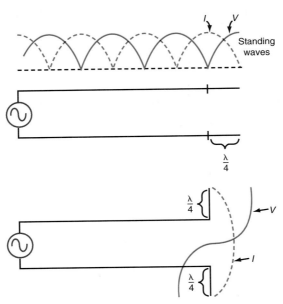

Fig. 14-9 Standing waves on an open transmission line and an antenna.

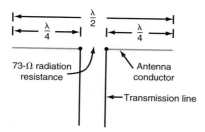

FIG. 14-10 The dipole antenna.

14-2 COMMON ANTENNA TYPES

All of the most common types of antennas used in the communications industry are based on a basic dipole, and most of them are some modified form of the half-wavelength dipole discussed in the last section.

THE DIPOLE ANTENNA

One of the most widely used antenna types is the half-wave dipole shown in Fig. 14-10. This antenna is also formally known as the *Hertz antenna* after Heinrich Hertz, who first demonstrated the existence of electromagnetic waves. Also called a *doublet,* a dipole antenna is two pieces of wire, rod, or tubing that are one-quarter wavelength long at the operating resonant frequency. Wire dipoles are supported with glass, ceramic, or plastic insulators at the ends and middle, as shown in Fig. 14-11. Self-supporting dipoles are made from a stiff metal rod or tubing.

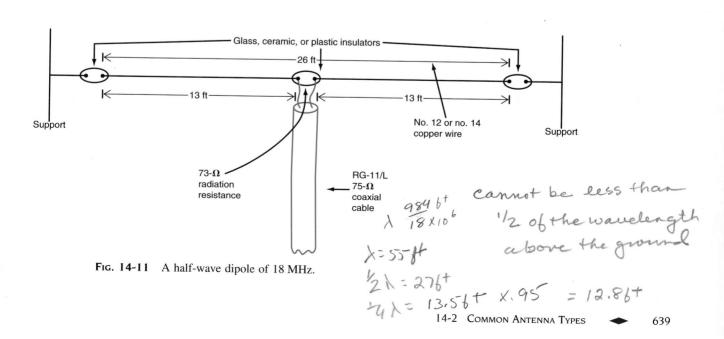

FIG. 14-11 A half-wave dipole of 18 MHz.

RADIATION RESISTANCE. The transmission line is connected at the center. The dipole has an impedance of 73 Ω at its center, which is the radiation resistance. At the resonant frequency, the antenna appears to be a pure resistance of 73 Ω. For maximum power transfer it is important that the impedance of the transmission line match the load. A 73-Ω coaxial cable like RG-59/U is a perfect transmission line for a dipole antenna. RG-11/U coaxial cable with an impedance of 75 Ω also provides an excellent match. When the radiation resistance of the antenna matches the characteristic impedance of the transmission line, the SWR is minimum and maximum power reaches the antenna.

The radiation resistance of the dipole is ideally 73 Ω when the conductor is infinitely thin and the antenna is in free space. Its actual impedance varies depending on conductor thickness, the ratio of diameter to length, and the proximity of the dipole to other objects, especially the earth.

As the conductor thickness increases with respect to the length of the antenna, the radiation resistance decreases. A typical length to diameter ratio is about 10,000 for a wire antenna, for a radiation resistance of about 65 Ω instead of 73. The resistance drops gradually as the diameter increases. With a large tubing conductor, the resistance can drop to as low as 55 Ω.

The graph in Fig. 14-12 shows how radiation resistance is affected by the height of a dipole above the ground. Curves for both horizontally and vertically mounted antennas are given. The resistance varies above and below an average of about 73 Ω, depending on height in wavelengths. The higher the antenna, the less effect the earth and surrounding objects have on it, and the closer the radiation resistance is to the theoretical ideal. Given that radiation resistance is affected by several factors, it often varies from 73 Ω. Nevertheless, a 75-Ω coaxial cable provides a good match, as does 50-Ω cable at lower heights with thicker conductors.

DIPOLE LENGTH. An antenna is a frequency-sensitive device. In Chap. 13, you learned that the formula $\lambda = 984/f$ can be used to calculate one wavelength at a specific frequency and $\lambda = 492/f$ can thus be used to calculate one-half wavelength. For example, one-half wavelength at 122 MHz is $492/122 = 4.033$ ft.

As it turns out, to get the dipole to resonate at the frequency of operation, the physical length must be somewhat shorter than the half wavelength computed with the expression given above, since actual length is related to the ratio of length to diameter,

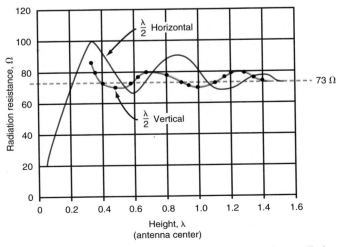

FIG. 14-12 The effect of dipole height above ground on radiation resistance.

conductor shape, Q, the dielectric (when the material is other than air), and a condition known as end effect. *End effect* is a phenomenon caused by any support insulators used at the ends of the wire antenna and has the effect of adding a capacitance to the end of each wire. At frequencies up to about 30 MHz, end effect shortens the antenna by about 5 percent. Thus the actual antenna length is only about 95 percent of the computed length. The formula must be modified as follows:

$$L = \frac{492 \times 0.95}{f} = \frac{467.4}{f}$$

where L = length of half-wave dipole antenna

For half-wave dipole wire antennas used below 30 MHz, the formula $L = 468/f$ provides a ballpark figure, so to speak. Minor adjustments in length can then be made to fine-tune the antenna to the center of the desired frequency range.

For example, an antenna for a frequency of 27 MHz would have a length of 468/27 = 17.333 ft. To create a half-wave dipole, two 8.666-ft lengths of wire would be cut, probably 12 or 14 gauge copper wire. Physically, the antenna would be suspended between two points as high as possible off the ground (see Fig. 14-11). The wire conductors themselves would be connected to glass or ceramic insulators at each end and in the middle to provide good insulation between the antenna and its supports. The transmission line would be attached to the two conductors at the center insulator. The transmission line should leave the antenna at a right angle so that it does not interfere with the antenna's radiation.

At frequencies above 30 MHz, the conductor is usually thicker because thicker rods or tubing, rather than wire, is used. Using thicker materials also shortens the length by a factor of about 2 or 3 percent. Assuming a shortening factor of 3 percent, one-half wavelength would be $492 \times 0.97/f$.

ANTENNA RESONANCE. Because the antenna is a half wavelength at only one frequency, it acts like a resonant circuit. To the generator, the antenna looks like a series resonant circuit (see Fig. 14-13). The inductance represents the magnetic field, and the capacitance represents the electric field. The resistance is the radiation resistance. As always, this resistance varies depending on antenna conductor thickness and height.

If the signal applied to the antenna is such that the antenna is exactly one-half wavelength long, the equivalent circuit will be resonant and the inductive reactance will cancel the capacitive reactance. Only the effect of the radiation resistance will be present, and the signal will radiate.

If the frequency of operation and the antenna length do not match, the equivalent circuit will not be resonant. Instead, like any resonant circuit, it will have a complex impedance made up of resistive and reactive components. If the frequency of operation is too low, the antenna will be too short and the equivalent impedance will be capacitive because the capacitive reactance is higher at the lower frequency. If the

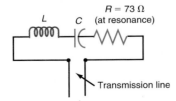

FIG. 14-13   The equivalent circuit of a dipole.

frequency of operation is too high, the antenna will be too long and the equivalent impedance will be inductive because the inductive reactance is higher at the higher frequency.

If the dipole is used at a frequency different from its design frequency, the antenna impedance no longer matches the transmission line impedance, so the SWR rises and power is lost. However, if the frequency of operation is close to that for which the antenna was designed, the mismatch will not be great and the antenna will work satisfactorily despite the higher SWR.

ANTENNA Q AND BANDWIDTH. The bandwidth of an antenna is determined by the frequency of operation and the Q of the antenna according to the familiar relationship $BW = f_r/Q$. Although it is difficult to calculate the exact Q for an antenna, as the above relationship shows, the higher the Q, the narrower the bandwidth. Lowering Q widens bandwidth. In resonant circuits a high Q (> 10) is usually desirable, because it makes the circuit more selective. For an antenna, low Q, and hence wider bandwidth, is desirable so that the antenna can operate over a wider range of frequencies with reasonable SWR. As a rule of thumb, any SWR below $2:1$ is considered good in practical antenna work. Modern communications transceivers rarely operate at just one frequency; typically they operate on a selected channel inside a broader band of frequencies. Further, the transmitter is modulated, so there are sidebands. If the antenna has too high a Q and its bandwidth is too narrow, the SWR will be higher than $2:1$ and sideband clipping can occur.

The Q and thus the bandwidth of an antenna are determined primarily by the ratio of the length of the conductor to the diameter of the conductor. When thin wire is used as the conductor, this ratio is very high, usually in the 10,000 to 30,000 range, resulting in high Q and narrow bandwidth. A length-to-diameter ratio of 25,000 results in a Q of about 14.

If the antenna conductors are made of larger-diameter wire or tubing, the length-to-diameter ratio and Q decrease, resulting in a wider bandwidth. A ratio of 1200 results in a Q of about 8.

When larger-diameter conductors are used to construct an antenna, the larger plate area causes the inductance of the conductor to decrease and the capacitance to increase. The L/C ratio is reduced for a given resonant frequency. Lowering L lowers the inductive reactance, which directly affects Q. Since $Q = X_L/R$, and $BW = f_r/Q$, lowering X_L reduces Q and increases bandwidth.

At UHF and microwave frequencies, antennas are typically made of short, fat conductors, such as tubing. It is not uncommon to see conductors as large as 0.5 in. in diameter. The result is wider bandwidth.

Bandwidth is sometimes expressed as a percentage of the resonant frequency of the antenna. A small percentage means a higher Q, and a narrower bandwidth means a lower percentage. A typical wire antenna has a bandwidth in the range of 3 to 6 percent of the resonant frequency. If thicker conductors are used, this percentage can be increased to the 7 to 10 percent range, which gives lower Q and wider bandwidth.

For example, if the bandwidth of a 24-MHz dipole antenna is given as 4 percent, the bandwidth can be calculated as $0.04 \times 24 = 0.96$ MHz (960 kHz). The operating range of this antenna, then, is the 960-kHz bandwidth centered on 24 MHz. This gives upper and lower frequency limits of 24 MHz ±480 kHz or one-half the bandwidth. The operating range is 23.52 to 24.48 MHz, where the antenna is still close to resonance.

The Q and bandwidth of an antenna are also affected by other factors. In array-type antennas with many conductors, Q is affected by the number of conductors used and their spacing to the dipole. These antennas usually have high Qs and, thus, narrow bandwidths, and so off-frequency operation produces greater changes in SWR than it does with lower-Q antennas.

CONICAL ANTENNAS. Another common way to increase bandwidth is to use a version of the dipole antenna known as the *conical antenna* [Fig. 14-14(a)]. Figure 14-14(b) shows a flat, broadside view of a conical antenna. The overall length of the antenna is 0.73λ or $0.73(984)/f = 718.32/f$. This is longer than the traditional half wavelength of a dipole antenna, but the physical shape changes the necessary dimensions for resonance. The cones are shaped such that the shaded area is equal to the unshaded area. When this is done, the distance between the boundaries marked by A and B in Fig. 14-14(b) is one-half wavelength, less about 5 percent, or approximately $468/f$, where f is in megahertz.

The center radiation resistance of a conical antenna is much higher than the 73 Ω usually found when straight-wire or tubing conductors are used. This center impedance is given by $Z = 120 \ln(\theta/2)$, where Z is the center radiation resistance at resonance and θ is the angle associated with the cone [see Fig. 14-14(b)]. For an angle of 30°, the center impedance is thus $Z = 120 \ln(30/2) = 120 \ln 15 = 120(2.7) = 325\ \Omega$. This is a reasonably good match to 300-Ω twin-lead transmission line.

To use coaxial cable, which is usually desirable, some kind of impedance-matching network is needed to transform the high center impedance to the 50 or 75 Ω characteristic of most coaxial cables.

Cones are difficult to make and expensive, and a popular and equally effective variation of the conical antenna is the *bow-tie antenna*. A two-dimensional cone is a

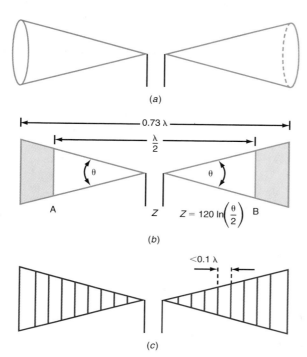

FIG. 14-14 The conical dipole and its variation. (*a*) Conical antenna. (*b*) Broadside view of conical dipole antenna (bow-tie antenna) showing dimensions. (*c*) Open-grill bow-tie antenna.

triangle; therefore, a flat version of the conical antenna looks like two triangles or a bow tie. One bow-tie version of the conical antenna in Fig. 14-14(b) would have the same shape and dimensions, but it would be made of flat aluminum. Bow-tie antennas can also be made of a grill-work of conductors, instead of a flat plate, as shown in Fig. 14-14(c); this configuration reduces wind resistance. If the spacing between the conductors is made less than 0.1 wavelength at the highest operating frequency, the antenna appears to be a solid conductor to the transmitter or receiver.

The primary advantage of conical antennas is their tremendous bandwidth: they can maintain a constant impedance and gain over a 4:1 frequency range. The length of the antenna is computed using the center frequency of the range to be covered. For example, an antenna to cover the 4:1 range at 250 MHz to 1 GHz (1000 MHz) would be cut for a center frequency of (1000 + 250)/2 = 1250/2 = 625 MHz.

DIPOLE POLARIZATION. Most half-wave dipole antennas are mounted horizontally to the earth. This makes the electric field horizontal to the earth; therefore, the antenna is horizontally polarized. Horizontal mounting is preferred at the lower frequencies (<30 MHz) because the physical construction, mounting, and support are easier. This type of mounting also makes it easier to attach the transmission line and route it to the transmitter or receiver.

A dipole antenna can also be mounted vertically, in which case the electric field will be perpendicular to the earth, making the polarization vertical. Vertical mounting is common at the higher frequencies (VHF and UHF), where the antennas are shorter and made of self-supporting tubing.

RADIATION PATTERN AND DIRECTIVITY. The *radiation pattern* of any antenna is the shape of the electromagnetic energy radiated from or received by that antenna. Most antennas have directional characteristics that cause them to radiate or receive energy in a specific direction. Typically that radiation is concentrated in a pattern that has a recognizable geometric shape.

The radiation pattern of a half-wave dipole has the shape of a doughnut. Figure 14-15 shows the pattern with half the doughnut cut away. The dipole is at the center hole of the doughnut, while the doughnut itself represents the radiated energy. To an

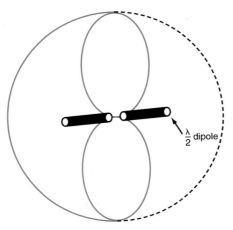

$\frac{\lambda}{2}$ dipole

FIG. 14-15 Three-dimensional pattern of a half-wave dipole.

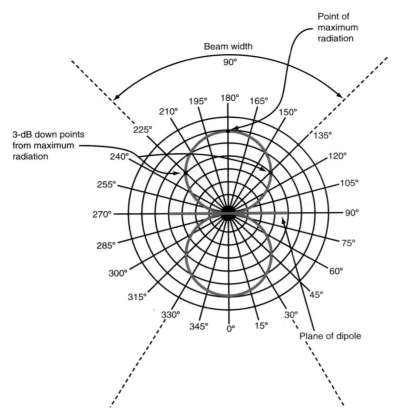

FIG. 14-16 Horizontal radiation pattern of a half-wave dipole.

observer looking down on the top of the dipole, the radiation pattern would appear to be a figure 8, as shown in Fig. 14-16. This horizontal radiation pattern is plotted on a polar coordinate graph in the figure. The center of the antenna is assumed to be at the center of the graph. The dipole is assumed to be aligned with the 90° to 270° degree axis. As shown, the maximum amount of energy is radiated at right angles to the dipole, at 0° and 180°. For that reason, a dipole is what is known as a *directional antenna.* For optimum transmission and reception, the antenna should be aligned broadside to the signal destination or source. For optimal signal transmission, the transmitting and receiving antennas must be parallel to one another.

Whenever a dipole receiving antenna is pointed toward a transmitter, or vice versa, it must be broadside to the direction of the transmitter. If the antenna is at some other angle, then maximum signal will not be received. As can be seen from the radiation pattern in Fig. 14-16, if the end of the receiving antenna is pointed directly at the transmitting antenna, no signal is received. As indicated earlier, this zero-signal condition could not occur in practice, because the radiated wave would undergo some shifts during propagation and, therefore, some minimal signal would be received from the ends of the antenna.

The measure of an antenna's directivity is *beamwidth,* the angle of the radiation pattern over which a transmitter's energy is directed or received. Beamwidth is measured on an antenna's radiation pattern. The concentric circles extending outward from the pattern in Fig. 14-16 indicate the relative strength of the signal as it moves away from the antenna. The beamwidth is measured between the points on the radiation curve that are 3 dB down from the maximum amplitude of the curve. As stated previously, the maximum amplitude of the pattern occurs at 0° and 180°. The 3-dB down points are 70.7 percent of the maximum. The angle formed with two lines extending from the center of the curve to these 3-dB points is the beamwidth. In this example, the beamwidth is 90°. The smaller the beamwidth angle, the more directional the antenna.

ANTENNA GAIN. Gain was previously defined as the output of an electronic circuit or device divided by the input. Obviously, passive devices like antennas cannot have gain in this sense. The power radiated by an antenna can never be greater than the input power. However, a directional antenna can radiate more power *in a given direction* than a nondirectional antenna, and in this "favored" direction, it acts as if it had gain. Antenna gain of this type is expressed as the ratio of the *effective radiated* output power P_{out} to the input power P_{in}. Effective radiated power is the actual power that would have to be radiated by a reference antenna (usually a nondirectional or dipole antenna) in order to produce the same signal strength at the receiver that the actual antenna produces. Antenna gain is usually expressed in decibels:

$$dB = 10 \log \frac{P_{out}}{P_{in}}$$

The power radiated by an antenna with directivity and therefore gain is called the *effective radiated power (ERP)*. The ERP is calculated by multiplying the transmitter power fed to the antenna (P_t) by the power gain (A_p) of the antenna:

$$ERP = A_p P_t$$

To calculate ERP, you must convert from decibels to the power ratio or gain.

The gain of an antenna is usually expressed in reference to either the dipole or an isotropic radiator. An *isotropic radiator* is a theoretical point source of electromagnetic energy. The E and H fields radiate out in all directions from the point source, and at any given distance from the point source, the fields form a sphere. To visualize this, think of a light bulb at the center of a large world globe and the light that illuminates the inside of the sphere as the electromagnetic energy.

In what is known as the *near field* of the antenna, defined as the part of the field less than 10 wavelengths from the antenna at the operating frequency, a portion of the surface area on the sphere looks something like that shown in Fig. 14-17. In the *far field*, 10 or more wavelengths *distant* from the source, the sphere is so large that a small area appears to be flat rather than curved, much as a small area of the earth appears flat. Most far-field analysis of antennas is done by assuming a flat surface area of radiation with the electric and magnetic fields at right angles to one another.

In reality, no practical antennas radiate isotropically; instead, the radiation is concentrated into a specific pattern, as seen in Figs. 14-15 and 14-16. This concentration of electromagnetic energy has the effect of increasing the radiation power over a surface area of a given size. In other words, antenna directivity gives the antenna gain over an isotropic radiator. The mathematics involved in determining the power gain of a dipole over an isotropic source is beyond the scope of this book. As it turns out, this gain is 1.64; in decibels, 10 log 1.64 = 2.15 dB.

Most formulas for antenna gain are expressed in terms of gain in decibels over a dipole. To compute the gain of an antenna with respect to an isotropic radiator, add 2.15 dB to the gain over the dipole. In general, the more concentrated an antenna's energy, the higher the gain. The effect is as if the transmitter power were actually increased by the antenna gain and applied to a dipole.

Antennas do not amplify they just can redirect energy (signal)

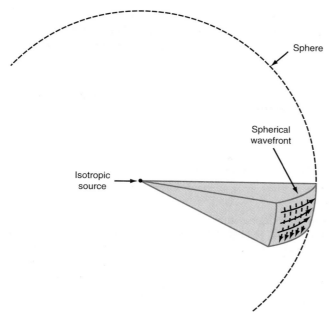

FIG. 14-17   An isotropic source and its spherical wavefront.

FOLDED DIPOLES.   A popular variation of the half-wave dipole is the folded dipole shown in Fig. 14-18(*a*). Like the standard dipole, it is one-half wavelength long. However, it consists of two parallel conductors connected at the ends with one side open at the center for connection to the transmission line. The impedance of this popular antenna is 300 Ω, making it an excellent match for the widely available 300-Ω twin lead. The spacing between the two parallel conductors is not critical, although it is typically inversely proportional to the frequency. For very high-frequency antennas, the spacing is less than 1 in; for low-frequency antennas, the spacing may be 2 or 3 in.

The radiation pattern and gain of a folded dipole are the same as those of a standard dipole. However, folded dipoles usually offer greater bandwidth. The radiation resistance impedance can be changed by varying the size of the conductors and the spacing. Three-element folded dipoles provide even greater bandwidth.

An easy way to make a folded dipole antenna is to construct it entirely of 300-Ω twin lead. A piece of twin lead is cut to a length of one-half wavelength and the two ends soldered together [Fig. 14-18(*b*)]. When calculating the half wavelength, the

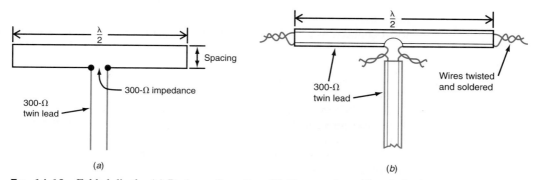

FIG. 14-18   Folded dipole. (*a*) Basic configuration. (*b*) Construction with twin lead.

velocity factor of twin lead (0.8) must be included in the formula. For example, the length of an antenna cut for 100 MHz (the approximate center of the FM broadcast band) is $\lambda/2 = 492/f = 492/100 = 4.92$ ft. Taking into account the velocity factor of twin lead, the final length is $4.92 \times 0.8 = 3.936$ ft.

The center of one conductor in the twin lead is then cut open and a 300-Ω twin lead transmission line is soldered to the two wires. The result is an effective, low-cost antenna that can be used for both transmitting and receiving purposes. Such antennas are commonly used for both TV and FM radio reception.

THE MARCONI GROUND-PLANE VERTICAL ANTENNA

Another widely used antenna, the quarter-wave vertical antenna, is like a vertically mounted dipole antenna. However, it offers major advantages because it is half the length of a dipole antenna.

RADIATION PATTERN. Most half-wave dipole antennas are mounted horizontally, but they can also be mounted vertically. The radiation pattern of a vertically polarized dipole antenna is still doughnut-shaped, but the radiation pattern, as seen from above the antenna is a perfect circle. Such antennas, which transmit an equal amount of energy in the horizontal direction, are called *omnidirectional* antennas. Because of the doughnut shape, the vertical radiation is zero and the radiation at any angle from the horizontal is greatly diminished.

The same effect can be achieved with a quarter-wavelength antenna. Figure 14-19 is a side view of a vertical dipole showing the doughnut-shaped radiation pattern. Half of the pattern is below the surface of the earth. This is called a *vertical radiation pattern.*

Vertical polarization and omnidirectional characteristics can also be achieved using a quarter-wavelength vertical radiator, as in Fig. 14-20(a). This antenna is known as a *Marconi* or *ground-plane antenna.* It is usually fed with coaxial cable; the center conductor is connected to the vertical radiator and the shield is connected to earth ground. With this arrangement, the earth acts as a type of electrical "mirror," effectively providing the other quarter wavelength of the antenna and making it the equivalent of a vertical dipole. The result is a vertically polarized omnidirectional antenna.

GROUND PLANE, RADIALS, AND COUNTERPOISE. The effectiveness of a vertically polarized omnidirectional antenna depends on making good electrical contact with the earth. This can be tricky. Sometimes, a reasonable ground can be obtained by driving a copper rod 5 to 15 ft long into the earth. If the earth is dry and has high resistance due to its content, however, even a ground rod is insufficient. Once a good electrical connection to the earth has been made, the earth becomes what is known as a

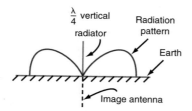

FIG. 14-19 Side view of the radiation pattern of a vertical dipole.

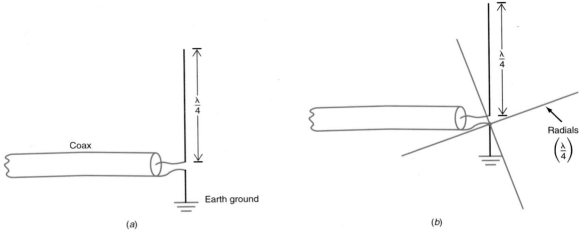

Fig. 14-20 Ground plane antenna. (*a*) Quarter-wavelength vertical antenna. (*b*) Using radials as a ground plane.

ground plane. If a good electrical connection (low resistance) cannot be made to the earth, then an artificial ground plane can be constructed of several quarter-wavelength wires laid horizontally on the ground or buried in the earth (Fig. 14-20(*b*). Four wires are usually sufficient, but in some antenna systems, more are used. These horizontal wires at the base of the antenna are referred to as *radials*. The greater the number of radials, the better the ground and the better the radiation. The entire ground-plane collection of radials is often referred to as a *counterpoise*.

At very high frequencies, when antennas are short, any large, flat metallic surface can serve as an effective ground plane. For example, vertical antennas are widely used on cars, trucks, boats, and other vehicles. The metallic roof of a car makes a superior ground plane for VHF and UHF antennas. In any case, the ground plane must be large enough so that it has a radius of greater than one-quarter wavelength at the lowest frequency of operation.

RADIATION RESISTANCE. The impedance of a vertical ground-plane antenna is exactly one-half the impedance of the dipole, or approximately 36.5 Ω. Of course, this impedance varies depending on the height above ground, the length/diameter ratio of the conductor, and the presence of surrounding objects. The actual impedance can drop to less than 20 Ω for a thick conductor very close to the ground.

Since there is no such thing as 36.5-Ω coaxial cable, 50-Ω coaxial cable is commonly used to feed power to the ground-plane antenna. This represents a mismatch. However, the SWR, 50/36.5 = 1.39, is relatively low and does not cause any significant power loss.

One way to adjust the antenna's impedance is to use "drooping" radials, as shown in Fig. 14-21. At some angle depending upon the height of the antenna above the ground, the antenna's impedance will be near 50 Ω, making it a nearly perfect match for most 50-Ω coaxial cable.

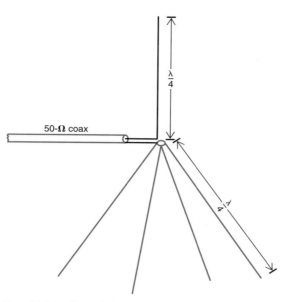

FIG. 14-21 Drooping radials adjust the antenna's impedance to near 50 Ω.

ANTENNA LENGTH.   In addition to its vertical polarization and omnidirectional characteristics, another major benefit of the quarter-wavelength vertical antenna is its length. it is one-half the length of a standard dipole, which represents a significant savings at lower radio frequencies. For example, a half-wave length antenna for a frequency of 2 MHz would have to be $468/f = 468/2 = 234$ ft. Constructing a 234-ft vertical antenna would present a major structural problem, as it would require a support at least that long. An alternative is to use a quarter-wave length vertical antenna, the length of which would only have to be $234/f = 117$ ft. Most low-frequency transmitting antennas use the quarter-wavelength of vertical configuration for this reason. AM broadcast stations in the 535- to 1635-kHz range use quarter-wavelength vertical antennas because they are short, inexpensive, and not visually obtrusive. Additionally, they provide an equal amount of radiation in all directions, which is usually ideal for broadcasting.

For many applications, for example with portable or mobile equipment, it is not possible to make the antenna a full quarter wavelength long. A cordless telephone operating in the 46- to 49-MHz range would have a quarter wavelength of $234/f = 234/46 = 5.1$ ft. A 5-ft whip antenna, or even a 5-ft telescoping antenna, would be impractical for a device you have to hold up to your ear. And a quarter-wavelength vertical antenna for a 27-MHz CB walkie-talkie would have to be an absurdly long $234/27 = 8.7$ ft!

To overcome this problem, much shorter antennas are used and lumped electrical components are added to compensate for the shortening. When a vertical antenna is made less than one-quarter wavelength, the practical effect is a decreased inductance. The antenna no longer resonates at the desired operating frequency, but at a higher frequency. To compensate for this, a series inductor, called a *loading coil,* is connected in series with the antenna coil (Fig. 14-22). The loading coil brings the antenna back into resonance at the desired frequency. The coil is sometimes mounted external to the equipment at the base of the antenna so that it can radiate along with the vertical rod. The coil can also be contained inside a hand-held unit, as in a cordless telephone.

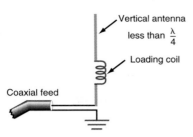

Fig. 14-22 Using a base leading coil to increase effective antenna length.

In some cases, the loading coil is placed at the center of the vertical conductor (see Fig. 14-23). CB antennas and cellular telephone antennas use this method. The CB antennas have a large coil usually enclosed inside a protective housing. The shorter cellular telephone antennas use a built-in self-supporting coil that looks like and serves the dual purpose of a flexible spring.

In both types of inductive vertical antennas, the inductor can be made variable, so that the antenna can be tuned, or the inductor can be made larger than needed and a capacitor connected in series to reduce the overall capacitance and tune the coil to resonance.

Another approach to using a shortened antenna is to increase the effective capacitance represented by the antenna. One way to do this is to add additional conductors at the top of the antenna, as shown in Fig. 14-24. Sometimes referred to as a *top hat,* this structure increases the capacitance to surrounding items bringing the antenna back into resonance. Obviously, such an arrangement is too top-heavy and inconvenient for portable and mobile antennas. However, it is sometimes used in larger fixed antennas at lower frequencies.

One of the most popular variations of the quarter-wavelength vertical antenna is the ⅝-wavelength vertical antenna (5 λ/8, or 0.625 λ). Like a quarter-wavelength ground-plane antenna, the ⅝-wavelength vertical antenna is fed at the base with coaxial cable and has four or more quarter-wavelength radials or the equivalent (i.e., body of a car).

A ⅝-wavelength vertical antenna is one-eighth wavelength longer than one-half wavelength. This additional length gives such an antenna about a 3-dB gain over a basic dipole and quarter-wavelength vertical antenna. The gain comes from concentrating the radiation into a narrower vertical radiation pattern at a lower angle to the horizon. Thus ⅝-wavelength vertical antennas are ideal for long-distance communications.

Since a ⅝-wavelength antenna is not some integer multiple of a one-quarter wavelength, it appears too long to the transmission line; that is, it looks like a capacitive circuit. To compensate for this, a series inductor is connected between the antenna and

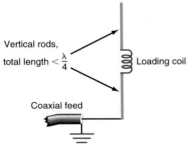

Fig. 14-23 A center-loaded shortened vertical.

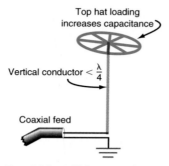

Top hat loading
increases capacitance

Vertical conductor $< \frac{\lambda}{4}$

Coaxial feed

FIG. 14-24 Using a top hat capacitive load to lower the resonant frequency of a shortened vertical antenna.

the transmission line, making the antenna a very close match to a 50-Ω coaxial transmission line. The arrangement is similar to that shown in Fig. 14-22.

The ⅝-wavelength antenna is widely used in the VHF and UHF bands, where its length is not a problem. It is very useful in mobile radio installations, where omnidirectional antennas are necessary but additional gain is needed for reliable transmission over longer distances.

Example 14-1

Calculate the length of the following antennas and state their radiation resistance at 310 MHz: (a) dipole; (b) folded dipole (twin lead; $Z = 300\ \Omega$, velocity factor $= 0.8$); (c) bow tie ($\theta = 35°$, 0.73 λ); (d) ground plane (vertical).

a. $\dfrac{\lambda}{2} = \dfrac{492 \times 0.97}{f} = \dfrac{477.25}{310} = 1.54\ \text{ft}$

Radiation resistance $= 73\ \Omega$

b. $\dfrac{\lambda}{2} = \dfrac{492}{f} = \dfrac{492}{310} = 1.587\ \text{ft}$

Taking the velocity factor into account, $0.8 \times 1.587 = 1.27\ \text{ft}$

c. $\lambda = \dfrac{984}{f} = \dfrac{984}{310} = 3.16\ \text{ft}$

$0.73\lambda = 3.16(0.73) = 2.3\ \text{ft}$

$Z = 120\ \ln \dfrac{\theta}{2} = 120\ \ln \dfrac{35}{2}$

$\quad = 120\ \ln 17.5 = 120(2.862)$

$\quad = 343.5\ \Omega$

d. $\dfrac{\lambda}{4} = \dfrac{234}{f} = 0.755\ \text{ft}$

$Z = 36.5\ \Omega$

GAIN

When a highly directive antenna is used, all the transmitted power is focused in one direction. Because the power is concentrated into a small beam, the effect is as if the antenna has amplified the transmitted signal. Directivity, because it focuses the power, causes the antenna to exhibit gain, which is one form of amplification. An antenna cannot, of course, actually amplify a signal; however, because it can focus the energy in a single direction, the effect is *as if* the amount of radiated power were substantially higher than the power output of the transmitter. The antenna has power gain.

The power gain of an antenna can be expressed as the ratio of the power transmitted (P_{trans}) to the input power of the antenna (P_{in}). Usually, however, power gain is expressed in decibels:

$$dB = 10 \log \frac{P_{trans}}{P_{in}}$$

The total amount of power radiated by the antenna, the ERP, is, as stated previously, the power applied to the antenna multiplied by the antenna gain. Power gains of 10 or more are easily achieved, especially at the higher RFs. This means that a 100-W transmitter can be made to perform like a 1000-W transmitter when applied to an antenna with gain.

RELATIONSHIP BETWEEN DIRECTIVITY AND GAIN

The relationship between the gain and the directivity of an antenna is expressed mathematically by the formula

$$B = \frac{203}{(\sqrt{10})^x}$$

where B = beamwidth of antenna, degrees
x = antenna power gain in decibels divided by 10 (x = dB/10)

The beamwidth is measured at the 3-dB down points on the radiation pattern. It assumes a symmetrical major lobe.

For example, the beamwidth of an antenna with a gain of 15 dB over a dipole is calculated as follows (x = dB/10 = 15/10 = 1.5):

$$B = \frac{203}{(\sqrt{10})^{1.5}} = \frac{203}{3.162^{1.5}} = \frac{203}{5.62} = 36.1°$$

It is possible to solve for the gain given the beamwidth by rearranging the formula and using logarithms:

$$x = 2 \log \frac{203}{B}$$

The beamwidth of an antenna of unknown gain can be measured in the field, and then the gain can be calculated. Assume, for example, a measured -3-dB beamwidth of $7°$. The gain is

$$x = 2 \log \frac{203}{7} = 2 \log 29 = 2(1.462) = 2.925$$

Since $x = \text{dB}/10$, $\text{dB} = 10x$. Therefore the gain in decibels is $2.925 \times 10 = 29.25$ db.

To create an antenna with directivity and gain, two or more antenna elements are combined to form an *array*. Two basic types of antenna arrays are used to achieve gain and directivity: parasitic arrays and driven arrays.

PARASITIC ARRAYS

A *parasitic array* consists of a basic antenna connected to a transmission line plus one or more additional conductors that are not connected to the transmission line. These extra conductors are referred to as *parasitic elements* and the antenna itself is referred to as the *driven element*. Typically the driven element is a half-wave dipole or some variation. The parasitic elements are slightly longer than and slightly less than one-half wavelength long. These parasitic elements are placed in parallel with and near the driven elements. A common arrangement is illustrated in Fig. 14-26. The elements of the antenna are all mounted on a common boom. The boom does not have to be an insulator. Because there is a voltage null at the center of a half-wavelength conductor at the resonant frequency, there is no potential difference between the elements so they can all be connected to a conducting boom with no undesirable effect. In other words, the elements are not "shorted together."

The *reflector,* a parasitic element which is typically about 5 percent longer than the half-wave dipole-driven element, is spaced from the driven element by a distance of 0.15 to 0.25 λ. When the signal radiated from the dipole reaches the reflector, it induces a voltage into the reflector and the reflector produces some radiation of its own. Because of the spacing, the reflector's radiation is mostly in phase with the radiation

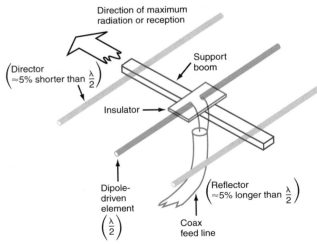

FIG. 14-26 A parasitic array known as a Yagi antenna.

of the driven element. As a result, the reflected signal is added to the dipole signal, creating a stronger, more highly focused beam in the direction of the driven element. The reflector minimizes the radiation to the right of the driven element and reinforces the radiation to the left of the driven element (see Fig. 14-26).

Another kind of parasitic element is a director. A *director* is approximately 4 percent shorter than the half-wave dipole driven element and is mounted in front of the driven element. The directors are placed in front of the driven element and spaced by some distance between approximately one-tenth to two-tenths of a wavelength from the driven element. The signal from the driven element causes a voltage to be induced into the director. The signal radiated by the director then adds in phase to that from the driven element. The result is increased focusing of the signal, a narrower beamwidth, and a higher antenna gain in the direction of the director. The overall radiation pattern of the antenna in Fig. 14-16 is very similar to that shown in Fig. 14-25.

An antenna made up of a driven element and one or more parasitic elements is generally referred to as a *Yagi antenna,* after one of its inventors. The antenna elements are usually made of aluminum tubing and mounted on an aluminum cross member or boom. Since the centers of the parasitic elements are neutral electrically, these elements can be connected directly to the boom. For the best lightning protection, the boom can then be connected to a metal mast and electrical ground. An antenna configured this way is often referred to as a *beam antenna* because it is highly directional and has very high gain. Gains of 3 to 15 dB are possible with beam angles of 20°–40°. The three-element Yagi antenna shown in Fig. 14-26 has a gain of about 8 dB when compared to a half-wave dipole. The simplest Yagi is a driven element and a reflector with a gain of about 3 dB over a dipole. Most Yagis have a driven element, a reflector, and from 1 to 20 directors. The greater the number of directors, the higher the gain and the narrower the beam angle.

Additional gain and directivity can be obtained by combining two or more Yagis to form an array. Two examples are shown in Fig. 14-27. In Fig. 14-27(*a*), two nine-element Yagis are stacked. The spacing determines overall gain and directivity. In Fig. 14-27(*b*), two 9-element Yagis are mounted side by side in the same plane. The gain and beamwidth depend on the spacing between the two arrays and how the transmission line is connected.

The drive impedance of a Yagi varies widely with the number of elements and the spacing. The parasitic elements greatly lower the impedance of the driven element, making it less than 10 Ω in some arrangements. Typically, some kind of impedance-matching circuit or mechanism must be used to attain a reasonable match to 50-Ω coaxial cable, which is the most commonly used feed line.

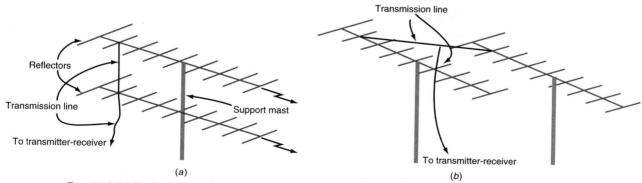

FIG. 14-27 Yagi arrays. (*a*) Stacked. (*b*) Side by side.

Despite the focusing action of the reflector and director in a Yagi, so that most power is radiated in the forward direction, a small amount is lost to the rear, making the Yagi a less-than-perfect directional antenna. Thus, in addition to the gain and beamwidth, another specification of a Yagi is the ratio of the power radiated in the forward direction to the power radiated in the backward direction, or the *front-to-back (F/B) ratio:*

$$\text{F/B} = 10 \log \frac{P_f}{P_b} \text{ dB}$$

where P_f = forward power
$\quad\quad P_b$ = backward power

Relative values of forward and backward power are determined by estimating the sizes of the loops in the radiation pattern for the antenna of interest. When the radiation patterns are plotted in decibels rather than in terms of power, the F/B ratio is simply the difference between the maximum forward value and the maximum rearward value, in decibels.

By varying the number of parasitic elements and their spacing, it is possible to maximize the F/B ratio. Of course, varying the number of elements and their spacing also affects the forward gain. However, maximum gain does not occur with the same conditions required to achieve the maximum F/B ratio. Most Yagis are designed to maximize F/B ratio rather than gain, thus minimizing the radiation and reception from the rear of the antenna.

Yagis are widely used communications antennas because of their directivity and gain. At one time they were widely used for TV reception, but since they are tuned to only one frequency, they are not good for reception or transmission over a wide frequency range. Amateur radio operators are major users of beam antennas. And many other communications services use them because of their excellent performance and low cost.

Beam antennas like Yagis are used mainly at VHF and UHF. For example, at a frequency of 450 MHz, the elements of a Yagi are only about 1 ft long, making the antenna relatively small and easy to handle. The lower the frequency, the larger the elements and the longer the boom. In general, such antennas are only practical above frequencies of about 15 MHz. At 15 MHz, the elements are in excess of 35 ft long. Although antennas this long are difficult to work with, they are still widely used in some communications services.

DRIVEN ARRAYS

The other major type of directional antenna is the *driven array,* an antenna that has two or more driven elements. Each element receives RF energy from the transmission line, and different arrangements of the elements produce different degrees of directivity and gain. The three basic types of driven arrays are the collinear, the broadside, and the end-fire. A fourth type is the wide-bandwidth log-periodic antenna.

COLLINEAR ANTENNAS. *Collinear antennas* usually consist of two or more half-wave dipoles mounted end to end as shown in Fig. 14-28. The lengths of the transmission lines connecting the various driven elements are carefully selected so that

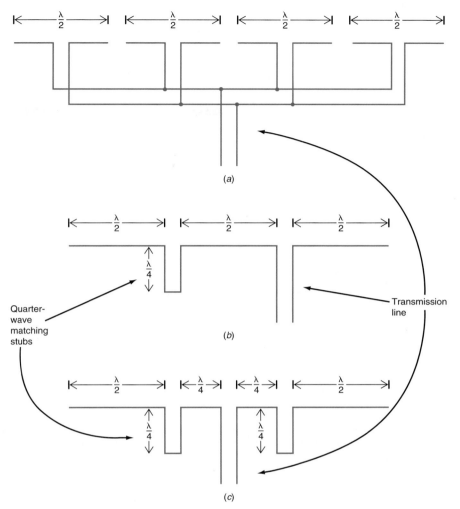

Fig. 14-28 Types of collinear antennas.

the energy reaching each antenna is in phase with all other antennas. With this configuration, the individual antenna signals combine, producing a more focused beam. Three different transmission-line connections are shown in the figure. Like dipole antennas, collinear antennas have a bidirectional radiation pattern, but the two beamwidths are much narrower, providing greater directivity and gain. With four or more half-wave elements, minor lobes begin to appear. A typical collinear pattern is shown in Fig. 14-29.

The collinear antennas in Fig. 14-28(*b*) and (*c*) use half-wave sections separated by shorted quarter-wave matching stubs which ensure that the signals radiated by each half-wave section are in phase. The greater the number of half-wave sections used, the higher the gain and the narrower the beamwidth.

Collinear antennas are generally used only on VHF and UHF bands because their length becomes prohibitive at the lower frequencies. At high frequencies, collinears are usually mounted vertically to provide an omnidirectional antenna with gain.

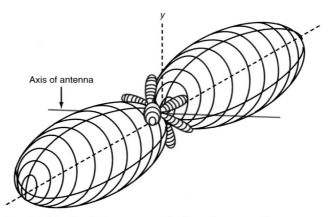

FIG. 14-29 Radiation pattern of a four-element collinear.

BROADSIDE ANTENNAS. A *broadside* array is, essentially, a stacked collinear antenna consisting of half-wave dipoles spaced from one another by one-half wavelengths, as shown in Fig. 14-30. Two or more elements can be combined. Each is connected to the other and to the transmission line. The crossover transmission line ensures the correct signal phasing. The resulting antenna produces a highly directional radiation pattern, not in the line of the elements, as in a Yagi, but broadside or perpendicular to the plane of the array. Like the collinear, the broadside is bidirectional in radiation, but the radiation pattern has a very narrow beamwidth and high gain. Its radiation pattern also resembles that of the collinear, as shown in Fig. 14-29. The radiation pattern is at a right angle to the plane of the driven elements.

END-FIRE ANTENNAS. The end-fire array shown in Fig. 14-31(*a*) uses two half-wave dipoles spaced one-half wavelength apart. Both elements are driven by the transmission line. The antenna has a bidirectional radiation pattern, but with narrower beam-

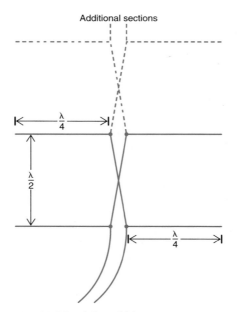

FIG. 14-30 A broadside array.

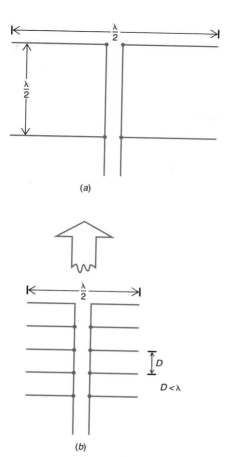

Fig. 14-31 End-fire antennas. (*a*) Bidirectional. (*b*) Unidirectional.

widths and lower gain. The radiation is in the plane of the driven elements as in a Yagi. The end of the fire array in Fig. 14-31(*b*) uses five driven elements spaced some fraction of a wavelength (*D*) apart. By careful selection of the optimal number of elements with the appropriately related spacing, a highly unidirectional antenna is created. The spacing causes the lobe in one direction to be canceled so that it adds to the other lobe, creating high gain and directivity in one direction.

Log-Periodic Antennas. A special type of driven array is the *wide-bandwidth log-periodic antenna* (Fig. 14-32). The lengths of the driven elements vary from long to short and are related logarithmically. The longest element has a length of one-half wavelength at the lowest frequency to be covered, and the shortest element is one-half wavelength at the higher frequency. The spacing is also variable. Each element is fed with a special short transmission-line segment to properly phase the signal. The transmission line is attached at the smallest element. The driving impedance ranges from about 200 to 800 Ω, and depends on the length-to-diameter ratio of the driven element. The result is a highly directional antenna with excellent gain.

The great advantage of the log-periodic antenna over a Yagi or other array is its very wide bandwidth. Most Yagis and other driven arrays are designed for specific frequency or a narrow band of frequencies. The lengths of the elements set the operating frequency. When more than one frequency is to be used, of course, multiple antennas

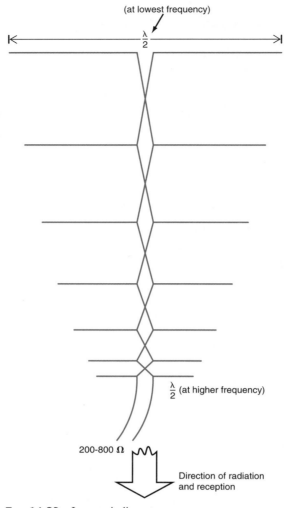

(at lowest frequency)

$\frac{\lambda}{2}$

$\frac{\lambda}{2}$ (at higher frequency)

200-800 Ω

Direction of radiation
and reception

Fig. 14-32   Log-periodic antenna.

can be used. The log-periodic antenna can achieve a 4:1 frequency range, giving it a very wide bandwidth. The driving impedance remains constant over this range. Most TV antennas in use today are of the log-periodic variety so that they can provide high gain and directivity on both VHF and UHF TV channels. Log-periodic antennas are also used in other two-way communications systems where multiple frequencies must be covered.

IMPEDANCE MATCHING

One of the most critical aspects of any antenna system is ensuring maximum power transfer from the transmitter to the antenna. An important part of this, of course, is the transmission line itself. When the characteristic impedance of the transmission line matches the output impedance of the transmitter and the impedance of the antenna itself, the SWR will be 1:1 and maximum power transfer will take place.

Example 14-2

An antenna has a gain of 14 dB. It is fed by an RG-8/U transmission line 250 ft long whose attenuation is 3.6 dB per 100 ft at 220 MHz. The transmitter output is 50 W. Calculate (*a*) transmission line loss, and (*b*) effective radiated power.

a. Total line attenuation:

$$3.6 \text{ dB/100 ft}$$
$$250/100 = 2.5 \text{ hundreds}$$
$$2.5 \times 3.6 = -9 \text{ dB} \qquad \text{(minus dB indicates a loss)}$$
$$\text{dB} = 10 \log P$$

$$P = \log^{-1} \frac{\text{dB}}{10} = \log^{-1} -0.9 = 0.126$$

$$P = \frac{P_{\text{out}}}{P_{\text{in}}}$$

where P_{in} = power into transmission line
P_{out} = power out of transmission line to antenna

$$\frac{P_{\text{out}}}{P_{\text{in}}} = 0.126$$
$$P_{\text{in}} = 50 \text{ W}$$
$$P_{\text{out}} = 0.126(50) = 6.3 \text{ W}$$

b. Gain = 14 dB

$$\text{Power ratio} = \log^{-1} \frac{\text{dB}}{10}$$
$$= \log^{-1} 1.4$$
$$= 25.1$$
$$\text{ERP} = \text{power to antenna} \times \text{power ratio} = 6.3(25.1) = 158.2 \text{ W}$$

The best way to prevent a mismatch between antenna and transmission line is correct design. In practice, when mismatches do occur, some corrections are possible. One solution is to tune the antenna, usually by adjusting its length to minimize SWR. Another is to insert an impedance-matching circuit or antenna tuner between the transmitter and the transmission line, such as the balun or *LC* L, T, or π networks described previously. These circuits can make the impedances equal so that no mismatch occurs, or at least the mismatch is minimized. The ideal is an SWR of 1, but any SWR value below 2 is usually acceptable.

Today, most transmitters and receivers are designed to have an antenna impedance of 50 Ω. Receivers must see an antenna system including transmission line that looks like a generator with a 50-Ω resistive impedance. Transmitters must see an antenna including transmission line with a 50-Ω resistive impedance over the desired operating frequency range if the SWR is to be close to 1 and maximum power is to be transferred to the antenna.

For low-frequency applications (< 30 MHz), using wire antennas high above the ground, the 50-Ω coaxial cable is a reasonable match to the antenna and no further action need be taken. However, if other antenna designs are used, or if the feed-point impedance is greatly different from that of a 50-Ω coaxial cable, some means must be used to ensure that the impedances are matched. For example, most Yagis and other multiple-element antennas have an impedance that is not even close to 50 Ω. Further, an antenna that is not the correct length for the desired frequency will have a large reactive component that can severely affect SWR and power output.

In situations in which a perfect match between antenna, transmission line, and transmitter is not possible, special techniques collectively referred to as *antenna tuning* or *antenna matching* are used to maximize power output and input. Most of these techniques are aimed at impedance matching, that is, making one impedance look like another through the use of tuned circuits or other devices. Several of the most popular methods are described in the following sections.

Q SECTIONS. A *Q section*, or *matching stub*, is a quarter wavelength of coaxial or balanced transmission line of a specific impedance that is connected between a load and a source for the purpose of matching impedances (see Fig. 14-33). A quarter-wavelength transmission line can be used to make one impedance look like another according to the relationship

$$Z_Q = \sqrt{Z_1 Z_0}$$

where Z_Q = characteristic impedance of quarter-wave matching stub or Q section
Z_0 = characteristic impedance of transmission line or transmitter at input of Q section
Z_L = impedance of load

Usually Z_L is the antenna feed-point impedance.

For example, suppose it is desired to use a Q section to match a standard 73-Ω coaxial transmission line to the 36.5-Ω impedance of a quarter-wave vertical antenna. Using the equation, we have

$$Z_Q = \sqrt{36.5 \times 73} = \sqrt{2664.5} = 51.6$$

This tells us that a quarter-wavelength section of 50-Ω coaxial cable will make the 73-Ω transmission line look like the 36.5-Ω antenna or vice versa. In this way maximum power is transferred.

When open-wire balanced transmission lines are used, a special balanced line can be designed and built to achieve the desired impedance. If it is desirable to use standard coaxial cable, then standard available impedances must be used. This means that

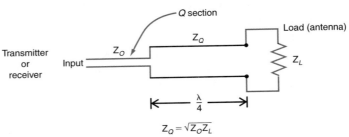

$$Z_Q = \sqrt{Z_0 Z_L}$$

FIG. 14-33 A quarter-wavelength matching stub or Q section.

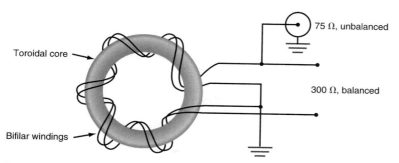

75 Ω, unbalanced

300 Ω, balanced

Toroidal core

Bifilar windings

FIG. 14-34 A bifilar toroidal balun for impedance matching.

the Q section will have to be 50, 75, or 93 Ω, since these are the only three standard values commonly available.

In using Q-section techniques, it is important to take into account the velocity factor of the cable used to make the Q section. For example, if a Q section of 93 Ω is needed, then RG-62/U coaxial cable can be used. If the operating frequency is 24 MHz, the length of one quarter wave is $246/f = 246/24 = 10.25$ ft. The velocity factor for this cable is 0.86, and so the correct length for the cable is $10.25 \times 0.86 = 8.815$ ft.

Two or more Q sections can be used in series to achieve the desired match, with each section performing an impedance match between its input and output impedances.

BALUNS. Another commonly used impedance-matching technique makes use of a balun, a type of transformer used to match impedances. Most baluns are made of a ferrite core, either a toroid or rod, and windings of copper wire. Baluns have a very wide bandwidth and, therefore, are essentially independent of frequency. Baluns can be created for producing impedance matching ratios of 4:1, 9:1, and 16:1. Some baluns have a 1:1 impedance ratio; the sole job of this type of balun is to convert between a balanced and unbalanced condition with no phase reversal.

Figure 14-34 shows a 4:1 balun made with bifilar windings on a toroid core. *Bifilar windings,* in which two parallel wires are wound together as if one around the core, provide maximum coupling between wires and core. The connections of the wires are such as to provide a balanced to unbalanced transformation. A common example of such a balun is one that provides a 75-Ω unbalanced impedance on one end and a 300-Ω balanced impedance on the other. Baluns are fully bidirectional; that is, either end can be used as the input or the output. Such baluns have excellent broadband capabilities and work over a very wide frequency range.

Baluns can also be constructed from coaxial cable. A half-wavelength coaxial cable is connected between the antenna and the feed line, shown in Fig. 14-35. When the length of the half-wave section for such a configuration is calculated, the velocity factor of the coaxial cable must be taken into account. Baluns of this type provide a 4:1 impedance transformation. For example, they can easily convert a 300-Ω antenna to a 75-Ω load and vice versa.

ANTENNA TUNERS. When baluns and matching sections cannot do the job, antenna tuners are used. An *antenna tuner* is a variable inductor, one or more variable capacitors, or a

DID YOU KNOW?

Using an antenna tuner at the transmitter only tricks the transmitter into seeing a low SWR, when in actuality the SWR on the line between the tuner and antenna is still high.

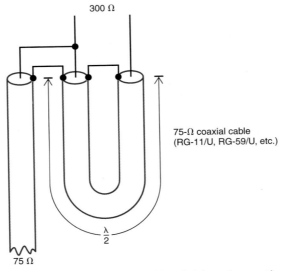

300 Ω

75-Ω coaxial cable
(RG-11/U, RG-59/U, etc.)

$\frac{\lambda}{2}$

75 Ω

FIG. 14-35 A coax balun with a 4:1 impedance ratio.

combination of these components connected in various configurations. L, T, and π networks are all widely used. The inductor and capacitor values are adjusted until the SWR indicates that the impedances match.

It is important to mention that using an antenna tuner at the transmitter only "tricks" the transmitter into seeing a low SWR. In reality, the SWR on the line between the tuner and the antenna is still high.

One popular variation of the antenna tuner is a *transmatch circuit,* which uses a coil and three capacitors to tune the antenna for optimal SWR (see Fig. 14-36). Capacitors C_2 and C_3 are ganged together and tuned simultaneously. Transmatch circuits can be made to work over a wide frequency range, typically from about 2 to 30 MHz. A similar circuit can be made using lower values of inductance and capacitance for matching VHF antennas. At UHF and microwave frequencies, other forms of impedance matching are used.

To use this circuit, an SWR meter is usually connected in series with the antenna transmission line at the transmatch output. L_1 is then adjusted for minimum SWR with the transmitter feeding power to the antenna. Then C_1 and C_2–C_3 are adjusted for minimum SWR. This procedure is repeated several times, alternately adjusting the coil and capacitors until the lowest SWR is reached.

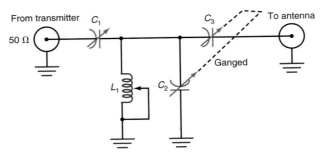

FIG. 14-36 An antenna tuner.

Example 14-3

Calculate the length of the impedance-matching section needed for a Q section to match a 50-Ω transmitter output to a Yagi with a feed impedance of 172 Ω. The operating frequency is 460 MHz.

$$Z_Q = \sqrt{Z_0 Z_l} = \sqrt{50(172)} = \sqrt{8600} = 92.74 \ \Omega.$$

92.74 Ω is a close match to RG-62 A/U coaxial cable with an impedance of 93 Ω (see Fig. 13-14). The velocity factor is 0.86

$$\frac{\lambda}{4} = \frac{246}{f} \times VF = \frac{246}{460} \times 0.86 = 0.5347 \times 0.86 = 0.46 \text{ ft } (5.5 \text{ in})$$

Example 14-4

Calculate the length of the impedance-matching section needed for a $\lambda/4$ balun to convert 75 Ω to 300 Ω. The operating frequency is 460 MHz. RG-59 A/U coaxial cable is the line of choice; the velocity factor of RG-59 A/U cable is 0.66.

$$\frac{\lambda}{4} = 0.5347 \times 0.66 = 0.353 \text{ ft } (4.23 \text{ in})$$

14-3 RADIO-WAVE PROPAGATION

Once a radio signal has been radiated by an antenna, it travels or propagates through space and ultimately reaches the receiving antenna. The energy level of the signal decreases rapidly with distance from the transmitting antenna. The electromagnetic wave is also affected by objects that it encounters along the way such as trees, buildings, and other large structures. In addition, the path that an electromagnetic signal takes to a receiving antenna depends upon many factors, including the frequency of the signal, atmospheric conditions, and the time of day. All these factors can be taken into account to predict the propagation of radio waves from transmitter to receiver.

THE OPTICAL CHARACTERISTICS OF RADIO WAVES

Radio waves act very much like light waves. Light waves can be reflected, refracted, diffracted, and focused by other objects. The focusing of waves by antennas to make them more concentrated in a desired direction is comparable to a lens focusing light waves into a narrower beam. Understanding the optical nature of radio waves gives a better feel for how they are propagated over long distances.

REFLECTION. Light waves are reflected by a mirror. Any conducting surface looks like a mirror to a radio wave, and so radio waves are reflected by any conducting surface they encounter along a propagation path. All metallic objects reflect radio waves, especially if the metallic object is at least one-half wavelength at the frequency of operation. Any metallic object on a transmission path, such as building parts, water towers, automobiles, airplanes, and even power lines, cause some reflection. Reflection is also produced by other partially conductive surfaces, such as the earth and bodies of water.

Radio-wave reflection follows the principles of light-wave reflection. That is, the angle of reflection is equal to the angle of incidence, as shown in Fig. 14-37. The radio wave is shown as a wavefront. To simplify the drawing, only the electrical lines of force, designated by the arrows, are shown. The angle of incidence is the angle between the incoming line of the wave and a perpendicular line to the reflecting surface. The angle of reflection is the angle between the reflected wave and the perpendicular line.

A perfect conductor would cause total reflection: all the wave energy striking the surface would be reflected. Since there are no perfect conductors in the real world, the reflection is never complete. But if the reflecting surface is a good conductor, such as copper or aluminum, and is large enough, most of the wave is reflected. Poorer conductors simply absorb some of the wave energy. In some cases, the wave penetrates the reflecting surface completely.

As the figure shows, the direction of the electric field approaching the reflecting surface is reversed from that leaving the surface. The reflection process reverses the polarity of a wave. This is equivalent to a 180° phase shift.

REFRACTION. Refraction is the bending of a wave due to the physical makeup of the medium through which the wave passes. The speed of a radio wave, like the speed of light, is approximately 300,000,000 m/s (186,400 mi/s) in free space, that is, in a vacuum or the air. When light passes through another medium, such as water or glass, it slows down. The slowing down as light enters or exits a different medium causes the light waves to bend.

The same thing happens to a radio wave. As a radio wave travels through free space, it encounters air of different densities, the density depending on the degree of ionization (caused by an overall gain or loss of electrons). This change of air density causes the wave to be bent.

The degree of bending depends on the index of refraction of a medium (n), obtained by dividing the speed of a light (or radio) wave in a vacuum and the speed of a

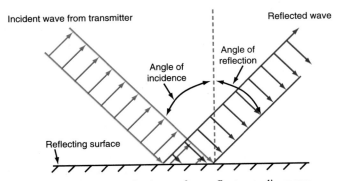

FIG. 14-37 How a conductive surface reflects a radio wave.

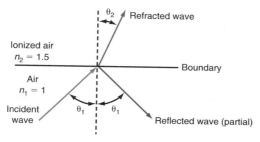

Fig. 14-38 How a change in the index of refraction causes bending of a radio wave.

light (or radio) wave in the medium that causes the wave to be bent. Since the speed of a wave in a vacuum is almost the same as the speed of a wave in air, the index of refraction for air is very close to 1. The index of refraction for any other medium will be greater than 1, how much greater depending upon how much the wavespeed is slowed.

Figure 14-38 shows how a wave is refracted. The incident wave from a transmitter travels through air where it meets a region of ionized air which causes the speed of propagation to slow. The incident wave has an angle of θ_1 to a perpendicular to the boundary line between air and the ionized air. The bent (refracted) wave passes through the ionized air; it now takes a different direction, however, which has an angle of θ_2 with respect to the perpendicular.

The relationship between the angles and the indices of refraction are given by a formula known as *Snell's law:*

$$n_1 \sin \theta_1 = n_2 \sin \theta_2$$

where n_1 = index of refraction of initial medium
n_2 = index of refraction of medium into which wave passes
θ_1 = angle of incidence
θ_2 = angle of refraction

Note that there will also be some reflection from the boundary between the two media because the ionization causes the air to be a partial conductor. However, this reflection is not total; a great deal of the energy passes into the ionized region.

DIFFRACTION. Remember that light and radio waves travel in a straight line. If an obstacle appears between a transmitter and receiver, some of the signal is blocked, creating what is known as a *shadow zone* [see Fig. 14-39(a)]. A receiver located in the shadow zone cannot receive a complete signal. However, some signal usually gets through due to the phenomenon of *diffraction,* the bending of waves around an object. Diffraction is explained by what is known as *Huygens' principle.* Huygens' principle is based on the assumption that all electromagnetic waves, light as well as radio waves, radiate as spherical wavefronts from a source. Each point on a wavefront at any given time can be considered as a point source for additional spherical waves. When the waves encounter an obstacle, they pass around it, above it, and on either side. As the wavefront passes the object, the point sources of waves at the edge of the obstacle create additional spherical waves that penetrate and fill in the shadow zone. This phenomenon, sometimes called knife-edge diffraction, is illustrated in Fig. 14-39(b).

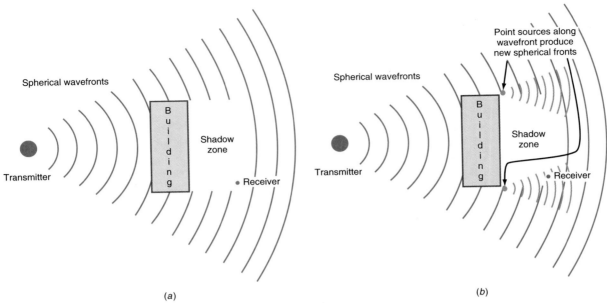

Fig. 14-39 Diffraction causes waves to bend around obstacles.

RADIO-WAVE PROPAGATION THROUGH SPACE

The three basic paths that a radio signal can take through space are the ground wave, the sky wave, and the space wave.

GROUND WAVES. *Ground* or *surface waves* leave an antenna and remain close to the earth (see Fig. 14-40). Ground waves actually follow the curvature of the earth and can, therefore, travel at distances beyond the horizon. Ground waves must have vertical polarization to be propagated from an antenna. Horizontally polarized waves are absorbed or shorted by the earth.

Ground-wave propagation is strongest at the low- and medium-frequency ranges. That is, ground waves are the main signal path for radio signals in the 30-kHz to 3-MHz range. The signals can propagate for hundreds and sometimes thousands of miles at these low frequencies. AM broadcast signals are propagated primarily by ground waves during the day and by sky waves at night.

The conductivity of the earth determines how well ground waves are propagated. The better the conductivity, the less the attenuation and the greater the distance the waves can travel. The best propagation of ground waves occurs over salt water because the water is an excellent conductor. Conductivity is usually lowest in low-moisture areas such as deserts.

At frequencies beyond 3 MHz, the earth begins to attenuate radio signals. Objects on the earth and features of the terrain become the same order of magnitude in size as the wavelength of the signal and thus absorb or adversely affect the signal. For this reason, the ground-wave propagation of signals above 3 MHz is insignificant except within several miles of the transmitting antenna.

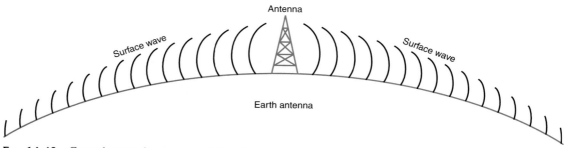

FIG. 14-40 Ground- or surface-wave radiation from an antenna.

SKY WAVES. Sky-wave signals are radiated by the antenna into the upper atmosphere, where they are bent back to earth. This bending of the signal is caused by refraction in a region of the upper atmosphere known as the *ionosphere* (see Fig. 14-41). Ultraviolet radiation from the sun causes the upper atmosphere to *ionize*, that is, to become electrically charged. The atoms take on or lose electrons, becoming positive or negative ions. Free electrons are also present. At its lowest point, the ionosphere is approximately 30 mi (50 km) above the earth and extends as far as 250 mi (400 km) from the earth. The ionosphere is generally considered to be divided into three layers, the D layer, the E layer, and the F layer; the F layer is subdivided into the F_1 and F_2 layers.

The D and E layers, the farthest from the sun, are weakly ionized. They exist only during daylight hours, during which they tend to absorb radio signals in the medium-frequency range from 300 kHz to 3 MHz.

The F_1 and F_2 layers, the closest to the sun, are the most highly ionized and have the most effect on radio signals. The F layers exist during both the day and at night.

The primary effect of the F layer is to cause refraction of radio signals when they cross the boundaries between layers of the ionosphere with different levels of ionization. When a radio signal goes into the ionosphere, the different levels of ionization cause the radio waves to be gradually bent. The direction of bending depends on the angle at which the radio wave enters the ionosphere and the different degrees of ionization of the layers as determined by Snell's law.

Figure 14-41 shows the effects of refraction with different angles of radio signals entering the ionosphere. When the angle is large with respect to the earth, the radio signals are bent slightly, pass on through the ionosphere, and are lost in space. Radiation directly vertical from the antenna, or 90° with respect to the earth, pass through the ionosphere. As the angle of radiation decreases from the vertical, some signals continue to pass through the ionosphere. But at some critical angle, which varies with signal frequency, the waves begin to be refracted back to the earth. The smaller the angle with respect to the earth, the more likely it is that the waves will be refracted and sent back to earth. This effect is so pronounced that it actually appears as though the radio wave has been reflected by the ionosphere.

In general, the higher the frequency, the smaller the radiation angle required for refraction to occur. At very high frequencies, essentially those above about 50 MHz, refraction seldom occurs regardless of the angle. VHF, UHF, and microwave signals usually pass through the ionosphere without bending. However, during a period of sunspot activity,

> ### DID YOU KNOW?
>
> AM broadcast signals are propagated primarily by ground waves during the day and by sky waves at night.

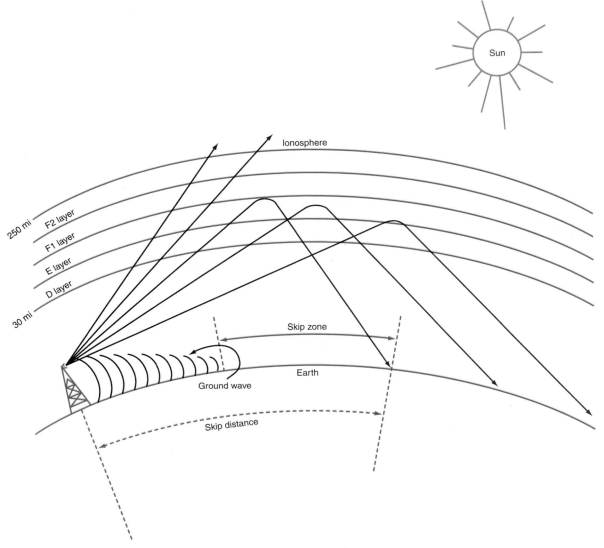

Fig. 14-41 Sky-wave propagation.

or other unusual electromagnetic phenomena, VHF and even UHF waves may be refracted by the ionosphere.

Reflected radio waves are sent back to earth with minimum signal loss. The result is that the signal is propagated over an extremely long distance. This effect is most pronounced in the 3- to 30-MHz or shortwave range, which permits extremely long distance communications.

In some cases, the signal reflected back from the ionosphere strikes the earth, is reflected back up to the ionosphere, and is re-reflected, back to earth. This phenomenon is known as *multiple-skip* or *multiple-hop* transmission. For strong signals and ideal ionospheric conditions, as many as 20 hops are possible. Multiple-hop transmission can extend the communications range by many thousands of miles. The maximum distance of a single hop is about 2000 miles, but with multiple hops, transmissions all the way around the world are possible.

The distance from the transmitting antenna to the point on earth where the first refracted signal strikes the earth to be reflected is referred to as the *skip distance* (see Fig. 14-41). If a receiver lies in that area between the place where the ground wave is fully attenuated and the point of first reflection from the earth, no signal will be received. This area is called the *skip zone*.

SPACE WAVES. The third method of radio-signal propagation is by *direct waves,* or *space waves*. A direct wave travels in a straight line directly from the transmitting antenna to the receiving antenna. Direct-wave radio signaling is often referred to as *line-of-sight communications*. Direct or space waves are not refracted, nor do they follow the curvature of the earth.

Because of their straight-line nature, direct-wave at signals travel horizontally from the transmitting antenna until they reach the horizon, at which point they are blocked, as shown in Fig. 14-42. If a direct-wave signal is to be received beyond the horizon, the receiving must be high enough to intercept it.

Obviously, the practical transmitting distance with direct waves is a function of the height of the transmitting and receiving antennas. The formula for computing the distance between a transmitting antenna and the horizon is

$$d = \sqrt{2h_t}$$

where h_t = height of transmitting antenna, ft
d = distance from transmitter to horizon, mi

This is called the *radio horizon*.

To find the practical transmission distance (D) for straight-wave transmissions, the height of the receiving antenna must be included in the calculations:

$$D = \sqrt{2h_t} + \sqrt{2h_r}$$

where h_r = height of receiving antenna, ft

For example, if a transmitting antenna is 350 ft high and the receiving antenna is 25 ft high, the longest practical transmission distance is

$$D = \sqrt{2(350)} + \sqrt{2(25)} = \sqrt{700} + \sqrt{50} = 26.46 + 7.07 = 33.53 \text{ mi}$$

Line-of-sight communications is characteristic of most radio signals with a frequency above approximately 30 MHz, particularly VHF, UHF, and microwave signals. Such signals pass through the ionosphere and are not bent. Transmission distances at

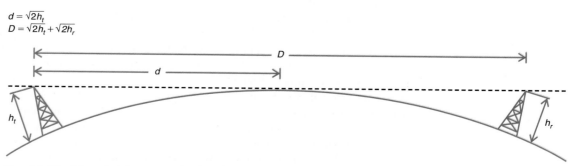

FIG. 14-42 Line-of-sight communications by direct or space waves.

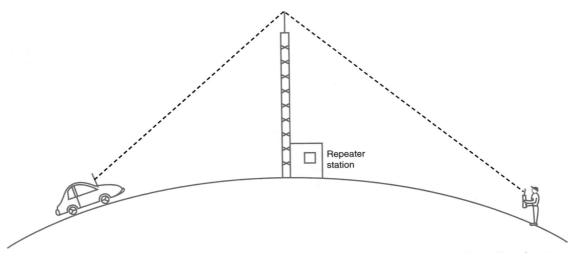

Fig. 14-43 How a repeater extends the communications distance of mobile radio units at VHF and UHF.

those frequencies are extremely limited, and it is obvious why very high transmitting antennas must be used for FM and TV broadcasts. The antennas for transmitters and receivers operating at the very high frequencies are typically located on top of tall buildings or on mountains, which greatly increases the range of transmission and reception.

To extend the communications distance at VHF, UHF, and microwave frequencies, special techniques have been adopted. The most important of these is the use of repeater stations (see Fig. 14-43). A repeater is a combination of a receiver and a transmitter operating on separate frequencies. The receiver picks up a signal from a remote transmitter, amplifies it, and retransmits it (on another frequency) to a remote receiver. Usually the repeater is located between the transmitting and receiving stations, and therefore, extends the communications distance. Repeaters have extremely sensitive receivers and high-power transmitters, and their antennas are located at high points.

Repeaters are widely used to increase the communications range for mobile and hand-held radio units, the antennas for which are naturally not very high off the ground. The limited transmitting and receiving range of such units can be extended considerably by operating them through a repeater located at some high point.

In high-activity areas, a repeater used for mobile units will become overloaded when too many users try to access it at the same time. When that happens, some users have to wait until free time becomes available, continuing to call until they get through. Such access delays are only a nuisance in some cases, but are clearly not acceptable when emergency services are unable to get through.

Although multiple repeaters can be used to ease overcrowding, they are often inadequate because communications activity is not equally distributed among them. This problem is solved by using *trunked repeater systems* where two or more repeaters are under the control of a computer system that can transfer a user from an assigned but busy repeater to another, available repeater. Thus the communications load is spread around between several repeaters.

Repeaters can also be used in series, as shown in Fig. 14-44. Each repeater contains a receiver and a transmitter. The original signal is picked up, amplified, and retransmitted on a different frequency to a second repeater, which repeats the process. Typically, such relay stations are located 20 to 60 mi apart, mostly at high elevations to ensure reliable communications over very long distances. Microwave relay stations are used by many telephone companies for long-distance communications.

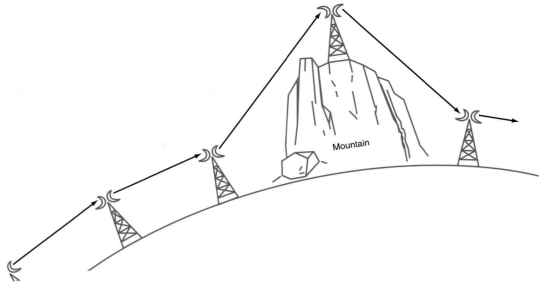

Fig. 14-44 Using repeater stations to increase communications distances at microwave frequencies.

The "ultimate" repeater is, of course, a communications satellite. Most communications satellites are located in a geostationary orbit 22,500 mi above the equator. Since at that distance, it takes exactly 24 h to rotate around the earth, communications satellites appear stationary. They act as fixed repeater stations. Signals sent to a satellite are amplified and retransmitted back to earth long distances away. The receiver-transmitter combination within the satellite is known as a *transponder.* Most satellites have many transponders, so that multiple signals can be relayed, making possible worldwide communications at microwave frequencies. (See Chap. 16.)

CALCULATING RECEIVED POWER

A transmitted signal is radiated at a specific power level. The output power of a transmitter can be accurately determined by calculation or measurement. That power level is increased if the antenna has gain due to improved directivity. As a signal leaves an antenna, it immediately begins to become attenuated. Basically, the degree of attenuation is proportional to the square of the distance between the transmitter and receiver. As discussed previously, other factors also affect attenuation. Ground-wave signals are greatly attenuated by objects on the earth, which block the signals and reduce their level at the receiver. In sky-wave propagation, the ionospheric conditions and the number of hops determine signal level at the receiver, with each hop further reducing signal level.

Despite these factors, it is possible to predict the approximate power level at a receiver, and such calculations are quite accurate for the short distances characteristic of direct- or space-wave transmission.

In analyzing the transmission of radio waves, it is often useful to start with the concept of an isotropic radiator, or point source of radio waves. That is, the signal radiates spherically in all directions.

The power density at a given distance from an isotropic radiator is predicted by the formula

$$P_d = \frac{P_t}{4\pi d^2}$$

where P_d = power density of signal, W/m^2
d = distance from point source, m
P_t = total transmitted power, W

The distance d is really the radius of an imaginary sphere, enclosing the source and $4\pi d^2$ is the area of that sphere at any given distance. However, since practical antennas are not purely isotropic sources, this formula must be modified somewhat. For example, if the transmitting antenna is a dipole, the dipole has a gain of 1.64 (or 2.15 dB) over an isotropic source, so the result must be multiplied by 1.64.

Suppose a transmitter puts a 50-W signal into a dipole antenna. The power density of the signal at a distance of 30 mi (4.83 km, or 48,300 m) is

$$P_d = \frac{1.64P_t}{4\pi d^2} = \frac{1.64(50)}{4(3.1416)(48{,}300)^2}$$
$$= 3 \times 10^{-9} = 3 \text{ nW/m}^2$$

Knowing the power density at a given distance is not a particularly useful thing. However, the formula for power at a given distance can be expanded to derive a general formula for computing the actual power value of a signal at a receiving antenna, as shown on the next page.

The expansive use of telephone networks has resulted in integratively designed transmission of telephone and data communications. The antennas are covered with a fiberglass material to protect them from the weather. The covering material does not impede the RF signals.

$$P_r = \frac{P_t G_t G_r \lambda^2}{16\pi^2 d^2}$$

where λ = signal wavelength, m

d = distance from transmitter, m

P_r and P_t = received and transmitted power, respectively

G_r and G_t = receiver and transmitter antenna gains expressed as a power ratio and referenced to an isotropic source

If the gains are those in reference to a dipole, each must be converted to a power ratio and multiplied by 1.64 before being used in the formula.

This formula is normally used only for ground-wave, direct-wave, or space-wave calculations. It is not used for sky-wave signal predictions because the refraction and reflection that occur make predictions highly inaccurate.

As an example, assume that a transmitter is operating at 150 MHz with a power of 3 W into a quarter-wavelength vertical antenna. The receiver, which is 20 mi (32.2 km, or 32,200 m) away, has an antenna with a gain of 8 dB. What is the received power?

The wavelength at 150 MHz is $\lambda = 300/f = 300/150 = 2$ m. The gain of the quarter-wave vertical transmitting antenna is the same as that for a dipole. With a dipole gain of 1, we must multiply this by 1.64 to get the gain over an isotropic source.

The gain of the receiving antenna is 8 dB. The gain is usually expressed as the gain over a dipole. To convert to gain with respect to an isotropic source, we add 2.15 dB. This is the same as multiplying the power ratio represented by the gain in decibels by 1.64. The result is $8 + 2.15 = 10.15$ dB.

This must now be converted to an actual power ratio. Since dB $= 10 \log (P_{\text{out}}/P_{\text{in}})$, where P_{in} and P_{out} are the input and output power of the antenna,

$$\frac{P_{\text{out}}}{P_{\text{in}}} = \log^{-1}\left(\frac{\text{dB}}{10}\right) = \log^{-1}\left(\frac{10.15}{10}\right) = \log^{-1}(1.015) = 10.35$$

The received power can now be calculated.

$$P_r = \frac{3(1.64)(10.35)(2)^2}{16(9.87)(32200)^2} = \frac{203.7}{16.37 \times 10^{-10}} = 1.24 \times 10^{-9} \text{ W or } 1.24 \text{ nW}$$

If the receiver antenna, transmission line, and front-end input impedance are 50 Ω, we can calculate the input voltage given this input power. Since $P = V^2/R$, $V = \sqrt{PR}$. Substituting into $V = \sqrt{PR}$,

$$V = \sqrt{(1.24 \times 10^{-9})(50)} = 78.7 \times 10^{-6} = 76.7 \ \mu\text{V}$$

This is a relatively strong signal; most good narrowband receivers can generate full intelligible output with 1 μV or less.

PATH ATTENUATION. Another way to predict received power is to estimate total power attenuation over a transmission path. This attenuation in decibels is given by the expression

$$\text{dB loss} = 37 \text{ dB} + 20 \log f + 20 \log d$$

where f = frequency of operation, MHz

d = distance traveled, mi

The distance can also be given in kilometers, in which case the 37-dB figure must be changed to 32.4 dB. Isotropic antennas are assumed.

The attenuation over a 20-mi path at a frequency of 150 MHz is

$$\text{dB loss} = 37 + 20 \log (150) + 20 \log (20) = 37 + 43.52 + 26 = 106.52$$

The dB loss formula, then, tells us that for every doubling of the distance between transmitter and receiver, the attenuation increases about 6 dB.

Example 14-5

A 275-ft-high transmitting antenna has a gain of 12 dB over a dipole. The receiving antenna, which is 60 ft high, has a gain of 3 dB. The transmitter power is 100 W at 224 MHz. Calculate (a) the maximum transmitting distance and (b) the received power at the distance calculated in part (a). (There are 1.61 km per mile.)

a. $D = \sqrt{2h_t} + \sqrt{2h_r}$
$\quad = \sqrt{2(275)} + \sqrt{2(60)} = \sqrt{550} + \sqrt{120}$
$\quad = 23.45 + 10.95 = 34.4 \text{ mi}$
$34.4 \times 1.61 = 55.4 \text{ km}$
$55.4 \text{ km} \times 1000 = 55,400 \text{ m}$

b. $P_r = \dfrac{P_t G_t G_r \lambda^2}{16\pi^2 d^2}$

$P_t = 100 \text{ W} \qquad G_t = 12 \text{ dB over dipole}$
$\qquad\qquad\qquad\qquad = 12 \text{ dB} + 2.15 \text{ dB}$
$\qquad\qquad\qquad\qquad = 14.15 \text{ dB}$

$\text{dB} = 10 \log \dfrac{P_{\text{out}}}{P_{\text{in}}}$

$\dfrac{P_{\text{out}}}{P_{\text{in}}} = \log^{-1}(\text{dB}/10) = \log^{-1}(14.15/10) = \log^{-1}(1.415)$
$\qquad = 26$

$G_t = 26$
$G_r = 3 \text{ dB} + 2.15 \text{ dB} = 5.15 \text{ dB}$

$\dfrac{P_{\text{out}}}{P_{\text{in}}} = \log^{-1}(5.15/10) = \log^{-1}(0.515) = 3.27$

$G_r = 3.27$

$\lambda = \dfrac{300}{f} = \dfrac{300}{224} = 1.34 \text{ m}$

$\lambda^2 = (1.34)^2 = 1.8$

$P_r = \dfrac{100(26)(3.27)(1.8)}{16(3.14)^2(55,400)^2}$

$\quad = \dfrac{15303.6}{4.85 \times 10^{11}} = 3.16 \times 10^{-9} = 3.16 \text{ mW}$

SUMMARY

Radio signals are electromagnetic waves. Signal voltage applied to an antenna creates an electric field and also causes current to flow, producing a magnetic field. The electric and magnetic fields propagate through space over very long distances at the speed of light. The electric and magnetic fields are perpendicular to one another and perpendicular to the direction of propagation of the wave.

If the electric field of a signal is parallel to the earth, the electromagnetic wave is said to be horizontally polarized; if it is perpendicular to the earth, the wave is vertically polarized. For optimal transmission and reception, both the transmitting and receiving antennas must be of the same polarization.

Antennas are some fraction of the signal wavelength, most commonly a half or quarter wavelength. A half-wavelength antenna fed at its center by the transmission line is called a dipole. Most antennas are some modified form of a basic half-wavelength dipole. Two common variations are the folded dipole antenna and the Marconi ground-plane vertical antenna. Each type offers different impedance values, directional radiation patterns, and gain.

To create an antenna with directivity and gain, two or more antenna elements are combined to form an array. The two basic array types are parasitic arrays, consisting of parasitic elements such as reflectors and directors, and driven arrays. Driven arrays are subdivided into broadside, collinear, end-fire, and log-periodic arrays. Yagi antennas are made up of a driven element and one or more parasitic elements.

For a given application, the characteristic impedance of a transmission line must match the output impedance of the transmitter and the impedance of the antenna itself. Mismatches cause an unfavorable standing-wave ratio and a loss of power. Antenna length, bandwidth, and Q are two characteristics that can be adjusted, through design, to produce the best match. Specialized components, such as Q sections, baluns, and tuners, can also be used.

The three basic types of radio-wave propagation paths are ground waves, sky waves, and space waves (direct waves). Ground waves, which are vertically polarized, follow the curvature of the earth and are strongest at medium and low frequencies. Sky-wave signals are radiated by an antenna into the upper atmosphere, where they are refracted in the ionosphere and sent back to earth. Direct waves travel in a straight line from the transmitting antenna to the receiving antenna, and are limited to what is known as line-of-sight transmission. To extend communications distances at VHF, UHF, and microwave frequencies, repeater stations can be used.

For direct- and space-wave transmissions over relatively short distances, it is possible to use simple formulas to predict attenuation given distance between transmitter and receiver, gain, and power figures.

KEY TERMS

Antenna (aerial)
Antenna array
Antenna bandwidth
Antenna directivity
Antenna polarization
Antenna Q

Antenna radiation pattern
Antenna reciprocity
Antenna resonance
Antenna tuners
Beam antenna
Beamwidth

Bow-tie antenna
Broadside array
Circular polarization
Collinear antenna
Communications satellite
Conductor

Conical antenna
Counterpoise
D layer
Diffraction
Dipole
Directivity
Director
Driven array
Drooping radial
E layer
Electric field
Electromagnetic field
End-fire array
F layer
Folded dipole
Friis formula
Front-to-back (F/B) ratio
Gain
Ground plane
Ground wave

Half-wave dipole antenna
Huygens' principle
Impedance matching
Index of refraction
Ionosphere
Left-hand circular polariza-
    tion (LHCP)
Line-of-sight communication
Loading coil
Log-periodic array
Magnetic field
Marconi ground-plane
    vertical
Minor lobe
Near field
Omnidirectional antenna
Parasitic array
Path attenuation
Polarization
Q section (matching stub)

Quarter-wavelength vertical
    antenna
Radial
Radiation resistance
Radio horizon
Radio wave
Reflection
Reflector
Refraction
Repeater
Right-hand circular polar-
    ization (PHCP)
Signal power density
Skip distance
Sky wave
Snell's law
Space wave
Transmatch circuit
Unidirectional antenna
Yagi antenna

QUESTIONS

1. What is the basic makeup of a radio wave?
2. What is the name of the mathematical expressions that describe the behavior of electromagnetic waves? Give a general statement regarding what these expressions state.
3. Describe the relationship between the types of fields that make up a radio wave. What is the orientation of these fields with respect to the transmission direction of the wave?
4. What part of a radio wave determines its polarization with respect to the earth's surface?
5. What is the polarization of a radio wave with a magnetic field that is horizontal to the earth?
6. Define what is meant by antenna reciprocity.
7. What is the name of the antenna type upon which most other antennas are based? What is the length of this antenna in terms of wavelength at the operating frequency?
8. What is the most commonly used medium for connecting an antenna to a transmitter or receiver?
9. What is the length of a dipole?
10. State the theoretical radiation resistance or drive impedance of a dipole.
11. What factors affect the impedance of a dipole and how?
12. What is the equivalent circuit of a dipole at resonance, above resonance, and below resonance?
13. What factors influence antenna bandwidth?
14. Describe the horizontal radiation pattern of a dipole. What is the three-dimensional shape of a dipole radiation pattern?
15. Define what is meant by directivity in an antenna.
16. What unit is used to state and measure an antenna's directivity?
17. Define antenna gain. What gives an antenna gain?
18. What is an isotropic radiator?
19. What gain does a dipole have over an isotropic radiator?
20. State two ways that antenna gain is expressed.
21. Define ERP.
22. What is a folded dipole? What are its advantages over a standard dipole?

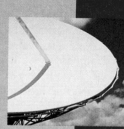

23. What two names are used to refer to a quarter-wavelength vertical antenna?
24. What is the horizontal radiation pattern of a quarter-wavelength vertical antenna?
25. What is the term used to describe an antenna that radiates equally well in all horizontal directions?
26. What must a quarter-wavelength vertical antenna have for proper operation, in addition to a vertical conductor radiator?
27. What is the name of a group of conductors connected to the transmission line at the base of some vertical antennas?
28. Name the most commonly used type of transmission line for quarter-wavelength vertical antennas.
29. What term is used to describe the directivity of an antenna that transmits or receives equally well in two opposite directions?
30. What is a unidirectional antenna?
31. Why does a directional antenna amplify a signal?
32. What physical attributes must an antenna have to exhibit gain and directivity?
33. Name the two basic classes of antenna arrays.
34. Name the three basic elements in a Yagi antenna.
35. Name the two parasitic elements in a Yagi beam antenna.
36. What is the upper limit on the number of directors that a Yagi can have?
37. What is the typical beamwidth range of a Yagi antenna?
38. State two factors that affect the beamwidth of a Yagi.
39. Define the term front-to-back ratio as it applies to a Yagi.
40. True or false. A Yagi can be operated with either horizontal or vertical polarization.
41. Which polarization is the most common for a Yagi?
42. List three kinds of driven arrays.
43. Describe the radiation pattern of the three most common types of driven arrays.
44. Do driven arrays have gain over a dipole?
45. What is the name of a popular wideband driven array with a unidirectional radiation pattern?
46. State one major benefit of a log-periodic antenna over other types of driven arrays.
47. What is the most common value of antenna input/output impedance for modern transmitters and receivers?
48. Why is impedance matching necessary in some antenna installations?
49. What is the name of a type of transformer used for impedance matching in antenna installations?
50. What do you call a quarter-wavelength section of transmission line that is used for impedance matching between antenna and transmission line?
51. What is the impedance matching ratio of a coaxial balun?
52. Describe how an antenna tuner works.
53. What is the usual procedure for adjusting an antenna tuner?
54. What are the requirements for the reflection of a radio signal?
55. Will a vertically polarized radio wave be received by a horizontally polarized antenna?
56. What is circular polarization? State the two types of circular polarization.
57. Will horizontally and vertically polarized antennas pick up circularly polarized signals?
58. What is refraction as it refers to radio waves? What causes refraction of radio waves?
59. What is diffraction as it refers to radio waves? Is diffraction harmful or advantageous in radio communications?
60. Name the three paths that a radio signal can take through space.
61. What two names are used to refer to a radio wave that propagates near the surface of the earth? Are they vertically or horizontally polarized?
62. What is the frequency range of signals that propagate best over the surface of the earth?
63. State the name of the radio wave that is refracted by the ionosphere.
64. What is the frequency range of signals that propagate best by ionospheric refraction?
65. What is the ionosphere? How is it different from atmospheric layers that are closer to the earth? Why does it differ?

66. What layer of the ionosphere has the greatest effect on a radio signal?
67. True or false. The ionosphere acts like a mirror, reflecting radio waves back to earth.
68. What phenomenon of propagation makes worldwide radio communications possible?
69. What factors determine whether a radio wave is refracted by the ionosphere or passes through to outer space?
70. What do you call a radio wave that propagates only over line-of-sight distances?
71. At what frequencies is line-of-sight transmission the only method of propagation?
72. How can the transmission distance of a UHF signal be increased?
73. What technique is used to increase transmission distances beyond the line-of-sight limit at VHF and above?
74. What is a repeater station and how does it work?

PROBLEMS

1. What is the approximate impedance of a dipole that is 0.1 λ above ground? 0.3 λ above ground? ◆
2. What is the length of a wire dipole antenna at 16 MHz?
3. A wire dipole antenna has length of 27 ft. What is its frequency of operation? What is its approximate bandwidth? ◆
4. At low frequencies (less than 30 MHz), what is the normal polarization of a dipole?
5. What is the most common transmission line used with low-frequency dipoles? ◆
6. The power applied to an antenna with a gain of 4 dB is 5 W. What is the ERP?
7. What is the length of a folded dipole made with a 300-Ω twin lead for a frequency of 216 MHz? ◆
8. Calculate the length of a quarter-wavelength vertical antenna at 450 MHz.
9. What is the length of the driven element in a Yagi at 290 MHz? ◆
10. Calculate the length of the coaxial loop used in a coaxial balun for a frequency of 227 MHz. Assume a velocity factor of 0.8.
11. What is the length in inches of a quarter-wavelength impedance-matching section of coaxial cable with a velocity factor of 0.66 at 162 MHz?
12. How far away is the radio horizon of an antenna 100 ft high?
13. What is the maximum line-of-sight distance between a paging antenna 250 ft high and a pager receiver 3.5 ft off the ground?
14. What is the path attenuation between transmitter and receiver at a frequency of 1.2 GHz and a distance of 11,000 mi?

CRITICAL THINKING

1. Of two dipoles, one using wire conductors and one using thin tubing conductors, which has a wider bandwidth?
2. The sun is at an angle of 30° to the horizon and causes a tall quarter-wavelength vertical antenna to cast a shadow 400 ft long. At what frequency does the antenna resonate?
3. Where are the two conductors of a coaxial transmission line connected in a quarter-wavelength vertical antenna installation?
4. Compare and contrast the operation of a quarter-wavelength vertical antenna to that of a half-wavelength dipole.
5. What is the feed impedance of a quarter-wavelength vertical antenna? What factors influence this impedance?
6. If a vertical antenna is too short, what can be done to bring it into resonance?

7. If a vertical antenna is too long for the desired frequency of operation, what can be done to make it resonant?

8. Refer to Fig. 14-45. What is the beamwidth of this antenna? Calculate its gain. Calculate the front-to-back ratio of this antenna in decibels. Note: This radiation pattern uses a logarithmic scale and is calibrated in decibels. The maximum radiation point on the major lobe at 0^0 is given the value of 0 dB. The -3-dB down points are clearly identified so that beam width can be determined. The center of the plot is -100 dB.

9. A transmitter has a directional antenna with a gain of 4 dB. The frequency is 72 MHz. The power into the antenna is 2 W. The receiver uses a quarter-wavelength ground plane. The distance between the transmitting and receiving antennas is 3 mi. What is the power at the receiver? What is the receiver input voltage with a 50-Ω input impedance?

10. A 915-MHz transmitter sends a signal to a receiver. The transmitter output power is 1 W. Both transmitter and receiver use quarter-wavelength vertical antennas. The receiver input impedance is 50 Ω. What is the maximum distance that can be achieved and still deliver 1 μV into the receiver front end?

11. What is the length of a ⅝ λ vertical antenna at 902 MHz?

12. A very low-power battery-operated transmitter at an inaccessible location is to send telemetry data to a data-collection center. The signal is often too weak for reliable data recovery. List some practical steps that can be taken to improve transmission success.

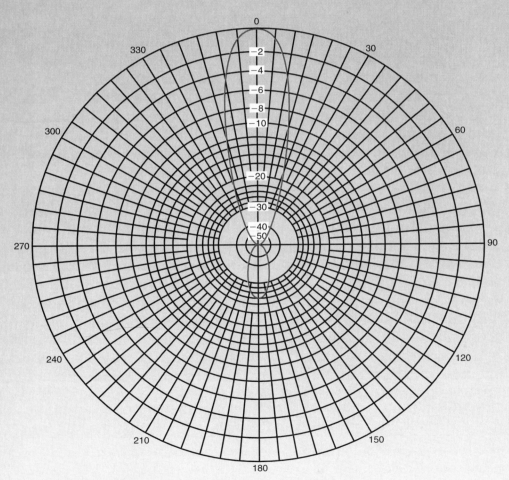

FIG. 14-45 Radiation pattern for critical thinking question 8.

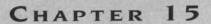

MICROWAVE COMMUNICATIONS

Objectives

After completing this chapter, you will be able to:

◆ *Explain* the reasons for the growing use of microwaves in communications applications.

◆ *Identify* the circuits that require the use of special microwave components.

◆ *Define* the term waveguide, *explain* how a waveguide works, and *calculate* the cutoff frequency of a waveguide.

◆ *Explain* the purpose and operation of direction couplers, circulators, isolators, T-sections, cavity resonators, and microwave vacuum tubes.

◆ *Describe* the operation of the major types of microwave diodes.

◆ *Name* five common types of microwave antennas and *calculate* the gain and beam width of horn and parabolic dish antennas.

◆ *Explain* the operation of each major section of a microwave relay station.

◆ *Explain* the basic concepts and operation of pulsed and Doppler radar systems.

As the use of electronic communications has increased over the years, the frequency spectrum normally used for radio signals has become extremely crowded. In addition, there has been an increasing need for more spectrum space to carry wider-bandwidth video and digital information. It is no secret that the electromagnetic frequency spectrum is a finite natural resource, one that is being used up very quickly. One of the primary solutions to this problem has been to move more radio communications higher in the spectrum. Initially, expansion was into the VHF and UHF ranges. Today, however, the primary expansion of radio communications services is into the microwave range, the 1- to 300-GHz range, which offers tremendous bandwidth for communications and other applications.

At these high frequencies standard electronic components do not work. Ordinary transistors do not amplify or oscillate at such high frequencies, and special transistors had to be developed. Many other special components were developed to amplify and process microwave signals. For microwave applications, microstrip and stripline transmission lines take the place of inductors, capacitors, and tuned circuits. Waveguides serve as transmission lines, and special tubes like magnetrons and traveling-wave tubes are used to obtain high power. Microwave semiconductor diodes are used for signal detection and mixing, frequency multiplication, attenuation and switching, and oscillation.

15-1 MICROWAVE CONCEPTS

Microwaves are the ultra-high, super-high, and extremely high frequencies directly above the lower frequency ranges where most radio communications now take place and below the optical frequencies that cover infrared, visible, and ultraviolet light. The outstanding benefits for radio communications of these extremely high frequencies and accompanying short wavelengths more than offset any problems connected with their use. Today, most new communications services and equipment use microwaves.

MICROWAVE FREQUENCIES AND BANDS

The practical microwave region is generally considered to extend from 1 to 30 GHz, although some definitions include frequencies up to 300 GHz. Microwave signals in the 1- to 30-GHz have wavelengths of 30 cm (about 1 ft) to 1 cm (or about 0.4 in.).

The microwave frequency spectrum is divided up into groups of frequencies, or bands, as shown in Fig. 15-1. Frequencies above 40 GHz are referred to as *millimeter* (*mm*) waves because their wavelength is only millimeters. Note that the submillimeter band overlaps part of the UHF band, which is 300 to 3000 MHz. Frequencies above 300 GHz are in the *submillimeter band.* Currently the only communications in either the millimeter or submillimeter ranges is for research and experimental activities.

THE BENEFITS OF MICROWAVES

Every electronic signal used in communications has a finite bandwidth. When a carrier is modulated by an information signal, sidebands are produced. The resulting signal occupies a certain amount of bandwidth, called a *channel,* in the radio frequency

Band Designation	Frequency Range, GHz
L	1–2
S	2–4
C	4–8
X	8–12
K_u	12–18
K	18–27
K_a	27–40
Millimeter	40–300
Submillimeter	> 300

Fig. 15-1 Microwave frequency bands.

spectrum. Channel center frequencies are assigned such that the signals using each channel do not overlap and interfere with signals in adjacent channels. As the number of communications signals and channels increase, more and more of the spectrum space is used up. Over the years as the need for electronic communication has increased, the number of radio communications stations has increased dramatically. As a result, the radio spectrum has become extremely crowded.

Use of the radio frequency spectrum is regulated by the federal government. In the United States, this job is assigned to the Federal Communications Commission (FCC). The FCC establishes various classes of radio communications and regulates the assignment of spectrum space. For example, in radio and TV broadcasting, certain areas of the spectrum are set aside and frequency assignments are given to stations. For two-way radio communications, other portions of the spectrum are used. The various classes of radio communications are assigned specific areas in the spectrum within which they can operate. Over the years, the available spectrum space, especially below 300 MHz, has essentially been used up. In many cases, communications services must share frequency assignments. In some areas, new licenses are no longer being granted because the spectrum space for that service is completely full. In spite of this, the demand for new electronic communications channels continues. The FCC must, on an on-going basis, evaluate users' needs and demands and reassign frequencies as necessary. Many compromises have been necessary.

Technological advances have helped solve some problems connected with overcrowding. For example, the selectivity of receivers has been improved so that adjacent channel interference is not as great. This permits stations to operate on more closely spaced frequencies.

On the transmitting side, new techniques have helped squeeze more signals into the same frequency spectrum. A classic example is the use of SSB, where only one sideband is used rather than two, thereby cutting the spectrum usage in half. Limiting the deviation of FM signals also helps to

DID YOU KNOW?

Most new communications services and equipment use microwaves.

As recently as the early 1990s, frequencies in the upper reaches of the radio spectrum had little value. In fact, there was a time when the Federal Communications Commission (FCC) issued licenses to use frequencies in the then useless microwave region free of charge. Today, however, with the development of new communications technology, these frequencies have found a valuable place in the field of electronic communications. The limitations which once caused the FCC to consider these frequencies useless have actually become the basis for their success. Signals in lower bands can pass straight through buildings and travel long distances, whereas these high-frequency signals are unable to pass through even small objects. However, they are able to carry huge amounts of information. These factors contribute to making the higher freqencies extremely valuable for use in wireless LANs. Because they cannot penetrate walls, the same frequencies may be used even in neighboring buildings without interference. Also, the large capacity at these frequencies makes them ideal for data transmission. (*Business Week,* March 11, 1996, pp. 87—88)

control bandwidth. In data communications new modulation techniques such as PSK and QAM have been used to narrow the required bandwidth of transmitted information or to transmit at higher speeds in narrower bandwidths. Digital compression methods also transmit more information through a narrow channel. Multiplexing techniques help put more signals or information into a given bandwidth.

The other major approach to solving the problem of spectrum crowding has been to move into the higher frequency ranges. Initially, the VHF and UHF bands were tapped. Today, most new communications services are assigned to the microwave region.

To give you some idea why more bandwidth is available at the higher frequencies, let's take an example. Consider a standard AM broadcast station operating on 1000 kHz. The station is permitted to use modulating frequencies up to 5 kHz, thus producing upper and lower sidebands 5 kHz above and below the carrier frequency, or 995 and 1005 kHz. This gives a maximum channel bandwidth of $1005 - 995 = 10$ kHz. This bandwidth represents $10/1000 = 0.01$ or 1 percent of the spectrum space at that frequency.

Now consider a microwave carrier frequency of 4 GHz. One percent of 4 GHz is $0.01 \times 4,000,000,000 = 40,000,000$ or 40 MHz. A bandwidth of 40 MHz is incredibly wide. In fact, it represents all of the low-frequency, medium-frequency, and high-frequency portions of the spectrum plus 10 MHz. This is the space that might be occupied by a 4-GHz carrier modulated by a 20-MHz information signal. Obviously, most information signals do not require that kind of bandwidth. A voice signal, for example, would take up only a tiny fraction of that. A 10-kHz AM signal represents only $10,000/4,000,000,000 = 0.00025$ percent of 4 GHz. Up to 4000 AM broadcast stations with 10-kHz bandwidths could be accommodated within the 40-MHz (1 percent) bandwidth.

Obviously then, the higher the frequency, the greater the bandwidth available for the transmission of information. This gives not only more space for individual stations, but also allows wide-bandwidth information signals such as video and high-speed digital data to be accommodated. The average TV signal has a bandwidth of approximately 6 MHz. It is impractical to transmit video signals on low frequencies because they use up entirely too much spectrum space. That is why most TV transmission is in the VHF and UHF ranges. There is even more space for video in the microwave region.

Wide bandwidth also makes it possible to use various multiplexing techniques to transmit more information. Multiplexed signals generally have wide bandwidths, but these can be easily handled in the microwave region. Finally, transmission of binary information often requires relatively wide bandwidths, and these are also easily transmitted on microwave frequencies.

Microwave radio transmission, then, offers an enormous amount of spectrum space for new communications services. Already the fight is on to acquire this space. Space in the microwave bands is being assigned by the FCC as fast as it is able to determine the best use for the spectrum.

HINTS AND HELPS

The great amount of spectrum space available in the microwave region makes multiplexing and the transmission of binary information, both which require wide bandwidths, more easily achievable.

THE DISADVANTAGES OF MICROWAVES

The higher the frequency, the more difficult it becomes to analyze electronic circuits. The analysis of electronic circuits at lower frequencies, say those below 30 MHz, is based upon current-voltage relationships (circuit analysis). Such relationships are simply not usable at microwave frequencies. Instead, most components and circuits are analyzed in terms of electric and magnetic fields (wave analysis). Thus techniques commonly used for analyzing antennas and transmission lines can also be used in designing microwave circuits. Measuring techniques are, of course, also different. In low-frequency electronics, currents and voltages are calculated. In microwave circuits, measurements are of electric and magnetic fields. Power measurements are more common than voltage and current measurements.

Another problem is that at microwave frequencies, conventional components become difficult, if not impossible, to implement. For example, a common resistor that looks like pure resistance at low frequencies does not exhibit the same characteristics at microwave frequencies. The short leads of a resistor, although they may be less than an inch, represent a significant amount of inductive reactance at very high frequencies. A small capacitance also exists between the leads. These small stray and distributed reactances are sometimes called *residuals*. Because of these effects, at microwave frequencies a simple resistor looks like a complex *RLC* circuit. This is also true of inductors and capacitors. Figure 15-2 shows equivalent circuits of components at microwave frequencies.

To physically realize resonant circuits at microwave frequencies, the values of inductance and capacitance must be smaller and smaller. Physical limits become a problem. Even a half-inch piece of wire represents a significant amount of inductance at microwave frequencies.

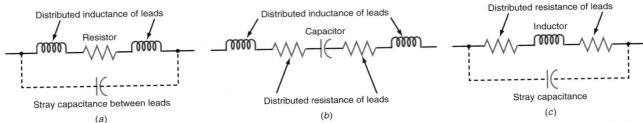

Fig. 15-2 Equivalent circuits of components at microwave frequencies. (*a*) Resistor. (*b*) Capacitor. (*c*) Inductor.

The solution is to use distributed circuit elements, such as transmission lines, rather than lumped components, at microwave frequencies. When transmission lines are cut to the appropriate length, they act as inductors, capacitors, and resonant circuits. Special versions of transmission lines known as striplines, microstrips, waveguides, and cavity resonators are widely used to implement tuned circuits and reactances.

In addition, because of inherent capacitances and inductances, conventional semiconductor devices such as diodes and transistors simply will not function as amplifiers, oscillators, or switches at microwave frequencies. Insufficient gain is available.

Another serious problem is transistor *transit time,* the amount of time it takes for the current carriers (holes or electrons) to move through a device. At low frequencies, transit times can be neglected, but at microwave frequencies, they are a high percentage of the actual signal period.

This problem has been solved by designing smaller and smaller microwave diodes, transistors, and ICs and using special materials such as gallium arsenide in which transit time is significantly less than in silicon. In addition, specialized components have been designed for microwave applications. This is particularly true for power amplification, where special vacuum tubes known as klystrons, magnetrons, and traveling-wave tubes are the primary components used for power amplification.

Another problem is that microwave signals, like light waves, travel in perfectly straight lines. This means that the communications distance is limited to line-of-sight range. Antennas must be very high for long-distance transmission. Microwave signals penetrate the ionosphere, so multiple-hop communications is not possible.

MICROWAVE COMMUNICATIONS SYSTEMS

Like any other communications system, a microwave communications system uses transmitters, receivers, and antennas. The same modulation and multiplexing techniques used at lower frequencies are also used in the microwave range. But the RF part of the equipment is physically different because of the special circuits and components that are used to implement the components.

TRANSMITTERS. Like any other transmitter, a microwave transmitter starts with a carrier generator and a series of amplifiers. It also includes a modulator followed by more stages of power amplification. The final power amplifier applies the signal to the transmission line and antenna. The carrier-generation and modulation stages of a microwave application are similar to those of lower-frequency transmitters. Only in the later power-amplification stages are special components used.

Figure 15-3 shows several ways that microwave transmitters are implemented. The special microwave stages and components are shaded. In the transmitter circuit shown in Fig. 15-3(a) a microwave frequency is first generated in the last multiplier stage. The operating frequency is 1680 MHz, where special microwave components and techniques must be used. Instead of tuned circuits made of loops of wire for inductors and discrete capacitors, microstrip transmission lines are used as tuned circuits and as impedance-matching circuits. One or more additional power amplifiers are then used to boost the signal to the desired power level. Both bipolar and MOSFET microwave power transistors are available that give power levels up to about 100 W or more. When FM is used, the remaining power amplifiers can also be class C, which pro-

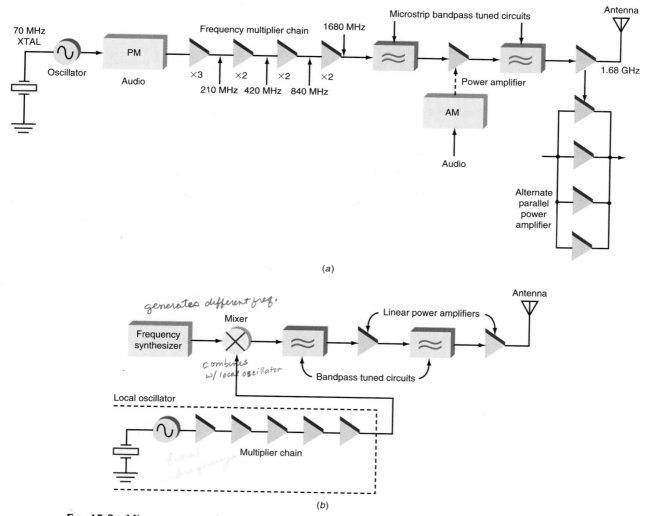

(a)

(b)

FIG. 15-3 Microwave transmitters. (a) Microwave transmitter using frequency multipliers to reach the microwave frequency. The shaded stages operate in the microwave region. (b) Microwave transmitter using up conversion with a mixer to achieve an output in the microwave range.

vides maximum efficiency. If more power is desired, several transistor power amplifiers can be paralleled, as in Fig. 15-3(a).

If AM is used in a circuit like that in Fig. 15-3(a), an amplitude modulator can be used to modulate one of the lower-power amplifier stages after the multiplier chain. When this is done, the remaining power-amplifier stages must be linear amplifiers to preserve the signal modulation.

For very high-power output levels—beyond several hundred watts—a special amplifier must be used, for example, the klystron.

Figure 15-3(b) shows another possible transmitter arrangement, in which a mixer is used to up-convert an initial carrier signal with or without modulation to the final microwave frequency.

The synthesizer output and a microwave local oscillator signal are applied to the mixer. The mixer then translates the signal up to the desired final microwave frequency. A conventional crystal oscillator using fifth-overtone VHF crystals followed by a chain of frequency multipliers can be used to develop the local oscillator frequency. Alternatively, one of several special microwave oscillators could be used, for example, a Gunn diode, a microwave semiconductor in a cavity resonator, or a dielectric resonator oscillator.

The output of the mixer is the desired final frequency at a relatively low power level, usually tens or hundreds of milliwatts at most. Linear power amplifiers are used to boost the signal to its final power level. At frequencies less than about 10 GHz, a microwave transistor can be used. At the higher frequencies, special microwave power tubes are used.

The tuned bandpass circuits shown in Figure 15-3(b) can be microstrip transmission lines when transistor circuits are used, or cavity resonators when the special microwave tubes are used.

Modulation could occur at several places in the circuit in Fig. 15-3(b). An indirect FM phase modulator might be used at the output of the frequency synthesizer; for some applications, a PSK modulator would be appropriate.

RECEIVERS. Microwave receivers, like low-frequency receivers, are the superheterodyne type. Their front ends are made up of microwave components. Most receivers use double conversion. A first down conversion gets the signal into the UHF or VHF range, where it can be more easily processed by standard methods. A second conversion reduces the frequency to an IF appropriate for the desired selectivity.

Figure 15-4 is a general block diagram of a double-conversion microwave receiver. The antenna is connected to a tuned circuit, which could be a cavity resonator or a mi-

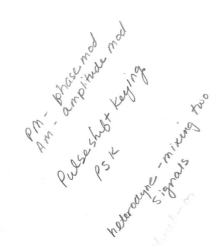

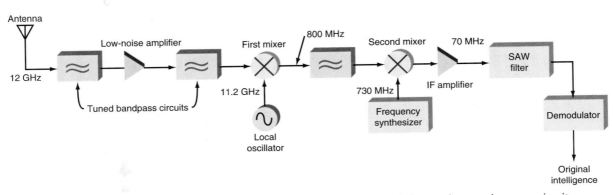

FIG. 15-4 A microwave receiver. The shaded areas denote microwave circuits.

crostrip or stripline tuned circuit. The signal is then applied to a special RF amplifier known as a *low-noise amplifier (LNA)*. Special low-noise transistors, usually gallium arsenide FET amplifiers, must be used to provide some initial amplification. Another tuned circuit connects the amplified input signal to the mixer. Most mixers are of the doubly balanced diode type, although some simple single-diode mixers are also used.

The local oscillator signal is applied to the mixer. The mixer output is usually within the UHF or VHF range. The 700- to 800-MHz range is typical. A microstrip tuned circuit selects out the difference signal, which in Fig. 15-4 is 12 GHz − 11.2 GHz = 0.8 GHz or 800 MHz.

The remainder of the receiver is typical of other superheterodynes. Note that the desired selectivity is obtained with an SAW filter, which is sometimes used to provide a specially shaped IF response. *Highest Q filter, Solid State, expensive   p 90   (Surface acoustic wave filter)*

TRANSMISSION LINES. The transmission line most commonly used in lower-frequency radio communications is coaxial cable. However, coaxial cable has very high attenuation at microwave frequencies, and conventional cable is unsuitable for carrying microwave signals except for very short runs, usually several feet or less.

Special microwave coaxial cable that can be used on the lower microwave bands—L, S, and C—is made of hard tubing rather than wire with an insulating cover and a flexible braid shield. The stiff inner conductor is separated from the outer tubing with spacers or washers, forming low-loss coaxial cable known as *hard line* cable. The insulation between the inner conductor and the outer tubing can be air; in some cases a gas such as nitrogen is pumped into the cable to minimize moisture build-up, which causes excessive power loss. This type of cable is used for long runs of transmission lines to an antenna on a tower. At higher microwave frequencies, C band and upward, a special hollow rectangular or circular pipe called *waveguide* is used for the transmission line (see Sec. 15-3).

> **DID YOU KNOW?**
>
> Special microwave coaxial cable called *hard line*, which is made of hard tubing rather than wire with an insulating cover, can be used on the lower microwave bands.

ANTENNAS. At low microwave frequencies, standard antenna types, including the simple dipole and the quarter-wave length vertical antenna, are still used. At these frequencies antenna sizes are very small; for example, the length of a half-wave dipole at 2 GHz is only about 3 in. A quarter-wave length vertical antenna for the center of the C band is only about 0.6 in long. At the higher frequencies, special antennas are generally used (see Sec. 15-6).

15-2 MICROWAVE TRANSISTOR AMPLIFIERS

Today, although vacuum tubes and microwave tubes like the klystron and magnetron are still used, especially for higher-power applications, most microwave systems use transistor amplifiers. Over the years, semiconductor manufacturers have learned to make transistors work at these higher frequencies. Special geometries are used to make bipolar transistors that provide both voltage and power gain at frequencies up to 10 GHz. Microwave FET transistors have also been created. The most important of these is the MESFET described in Chap. 8. The use of gallium arsenide rather than silicon has fur- *metal semicon*

gallium arsenide

ther increased the frequency capabilities of FETs. Both small-signal and power FETs are available to operate at frequencies up to about 10 GHz. Since most microwave communications activity takes place in the lower-frequency ranges (L, S, and C bands), transistors are the primary active components used.

In the following sections both discrete transistor types with microstrip tuned circuits and monolithic microwave integrated circuits (MMICs) are discussed.

MICROSTRIP TUNED CIRCUITS

Before specific amplifier types are introduced, it is important to examine the method by which tuned circuits are implemented in microwave amplifiers. Lumped components such as coils and capacitors are still used in some cases at the high UHF and low microwave frequencies (below about 2 GHz) to create resonant circuits, filters, or impedance-matching circuits. However, at higher frequencies, standard techniques for realizing such components become increasingly harder to implement. Instead, transmission lines, specifically microstrip, are used.

Chapter 13 described how transmission lines can be used as inductors and capacitors as well as series and parallel resonant circuits. These are readily implemented at the microwave frequencies because half- or quarter-wavelength transmission lines are only inches or some fraction thereof at those frequencies. Microstrip is preferred for reactive circuits at the higher frequencies because it is simpler and less expensive than stripline, but stripline is used where shielding is necessary. The tuned circuits are created using a copper pattern printed circuit board (PCB) upon which are mounted the transistors, ICs, and other components of the circuit.

Figure 15-5 shows several views of microstrip transmission line used for a reactive circuit. The printed circuit board (PCB) is usually made of G-10 or FR-4 fiberglass or a combination of fiberglass and Teflon. The bottom of the PCB is a thin solid copper sheet which serves as a ground plane and one side of the transmission line. The copper strip is the other conductor of the transmission line.

Figure 15-5(*a*) is a perspective view of a microstrip line; Fig. 15-5(*b*) is an end view; and Fig. 15-5(*c*) and Fig. 15-5(*d*) are side views. Both open and shorted segments of line can be used, although shorted segments are preferred because they do not radiate as much as open segments. Quarter-wave length sections are preferred because they are shorter and take up less space on the PCB. Figure 15-6 summarizes the open- and shorted-line possibilities for microstrip.

An important characteristic of microstrip is its impedance. As discussed in Chap. 13, the characteristic impedance of a transmission line depends on its physical characteristics, in this case, for example, on the width of the strip and the spacing between the strip and the copper ground plane, which is the thickness of the PCB material. The dielectric constant of the insulating material is also a factor. Most characteristic im-

Each of these microwave antennas located in Stephen Butte, Washington, is specifically designed for long-distance telephone signal transmission. They are covered to reduce weather-related maintenance problems.

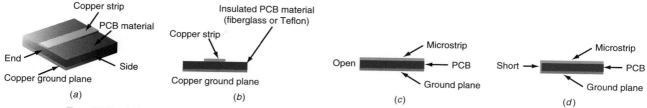

FIG. 15-5 Microstrip transmission line used for reactive circuits. (*a*) Perspective view. (*b*) Edge or end view. (*c*) Side view (open line). (*d*) Side view (shorted line).

pedances are less than 100 Ω; 50 Ω is the most common, followed closely by 75 Ω. Values higher than 100 Ω are used for cases when impedance-matching requirements demand it.

Quarter-wavelength transmission line can be used to make one type of component look like another. For example, in Fig. 15-7(*a*), the $\lambda/4$ microstrip line can make a resistor at one end look like a resistance of another value; specifically,

$$R_2 = \frac{Z_0^2}{R_1}$$

Here, R_1 is the resistance value of the resistor connected to one end of the line, and Z_0 is the characteristic impedance of the microstrip, such as 50 Ω. If R_1 is the characteristic impedance of the microstrip, such as 50 Ω. If R_1 is 150 Ω, then the other end of the line will have a value of R_2, or

$$50^2/150 = 2500/150 = 16.67 \ \Omega$$

This application is the same as the quarter-wavelength matching transformer or Q section discussed in Chap. 14. Recall that two impedances can be matched using a line of length $\lambda/4$ according to the relationship $Z_0 = \sqrt{R_1R_2}$, where Z_0 is the characteristic impedance of the quarter-wavelength line and while R_1 and R_2 are the input and output impedances to be matched.

LENGTH	SHORTED LINES	OPEN LINES
Less than $\lambda/4$	Inductor	Capacitor
$\lambda/4$	Parallel resonant or open circuit	Series resonant or short circuit
$>\lambda/4, <\lambda/2$	Capacitor	Inductor
$\lambda/2$	Series resonant circuit or short repeater	Parallel resonant circuit or open repeater

FIG. 15-6 Equivalent circuits of open and shorted microstrip lines.

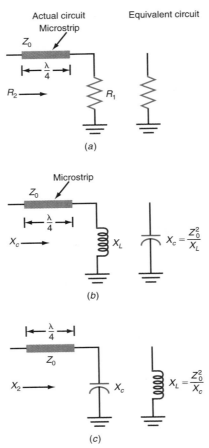

FIG. 15-7 How a quarter-wavelength microstrip can transform impedances and reactances.

A quarter-wavelength line can also make a capacitor look like an inductor or an inductance look like a capacitance [see Fig. 15-7(b) and (c)]. For example, a 75-Ω microstrip $\lambda/4$ long will make an inductive reactance of 30 Ω look like a capacitive reactance of

$$X_c = \frac{Z_0^{\,2}}{X_l} = \frac{75^2}{30} = \frac{5635}{30} = 187.5 \ \Omega$$

Figure 15-8 shows the physical configurations for equivalent coils and capacitors in microstrip form. The thin segment of microstrip shown in Fig. 15-8(a) acts like a series inductor. Figure 15-8(b) shows a short right-angle segment whose end is grounded; this microstrip acts like a parallel inductor. When series capacitance or capacitive coupling is needed, a small capacitor can be created by using the ends of microstrip lines as tiny capacitor plates separated by an air dielectric [Fig. 15-8(c)]. A shunt capacitor can be created using a short, fat segment of microstrip as in Fig. 15-8(d). As these general forms demonstrate, it is often possible to visualize or even draw the equivalent circuit of a microstrip amplifier by observing the pattern on the PCB.

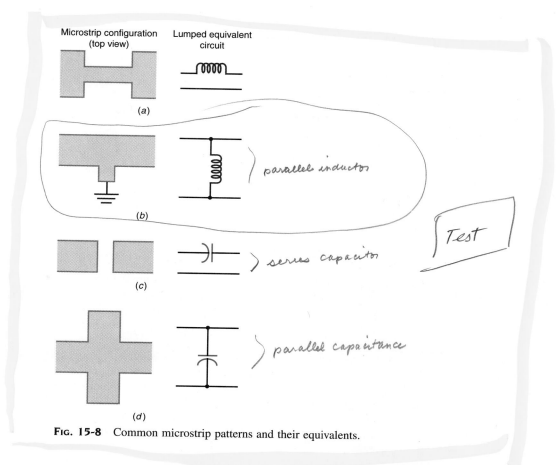

Microstrip configuration (top view) | **Lumped equivalent circuit**

(a)

(b) *parallel inductor*

(c) *series capacitor*

(d) *parallel capacitance*

Test

FIG. 15-8 Common microstrip patterns and their equivalents.

Microstrip can also be used to realize coupling from one circuit, as illustrated in Fig. 15-9. One microstrip line is simply placed parallel to another segment of microstrip. The degree of coupling between the two depends on the distance of separation and the length of the parallel segment. The closer the spacing and the longer the parallel run, the greater the coupling. There is always signal loss by such a coupling method, but it can be accurately controlled.

Although microstrip performs best when it is a straight line, 90° turns are often necessary on a PCB. When turns must be used, a straight right-angle turn, like that shown in Fig. 15-10(a) is forbidden because it acts like a low-pass filter across the line. A gradually curved line, like that shown in Fig. 15-10(b) (or Fig. 15-9) is preferred when the turn radius is much greater than the width of the line. An acceptable alternative method is shown in Fig. 15-10(c). The cut on the corner is critical. Note that the dimensions must be held to $\lambda/4$, which is one-quarter the width of the microstrip.

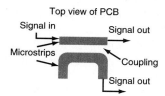

FIG. 15-9 Coupling between microstrips.

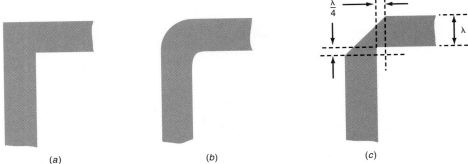

(a)   (b)   (c)

FIG. 15-10 Microstrip 90° turns. (*a*) Right angles like this should be avoided. (*b*) Gradual curves or turns are preferred. (*c*) This arrangement is also acceptable.

A special form of microstrip is the *hybrid ring* shown in Fig. 15-11. The total length of the microstrip ring is 1.5 λ. There are four taps or ports on the line, spaced at quarter-wave length intervals (λ/4), which can be used as inputs or outputs.

Now, a signal is applied to port 1, and some interesting things happen. Output signals appear at the ports 2 and 4, but their levels are at one-half the power of the input. Thus the circuit acts as a power divider to supply two signals of equal level to other circuits. There is no output at port 3. The effect of applying a signal at port 4 is similar. Equal half-power outputs appear at ports 1 and 3, but no signal appears at port 2.

If individual signals are applied simultaneously to ports 1 and 3, the output at port 2 will be their sum and the output at port 4 will be their difference.

The unique operation of the hybrid ring makes it very useful for splitting signals or combining them.

Microstrip can be used to create almost any tuned circuit necessary in an amplifier, including resonant circuits, filters, and impedance-matching networks. Figure 15-12(*a*) shows how a low-pass filter is implemented with microstrip sections. The component shown is a highly selective low-pass filter for use in the 1- to 3-GHz range depending on the exact dimensions. The lumped constant equivalent circuit is

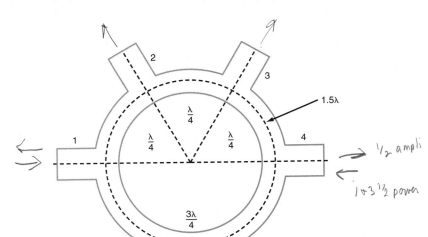

FIG. 15-11 A microstrip hybrid ring.

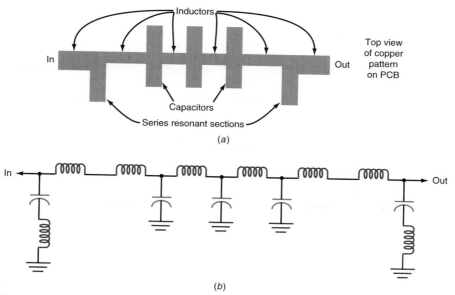

FIG. 15-12 A microstrip filter. (a) Microstrip low-pass filter. (b) Lumped constant equivalent circuit.

shown in Fig. 15-12(b). The transmission-line segments are formed on the PCB itself and connected end to end as required. The transistors or ICs are then soldered to the PCB along with any resistors or larger discrete components that may be needed.

Example 15-1

A quarter-wave Q-matching section made of microstrip is designed to match a source of 50 Ω to a load of 136 Ω at 5.8 GHz. The PCB dielectric constant ϵ_r is 2.4. Calculate (a) the required impedance of the microstrip and (b) its length.

a. $Z_Q = \sqrt{Z_{source} \times Z_{load}}$
   $= \sqrt{50(136)} = 82.46\ \Omega$

b. $\lambda = \dfrac{300}{f}$ MHz

   5.8 GHz = 5800 MHz

   $\lambda = \dfrac{300}{5800} = 0.0517$ m

   $\lambda/4 = 0.0517/4 = 0.12931$ m

   1 m = 38.37 in

   $\lambda/4 = 0.012931\ (39.37) = 0.51$ in

   Velocity of propagation $= \dfrac{1}{\sqrt{\epsilon_r}} = \dfrac{1}{\sqrt{24}} = 0.645$

   $\lambda/4 = 0.51$ in $(0.645) = 0.3286$ in

Microwave transistors, whether they are bipolar or FET types, operate just like other transistors. The primary differences between standard lower-frequency transistors and microwave types are internal geometry and packaging.

To reduce internal inductances and capacitances of transistor elements, special chip configurations, known as *geometries* are used that permit the transistors to operate at higher power levels and at the same time minimize distributed and stray inductances and capacitances.

Figure 15-13 shows several types of microwave transistors. Figure 15-13(*a*) and Fig. 15-13(*b*) are low-power small-signal microwave transistors: these types are either NPN silicon or gallium arsenide FET. Note the very short leads. Both packages are designed for surface mount directly to microstrip on the PCB. The transistor in Fig. 15-13(*b*) has four leads, usually two emitter (or source) leads plus the base (or gate) and the emitter (or drain). The two emitter leads in parallel ensure low inductance. Some transistors of this type have two base or two collector leads instead of two emitter leads.

Figure 15-13(*c*) shows an enhancement-mode power MOSFET. The short, fat leads are thick strips of copper that are soldered directly to the microstrip circuitry on the PCB. Figure 15-13(*d*) is an NPN power transistor with two emitter leads. The short, fat leads ensure low inductance and also permit high currents to be accommodated. The fat copper strips help to dissipate heat. The devices in Fig. 15-13(*c*) and (*d*) can handle power levels up to several hundred watts.

Transistors for small-signal amplification and oscillators are available for frequency up to about 30 GHz. For power amplification transistors are available for frequencies up to 20 GHz.

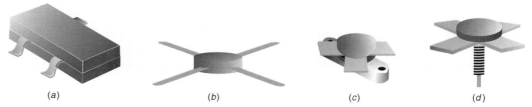

(*a*)   (*b*)   (*c*)   (*d*)

FIG. 15-13   Microwave transistors. (*a*) and (*b*) Low-power small signal. (*c*) FET power. (*d*) NPN bipolar power.

SMALL-SIGNAL AMPLIFIERS

A small-signal microwave amplifier can be made up of a single transistor or multiple transistors combined with a biasing circuit and any microstrip circuits or components as required. Most microwave amplifiers are of the tuned variety. That is, their bandwidth is set by the application and implemented by microstrip series or parallel tuned circuits, and then microstrip lines are used to perform the various impedance-matching duties required to get the amplifier to work.

Another type of small-signal microwave amplifier is a multistage IC, a variety of MMIC. Besides amplifiers, other MMICs available include mixers, switches and phase shifters.

TRANSISTOR AMPLIFIERS. Figure 15-14 shows a microwave amplifier for small signals. This type of amplifier is commonly used in the front end of a microwave receiver to provide initial amplification for the mixer. A low-noise transistor is used. The typical gain range is 10 to 25 dB.

Most microwave amplifiers are designed to have input and output impedances of 50 Ω. In the circuit shown in Fig. 15-14, the input can come from an antenna or another microwave circuit. The blocks labeled TL are microstrip sections that act as tuned circuits, inductors, or capacitors. An input microstrip, TL1, acts as an impedance-matching section, and TL2 and TL3 form a tuned circuit. TL4 is another impedance-matching section that matches the tuned circuit to the complex impedance of the base input. The tuned circuits set the bandwidth of the amplifier input.

A similar sequence of impedance-matching sections and tuned circuits is used in the collector to set the bandwidth and to match the transistor collector impedance to the output. C_2 and C_3 are variable capacitors that allow some tuning of the bandwidth.

All the other components are of the surface-mount chip type to keep lead inductances short. The microstrip segments TL9 and TL10 act as inductors, forming part of the substantial decoupling networks used on the base bias supply V_{BB} and the collector supply V_{CC}. The base supply voltage and the value of R_1 set the base bias, thus biasing the transistor into the linear region for class A amplification. RFCs are used in the supply leads to keep the RF out of the supply and to prevent feedback paths through the supply which can cause oscillation and instability in multistage circuits. Ferrite beads (FB) are used in the collector supply lead for further decoupling.

> **DID YOU KNOW?**
>
> Most microwave amplifiers are designed to have input and output impedances of 50 Ω.

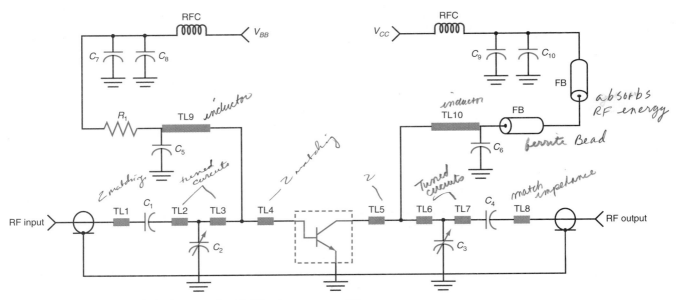

FIG. 15-14 A single-stage class A RF microwave amplifier.

MMIC AMPLIFIERS. A common MMIC amplifier is one that incorporates two or more stages of FET or bipolar transistors made on a common chip to form a multi-stage amplifier. The chip also incorporates resistors for biasing and small bypass capacitors. Physically, these devices look like the transistors in Fig. 15-13(*a*) and (*b*). They are soldered to a PCB containing microstrip circuits for impedance matching and tuning.

Another popular form of MMIC is the *hybrid circuit,* which combines an amplifier IC connected to microstrip circuits and discrete components of various types. All the components are formed on a tiny alumina substrate which serves both as a base and a place to form microstrip lines. Surface mount chip resistors, capacitors, transistors, and MMIC amplifiers are connected. The entire unit is packaged into a housing, usually metal for shielding, and connected to additional circuits on a PCB.

POWER AMPLIFIERS. A typical class A microwave power amplifier is shown in Fig. 15-15. The microstrip lines are used for impedance matching and tuning. Most microstrip circuits simulate L-type matching networks with a low-pass configuration. Small wire loop inductors and capacitors are used to form the decoupling networks to prevent feedback through the power supply, which would cause oscillation. The input and output impedances are 50 Ω. This particular stage operates at 1.2 GHz and provides an output power of 1.5 W. Typical power-supply voltages are 12, 24, and 28 V, but can go to 36 or 48 V for high-power applications.

Note that bias is not supplied by a resistive voltage divider. Instead, it usually comes from a separate bias-current circuit like that shown in Fig. 15-16, which is an ordinary constant-current source. Resistors R_1, R_2, and R_3 form a voltage divider to set the base voltage on Q_1. A voltage is developed across emitter resistor R_4. This voltage divided by R_4 gives the value of the current supplied by Q_1 to the transistor in the microwave amplifier. A diode in series with the voltage divider provides some temperature compensation for variations in the emitter-base voltage that occur in Q_1. R_1 is adjustable to set the bias current to the precise value for optimum power and minimum distortion. Most power amplifiers obtain their bias from constant-current sources; this provides a bias current that is relatively temperature-independent, which provides superior protection against damage.

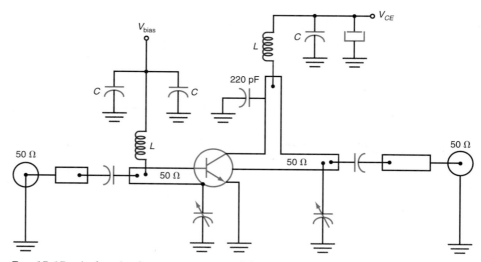

FIG. 15-15 A class A microwave power amplifier.

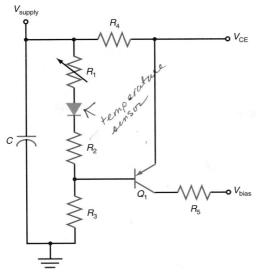

FIG. 15-16 A constant-current bias supply for a linear power amplifier.

The bias is applied to the base of the amplifier in Fig. 15-15 through a small inductor. This and the bypass capacitors keep the microwave energy output of the bias circuit.

The single-stage FET power amplifier shown in Fig. 15-17 can achieve a power output of 100 W in the high UHF and low microwave region. The transistor is an enhancement-mode FET; that means the FET does not conduct with drain voltage applied. A positive gate signal of 3 V or more must be applied to achieve conduction. The gate signal is supplied by a lower power driver stage, and the gate bias is supplied by a zener diode. The gate voltage is made adjustable with potentiometer

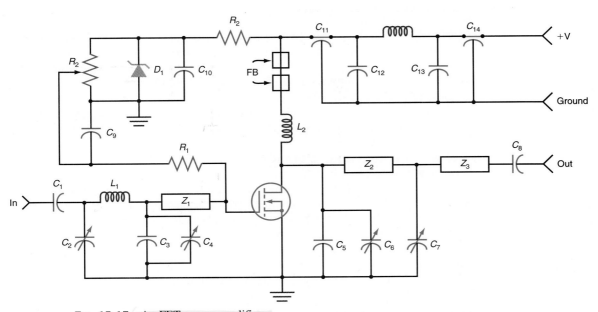

FIG. 15-17 An FET power amplifier.

R_2. This allows the bias to be set for linear, class B or class C operation. For linear class A operation, a quiescent value of drain current is set with the bias. This bias circuit can also be used to adjust the power output over a small range.

In the circuit shown in Fig. 15-17, a combination of LC tuned circuits and microstrips is used for tuning and impedance matching. L_1 is a tiny hairpin loop of heavy wire. L_2 is a larger inductor made of multiple turns of wire to form a high impedance for decoupling.

15-3 WAVEGUIDES AND CAVITY RESONATORS

Long parallel transmission lines, such as 300-Ω twin lead, radiate electromagnetic energy while transporting it from one place to another. As the frequency of operation gets higher, the amount of radiation from the line increases. At microwave frequencies, virtually all the energy is radiated; almost no energy ever reaches the end of the transmission line. Coaxial cable eliminates the radiation problem, but it has significant loss at the higher frequencies. For the lower microwave frequencies, special coaxial cables have been developed that can be used up to approximately 6 GHz if the length is kept short (less than 100 ft). Above this frequency, coaxial cable loss is too great. Short lengths of coaxial cable, several feet or less, can be used to interconnect pieces of equipment that are close together, but for longer runs other methods of transmission must be used. To minimize loss, special types of coaxial cables with large inner conductors and shields have been developed; however, in most cases these cables are rigid rather than flexible, which makes them difficult to use.

WAVEGUIDES

Most microwave energy transmission above 6 GHz is handled by *waveguides*, hollow metal conducting pipes designed to carry and constrain the electromagnetic waves of a microwave signal. Most waveguides are rectangular. Waveguides can be used to carry energy between pieces of equipment or over longer distances to carry transmitter power to an antenna or microwave signals from an antenna to a receiver.

Waveguides are made from copper, aluminum, or brass. These metals are extruded into long rectangular or circular pipes. Often the insides of waveguides are plated with silver to reduce resistance, keeping transmission losses to a very low level.

SIGNAL INJECTION AND EXTRACTION. A microwave signal to be carried by a waveguide is introduced into one end of the waveguide with an antennalike probe that creates an electromagnetic wave which propagates through the waveguide. The electric and magnetic fields associated with the signal bounce off the inside walls back and forth as the signal progresses down the waveguide. The waveguide totally contains the signal so that none escapes by radiation.

The probe shown in Fig. 15-18(a) is a quarter-wavelength vertical antenna at the signal frequency which is inserted in the waveguide one-quarter wavelength from the end, which is closed. The signal is usually coupled to the probe through a short coaxial cable and a connector. The probe sets up a vertically polarized electromagnetic wave in the waveguide, which is then propagated down the line. Because the probe is lo-

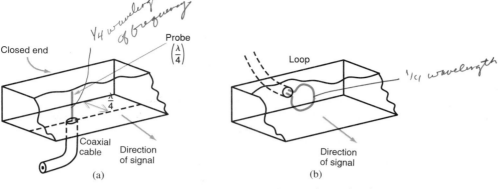

Handwritten annotations: "1/4 wavelength of frequency" (top left), "1/4 wavelength" (top right)

Closed end

Probe $\left(\dfrac{\lambda}{4}\right)$

$\dfrac{\lambda}{4}$

Coaxial cable

Direction of signal

(a)

Loop

Direction of signal

(b)

FIG. 15-18 Injecting a sine wave into a waveguide and extracting a signal.

cated one-quarter wavelength from the closed end of the waveguide, the signal from the probe is reflected from the closed end of the line and reflected back toward the open end. Over a quarter-wavelength distance, the reflected signal appears back at the probe in phase to aid the signal going in the opposite direction. Remember that a radio signal consists of both electric and magnetic fields at right angles to one another. The electric and magnetic fields established by the probe propagate down the waveguide at a right angle to the two fields of the radio signal. The position of the probe determines whether the signal is horizontally or vertically polarized. In Fig. 15-18(a), the electric field is vertical, so the polarization is vertical. The electric field begins propagating down the line and sets up charges on the line that in turn cause current to flow. This in turn generates a companion magnetic field at right angles to the electric field and the direction of propagation.

Loops can also be used to introduce a magnetic field into a waveguide. The loop in Fig. 15-18(b) is mounted in the closed end of the waveguide. Microwave energy applied through a short piece of coaxial cable causes a magnetic field to be set up in the loop. The magnetic field also establishes an electric field, which is then propagated down the waveguide.

Probes and loops can also be used to extract a signal from a waveguide. When the signal strikes a probe or a loop, a signal is induced which can then be fed to other circuitry through a short coaxial cable.

WAVEGUIDE SIZE AND FREQUENCY. Figure 15-19 shows the most important dimensions of a rectangular waveguide: the width (a) and the height (b). Note that these are the inside dimensions of the waveguide. The frequency of operation of a waveguide is determined by the size of a. This dimension is usually made equal to one-half wavelength, a bit below the lowest frequency of operation. This frequency is known as the *waveguide cutoff frequency*. At its cutoff frequency and below, a waveguide will not transmit energy. At frequencies above the cutoff frequency, a waveguide will propagate electromagnetic energy. A waveguide is essentially a high-pass filter with a cutoff frequency equal to

$$f_{co} = \frac{300}{2a}$$

where f_{co} is in megahertz and a is in meters.

For example, suppose it is desired to determine the cutoff frequency of a waveguide with an a dimension of 0.7 in.

DID YOU KNOW?

At higher microwave frequencies, above approximately 6–10 GHz, a special hollow rectangular or circular pipe called *waveguide* is used for the transmission line.

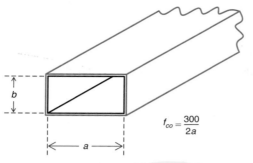

$$f_{co} = \frac{300}{2a}$$

Fig. 15-19 The dimensions of a waveguide determine its operating frequency range.

To convert 0.7 in to meters, multiply by 2.54 to get centimeters and divide by 100 to get meters; 0.7 in = 0.01778 m. Thus,

$$f_{co} = \frac{300}{2(0.01778)} = 8436 \text{ MHz} = 8,426 \text{ GHz}$$

Normally, the height of a waveguide is made equal to approximately one-half the *a* dimension, or 0.35 in. The actual size might be 0.4 in.

Example 15-2

A rectangular waveguide has a width of 0.65 in. and a height of 0.38 in. (a) What is the cutoff frequency? (b) What is a typical operating frequency for this waveguide?

a. $0.65 \times 2.54 = 1.651$ cm

$1.651/100 = 0.01651$ m

$$f_{co} = \frac{300}{2a} = \frac{300}{2(0.01651)} = 9085 \text{ MHz} \ (9.085 \text{ GHz})$$

b. $f_{co} = 0.7f$

$$f = \frac{f_{co}}{0.7} = 1.42 f_{co}$$

$f = 1.42(9.085) = 12.98$ GHz and above

Example 15-3

Would the rectangular waveguide in Example 15-2 operate in the C band?

C band is approximately 4 to 6 GHz. Since a waveguide acts as a high-pass filter with a cutoff of 9.085 GHz, it will not pass a C-band signal.

SIGNAL PROPAGATION. When a probe or loop launches energy into a waveguide, the signal enters the waveguide at an angle so that the electromagnetic fields bounce off the side walls of the waveguides as the signal propagates along the line. In Fig. 15-20(a), a vertical probe is generating a vertically polarized wave with a vertical electric field and a magnetic field at a right angle to the electric field. The electric field is at a right angle to the direction of wave propagation, so it is called a *transverse electric (TE)* field. Figure 15-20(b) shows how a loop would set up the same signal. In this case, the magnetic field is transverse to the direction of propagation, so it is called a *transverse magnetic (TM)* field.

The angles of incidence and reflection depend on operating frequency (see Fig. 15-21). At high frequencies, the angle is large and therefore the path between the opposite walls is relatively long, as shown in Fig. 15-21(a). As the operating frequency decreases, the angle also decreases and the path between the sides shortens. When the operating frequency reaches the cutoff frequency of the waveguide, the signal simply bounces back and forth between the side walls of the waveguide. No energy is propagated.

Whenever a microwave signal is launched into a waveguide by a probe or loop, electric and magnetic fields are created in various patterns depending upon method of energy coupling, frequency of operation, and size of waveguide. Figure 15-22 shows typical fields in a waveguide. In the end view, the lines represent the electric field (E) lines. The dots represent the magnetic field (H) lines. In the top view the dashed lines represent the H field and the ✕s and dots represent the E field. The ✕ means the line is going into the page; the • means the line is coming out of the page.

The pattern of the electromagnetic fields within a waveguide takes many forms. Each form is called an operating mode. As indicated earlier, the magnetic or the electric field must be perpendicular to the direction of propagation of the wave. In the TE mode, the electric field exists across the guide and no E lines extend lengthwise along the guide. In the TM mode, the H lines form loops in planes perpendicular to the walls of the guide and no part of an H line is lengthwise along the guide.

Subscript numbers are used along with the TE and TM designations to further describe the E and H field patterns. A typical designation is $TE_{0,1}$. The first number indicates the number of half-wavelength patterns of transverse lines that exist along the short dimension of the guide through the center of the cross section. Transverse lines are those perpendicular to the walls of the guide. The second number indicates the number of transverse half-wavelength patterns that exist along the long dimension of the guide through the center of the cross section. If there is no change in the field intensity of one dimension, a zero is used.

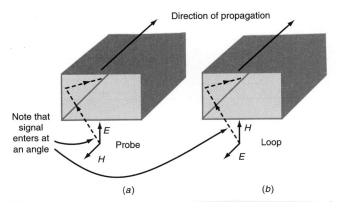

Direction of propagation

Note that signal enters at an angle

E
Probe
H

H
Loop
E

(a)　　　　　(b)

FIG. 15-20 (a) Transverse electric (TE). (b) Transverse magnetic (TM).

electric field in vertical

magnetic field vertical

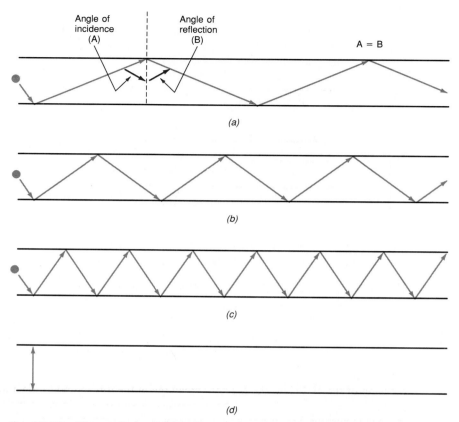

FIG. 15-21 Wave paths in a waveguide at various frequencies. (a) High frequency. (b) Medium frequency. (c) Low frequency. (d) At cutoff frequency.

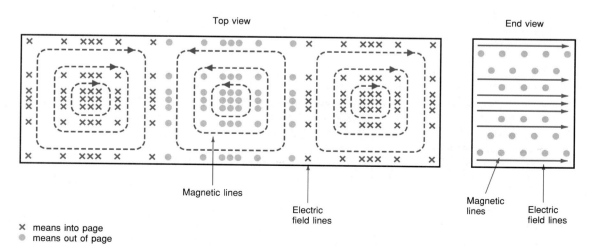

FIG. 15-22 Electric (E) and magnetic (H) fields in a rectangular waveguide.

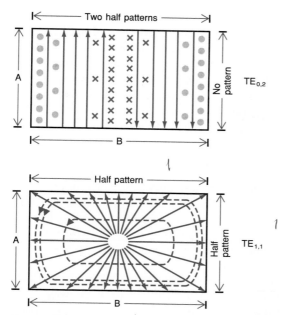

FIG. 15-23 Other waveguide operating modes.

The waveguide in Fig. 15-22 is TE because the E lines are perpendicular (transverse) to the sides of the guide. Looking at the end view in Fig. 15-22 along the short length, there is no field intensity change, so the first subscript is 0. Along the long dimension of the end view, the E lines spread out at the top and bottom but are close together in the center. The actual field intensity is sinusoidal—zero at the ends, maximum in the center. This is one-half of a sine-wave variation, and so the second subscript is 1. The mode of the line in Fig. 15-22 is therefore $TE_{0,1}$. This, by the way, is the main or dominant mode of most rectangular waveguides. Many other patterns are possible, such as the two shown in Fig. 15-23.

WAVEGUIDE HARDWARE AND ACCESSORIES

In some ways, waveguides have more in common with plumbing equipment than they do with the standard transmission lines used in radio communications. And, like plumbing, waveguides have a variety of special parts, such as couplers, turns, joints, rotary connections, and terminations. Most waveguides and their fittings are precision-made so that the dimensions match perfectly. Any mismatch in dimensions or misalignment of pieces that fit together will introduce significant losses and reflections. Waveguides are available in a variety of standard lengths which are interconnected to form a path between a microwave generator and its ultimate destination.

CONNECTION JOINTS. Figure 15-24 shows a choke joint which is used to interconnect two sections of waveguide. It consists of two flanges connected to the waveguide at the center. The right-hand flange is flat, and the one at the left is slotted a quarter-wavelength deep at a distance of a quarter wavelength from the point at which

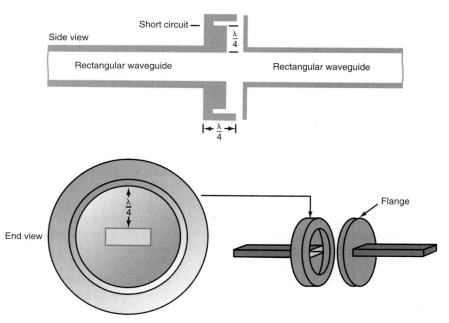
the walls of the guide are joined. The quarter waves together become a half wave and reflect a short circuit at the place where the walls are joined together. Electrically, this creates a short circuit at the junction of the two waveguides. The two sections can actually be separated as much as a tenth of a wavelength without excessive loss of energy at the joint. This separation allows room to seal the interior of the waveguide with a rubber gasket for pressurization. The choke joint effectively keeps the RF inside the waveguide. And it introduces minimum loss, 0.03 dB or less.

CURVED SECTIONS. Special curved waveguide sections are available for making 90° bends. Curved sections introduce reflections and power loss, but these are kept small by proper design. When the radius of the curved section is greater than two wavelengths at the signal frequency, losses are minimized. Figure 15-25 shows several different configurations. Flanges are used to connect curve sections with straight runs.

T SECTIONS. It is occasionally necessary to split or combine two or more sources of microwave power. This is done with *T sections* or *T junctions* (see Fig. 15-26). The T can be formed on the short or long side of the waveguide. If the junction is formed on the short side, it is called a *shunt T.* If the junction is formed on the long side, it is called a *series T.* Each T section has three ports which can be used as inputs or outputs.

If a signal is propagated along a waveguide connected to the C port of a shunt T like that in Fig. 15-26(*a*), equal amounts of the signal will appear in phase at output ports A and B. The power level of the signals at A and B is one-half the input power. Such a device is called a *power divider.*

FIG. 15-24 A choke joint permits sections of waveguide to be interconnected with minimum loss and radiation.

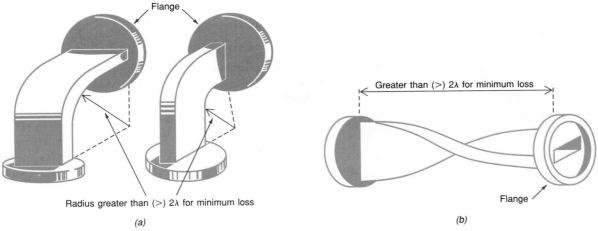

Fig. 15-25 Curved waveguide sections. (*a*) Right-angle bends. (*b*) Twisted section.

The shunt T can also combine signals. If the signal input to A is of the same phase as the signal input to B, they will be combined at the output port C. The output power is the sum of the individual powers. This is called a *power combiner.*

In the series T shown in Fig. 15-26(*b*), a signal entering port D will be split into two half-power signals which appear at ports A and B but which are 180° out of phase with one another.

HYBRID Ts. A special device called a *hybrid T* can be formed by combining the series and shunt T sections (see Fig. 15-27). Sometimes referred to as a *magic T,* this device is used as a duplexer to permit simultaneous use of a single antenna by both a transmitter and a receiver. The antenna is connected to port B. Any received signal is passed to port D, which is connected to the receiver front end. Port A is terminated and not used. The received signal will not enter port C, which is connected to the transmitter.

A transmitted signal will pass through to the antenna at port B, but will not enter port D to the receiver. If a transmitter and receiver operate on the same frequency or near the same frequencies, and the transmitter output signal is at a high power level, some means must be used to prevent the power from entering the receiver and causing damage. The hybrid T can be used for this purpose.

TERMINATIONS. In many cases it is necessary to terminate the end of an unused port of a waveguide. For example, port A in Fig. 15-27 would be connected to a load of the correct impedance to prevent high reflections, unfavorable SWR, and loss. If the

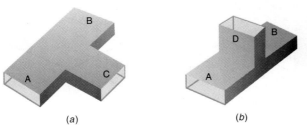

Fig. 15-26 Waveguide T sections. (*a*) Shunt T. (*b*) Series T.

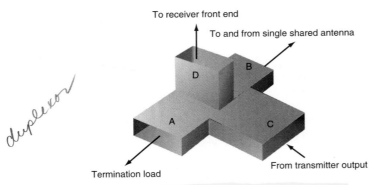

duplexor

FIG. 15-27 A hybrid or magic T used as a duplexer.

length of the line is some multiple of one-quarter or one-half wavelength, it may be possible to open or short the waveguide. In most cases this is not possible, and other means of termination are used.

One approach to termination is to insert a pyramid-shaped metallic section in the end of the line, as shown in Fig. 15-28(*a*). The taper provides a correct match. Usually the tapered section is movable so that the termination can be adjusted for minimum SWR.

It is also possible simply to fill the end of the line with a powdered graphite resistive material, as shown in Fig. 15-28(*b*). This absorbs the signal and dissipates it as heat so that no reflections occur.

Termination can be accomplished by using a resistive material shaped like a triangle or wedge at the end of a closed line [see Fig. 15-28(*c*)]. The tapered resistive element is oriented to match the orientation of the electric field in the guide. The magnetic component of the wave induces a voltage into the tapered resistive material and current flows. Thus the signal is absorbed and dissipated as heat.

DIRECTIONAL COUPLERS. One of the most commonly used waveguide components is the directional coupler. *Directional couplers* are used to facilitate the measurement of microwave power in a waveguide and the SWR. They can also be used to tap off a small portion of a high-power microwave signal to be sent to another piece of equipment.

The directional coupler in Fig. 15-29 is simply a short segment of waveguide with coupling joints that are designed to be inserted into a longer run of waveguide between a transmitter and an antenna or between some source and a load. A similar section of

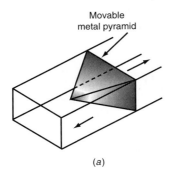

(*a*)

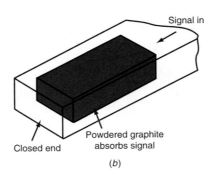

(*b*)

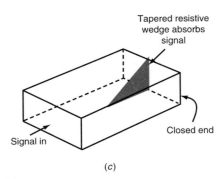

(*c*)

FIG. 15-28 Terminating matched loads for waveguides.

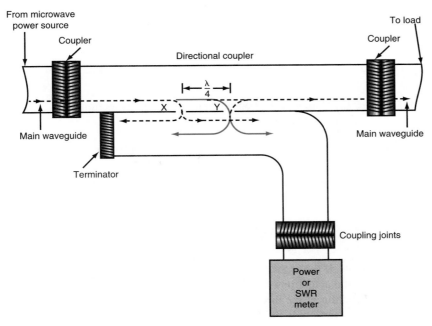

Fig. 15-29 Directional coupler.

waveguide is physically attached to this short segment of line. It is terminated at one end and the other end is bent away at a 90° angle. The bent section is designed to attach to a microwave power meter or an SWR meter. The bent section is coupled to the straight section by two holes (X and Y) which are a quarter wavelength apart at the frequency of operation. Some of the microwave energy passes through the holes in the straight section into the bent section. The amount of coupling between the two sections depends on the size of the holes.

If energy from the source at the left moves through the waveguide from left to right, microwave energy passes through hole X into the bent section. The energy splits in half. Part of the energy goes to the left, where it is absorbed by the terminator; the other half moves to the right toward hole Y. This is indicated by the dashed lines.

Additional energy from the straight section enters hole Y. It, too, splits into two equal components, one going to the left and the other going to the right. These components are indicated by the solid lines. The energy moving from hole Y to the left toward hole X in the curved section cancels the energy moving from hole X to Y. The signal moving from hole Y to hole X travels a total distance of one-half wavelength or 180°, so it is in phase with the signal at hole X. But between Y and X the signals are exactly out of phase and equal in amplitude, so they cancel. Any small residual signal moving to the left is absorbed by the terminator. The remaining signal moves on to the power or SWR meter.

The term *directional coupler* derives from the operation of the device. A portion of the energy of signals moving from left to right will be sampled and measured. Any signal entering from the right moving to the left will simply be absorbed by the terminator; none will pass on to the meter.

The amount of signal energy coupled into the bend section depends on the size of the holes. Usually only a very small portion of the signal, less than 1 percent, is extracted or sampled. Thus the primary signal is not materially attenuated. However, a sufficient amount of signal is present to be measured. The exact amount of signal extracted is determined by the coupling factor (C), which is determined by the familiar formula for power ratio:

$$C = 10 \log \frac{P_{\text{in}}}{P_{\text{out}}} \qquad \text{dB}$$

where P_{in} = amount of power applied to straight section
P_{out} = signal power going to power meter

Most directional couplers are available with a fixed coupling factor, usually 10, 20, or 30 dB.

The amount of power that is actually measured can be calculated by rearranging the power ratio formula:

$$P_{\text{out}} = \frac{P_{\text{in}}}{\log^{-1}(C/10)} = \frac{P_{\text{in}}}{10^{C/10}}$$

For example, if the input power is 2000 W to a directional coupler with a coupling factor of 30, the actual power to the meter is $2000/10^3 = 2000/1000 = 2$ W.

A variation of the directional coupler is the *bidirectional coupler*. Using two sections of line to sample the energy in both directions allows determination of the SWR. One segment of line samples the incident or forward power while the other is set up to sample the reflected energy in the opposite direction.

CAVITY RESONATORS

capacitive/inductive network   narrow band of frequencies

high frequency range

A *cavity resonator* is a waveguide-like device that acts like a high-Q parallel resonant circuit. A simple cavity resonator can be formed with a short piece of waveguide one-half wavelength long, as illustrated in Fig. 15-30(a). The ends are closed. Energy is coupled into the cavity with a coaxial probe at the center, as shown in the side view in Fig. 15-30(b). When microwave energy is injected into the cavity, the signal bounces off the shorted ends of the waveguide and reflects back toward the probe. Because the probe is located a quarter wavelength from each shorted end, the reflected signal reinforces the signal at the probe. The result is that the signal bounces back and forth off the shorted ends. If the signal is removed, the wave continues to bounce back and forth until losses cause it to die out.

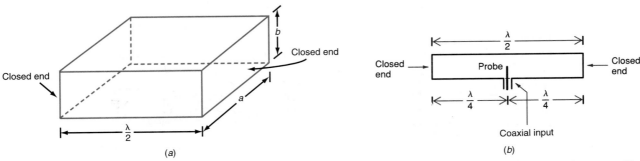

FIG. 15-30 Cavity resonator made with waveguide. (a) $\lambda/2$ section of rectangular waveguide used as a cavity resonator. (b) Side view of cavity resonator showing coupling of energy by a probe.

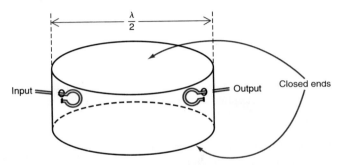

Fig. 15-31 Circular resonant cavity with input-output loops.

This effect is pronounced at a frequency where the length of the waveguide is exactly one-half wavelength. At that frequency, the cavity is said to resonate and acts like a parallel resonant circuit. A brief burst of energy applied to the probe will make the cavity oscillate; the oscillation continues until losses cause it to die out. Cavities such as this have extremely high Q, as high as 30,000. For this reason, they are commonly used to create resonant circuits and filters at microwave frequencies.

A cavity can also be formed by using a short section of circular waveguide such as that shown in Fig. 15-31. With this shape, the diameter should be one-half wavelength at the operating frequency. Other cavity shapes are also possible.

Typically, cavities are specially designed components. Often, they are hollowed-out sections in a block of metal that have been machined to very precise dimensions for specific frequencies. The internal walls of the cavity are often plated with silver or some other low-loss material to ensure minimum loss and maximum Q.

Some cavities are also tunable. One wall of the cavity is made movable, as shown in Fig. 15-32(a). An adjustment screw moves the end wall in and out to adjust the resonant frequency. The smaller the cavity, the higher the operating frequency. Cavities can also be tuned with adjustable plugs in the side of the cavity, as in Fig. 15-32(b). As the plug is screwed in, more of it intrudes into the cavity and the operating frequency goes up.

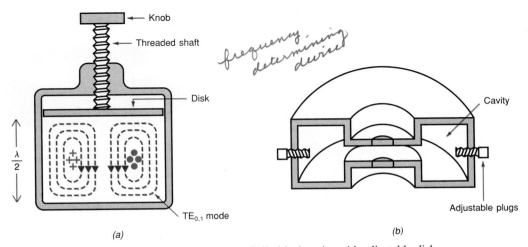

Fig. 15-32 Adjustable or tunable cavities. (a) Cylindrical cavity with adjustable disk. (b) Cavity with adjustable plugs.

A circulator is a three-port microwave device used for coupling energy in only one direction around a closed loop. A schematic diagram of a circulator is shown in Fig. 15-33(a). Microwave energy applied to port 1 is passed to port 2 with only minor attenuation; however, the signal will be greatly attenuated on its way to port 3. The loss from port 1 to port 3 is usually 20 dB or more. A signal applied to port 2 will be passed with little attenuation to port 3, but little or none will reach port 1.

The primary application of a circulator is as a *diplexer*, which allows a single antenna to be shared by a transmitter and receiver. As shown in Fig. 15-33(a), the signal from the transmitter is applied to port 1 and passed to the antenna, which is connected to port 2. The transmitted signal does not reach the receiver connected to port 3. The received signal from the antenna comes into port 2 and is sent to port 3, but is not fed into the transmitter on port 1.

A common way of making a circulator is to create a microstrip pattern that looks like a Y, as shown in Fig. 15-33(b). The three ports are 120° apart. These ports, which are formed on PCBs, can be connected to coaxial connectors.

On top of the Y junction, and sometimes on the bottom, is a ferrite disk. Ferrites are ceramics made of compounds such as $BaFe_2O_3$. They have magnetic properties like iron or steel, so they support magnetic fields, but they do not cause the induction of eddy currents. The iron is often mixed with zinc or manganese. A widely used type of ferrite is yttrium-iron-garnet (YIG). A permanent magnet is then placed such that it forces magnetic lines of force perpendicular to the device through the ferrite. This strong magnetic field interacts with the magnetic fields produced by any input signals, creating the characteristic action of the circulator.

The type of circulator described here is for low and medium power levels. For high-power applications, a circulator made with a waveguide-like structure containing ferrite material is used.

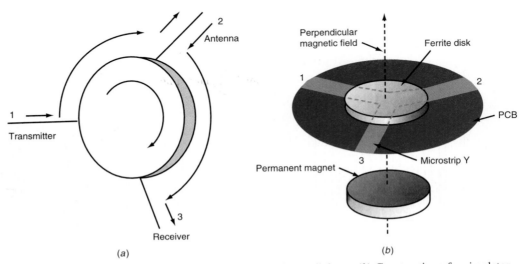

FIG. 15-33 Circulator. (a) Circulator used as a diplexer. (b) Construction of a circulator.

ISOLATORS

Isolators are variations of circulators, but they have one input and one output. That is, they are configured like the circulator diagrammed in Fig. 15-33, but only ports 1 and 2 are used. An input to port 1 is passed to port 2 with little attenuation. Any reflected energy, as would occur with an SWR higher than 1, is not coupled back into port 1. Isolators are often used in situations where a mismatch, or the lack of a proper load, could cause reflection so large as to damage the source.

15-4 MICROWAVE SEMICONDUCTOR DIODES

SMALL-SIGNAL DIODES

Diodes used for signal detection and mixing are the most common microwave semiconductor devices. Two types are widely used: the point-contact diode and the Schottky barrier or hot-carrier diode.

The typical semiconductor diode is a junction formed of P- and N-type semiconductor materials. Because of the relatively large surface area of the junction, diodes exhibit a high capacitance which prevents normal operation at microwave frequencies. For this reason, standard PN junction diodes are not used in the microwave region.

POINT-CONTACT DIODES. Perhaps the oldest microwave semiconductor device is the *point-contact diode,* also called a *crystal diode.* A point-contact diode is nothing more than a piece of semiconductor material and a fine wire which makes contact with the semiconductor material. Because the wire makes contact with the semiconductor over a very small surface area, the capacitance is extremely low. Current flows easily from the cathode, the fine wire, to the anode, the semiconductor material. However, current does not flow easily in the opposite direction.

Most early point-contact diodes used germanium as the semiconductor material, but today these devices are made of P-type silicon with a fine tungsten wire as the cathode (Fig. 15-34). The forward threshold voltage is extremely low.

Point-contact diodes are ideal for small-signal applications. They are widely used in microwave mixers and detectors and in microwave power measurement equipment. They are extremely delicate and cannot withstand high power. They are also easily damaged, and therefore must be used in such a way to minimize shock and vibration.

HOT CARRIER DIODES. For the most part, point-contact diodes have been replaced by *Schottky* diodes, sometimes referred to as *hot carrier diodes.* Most Schottky diodes are made with N-type silicon on which has been deposited a thin metal layer. Gallium arsenide is also used. The semiconductor forms the cathode, and the metal forms

> *very fragile*
>
> *mixer — non-linear (diodes)*

> **HINTS AND HELPS**
>
> Although they are widely used in microwave mixers and detectors and in microwave power measurement equipment, point-contact diodes are extremely delicate and cannot withstand high power.

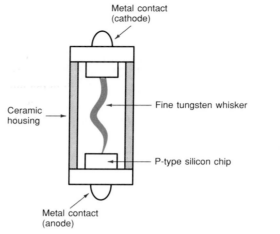

Metal contact
(cathode)

Ceramic
housing

Fine tungsten whisker

P-type silicon chip

Metal contact
(anode)

FIG. 15-34 A point-contact diode.

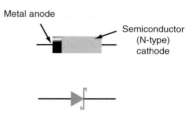

Metal anode

Semiconductor
(N-type)
cathode

FIG. 15-35 Hot carrier or Schottky diode.

the anode. The structure and schematic symbol for a Schottky diode are shown in Fig. 15-35. Typical anode materials are nickel chromium and aluminum, although other metals, for example, gold, are also used.

Like the point-contact diode, the Schottky diode is extremely small and therefore has a tiny junction capacitance. It also has a low bias threshold voltage. It conducts as a forward bias of 0.2 to 0.3 V, whereas the silicon junction diode conducts at 0.6 V. Higher voltage drops across PN silicon diodes result from bulk resistance effects. Schottky diodes are ideal for mixing, signal detection, and other low-level signal operations. They are widely used in balanced modulators and mixers. Because of their very-high-frequency response, Schottky diodes are also used as fast switches at microwave frequencies.

The most important use of microwave diodes is as mixers. Microwave diodes are usually installed as part of a waveguide or cavity resonator tuned to the incoming signal frequency. A local-oscillator signal is injected with a probe or loop. The diode mixes the two signals and produces the sum and difference output frequencies. The difference frequency is usually selected with another cavity resonator or with a low-frequency LC tuned circuit. In a microwave circuit the mixer is usually the input circuit. This is because it is desirable to convert the microwave signal down to a lower frequency level as early as possible, at a point where amplification and demodulation can take place with simpler, more conventional electronic circuits.

FREQUENCY-MULTIPLIER DIODES

Microwave diodes designed primarily for frequency-multiplier service include varactor diodes and step-recovery diodes.

VARACTOR DIODES. A *varactor diode* is basically a voltage variable capacitor. When a reverse bias is applied to the diode, it acts like a capacitor. Its capacitance depends upon the value of the reverse bias. Varactor diodes made with gallium arsenide are optimized for use at microwave frequencies. Their main application in microwave circuits is as frequency multipliers.

Figure 15-36 shows a varactor frequency-multiplier circuit. When an input signal is applied across the diode, it alternately conducts and cuts off. The result is a nonlinear or distorted output containing many harmonics. By using a tuned circuit in the output, the desired harmonic is selected and others are rejected. Since the lower harmonics produce the greatest amount of energy, varactor multipliers are usually used only for doubling and tripling operations. In Fig. 15-36, the input tuned circuit L_1–C_2 resonates at the input frequency (f_{in}), and the output tuned circuit L_2–C_3 resonates at two or three times the input frequency as desired. In practice, the tuned circuits are not actually made up of individual inductors or capacitors. Instead, they are microstrip, stripline, or cavity resonators. C_1 and C_4 are used for impedance matching.

A varactor frequency multiplier does not have gain like a class C amplifier used as a multiplier. In fact, a varactor introduces a signal power loss. However, it is a relatively efficient circuit, and the output can be as high as 80 percent of the input. Typical efficiencies are in the 50 to 80 percent range. No external source of power is required for this circuit; only the RF input power is required for proper operation. Outputs up to 50 W are obtainable with special high-power varactors.

Varactors are used in those applications where it is difficult to generate microwave signals. Usually it is a lot easier to generate a VHF or UHF signal and then use a series of frequency multipliers to put it into the desired microwave region. Varactor diodes are available for producing relatively high power outputs at frequencies up to 100 GHz.

STEP-RECOVERY DIODES. Another diode used in microwave frequency-multiplier circuits like that in Fig. 15-36 is the *step-recovery diode* or *snap-off varactor*. It is a PN-junction diode made with gallium arsenide or silicon. When it is forward-biased, it conducts like any diode, but a charge is stored in the depletion layer. When reverse bias is applied, the charge keeps the diode on momentarily. Then the diode turns off abruptly. This snap-off produces an extremely high-intensity reverse current pulse with a duration of about 100 ps (1 ps $= 10^{-12}$ s). It is extremely rich in harmonics. Even the higher harmonics are of relatively high amplitude.

Step-recovery diodes can also be used circuits like that in Fig. 15-36 to produce multipliers with power ratings up to 5 and 10. Power ratings of 50 W can be obtained. Operating frequencies up to 100 GHz are possible with an efficiency of 80 percent or better.

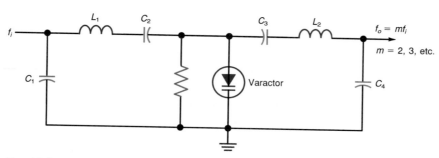

FIG. 15-36 A varactor frequency multiplier.

(handwritten margin notes: current 180° out of phase with voltage / forward bias — resistance goes down — reverse bias — resistance goes up)

Three types of diodes other than the tunnel diode that can oscillate due to negative-resistance characteristics are the Gunn diodes and IMPATT and TRAPATT diodes.

GUNN DIODES. *Gunn diodes,* also called *transferred-electron devices* (TEDs), are not diodes in the usual sense because they do not have junctions. A Gunn diode is a thin piece of N-type gallium arsenide (GaAs) or indium phosphide (InP) semiconductor which forms a special resistor when voltage is applied to it. This device exhibits a negative-resistance characteristic. That is, over some voltage range, an increase in voltage results in a decrease in current and vice versa, just the opposite of Ohm's law. When so biased, the time it takes for electrons to flow across the material is such that the current is 180° out of phase with the applied voltage. If a Gunn diode so biased is connected to a cavity resonant near the frequency determined by the electron transit time, the resulting combination will oscillate. The Gunn diode, therefore, is used primarily as a microwave oscillator.

Gunn diodes are available that will oscillate at frequencies up to about 50 GHz. In the lower microwave range, power outputs up to several watts are possible. The thickness of the semiconductor determines the frequency of oscillation. However, if the cavity is made variable, the Gunn oscillator frequency can be adjusted over a narrow range.

IMPATT AND TRAPATT DIODES. Two other microwave diodes widely used as oscillators are the *IMPATT* and *TRAPATT* diodes. Both are PN-junction diodes made of silicon, GaAs, or InP. They are designed to operate with a high reverse bias that causes them to avalanche or break down. A high current flows. Over a narrow range, a negative-resistance characteristic is produced that causes oscillation when the diode is mounted in a cavity and properly biased. IMPATT diodes are available with power ratings up to about 25 W to frequencies as high as about 30 GHz. IMPATT diodes are preferred over Gunn diodes if higher power is required. Their primary disadvantages are their higher noise level and the higher operating voltages.

PIN DIODES

A PIN diode is a special PN-junction diode with an I (intrinsic) layer between the P and N sections as shown in Fig. 15-37(*a*). The P and N layers are usually silicon, although GaAs is sometimes used. In practice, the I layer is a very lightly doped N-type semiconductor.

At frequencies less than about 100 MHz, the PIN diode acts just like any other PN junction diode. At higher frequencies, it acts like a variable resistor or like a switch. When the bias is zero or reverse, the diode acts like a high value of resistance, 5 Ω and higher. If a forward bias is applied, the diode resistance drops to a very low level, typically a few ohms or less. By varying the amount of forward bias, the value of the resistance can be varied over a linear range. The characteristic curve for a PIN diode is shown in Fig. 15-37.

PIN diodes are used as switches in microwave circuits. A typical application is to connect a PIN diode across the output of a microwave transmission line like microstrip

(handwritten margin notes: 5 mΩ → / PIN diodes used in radar)

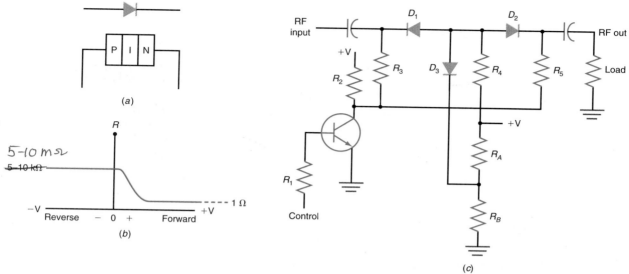

FIG. 15-37 A PIN diode. (*a*) Construction. (*b*) Characteristic curve. (*c*) Tee switch.

or stripline. When the diode is reverse-biased, it acts as a very high resistance and has little effect on the normally much lower characteristic impedance of the transmission line. When the diode is forward-biased, it shorts the line, creating almost total reflection. PIN diodes are widely used to switch sections of quarter- or half-wavelength transmission lines to provide varying phase shifts in a circuit.

A popular switching circuit is the tee configuration shown in Fig. 15-37(*c*), which is a combination of two series switches and a shunt switch. It has excellent isolation between input and output. A simple saturated transistor switch is used for control. When the control input is binary 0, the transistor is OFF, so positive voltage is applied to the cathodes of diodes D_1 and D_2. D_3 conducts; because of the voltage divider, which is made up of R_A and R_B, the cathode of D_3 is at a lower positive voltage level than its anode. Under this condition, the switch is OFF, so no signal reaches the load.

When the control input is positive, or binary 1, the transistor conducts, pulling the cathodes of D_1 and D_2 to ground through resistors R_3 and R_5. The anodes of D_1 and D_2 are at positive voltage through resistor R_4, so these diodes conduct. D_3 is cut off. Thus the signal passes to the load.

PIN diodes are also sometimes used for their variable-resistance characteristics. Varying their bias to change resistance permits variable-voltage attenuator circuits to be created. PIN diodes can also be used as amplitude modulators (see Chap. 4).

15-5 MICROWAVE TUBES

Before transistors were invented, all electronic circuits were implemented with vacuum tubes. Tubes are devices used for controlling a large current with a small voltage to produce amplification, oscillation, switching, and other operations. Today, vacuum tubes are used only for special applications. The cathode-ray tube (CRT) used in TV sets, computer monitors, oscilloscopes, spectrum analyzers, and other display devices is a special form of vacuum tube which shows no signs of obsolescence. LCDs offer the only alternative to the CRT, and their quality has increased significantly over the years.

Vacuum tubes are also still found in microwave equipment. This is particularly true in microwave transmitters used for producing high output power. Both bipolar field-effect transistors can produce power in the microwave region up to approximately several hundred watts, but many applications require more power. Radio transmitters in the UHF and low microwave bands use standard vacuum tubes designed for power amplification. These can produce power levels of up to several thousand watts. At higher microwave frequencies (above 2 GHz) special tubes are used. In satellites and their earth stations, TV stations, and in some military equipment such as radar, very high output powers are needed. Special microwave tubes developed during World War II, the klystron, the magnetron, and the traveling wave tube, are still widely used for such microwave power amplification.

KLYSTRONS

A *klystron* is a microwave vacuum tube using cavity resonators to produce velocity modulation of an electron beam which produces amplification. Figure 15-38 is a schematic diagram of a two-cavity klystron amplifier. The vacuum tube itself consists of a cathode which is heated by a filament. At a very high temperature, the cathode emits electrons. These negative electrons are attracted by a plate or collector, which is biased with a high positive voltage. Thus current flow is established between the cathode and the collector inside the evacuated tube.

The electrons emitted by the cathode are focused into a very narrow stream by using electrostatic and electromagnetic focusing techniques. In *electrostatic focusing,* special elements called focusing plates to which have been applied high voltages force the electrons into a narrow beam. Electromagnetic focusing makes use of coils around the tube through which current is passed to produce a magnetic field. This magnetic field helps focus the electrons into a narrow beam.

The sharply focused beam of electrons is then forced to pass through the centers of two cavity resonators that surround the open center cavity. The microwave signal to be amplified is applied to the lower cavity through a coupling loop. This sets up electric and magnetic fields in the cavity which cause the electrons to speed up and slow down as they pass through the cavity. On one-half cycle, the electrons are speeded up; on the next half cycle of the input, they are slowed down. The effect of this is to create bunches of electrons that are one-half wavelength apart in the drift space between the cavities. This speeding up and slowing down of the electron beam is known as *velocity modulation*. Since the input cavity produces bunches of electrons, it is commonly referred to as the *buncher cavity*.

Since the bunched electrons are attracted by the positive collector, they move on through the tube, eventually passing through the center of another cavity known as the catcher cavity. Because the bunched electrons move toward the collector in clouds of alternately dense and sparse areas, the electron beam can be referred to as a *density-modulated* beam.

As the bunches of electrons pass through the catcher cavity, the cavity is excited into oscillation at the resonant frequency. Thus the DC energy in the electron beam is converted into RF energy at the cavity frequency and amplification occurs. The output is extracted from the catcher cavity with a loop.

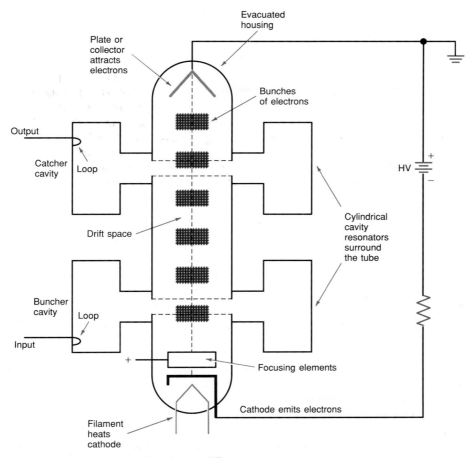

Fɪɢ. 15-38 A two-cavity klystron amplifier.

Klystrons are also constructed with additional cavities between the buncher and catcher cavities. These intermediate cavities produce further bunching which causes increased amplification of the signal. If the buncher cavities are tuned off center frequency from the input and output cavities, the effect is to broaden the bandwidth of the tube. The frequency of operation of a klystron is set by the sizes of the input and output cavities. Since cavities typically have high Qs, their bandwidth is limited. By lowering the Qs of the cavities and by introducing intermediate cavities, wider bandwidth operation can be achieved.

Klystrons are no longer widely used in most microwave equipment. Gunn diodes have replaced the smaller reflex klystrons in signal-generating applications because they are smaller and lower in cost, and do not require high DC supply voltages. The larger multi-cavity klystrons are being replaced by traveling wave tubes in high-power applications.

MAGNETRONS

Another widely used microwave tube is the *magnetron,* a combination of a simple diode vacuum tube with built-in cavity resonators and an extremely powerful permanent magnet. The typical magnetron assembly shown in Fig. 15-39 consists of a circular anode

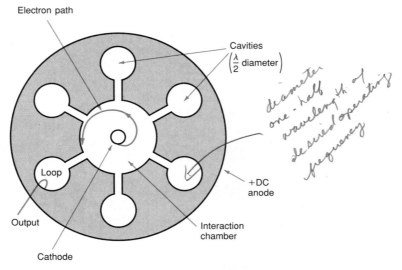

Electron path

Cavities
$\left(\frac{\lambda}{2}\text{ diameter}\right)$

+DC anode

Loop

Output

Cathode

Interaction chamber

diameter one-half wavelength of desired operating frequency

Note: Magnetic field is perpendicular to the page.

FIG. 15-39   A magnetron tube used as an oscillator.

into which has been machined an even number of resonant cavities. The diameter of each cavity is equal to a one-half wavelength at the desired operating frequency. The anode is usually made of copper and is connected to a high-voltage positive DC.

In the center of the anode, called the interaction chamber, is a circular cathode that emits electrons when heated. In a normal diode vacuum tube, the electrons would flow directly from the cathode straight to the anode, causing a high current to flow. In a magnetron tube, however, the direction of the electrons is modified because the tube is surrounded by a strong magnetic field. The field is usually supplied by a C-shaped permanent magnet centered over the interaction chamber. In the figure, the field is labeled "perpendicular to the page"; this means that the lines of force could be coming out of the page or going into the page depending upon the construction of the tube.

The magnetic fields of the moving electrons interact with the strong field supplied by the magnet. The result is that the path for electron flow from the cathode is not directly to the anode, but instead, is curved. By properly adjusting the anode voltage and the strength of the magnetic field, the electrons can be made to bend such that they rarely reach the anode and cause current flow. The path becomes circular loops, as illustrated in Fig. 15-39. Eventually, the electrons do reach the anode and cause current flow. By adjusting the DC anode voltage and the strength of the magnetic field, the electron path is made circular. In making their circular passes in the interaction chamber, the electrons excite the resonant cavities into oscillation. A magnetron, therefore, is an oscillator, not an amplifier. A takeoff loop in one cavity provides the output.

Magnetrons are capable of developing extremely high levels of microwave power. Thousands and even millions of watts of power can be produced by a magnetron. When operated in a pulsed mode, magnetrons can generate several megawatts of power in the microwave region. Pulsed magnetrons are commonly used in radar systems. Continuous-wave magnetrons are also used and can generate hundreds and even thousands of watts of power. A typical application for a continuous-wave magnetron is for heating purposes in microwave ovens.

One of the most versatile microwave RF power amplifiers is the *traveling-wave tube* (TWT), which can generate hundreds and even thousands of watts of microwave power. The main virtue of the TWT is an extremely wide bandwidth. It is not resonant at a single frequency.

Figure 15-40 shows the basic structure of a traveling-wave tube. It consists of a cathode and filament heater plus an anode that is biased positively to accelerate the electron beam forward and to focus it into a narrow beam. The electrons are attracted by a positive plate called the *collector* to which is applied a very high DC voltage. Traveling-wave tubes can be anywhere from 1 ft to several feet. In any case, the length of the tube is usually many wavelengths at the operating frequency. Permanent magnets or electromagnets surround the tube, keeping the electrons tightly focused into a narrow beam.

Encircling the length of a traveling-wave tube is a helix or coil. The electron beam passes through the axis of the helix. The microwave signal to be amplified is applied to the end of the helix near the cathode, and the output is taken from the end of the helix near the collector. The purpose of the helix is to provide a path for the RF signal that will slow down its propagation. The propagation of the RF signal along the helix is made approximately equal to the velocity of the electron beam from cathode to collector. The helix is configured such that the wave traveling along it is slightly slower than that of the electron beam.

The passage of the microwave signal down the helix produces electric and magnetic fields that interact with the electron beam. The effect on the electron beam is similar to that in a klystron. The electromagnetic field produced by the helix causes the electrons to be speeded up and slowed down. This produces velocity modulation of the beam, which in turn produces density modulation. Density modulation, of course, causes bunches of electrons to group together one wavelength apart. These bunches of electrons travel down the length of the tube toward the collector. Since the density-modulated electron beam is essentially in step with the electromagnetic wave travel-

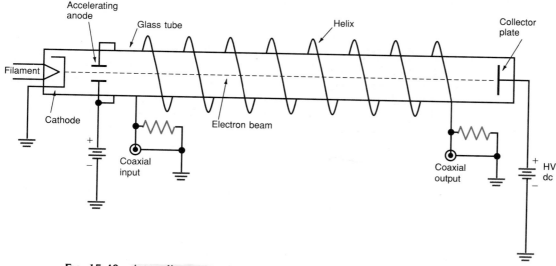

FIG. 15-40 A traveling-wave tube (TWT).

ing down the helix, the electron bunches induce voltages into the helix which reinforce the existing voltage. The result is that the strength of the electromagnetic field on the helix increases as the wave travels down the tube toward the collector. At the end of the helix, the signal is considerably amplified. Coaxial-cable of waveguide structures are used to extract energy from the helix.

Traveling-wave tubes can be made to amplify signals in a range from UHF to hundreds of gigahertz. Most TWTs have a frequency range of approximately 2:1 in the desired segment of the microwave region to be amplified. TWTs can be used in both continuous and pulsed modes of operation. One of the most common applications of TWTs is as power amplifiers in satellite transponders.

MISCELLANEOUS MICROWAVE TUBES

In a *backward wave oscillator* (*BWO*), a variation of the TWT, the wave travels from the anode end of the tube back toward the electron gun, where it is extracted. BWOs can generate up to hundreds of watts of microwave power in the 20- to 80-GHz range. The operating frequency of the BWO can easily be tuned by varying the collector voltage.

Gyrotrons, which are built and operated like klystrons, are used for amplification at microwave frequencies above 30 GHz into the millimeter-wave range. They can also be connected to operate as oscillators. Gyrotrons are the only devices currently available for power amplification and signal generation in the millimeter-wave range.

A *crossed-field amplifier* (*CFA*) is similar to a TWT. Its gain is lower but is somewhat more efficient. For a given power level, the operating voltage of a CFA is usually lower than that of a TWT. The bandwidth is about 20 to 60 percent of the design frequency. Power levels up to several megawatts in a pulse mode can be achieved.

15-6 MICROWAVE ANTENNAS

All the antennas we discussed in Chap. 14 can also be used at microwave frequencies. However, these antennas will be extremely small. At 5 GHz a half-wave dipole is slightly less than 1 in long and a quarter-wave vertical is slightly less than 0.5 in long. These antennas can, or course, radiate microwave signals, but inefficiently. Because of the line-of-sight transmission of microwave signals, highly directive antennas are preferred because they do not waste the radiated energy and because they provide an increase in gain, which helps offset the noise problems at microwave frequencies. For these important reasons, special high-gain, highly directive antennas are normally used in microwave applications.

LOW-FREQUENCY ANTENNAS

At low microwave frequencies, less than 2 GHz, standard antennas are commonly used, including the dipole and its variations such as the bow-tie, the Yagi, and the ground-plane antenna. Another variation is the corner reflector shown in Fig. 15-41. This an-

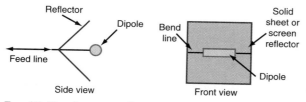

Side view
Front view

FIG. 15-41 A corner reflector used with a dipole for low microwave frequencies.

more directivity
narrow beam
higher gain

tenna is a fat, wide-bandwidth, half-wave dipole fed with low-loss coaxial cable. Behind the dipole is a reflector made of solid sheet metal, a group of closely spaced horizontal rods, or a fine-mesh screen material to reduce wind resistance. This arrangement gives better reflection than a simple rod reflector as used in a Yagi, so the gain is higher.

The angle of the reflector is usually 45°, 60°, or 90°. The spacing between the dipole and the corner of the reflector is usually in the range from about 0.25 to 0.75 wavelength. Within that spacing range, the gain varies only about 1.5 dB. However, the feed-point impedance of the dipole varies considerably with spacing. The spacing is usually adjusted for the best impedance match. It is common to use 50- or 75-Ω coaxial cable, which is relatively easy to match. A folded dipole can also be used if the antenna is operated in the UHF or VHF ranges. The overall gain of a corner reflector antenna is 10 to 15 dB. Higher gains can be obtained with a parabolic reflector, but corner reflectors are easier to make and much less expensive.

HINTS AND HELPS

Special high-gain, highly directive antennas are used in microwave applications because they do not waste the radiated energy. They also provide an increase in gain, which helps offset the noise problems at these frequencies.

HORN ANTENNAS

As discussed previously, waveguides are the most predominant type of transmission line used with microwave signals. Below approximately 6 GHz, special coaxial cables can be used effectively if the distances between the antenna and the receiver or transmitter are less than 50 ft. In most microwave systems, waveguides are preferred because of their low loss. Microwave antennas, therefore, must be some extension of or compatible with a waveguide. Waveguides are, of course, inefficient radiators if simply left open at the end. The problem with using a waveguide as a radiator is that it provides a poor impedance match with free space, and the mismatch results in standing waves and reflected power. The result is tremendous power loss of the radiated signal. This mismatch can be offset by simply flaring the end of the waveguide to create a *horn antenna,* as shown in Fig. 15-42. The longer and more gradual the flair, the better the impedance match and the lower the loss. Horn antennas have excellent gain and directivity. The longer the horn, the greater its gain and directivity.

Different kinds of horn antennas can be created by flaring the end of the waveguide in different ways. For example, flaring the waveguide in only one dimension creates a sectoral horn, as shown in Fig. 15-43(*a*) and (*b*). Two of the sides of the horn remain parallel with the sides of the waveguide, and the other dimension is flared. Flaring both dimensions of the horn produces a pyramidal horn, as shown in Fig. 15-43(*c*). If a circular waveguide is used, the flair produces a conical horn, as in Fig. 15-43(*d*).

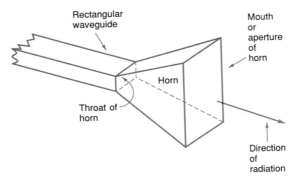

FIG. 15-42 Basic horn antenna.

The gain and directivity of a horn are a direct function of its dimensions; the most important dimensions are horn length, aperture area, and flare angle (see Fig. 15-44).

The length of a typical horn is usually 2 to 15 λ at the operating frequency. Assuming an operating frequency of 10 GHz, the length of 1 λ is 300/f = 300/10,000 = 0.03 m, where f is in megahertz. A length of 0.03 m equals 3 cm; 1 in equals 2.54 cm, so a wavelength at 10 GHz is 3/2.54 = 1.18 in Thus a typical horn at 10 GHz could be anywhere from about 2½ to 18 in long. Longer horns are, of course, more difficult to mount and work with, but provide higher gain and better directivity.

The aperture is the area of the rectangle formed by the opening of the horn and is simply the product of the height and width of the horn, as shown in Fig. 15-44. The greater this area, the higher the gain and directivity. The flare angle also affects gain and directivity. Typical flare angles vary from about 20° to 60°. Obviously, all these dimensions are

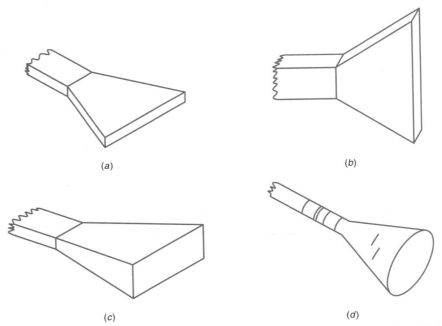

(a)

(b)

(c)

(d)

FIG. 15-43 Types of horn antennas. (a) Sectoral horn. (b) Sectoral horn. (c) Pyramidal horn. (d) Conical horn.

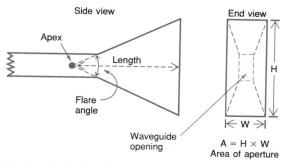

Fig. 15-44 Dimensions of a horn.

interrelated. For example, increasing the flare angle increases the aperture area, and for a given aperture area, decreasing the length increases the flare angle. Any of these dimensions can be adjusted to achieve a desired design objective.

BEAMWIDTH. Remember that the directivity of an antenna is measured in terms of beamwidth, the angle formed by extending lines from the center of the antenna response curve to the 3-dB down points. In the example in Fig. 15-45, the beamwidth is approximately 30°. Horn antennas typically have a beam angle somewhere in the 10° to 60° range.

The signal radiated from an antenna is three-dimensional. The directivity patterns indicate the horizontal radiation pattern of the antenna. The antenna also has a vertical radiation pattern. Figure 15-46 is a typical plot of the vertical radiation pattern of a horn antenna. On pyramidal and circular horns, the vertical beamwidth is usually about the same angle as the horizontal beamwidth. This is not true for sectoral horns.

The horizontal beamwidth B of a pyramidal horn is computed using the expression

$$B = \frac{80}{w/\lambda}$$

where w = horn width
λ = wavelength of operating frequency

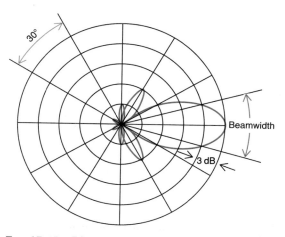

Fig. 15-45 Directivity of an antenna as measured by beamwidth.

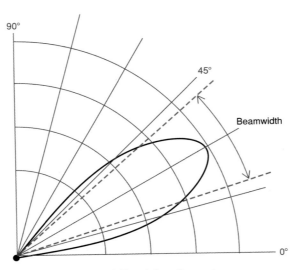

Fig. 15-46 Vertical directivity of an antenna.

As an example, assume an operating frequency of 10 GHz (10,000 MHz), which gives a wavelength of 0.03 m, as computed earlier. If the pyramidal horn is 10 cm high and 12 cm wide, the beamwidth is $80/(0.12/0.03) = 80/4 = 20°$.

GAIN. The gain of a pyramidal horn can also be computed from its dimensions. The approximate power gain of a pyramidal horn antenna is

$$G = 4\pi\frac{KA}{\lambda^2} \qquad \text{decibels}$$

where A = aperture of horn, m^2
λ = wavelength, m
K = constant derived from how uniformly the phase and amplitude of the electromagnetic fields are distributed across the aperture

The typical values of K are in the 0.5 to 0.6 range.

Let us take as an example the previously described horn (height = 10 cm; width = 12 cm). The aperture area is

$$A = \text{height} \times \text{width} = 10 \times 12 = 120 \text{ cm}^2 \text{ or } 0.012 \text{ m}^2$$

The operating frequency is 10 GHz, so 1 λ = 0.03 m or 3 cm. The gain is

$$G = \frac{4(3.14)(0.5)(0.012)}{(0.03)^2} = \frac{0.07536}{0.0009} = 83.7$$

This is the power ratio P. To find the gain in decibels, the standard power formula is used:

$$dB = 10 \log P$$

where P is the power ratio or gain. Here,

$$dB = 10 \log 83.7$$
$$= 10(1.923)$$
$$= 10.23$$

This is the power gain of the horn over a standard half-wave dipole or quarter-wave vertical.

BANDWIDTH. Most antennas have a narrow bandwidth because they are resonant at only a single frequency. Their dimensions determine the frequency of operation. Bandwidth is an important consideration at microwave frequencies because the spectrum transmitted on the microwave carrier is usually very wide so that a considerable amount of information can be carried. Horns are essentially nonresonant or aperiodic, which means they can operate over a wide frequency range. The bandwidth of a typical horn antenna is approximately 10 percent of the operating frequency. The bandwidth of a horn at 10 GHz is approximately 1 GHz. This is an enormous bandwidth—plenty wide enough to accommodate almost any kind of complex modulating signal.

PARABOLIC ANTENNAS

Horn antennas are used by themselves in many microwave applications. When higher gain and directivity are desirable, they can easily be obtained by using a horn in conjunction with a parabolic reflector. A *parabolic reflector* is a large dish-shaped structure made of metal or screen mesh. The energy radiated by the horn is pointed at the reflector, which focuses the radiated energy into a narrow beam and reflects it toward its destination. Because of the unique parabolic shape, the electromagnetic waves are narrowed into an extremely small beam. Beamwidths of only a few degrees are typical with parabolic reflectors. Of course, such narrow beamwidths also represent extremely high gains.

A parabola is a common geometric figure. A key dimension of a parabola is a line drawn from its center at point Z to a point on the axis labeled F, which is the focal point. The ends of a parabola theoretically extend outward for an infinite distance, but for practical applications they are limited. In the figure, the limits are shown by the dashed vertical line; the end points are labeled X and Y.

The distance between a parabola's focal point and any point on the parabola and then to the vertical dashed line is a constant value. For example, the sum of lines FA and AB is equal to the sum of lines FC and FD. This effect causes a parabolic surface to collimate electromagnetic waves into a narrow beam of energy. An antenna placed at the focal point F will radiate waves from the parabola in parallel lines. If used as a receiver, the parabola will pick up the electromagnetic waves and reflect them to the antenna located at the focal point.

The key thing to remember about a parabolic reflector is that it is not two-dimensional. If a parabola is rotated about its axis, a three-dimensional dish-shaped structure results. This is called a paraboloid.

Figure 15-47 shows how a parabolic reflector is used in conjunction with a conical horn antenna for both transmission and reception. The horn antenna is placed at the focal point. In transmitting, the horn radiates the signal toward the reflector, which bounces the waves off the reflector and collimates them into a narrow parallel beam. When used for receiving, the reflector picks up the electromagnetic signal and bounces the waves toward the antenna at the focal point. The result is an extremely high-gain, narrow-beamwidth antenna.

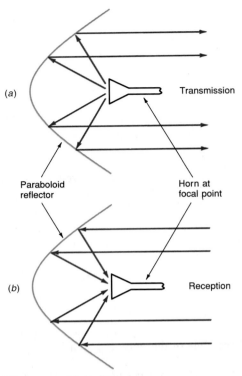

Fig. 15-47 Sending and receiving with a parabolic reflector antenna.

Note that any common antenna type (e.g., a dipole) can be used with a parabolic reflector to achieve the effects just described.

GAIN. The gain of a parabolic antenna is directly proportional to the aperture of the parabola. The aperture, the area of the outer circle of the parabola, is

$$A = \pi R^2$$

The gain of a parabolic antenna is given by the simple expression

$$G = 6\left(\frac{D}{\lambda}\right)^2$$

where G = gain expressed as a power ratio
D = diameter of dish, m
λ = wavelength, m

Most parabolic reflectors are designed so that the diameter is no less than 1 λ at the lowest operating frequency. However, the diameter can be as much as 10 λ if greater gain and directivity are required.

For example, the power gain of a 5-m-diameter dish at 10 GHz ($\lambda = 0.03$, as computer earlier) is

$$G = 6(5/0.03)^2$$
$$= 6(166.67)^2$$
$$= 6(27778.9) = 166,673$$

Expressed in decibels,

$$dB = 10 \log 166673 = 10 \,(5.22) = 52.2$$

BEAMWIDTH. The beamwidth of a parabolic reflector is inversely proportional to the diameter. It is given by the expression

$$B = \frac{58}{D/\lambda}$$

The beamwidth of our 5-m, 10-GHz antenna is $58(5/0.03) = 5/166.67 = 0.348°$.

With a beamwidth of less than $0.5°$, the signal radiated from a parabolic reflector is a pencil-thin beam that must be pointed with great accuracy in order for the signal to be picked up. Usually both the transmitting and receiving antennas are of a parabolic design and have extremely narrow beamwidths. For that reason, they must be accurately pointed if contact is to be made. Despite the fact that the beam from a parabolic reflector spreads out and grows in size with distance, directivity is good and gain is high. The precise directivity helps to prevent interference from signals coming in at angles outside of the beamwidth.

FEED METHODS. Keep in mind that the parabolic dish is not the antenna, only a part of it. The antenna is the horn at the focal point. There are many physical arrangements used in positioning the horn. One of the most common is the seemingly awkward configuration shown in Fig. 15-48. The waveguide feeds through the center of the parabolic dish and is curved around so that the horn is positioned exactly at the focal point.

Another popular method of feeding a parabolic antenna is shown in Fig. 15-49. Here the horn antenna is positioned at the center of the parabolic reflector. At the focal point is another small reflector with either a parabolic or hyperbolic shape. The electromagnetic radiation from the horn strikes the small reflector, which then reflects the energy toward the large dish which in turn radiates the signal in parallel beams. This arrangement is known as a *Cassegrain feed.*

The Cassegrain feed has several advantages over the feed arrangement in Fig. 15-48. The first is that the waveguide transmission line is shorter. In addition, the radical bends in the waveguide are eliminated. Both add up to less signal attenuation. The noise figure is also improved somewhat. Most large earth-station antennas use a Cassegrain feed arrangement.

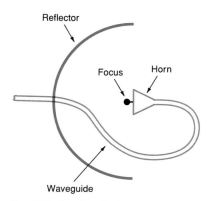

FIG. 15-48 Standard waveguide and horn feed.

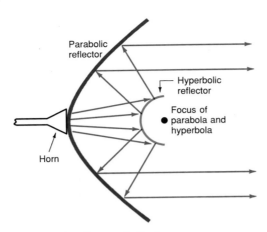

Fig. 15-49 Cassegrain feed.

Many other feed arrangements have been developed for parabolic reflectors. In those antennas used for satellite TV reception, a waveguide is not used. Instead, a horn antenna mounted at the focal point is usually fed with large microwave coaxial cable. In other large antenna systems, various mechanical arrangements are used to permit the antenna to be rotated or its position otherwise physically changed. Many earth-station antennas must be set up so that their azimuth and elevation can be changed to ensure proper orientation for the receiving antenna. This is particularly true of antennas used in satellite communications systems.

Example 15-4

A parabolic reflector antenna has a diameter of 5 ft. Calculate (*a*) the lowest possible operating frequency, (*b*) gain at 15 GHz, and (*c*) beamwidth at 15 GHz. (The lowest operating frequency occurs where the dish diameter is 1 λ.)

a. $\lambda = \dfrac{984}{f_{MHz}}$

$= 5 \text{ ft}$

$f_{MHz} = \dfrac{984}{5} = 196.8 \text{ MHz}$

b. There are 3.28 ft per meter, so the diameter is 5/3.28 = 1.524 m.

15 GHz = 15,000 MHz

$\lambda = 300/f_{MHz} = 300/15,000 = 0.02 \text{ m}$

$G = 6\left(\dfrac{D}{\lambda}\right)^2 = 6\left(\dfrac{1.524}{0.02}\right)^2 = 34838.6$ (power density ratio)

dB = 10 log 34838.6 = 45.42

c. $B = \dfrac{58}{\dfrac{D}{\lambda}} = \dfrac{58\,\lambda}{D} = \dfrac{58(0.02)}{1.525} = 0.76°$

Some sophisticated systems use multiple horns on a single reflector. Such multiple feeds permit several signals on different frequencies to be either radiated or received with a single large reflecting structure.

HELICAL ANTENNAS

A *helical antenna,* as its name suggests, is a wire helix (Fig. 15-50). A center insulating support is used to hold heavy wire or tubing formed into a circular coil or helix. The diameter of the helix is typically one-third wavelength, and the spacing between turns is approximately one-quarter wavelength. Most helical antennas use from 6 to 8 turns. A circular or square ground-plane antenna or reflector is used behind the helix. Figure 15-50 shows a coaxial feed line. Helical antennas are widely used at VHF and UHF frequency ranges.

The gain of a helical antenna is typically in the 12- to 20-dB range, and beam widths vary from approximately 12° to 45°. Although these values do not compare favorably with those obtainable with horns and parabolic reflectors, helical antennas are favored in many applications because of their simplicity and low cost.

Most antennas transmit either a vertically or horizontally polarized electromagnetic field. With a helical antenna, however, the electromagnetic field is caused to rotate. This is known as *circular polarization.* Either right-hand (clockwise) and left-hand (counterclockwise) circular polarization can be produced, depending on the direction of winding of the helix. Because of the rotating nature of the magnetic field, a circularly polarized signal can easily be received by either a horizontally or vertically polarized receiving antenna. A helical receiving antenna can also easily receive horizontally or vertically polarized signals. Note, however, that a right-hand circularly polarized signal will not be picked up by a left-hand circularly polarized antenna, and vice versa. Therefore, helical antennas used at both transmitting and receiving ends of a communications link must both have the same polarization.

To obtain greater gain and narrower beamwidth, several helical antennas can be used in an array with a common reflector. A popular arrangement is a group of four helical antennas.

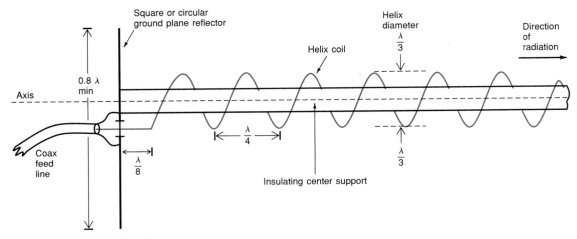

FIG. 15-50 The helical antenna.

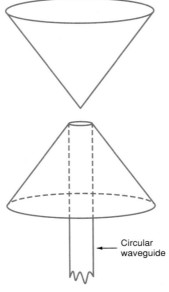

FIG. 15-51 The omnidirectional bicone antenna.

BICONE ANTENNAS

Most microwave antennas are highly directional. But in some applications an omnidirectional antenna may be required. One of the most widely used omnidirectional microwave antennas is the *bicone* (Fig. 15-51). The signals are fed into bicone antennas through a circular waveguide ending in a flared cone. The upper cone acts as a reflector, causing the signal to be radiated equally in all directions with a very narrow vertical beamwidth.

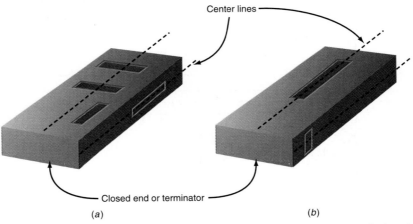

FIG. 15-52 Slot antennas on a waveguide. (*a*) Radiating slots. (*b*) Nonradiating slots.

Fig. 15-53 A slot antenna array.

Slot Antennas

A *slot antenna* is a radiator made by cutting a half-wavelength slot in a conducting sheet of metal or into the side or top of a waveguide. The basic slot antenna is made by cutting a half-wavelength slot in a large metal sheet. It has the same characteristics as a standard dipole antenna, as long as the metal sheet is very large compared to 1 λ at the operating frequency. A more common way of making a slot antenna is shown in Fig. 15-52. The slot must be one-half wavelength long at the operating frequency. Figure 15-52(*a*) shows how the slots must be positioned on the waveguide to radiate. If the slots are positioned on the center lines of the waveguide sides, as in Fig. 15-52(*b*), they will not radiate.

Several slots can be cut into the same waveguide to create a slot antenna array (Fig. 15-53). Slot arrays, which are equivalent to driven arrays with many elements, have better gain and better directivity than single-slot antennas.

Slot antennas are widely used on high-speed aircraft. External antennas would be torn off at such high speeds, or would slow the aircraft. The slot antenna can be integrated into the metallic skin of the aircraft. The slot itself is filled in with an insulating material to create a smooth skin surface.

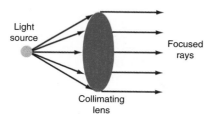

Fig. 15-54 How a lens focuses light rays.

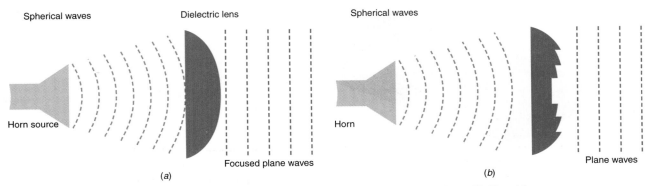

Labels on image (a): Spherical waves, Dielectric lens, Horn source, Focused plane waves

Labels on image (b): Spherical waves, Horn, Plane waves

FIG. 15-55 Lens antenna operations. (*a*) Dielectric lens. (*b*) Zoned lens.

DIELECTRIC (LENS) ANTENNAS

As discussed previously, radio waves, like light waves, can be reflected, refracted, diffracted, and otherwise manipulated. This is especially true of microwaves, which are close in frequency to light. Thus a microwave antenna can be created by constructing a device that serves as a lens for microwaves just as glass or plastic can serve as a lens for light waves. These *dielectric* or *lens* antennas use a special dielectric material to collimate or focus the microwaves from a source into a narrow beam. Figure 15-54 shows how a lens concentrates light rays from a source into a focused narrow beam. A dielectric lens antenna operates in a similar way.

An example of a lens antenna is one used in the millimeter-wave range. The microwave energy is coupled to a horn antenna through a waveguide. A dielectric lens is placed over the end of the horn, which focuses the waves into a narrower beam with greater gain and directivity. In technical terms, the lens takes the microwaves from a source with a spherical wavefront (e.g., a horn antenna) and concentrates them into a plane wavefront. A lens like that shown in Fig. 15-55(*a*) can be used. The shape of the lens ensures that all of the entering waves with a spherical wavefront are put into phase at the output to create the concentrated plane wavefront. However, the lens in Fig. 15-55(*a*) will work only when it is very thick at the center. This creates great signal loss, especially at the lower microwave frequencies. To get around this problem, a stepped or zoned lens, such as the one shown in Fig. 15-55(*b*), can be used. The spherical wavefront is still converted into a focused plane wavefront, but the thinner lens causes less attenuation.

Lens antennas are usually made of polystyrene or some other plastic, although other types of dielectric can be used. They are rarely used at the lower microwave frequencies. Their main use is in the millimeter range above 40 GHz.

PATCH ANTENNAS

Patch antennas are made with microstrip on PCBs. The antenna is a circular or rectangular area of copper separated from the ground plane on the bottom of the board by the thickness of the PCB's insulating material (see Fig. 15-56). The width of the rectangular antenna is approximately one-half wavelength, and the diameter of the circular antenna is about 0.55 to 0.59 wavelength. In both cases the exact dimensions depend upon the dielectric constant and the thickness of the PCB material. The most commonly used PC board material for patch antennas is a Teflon-fiberglass combination.

The feed method for patch antennas can be either coaxial or edge. With the coaxial method, the center conductor of a coaxial cable is attached somewhere between the center and the edge of the patch and the coaxial shield is attached to the ground plane [Fig. 15-56(a)]. If the antenna is fed at the edge, a length of microstrip is connected from the source to the edge, as shown in Fig. 15-56(b). The impedance of the edge feed is about 120 Ω. A quarter-wave Q section can be used to match this impedance to the 50-Ω impedance that is characteristic of most circuits. When coaxial feed is used, the impedance is zero at the center of the antenna and increases to 120 Ω at the edge. Correctly positioning the coaxial-cable center on the patch [dimension x in Fig. 15-56(a)] allows extremely accurate impedance matching.

Patch antennas are small, inexpensive, and easy to construct. In many applications they can simply be integrated on the PCB with the transmitter or receiver. A disadvantage of patch antennas is their narrow bandwidth, which is usually no more than about 5 percent of the resonant frequency with circular patches, and up to 10

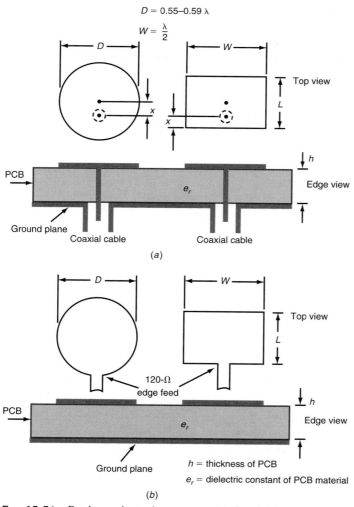

FIG. 15-56 Patch or microstrip antennas. (a) Coaxial feed. (b) Edge feed.

percent with rectangular patches. The bandwidth is directly related to the thickness of the PCB material, that is, the distance between the antenna and the ground plane [*h* in Fig. 15-56(*b*)]. The greater the thickness of the PCB dielectric, the greater the bandwidth.

The radiation pattern of a patch antenna is approximately circular in the direction opposite to that of the ground plane.

PHASED ARRAYS

A *phased array* is an antenna system made up of a large group of similar antennas on a common plane. Patch antennas on a common PCB can be used, or separate antennas like dipoles can be physically mounted together in a plane. The antennas are driven by transmission lines that incorporate impedance matching, power splitting, and phase-shift circuits. The basic purpose of an array is to improve gain and directivity. Arrays also offer better control of directivity, since individual antennas in an array can be turned off or on, or driven through different phase shifters. The result is that the array can be "steered"—that is, its radiation pattern can be pointed over a wide range of different directions—without physically moving the antenna, as is necessary with Yagis or parabolic dish antennas.

There are two common arrangements for phased arrays. In one configuration, the multiple antennas are driven by a common transmitter or feed a common receiver. Another approach is to have a low-power transmitter amplifier or low-noise receiver amplifier associated with each dipole or patch in the array. In both cases, the switching and phase shifting are under the control of a microprocessor or computer. Different programs in the processor select the gain, directivity, and other factors as required by the application.

Most phased arrays are used in radar systems, but they are finding applications in some specialized communications systems and in satellites.

15-7 MICROWAVE APPLICATIONS

The communications applications in which microwaves are most widely used today are telephone communications and radar. However, there are many other significant uses of microwave frequencies in communications. For example, TV stations use microwave relay links instead of coaxial cables to transmit TV signals over long distances, and cable TV networks use satellite communications to transmit programs from one location to another. Communications with satellites, deep-space probes, and other spacecraft is usually done by microwave transmission because microwave signals are not reflected or absorbed by the ionosphere as are many lower-frequency signals. Electromagnetic radiation from the stars is also primarily in the microwave region, only sensitive radio receivers and large antennas operating in the microwave region are used to map outer space with far greater precision than could be achieved with optical telescopes. Finally, microwaves are also used for heating—in the kitchen (microwave ovens), in medical practice (diathermy machines used to heat muscles and tissues without causing skin damage), and in industry.

The rest of this chapter is devoted to a discussion of microwave relay stations and radar. Satellites are covered in detail in Chap. 16.

Microwave radio repeater or relay stations are used by telephone companies to carry telephone calls over long distances. As discussed in Chap. 10, the signals are transmitted using various multiplexing techniques. Microwave relay or repeater stations are an alternative to using standard telephone wiring (twisted pair) or fiber-optic cable. Microwave radio is used in locations where it is difficult or too expensive to string cable of any kind, for example, mountainous areas and deserts.

Telephone companies have terminal equipment where signals originate and terminate. This equipment communicates with a string of repeater stations along the path from source to destination. Each repeater station contains both a receiver and a transmitter, and can send and receive signals in both directions at the same time.

These systems operate in assigned bands with frequencies near 2, 4, 6, 11, 13, and 18 GHz. The distance between repeaters varies with both terrain and frequency of operation. Maximum repeater separation is usually about 40 to 60 km (25 to 37 mi) at the lower frequencies. At higher frequencies, lower power is used, and the path loss is much greater, especially in bad weather with rain or fog. Therefore, the spacing between repeaters is much less, typically 25 km (15 mi) or so. The antennas are parabolic dishes mounted on high towers; where possible, the towers are located on hills, mountain tops, or even high buildings to increase the transmitting distance.

TERMINAL EQUIPMENT. Figure 15-57 shows the terminal equipment used for a microwave relay system. A multiplexed signal containing dozens, hundreds, or even thousands of calls is used to modulate a carrier of 70 MHz. FM is used with analog signals, and BPSK is used with digital signals. The signal is amplified in an IF amplifier and sent to a mixer which is used as an up converter to translate the signal up to the desired microwave operating frequency. The local oscillator frequency adds to the 70-MHz IF to create the final operating frequency. The resulting signal is then am-

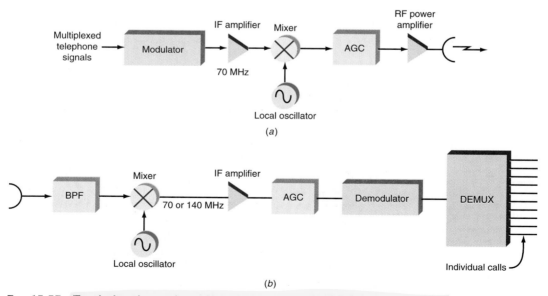

FIG. 15-57 Terminal equipment in a microwave relay system. (*a*) Terminal transmitter. (*b*) Terminal receiver.

plified by class C power amplifiers and applied to the antenna, which is usually a horn feeding a parabolic dish.

At the receiving end, the antenna picks up the signal and feeds it to a mixer by way of a tuned circuit. The tuned circuit can be a microstrip or a cavity resonator bandpass filter. The mixer is usually a hot carrier diode. The mixer, also driven by a local oscillator, down-converts the signal to an IF; the IF is usually 70 MHz, although 140 MHz is used in some systems. The signal passes through an IF amplifier where the desired receiver selectivity and bandwidth are obtained. As in most superheterodyne receivers, AGC is used. Other processing techniques may also be used such as equalization, which compensates for phase shifts and frequency-response problems. Finally, the signal is demodulated. The recovered signal, a composite of many phone calls, is sent to a demultiplexer, where the signals are individually recovered.

ANALOG REPEATER STATIONS. *Analog repeater stations* are used to transmit and receive multiplexed analog telephone calls through frequency-division multiplexing (Chap. 10). Figure 15-58 is a block diagram of an analog repeater station. Two antennas are used, one pointing in each direction of the path. A circulator used as a diplexer permits both antennas to simultaneously transmit and receive.

A received signal is band-limited by a tuned circuit such as a bandpass filter, then sent to a mixer where it is down-converted to an IF of 70 or 140 MHz. It is then amplified by an IF amplifier with AGC. The signal is *not* demodulated. (The points in the system marked X are those at which demodulation would be used.) The amplified signal with the modulation is then fed to a mixer that up-converts it to the outgoing microwave frequency, which is different from the received microwave signal. The same frequency cannot be used because the transmitter would interfere with the receiver. The signal is then sent to a power amplifier before it is applied to the antenna. The power output is typically 2 to 5 W, although higher-power transmitters are sometimes used. The dish antennas boost the effective radiated power due to their gain and directivity. Note that the repeater system has two receive-transmit paths, one for each direction.

DIGITAL REGENERATIVE REPEATER STATIONS. A different type of repeater station is used when digital telephone calls are transmitted. Telephone calls are commonly digitized and transmitted by time-division multiplexing techniques, such as the T-1 system (also known as DS-1) discussed in Chap. 11. The T-1 signal contains 24 channels

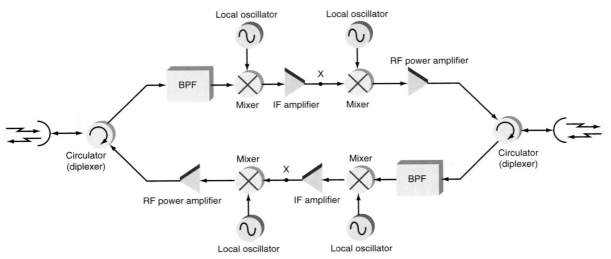

FIG. 15-58 A microwave relay station for analog signals.

digitized at an 8-kHz rate using 8-bit words. The final resulting bit stream of serial data is transmitted at a rate of 1.544 Mbps.

Larger, faster digital multiplexing schemes are becoming more widely used by telephone companies. The DS-2 system transmits 96 channels at 6.312 Mbps, the DS-3 system transmits 672 channels at 44.736 Mbps, and the DS-4 system transmits 4032 channels at 274.176 Mbps. These systems often use fiber-optic cable, but the signals can also be sent via microwave radio.

Typically, the only difference between a digital repeater station and an analog repeater station is inclusion of a demodulator and modulator section between the IF amplifier and the up converter. This circuitry is added at the points designated X in Fig. 15-58.

RADAR

The electronic communications system known as radar (*radio detection and ranging*) is based on the principle that high-frequency RF signals are reflected by conductive targets. The usual targets are airplanes, missiles, ships, and automobiles. In a radar system, a signal is transmitted toward the target. The reflected signal is picked up by a receiver in the radar unit. The reflected or return radio signal is called an *echo*. The radar unit can then determine the distance to the target (range), its direction (azimuth), and in some cases, its elevation (distance above the horizon).

The ability of radar to determine the distance between a remote object and the radar unit is dependent upon knowing the exact speed of radio-signal transmission. In most radar applications, nautical miles are used instead of statute miles to express transmission speeds. A nautical mile is equal to 6076 feet. The speed of a radio signal is 162,000 nautical miles per second. (Sometimes, a special unit known as a radar mile is used. A radar mile is equal to 6000 ft.) It takes a radio signal 5.375 μs to travel 1 mi and 6.18 μs to travel 1 nautical mile.

A radar signal must travel twice the distance between the radar unit and the remote target. The signal is transmitted, a finite time passes before the signal reaches the target and is reflected, and the signal then travels an equal distance back to the source. If an object is exactly 1 nautical mile away, the signal takes 6.18 μs to reach the target and 6.18 μs to return. The total elapsed time from the instant of initial transmission to the reception of the echo is 12.36 μs.

The distance to a remote target is calculated using the expression

$$D = \frac{T}{12.36}$$

where D = distance between radar unit and remote object, nautical miles
T = total time between transmission and reception of signal, μs

In short-distance applications, the yard is the common unit of distance measurement. A radio signal travels 328 yards per microsecond, so the distance to an object in yards is computed using the expression

$$D = \frac{328T}{2} = 164T$$

A measured time of 5.6 μs corresponds to a distance of 164(5.6) = 918.4 yd.

In order to obtain a strong reflection or echo from a distant object, the wavelength of the radar signal should be small compared to the size of the object being observed. If the wavelength of the radar signal is long with respect to the distant object, only a small amount of energy will be reflected. At higher frequencies, the wavelength is shorter and, therefore, the reflected energy is greater. For optimal reflection, the size of the target should be one-quarter wavelength or more at the transmitted frequency.

The shorter the wavelength of the signal compared to the observed object, the higher the resolution or definition of the remote object. In most cases, it is necessary only to detect the presence of a remote object. But if very short wavelengths are used, in many cases the actual shape of an object can be clearly determined.

The term *cross section* is often used with reference to a radar target. If a target is at least 10 times larger than 1 λ of the radar signal detecting it, the cross section is constant. The cross section of a target, a measure of the area of the target "illuminated" by the radar signal, is given in square meters. A target's cross section is determined by the size of the object, the unique geometry of the target, the viewing angle, and the position.

The larger the cross section, the greater the reflected signal power and the greater the distance of detection and the higher the probability of the signal being greater than the noise. Another factor influencing the return-signal strength is the material of the target. Metal returns the greatest signal; other materials can also reflect radar waves, but not as effectively. Some objects will absorb radar waves, making the reflected signal very small.

The F117, the U.S. Air Force's stealth fighter, has a large physical cross section (about 1), but it is designed to deflect and absorb any radar signal aimed at it. All the plane's surfaces are at an angle such that radio signals striking them are not reflected directly back to the radar unit. Instead, they are reflected off at an angle, and little if any reflected energy is received. The surface of the aircraft is also coated with a material that absorbs radio waves. This combination gives the F117 an effective cross section the size of a small bird.

All of the most important factors affecting the amount of received signal reflected from a target are summed up in what is called the *radar equation:*

$$P_r = \frac{P_t G \sigma A_e}{(4\pi)^2 R^4}$$

where P_r = received power
G = antenna gain (product of transmitting and receiving gains)
σ = cross section of target
A_e = effective area of receiving antenna (dish area)
R = range or distance to target

Most of the relationships between the variables are obvious. However, given the fact that the received power is inversely proportional to the fourth power of the distance to the target, it is not surprising that radar range is generally so limited. Keep in mind that the radar equation does not take into account the S/N ratio. Very low-noise receiver front ends are essential for target acquisition.

Since radar uses microwave frequencies, line-of-sight communications results. In other words, radar cannot detect objects beyond the horizon. Objects do not have to be physically visible, but must be within line-of-sight radio distance in order for detection to occur.

The relationship between range, azimuth, and elevation can be expressed by a right triangle, as shown in Fig. 15-59. Assume that the radar is land-based and used to detect aircraft. The distance between the radar unit and the remote airplane is the hypotenuse of the right triangle. The angle of elevation is the angle between the hypotenuse and the baseline, which is a line tangent to the surface of the earth at the radar location. The altitude is defined by the angle of elevation. The greater the angle of elevation, the greater the altitude. Knowing the range and the angle of elevation allows the altitude to be computed using standard trigonometric techniques.

Another important factor in locating a distant object is knowing its direction with respect to the radar set. If the radar station is fixed and land-based, the direction (bearing or azimuth) of the remote object is usually given as a compass direction in degrees. Recall that true north is 0 or 360°, east is 90°, south is 180°, and west is 270°. If the radar unit is located in a moving vehicle, such as an airplane or ship, the azimuth is given as a relative bearing with respect to the forward direction of the vehicle. Straight ahead is 0° or 360°, directly to the right is 90°, directly behind is 180°, and directly to the left is 270°.

The ability of a radar unit to determine the direction of a remote object requires the use of a highly directional antenna. An antenna with an extremely narrow beamwidth will receive signals only over a narrow angle. The narrower the beamwidth of the antenna, the more precisely the actual bearing can be determined.

Since most radar systems operate in the microwave region, highly directional antennas are easily obtained. Horns with parabolic reflectors are the most common, and beam widths of less than 1° are readily attainable. These highly directional antennas are continuously rotated 360°. The same antenna is used for transmitting the original signal and receiving the reflected signal.

Circuits within the radar unit are calibrated so that the direction in which the antenna is pointing is accurately known. When the echo is received, it is compared to the calibrated values and the precise direction determined.

The ability of a radar unit to determine the altitude of a remote target depends on the vertical beamwidth of the radar antenna. The radar antenna may scan vertically while measuring the distance of the object during the scan. When the object is detected, the vertical elevation of the antenna is noted and the actual altitude is then computed.

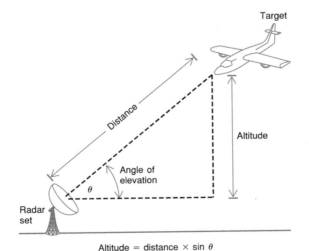

Altitude = distance × sin θ

FIG. 15-59 The trigonometry of radar.

Example 15-5

A radar set detects the presence of an aircraft. The time between the radiated and received pulses is 9.2 μs. The antenna is set to an angle of elevation of 20°. Determine (*a*) the line-of-sight distance to the aircraft in statute miles and (*b*) the altitude of the aircraft.

a. D (nautical miles) = $T/12.36$ = 9.2/12.36 = 0.744
    1 statute mile = 5280 ft
    1 nautical mile = 6076 ft
    $\dfrac{5280}{6076}$ = 0.87 statute mile per nautical mile
    D (statute miles) = 0.741(0.87) = 0.644

b. $A = D \sin \theta$
    where θ = angle of elevation
    A = 0.644 sin 20
       = 0.644(0.342) = 0.22 mile
       = 0.22(5280 ft) = 1163 ft

PULSED RADAR. There are two basic types of radar systems: pulsed and continuous wave (CW). There are also numerous variations of each. By far the most commonly used radar system is the pulsed type. Signals are transmitted in short bursts or pulses, as shown in Fig. 15-60. The duration or width (W) of the pulse is very short and, depending upon the application, can be anywhere from less than a microsecond to several microseconds. The time between transmitted pulses is known as the *pulse repetition time* (*PRT*). If the PRT is known, the *pulse repetition frequency* (*PRF*) can be determined using the formula

$$PRF = \frac{1}{PRT}$$

For example, if the pulse repetition time is 150 μs, the PRF is $1/150 \times 10^{-6}$ = 666.7 kHz.

The ratio of the pulse width to the PRT is known as the *duty cycle.* The duty cycle is normally expressed as a percentage:

$$\text{Duty cycle} = W \times \frac{100}{PRT}$$

For example, a pulse width of 7 μs with a PRT of 280 μs produces a duty cycle of $7 \times 100/280$ = 2.5 percent.

It is during the interval between the end of the transmitted pulse and the beginning of the next pulse in sequence in Fig. 15-60 that the echo is received.

The duration of the transmitted pulse and the PRT are extremely critical in determining the performance of a radar system. Very short-range radars have narrow pulses and short pulse repetition times. If the target is only a short distance away, the echo travel time will be relatively short. In short-range radars, the pulse width is made narrow to ensure that the pulse is terminated before the echo of the target is received. If the pulse is too long, the return signal may be masked or blanked by the transmitted

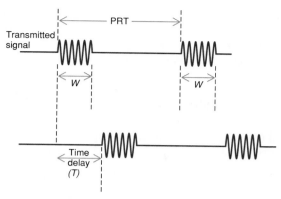

FIG. 15-60 Pulsed radar signals.

pulse. Long-range radars typically have a longer pulse repetition time because it takes longer for the echo to return. This also permits a longer burst of energy to be transmitted, ensuring a stronger return.

If the PRT is too short relative to the distance of the target, the echo may not return during the time interval between two successive pulses, but after the second transmitted pulse. This is known as a *double range* or *second return echo*. Naturally, such echoes lead to imprecise distance measurements.

CONTINUOUS WAVE (CW) RADAR. In *continuous-wave (CW)* radar, a constant-amplitude continuous microwave sine wave is transmitted. The echo is also a constant-amplitude microwave sine wave of the same frequency, but of lower amplitude and obviously shifted in phase. The question is, "How is such a signal used to determine target characteristics?"

The answer has to do with the object itself. In most cases, the target is moving with respect to the radar unit. The reflected signal from moving airplane, ship, missile, or automobile undergoes a frequency change. It is this frequency change between the transmitted signal and the returned signal that is used to determine the speed of the target.

The frequency shift that occurs when there is relative motion between the transmitting station and a remote target is known as the *Doppler effect*. A familiar example of the Doppler effect is the fixed-frequency sound waves emitted by an automobile horn. If the horn sounds when the car is stationery, you will hear a single tone. However, if the car is moving toward you while the horn is on, you will experience a tone of continuously increasing frequency. As the car moves closer to you, the sound waves are compressed, creating the effect of a higher-frequency signal. If the car is moving away from you with its horn on, you will experience a continually decreasing frequency. As the car moves away, the sound waves are stretched out, creating the effect of a lower-frequency signal. This same effect works on both radio and light waves.

In a Doppler system, the transmitter sends out a continuous frequency signal. If the frequency difference between a transmitted signal and a reflected signal is known, the relative speed between the radar unit and the observed object can be determined using the formula

$$V = \frac{f\lambda}{1.03}$$

where f = frequency difference between transmitted and reflected signals, Hz
λ = wavelength of transmitted signal, m
V = relative velocity between the two objects, mph

Assume, for example, a frequency shift of 1500 Hz at a frequency of 10 GHz. A frequency of 10 GHz represents a wavelength of 300/*f* (in MHz) = 300/10,000 = 0.03 m. The speed is therefore (1500)(0.03)/1.03 = 43.7 mph.

In CW radar, it is the Doppler effect that provides frequency modulation of the reflected carrier. In order for there to be a frequency change, the observed object must be moving toward or away from the radar unit. If the observed object moves parallel to the radar unit, there is no relative motion between the two and no frequency modulation occurs.

The greatest value of CW radar is its ability to measure the speed of distant objects. Police radar units use CW Doppler radar for measuring the speed of cars and trucks.

Some radar systems combine both pulse and Doppler techniques to improve performance and measurement capabilities. One such system evaluates successive echoes to determine phase shifts that indicate when a target is moving. Such radars are said to incorporate *moving target indication* (*MTI*). Through a variety of special signal-processing techniques, multiple moving targets can be distinguished not only from one another, but from fixed targets as well.

BLOCK DIAGRAM ANALYSIS. Figure 15-61 is a block diagram of a typical pulsed radar unit. There are four basic subsystems: the antenna, the transmitter, the receiver, and the display unit.

The transmitter in a pulsed radar system invariably uses a magnetron. Recall that a magnetron is a special high-power vacuum tube oscillator that operates in the microwave region. The cavity size of the magnetron sets the operating frequency. A master timing generator develops the basic pulses used for triggering the magnetron. The timing generator sets the pulse duration, the PRT, and the duty cycle. The pulses from the timing network trigger the magnetron into oscillation and it emits short bursts of microwave energy. Magnetrons are capable of extremely high powers, especially when operated on a pulse basis. Continuous average power may be low but when pulsed, magnetrons can produce many megawatts of power for the short duration required by the application. This helps ensure a large reflection. Klystrons and TWTs are commonly used in CW Doppler radars. Low-power radars such as those used by the police for speed detection use Gunn diodes.

In the diagram in Fig. 15-61, you can see that the transmitter output is passed through a circulator and then applied to the antenna. The circulator is really a type of duplexer, which allows the transmitter and receiver to share a single antenna and prevents the high-power transmitted signal from getting into the receiver and damaging it.

A radar duplexer is a waveguide assembly containing special devices that prevent interference between the transmitter and receiver. The most commonly used device is a spark gap tube. Spark gap tubes are either *TR* (*transmit-receive*) or *anti-transmit-receive* (*ATR*) types. TR tubes prevent transmitter power from reaching the receiver. When RF energy from the transmitter is detected, the spark gap breaks down, creating a short circuit for the RF energy. ATR tubes effectively disconnect the transmitter from the circuit during the receive interval. The TR and ATR tubes, when combined with the appropriate quarter- and half-wavelength waveguides, provide effective isolation between transmitter and receiver. In low-power radars, PIN diodes or a standard circulator can be used for this purpose.

The antenna system is typically a horn with a parabolic reflector that produces a very narrow beamwidth. A special waveguide assembly with a rotating joint allows the waveguide and horn antenna to be rotated continuously over 360°.

The same antenna is also used for reception. During the pulse OFF time, the received signal passes through the antenna, the associated waveguide, and either the duplexer or circulator to the receiver. The receiver is a standard high-gain superhetero-

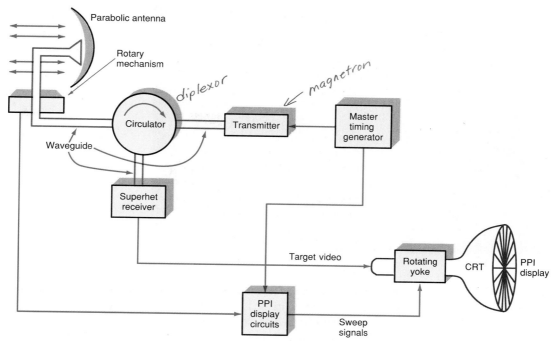

Fig. 15-61 General block diagram of a pulsed radar system.

dyne type. The signal is usually fed directly to a mixer in most systems, although some radars use an RF amplifier. The mixer is typically a single diode or diode bridge assembly. The local oscillator feeds a signal to the mixer at the appropriate frequency so that an IF is developed. Both klystrons and Gunn diodes are used in local oscillator applications. The IF amplifiers provide very high gain prior to demodulation. Some radar receivers use double conversion.

The demodulator in a pulsed radar system is typically just a diode detector, since only the pulses must be detected. In Doppler and other more complex radars, some form of frequency- and phase-sensitive demodulator is used. Phase-locked loops are common in this application. The output of the demodulator is fed to a video amplifier which creates signals that are ultimately displayed.

The display in most radar systems is a CRT. Various display formats can be used. The most common type of CRT display is known as a *type P* display or *plan position indicator (PPI)*. PPI displays show both the range and the azimuth of a target. The center of the display is assumed to be the location of the radar unit. Concentric circles indicate the range. The azimuth or direction is indicated by the position of the reflected target on the screen with respect to the vertical radius line. The targets show up as lighted blips on the screen.

The PPI display is developed by a sophisticated scanning system tied in to antenna rotation. As the antenna is rotated by a motor, an encoder mechanism sends signals to the PPI control circuits that designate the azimuth or direction of the antenna. At the same time, the horizontal and vertical deflection coils (the *yoke*) around the neck of the CRT rotate in synchronism with the antenna. The electron beam in the CRT is swept from the center of the screen out to the edge. The sweep begins at the instant of pulse transmission and sweeps outward from the center. The beginning of each transmitted pulse begins another sweep of the electron beam. As the deflection yoke rotates, the beam moves in such a way that it appears that a radius from the center to the edge is

Lightweight, handheld portable radar systems with compact electronic circuitry now enable police departments to improve law enforcement.

continuously rotating. Target reflections appear as lighted blips on the screen. Calibrations on the screen in the form of graticule markings or superimposed electron beam patterns permit distance and azimuth values to be read directly.

Some PPI radars scan only a narrow range of azimuth rather than the entire 360°. Airplanes with radars in the nose, for example, only scan a 90° to 180° range forward.

An important high-tech type of radar known as *phased array radar* provides greater flexibility in scanning narrow sectors and tracking multiple targets. Instead of a single horn and parabolic reflector, multiple dipoles or patch antennas are used. Slots in a waveguide are also used. A half-wave dipole at microwave frequencies is very short. Therefore, many of them can be mounted together in a matrix or array. The result is a special collinear array above a reflecting surface with very high gain. By using a system of separate feed lines and a variable phase shifter for each antenna, the beam width and directivity can be controlled electronically. This permits rapid scanning and on-the-fly adjustment of directivity. Phase array radars eliminate the mechanical systems needed for conventional radars.

RADAR APPLICATIONS. One of the most important uses of radar is in weapons defense systems and in safety and navigation systems. Search radars are used to locate enemy missiles, planes, and ships. Tracking radars are used on missiles and planes to locate and zero in on targets. Radars are widely used on planes and ships for navigating blind in fog or bad weather. Radars help ground controllers locate and identify nearby planes. Special radars assist planes in landing in bad weather when visibility is near zero.

Radars are also used as altimeters to measure height. High-frequency radars can actually be used to plot or map the terrain in an area. Special terrain-following radars allow high-speed jets to fly very close to the ground to avoid detection by enemy radar.

In civilian applications, radars are used on boats of all sizes for navigation in bad weather. The police use radar to catch speeders. Small handheld Doppler radar units can also be used in sporting events—to time race cars or determine the speed of a pitched baseball or tennis serve. Finally, ground and satellite based radars are widely used to track clouds, storms, and other phenomena for the purpose of weather forecasting.

SUMMARY

Microwaves occupy the 1- to 300-GHz frequency range. The best-known microwave applications are telephone systems (microwave relay systems), radar, satellite communications, and heating. Today, most new communications services are assigned to the microwave region.

The current-voltage relationships that exist at low frequencies are not usable at microwave frequencies. At these frequencies even a half-inch piece of wire represents a significant amount of inductance. Instead of lumped components, distributed circuit elements are used. Microstrip can be used to create almost any tuned circuit for microwave transmission, including resonant circuits, filters, and impedance-matching networks.

Most microwave energy transmission above 6 GHz is handled by waveguides, hollow metal tubes designed specifically for those high frequencies. Waveguides have a variety of special parts, such as directional couplers, turns, joints, rotary connections, and terminations. Other special microwave components are cavity resonators (waveguide-like devices that act like high-Q parallel resonant circuits), circulators (diplexers that allow a single antenna to be shared by a transmitter and receiver), and isolators.

Diodes that have been specially designed for microwave use are point-contact diodes, Schottky (hot carrier) diodes, tunnel diodes, varactor diodes, step-recovery diodes, and oscillators (Gunn diodes, and IMPATT and TRAPATT diodes). Tubes used for microwave power amplification are the klystron, the magnetron, and the traveling-wave tube. Other microwave devices are the backward-wave oscillator, the gyrotron, and the cross-field amplifier.

The most widely used microwave antenna is the horn, a flared waveguide with high gain and customizable directivity. Horns are often used in combination with parabolic reflectors, the familiar dish antennas associated with TV and satellite transmissions. Other types of microwave antenna are the helical antenna, the slot, the dielectric lens antenna, and the patch. Patch antennas are made with microstrip on printed circuit boards (PCBs). A phased array is an antenna system made up of a large group of similar antennas on a common plane. Patch antennas on a common PCB can be used, or separate antennas like dipoles can be physically mounted together in a plane.

Parabolic dishes used in repeater systems are mounted on high towers; when feasible, the towers are located on hills, mountain tops, or even high buildings to increase the transmitting distance. Microwave repeater systems operate in assigned bands with frequencies near 2, 4, 6, 11, 13, and 18 GHz. The distance between repeaters varies with both the terrain and the frequency of operation.

Radar, a microwave-frequency communications system used to detect objects at a distance that cannot be observed visually, is based on the principle that high-frequency RF signals are reflected by conductive targets. The two basic types of radar are pulsed and continuous wave (Doppler).

KEY TERMS

Analog repeater station
Anti-transmit-receive tube
Backward wave oscillator
Bicone antenna

Cassegrain feed
Cavity resonator
Channel
Chip geometry

Choke joint
Circulator
Continuous-wave radar
Crossed field amplifier

Crystal oscillator
Dielectric lens antenna
Digital regenerative
   repeater station
Diode
Directional coupler
Doppler effect
Frequency-multiplier diode
Frequency spectrum
Gunn diode
Gyrotron
Helical antenna
Horn antenna
Hot carrier diode
Hybrid circuit
Hybrid ring
Hybrid T
IMPATT diode
Interaction chamber
Isolator
Klystron
Low-noise amplifier
Magnetron
Microstrip
Microstrip tuned circuit
Microwave

Microwave antenna
Microwave receiver
Microwave relay station
Microwave transistor
Microwave transistor
   amplifier
Microwave transmitter
Microwave tubes
Millimeter wave
Mixer
Monolithic microwave
   integrated circuit
Multistage integrated
   circuit amplifier
Oscillator diode
Parabolic antenna
Patch antenna
Phased array
Phased array radar
PIN diode
Plan-position indicator
Point-contact diode
Pulsed radar
Quarter-wavelength trans-
   mission line
Radar

Radar duplexer
Schottky diode
Semiconductor diode
Signal propagation
Single-pole double-throw
   switch
Slot antenna
Small-signal diodes
Small-signal microwave
   amplifier
Step-recovery diode
Stripline
Submillimeter wave
T section
Termination
Transistor amplifier
Transmit-receive tube
Transverse electric mode
Transverse magnetic mode
TRAPATT diode
Traveling-wave tube
Type-P display
Vacuum tube
Varactor diode
Waveguide
Waveguide cutoff frequency

QUESTIONS

1. What is the microwave frequency range?
2. What is the primary advantage of using microwave frequencies?
3. List seven reasons why microwaves are more difficult to work with than lower-frequency signals.
4. What is the designation of the lowest-frequency microwave band?
5. What is the name for microwaves above 40 GHz?
6. Name four techniques that have helped squeeze more signals into a given spectrum space.
7. Why won't conventional transistors work at microwave frequencies?
8. Name two common methods for generating a microwave signal with conventional lower-frequency (VHF and UHF) components.
9. Name the parts of a microwave receiver and a microwave transmitter that require special microwave components.
10. Name the two types of transmission lines made of PCB material used to produce circuits and components in the microwave region. Which is the most commonly used and why?
11. List three common circuits or functions performed by microstrip in microwave equipment.
12. What is the name of the microstrip device that has four ports and allows multiple circuits and devices to be interconnected without interfering with one another?

13. How are microwave transistors different from lower-frequency transistors?
14. What semiconductor material is used instead of silicon to make transistors and ICs perform satisfactorily in the microwave region?
15. How are microstrip lines used in microwave amplifiers?
16. Up to what frequency are microwave transistors and ICs available?
17. How are microwave linear amplifiers biased and why?
18. What is the main disadvantage of using coaxial cable at microwave frequencies?
19. Coaxial cable is not used beyond a certain frequency. What is that frequency?
20. Name the special coaxial cable made for microwave applications. What is its main disadvantage?
21. Name the two ways to couple or extract energy from a waveguide.
22. A waveguide acts like what kind of filter?
23. True or false. A waveguide with a cutoff frequency of 6.35 GHz will pass a signal of 7.5 GHz.
24. The magnetic and electric fields in a waveguide are designated by what letters?
25. State the basic operating mode of most waveguides.
26. What do you call the coupling mechanism used to attach two sections of a waveguide?
27. Name the two types of T sections of waveguide and briefly tell how each is used.
28. Explain how a hybrid T works and give a common application.
29. What devices are used to terminate a waveguide? Why is termination necessary?
30. State the main purpose of a directional coupler.
31. What useful device is created by a half-wavelength section of waveguide shorted at both ends?
32. A cavity resonator acts like what type of circuit?
33. How are resonant cavities used in microwave circuits?
34. What characteristic of a cavity can be adjusted by mechanically varying the cavity's dimensions?
35. What is a circulator?
36. What is an isolator?
37. Describe the anode and cathode of a point-contact diode.
38. What are two names for a microwave diode with an N-type silicon cathode and a metal anode forming a junction?
39. Name the most common application of microwave signal diodes.
40. What is the name of a microwave diode that acts like a variable-voltage capacitor? What is its main application?
41. Name two types of diodes widely used as frequency multipliers in the microwave region? Do such multiplier circuits amplify?
42. Name two other diodes used as microwave diode oscillators.
43. Name two ways that a PIN diode is used at microwave frequencies.
44. Does an IMPATT diode operate with forward or reverse bias?
45. True or false. Klystrons can be used as amplifiers or oscillators.
46. True or false. A magnetron operates as an amplifier.
47. True or false. A TWT is an amplifier.
48. The cavities in a klystron produce what kind of modulation of the electron beam?
49. State at which points in a klystron amplifier the input is applied and the output taken from.
50. Low-power klystrons are being replaced by what semiconductor device? High-power klystrons are being replaced by what other device?
51. What component of a magnetron causes the electrons to travel in circular paths?
52. State two major applications of magnetrons.
53. What is the main application of a TWT?
54. How is density modulation of the electron beam in a TWT achieved?
55. What is the main benefit of a TWT over a klystron?
56. What is the most commonly used microwave antenna?
57. How does increasing the length of a horn affect its gain and beamwidth?
58. What is the typical beamwidth range of a horn?

59. Describe the bandwidth of horns.
60. State the geometric shape of a commonly used microwave antenna reflector.
61. In order for a dish reflector to work, how must the antenna be positioned?
62. What type of antenna is most commonly used with a dish reflector?
63. What do you call the feed arrangement of a dish with a horn at its center and a small reflector at the focal point? What is its purpose?
64. Name two benefits of a helical antenna.
65. What is the gain range of a helical antenna?
66. What is the typical beamwidth range of a helical antenna?
67. What is the type of polarization radiated by a helical antenna?
68. What does RHCP mean? LHCP?
69. True or false. A helical antenna can receive either vertically or horizontally polarized signals.
70. True or false. An RHCP antenna can receive a signal from an LHCP antenna.
71. Name a popular omnidirectional microwave antenna.
72. What is the typical length of a slot antenna?
73. Where are slot antennas most often used?
74. How is a slot antenna driven?
75. What basic function is performed by a lens antenna?
76. What is the name given to antennas made with microstrip?
77. Name the two common shapes of microstrip antennas.
78. Name two ways that microstrip antennas are fed.
79. What is the purpose of a microwave relay station? What is another name for this station?
80. In what application are microwave relay stations most commonly used?
81. Name the two types of microwave relay stations and describe the signals transmitted by each.
82. Microwave relay stations are an alternative to what other types of communications?
83. List four popular uses for microwaves in communications.
84. Knowing the speed of radio signals allows radar to determine what characteristic of a target?
85. What characteristic of a radar antenna determines the azimuth of a target?
86. What characteristic of a target determines the amount of reflected signal?
87. Radars operate in what microwave bands?
88. True or false. Both transmitter and receiver in a radar share the same antenna.
89. Name the two main types of radar.
90. What is the name given to a frequency shift of sound, radio, or light waves that occurs as the result of the relative motion between objects?
91. True or false. Distance can be measured with CW radar.
92. What is the key component in a radar transmitter?
93. What are TR and ATR tubes? Why are they used?
94. What is the output display in most radar sets?
95. Name three major military and four civilian uses of radar.

PROBLEMS

1. The TV channels 2–13 occupy a total bandwidth of about 72 MHz. At a frequency of 200 MHz, what is the bandwidth percentage? The bandwidth percentage at 20 GHz? ◆
2. A quarter-wavelength piece of 75-Ω microstrip line has a capacitive load of $X_c = 22$ Ω. What impedance does a driving generator see?
3. What is the length of a quarter-wavelength microstrip at 2.2 GHz? (The dielectric constant of the PCB is 4.5.) ◆

4. A rectangular waveguide has a width of 1.4 in. and a height of 0.8 in. What is the cutoff frequency of the waveguide?

5. A 20-dB directional coupler is installed in a waveguide carrying a 5-W 6-GHz signal. What is the output of the coupler? ◆

6. What is the length of a dipole at 2.4 GHz?

7. A horn antenna is 5 λ long at 7 GHz. What is that length in inches? ◆

8. A horn aperture is 6 cm wide and 4 cm high. The operating frequency is 18 GHz. Compute the beam angle and the gain. Assume $K = 0.5$.

9. A parabolic reflector antenna has a diameter of 6 m. The frequency of operation is 20 GHz. What are the gain and beamwidth? ◆

10. What is the approximate lowest frequency of operation of a dish 18 in. in diameter?

11. State the speed of a radar signal in microseconds per nautical mile.

12. Convert 22 standard statute miles to nautical miles.

13. The elapsed time between radiating a radar signal and receiving its echo is 23.9 μs. What is the target distance in nautical miles?

14. How many microseconds is the time period between the transmitted signal and receipt of an echo from a target 1800 yards away?

15. For optimum reflection from a target, what should the size of the target be in wavelengths of the radar signal?

16. A CW radar operates at a frequency of 9 GHz. A frequency shift of 35 kHz is produced by a moving target. What is its speed?

CRITICAL THINKING

1. If microwave power transistors are limited to, say, an upper power limit of 100 W each, explain how you could build a microwave power amplifier with a power rating of 1 kW.

2. A NASA deep space probe operates on 8.4 GHz. Its location is past Mars heading into outer space. Explain in detail how you would go about maintaining and/or improving signal strength at the main receiver in California.

3. Explain how the speed of an aircraft can be measured using pulse radar techniques.

Preventive maintenance is a necessity for this covered traffic control radar dome tower.

CHAPTER 16

SATELLITE COMMUNICATIONS

Objectives

After completing this chapter, you will be able to:

◆ Define the terms *posigrade, retrograde, geocenter, apogee, perigee, ascending, descending, period, angle of inclination, geosynchronous, latitude, longitude,* and *meridian.*

◆ State the operative physical principles of launching a satellite and maintaining its orbit.

◆ Draw a block diagram of the communications system in a communications satellite, give its name, and explain how it works.

◆ List the six main subsystems of a satellite.

◆ Draw a block diagram of a satellite ground station, identifying the five main subsystems and explaining the operation of each.

◆ Name three common applications for satellites and state which is the most common.

◆ Explain the concept and operation of the Global Positioning System. Draw a block diagram of a GPS receiver and explain the function of each component.

A *satellite* is a physical object that orbits, or rotates, around some celestial body. Satellites occur in nature, and our own solar system is a perfect example. The earth and other planets are satellites rotating around the sun. The moon is a satellite to the earth. A balance between the inertia of the rotating satellite at high speed and gravitational pull of the orbited body keeps the satellite in place.

Manufactured satellites are launched and orbited for a variety of purposes. The most common application is communications where the satellite is used as a repeater. In this chapter, we introduce satellite concepts and discuss how satellites are identified and explained. The operation of a satellite ground station is summarized. Typical satellite applications are reviewed with particular emphasis on the Global Positioning System, a worldwide satellite-based navigational system.

16-1 SATELLITE ORBITS

The ability to launch a satellite and keep it in orbit depends upon following well-known physical and mathematical laws that are referred to collectively as *orbital dynamics*. In the first section we will introduce you to these principles before discussing the physical components of a satellite and how it is used in various communications applications.

THE PRINCIPLES OF SATELLITE ORBITS AND POSITIONING

If a satellite were launched vertically from the earth and then released, it would fall back to earth because of gravity. In order for the satellite to go into orbit around the earth, it must have some forward motion. For that reason, when the satellite is launched, it is given both vertical and forward motion. The forward motion produces inertia, which tends to keep the satellite moving in a straight line. However, gravity tends to pull the satellite toward the earth. The inertia of the satellite is equalized by the earth's gravitational pull. The satellite constantly changes its direction from a straight line to a curved line to rotate about the earth.

If a satellite's velocity is too high, it will overcome the earth's pull and go out into space. It takes an escape velocity of approximately 25,000 mi/h to cause a spacecraft to break the gravitational pull of the earth. At lower speeds gravity constantly pulls the satellite toward the earth. The goal is to give the satellite acceleration and speed that will exactly balance the gravitational pull.

The closer the satellite is to earth, the stronger the effect of the earth's gravitational pull. So in low orbits, the satellite must travel faster to avoid falling back to earth. The lowest practical earth orbit is approximately 100 mi. At this height, the satellite's speed must be about 17,500 mi/h to keep the satellite in orbit. At this speed, the satellite orbits the earth in approximately 1½ h. Communications satellites are usually much farther from earth. A typical distance is 22,300 mi. A satellite need travel only about 6800 mi/h in order to stay in orbit at that distance. At this speed, the satellite rotates the earth in approximately 24 h, the earth's own rotational time.

There is more to a satellite orbit than just the velocity and gravitational pull. The satellite is also affected by the gravitational pull of the moon and sun.

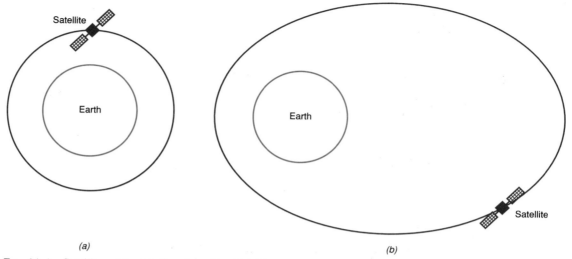

FIG. 16-1 Satellite orbits. (*a*) Circular orbit. (*b*) Elliptical orbit.

A satellite rotates around the earth in either a circular or an elliptical path as shown in Fig. 16-1. Circles and ellipses are geometric figures that can be accurately described mathematically. Because the orbit is either circular or elliptical, it is possible to calculate the position of a satellite at any given time.

A satellite rotates in an orbit that forms a plane that passes through the center of gravity of the earth called the *geocenter* (Fig. 16-2). In addition, the direction of satellite rotation may be either in the same direction as the earth's rotation or against the direction of earth's rotation. In the former case, the orbit is said to be *posigrade,* and in the latter case, *retrograde.* Most orbits are posigrade. In a circular orbit the speed of rotation is constant. However, in an elliptical orbit the speed changes depending

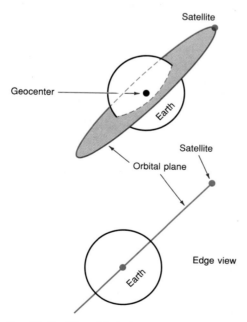

FIG. 16-2 The orbital plane passes through the geocenter.

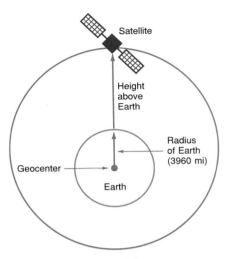

Fig. 16-3   Satellite height.

upon the height of the satellite above the earth. Naturally the speed of the satellite is higher when it is close to the earth than when it is far away.

SATELLITE HEIGHT.   In a circular orbit, the height is simply the distance of the satellite from the earth. However, in geometrical calculations, the height is really the distance between the center of the earth and the satellite. In other words, that distance includes the radius of the earth, which is generally considered to be about 3960 mi (or 6370 km). A satellite that is 5000 mi above the earth in circular orbit is 3960 + 5000, or 8960, mi from the center of the earth (see Fig. 16-3).

When the satellite is in an elliptical orbit, the center of the earth is one of the focal points of the ellipse (refer to Fig. 16-4). In this case, the distance of the satellite from the earth varies depending upon its position. Typically the two points of greatest interest are the highest point above the earth (the *apogee*) and the lowest point (the

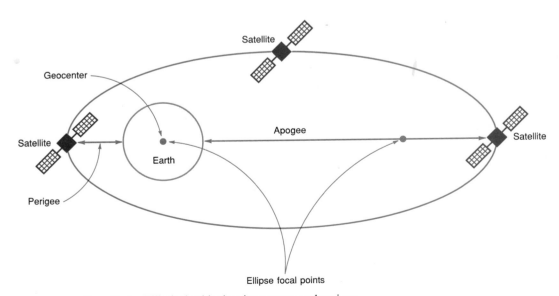

Fig. 16-4   Elliptical orbit showing apogee and perigee.

perigee). The apogee and perigee distances typically are measured from the geocenter of the earth.

SATELLITE SPEED.   As indicated earlier, the speed varies depending upon the distance of the satellite from the earth. For a circular orbit the speed is constant, but for an elliptical orbit the speed varies depending upon the height. Low earth satellites of about 100 mi height have a speed in the neighborhood of 17,500 mi/h. Very high satellites such as communications satellites, which are approximately 22,300 mi out, rotate much more slowly, a typical speed being in the neighborhood of 6800 mi/h.

SATELLITE PERIOD.   The *period* is the time it takes for a satellite to complete one orbit. It is also called the *sidereal period*. A sidereal orbit uses some external fixed or apparently motionless object such as the sun or star for reference in determining a sidereal period. The reason for using a fixed reference point is that while the satellite is rotating around the earth, the earth itself is rotating.

Another method of expressing the time for one orbit is the revolution or synodic period. One *revolution* is the period of time that elapses between the successive passes of the satellite over a given meridian of earth longitude. Naturally, the synodic and sidereal periods differ from one another because of the earth's rotation. The amount of time difference is determined by the height of the orbit, the angle of the plane of the orbit, and whether the satellite is in a posigrade or retrograde orbit. Period is generally expressed in hours. Typical rotational periods range from about 1½ h for a 100-mi height to 24 h for a 22,300-mi height.

ANGLE OF INCLINATION.   The *angle of inclination* of a satellite orbit is the angle formed between the line that passes through the center of the earth and the north pole and a line that passes through the center of the earth but that is also perpendicular to the orbital plane. This angle is illustrated in Fig. 16-5(*a*). Satellite orbits can have inclination angles of 0 through 90°.

Another definition of *inclination* is the angle between the equatorial plane and the satellite orbital plane as the satellite enters the northern hemisphere. This definition may be a little bit easier to understand, but it means the same thing as the previous definition.

When the angle of inclination is 0°, the satellite will be directly above the equator. When the angle of inclination is 90°, the satellite will pass over both the north and south poles once for each orbit. Orbits with 0° inclination are generally called *equatorial orbits* while satellites with orbits with inclinations of 90° are referred to as *polar orbits*.

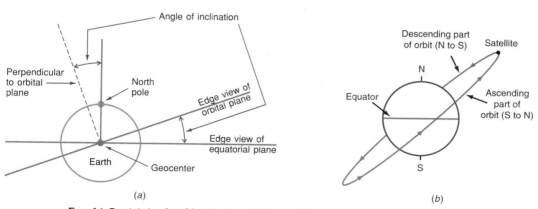

FIG. 16-5   (*a*) Angle of inclination. (*b*) Ascending and descending orbits.

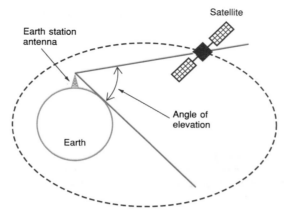

FIG. 16-6 Angle of elevation.

When the satellite has an angle of inclination, the orbit is said to be either *ascending* or *descending*. As the satellite moves from south to north and crosses the equator, the orbit is ascending. When the satellite goes from north to south across the equator, the orbit is descending [see Fig. 16-5(*b*)].

ANGLE OF ELEVATION. The *angle of elevation* of a satellite is that angle that appears between the line from the earth station's antenna to the satellite and the line between the earth station's antenna and the earth's horizon (see Fig. 16-6). If the angle of elevation is too small, the signals between the earth station and the satellite have to pass through much more of the earth's atmosphere. Because of the very low powers used and the high absorption of the earth's atmosphere, it is desirable to minimize the amount of time that the signals spend in the atmosphere. Noise in the atmosphere also contributes to poor performance. The lower the angle of radiation, the more time that this signal spends in the atmosphere. The minimum practical angle of elevation for good satellite performance is generally 5°. The higher the angle of elevation, the better.

GEOSYNCHRONOUS ORBITS. To use a satellite for communications relay or repeater purposes, the ground station antenna must be able to follow or track the satellite as it passes overhead. Depending upon the height and speed of the satellite, the earth station will be able to use it only for communications purposes for that short period of time when it is visible. The earth station antenna will track the satellite from horizon to horizon. But at some point, the satellite will disappear around the other side of the earth. At this time, it can no longer support communications.

One solution to this problem is to launch a satellite with a very long elliptical orbit so that the earth station can "see" the apogee. In this way the satellite stays in view of the earth station for most of its orbit and is useful for communications for a longer period of time. It is only that short duration when the satellite disappears on the other side of the earth (perigee) that it cannot be used.

The intermittent communications caused by these orbital characteristics are highly undesirable in many communications applications. One way to reduce interruptions is to use more than one satellite. Typically three satellites, if properly spaced in the correct orbits, can provide continuous communications at all times. However, multiple tracking stations and complex signal switching or "hand-off" systems between stations are required. Maintaining these stations is expensive and inconvenient.

The solution is to launch a synchronous or geostationary satellite. This satellite orbits the earth around the equator at a distance of 22,300 mi (or 35,860 km). A satellite at that distance rotates around the earth in exactly 24 h. In other words, the satellite rotates in exact synchronism with the earth. For that reason, it appears to be fixed or stationary, thus the terms *synchronous, geosynchronous,* or *geostationary* orbit. Since the satellite remains apparently fixed, no special earth station tracking antennas are required. The antenna is simply pointed at the satellite and remains in a fixed position. With this arrangement, continuous communications are possible. Most communications satellites in use today are of the geosynchronous variety. Approximately 40 percent of the earth's surface can be "seen" or accessed from such a satellite. Users inside that area can use the satellite for communications.

POSITION COORDINATES IN LATITUDE AND LONGITUDE. In order to use a satellite, you must be able to locate its position in space. That position is usually predetermined by the design of the satellite and is achieved during the initial launch and subsequent position adjustments. Once the position is known, the earth station antenna can be pointed at the satellite for optimum transmission and reception. For geosynchronous satellites, the earth station antenna can be adjusted once, and it will remain in that position except for occasional minor adjustments. The position of other satellites above the earth will vary depending upon their orbital characteristics. To use these satellites, special tracking systems must be used. A tracking system is essentially an antenna whose position can be changed in order to follow the satellite across the sky. In order to maintain optimum transmission and reception, the antenna must be continually pointed at the satellite as it rotates. In this section, we discuss methods of locating and tracking satellites.

The location of a satellite is generally specified in terms of latitude and longitude, just as a point on earth would be described. The satellite location is specified by a point on the surface of the earth directly below the satellite. This point is known as the *subsatellite point (SSP)*. The subsatellite point is then located using conventional latitude and longitude designations.

Latitude and longitude form a system for locating any given point on the surface of the earth. This system is widely used for navigational purposes. If you have ever studied a globe of the earth, you have seen the lines of latitude and longitude. The lines of longitude, or *meridians,* are those drawn on the surface of the earth between the north and south poles. Lines of latitude are those drawn on the surface of the earth from east to west parallel to the equator. The centerline of latitude is the equator, which separates the earth into the northern and southern hemispheres.

Latitude is defined as the angle between the line drawn from a given point on the surface of the earth to the point at the center of the earth called the *geocenter* and the line between the geocenter and the equator (see Fig. 16-7). Zero degree latitude is at the equator while 90° latitude is at either the north or south pole. Usually an N or an S is added to the latitude angle in order to designate whether the point is in the northern or southern hemisphere.

A line drawn on the surface of the earth between the north and south poles is generally referred to as a *meridian.* A special meridian called the *prime meridian* is used as a

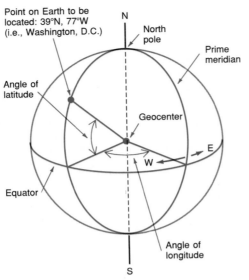

Fig. 16-7 Tracking and navigation by latitude and longitude.

reference point for measuring longitude. It is the line on the surface of the earth, drawn between the north and south poles, that passes through Greenwich, England. The longitude of a given point is the angle between the line connected to the geocenter of the earth to the point where the prime meridian and equator and the meridian containing the given point of interest intersect (see Fig. 16-7). The designation east or west is usually added to the longitude angle to indicate whether the angle is being measured to the east or west of the prime meridian. As an example, the location of Washington, D.C., is given by the latitude and longitude of 39° north and 77° west.

To show how latitude and longitude are used to locate a satellite, refer to Fig. 16-8. This figure shows some of the many geosynchronous communications satellites serving the United States and other parts of North America. Since geosynchronous satellites rotate around the equator, their subsatellite point is on the equator. For that reason, all geosynchronous satellites have a latitude of 0°.

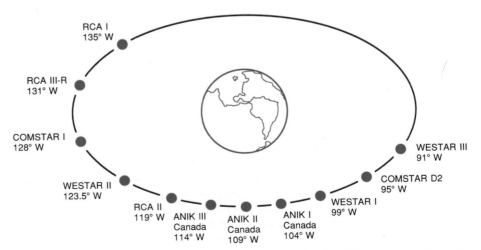

Fig. 16-8 The location of some of the many communications satellites in geosynchronous orbit.

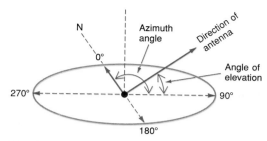

Fig. 16-9 Azimuth and elevation: Azimuth = 90°; elevation = 40°.

Azimuth and Elevation. Knowing the location of the satellite is insufficient information for most earth stations that must communicate with the satellite. The earth station really needs to know the azimuth and elevation settings of its antenna to intercept the satellite. Most earth station satellite antennas are highly directional and must be accurately positioned to "hit" the satellite. The azimuth and elevation designations in degrees tell where to point the antenna (see Fig. 16-9). *Azimuth* refers to the direction where north is equal to 0°. The *azimuth angle* is measured clockwise with respect to north. The *angle of elevation* is the angle between the horizontal plane and the pointing direction of the antenna.

Once the azimuth and elevation are known, the earth station antenna can be pointed in that direction. For a geosynchronous satellite, the antenna will simply remain in that position. For any other satellite, the antenna must be moved as the satellite passes overhead.

For geosynchronous satellites, the angles of azimuth and elevation are relatively easy to determine. Because geosynchronous satellites are fixed in position over the equator, special formulas and techniques have been developed to permit easy determination of azimuth and elevation for any geosynchronous satellite for any point on earth. Example 16-1 shows how calculations are made.

Example 16-1

A satellite earth station is to be located at longitude 95° west, latitude 30° north. The satellite is located at 121° west longitude in geosynchronous orbit. Determine the approximate azimuth and elevation settings for the antenna.

a. Difference between longitude of satellite and longitude of location:

$$121 - 95 = 26°$$

b. Locate longitude difference (relative longitude) and latitude.

c. Determine the elevation from the curves:

$$45°$$

d. Determine the radial position:

$$137°$$

e. Calculate the actual azimuth:

$$360 - 137 = 223°$$

16-2 SATELLITE COMMUNICATIONS SYSTEMS

Communications satellites are not originators of information to be transmitted. Although some other types of satellites generate the information to be transmitted, communications satellites do not. Instead, these satellites are relay stations for earth sources. If a transmitting station cannot communicate directly with one or more receiving stations because of line-of-sight restrictions, a satellite can be used. The transmitting station sends the information to the satellite, which in turn retransmits it to the receiving stations. The satellite in this application is what is generally known as a *repeater*.

REPEATERS AND TRANSPONDERS

Figure 16-10 shows the basic operation of a communications satellite. An earth station transmits information to the satellite. The satellite contains a receiver that picks up the transmitted signal, amplifies it, and translates it on another frequency. The signal on the new frequency is then retransmitted to the receiving stations back on earth. The original signal being transmitted from the earth station to the satellite is called the *uplink,* and the retransmitted signal from the satellite to the receiving stations is called the *downlink.* Usually the downlink frequency is lower than the uplink frequency. A typical uplink frequency is 6 GHz, and a common downlink frequency is 4 GHz.

The transmitter-receiver combination in the satellite is known as a *transponder.* The basic function of a transponder is amplification and frequency translation (see Fig. 16-11). The reason for frequency translation is that the transponder cannot transmit and receive on the same frequency. The transmitter's strong signal would overload, or

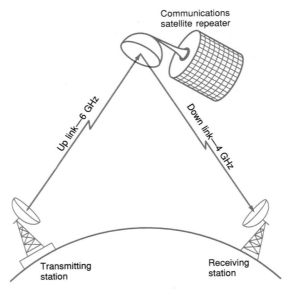

FIG. 16-10 Using a satellite as a microwave relay link.

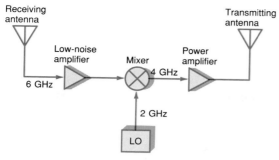

Fig. 16-11 A satellite transponder.

"desensitize," the receiver and block out the very small uplink signal, thereby prohibiting any communications. Widely spaced transmit and receive frequencies prevent interference.

Transponders are also wide bandwidth units so that they can receive and retransmit more than one signal. Any earth station signal within the receiver's bandwidth will be amplified, translated, and retransmitted on a different frequency.

Although the typical transponder has a wide bandwidth, it is used with only one uplink or downlink signal to minimize interference and improve communications reliability. In order to be economically feasible, a satellite must be capable of handling several channels. As a result, most satellites contain multiple transponders, each operating at a different frequency. A typical communications satellite has 12, 24, or more transponders. Each transponder represents an individual communications channel. Various multiplexing schemes are used so that each channel can carry multiple information transmissions.

FREQUENCY ALLOCATIONS. Most communications satellites operate in the microwave frequency spectrum. However, there are some exceptions. For example, many military satellites operate in the 200 to 400 VHF/UHF range. Also, the amateur radio OSCAR satellites operate in the VHF/UHF range. VHF, UHF, and microwave signals penetrate the ionosphere with little or no attenuation and are not refracted back to earth as are lower-frequency signals in the 3- to 30-MHz range.

The microwave spectrum is divided up into frequency bands that have been allocated to satellites as well as other communications services such as radar. These frequency bands are generally designated by a letter of the alphabet. Figure 16-12 shows the various frequency bands used in satellite communications.

The most widely used satellite communications band is the C band. The uplink frequencies are in the 5.925- to 6.425-GHz range. In any general discussion of the C band, the uplink is generally said to be 6 GHz. The downlink is in the 3.7- to 4.2-GHz range. But again, in any general discussion of the C band, the downlink is nominally said to be 4 GHz. Occasionally, the C is referred to by the designation 6/4 GHz, where the uplink frequency is given first.

DID YOU KNOW?

Amateur radio operators are probably called hams because *HAM* was the call sign used by an early wireless station operated by three amateurs named <u>H</u>yman, <u>A</u>lmy, and <u>M</u>urray. (Shrader, *Electronic Communication*, 6th ed., Glencoe/McGraw-Hill, 1996)

FREQUENCY	BAND
225–390 MHz	P
350–530 MHz	J
1530–2700 MHz	L
2500–2700 MHz	S
3400–6425 MHz	C
7250–8400 MHz	X
10.95–14.5 GHz	Ku
17.7–21.2 GHz	Kc
27.5–31 GHz	K
36–46 GHz	Q
46–56 GHz	V
56–100 GHz	W

FIG. 16-12 Frequency bands used in satellite communications.

Although most satellite communications activity takes place in the C band, there is a steady move toward the higher frequencies. Currently, the Ku band is receiving the most attention. The uplinks are in the 14- to 14.5-GHz range while the downlinks are from 11.7 to 12.2 GHz. You will see the Ku band designated as 14/12 GHz.

Most new communications satellites will operate in this band. This upward shift in frequency is happening because the C band is becoming overcrowded. There are many communications satellites in orbit now, most of them operating in the C band. However, there is some difficulty with interference because of the heavy usage. The only way this interference will be minimized is to shift all future satellite communications to higher frequencies. Naturally, the electronic equipment achieving these higher frequencies is more complex and expensive. Yet, the crowding and interference problems cannot be resolved in any other way. Further, for a given antenna size, the gain is higher in the Ku band than in the C band. This can improve communications reliability while decreasing antenna size and cost.

Two other bands of interest are the X and L bands. The military uses the X band for its satellites and radar. The L band is used for navigation as well as marine and aeronautical communications and radar.

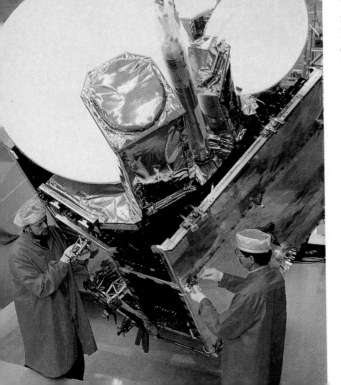

Technicians make final adjustments on a satellite assembly in a clean room. Satellites have dramatically improved worldwide communications.

Recall the frequencies designated for the C band uplink and downlink. These are 5925–6425 and 3700–4200 MHz, respectively. You can see that the bandwidth between the upper and lower limits is 500 MHz. This is an incredibly wide band, capable of carrying an enormous number of signals. In fact, 500 MHz covers all the radio spectrum so well known from VLF through VHF and beyond. Most communications satellites are designed to take advantage of this full bandwidth. This allows them to carry the maximum possible number of communications channels. Of course, this extremely wide bandwidth is one of the major reasons why microwave frequencies are so useful in communications. Not only can many communications channels be supported but also very high speed digital data requiring a wide bandwidth.

The transponder receiver "looks at" the entire 500-MHz bandwidth and picks up any transmission there. However, the input is "channelized" because the earth stations operate on selected frequencies or channels. The 500-MHz bandwidth is typically divided into 12 separate transmit channels, each 36 MHz wide. There are 4-MHz guardbands between channels that are used to minimize adjacent channel interference (see Fig. 16-13). Note the center frequency for each channel. Remember that the uplink frequencies are translated by frequency conversion to the downlink channel. In both cases, the total bandwidth (500 MHz) and channel bandwidth (36 MHz) are the same. Onboard the satellite, a separate transponder is allocated to each of the 12 channels.

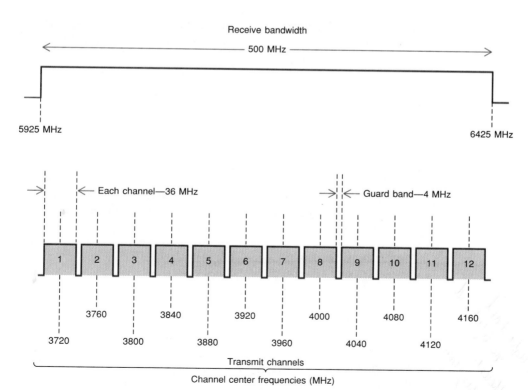

FIG. 16-13 Receive and transmit bandwidths in a C-band communications satellite.

Although 36 MHz seems narrow compared to 500 MHz, each transponder bandwidth is capable of carrying an enormous amount of information. For example, one typical transponder can handle up to 1000 one-way analog telephone conversations as well as one full-color TV channel. Each transponder channel can also carry high-speed digital data. Using certain types of modulation, a standard 36-MHz bandwidth transponder can handle digital data at rates of up to 60 megabits per second.

Example 16-2

A satellite transponder operates in the C band (see Fig. 16-13). Assume a local oscillator frequency of 2 GHz.

a. What is the uplink receiver frequency if the downlink transmitter is on channel 4?

The downlink frequency of channel 4 is 3840 MHz (Fig. 16-13). The downlink frequency is the difference between the uplink frequency (f_u) and the local oscillator frequency f_{LO}:

$$f_d = f_u - f_{LO}$$

Therefore:

$$f_u = f_d + f_{LO}$$
$$f_u = 3840 + 2000 = 5840 \text{ MHz} \qquad (\text{or } 5.84 \text{ GHz})$$

b. What is the maximum theoretical data rate if one transponder is used for binary transmission?

The bandwidth of one transponder channel is 36 MHz. For binary transmission, the maximum theoretical data rate or channel capacity (C) for a given bandwidth (B) is:

$$C = 2B$$
$$C = 2(36) = 72 \text{ Mb/s}$$

Although the transponders are quite capable, they nevertheless rapidly become overloaded with traffic. Further, at times there is more traffic than there are transponders to handle it. For that reason, numerous techniques have been developed to effectively increase the bandwidth and signal-carrying capacity of the satellite. Two of these techniques are known as *frequency reuse* and *spatial isolation.*

FREQUENCY REUSE. One system for effectively doubling the bandwidth and information-carrying capacity of a satellite is known as *frequency reuse.* In this system, a communications satellite is provided with two identical sets of 12 transponders. The first channel in one transponder operates on the same channel as the first transponder in the other set, and so on. With this arrangement, the two sets of transponders transmit in the same frequency spectrum and, therefore, appear to interfere with one another. However, this is not the case. The two systems, while operating on exactly the same frequencies, are isolated from one another by the use of special antenna techniques.

All satellite communication systems consist of two basic parts, the satellite or space-craft and two or more earth stations. The satellite performs the function of a radio repeater or relay station. Two or more earth stations may communicate with one another through the satellite rather than directly point-to-point on the earth.

The heart of a communications satellite is the communications subsystem. This is a set of transponders that receive the uplink signals and retransmit them to earth. A transponder is a repeater that implements a wideband communications channel that can carry many simultaneous communications transmissions.

The transponders are supported by a variety of additional "housekeeping" subsystems. These include the power subsystem, the telemetry tracking and command

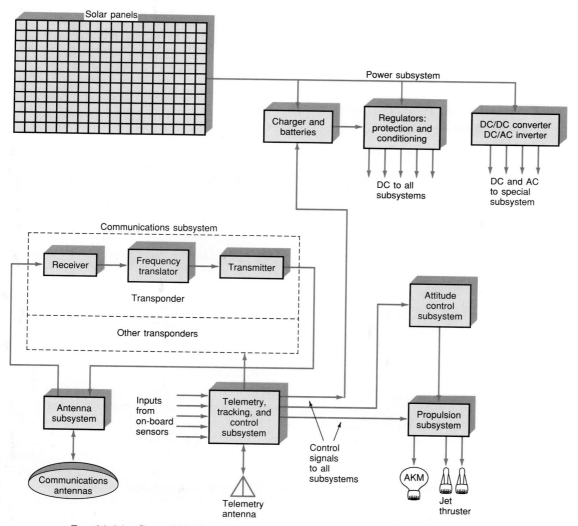

FIG. 16-14 General block diagram of a communications satellite.

subsystems, the antennas, and the propulsion and attitude stabilization subsystems. These are essential to the self-sustaining nature of the satellite.

Figure 16-14 is a general block diagram of a satellite. All the major subsystems are illustrated. The solar panels supply the electric power for the spacecraft. They drive regulators that distribute DC power to all other subsystems. And they charge the batteries that operate the satellite during eclipse periods. AC to DC converters and DC to AC inverters are used to supply special voltages to some subsystems.

The communications subsystem consists of multiple transponders. These receive the uplink signals, amplify them, translate them in frequency, and amplify them again for retransmission as downlink signals. The transponders share a common antenna subsystem for both reception and transmission.

The telemetry, tracking, and command subsystem monitors onboard conditions such as temperature and battery voltage and transmits this data back to a ground station for analysis. The ground station may then issue orders to the satellite by transmitting a signal to the command subsystem, which then is used to control many spacecraft functions such as firing the jet thrusters.

The jet thrusters and the *apogee kick motor* (*AKM*) are part of the propulsion subsystem. They are controlled by commands from the ground.

The attitude control subsystem provides stabilization in orbit and senses changes in orientation. It fires the jet thrusters to perform attitude adjustment and station-keeping maneuvers that keep the satellite in its assigned orbital position.

COMMUNICATIONS SUBSYSTEMS

The main payload on a communications satellite, of course, is the communication subsystem that performs the function of a repeater or relay station. An earth station takes the signals to be transmitted, known as *baseband signals,* and frequency modulates a microwave carrier. The three most common baseband signals are voice, video, or computer data. These uplink signals are then amplified, translated in frequency, and retransmitted on the downlink to one or more earth stations. The component that performs this function is known as a *transponder.* Most modern communications satellites contain at least 12 transponders. More advanced satellites contain many more. These transponders operate in the microwave frequency range.

The basic purpose of a transponder is simply to rejuvenate the uplink signal and retransmit it over the downlink. In this role the transponder simply performs the function of an amplifier. By the time the uplink's signal reaches the satellite, it is extremely weak. Therefore, it must be amplified before it can be retransmitted to the receiving earth station.

However, transponders are more than just amplifiers. An amplifier is a circuit that takes a signal and does nothing more than increase the voltage, current, or power level of that signal without changing its frequency or content. Such a transponder then would literally consist of a receiver and a transmitter that would operate on the same frequency. Because of the close proximity of the transmitter and the receiver in the satellite, the high transmitter output power for the downlink would be picked up by that satellite receiver. Naturally, the uplink signal would be totally obliterated. Further, the transmitter output fed back into the receiver input would cause oscillation.

To avoid this problem, the receiver and transmitter in the satellite transponder are designed to operate at separate frequencies. In this way, they will not interfere with one another. The frequency spacing is made as wide as practical to prevent oscillation and to minimize the effect of the transmitter desensitizing the receiver. In many repeaters, even though the receive and transmit frequencies are different, the high output power of the transmitter can still affect the sensitive receiver input circuits and, in effect, desensitize them, making them less sensitive in receiving the weak uplink signals. The wider the frequency spacing between transmitter and receiver, the less this desensitizing problem is.

In typical satellites, the input and output frequencies are separated by huge amounts. At C band frequencies, the uplink signal is in the 6-GHz range while the downlink signal is in the 4-GHz range. This 2-GHz spacing is sufficient to eliminate most problems. However, to ensure maximum sensitivity and minimum interference between uplink and downlink signals, the transponder contains numerous filters that not only provide channelization but also help to eliminate interference from external signals regardless of their source.

Three basic transponder configurations are used in communications satellites. They are all essentially minor variations of one another, but each has its advantages and disadvantages. These are the single-conversion, double-conversion, and regenerative transponders.

A *single-conversion transponder* uses a single mixer to translate the uplink signal to the downlink frequency. A *dual-conversion transponder* makes the frequency translation in two steps with two mixers. No demodulation occurs. A regenerative repeater demodulates the uplink signal after the frequency is translated to some lower intermediate frequency. The recovered baseband signal is then used to modulate the downlink signal.

MULTICHANNEL CONFIGURATIONS. Virtually all modern communications satellites contain multiple transponders. This permits many more signals to be received and transmitted. A typical commercial communications satellite contains 12 transponders,

24 if frequency reuse is incorporated. Military satellites often contain fewer transponders, whereas the newer, larger commercial satellites have provisions for up to 50 channels. Each transponder operates on a separate frequency, but its bandwidth is wide enough to carry multiple channels of voice, video, and digital information.

There are two basic multichannel architectures in use in communications satellites. One is a broadband system, and the other is a fully channelized system.

BROADBAND SYSTEM. As indicated earlier, a typical communications satellite spectrum is 500 MHz wide. This is typically divided into 12 separate channels, each with a bandwidth of 36 MHz. The center frequency spacing between adjacent channels is 40 MHz, thereby providing a 4-MHz spacing between channels to minimize adjacent channel interference. Refer to Fig. 16-13 for details. A wide bandwidth repeater (Fig. 16-15) is designed to receive any signal transmitted within the 500-MHz total bandwidth.

The receive antenna is connected to a low noise amplifier as in every transponder. Very wideband tuned circuits are used so that the entire 500-MHz bandwidth is received and amplified. A low-noise amplifier (LNA), usually a GaAsFET, provides gain. A mixer translates all incoming signals into their equivalent lower downlink frequencies. In a C-band communications satellite, the incoming signals are located between 5.925 and 6.425 MHz. A local oscillator operating at the frequency of 2.225 GHz is used to translate the inputs to the 3.7- to 4.2-GHz range. A wideband amplifier following the mixer amplifies this entire spectrum.

The channelization process occurs in the remainder of the transponder. For example, in a 12-channel satellite, 12 bandpass filters, each centered on one of the 12 channels, are used to separate all the various received signals. Figure 16-13 shows the 12 basic channels with their center frequencies, each having a bandwidth of 36 MHz. The bandpass filters separate out the unwanted mixer output signals and retain only the difference signals. Then, individual *high-power amplifiers* (*HPAs*) are used to increase signal level. These are usually *traveling wave tubes* (*TWTs*). The output of each TWT amplifier is again filtered to minimize harmonic and intermodulation distortion

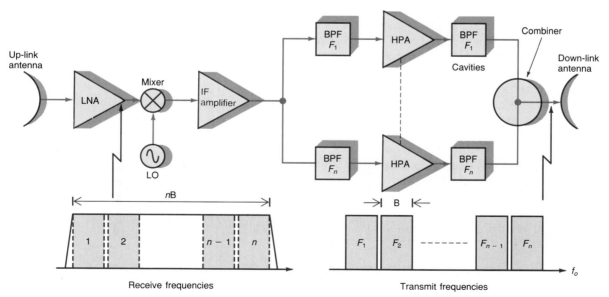

FIG. 16-15 A broadband multiple-channel repeater.

problems. These filters are usually part of a larger assembly known as a *multiplexer*, or *combiner*. This is a waveguide-cavity resonator assembly that filters and combines all of the signals for application to a single antenna.

It is logical to assume that if the receive function can be accomplished by wideband amplifier and mixer circuits, then it must be possible to provide the transmit function in the same way. However, it is generally not possible to generate very high output power over such wide bandwidth. The fact is that no components and circuits can do this well. The high-power amplifiers in most transponders are traveling-wave tubes that inherently have limited bandwidth. They operate well over a small range but cannot deal with the entire 500-MHz bandwidth allocated to a satellite. Therefore, in order to achieve the high-power levels, the channelization process is used.

POWER SUBSYSTEM

Today virtually every satellite uses solar panels for its basic power source. Solar panels are large arrays of photocells connected in various series and parallel circuits to create a powerful source of direct current. Early solar panels could generate hundreds of watts of power. Today huge solar panels are capable of generating many kilowatts. A key requirement is that the solar panels always be pointed toward the sun. There are two basic satellite configurations. In cylindrical-shaped satellites, the solar cells surround the entire unit, and, therefore, some portion of them is always exposed to sunlight. In body-stabilized, or three-axis, satellites, individual solar panels are manipulated with various controls to ensure that they are correctly oriented with respect to the sun.

Solar panels generate a direct current that is used to operate the various components of the satellite. However, the DC power is typically used to charge nickel cadmium batteries that act as a buffer. When a satellite goes into an eclipse or when the solar panels are not properly positioned, the batteries take over temporarily and keep the satellite operating. The batteries are not large enough to power the satellite for a long period of time; they are used as a backup system for eclipses, initial satellite orientation and stabilization, or emergency conditions.

The basic DC voltage from the solar panels is conditioned in various ways. For example, it is typically passed through voltage regulator circuits before being used to power individual electronic circuits. Occasionally, voltages higher than those produced by the solar panels must also be generated. For example, the traveling-wave tube (TWT) amplifiers in most communications transponders require thousands of volts for proper operation. Special DC-to-DC converters are used to translate the lower DC voltage of the solar panels to the higher DC voltage required by the TWTs.

TELEMETRY, COMMAND, AND CONTROL SUBSYSTEMS

All satellites have a telemetry, command, and control (TC&C) subsystem that allows a ground station to monitor and control conditions in the satellite. The telemetry system is used to report the status of the onboard subsystems to the ground station (see

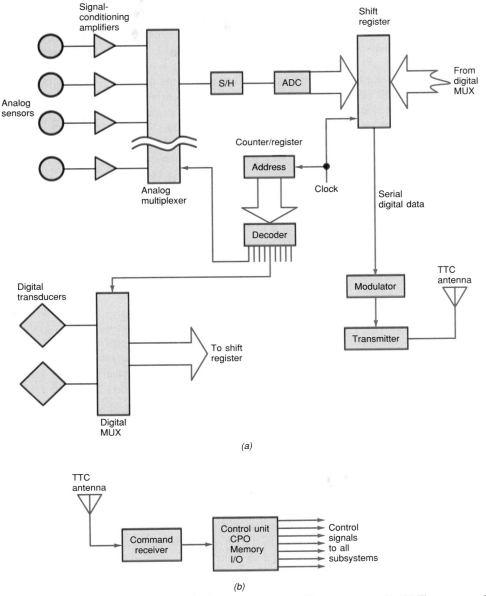

(a)

(b)

FIG. 16-16 (a) General block diagram of a satellite telemetry unit. (b) The command receiver and controller.

Fig. 16-16). The telemetry system typically consists of various electronic sensors for measuring temperatures, radiation levels, power supply voltages, and other key operating characteristics. Both analog and digital sensors may be used. The sensors are selected by a multiplexer and then converted to a digital signal, which then modulates an internal transmitter. This transmitter sends the telemetry information back to the earth station where it is recorded and monitored. With this information, the ground station then determines the operational status of the satellite at all times.

A command and control system permits the ground station to control the satellite. Typically the satellite contains a command receiver that receives control signals from an earth station transmitter. The control signals are made up of various digital codes that tell the satellite what to do. Various commands may initiate a telemetry sequence, activate thrusters for attitude correction, reorient an antenna, or perform other operations as required by the special equipment specific to the mission. Usually, the control signals are processed by an onboard computer.

Most satellites contain a small digital computer, usually microprocessor based, that acts as a central control unit for the entire satellite. The computer contains a built-in ROM with a master control program that operates the computer and causes all other subsystems to be operated as required. The command and control receiver typically takes the command codes that it receives from the ground station and passes them on to the computer, which then carries out the desired action.

The computer may also be used to make necessary computations and decisions. Information collected from the telemetry system may be first processed by the computer before it is sent to the ground station. The memory of the computer may also be used to store data temporarily prior to processing or prior to being transmitted back to earth. The computer may also serve as an event timer or clock. Thus the computer is a versatile control element that can be reprogrammed via the command and control system to carry out any additional functions that may be required, particularly those that were not properly anticipated by those designing the mission.

APPLICATIONS SUBSYSTEMS

The applications subsystem is made up of the special components that enable the satellite to fulfill its intended purpose. For a communications satellite, this subsystem is made up of the transponders.

An observation satellite such as those used for intelligence gathering or weather monitoring may use TV cameras or infrared sensors to pick up various conditions on earth and in the atmosphere. This information is then transmitted back to earth by a special transmitter designed for this purpose. There are many variations of this subsystem depending upon the use. The *Global Positioning System* satellite system is an example of a subsystem, the application payload for which is used for navigation. This system will be discussed in detail later in this chapter.

16-4 GROUND STATIONS

The ground station, or earth station, is the terrestrial base of the system. The ground station communicates with the satellite to carry out the designated mission. The earth station may be located at the end user's facilities or may be located remotely with ground-based intercommunication links between the earth station and the end user. In the early days of satellite systems, earth stations were typically placed in remote country locations. Because of their enormous antennas and other critical requirements, it was not practical to locate them in downtown or suburban areas. Today earth stations are still complex, but their antennas are smaller. Many earth stations are now located on top of tall buildings or in other urban areas directly where the end user resides. This offers the advantage of eliminating complex intercommunications systems between the earth station and the end user.

Like the satellite, the earth station is made up of a number of different subsystems. The subsystems, in fact, generally correspond to those onboard the satellite but larger and much more complex. Further, several additional subsystems exist at earth stations that would not be appropriate in a satellite.

An earth station consists of five major subsystems: the antenna subsystem, the receive subsystem, the transmit subsystem, the *ground control equipment* (*GCE*) subsystem, and the power subsystem (see Fig. 16-17). Not shown here are the telemetry, control, and instrumentation subsystems.

The power subsystem furnishes all the power to the other equipment. The primary source of power is the standard AC power lines. This subsystem operates power supplies that distribute a variety of DC voltages to the other equipment. The power sub-

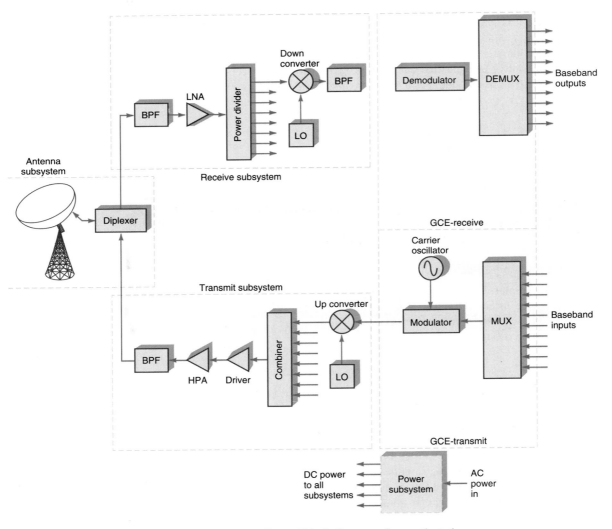

FIG. 16-17 General block diagram of an earth station.

system also consists of emergency power sources such as diesel generators, batteries, and inverters to ensure continuous operation during power failures.

As shown in Fig. 16-17, the antenna subsystem usually includes a *diplexer,* that is, a waveguide assembly that permits both the transmitter and receiver to use the same antenna. The diplexer feeds a bandpass filter in the receiver section that ensures that only the received frequencies pass through to the sensitive receiving circuits. This bandpass filter blocks the high-power transmit signal that can occur simultaneously with reception. This prevents overload and damage to the receiver.

The output of the bandpass filter feeds a low noise amplifier (LNA). This low noise amplifier drives a power divider. This is a waveguidelike assembly that splits the received signal into smaller but equal power signals. The power divider feeds several down converters. These are standard mixers fed by local oscillators (LOs) that translate the received signals down to an intermediate frequency, usually of 70 MHz. A bandpass filter ensures the selection of the proper sidebands out of the down converter.

The IF signal containing the data is then sent to the GCE receive equipment where it is demodulated and fed to a demultiplexer where the original signals are finally obtained. The demultiplexer outputs are usually the baseband or original communications signals. In actual systems, several levels of demodulation and demultiplexing may have to take place to obtain the original signals. In the GCE transmit subsystem, the baseband signals such as telephone conversations are applied to a multiplexer that permits multiple signals to be carried on a single channel.

ANTENNA SUBSYSTEMS

All earth stations have a relatively large parabolic dish antenna that is used for sending and receiving signals to and from the satellite. Early satellites had very low power transmitters, and, therefore, the signals received on earth were extremely small. Huge high-gain antennas were required to pick up minute signals from the satellite. The earth station dishes were 80 to 100 ft or more in diameter. Antennas of this size are still used in some satellite systems today, and even larger antennas have been used for deep space probes.

Modern satellites now transmit with much more power. Advances have also been made in receiver components and circuitry. For that reason, smaller earth station antennas are now practical. In some applications, antennas as small as 18 in in diameter can be used.

Typically the same antenna is used for both transmitting and receiving. A diplexer, a special coupling device for microwave signals, is used to permit a single antenna to be used for multiple transmitters and/or receivers. In some applications, a separate antenna is used for telemetry and control functions.

The antenna in an earth station must also be steerable. That is, it must be possible to adjust its azimuth and elevation so that the antenna can be properly aligned with the satellite. Earth stations supporting geosynchronous satellites can generally be fixed in position, however. Azimuth and elevation adjustments are necessary to initially pinpoint the satellite and to permit minor adjustments over the satellite's life.

DID YOU KNOW?

Because of the low-power transmitters on early satellites, huge high-gain antennas 80 to 100 ft or more in diameter were required to pick up signals. Antennas of this size are still used in some satellite systems and for deep space probes.

The downlink is the receive subsystem of the earth station. It usually consists of very low noise preamplifiers that take the small signal received from the satellite and amplify it to a level suitable for further processing. The signal is then demodulated and sent on to other parts of the communications system.

RECEIVER CIRCUITS. The receive subsystem consists of the low noise amplifier, down converters, and related components. The purpose of the receive subsystem is to amplify the downlink satellite signal and translate it to a suitable intermediate frequency. From that point, the IF signal is demodulated and demultiplexed as necessary to generate the original baseband signals.

Refer to the general block diagram of the receive subsystem shown in Fig. 16-29. The output from the diplexer or circulator in the antenna subsystem feeds the RF signal to a *bandpass filter (BPF)*. The purpose of this filter is to pass only the receive signal and to block any transmitted energy. While the circulator or diplexer generally provides sufficient isolation between the received and transmitted signals, the filter provides additional protection to the receiver circuits.

The filter output is applied to a low noise amplifier (LNA). This provides initial amplification of the downlink signal. Typically several stages of amplification are used to raise the tiny downlink signal to an amplitude suitable for use with the remaining circuits.

The output of the LNA is fed to a down-converter circuit consisting of a mixer and a local oscillator. The down converter translates the RF signal to an intermediate frequency where it will be dealt with further by the ground communications equipment.

The amplified signal from the low noise amplifier is fed to a power divider circuit or branching network. This is either a waveguidelike device or in some cases a special stripline assembly that divides a single signal into multiple signals. The signals are identical to the received signal except that their amplitude is lower. The power dividers simply provide a means of allocating some of the amplified received signal to each of the independent down-converter circuits. It also provides necessary impedance matching.

The next stage in a satellite earth station receiver is a down converter which translates the microwave signal down to one or more intermediate frequencies. The IF signal containing the original modulation is then sent to the ground control equipment where it is demodulated and demultiplexed.

There are basically two types of down converters used in earth station receivers, single conversion and double conversion. In practice, single-conversion down converters are rarely used because of image rejection problems and the difficulty of tuning or channel selection. To solve these problems, dual-conversion down converters are normally used.

Figure 16-18(*a*) shows a typical dual-conversion down converter. The input bandpass filter passes the entire 500-MHz satellite signal. This is fed to a mixer along with a local oscillator. The output of the mixer is an IF signal, usually 770 MHz. This is passed through a bandpass filter at that frequency with a bandwidth of 36 MHz.

The signal is then applied to another mixer. When combined with the local oscillator frequency, the mixer output is the standard 70-MHz IF value. An IF of 140 MHz is used in some systems. A 36-MHz-wide bandpass filter positioned after the mixer passes the desired channel.

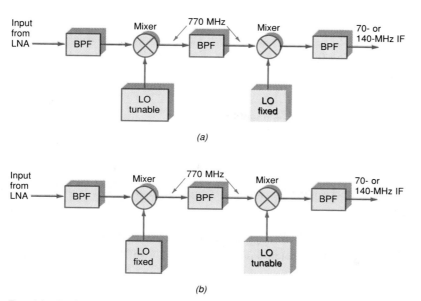

Fig. 16-18 Dual-conversion down converters. (*a*) RF tuning. (*b*) IF tuning.

In dual-conversion down converters, two different tuning or channel selection arrangements are used. One is referred to as *RF tuning* while the other is referred to as *IF tuning*. In RF tuning, shown in Fig. 16-18(*a*), the first local oscillator is made adjustable. Typically a frequency synthesizer is used in this application. The frequency synthesizer generates a highly stable signal at selected frequency increments. The frequency synthesizer is set to a frequency that will select a desired channel. In RF tuning, the second local oscillator has a fixed frequency to achieve the final conversion.

In IF tuning, the first local oscillator is fixed in frequency while the second is made tunable. Again, a frequency synthesizer is normally used for the tunable local oscillator [see Fig. 16-18(*b*)].

Example 16-3

Refer to Fig. 16-18(*b*). If the earth station downlink signal received is at $f_s = 4.08$ GHz, what are the local oscillator frequencies (f_{LO}) needed to achieve IFs of 770 and 140 MHz?

Assume LO frequencies lower than the received signals. First IF
$$f_{IF} = f_s - f_{LO}$$
$$f_{LO} = f_s - f_{IF} = 4080 - 770 = 3310 \text{ MHz}$$
Second IF
$$f_{LO} = f_s - f_{IF}$$
$$f_s = 770 \text{ MHz}$$
$$f_{LO} = 770 - 140 = 630 \text{ MHz}$$

RECEIVER GROUND CONTROL EQUIPMENT. The receiver ground control equipment (GCE) consists of one or more racks of equipment used for demodulating and demultiplexing the received signals. The down converters provide initial channelization by transponder, while the demodulators and demultiplexing equipment process the 70-MHz IF signal into the original baseband signals. Other intermediate signals may be developed as required by the application.

The outputs from the down converters are usually made available on a patch panel of coaxial cable connectors. These are interconnected via coaxial cables to the demodulators. The demodulators are typically packaged in a thin narrow vertical module that plugs into a chassis in a rack. Many of these demodulators are provided. They are all identical in that they demodulate the IF signal. In FDM systems, each demodulator is an FM detector. The most commonly used type is the phase-locked loop discriminator. Equalization and de-emphasis are also taken care of in the demodulator.

In systems using TDM, the demodulators are typically those used to detect four-phase, or quadrature, PSK at 60 or 120 megabits per second. The IF is usually at 140 MHz. Again, a patch panel between the down converters and the demodulators permits flexible interconnection to provide any desired configuration. When video data are transmitted, the output of the FM demodulator is the baseband video signal, which can then be transmitted by cable or used on the premises as required.

If the received signals are telephone calls, then the demodulator outputs are sent to demultiplexing circuits. Again in many systems, a patch panel between the demodulator outputs and the demultiplexer inputs is provided to make various connections as may be required. In FDM systems, standard frequency division multiplexing is used. This consists of additional single-sideband (SSB) demodulators and filters. Depending upon the number of signals multiplexed, several levels of channel filters and SSB demodulators may be required to generate the original baseband voice signals. Once this is done, the signal may be transmitted over the standard telephone system network as required.

In TDM systems, time division demultiplexing equipment is used to reassemble the original transmitted data. The original baseband digital signals may be developed in some cases, or in others these signals are used with modems as required for interconnecting the earth station with the computer that will process the data.

Passenger cars can now be equipped with a dashboard-mounted computerized map along with a dual-mode route guidance system.

TRANSMITTER SUBSYSTEMS

The uplink is the transmitting subsystem of the earth station. It consists of all the electronic equipment that takes the signal to be transmitted, amplifies it, and sends it to the antenna. In a communications system, the signals to be sent to the satellite might be TV programs, multiple telephone calls, or digital data from a computer. These signals are used to modulate the carrier which is then amplified by a large traveling wave tube or klystron amplifier. Such amplifiers generally generate many hundreds of watts of output power. This is sent to the antenna by way of microwave waveguides, combiners, and diplexers.

The transmit subsystem consists of two basic parts, the up converters and the power amplifiers. The up converters translate the baseband signals modulated on to carriers up to the final uplink microwave frequencies. The power amplifiers generate the high-power signals that are applied to the antenna. The modulated carriers are created in the GCE transmit equipment.

TRANSMIT GROUND CONTROL EQUIPMENT.
The transmit subsystem begins with the baseband signals. These are first fed to a multiplexer if multiple signals are to be carried by a single transponder. Telephone calls are a good example. Frequency or time division multiplexers are used to assemble the composite signal. The multiplexer output is then fed to a modulator. In analog systems, a wideband frequency modulator is normally used. It operates at a carrier frequency of 70 MHz with a maximum deviation of ±18 MHz. Video signals are fed directly to the modulator; they are not multiplexed.

In digital systems, analog signals are first digitized with PCM converters. The resulting serial digital output is then used to modulate a QPSK modulator.

TRANSMITTER CIRCUITS.
Once the modulated IF signals have been generated, up conversion and amplification will take place prior to transmission. Individual up converters are connected to each modulator output. Each up converter is driven by a frequency synthesizer that allows selection of the final transmitting frequency. The frequency synthesizer selects the transponder it will use in the satellite. The synthesizers are ordinarily adjustable in 1-kHz increments so that any up converter can be set to any channel frequency or transponder.

As in down converters, most modern up converters use dual conversion. Both RF and IF tuning are used (Fig. 16-19). With IF tuning, a tunable carrier from a frequency synthesizer is applied to the mixer to convert the modulated signal to an intermediate IF level, usually 700 MHz. Another mixer fed by a fixed frequency local oscillator (LO) then performs the final up conversion to the transmitted frequency.

In RF tuning, a mixer fed by a fixed frequency local oscillator performs an initial up conversion to 700 MHz. Then a sophisticated RF frequency synthesizer applied to a second mixer provides up conversion to the final microwave frequency.

In some systems, all the IF signals at 70 or 140 MHz are combined prior to the up conversion. In this system, different carrier frequencies are used on each of the modulators to provide the desired channelization. The carrier frequencies when translated by the up converter will translate to the individual transponder center frequencies. A special IF combiner circuit mixes all of the signals linearly and applies them to a single up converter. This up converter creates the final microwave signal.

In most systems, however, individual up converters are used on each modulated channel. At the output of the up converters, all the signals are combined in a microwave combiner which produces a single output signal that will be fed to the final amplifiers. This arrangement is illustrated in Fig. 16-20.

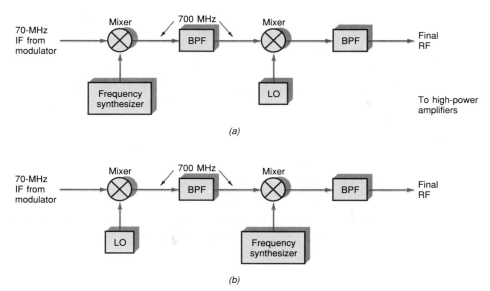

(a)

(b)

FIG. 16-19 Typical up-converter circuits. (*a*) IF tuning. (*b*) RF tuning.

The final combined signal to be transmitted to the satellite appears at the output of the RF combiner. But it must first be amplified considerably before being sent to the antenna. This is done by the power amplifier. The power amplifier usually begins with an initial stage called the *intermediate power amplifier* (*IPA*). This provides sufficient drive to the final high-power amplifier (HPA). Note also in Fig. 16-20 that redundant amplifiers are used. The main IPA and HPA are used until a failure occurs. Switches (SW) automatically disconnect the defective main amplifier and connect the

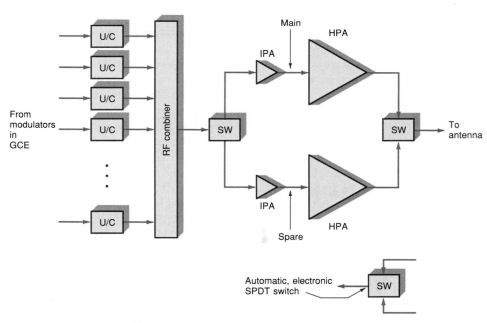

FIG. 16-20 An RF combiner and power amplifiers.

spare to ensure continuous operation. The amplified signal is then sent to the antenna via a waveguide, the diplexer, and a filter.

There are three types of power amplifiers used in earth stations: transistor, traveling wave tube (TWT), and klystron. Transistor power amplifiers are used in small and medium earth stations with low power. Powers up to 10 W are common. Normally, power *gallium arsenide field-effect transistors (GaAs MESFETs)* are used in this application. In some instances, a chain of varactor or IMPATT diodes is used to create a string of frequency multipliers which translate an RF power signal generated by a bipolar transistor at a lower frequency to the desired final operating frequency. Again, powers of up to 10 W are achievable with these systems. The newer GaAs MESFETs power devices, when operated in parallel, can achieve powers of up to approximately 40 W. Further improvements are expected.

Most medium- and high-power earth stations use either TWTs or klystrons for the power amplifiers. There are two typical power ranges, one in the 200- to 400-watt range and the other in the 2- to 3-kW range. The amount of power used depends upon the location of the station and its antenna size. Satellite transponder characteristics also influence the power required by the earth station. As improvements have been made in low noise amplifiers and satellites have been able to carry higher-power transponders, earth station transmitter power requirements have greatly decreased. This is also true for antenna sizes.

POWER SUBSYSTEMS

Most earth stations receive their power from the normal AC mains. Standard power supplies convert the AC to the DC voltages required to operate all subsystems. However, most earth stations have backup power systems. Satellite systems, particularly those used for reliable communications of telephone conversations, TV programs, computer data, and so on, must not go down. The backup power system takes over if an AC power failure occurs. The backup power system may consist of a diesel engine driving an AC generator. When AC power fails, an automatic system starts the diesel engine. The generator creates the equivalent AC power, which is automatically switched to the system. Smaller systems may use *uninterruptible power supplies (UPS)*, which derive their main power from batteries. Large battery arrays drive DC-to-AC inverters which produce the AC voltages for the system. Uninterruptible power supplies are not suitable for long power failures and interruptions because the batteries quickly become exhausted. However, for short interruptions of power, less than an hour, they are adequate.

TELEMETRY AND CONTROL SUBSYSTEMS

The telemetry equipment consists of a receiver and the recorders and indicators that display the telemetry signals. The signal may be received by the main antenna or a separate telemetry antenna. A separate receiver on a frequency different from the communications channels is used for telemetry purposes. The telemetry signals from the various sensors and transducers in the satellite are multiplexed onto a single carrier and are sent to the earth station. The earth station receiver demodulates and demultiplexes the telemetry signals into the individual outputs. These are then recorded and sent to various indicators, such as strip chart recorders, meters, and digital displays.

Signals may be in digital form or converted to digital. They can be sent to a computer where they can be further processed and stored. Mathematical operations on the signals may be necessary. For example, the number received from the satellite representing temperature may have to be processed by some mathematical formula to obtain the actual output temperature in Celsius or Fahrenheit. The computer can also generate and format printed reports as required.

The control subsystem permits the ground station to control the satellite. This system usually contains a computer for entering the commands that modulate a carrier that is amplified and fed to the main antenna. The command signals can make adjustments in the satellite attitude, turn transponders off and on, actuate redundant circuits if the circuits they are backing up fail, and so on.

In some satellite systems where communications is not the main function, some instrumentation may be a part of the ground station. *Instrumentation* is a general term for all the electronic equipment used to deal with the information transmitted back to the earth station. It may consist of demodulators and demultiplexers, amplifiers, filters, A/D converters, or signal processors. The instrumentation subsystem is in effect an extension of the telemetry system. Besides relaying information about the satellite itself, the telemetry system may also be used to send information back related to various scientific experiments being conducted on the satellite. In satellites used for surveillance, the instrumentation may be such that it can deal with television signals sent back from an onboard TV camera. The possibilities are extensive depending upon the actual satellite mission.

16-5 SATELLITE APPLICATIONS

Every satellite is designed to perform some specific task. Its predetermined application specifies the kind of equipment it must have onboard and its orbit. While the emphasis on satellites in this chapter is communications, satellites are useful for many other purposes.

COMMUNICATIONS SATELLITES

The main application for satellites today is communications. Satellites used for this purpose act as relay stations in the sky. They permit reliable long-distance communications worldwide. They solve many of the growing communications needs of industry and government. Communications applications will continue to dominate this industry.

The primary use of communications satellites is long-distance telephone service. Satellites greatly simplify long-distance calls not only within but also outside the United States.

Another major communications application is TV. For years, TV signals have been transmitted through satellites for redistribution. Because of the very high frequency signals involved in TV transmission, other long-distance transmission methods are not technically or economically feasible. Special coaxial cables and fiber-optic cables as well as microwave relay links have been used to transmit TV signals from one place to another. However, with today's communications satellites, TV signals can be transmitted easily from one place to another. All the major TV networks and cable TV companies rely on communications satellites for TV signal distribution.

Consumers have discovered that with an appropriate home satellite antenna and receiver, they can pick up these "free" TV signals. Satellite TV transmissions are intended for use within the TV industry so that the networks and cable TV companies can distribute their signals. However, such signals can be picked up by anyone with an appropriate microwave receiver and antenna. An entire industry has grown up around receiving these network distribution TV signals at home. Millions of home satellite TV receivers are in use, and those who have them enjoy the broadest range of TV programming possible. These signals contain virtually everything that exists on the standard TV channels, much of which never reaches the public. The programs being transmitted are often uncut and unedited.

The so-called premium TV signals like the movie channels that cable companies distribute for pay are no longer available directly to the public. Sources like HBO, Showtime, and others encrypt their signals so that only the cable companies can receive them. However, home satellite receivers can be equipped with decoders, and the premium signals can be had by subscription for a price.

A more recent satellite TV service is the *Direct Broadcast Satellite* (*DBS*) service. This is a TV signal distribution system designed to distribute signals directly to consumers. Special broad U.S. coverage satellites with high power transmit cable-TV-like services direct to homes equipped with the special DBS receivers. These newer satellites work at the higher microwave frequencies (K band) with higher power so that only very small antennas are needed. Older satellite systems (C band) still require large dish antennas 6 ft or more in diameter. The newer DBS system allows dishes as small as 18 in in diameter to be used reliably. The newer low noise GaAs FET receiver front-ends make this possible.

Communications applications will continue to dominate the satellite field. An application under development is satellite-based cellular telephone service. Current cellular telephone systems (to be covered in Chap. 17) rely upon many low-power, ground-based cells to act as intermediaries between the standard telephone system and the millions of roving cellular telephones. The proposed new systems use low earth orbit satellites to perform the relay services to the main telephone system or to make connection directly between any two cellular telephones using the system.

While nearly a dozen different satellite cellular telephone systems have been proposed, the most advanced system is that developed by Motorola. Known as *IRIDIUM,* it uses a constellation of 66 satellites in six polar orbits with 11 satellites per orbit 420 mi above the earth (refer to Fig. 16-21). The satellites operate in the L band over the frequency range of 1.61 to 1.6265 GHz. The satellites communicate with ground stations called *gateways* that connect the system to the public switched telephone network. The satellites also communicate among themselves. Both gateway and inter-satellite communications take place over K_a-band frequencies. The system provides truly global coverage between any two handheld cellular telephones or between one of the cellular telephones and any other telephone on earth.

Each satellite in the system transmits back to earth, creating 48 spot beams that in effect form moving cell areas of coverage. The spot beams provide spatial separation to permit more than one satellite to use the same frequencies for

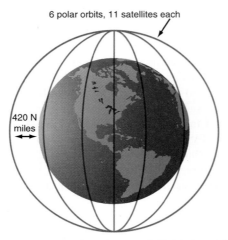

Fig. 16-21   The Motorola IRIDIUM satellite cellular telephone system.

simultaneous communications. Frequency reuse makes efficient use of a small number of frequencies.

Transmission is by digital techniques. A special form of QPSK is used for modulation. *Time division multiple access* (*TDMA*), a special form of time division multiplexing, is used to provide multiple voice channels per satellite.

In addition to voice communications, IRIDIUM will also be able to provide a whole spectrum of other communications services including:

1. *Data Communications*   Data rates of up to 2400 bps for duplex E-mail and other computer communication.
2. *Fax*   Two-way facsimile at 2400 bps.
3. *Paging*   Global paging to receivers with a two-line alphanumeric display.
4. *Radio Determination Services (RDSs)*   A subsystem that permits satellites to locate transceivers on earth. Accuracy is expected to be within 3 miles.

The IRIDIUM system is expected to be fully operational by 2000.

SURVEILLANCE SATELLITES

Another application of satellites is surveillance or observation. From their vantage point high in the sky, satellites can look at the earth and transmit what they see back to ground stations for a wide variety of purposes. For example, military satellites are used to perform reconnaissance. Onboard cameras take photographs that can later be ejected from the satellite and brought back to earth for recovery. TV cameras can take pictures and send them back to earth as electrical signals. Infrared sensors detect heat sources. Small radars can profile earth features.

Intelligence satellites collect information about enemies and potential enemies. They permit monitoring for the purpose of proving other countries' compliance with nuclear test ban and missile stockpile treaties.

There are many different kinds of observation satellites. One special type is the meteorological, or weather, satellite. These satellites photograph cloud cover and send back to earth pictures that are used for determining and predicting the weather. Geodetic satellites photograph the earth for the purpose of creating more accurate and more detailed maps.

NAVIGATION SATELLITES

A third applications area is navigation. Electronic systems have been used for years to provide accurate position information to ships, airplanes, and land-based vehicles. Loran and Omega are well-known systems used in marine navigation. Loran is a highly accurate system that uses coastal stations along all U.S. water borders. Over 50 stations transmit on a frequency of about 100 kHz in a range that is very limited.

The Omega system uses signals in the 10- to 13-kHz range, and with 8 stations it can cover the world. In both cases accuracy is limited to within the 1000- to 2000-ft range. Yet this is good accuracy for general navigation.

Aircraft use the *VOR/DME Tacan system* of navigation. *VOR* means *VHF omnidirectional ranging* and *DME* means *distance measuring equipment. Tacan* refers to *Tactical Air Navigation,* which is a similar but higher-frequency, higher-accuracy version of the VOR/DME used by the military. VOR uses hundreds of beacon stations transmitting a rotating narrow beam signal that is used to vector planes from one VOR station to another along common air traffic routes. DME permits accurate distance measurements between stations. VOR transmits in the 108- to 118-MHz range while Tacan operates in the 960- to 1215-MHz range. Typical accuracy is in the 200- to 600-ft range. Obviously, use is limited to the United States.

Satellite systems overcome some of the problems with land-based navigation systems. First, they can provide global coverage so that planes and ships can navigate anywhere on earth. By using multiple satellites, global coverage can be obtained, thus overcoming the limitations of earth-based systems that are confined in range by signal propagation conditions.

By using satellite systems, very high frequency signals can be used. High-frequency signals give better resolution and accuracy in positioning. Finally, satellite systems can provide altitude positioning data that other systems cannot.

The first successful satellite navigation was the Transit system built and deployed by the Navy. It uses a system of satellites in polar "bird cage" orbits 580 nautical miles above the earth (see Fig. 16-22). (One nautical mile equals 1.15 statute miles.) Five or six Transit satellites are orbiting the earth at any given time. Each satellite transmits continuous carrier signals on frequencies near 150 and 400 MHz. Receivers on earth, known as SatNav receivers, pick up the signals and use Doppler measurements to determine the speed of the satellite with respect to the receiver.

Doppler refers to a shift in frequency of a wave that occurs due to its movement from one place to another. Assume that a continuous sine wave carrier is transmitted between two fixed points. The receiver will pick up the transmitted signal at exactly the frequency it was transmitted. But now imagine that the transmitter moves toward the still fixed receiver. The forward movement has the effect of compress-

DID YOU KNOW?

Satellite navigation systems can provide global coverage unavailable with land-based systems satellites. This enables ships and planes to navigate anywhere on earth.

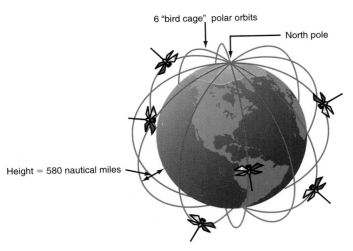

6 "bird cage" polar orbits

North pole

Height = 580 nautical miles

Fig. 16-22 The Transit satellite navigation system.

ing the waves as they travel toward the receiver. The received signal is therefore higher in frequency. If the transmitter is moving away from the receiver, the signal waves are stretched out, making the receiver pick up lower-frequency signals.

This change in frequency due to motion between transmitter and receiver is called the *Doppler shift.* By detecting and accurately measuring the amount of the Doppler shift, the velocity of movement can be precisely determined. This permits speed measurements, which can be translated into position information.

In the Transit system, the SatNav receiver measures the Doppler shift during the time a satellite transverses the sky from horizon to horizon. Then, since the exact ground path of the satellite is known from codes transmitted by the satellite, the position of the receiver can be calculated from the Doppler data.

Transit is widely used by the Navy and many ships and smaller boats for global navigation. Its accuracy is within about 1500 ft. The Transit system has been in use since the late 1960s, and civilian use was permitted in 1973. Today, Transit is being phased out in favor of another better satellite navigation system known as the *Global Positioning System* (*GPS*). The next and last section of this chapter covers GPS in detail.

16-6 GLOBAL POSITIONING SYSTEM

The *Global Positioning System* (*GPS*), also known as *Navstar*, is a satellite-based navigation system that can be used by anyone with an appropriate receiver to pinpoint his or her location on earth. The array of GPS satellites transmits highly accurate, time-coded information that permits a receiver to calculate its exact location in terms of the latitude and longitude on earth as well as the altitude above sea level.

GPS was developed by the U.S. Air Force for the Department of Defense as a continuous global radio navigation system that all elements of the military services would use for precision navigation. Development was started in 1973, and by 1994, the system was fully operational.

The GPS Navstar system is an *open navigation system;* that is, anyone with a GPS receiver can use it. The system is designed, however, to provide a base navigation sys-

Navigational computers, such as this Magellan handheld device, can pinpoint exact positions on the surface of the Earth using a GPS.

tem with an accuracy to within 100 m. A supplementary part of the system accessible only to the military provides much greater precision, to within 10 to 20 m. The base system is highly accurate and more so than virtually any other electronic navigation system in existence. As a result, GPS is gradually replacing older military systems such as the Navy's Transit satellite navigation system and civilian land-based systems such as Loran.

The GPS system is an excellent example of a modern satellite-based system and the high-technology communications techniques used to implement it. Learning about GPS is an excellent way to bring together and illustrate the complex concepts presented in this chapter and previous chapters.

The GPS system consists of three major segments: the space segment, the control segment, and the user segment.

SPACE SEGMENT

The space segment is the constellation of satellites orbiting above the earth that contain transmitters which send highly accurate timing information to GPS receivers on earth. The receivers in the user segment themselves may be used on land, sea, or air.

The fully implemented GPS system consists of 21 main operational satellites plus 3 active spare satellites (see Fig. 16-23). The satellites are arranged in six orbits, each orbit containing 3 or 4 satellites. The orbital planes form a 55° angle with the equator. The satellites orbit at a height of 10,898 nautical miles above the earth (20,200 km). The orbital period for each satellite is approximately 12 h (11:58).

Each of the orbiting satellites contains four highly accurate atomic clocks. These provide precision timing pulses used to generate a unique binary code, that is, a pseudo random code identifying the specific satellite in the constellations, which is transmitted to earth. The satellite also transmits a set of digitally coded ephemeris data that completely defines its precise orbit. The term *ephemeris* is normally associated with specifying the location of a celestial body. Tables have been computed so that it is possible to pinpoint the location of planets and other astronomical bodies at precise locations and times. Ephemeris data can also be computed for any orbiting body such as a

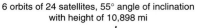

6 orbits of 24 satellites, 55° angle of inclination
with height of 10,898 mi

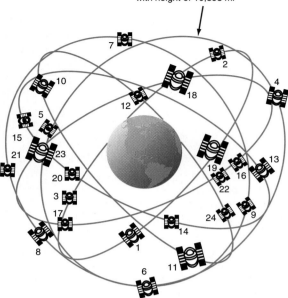

FIG. 16-23 The GPS space segment.

satellite. This data tells where the satellite is at any given time, and its location can be specified in terms of the satellite ground track in precise latitude and longitude measurements. This ephemeris information is coded and transmitted from the satellite providing an accurate indication of the exact position of the satellite above the earth at any given time. The satellite's ephemeris data is updated once per day by the ground control station to ensure accuracy.

A GPS receiver on earth is designed to pick up signals from three, four, or more satellites simultaneously. The receiver decodes the information and, using the time and ephemeris data, calculates the exact position of the receiver. The receiver contains a high-speed, floating-point microcomputer that performs the necessary calculations. The output of the receiver is a decimal display of latitude and longitude as well as altitude. Readings from only three satellites are necessary for latitude and longitude information only. A fourth satellite reading is required in order to compute altitude.

THE CONTROL SEGMENT AND ATOMIC CLOCKS

The *control segment of the GPS system* refers to the various ground stations that monitor the satellites and provide control and update information. The master control station is operated by the U.S. Air Force in Colorado Springs. Additional monitoring and control stations are located in Hawaii, Kwajalein, Diego Garcia, and the Ascension Islands. These four monitoring stations in the islands are unmanned. They constantly monitor the satellites and collect range information from each. The positions of these monitoring stations are accurately known. The information is sent back to the master control station in Colorado where all the information is collected and position data on each satellite calculated. The master control station then transmits new ephemeris and clock data to each satellite on the S-band uplink once per day. This data updates the NAV-msg.

The telemetry data transmitted as part of the NAV-msg is also received by the ground control station to keep track of the health and status of each receiver. The uplink S-band control system allows the ground station to do some station keeping to correct the satellite position as may be needed. Positioning is accomplished with the hydrazene thrusters.

ATOMIC CLOCKS. The precision timing signals are derived from atomic clocks. Most digital systems derive their timing information from a precision crystal oscillator called a *clock*. Crystal oscillators, while precise and stable, simply do not have the necessary precision and stability for the GPS system. Timing data must be extremely precise to provide accurate navigation information.

Atomic clocks are electronic oscillators that use the oscillating energy of a gas to provide a stable operating frequency. Certain chemicals have atoms that can oscillate between low and high energy levels. This frequency of oscillation is extremely precise and stable.

Cesium and rubidium are used in atomic clocks. In gaseous form, they are irradiated by electromagnetic energy at a frequency near their oscillating point. For cesium this is 9,192,631,770 Hz; for rubidium, it is 6,834,682,613 Hz. This signal is generated by a quartz crystal controlled oscillator whose frequency can be adjusted by the application of a control voltage applied to a varactor. The oscillation signal from the cesium (or rubidium) gas is detected and converted into a control signal that operates the oscillator. An error detector determines the difference between the crystal frequency and the cesium oscillations and generates the control voltage to set the crystal oscillator precisely. The output of the crystal oscillator is then used as the accurate timing signal. Frequency dividers and phase-locked loops are provided to generate the lower frequency signals to be used. These signals operate the C/A- and P-code generators.

The GPS satellite contains two cesium clocks and two rubidium clocks. Only one of them is used at any given time. The clocks are kept fully operational so that, if one fails, another can be switched in immediately.

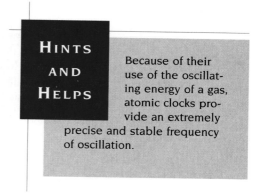

HINTS AND HELPS

Because of their use of the oscillating energy of a gas, atomic clocks provide an extremely precise and stable frequency of oscillation.

GPS RECEIVERS

A *GPS receiver* is a complex superheterodyne microwave receiver designed to pick up the GPS signals, decode them, and then compute the actual location of the receiver. The output is usually an LCD display giving latitude, longitude, and altitude information.

There are many different types of GPS receivers. Over 40 manufacturers now make some form of GPS receiver. The larger and more sophisticated units are used in military vehicles. There are also sophisticated civilian receivers for use in various kinds of precision applications like surveying and mapmaking. Different models are available for use in aircraft, ships, and trucks. Handheld units are also available.

The most widely used GPS receiver is the popular handheld portable type, not much larger than an oversized handheld calculator. Most of the circuitry to make a GPS receiver has been reduced to integrated circuit form, thereby permitting an entire receiver to be contained in an extremely small, portable, battery-operated unit. Keep in mind that all GPS receivers are not only superheterodyne communications receivers but also sophisticated computers. A considerable amount of high-level mathematics must be carried out to compute the receiver position from the received data.

Figure 16-24 is a general block diagram of a GPS receiver, typical of the simpler, low-cost, handheld units on the market. The receiver consists of the antenna, the RF/IF section, the frequency standard clock oscillator, and a frequency synthesizer that provides local oscillator signals as well as clock and timing signals for the rest of the receiver.

Other sections of the receiver include a *digital signal processor,* a control microcomputer along with its related RAM and ROM. Interface circuits provide connection to the LCD display.

The antenna system is a type of patch antenna made on a printed circuit board. It is designed to receive *right-hand circularly polarized* (*RHCP*) signals from the GPS satellites. In the handheld units, the antenna is part of the single physical structure and is connected directly to the receiver front end. In some larger and more complex GPS receivers, the antenna is a separate unit and may be mounted at a high clear point and connected to the receiver with a coaxial cable.

The receiver can determine its exact position only by computing the position information obtained from four satellites. The receiver picks up signals from four satellites simultaneously (see Fig. 16-25). R1 through R4 are the ranges to the satellites from the yacht. Because spread spectrum is used, the receiver ignores all of the signals except the one whose pseudo code has been entered and used to obtain lock. Once the receiver locks on the one satellite, all the information is extracted from the satellite. Then the C/A code is switched to another satellite within view and the process is repeated. In other words, the receiver performs a time multiplexing operation on the four satellites within view of the receiver. The data is extracted from each of the four satellites and stored in the receiver's memory.

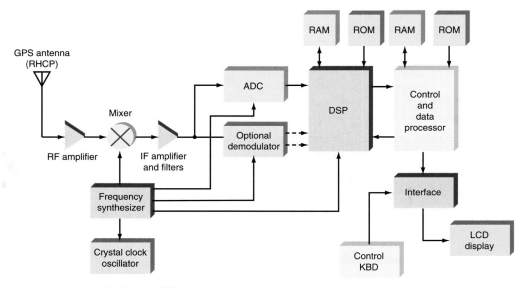

Fɪɢ. 16-24 A GPS receiver.

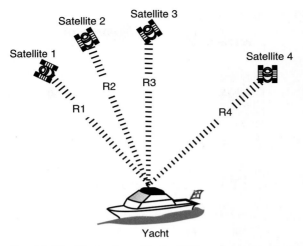

Fig. 16-25 Navigating with the GPS.

Once all the data has been accumulated, the high-speed control and data microprocessor in the receiver performs the final calculations. The microprocessor is typically a 16-bit unit with floating-point capability. Floating-point numbers must be used in order to provide the precision of calculation for accurate location.

The computations performed by the receiver are given below. First the receiver calculates the ranges to each of the four satellites, R1 through R4. These are obtained by measuring the time shift between the received pulses which is the delay in transmission between the satellite and the receiver. These are the T_1 through T_4 times given below. Multiplying these by the speed of light (C) gives the range in meters between the satellite and the receiver. These four range values are used in the final calculations:

$$R1 = C \times T_1$$
$$R2 = C \times T_2$$
$$R3 = C \times T_3$$
$$R4 = C \times T_4$$

The basic ranging calculation is the solution to four simultaneous equations, as indicated in Fig. 16-26. The X, Y, and Z values are derived from the NAV-msg data transmitted by each of the satellites. The C_B is the clock bias. Since there is a difference

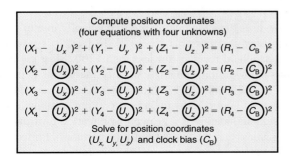

Fig. 16-26 The equations solved by the GPS receiver.

Communications technicians must be well trained and up to date with the latest technologies to work in this satellite network data processing center. The instrumentation shown in this photograph is used to track a positioning satellite.

between the clock frequency in the receiver and the clock in the satellites, there will be some difference called the *bias*. However, by factoring the clock error into the equations, it will be canceled out.

The goal of the microprocessor then is to solve for the user position, designated U_X, U_Y, and U_Z in the equations. Once the calculation has been made, the microprocessor converts that information into the latitude, longitude, and altitude data, which are displayed on the LCD screen. The receiver display also shows the time of day, which is highly precise as it is derived from the atomic clocks in the satellites.

GPS Applications

The primary application of GPS is military navigation and related applications. GPS is being used by all services for ships, aircraft of all sorts, and ground troops. The GPS was first widely used in Operation Desert Storm during the Persian Gulf War with Iraq in 1991. It was an extremely good test and very successful. Now that GPS is fully operational, its uses in the military will continue to grow.

Civilian uses have also increased dramatically due to the availability of many low-cost portable receivers. In fact, it is now possible to purchase a handheld receiver for less than $200. Most civilian applications involve navigation, which is usually marine or aviation related. Hikers and campers and other outdoors sports enthusiasts also use GPS.

Commercial applications include surveying, mapmaking, and construction. Vehicle location is a growing application for trucking and delivery companies, taxi, bus, and train transportation. Police, fire, ambulance, and forest services also use GPS. Private citizens will begin using GPS-based systems that will be installed in cars to provide a continuous readout of current vehicle location.

DID YOU KNOW?

The GPS was first widely used in Operation Desert Storm with much success. Nonmilitary applications of this system involve navigation, surveying, mapmaking, and vehicle tracking devices.

SUMMARY

A *satellite* is a natural or human-made physical object that orbits, or rotates, around some celestial body. Satellites are kept in orbit by programming them in accordance with physical and mathematical laws collectively referred to as *orbital dynamics*.

In order for a satellite to go into orbit around the earth, it must have some forward motion. Thus, when it is launched, it is given both vertical and forward motion. The forward motion produces inertia, which tends to keep the satellite moving in a straight line. However, gravity tends to pull the satellite toward earth. The inertia of the satellite is equalized by the earth's gravitational pull. The satellite constantly changes its direction from a straight line to a curved line to rotate about the earth. Ground stations receive signals from the satellite that give the exact position of the satellite. Corrections to the orbital path can be made from the ground stations.

Communications satellites are not originators of information; rather, they are *repeaters*. A transmitting station sends the information to the satellite, which in turn retransmits it to the receiving stations.

Every satellite is designed to perform some specific task. Its predetermined application specifies the kind of equipment it must have onboard and its orbit. The main application for satellites today is communications. Satellites used for this purpose act as relay stations in the sky. They permit reliable long-distance communications worldwide. The primary use of communications satellites is long-distance telephone calls. Another major application is television signal transmission. Still another is cellular phone transmission. Satellites are also used for military surveillance, meteorological observations, and the monitoring of natural resources and pollution.

Navigation of ships, airplanes, and land-based vehicles is another common use of satellites. Developed by the U.S. Air Force, the *Global Positioning System* (*GPS*), also known as *Navstar*, is a satellite-based navigation system that can be used by anyone with an appropriate receiver to pinpoint his or her location on earth. The GPS is being used increasingly by civilians for navigation assistance in, for example, hiking or boating. Other commercial uses are for mapmaking and surveying, transportation, and police and fire protection.

KEY TERMS

Angle of elevation
Angle of inclination
Antenna subsystem
Apogee
Atomic clocks
Azimuth
Bandwidth
C band
Centripetal acceleration
Channelization process
Communications satellite
Direct Broadcast Satellite
 (DBS) service

Distance measuring equipment (DME)
Doppler shift
Dual-conversion trans-ponder
Downlink
Frequency reuse
Geocenter
Global Positioning System
 (GPS)
GPS receiver
Ground control equipment
 (GCE)
Ground station

High-power amplifier
 (HPA)
Intermediate-power amplifier (IPA)
IRIDIUM
Klystron
Latitude
Navigation satellites
Orbit
Perigee
Posigrade orbit
Power subsystem
Prime meridian

REVIEW

Receive subsystem
Receiver circuits
Regenerative repeater
Repeater
Retrograde orbit
Satellite
Satellite period
Satellite-based cellular
  telephone service
SatNav receivers
Sidereal period
Single-conversion
  transponder

Solar panels
Subsatellite point
  (SSP)
Surveillance satellites
Synchronous (geostation-
  ary) satellite
Tactical Air Navigation
  (Tacan)
Telemetry, command, and
  control subsystem
  (TC&C)
Transmit subsystem
Transmitter circuits

Transponder
Traveling wave tube
  (TWT)
TV transmission
Uplink
VHF omnidirectional
  ranging
  (VOR)
Y code

QUESTIONS

1. A satellite that rotates in the same direction as the earth's rotation is said to be in a:
   a. Posigrade orbit
   b. Retrograde orbit
2. What is the geometric shape of a noncircular orbit?
3. What is the name of the center of gravity of the earth?
4. State the name for the time for one orbit.
5. Define *ascending orbit* and *descending orbit*.
6. State the effects on a satellite signal if the angle of elevation is too low.
7. What do you call a satellite that rotates around the equator 22,300 mi from the earth?
8. Why are small jet thrusters on a satellite fired occasionally?
9. What is the name of the point on the earth directly below a satellite?
10. Name the two angles used to point a ground station antenna toward a satellite.
11. What is the basic function and purpose of a communications satellite? Why are satellites used instead of standard earth-bound radio?
12. What element of a satellite system transmits an uplink signal to the satellite?
13. What do you call the signal path from a satellite to a ground station? What is the name of the signal path from the satellite to ground?
14. State the name of the basic communications electronics unit on a satellite. Explain the purpose and operation of the four main elements.
15. What is the most common operational frequency range of most communications satellites?
16. One of the most popular satellite frequency ranges is 4 to 6 GHz. What letter name is given to that range?

17. Military satellites often operate in which frequency band?
18. What is the frequency range of the Ku band?
19. What is the bandwidth of a typical satellite transponder?
20. A typical C-band transponder can carry how many channels? What is the bandwidth of each?
21. Name three common baseband signals handled by a satellite.
22. How is power amplification achieved in a transponder?
23. What are two common transponder intermediate frequencies?
24. What is the name of the circuit that provides channelization in a transponder?
25. What is the main power supply in a satellite?
26. What is used to power the satellite during an eclipse?
27. State the purpose of the TC&C subsystem on a satellite.
28. The signals to be communicated by the earth station to the satellite are known as _____ signals.
29. What does the receiver in a ground station do to the downlink signal before demodulation and demultiplexing?
30. What kind of circuit is often used to replace local oscillators for channel selection in earth station transmitters and receivers?
31. Name the three main types of power amplifiers used in earth stations.

PROBLEMS

1. What is the angle of inclination of a satellite that orbits over the equator? ◆
2. What standard navigation coordinates are used to locate a satellite in space?
3. How are satellite attitude adjustments made from the ground station? ◆
4. How do GPS receivers distinguish between the different satellite signals all transmitted on the same frequencies?
5. How is satellite position data transmitted by the satellite? What is the data rate? ◆
6. How is the speed of the satellite measured in the receiver?
7. Explain the need for and concept behind differential GPS. ◆

CRITICAL THINKING

1. How is a satellite kept in orbit? Explain the balancing forces.
2. Define the frequency and format of the C/A- and P-code signals. What is the Y code, and why is it needed?
3. Assume a satellite communications system based upon a highly elliptical polar orbit approximately centered over the Atlantic Ocean. Communications is to be maintained between a U.S. station and one in the United Kingdom. The frequency of operation is 435 MHz uplink and 145 MHz downlink. Discuss the type of antennas that might be used for both uplink and downlink. What are the implications, pros and cons, of each type? Will tracking or positioning equipment be needed and if so for which types of antennas? Will communications ever be interrupted and if so when?
4. Explain how you can find where you are on the earth if you have a GPS receiver that gives outputs in latitude and longitude.
5. Frequency reuse in a satellite allows two transponders to share a common frequency because the signals are kept from interfering with one another by using different antenna polarizations. How else could frequency reuse be achieved in a given area on earth?

CHAPTER 17

TELECOMMUNICATIONS SYSTEMS

Objectives

After completing this chapter, you will be able to:

◆ *Name* and *describe* the components in conventional and electronic telephones.

◆ *Describe* the characteristics of the various signals used in telephone communications.

◆ *State* the general operation of a cordless telephone.

◆ *Describe* the operation of a PBX.

◆ *Explain* the general hierarchy of signal transmission within the telephone system.

◆ *Explain* the basic operation of the cellular telephone system.

◆ *Explain* the operation of a facsimile machine.

◆ *Explain* the operation of a paging system.

◆ *Draw* a block diagram of a basic paging receiver and explain its operation.

◆ *Define* ISDN and explain its operation.

The telephone system is the largest and most complex electronic communications system in the world. It uses just about every type of electronic communications technique available including virtually all of those described in this book. The telephone communications system is so large and widely used that no text on electronic communications would be complete without a discussion of it. On the other hand, because the system is so large and complex, space is simply not available to cover it in great depth. However, a general discussion of the telephone system will better prepare readers for those communications applications that use the telephone system.

The approach taken in this chapter is to introduce the parts of the telephone system most commonly in use and accessible. The basic operation of standard and electronic telephones and the operation of cordless telephones are covered. Next, the telephone transmission system is briefly introduced and its operation described.

Although the primary purpose of the telephone system is to provide voice communications between individuals, it is also widely used for many other purposes. These include facsimile transmission and computer data transmission, as well as wireless transmission used in cellular and paging telephone systems. Data transmission via modem was discussed in Chap. 11, so it will not be repeated here. However, data transmission by the *integrated services digital network (ISDN) system,* which is a part of the telephone system, will be discussed.

17-1 TELEPHONES

The original telephone system was designed for full-duplex analog communications of voice signals. Today, the telephone system is still primarily analog in nature, but it employs a considerable number of digital techniques, not only in signal transmission but also in control operations.

The telephone system permits any telephone to connect with any other telephone in the world. This means that each telephone must have a unique identification code—the 10-digit telephone number assigned to each telephone. The telephone system provides a means of recognizing each individual number and switching systems that can connect any two telephones.

THE LOCAL LOOP

DID YOU KNOW?

The modern telephone system is still primarily analog in nature, although it employs a considerable number of digital techniques.

Standard telephones are connected to the telephone system by way of a two-wire, twisted-pair cable that terminates at the local exchange or central office. As many as 10,000 telephone lines can be connected to a single central office (see Fig. 17-1). The connections from the central office go to the "telephone system" represented in Fig. 17-1 by the large cloud. This part of the system, which is mainly long distance, will be described in the next section. A call originating at telephone A will pass through the central office and then into the main system, where it is transmitted via one of many different routes to the central office connected to the desired location designated as B in Fig. 17-1. The connection between nearby local exchanges is direct rather than long distance.

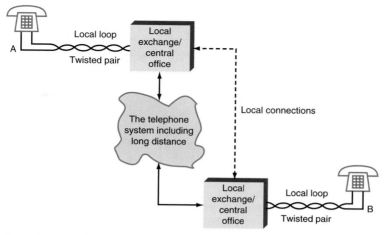

Fig. 17-1 The basic telephone system.

The two-wire, twisted-pair connection between the telephone and the central office is referred to as the *local loop* or *subscriber loop*. The circuits in the telephone and at the central office form a complete electric circuit, or loop. This single circuit is analog in nature and carries both DC and AC signals. The DC power for operating the telephone is generated at the central office and supplied to each telephone over the local loop. The AC voice signals are transmitted along with the DC power. Despite the fact that only two wires are involved, full-duplex operation, that is, simultaneous send and receive, is possible. All dialing and signaling operations are also carried on this single twisted pair.

The basic telephone set and the local loop, standard and electronic telephone sets, and cordless telephones will be discussed in this section.

THE TELEPHONE SET

A basic telephone or telephone set is an analog baseband transceiver. It has a handset which contains a microphone and a speaker, better known as a *transmitter* and a *receiver.* It also contains a ringer and a dialing mechanism. Overall, the telephone set fulfills the following basic functions:

The receive mode provides:

1. An incoming signal that rings a bell or produces an audio tone indicating that a call is being received.
2. A signal to the telephone system indicating that the signal has been answered.
3. Transducers to convert voice into electrical signals and electrical signals into voice.

The transmit mode:

1. Indicates to the telephone system that a call is to be made when the handset is lifted.
2. Indicates that the telephone system is ready to use by generating a signal called the *dial tone.*

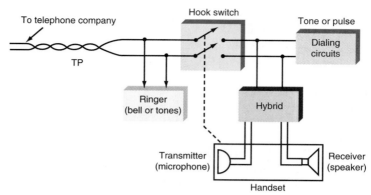

FIG. 17-2 Block diagram of a basic telephone.

3. Provides a way of transmitting the telephone number to be called to the telephone system.
4. Receives an indication that the call is being made by receiving a ringing tone.
5. Provides a means of receiving a special tone indicating that the called line is busy.
6. Provides a means of signaling the telephone system that the call is complete.

All telephone sets provide these basic functions. Some of the more advanced electronic telephones have other features such as multiple line selection, hold, and speaker phone.

Figure 17-2 is a basic block diagram of a telephone set. The function of each block is described below. Detailed circuits for each of the blocks and their operation will be described later when the standard and electronic telephones are discussed in detail.

RINGER. The *ringer* is either a bell or an electronic oscillator connected to a speaker. It is continuously connected to the twisted pair of the local loop back to the central office. When an incoming call is received, a signal from the central office will cause the bell or ringer to produce a tone.

SWITCH HOOK. A *switch hook* is a double-pole mechanical switch that is usually controlled by a mechanism actuated by the telephone handset. When the handset is "on the hook," the hook switch is open, thereby isolating all the telephone circuitry from the central office local loop. When a call is to be made or to be received, the handset is taken off the hook. This closes the switch and connects the telephone circuitry to the local loop. The direct current from the central office is then connected to the telephone, closing its circuits to operate.

DIALING CIRCUITS. The *dialing circuits* provide a way for entering the telephone number to be called. In older telephones, a pulse dialing system was used. A rotary dial connected to a switch produced a number of ON-OFF pulses corresponding to the digit dialed. These ON-OFF pulses formed a simple binary code for signaling the central office.

In most modern telephones, a tone dialing system is used. Known as the *dual-tone multifrequency (DTMF) system,* this dialing method uses a number of pushbuttons that generate pairs of audio tones that indicate the digits called.

Whether pulse dialing or tone dialing is used, circuits in the central office recognize both types of signals and make the proper connections to the dialed telephone.

HANDSET. This unit contains a microphone for the transmitter and a speaker or receiver. When you speak into the transmitter, it generates an electrical signal representing your voice. When a received electrical voice signal occurs on the line, the receiver translates it into sound waves. The transmitter and receiver are independent units, and each has two wires connecting to the telephone circuit. Both of these connect to a special device known as the *hybrid.*

HYBRID. The *hybrid* is a special transformer used to convert signals from the four wires from the transmitter and receiver into a signal suitable for a single two-line pair to the local loop. The hybrid permits *full-duplex,* that is, simultaneous send and receive, analog communications on the two-wire line. The hybrid also provides a side tone from the transmitter to the receiver so that the speaker can hear his or her voice in the receiver. This feedback permits automatic voice-level adjustment.

THE STANDARD TELEPHONE AND LOCAL LOOP

Figure 17-3 is a simplified schematic diagram of a conventional telephone and the local loop connections back to the central office. The circuitry at the central office will be discussed in more detail later. For now, note that the central office applies a DC voltage over the twisted-pair line to the telephone. This DC voltage is approximately -48 V with respect to ground in the open-circuit condition. When a subscriber picks up the telephone, the switch hook closes, connecting the circuitry to the telephone line. The load represented by the telephone circuitry causes current to flow in the local loop and the voltage inside the telephone to drop to approximately 5 to 6 V.

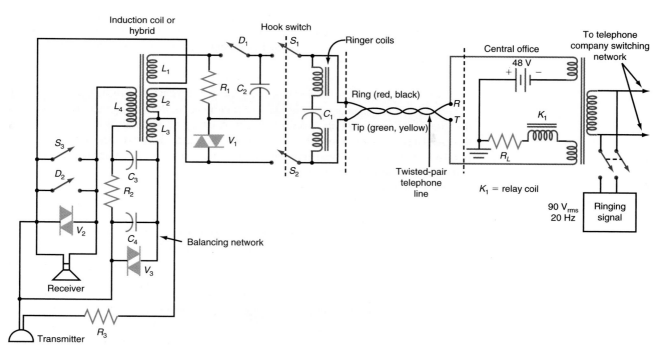

FIG. 17-3 Standard telephone circuit diagram showing connection to central office.

The amount of current flowing in the local loop depends upon a number of factors. The DC voltage supplied by the central office may not be exactly −48 V. It can, in fact, vary several volts above or below the 48-V normal value.

As Fig. 17-3 shows, the central office also inserts some resistance R_L to limit the total current flow if a short circuit should occur on the line. This resistance can range from about 350 to 800 Ω. In Fig. 17-3, the total resistance is approximately 400 Ω.

The resistance of the telephone itself also varies over a relatively wide range. It can be as low as 100 Ω and as high as 400 Ω, depending upon the circuitry. The resistance varies because of the resistance of the transmitter element and because of the variable resistors called *varistors* used in the circuit to provide automatic adjustment of line level.

The local loop resistance depends considerably on the length of the twisted pair between the telephone and the central office. Although the resistance of copper wire in the twisted pair is relatively low, the length of the wire between the telephone and the central office can be many miles long. Thus the resistance of the local loop can be anywhere from 1000 to 1800 Ω, depending upon the distance.

Finally, the frequency response of the local loop is approximately 300 to 3400 Hz. This is sufficient to pass voice frequencies that produce full intelligibility. An unloaded twisted pair has an upper cutoff frequency of about 4000 Hz. But this cutoff varies considerably depending upon the overall length of the cable. When long runs of cable are used, special loading coils are inserted into the line to compensate for excessive rolloff at the higher frequencies.

The wires in Fig. 17-3 end at terminals on the telephone labeled *tip* and *ring*. These designations refer to the plug used to connect telephones to one another at the central office. At one time, large groups of telephone operators at the central office used plugs and jacks at a switchboard to connect one telephone to another manually. The jack is shown in Fig. 17-4. The tip and the ring are metallic contacts that are attached to the wires in the cable. They touch spring-loaded contacts in the jack to make the connection. Although such plugs and jacks are no longer used, the tip and ring designation is still used to refer to the two wires of the local loop.

The wires are also usually color-coded red and green. The tip wire is green and is usually connected to ground; the ring wire is red. Many telephone cables into a home or office also contain a second twisted pair if a separate telephone line is to be installed. These wires are usually color-coded black and yellow. Black and yellow correspond to ring and tip, respectively, where yellow is ground. Other color combinations are used in telephone wiring.

Now let's take a more detailed look at the operation of the various circuits in the telephone. The local loop circuitry will be discussed later.

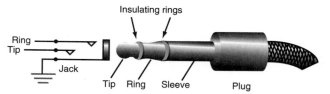

FIG. 17-4 Tip and ring designation on an old plug and jack.

RINGER. In Fig. 17-3, the circuitry connected directly to the tip and ring local loop wires is the ringer. The ringer in most older telephones is an electromechanical bell. A pair of electromagnetic coils is used to operate a small hammer that alternately strikes two small metallic bells. Each bell is made of different materials so that each produces a slightly different tone. When an incoming call is received, a voltage from the central office operates the electromagnetic coils which, in turn, operate the hammer to ring the bells. The bells make the familiar tone produced by most standard telephones.

In Fig. 17-3, the ringing coils are connected in series with a capacitor C_1. This allows the AC ringing voltage to be applied to the coils but blocks the 48 V of direct current, thus minimizing the current drain on the 48 V of power supplied at the central office.

The ringing voltage supplied by the central office is a sine wave of approximately 90 V rms at a frequency of about 20 Hz. These are the nominal values, because the actual ringing voltage can vary from approximately 80 to 100 V rms with a frequency somewhere in the 15- to 30-Hz range. This AC signal is supplied by a generator at the central office. The ringing voltage is applied in series with the −48-Vdc signal from the central office power supply. The ringing signal is connected to the local loop line by way of a transformer T_1. The transformer couples the ringing signal into its secondary winding where it appears in series with the 48-Vdc supply voltage.

The standard ringing sequence is shown in Fig. 17-5. In U.S. telephones, the ringing voltage occurs for 1 s followed by a 3-s interval. Telephones in other parts of the world use different ringing sequences. For example, in the United Kingdom, the standard ring sequence is a higher-frequency tone occurring more frequently, and it consists of two ringing pulses 400 ms long, separated by 200 ms. This is followed by a 2-s interval of quiet before the tone sequence repeats.

TRANSMITTER. The transmitter is the microphone into which you speak during a telephone call. In a standard telephone, this microphone uses a carbon element that

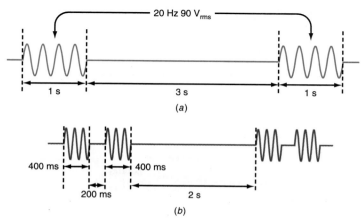

FIG. 17-5 Telephone ringing sequence. (*a*) United States and Europe. (*b*) United Kingdom.

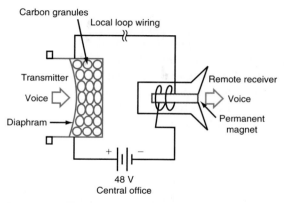

Carbon granules
Local loop wiring
Transmitter
Voice
Diaphram
Remote receiver
Voice
Permanent magnet
+ −
48 V
Central office

Fig. 17-6 The transmitter and receiver in a telephone.

effectively translates acoustical vibrations into resistance changes. The resistance changes, in turn, produce current variations in the local loop representing the speaker's voice. A DC voltage must be applied to the transmitter so that current flows through it during operation. The 48 V from the central office is used in this case to operate the transmitter.

Figure 17-6 is a simplified diagram showing how a telephone transmitter works. The basic transmitter element is a small module containing fine carbon granules. One side of the module is a flexible diaphragm. Whenever you speak, you create acoustic energy in the form of the movement of air. The air molecules move in accordance with your voice frequency. The acoustic energy from the voice reaches the diaphragm and causes it to vibrate in accordance with the speech. An outward acoustic pressure wave causes the carbon granules to be compressed. Pushing the carbon granules closer together causes the overall resistance of the element to decrease. When the acoustic energy moves in the opposite direction, the carbon granules expand outward. Since they are less tightly compressed, their resistance increases. The transmitter element is in series with the telephone circuit, which includes the 48-V central office battery and the speaker in the remote handset. Speaking into the transmitter causes the current flow in the circuit to vary in accordance with the voice signal. The resulting AC voice signal produced on the telephone line is approximately 1 to 2 V rms.

RECEIVER. The receiver, or earpiece, is basically a small permanent magnet speaker. A diaphragm is physically attached to a coil which rests inside a permanent magnet. Whenever a voice signal comes down a telephone line, it develops a current in the receiver coil. The coil produces a magnetic field that interacts with the permanent magnet field. The result is vibration of the diaphragm in the receiver, which converts the electrical signal into the acoustic energy that supplies the voice to the ear. As it comes in over the local loop lines, the voice signal has an amplitude of approximately 0.5 to 1 V rms.

HYBRID. The hybrid is a transformerlike device that is used to simultaneously transmit and receive on a single pair of wires. The hybrid, which is also sometimes referred to as an *induction coil,* is really several transformers combined into a single unit. The windings on the transformers are connected in such a way that signals produced by the transmitter are put on the two-wire local loop but do not occur in the receiver. In the same way, the transformer windings permit a signal to be sent to the receiver, but the resulting voltage is not applied to the transmitter.

In practice, the hybrid windings are set up so that a small amount of the voice signal produced by the transmitter does occur in the receiver. This provides feedback to the speaker so that he or she may speak with normal loudness. The feedback from the transmitter to the receiver is referred to as the *side tone*. If the side tone was not provided, there would be no signal in the receiver and the person speaking would have the sensation that the telephone was dead. By hearing his or her own voice in the receiver at a moderate level, the caller can speak at a normal level. Without the side tone, the speaker tends to speak more loudly, which is unnecessary.

AUTOMATIC LEVEL ADJUSTMENT. Because of the wide variation in the different loop lengths of the two telephones connected to one another, the circuit resistances will vary considerably, thereby causing a wide variation in the transmitted and received voice signal levels. All telephones contain some type of component or circuit that provides automatic voice-level adjustment so that the signal levels are approximately the same regardless of the loop lengths. In the standard telephone, this automatic loop length adjustment is handled by components called *varistors*. These are labeled V_1, V_2, and V_3 in Fig. 17-3.

A varistor is a nonlinear resistance element whose resistance changes depending upon the amount of current passing through it. When the current passing through the varistor increases, its resistance decreases. A decrease in current causes the resistance to increase.

The varistors are usually connected across the line. In Fig. 17-3, varistor V_1 is connected in series with a resistor R_1. This varistor automatically shunts some of the current away from the transmitter and the receiver. If the loop is long, the current will be relatively low and the voltage at the telephone will be low. This causes the resistance of the varistor to increase, thus shunting less current away from the transmitter and receiver. On short local loops, the current will be high and the voltage at the telephone will be high. This causes the varistor resistance to decrease; thus more current is shunted away from the transmitter and receiver. The result is a relatively constant level of transmitted or received speech.

Note that a second varistor V_3 is used in the balancing network. The balancing network (C_3, C_4, R_2) works in conjunction with the hybrid to provide the side tone discussed earlier. The varistor adjusts the level of the side tone automatically.

PULSE DIALING. The term *dialing* is used to describe the process of entering a telephone number to be called. In older telephones, a rotary dial was used. In more modern telephones, pushbuttons that generate electronic tones are used for "dialing."

The use of a rotary dialing mechanism produces what is known as *pulse dialing*. Rotating the dial and releasing it cause a switch contact to open and close at a fixed rate, producing current pulses in the local loop. These current pulses are detected by the central office and used to operate the switches that connect the dialing telephone to the called telephone.

Figure 17-7 shows the process of dialing the number 5. Five ON-OFF pulses are produced. This is followed by a variable time interval until the next digit is dialed.

Referring to Fig. 17-3, you can see that the pulse-dialed switch D_1 is connected in series with the telephone circuit. This switch contact is normally closed when the dial is not in use. It remains closed as the dial is rotated in the clockwise direction. When the dial is released, the switch opens and closes at a constant rate as indicated earlier, opening and closing the circuit and switching the loop current off and on.

Also during the dialing process, switch D_2, which is also mounted on the dialing mechanism, closes. This short circuits the receiver to prevent the current pulses from producing loud clicks in the receiver.

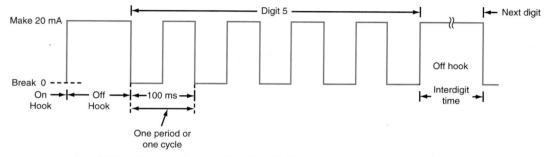

Fig. 17-7 Current pulses produced by dialing.

TONE DIALING. Although some dial telephones are still in use and all central offices can accommodate them, most modern telephones use a dialing system known as *TouchTone*. It uses pairs of audio tones to create signals representing the numbers to be dialed. This dialing system is referred to as the *dual-tone multifrequency (DTMF) system.*

A typical DTMF keyboard on a telephone is shown in Fig. 17-8. Most telephones use a standard keypad with 12 buttons or switches for the numbers 0 through 9 and the special symbols * and #. The DTMF system also accommodates 4 additional keys for special applications.

In Fig. 17-8 numbers represent audio frequencies associated with each row and column of pushbuttons. For example, the upper horizontal row containing the keys for 1, 2, and 3 is labeled 697, meaning that when any one of these three keys is depressed, a sine wave of 697 Hz is produced. Each of the four horizontal rows produces a different frequency. The horizontal rows generate what is generally known as the *low group of frequencies.*

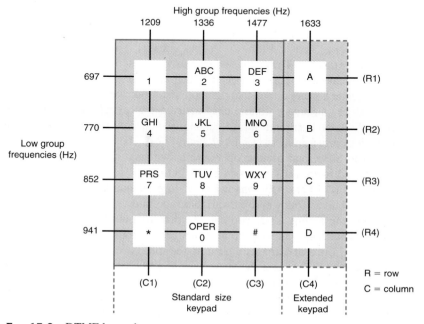

Fig. 17-8 DTMF keypad.

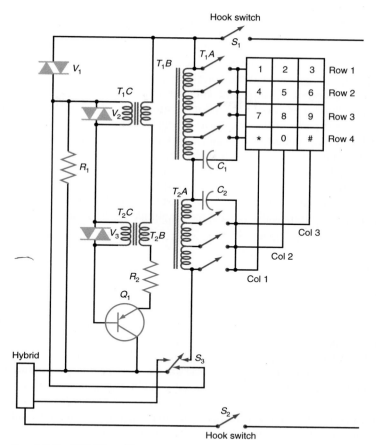

Fig. 17-9 DTMF oscillator.

A higher group of frequencies is associated with the vertical columns of keys. For example, the keys for the numbers 2, 5, 8, and 0 produce a frequency of 1336 Hz when depressed.

If the number 2 is depressed, two sine waves are generated simultaneously, one at 697 Hz and the other at 1336 Hz. These two tones are linearly mixed. This combination produces a unique sound and is easily detected and recognized at the central office as the signal representing the dialed digit 2. The tolerance on the generated frequencies is usually within ±1.5 percent.

The earliest TouchTone telephones used a simple single transistor oscillator to generate two tones simultaneously. A typical circuit is shown in Fig. 17-9. The oscillator circuit consists of two LC–tuned circuits. The inductance is provided by tapped transformer windings. Winding A of transformer T_1 is the inductance for the low-frequency tones, and winding A on transformer T_2 provides the inductance for the high frequencies. The A windings are tapped to provide different inductance values for each of the various frequencies. A fixed capacitor C_1 resonates with the selected winding on T_1A, and capacitor C_2 resonates with the selected winding on T_2A. Whenever one of the pushbuttons on the keypad is depressed, one of the switches in the low-frequency group and one of the switches in the high-frequency group are closed simultaneously.

Referring to Fig. 17-9, note switch S_3. It is also actuated when any one of the keys on the keypad is depressed. A slight depression of the key establishes connection to the two resonant circuits as indicated above. During this time, the contacts on S_3 remain close so that direct current flows through windings T_1A and T_2A. Further depression of the switch causes the contacts S_3 to open, breaking the current through windings T_1A and T_2A. Thus the two LC–resonant circuits are shocked into oscillation and two damped sine waves at the correct frequencies are generated.

The depression of S_3 causes the collector of transistor Q_1 to be connected to the DC voltage. Thus an electronic oscillator circuit is formed. Transformer windings T_1C and T_2C transmit some of the sine wave energy back to the base circuit of the transistor with the correct polarity to sustain continuous oscillation. Transformer windings T_1B and T_2B are connected in series with the emitter of the transistor. The resulting two-tone signal is placed on the local loop where it will be received and detected by the central office. Some of the tone is also fed back through the hybrid to the receiver so that the calling party can hear it. The varistors V_2 and V_3 across windings T_1C and T_2C are used to automatically control the amplitude of the signal depending upon the amount of voltage on the local loop.

ELECTRONIC TELEPHONES

When solid-state circuits came along in the late 1950s, an electronic telephone became possible and practical. Today, most new telephones are electronic, and they use integrated circuit technology.

The development of the microprocessor has also affected telephone design. Although simple electronic telephones do not contain a microprocessor, most multiple-line and full-featured telephones do. A built-in microprocessor permits automatic control of the telephone's functions and provides features such as telephone number storage and automatic dialing and redialing that are not possible in conventional telephones.

The variety of available electronic telephones is immense. The next section will give you an example of a sophisticated microprocessor-based electronic telephone.

TYPICAL IC ELECTRONIC TELEPHONE. The major components of an electronic telephone circuit are shown in Fig. 17-10. Most of the functions are implemented with circuits contained within a single IC.

In Fig. 17-10, note that the TouchTone keypad drives a DTMF tone generator circuit. An external crystal or ceramic resonator provides an accurate frequency reference for generating the dual dialing tones.

The tone ringer is driven by the 20-Hz ringing signal from the phone line and drives a piezoelectric sound element.

The IC also contains a built-in line voltage regulator. It takes the DC voltage from the local loop and stabilizes it to provide a constant voltage to the internal electronic circuits. An external zener diode and transistor provide bias to the electret microphone.

The internal speech network contains a number of amplifiers and related circuits that fully duplicate the function of a hybrid in a standard telephone. This IC also contains a microcomputer interface. The box labeled MPU is a single-chip microprocessing unit. Although it is not necessary to use a microprocessor, if automatic dialing and other functions are implemented, this circuit is capable of accommodating them.

Finally, note the bridge rectifier and hook switch circuit. The twisted pair from the local loop is connected to the tip and ring connections. Both the 48-Vdc and 20-Hz ring voltages will be applied to this bridge

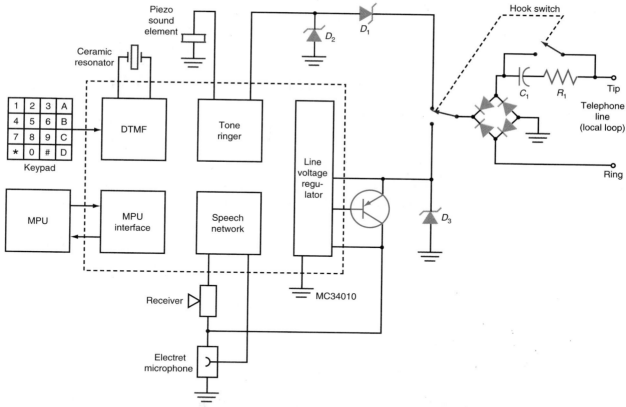

FIG. 17-10 Single-chip electronic telephone.

rectifier. For direct current, the bridge rectifier provides polarity protection for the circuit, ensuring that the bridge output voltage is always a positive voltage. When the AC ringing voltage is applied, the bridge rectifier rectifies it into a pulsating DC voltage. The hook switch is shown with the telephone on the hook or in the "hung-up" position. Thus the DC voltage is not connected to the circuit at this time. However, the DC ringing voltage will be coupled through the resistor and capacitor to the bridge, where it will be rectified and applied to the two zener diodes D_1 and D_2 that drive the tone ringer circuit.

When the telephone is taken off the hook, the hook switch closes, providing a DC path around the resistor and capacitor R_1 and C_1. The path to the tone ringer is broken, and the output of the bridge rectifier is connected to zener diode D_3 and the line voltage regulator. Thus the circuits inside the IC are powered up and calls may be received or made.

The tone ringer circuit is shown in Fig. 17-11. This IC is the Motorola MC34010 which is representative of a typical single-chip telephone IC. When an incoming call is being received, the central office applies the 20-Hz sine wave to the local loop. This passes through capacitor C_1, and R_1 is applied to the bridge rectifier circuit. The output of the bridge rectifier is applied to two zener diodes D_1 and D_2. D_1 provides a threshold level that must be exceeded before the tone ringer circuits are enabled. A DC voltage developed across D_2 and filtered by R_1 and C_1 is applied to the tone ringer input circuits, powering an internal bias circuit generating 21 Vdc. This is applied to an internal threshold detector circuit. When the ringing voltage reaches a specific level, it

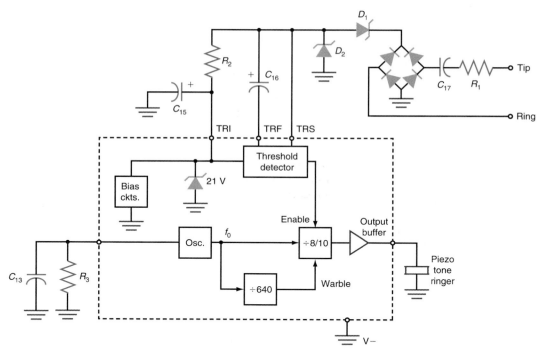

Fig. 17-11 Tone ringer circuit.

triggers and powers up the remaining circuits. Because there is a threshold to be exceeded, false ringing will not occur when stray signals appear on the line or when the DC voltage from the loop drops to a low level.

The ringing tone is generated by an internal oscillator circuit, the frequency of which is set to 8 kHz by external components C_{13} and R_3. The 8-kHz signal is applied to a frequency divider which can divide by 8 or by 10. The division ratio is determined by the warble which is generated by the oscillator and a circuit which is divided by 640. This produces a 12.5-Hz warble square wave that alternates the frequency divider between divide by 8 and divide by 10. As a result, the frequency divider output is a rectangular wave that switches between 800 and 1000 Hz at a 12.5-Hz rate. The signal is applied to an output buffer amplifier that drives an external piezoelectric tone ringer. This type of tone ringer is a transducer that reproduces the signal applied to it. It is unlike the previously described tone ringer that produces a fixed resonant tone output in the 2- to 3-kHz range.

Details of the DTMF dialing tone generator are shown in Fig. 17-12. The row and column connections to the TouchTone keypad are connected to the keypad comparator and logic circuits. The output of the logic circuits goes to program two counter-circuits, a column counter and a row counter. These counters generate 8-bit binary numbers that drive two digital-to-analog (D/A) converters. As the counters are stepped, they generate a binary number sequence that, when applied to the D/A converter, produces a stair-step approximation of the sine wave at a frequency related to the number key pressed. The column counter and its D/A converter generate one tone, and the row counter and its D/A converter generates the second tone. The two tones are combined at the output of the D/A converters and applied to the output buffer amplifier. The signal is then connected to the telephone line.

Timing for the circuitry is provided by a built-in oscillator that runs at 500 kHz. This frequency is set by an external ceramic resonator and capacitors C_1 and C_2.

Details of the speech network are shown in Fig. 17-13. This collection of amplifiers and other circuits duplicates the function of the hybrid transformer and the related level control circuits in a conventional telephone.

The DC voltage from the local loop through the bridge rectifier is applied to C_{12} and zener diode D_3, providing a stabilized DC voltage for the internal regulator circuit. The DC voltage from transistor T_1 and capacitor C_9 biases the electret microphone in the telephone handset. This bias voltage at input pin TXO also biases the transmit amplifier. The voice signal from the microphone is coupled through capacitor C_5 and R_{13}, C_4 and R_{12} to the transmit amplifier. The peak limiter circuit provides feedback to the transmit amplifier to reduce the gain and minimize distortion when very loud talkers use the telephone. The transmit amplifier output appears at the TXO pin and is applied to R_{10} and transistor T_1 to the telephone line. Thus T_1 acts as a final amplifier for the voice signal before being placed on the local loop.

The output of the transmit amplifier is inverted by a side tone amplifier and appears at the STA output. Here it is applied to an RC network and reduced in level. It is applied through C_7, C_8, and R_7 to the receive amplifier at input RXI. The receive amplifier output at RXO drives the small receiver speaker in the handset.

When an incoming voice signal is received, it is passed through R_5, C_7, C_8, and R_7 to the receive amplifier, where it again is passed to the speaker in the handset.

Finally, note that this circuit has a built-in mute control driven from the dialer circuit. Whenever dialing takes place, the mute control shuts off the transmit circuits and reduces the amplitude of the signal to the receive amplifier.

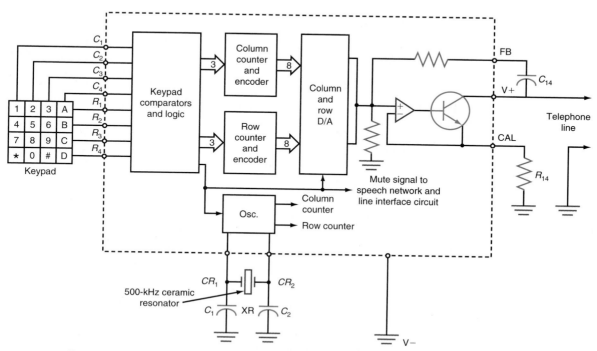

Fig. 17-12   DTMF dialing circuits.

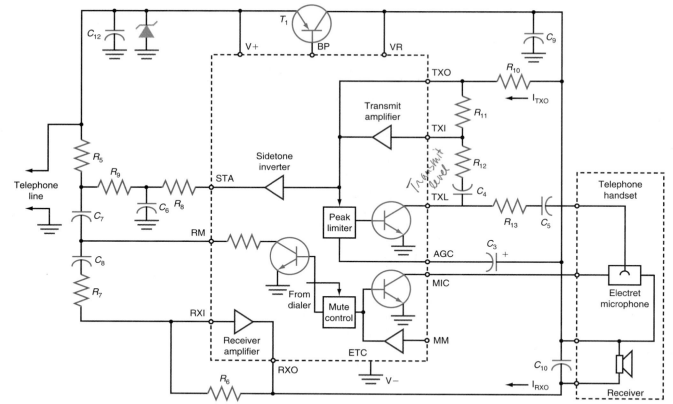

Fig. 17-13 Electronic speech circuits.

Microprocessor Control. The more advanced electronic telephones also contain a built-in microcontroller. Like any microcontroller, it consists of the CPU, a ROM where a control program is stored, a small amount of random access read-write memory, and I/O circuits. The microcontroller, usually a single-chip IC, may be directly connected to the telephone IC, or some type of intermediate interface circuit may be used.

The function performed by the microcomputer is primarily that of storing telephone numbers and automatically redialing. Many advanced telephones have the capability of storing 12 commonly called numbers. The user puts the telephone into a program mode and uses the TouchTone keypad to enter the most frequently dialed numbers. These are stored in the microcontroller's RAM. To automatically dial one of the numbers, a pushbutton on the front of the telephone is depressed. This may be one of the TouchTone pushbuttons, or it may be a separate set of pushbuttons provided for the purpose. When one of the pushbuttons is depressed, the microcontroller supplies a preprogrammed set of binary codes to the DTMF circuitry in the telephone IC. Thus the number is automatically dialed.

Line Interface. We have talked a great deal in general terms about how the telephone is connected to the local loop. Let's now examine the system more closely.

Most telephones are connected by way of a thin multiwire cable to a wall jack. A special connector on the cable, called an *RJ-11 modular connector,* plugs into the matching wall jack. Two local loops are available if needed.

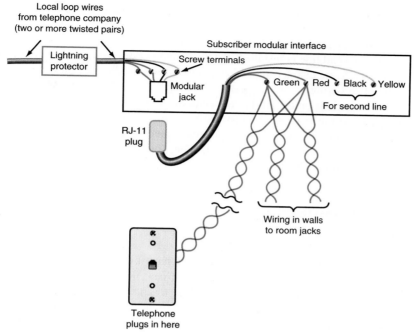

FIG. 17-14 Subscriber interface.

The wall jack is connected by way of wiring inside the walls to a central wiring point called the *subscriber interface*. Also known as the *wiring block* or *modular interface*, this is a small plastic housing containing all of the wiring that connects the line from the telephone company to all of the telephone wires in the house. Most houses and apartments are wired so that there is a wall jack in every room.

Figure 17-14 is a general diagram of the modular interface. The line from the telephone company usually passes through a protector which is for lightning protection. It then terminates at the interface box. An RJ-11 jack and plug are provided to connect to the rest of the wiring. This gives the telephone company a way to disconnect the incoming line from the rest of the house wiring to making testing and troubleshooting easier.

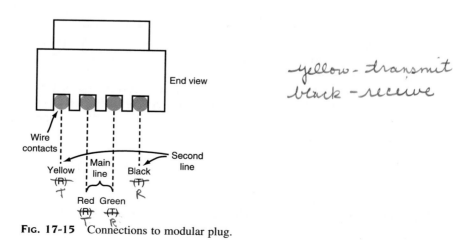

FIG. 17-15 Connections to modular plug.

All the wiring is made by way of screw terminals. For a single-line house, the green and red tip and ring connections terminate at the terminals, and all wiring to the room wall jacks is connected in parallel at these terminals.

If a second line is installed, the black and yellow wires, which are the tip and ring connections, are also terminated at screw terminals. They are then connected to the inside house wiring.

Connections on the RJ-11 connector are shown in Fig. 17-15 on the previous page. The red and green wires terminate at the two center connections while the black and yellow wires terminate at the two outside connections. Most telephone wire and RJ-11 connectors have four wires and connections, but there are exceptions. With four wires a two-line phone can be accommodated.

CORDLESS TELEPHONES

Virtually all offices and most homes now have two or more telephones, most homes and apartments have a standard telephone jack in every room. This permits a single phone to be moved easily from one place to another, and it permits multiple (extension) phones. However, the ultimate convenience is a cordless telephone, which uses two-way radio transmission and provides total portability. Today, many homes have a cordless unit.

CORDLESS TELEPHONE CONCEPTS. A cordless telephone is a full-duplex, two-way radio system made up of two units, the portable unit or handset and the base unit. The base unit is wired to the telephone line by way of a modular connector. It receives its power from the AC line. The base unit is a complete transceiver in that it contains a transmitter that sends the received audio signal to the portable unit and receives signals transmitted by the portable unit and retransmits them on the telephone line. It also contains a battery charger that rejuvenates the battery in the handheld unit.

The portable unit is also a full transceiver. It is battery powered and fully portable. This unit is designed to rest in the base unit where its battery can be recharged. Both units have an antenna.

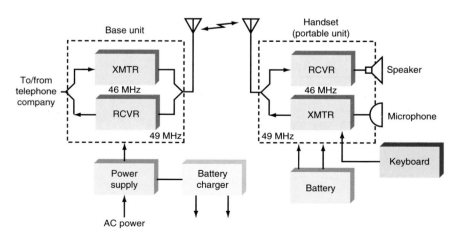

FIG. 17-16 Cordless telephone system.

The transceivers in both the portable and the base units use full-duplex operation. To achieve this, the transmitter and receiver must operate on different frequencies.

Figure 17-16 shows simplified block diagrams of the base and portable units of a typical cordless telephone. The base unit transmitter operates in the 46-MHz range and receives in the 49-MHz range. The portable unit has a receiver that operates in the 46-MHz range and a transmitter that operates in the 49-MHz range. These frequencies are far enough apart so that full simultaneous send and receive operations are possible at all times.

DID YOU KNOW?

The base unit and receiver unit in cordless telephones each contain a separate full-duplex, complete transceiver.

FREQUENCY ALLOCATIONS. The FCC has allocated 10 duplex channels for cordless telephone operation (see Fig. 17-17). Frequency modulation is used.

The earliest available cordless telephones used the 49-MHz transmitting frequencies but used frequencies in the 1.6- to 1.8-MHz band for the second channel. The base unit transmitter operated in this frequency range using amplitude modulation. The base station coupled the RF energy in this frequency range to the AC power line, which was used as the antenna. This system was excessively noisy. Because of the low power used, the signals were weak and a considerable amount of noise was picked up from nearby electrical equipment and strong AM broadcast stations operating in the area. Newer telephone units use the 46-MHz band for the base station transmit and portable unit receive frequencies.

The transmitters in both the base station and the receiver are required to operate on the specific frequencies indicated in Fig. 17-17. Crystal control is used to set the transmit and receive frequencies. Most cordless telephones operate on only one of the

CHANNEL	BASE TRANSMITTER (MHz)	HANDSET TRANSMITTER (MHz)
1	46.610	49.670
2	46.630	49.845
3	46.670	49.860
4	46.710	49.770
5	46.730	49.875
6	46.770	49.830
7	46.830	49.890
8	46.870	49.930
9	46.930	49.990
10	46.970	49.970

FIG. 17-17 FCC–assigned channels for cordless telephones.

channels indicated. More sophisticated cordless telephones may have two channels available so that the operating frequency can be changed if interference is occurring on one channel.

The FCC also regulates the transmitter power. These must meet rigid specifications regarding the maximum power output which cannot exceed 500 mW. This requirement ensures that the transmitting distance is limited so that nearby cordless telephones will not interfere with one another. The maximum usable operating range is usually up to approximately 100 ft between the base station and portable unit. Operation over distances as great as 1000 ft have been achieved with some cordless telephone, although not typically. The maximum distance depends considerably upon the environment in which the telephone is used. The transmitting distance is extremely limited when the units are used in a concrete and steel building. Operation in a typical home is reliable at distances of about 100 ft. In most cases it is difficult to predict the actual operating range, and performance varies from one part of the house to another.

Despite the relatively limited distance over which a cordless telephone operates, there is still potential for interference with other nearby telephones. Under some conditions, you could receive signals from a cordless telephone at a neighbor's home, or your transmitter could interfere with the telephone usage of a neighbor. This problem is avoided in most modern telephones by the use of specific audio signaling tones that enable the transmitter and receiver circuitry in the telephone. If the correct tones are not transmitted and received between the portable and base units, the telephone will not operate. As a result, a cordless telephone will typically not interfere with a nearby unit, nor will the phone of a neighbor interfere with your own. Such tone signaling provides the privacy and security that you ordinarily expect from standard telephones.

CORDLESS TELEPHONE SECURITY. Figure 17-18 shows how the tone signaling system operates in a cordless telephone to provide privacy and avoid interference from nearby cordless telephones. Simplified block diagrams of the base unit and the portable unit are shown.

Note that the base unit is connected directly to the local loop line. The ring detector circuit recognizes the standard 20-Hz ringing sine wave from the central office. It converts this into a signal that operates a ring signal generator. The ring signal generator is an oscillator that develops an audio tone in the 700- to 1500-Hz range. It is sent to the 46-MHz transmitter where it frequency modulates and transmits the tone to the portable unit.

The ring signal is received on the portable unit's 46-MHz receiver. The receiver demodulates the signal to recover the ring tone. A ring signal detector circuit in the portable unit is designed to recognize the tone generated by the base station. The ring signal detector is usually just a simple bandpass filter set to the correct frequency. If passed by the filter, the ring signal is sent to an audio amplifier where it operates a speaker to signal the user that an incoming call is being received.

To respond to a ringing signal, the user picks up and turns on the unit, usually with a pushbutton switch. This disconnects the ringing signal from the speaker. The internal 49-MHz transmitter is also turned on. A pilot signal generator in the portable unit generates a sine wave between 4 and 7 kHz. The pilot signal is well above the normal voice frequency range of 300 to 3 kHz and, therefore, will not interfere with it. Both the pilot tone and the voice signal are transmitted to the 49-MHz receiver in the base station.

Once the connection to the local loop is made, the dial tone from the central office passes through the relay to the hybrid transformer. The dial tone is then used to

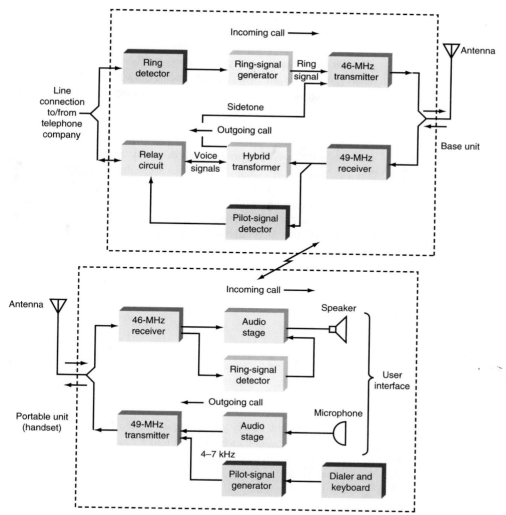

FIG. 17-18 Block diagram of cordless telephone showing calling and security circuits.

modulate the 46-MHz transmitter. This is picked up by the receiver in the portable unit and demodulated so that the dial tone is heard in the speaker.

Dialing now takes place. The portable unit has a built-in DTMF TouchTone dialing unit. These tones modulate the 49-MHz transmitter. The dialing signals are picked up by the base unit receiver and demodulated and fed to the hybrid. The hybrid places the dialing tones on the local loop. Once the called station answers, a full-duplex link has been established and normal telephone operation takes place.

ADVANCED CORDLESS TELEPHONES. Although cordless telephones work reliably, their range is limited, and signal quality is not quite up to standard telephone levels. Cordless telephones, particularly the portable units, are highly susceptible to all types of electrical noise and radio interference that may occur nearby and easily degrade the signal.

Because of this, a new higher-quality cordless telephone has been developed. It uses the *personal communications services (PCS) frequencies* of 902 to 928 MHz.

This is in the approximate same range as cellular telephones. This frequency band has been designated for a variety of wireless communications services including cordless telephones.

Cordless telephones and other wireless devices using this frequency range do not require an FCC license. Yet, high-power output of up to 1 W can be used. This significant increase in power over standard cordless phones permits greater distances to be covered more reliably. And because of the digital transmission of voice, the clarity is superior to that of ordinary analog telephones.

The 902- to 928-MHz cordless telephones also use spread spectrum modulation. In other words, the telephones are fully digital. Voice is digitized into a serial bit stream and used to modulate the carrier using standard BPSK techniques. By using spread spectrum techniques, either frequency hopping or direct sequence, many cordless telephones can easily share the same frequency band with little or no interference.

17-2 THE TELEPHONE SYSTEM

Most of us take telephone service for granted, as we do other so-called utilities, electric power for example. In the United States telephone service is excellent. But this is certainly not the case in many other countries in the world.

When we refer to the "telephone system," we are talking about the organizations and facilities involved in connecting your telephone to the called telephone regardless of where it might be in United States or anywhere else in the world. A number of different companies are involved in long-distance calls, although a single company is usually responsible for local calls in a given area. These companies make up the telephone system, and they design, build, maintain, and operate all the facilities and equipment used in providing universal telephone service. A vast array of equipment and technology is employed. Practically every conceivable type of electronic technology is used to implement worldwide telephone service.

The telephone, a small but relatively complex entity, is nothing compared to the massive system that backs it up. The telephone system can connect any two telephones in the world, and most people can only speculate on the method by which this connection takes place. It takes place on many levels and involves an incredible array of systems and technology. Obviously, it is difficult to describe such a massive system here. However, in this brief section, we will attempt to describe the technical complexities of interconnecting telephones, the central office and the subscriber line interface which connects each user to the telephone system, the hierarchy of interconnections within the telephone system, and the major elements and general operation of the telephone system. Long-distance operation and special telephone interconnection systems such as the PBX are also discussed.

SUBSCRIBER INTERFACE

Most telephones are connected to a local central office by way of the two-line, twisted-pair local loop cable. The central office contains all the equipment that operates the

The staff of this telephone network operational control center must be well trained in electronics and computer technology to operate a complex communication system. The network staff are currently routing long-distance telephone calls.

telephone and connects it to the "telephone system" that makes the connection to any other telephone.

Each telephone connected to the central office is provided with a group of basic circuits that power the telephone and provide all the basic functions such as ringing, dial tone, and dialing supervision. These circuits are collectively referred to as the *subscriber interface* or the *subscriber line interface circuits (SLIC)*. In older central office systems, the subscriber interface circuits used discrete components. Today, most functions of the subscriber line interface are implemented by one or perhaps two integrated circuits plus supporting equipment. The subscriber line interface is also referred to as the *line side interface*.

The SLIC provides seven basic functions generally referred to as *BORSCHT* (representing the first letters of the functions *battery, overvoltage protection, ringing, supervision, coding, hybrid,* and *test*). A general block diagram of the subscriber interface and BORSCHT functions is given in Fig. 17-19.

BATTERY. The subscriber line interface at the central office must provide a DC voltage to the subscriber to operate the telephone. In the United States, this is typically −48 Vdc with respect to ground. The actual voltage can be anything between approximately −20 and −80 V when the phone is on the hook, that is, disconnected. The voltage at the telephone drops to approximately 6 V when the phone is taken off the

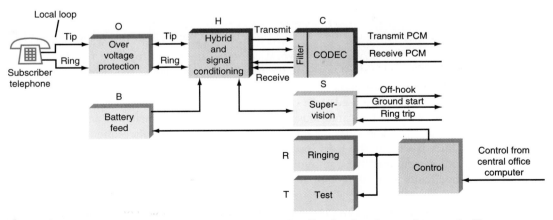

FIG. 17-19 BORSCHT functions in the subscriber line interface at the central office.

hook. The large differences between the on-hook and off-hook voltages has to do with the large voltage drop that occurs across the components in the telephone and the long local loop cable.

OVERVOLTAGE PROTECTION. The circuits and components that protect the subscriber line interface circuits from electrical damage are referred to collectively as *overvoltage protection*. The phone lines are vulnerable to many types of electrical problems. Lightning is by far the worst threat, although other hazards exist, including accidental connection to an electrical power line or some type of misconnection that would occur during installation. Induced disturbances from other sources of noise can also cause problems. Overvoltage protection ensures reliable telephone operation even under such conditions.

RINGING. When a specific telephone is receiving a call, the telephone local office must provide a ringing signal. As indicated earlier, this is commonly a 90-V rms AC signal at approximately 20 Hz. The SLIC must connect the ringing signal to the local loop when a call is received. This is usually done by closing relay contacts that connect the ringing signal to the line. The SLIC must also detect when the phone is picked up (off hook) so that the ringing signal can be disconnected.

SUPERVISION. *Supervision* refers to a group of functions within the subscriber line interface that monitor local loop conditions and provide various services. For example, the supervision circuits in the SLIC detect when a telephone is picked up to initiate a new call. A sensing circuit recognizes the off-hook condition and signals circuits within the SLIC to connect a dial tone. The caller then dials the desired number, which causes interconnection through the telephone system.

The supervision circuits continuously monitor the line during the telephone call. The circuits sense when the call is terminated and provide the connection of a busy signal if the called number is not available.

CODING. Coding is another name for analog-to-digital conversion and digital-to-analog conversion. Today, many telephone transmissions are made by way of serial digital data methods. The SLIC may contain codec that converts the analog voice signals into serial PCM format or convert received digital calls back into analog signals to be placed on the local loop. Transmission over trunk lines to other central offices or toll offices or for use in long-distance transmission is typically by digital PCM signals in modern systems.

HYBRID. Recall that in the telephone, a hybrid circuit (also known as a *two-wire* to *four-wire circuit*), usually a transformer, provides simultaneous two-way conversations on a single pair of wires. The hybrid combines the signal from the telephone transmitter with the received signal to the receiver on the single twisted pair. It keeps the signals separate within the telephone.

A hybrid is also used at the central office. It effectively translates the two-wire line to the subscriber back into four lines, two each for the transmitted and received signals. The hybrid provides separate transmit and receive signals. Although a single pair of lines is used in the local loop to the subscriber, all other connections to the telephone system treat the transmitted and received signals separately and have independent circuits for dealing with them along the way.

TEST. In order to check the status and quality of subscriber lines, the phone company often puts special test tones on the local loop and receives resulting tones in return. These can give information about the overall performance of the local loop. The SLIC provides a way to connect the test signals to the local loop and to receive the resulting signals for measurement.

The basic BORSCHT functions are usually divided into two groups, high voltage and low voltage. The high-voltage parts of the system are the battery feed, the overvoltage protection, the ringing circuits, and the test circuits. The low-voltage group includes the supervision, coding, and hybrid functions. In older systems, all the functions were implemented with discrete component circuits. Today, these functions are generally divided between two ICs, one for the high-voltage functions and the other for the low-voltage functions.

THE TELEPHONE HIERARCHY

Whenever you make a telephone call, your voice is connected through your local exchange to the telephone system. From there it passes through at least one other local exchange, which is connected to the telephone you are calling. Several other facilities may provide switching, multiplexing, and other services required to transmit your voice. The organization of this hierarchy in the United States is discussed in the next sections.

CENTRAL OFFICE. The central office or local exchange is the facility to which your telephone is directly connected by a twisted-pair cable. Also known as an *end office (EO)*, the local exchange can serve up to 10,000 subscribers, each of whom are identified by a four-digit number from 0000 through 9999 (the last four digits of the telephone number).

The local exchange also has an exchange number. These are the three additional digits that make up a telephone number. Obviously, there can be as many as 1000 exchanges with numbers from 000 through 999. These exchanges become part of an area code region, which is defined by an additional three-digit number. Each area code is fully contained within one of the geographical areas assigned to one of the seven Bell regional operating companies.

LONG-DISTANCE OPERATION. As indicated earlier, the United States is divided into seven telephone service regions. Service within each region is provided by one of the Regional Bell Operating Companies (RBOCs) that belong to one of the RBHCs. Each of the RBHCs are typically divided into smaller operating companies serving one portion of the designated area. In all, there are 22 Bell operating companies that make up the seven large RBHC corporations.

In addition to the Bell operating companies, there are many other independent telephone companies. One of the largest is General Telephone and Electronics (GTE), which provides telephone service primarily in limited geographical areas around the country where the Bell operating companies do not serve. GTE and other smaller companies often provide telephone service in remote rural areas. All these Bell operating companies and any independent companies are referred to as *local exchange carriers* or *local exchange companies (LECs)*.

The LECs provide telephone services to designated geographical areas referred to as *local access and transport areas (LATAs)*. The United States is divided into approximately 200 LATAs. The LATAs are defined within individual states making up the seven operating regions. The LECs provide the telephone service for the LATAs within their regions but do not provide long-distance service for the LATAs.

Long-distance service is provided by long-distance carriers known as *interexchange carriers (IXCs)*. The IXCs are the familiar long-distance carriers such as AT&T, MCI, and US Sprint. Long-distance carriers must be used for the interconnection for any inter-LATA connections. The LECs can provide telephone service within the LATAs that are part of their operating region, but links between LATAs within a region, even though they may be directly adjacent to one another, must be made through an IXC.

Each LATA contains a *serving*, or *point of presence (POP), office* that is used to provide the interconnections to the IXCs. The local exchanges communicate with one another via individual trunks. And all local exchanges connect to an LEC central office, which provides trunks to the POP. It is at the POP where the long-distance carriers can make their interface connections. The POPs must provide equal access for any long-distance carrier desiring to connect. Many POPs are connected to multiple IXCs, but in many areas, only one IXC serves a POP.

Figure 17-20 summarizes the hierarchy just discussed. Individual telephones within a LATA connect to the local exchange or central office by way of the two-wire local loop. The central offices within an LATA are connected to one another by trunks. These trunks may be standard baseband twisted-pair cables run underground or on telephone poles but they may also be coaxial cable, fiber-optic cable, or microwave radio links. In some areas, two or more central offices are located in the same building or physical facility. Trunk interconnections are usually made by cables.

The local exchanges are also connected to an LEC central office when a connection cannot be made between two local exchanges that are not directly trunked. The call passes from the local exchange to the LEC central office, where the connection is made to the other local exchange.

The LEC central office is also connected to the POP. Depending upon the organization of the LEC within the LATA, the LEC central office may contain the POP.

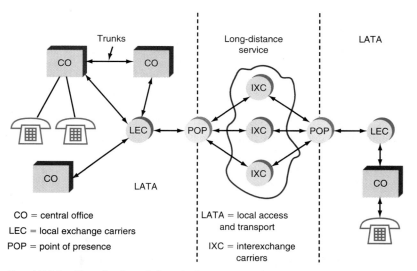

CO = central office
LEC = local exchange carriers
POP = point of presence

LATA = local access and transport
IXC = interexchange carriers

FIG. 17-20 Organization of the telephone system in the United States.

Note in Fig. 17-20 that the POP provides the connections to the long-distance carriers, or IXCs. The cloud represents the long-distance networks of the IXCs. The long-distance network connects to the remote POPs, which in turn are connected to other central offices and local exchanges.

Today, AT&T has implemented what it calls *dynamic nonhierarchical routing (DNHR)*. In most cases, the types of switching offices and trunk paths available make it possible to establish a long-distance connection with two or fewer switching centers.

Most other long-distance carriers have their own specific hierarchical arrangements. A variety of switching offices across the country are linked by trunks using fiber-optic cable or microwave relay links. Multiplexing techniques are used throughout to provide many simultaneous paths for telephone calls.

In all cases, the various central offices and routing centers provide switching services. The whole idea is to permit any one telephone to directly connect with any other specific telephone. The purpose of all the different levels in the telephone system hierarchy is to provide the interconnecting trunk lines as well as switching equipment that makes the desired interconnection.

PRIVATE TELEPHONE SYSTEM

Telephone service provided to companies or large organizations with many employees and many telephones is considerably different from basic local loop service provided for individuals. Depending upon the size of the organization, there may be dozens, hundreds, or even thousands of telephones required. It is simply not economical to provide each telephone in the organization its own separate local loop connection to the central office. It is also inefficient use of expensive facilities to use a remote central office for intercompany communications. For example, an individual in one office often needs to make an intercompany call to a person in another office, which may be only a few doors down the hall or a couple of floors away. Making this connection through the local exchange is wasteful.

This problem is solved by the use of private telephone systems within a company or organization. Private telephone systems implement telephone service amongst the telephones in the organization and provide one or more local loop connections to the central office. The two basic types of private telephone systems are known as *key systems* and *private branch exchanges*.

KEY SYSTEMS. *Key systems* are small telephone systems designed to serve from 2 to 50 user telephones within an organization. Commercially available systems usually have provisions for 6, 10, 12, or 50 telephones.

Simple key telephone systems are made up of the individual telephone units generally referred to as *stations,* all of which are connected to a central answering station. The central answering station is connected to one or more local loop lines known as *trunks* back to the local exchange. Most systems also contain a central electronic switching unit that makes all of the internal and external connections.

The telephone sets in a key system typically have a group of pushbuttons that allow each telephone to select two or more outgoing trunking lines. Phone calls are made in the usual way.

PRIVATE BRANCH EXCHANGE. A *private branch exchange,* or *PBX,* as it is known, is a private telephone system for larger organizations. Most PBXs are set up to handle

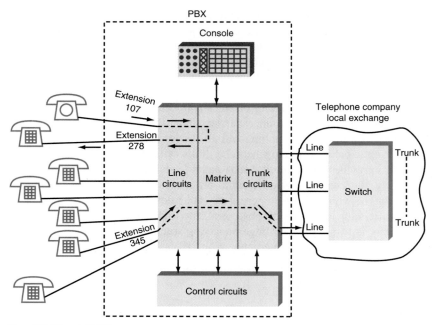

FIG. 17-21 A PBX.

50 or more telephone interconnections. They can handle thousands of individual telephones within an organization. Private branch exchanges are made by three primary companies: AT&T, Northern Telcom, and Rolm. These systems may also be referred to as *private automatic branch exchanges (PABXs)* or *computer branch exchanges (CPXs)*. Of the three terms, the expression *PBX* is the most widely known and used.

A PBX (see Fig. 17-21) is, in effect, a miniature complete telephone system. It provides baseband interconnections to all the telephones in an organization. All the telephones connect to a central switching system which makes intercompany connections as well as external connections to multiple trunk lines to the central office.

Like the key system, the PBX offers the advantage of efficiency and cost reduction when many telephones are required. Interoffice calls can be completed by the PBX system without accessing the local exchange. Further, it is more economical to limit the number of trunk lines to the central office, for all telephones in the organization will not be attempting to access an outside line at one time.

The modern PBX is usually fully automated by computer control. Although no operator is required, most large organizations have one or more operators who answer incoming telephone calls and route them appropriately with a control console. However, some PBXs are automated so that the individual user's telephone whose extension is the last four digits of the telephone number can be called directly from outside.

As you can see from Fig. 17-21, the PBX is made up of line circuits that are similar to the subscriber line interface circuits discussed earlier. The matrix is the electronic switch that connects any phone to any other phone in the system. It also permits conference calls. The trunk circuits interface to the local loop lines to the central office. All the circuits are under the control of a central computer dedicated to the operation of the PBX.

An alternative to the PBX is known as *Centrex*. This service, normally provided by the local telephone company, performs the function of a PBX but uses special equipment, and most of the switching is carried out by the local exchange switching equipment over special trunk lines. Its advantage over a standard PBX is that the high initial cost of PBX equipment can be avoided by leasing the Centrex equipment from the telephone company.

17-3 FACSIMILE

Facsimile, or *fax,* is an electronic system for transmitting graphical information by wire or radio. Facsimile is used to send printed material by scanning it and converting it into electronic signals that modulate a carrier to be transmitted over the telephone lines. Since modulation is involved, fax transmission can also take place by radio. With facsimile, documents such as letters, photographs, line drawings, or any printed information can be converted into an electrical signal and transmitted with conventional communications techniques. The components of a fax system are illustrated in Fig. 17-22.

Although facsimile is used to transmit pictures, it is not TV because it does not transmit sound messages or live scenes and motion. However, it does use scanning techniques that are in some way generally similar to those used in TV. A scanning process is used to break a printed document up into many horizontal scan lines which can be transmitted and reproduced serially.

HISTORY OF FACSIMILE

Like most other communications technologies, facsimile is not new. It was invented in 1842 by a Scottish inventor named Alexander Bain. Later, the original process was improved in England in the 1850s by Fredrick Bakewell. A German scientist, Arthur Korn, incorporated photoelectric scanning into the fax process in 1907, which made fax truly practical. In the 1920s, fax development was expedited, and commercial products were developed. The first radio fax transmissions were made in the 1930s.

The earliest users of fax were the newspapers and wire service companies which transmitted news information and photos by fax. It was also used by telegram companies, the government and military, and in many cases, commercial business organizations. Today fax is widely used to transmit weather photos from satellites to ground stations.

The fax machine has long been recognized as a major communications convenience. But its cost remained high until the 1970s, when major breakthroughs were

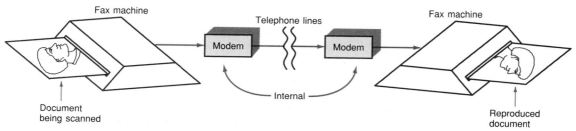

Fig. 17-22 Components of a facsimile system.

made in the development of semiconductor devices such as photosensors and micro-processors. The development of data compression techniques further improved transmission times. Today, the fax machine is just about as common in offices as the telephone and the personal computer.

HOW FACSIMILE WORKS

The early facsimile machines were electromechanical devices consisting of a rotary scanning drum and some form of printing mechanism. The typical facsimile machine was capable of both sending and receiving. Figure 17-23 shows the basic scanning mechanism. The document to be transmitted was wrapped around and tightly affixed to the drum. A motor rotated the drum through a gear train mechanism. The motor was designed to operate at an accurate speed so that the scanning rate was precise. A precision tuning fork oscillator was used to control the scanning drum motor for precise speed value.

Scanning of the document was done with a light and photocell arrangement. A lead screw driven by the gear train moved a scanning head consisting of a light source and a photocell. An incandescent light source, focused to a tiny point with a lens system, was used to scan the document. The lens was also used to focus the reflected light from on the document onto the photocell. As the light scanned the letters and numbers in a typed or printed document or the gray scale in a photograph, the photocell produced a varying electronic signal whose output amplitude was proportional to the amount of reflected light. This baseband signal was then used to amplitude or frequency modulate a carrier in the audio frequency range. This permitted the signal to be transmitted over the telephone lines.

Figure 17-24 shows how a printed letter might have been scanned. Assume the letter F is black on a white background. The output of a photodetector as it scans across line a is shown at A. The output voltage is high for white and low for black. The output of the photodetector is also shown for scan lines b and c. The output of the photodetector is used to modulate a carrier, and the resulting signal is put on the telephone line.

The resolution of the transmission is determined by the number of scan lines per vertical inch. The greater the number of lines scanned, the finer the detail transmitted and the higher the quality of reproduction. Older systems had a resolution of 96 lines per inch (LPI), and the new systems use 200 LPI.

On the receiving end, a demodulator recovered the original signal information, which was then applied to a stylus. The purpose of the stylus was to redraw the original information on a blank sheet of paper. A typical stylus converted the electrical signal into heat variations that burned the image into heat-sensitive paper. Other types of printing mechanisms were used.

In the printing process, a blank sheet of paper was wrapped around a cylindrical drum and rotated at the same speed as the transmitting drum. The drum rotation was precisely controlled by a precision tuning fork oscillator. The drum rotations in both the transmitter and receiver were synchronized so that the information would be reproduced simultaneously with transmission. Because of the frequency spectrum limitations of the telephone lines and radio channels, scanning speed was slow

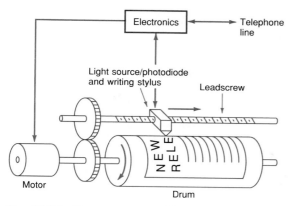

FIG. 17-23 Drum scanner used in early fax machines.

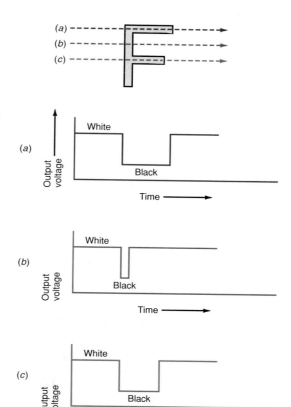

FIG. 17-24 Output of a photosensitive detector during different scans.

to ensure the production signals that were within the available bandwidth. Transmission times of many minutes per page were typical.

Today's modern fax machine is a high-tech electro-optical machine. Scanning is done electronically, and the scanned signal is converted into a binary signal. Then digital transmission with standard modem techniques is used.

Figure 17-25 is a block diagram of a modern fax machine. The transmission process begins with an image scanner that converts the document into hundreds of horizontal scan lines. Many different techniques are used, but they all incorporate a photo- (light-) sensitive device to convert light variations along one scanned line into an electrical voltage. The resulting signal is then processed in various ways to make the data smaller and thus faster to transmit. The resulting signal is sent to a modem where it modulates a carrier set to the middle of the telephone voice spectrum bandwidth. The signal is then transmitted to the receiving fax machine over the public-switched telephone network.

The receiving fax machine's modem demodulates the signal that is then processed to recover the original data. The data is decompressed and then sent to a printer, which reproduces the document. Since all fax machines can transmit as well as receive, they are referred to as *transceivers*. The transmission is half duplex because only one machine may transmit or receive at a time.

Most fax machines have a built-in telephone, and the printer can also be used as a copy machine. An embedded microcomputer handles all control and operation including paper handling.

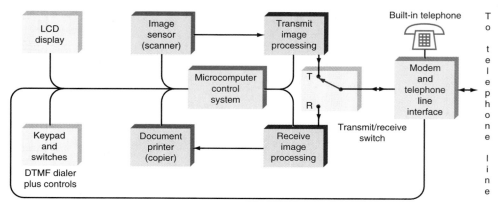

Fig. 17-25 Block diagram of modem fax machine.

IMAGE PROCESSING

Most fax machines use *charged coupled devices (CCDs)* for scanning. A CCD is a light-sensitive semiconductor device that converts varying light amplitudes into an electrical signal. The typical CCD device is made up of many tiny reverse-biased diodes that act like capacitors which are manufactured in a matrix on a silicon chip (see Fig. 17-26). The base forms one large plate of a capacitor which is electrically separated by a dielectric from many thousands of tiny capacitor plates as shown. When the CCD is exposed to light, the CCD capacitors charge to a value proportional to the light intensity. The capacitors are then scanned or sampled electronically to determine their charge. This creates an analog output signal that accurately depicts the image focused on the CCD.

A CCD is actually a device that breaks up any scene or picture into *individual picture elements,* or *pixels.* The greater the number of CCD capacitors, or pixels, the higher the resolution and the more faithfully a scene, photograph, or document may be reproduced. CCD devices are available with a matrix of many thousands of pixels, thereby permitting very high resolution picture transmission. CCDs are widely used in modern video cameras in place of the more delicate and more expensive vidicon tubes. In the video camera (camcorder), the lens focuses the entire scene on a CCD matrix. This same approach is used in some fax machines. In one type of fax machine, the document to be transmitted is placed face down as it might be in a copy machine. The document is then illuminated with brilliant light from a xenon or fluorescent bulb. A lens system focuses the reflected light on a CCD. The CCD is then scanned, and the resulting output is an analog signal whose amplitude is proportional to the amplitude of the reflected light.

In most desktop fax machines, the entire document is not focused on a single CCD. Instead, only a narrow portion of the document is lighted and examined as it is moved through the fax machine with rollers. A complex system of mirrors is used to focus the lighted area on the CCD (see Fig. 17-27).

The more modern fax machines use another type of scanning mechanism that does not use lenses. The scanning mechanism is an assembly made up of an LED array and a CCD array. These are arranged so that the entire width of a standard 8½ × 11 in page is scanned simultaneously one line at a time. The LED array illuminates a narrow portion of the document. The reflected light is picked up by the CCD scanner. A typical scanner has 2048 light sensors forming one scan line. Figure 17-28 shows a

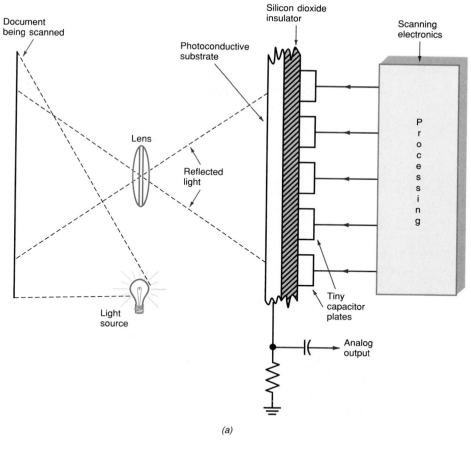

(a)

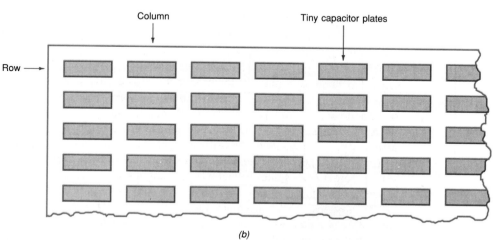

(b)

FIG. 17-26 A charge coupled device (CCD) is used to scan documents in modern fax machines. (a) Cross section. (b) Detail of capacitor matrix.

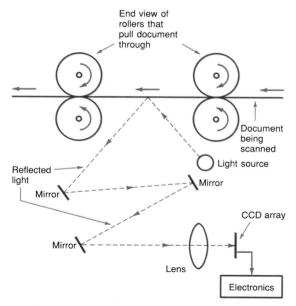

FIG. 17-27 Scanning mechanism in a fax machine.

side view of the scanning mechanism. The 2048 pixels of light are converted into voltages proportional to the light variations on one scanned line. These voltages are converted from a parallel format to a serial voltage signal. The resulting analog signal is amplified and sent to an AGC circuit and an S/H amplifier. The signal is then sent to an analog-to-digital converter where the light signals are translated into binary data words for transmission.

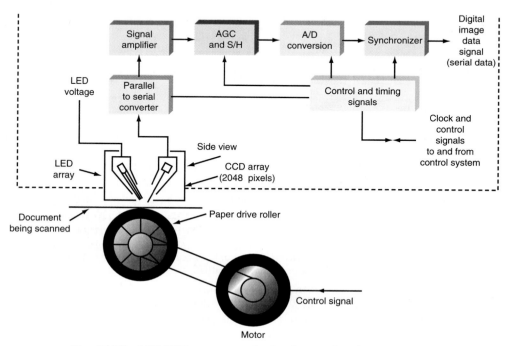

FIG. 17-28 LED/CCD scanner mechanism in a modern fax machine.

DATA COMPRESSION

An enormous amount of data is generated by scanning one page of a document. A typical $8\frac{1}{2} \times 11$ in page represents about 40,000 bytes of data. This can be shortened by a factor of 10 or more with data compression techniques. Furthermore, because of the narrow bandwidth of telephone lines, data rates are also limited. That is why it takes so long to transmit one page of data. Developments in high-speed modems have helped reduce the transmission time, but the most important developments are data compression techniques that reduce the overall amount of data, which significantly decreases the transmission time and telephone charges.

Data compression is a digital data processing technique that looks for redundancy in the transmitted signal. White space or continuous segments of the page that are the same shade produce continuous strings of data words that are the same. These can be eliminated and transmitted as a special digital code that is significantly faster to transmit. Other forms of data compression use various mathematical algorithms to reduce the amount of data to be transmitted.

HINTS AND HELPS

Because of the large amount of data generated through scanning, data compression techniques can be used to shorten facsimile transmission by a factor of 10 or more.

The data compression is carried out by a *digital signal processing (DSP) chip.* This DSP chip is a super high speed microprocessor with embedded ROM containing the compression program. The digital data from the A/D converter is passed through the DSP chip, from which comes a significantly shorter string of data that represents the scanned image. This is what is transmitted, and in far less time than the original data could be transmitted.

At the receiving end, the demodulated signal is decompressed. Again, this is done through a DSP chip especially programmed for this function. The original data signal is recovered and sent to the printer.

MODEMS

Every fax machine contains a built-in modem that is similar to a conventional data modem for computers. These modems are optimized for fax transmission and reception. And they follow international standards so that any fax machine can communicate with any other fax machine.

A number of different modulation schemes are used in fax systems. Analog fax systems use AM or FM. Digital fax uses PSK or QAM. To ensure compatibility between fax machines of different manufacturers, standards have been developed for speed, modulation methods, and resolution by the *International Telegraph and Telephone Consultative Committee,* better known by its French abbreviation *CCITT.* The CCITT is now known as the *ITU-T,* or *International Telecommunications Union.* The ITU-T fax standards are divided into four groups:

1. *Group 1 (G1 or GI):* Analog transmission using frequency modulation where white is 1300 Hz and black is 2100 Hz. Most North American equipment uses 1500 Hz for white and 2300 Hz for black. The scanning resolution is 96 lines per inch (LPI). Average transmission speed is 6 minutes per page ($8\frac{1}{2} \times 11$ in or A4 metric, size which is slightly longer than 11 in).

2. *Group 2 (G2 or GII):* Analog transmission using FM or vestigial sideband AM. The vestigial sideband AM uses a 2100-Hz carrier. The lower sideband plus part of the upper sideband are transmitted. Resolution is 96 LPI. Transmission speed is 3 min or less for an $8\frac{1}{2} \times 11$ in or A4 page.
3. *Group 3 (G3 or GIII):* Digital transmission using PCM black and white only or up to 32 shades of gray. PSK or QAM to achieve transmission speeds of up to 9600 baud. Resolution 200 LPI. Transmission speed of less than 1 minute per page with 15 to 30 s being typical.
4. *Group 4 (G4 or GIV):* Digital transmission, 56K b/s, resolution up to 400 LPI, and speed of transmission less than 5 s.

The older G1 and G2 machines are no longer used. The most common configuration is Group 3. Most G3 machines can also read the G2 format.

G4 machines are not yet widely used. They are designed to use digital transmission only with no modem over very wide band dedicated digital-grade telephone lines. G4 machines will become popular when the new integrated services digital network (ISDN) telephone system goes into use in the future. Both G3 and G4 formats also employ digital data compression methods that shorten the binary data stream considerably, thereby speeding up page transmission. This is important because shorter transmission times cut long-distance telephone charges and reduce operating costs.

FAX MACHINE OPERATION

Figure 17-29 is a simplified block diagram of the transmitting circuits in a modern G3 fax transceiver. The analog output from the CCD array is serialized and fed to an A/D converter which translates the continuously varying light intensity into a stream of binary numbers. Sixteen gray scale values between white and black are typical. The binary data is sent to a DSP digital data compression circuit as described earlier. The binary output in serial data format is used to modulate a carrier which is transmitted over the telephone lines. The techniques are similar to those employed in modems. Speeds of 2400/4800 and 7200/9600 baud are common. Most systems use some form of PSK or QAM to achieve very high data rates on voice-grade lines.

In the receiving portion of the fax machine, the received signal is demodulated and then sent to DSP circuits where the data compression is removed and the binary signals are restored to their original form. The signal is then applied to a printing mechanism. In most fax machines, thermal printers are used. Most thermal printers use a special heat-sensitive paper. The *printhead,* or *writing stylus* as it is sometimes called, contains tiny heating elements that are turned off and on by the received signal. A typical printhead has 2048 thick-film resistors that are heated individually to reproduce the transmitted pixels. The heat darkens the paper at the appropriate point to re-create the original image. Some newer thermal printers use a heating element printhead to melt ink on a special ribbon onto plain 20-lb bond paper.

In the high-priced machines, laser scanning of an electrosensitive drum, similar to the drum used in laser printers, produces output copies using the proven techniques of xerography.

The control logic in Fig. 17-29 is usually an embedded microcomputer. Besides all of the internal control functions it implements, it is used for "handshaking" between the two machines that will communicate. This ensures compatibility. Handshaking is usually carried out by exchanging different audio tones. The called machine responds with tones designating its capability. The calling machine compares this to its own

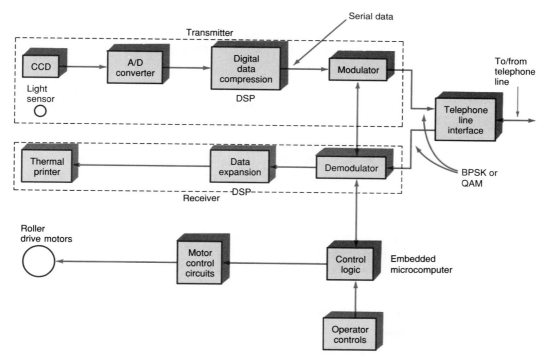

Fig. 17-29 Block diagram of a facsimile machine.

standards and then either initiates the transmission or terminates it because of incompatibility. If the transmission proceeds, the calling machine sends synchronizing signals to ensure that both machines start at the same time. The called machine acknowledges the receipt of the sync signal, and transmission begins. All the protocols for establishing communications and sending and receiving the data are standardized by the ITU-T. Transmission is half duplex.

As improvements have been made in picture resolution quality, transmission speed, and cost, facsimile machines have become much more popular. The units can be easily attached with standard RJ-11 modular connectors to any telephone system. In most business applications, the fax machine is typically dedicated to a single line. Most fax machines feature fully automatic operation with microprocessor-based control. A document can be sent to a fax machine automatically. The sending machine simply dials the receiving machine and initiates the transmission. The receiving machine answers the initial call and then reproduces the document before hanging up.

Most fax machines have a built-in telephone and are designed to share a single line with conventional voice transmission. The built-in telephone usually features TouchTone dialing and number memory plus automatic redial and other modern telephone features. Most fax machines also have automatic send and receive features for fully unattended operation. Smaller portable fax machines are available. Fax machines may also be used with standard cellular telephone systems in automobiles.

Another popular variation is the *fax modem,* an internal modem designed to be plugged into a personal computer. It allows standard modem operation for connection to online services and any desired remote computer usage. However, this device also contains the circuits for a fax modem but without the image scanning and printing.

A document produced with a word processor is usually stored in ASCII format as a file on disk. This file can be transmitted serially to the fax modem, which compresses it and sends it over the telephone lines. The receiving computer demodulates the signal, decompresses it, and stores it in RAM and then on disk. The resulting document can be read on the video monitor or printed by the usual means.

17-4 CELLULAR TELEPHONE SYSTEMS

A *cellular radio system* provides standard telephone operation by two-way radio at remote locations. Cellular radios or telephones are usually installed in cars or trucks but are also available in handheld models. Each cellular telephone permits the user to link up with the standard telephone system which permits calls to any part of the world.

The Bell Telephone Company division of AT&T developed the *cellular radio system* during the late 1970s and fully implemented it in the early 1980s. Today, cellular radio telephone service is available nationwide in most medium-size and large cities. It is known as the *advanced mobile phone service,* or *AMPS.*

CELLULAR CONCEPTS

The basic concept behind the cellular radio system is that rather than serving a given geographical area with a single transmitter and receiver, the system divides the service area into many smaller areas known as *cells,* as shown in Fig. 17-30. The typical cell covers only several square miles and contains its own receiver and low-power transmitter. The cell site is designed to reliably serve only vehicles in its small cell area.

Each cell is connected by telephone lines or a microwave radio relay link to a master control center known as the *mobile telephone switching office (MTSO).* The MTSO

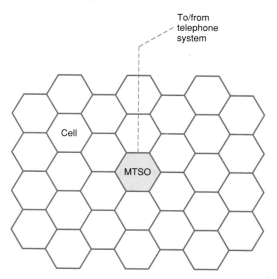

FIG. 17-30 The area served by a cellular telephone system is divided into small areas called *cells. Note:* Cells are shown as ideal hexagons, but in reality they will have circular or other geometric shapes. These areas may overlap, and the cells may be of different sizes.

controls all the cells and provides the interface between each cell and the main telephone office. As the vehicle containing the telephone passes through a cell, it is served by the cell transceiver. The telephone call is routed through the MTSO and to the standard telephone system. As the vehicle moves, the system automatically switches from one cell to the next. The receiver in each cell station continuously monitors the signal strength of the mobile unit. When the signal strength drops below a desired level, it automatically seeks a cell where the signal from the mobile unit is stronger. The computer at the MTSO causes the transmission from the vehicle to be switched from the weaker cell to the stronger cell. This is called a *handoff*. All of this takes place in a very short period of time and is completely unnoticeable to the user. The result is that optimum transmission and reception are obtained.

The cellular system operates in the 800- to 900-MHz range, previously reserved for the higher UHF TV channels 68 through 83, which were rarely used. Originally there were 666 30-kHz-wide full-duplex channels available for communications. Today, 832 channels are used. The cellular system also uses what is known as *frequency reuse,* which allows cells within the system to use the same frequency channel. Because the cells are physically small, low-power transmitters are used, and the cell sites use directional antennas, the signal does not stray beyond the cell boundaries. This allows other cells within the system to share the same frequency channel without interference. Frequency reuse tremendously increases the number of available channels. Obviously, to prevent interference, adjacent cells are not permitted to use the same channel.

Another feature of the cellular system is that different sizes of cells can be accommodated. In low-usage areas, the cells can be large. As the number of users increases, the large cells can be divided into smaller cells. This provides ease of expansion as the number of users grows. A typical system can serve up to about 50,000 subscribers, and 10,000 of them may use the system simultaneously.

A CELLULAR TELEPHONE UNIT

Figure 17-31 is a general block diagram of a cellular mobile radio unit. This applies to either the larger units designed to be mounted in a vehicle or the smaller handheld devices. The two types of units are identical in operation, but the larger units used in vehicles have a higher power transmitter and are inherently larger.

The unit consists of five major sections: transmitter, receiver, synthesizer, logic unit, and control unit. Mobile radios derive their operating power from the car battery. Portable units contain built-in rechargeable batteries. The transmitter and receiver share a single antenna. The sections are discussed below.

TRANSMITTER. The transmitter block diagram is shown in Fig. 17-32. It is a low-power FM unit operating in the frequency range of 825 to 845 MHz. There are 666 30-kHz transmit channels. Channel 1 is 825.03 MHz, channel 2 is 825.06 MHz, and so on up to channel 666 on 844.98 MHz. The carrier furnished by a frequency synthesizer is phase modulated by the voice signal. The phase modulator produces a deviation of ±12 kHz. Preemphasis is used to help minimize noise. The modulator output is translated up to the final transmitter frequency by a mixer whose second input

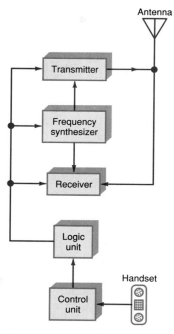

Antenna

Transmitter

Frequency
synthesizer

Receiver

Logic
unit

Handset

Control
unit

FIG. 17-31 General block diagram of a cellular radio.

also comes from the frequency synthesizer. The mixer output is fed to class C power amplifier stages where the output signal is developed. The final amplifier stage is designed to supply 3 W to the antenna of a vehicle-mounted unit but only about 500 MW in a handheld unit.

A unique feature of the higher-power transmitter is that its output power is controllable by the cell site and MTSO. Special control signals picked up by the receiver are sent to an *automatic power control (APC) circuit* that sets the transmitter to one of 8 power output levels. The APC circuit can introduce power attenuation in steps of 4 dB from 0 dB (3 W) to 28 dB (4.75 MW). This is done by controlling the supply voltage to one of the intermediate-power amplifier stages.

The output power of the transmitter is monitored internally by built-in circuits. A microstrip directional coupler taps off an accurate sample of the transmitter output power and rectifies it into a proportional DC signal. This signal is used in the APC cir-

Cellular telephones are now easily carried in a pocket or purse. This salesman is able to keep in constant contact with his clients and his office staff while making his daily sales calls.

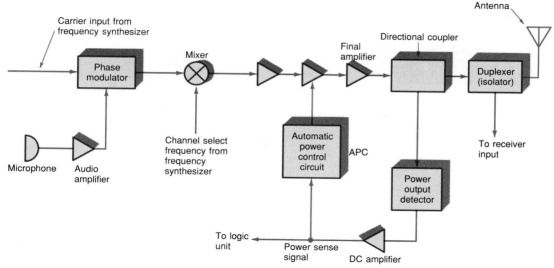

Fig. 17-32 Cellular transmitter.

cuit and is transmitted back to the cell site permitting the MTSO to know the present power level.

This automatic power control feature permits optimum cell site reception with minimal power. It also helps to minimize interference from other stations in the same or adjacent cells.

The transmitter output is fed to a duplexer circuit or isolator that allows the transmitter and receiver to share the same antenna. Since cellular telephone units use full-duplex operation, the transmitter and receiver will operate simultaneously. The transmit and receive frequencies are spaced 45 MHz apart to minimize interference. However, an isolator is still needed to keep transmitter power out of the sensitive receiver. The duplexer consists of two very sharp bandpass filters, one for the transmitter and one for the receiver. The transmitter output passes through this filter to the antenna.

RECEIVER.   The receiver is a dual-conversion superheterodyne (refer to Fig. 17-33). An RF amplifier boosts the level of the received cell site signal. The receiver frequency range is 870.03 to 889.98 MHz. There are 666 receive channels spaced 30 kHz apart. The first mixer translates the incoming signal down to a first IF of 82.2 MHz. Some receivers use a 45-MHz first IF. The local oscillator signal for the mixer is derived from the frequency synthesizer. The local oscillator frequency sets the receive channel. The signal passes through IF amplifiers and filters to the second mixer, which is driven by a crystal controlled local oscillator. The second IF is usually either 10.7 MHz or 455 kHz. The signal is then demodulated, de-emphasized, filtered, and amplified before being applied to the output speaker in the handset.

The output of the demodulator is also fed to other filter circuits that select out the control audio tones and digital control data stream sent by the cell site to set and control both the transmitter and the receiver. The demodulator output is also filtered into a DC level whose amplitude is proportional to the strength of the received signal. This is the *receive signal strength indicator (RSSI) signal* that is sent back to the cell site so that the MTSO can monitor the received signal from the cell and make decisions about switching to another cell.

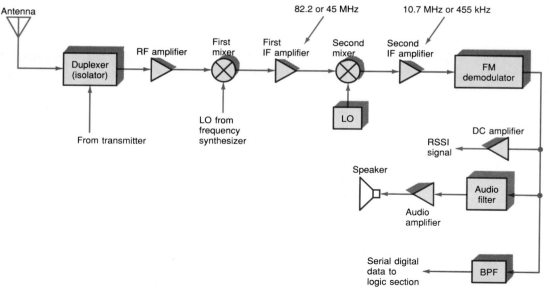

Fig. 17-33 Cellular receiver.

FREQUENCY SYNTHESIZER.   The frequency synthesizer section develops all the signals used by the transmitter and receiver (see Fig. 17-34). It uses standard PLL circuits and a mixer. A crystal controlled oscillator provides the reference for the PLLs. One PLL incorporates a VCO (number 2) whose output frequency is used as the local oscillator for the first mixer in the receiver. This signal is mixed with the output of a second PLL VCO to derive the transmitter output frequency.

As in other PLL circuits, the output VCO frequency is determined by the frequency division ratio of the divider in the feedback path between the VCO and the phase

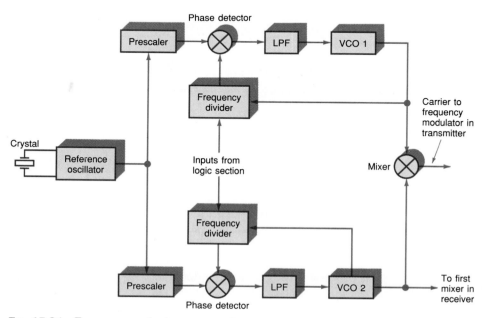

Fig. 17-34 Frequency synthesizer.

detector. In a cellular radio, this frequency division ratio is supplied by the MTSO via the cell site. When a mobile unit initiates or is to receive a call, the MTSO computer selects an unused channel. It then transmits a digitally coded signal to the receiver containing the frequency division ratios for the transmitter and receiver PLLs. This sets the transmit and receive channel frequencies.

LOGIC UNIT. The logic unit shown in Fig. 17-35 contains the master control circuitry for the cellular radio. It is made up of a microprocessor with both RAM and ROM plus additional circuitry used for interpreting signals from the MTSO/cell site and generating control signals for the transmitter and receiver.

All cellular radios contain a *programmable read-only memory (PROM) chip* called the *number assignment module (NAM)*. The NAM contains the *mobile identification number (MIN)*, which is the telephone number assigned to the unit. The NAM PROM is "burned" when the cellular radio is purchased and the MIN assigned. This chip allows the radio to identify itself when a call is initiated or when the radio is interrogated by the MTSO.

All cellular mobile radios are fully under the control of the MTSO through the cell site. The MTSO sends a serial digital data stream at 10K bps through the cell site to the radio to control the transmit and receive frequencies and transmitter power. The MTSO monitors the received cell signal strength at the cellular radio by way of the RSSI signal, and it monitors transmitter power level. These are transmitted back to the cell site and MTSO. Audio tones are also used for signaling purposes.

CONTROL UNIT. The control unit contains the handset with speaker and microphone. This may be a standard handset as used in a regular telephone on a mobile unit.

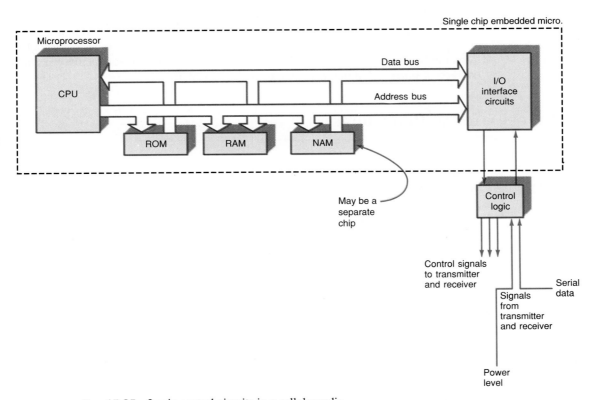

FIG. 17-35 Logic control circuits in a cellular radio.

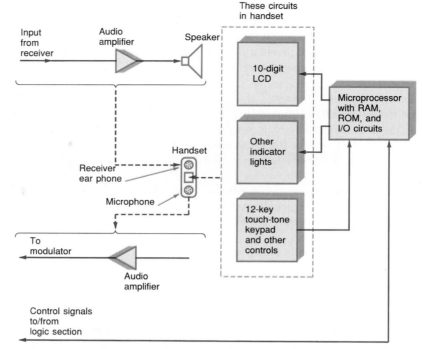

Fig. 17-36 Control unit with handset.

However, these circuits are built into the handheld units. The main control unit contains a complete TouchTone dialing circuit (see Fig. 17-36). The control unit is operated by a separate microprocessor that drives the LCD display and other indicators. It also implements all manual control functions. The microprocessor memory permits storage of often-called numbers and an auto-dial feature.

OPERATIONAL PROCEDURE

Described below is the sequence of operations that occur when a person initiates a cellular telephone call:

1. The operator applies power to the unit which turns on the transmitter and receiver.
2. The receiver seeks an open control channel. Twenty-one control or paging channels are used to establish initial contact with a cell and the MTSO. When contact is made, the cell site reads the NAM data and the MTSO computer verifies that it is a valid number.
3. The operator enters the number to be called via a keyboard.
4. The operator sends the telephone number to be called by pressing a send or call button.
5. The cell and MTSO search for an open channel and send the frequency data to a cellular transceiver.
6. The RSSI signal is read to determine the optimum cell selection. The transmitter power is adjusted.

7. Handshake signals are exchanged, signifying that contact has been established.
8. The MTSO calls the designated number. Conversation takes place.
9. If the mobile unit passes from one cell to another, the MTSO senses the RSSI signal and "hands off" the mobile signal from one cell to another to maintain maximum signal strength.
10. The call is terminated.

When a mobile unit is to receive a call, the MTSO and cell site transmit a call signal containing the MIN over a control channel. The transceiver monitors the control channels, identifies its MIN, and turns on. From that point on, the sequence beginning at step 5 above is similar.

Keep in mind that while most cellular telephones carry voice communications, like other telephones they can also carry data. Special modems are available that permit a personal computer to be connected to a cellular telephone for online communications. Even a fax machine can be connected to a cellular telephone.

FUTURE CELLULAR TELEPHONE SYSTEMS

The cellular telephone system has been far more successful than ever expected. When cellular service was first started during the mid-1980s, few expected it to be as commonly used as it is. Industry analysts predicted that fewer than 1 million cellular telephones would be in use by 1995. However, by 1995, that figure was exceeded by a factor of 10 or more.

Thanks to rapid technological developments, the cost and size of a cellular telephone have dropped considerably, from as high as $3000 per unit in the early days to less than $100 for a simple unit today. However, the cost of cellular service has remained high and in fact has increased because of the high cost of system expansion and maintenance. The resulting "sticker shock" has slowed growth.

Today, cellular telephone service is available in most areas of the United States and Europe. In addition, most systems are at full capacity. Although there is still room for some growth in most areas, the systems are reaching their limits of growth. This has mainly to do with the lack of available spectrum space for expansion.

New cellular systems are under development to solve the growth and expansion problem. The goals are to provide a wider range of cellular services to more consumers through the use of similar wireless technologies. These new systems are referred to as the *personal communications services (PCSs)* or *personal communications networks (PSNs)*. They will eventually replace cellular systems and provide not only wireless telephone service anywhere in the United States or Europe but also will permit easier connection of computers to remote online services, E-mail, and bulletin board services. Such systems are expected to replace the widespread paging services now in use. Paging systems are covered in the next section of this chapter.

The new cellular systems are expected to use the microwave bands. The FCC has set aside the 1.7- to 2.3-GHz band for new cellular telephone and other wireless services. All new systems will be digital. The current AMPS system

DID YOU KNOW?

Most cellular systems in the United States and Europe are at full capacity because of the lack of available spectrum space with the current technology.

Enhanced cellular telephone technology has lowered costs and contributed to the wider use of mobile telephones, not only for voice but also for fax and data from laptop computers.

is fully analog except for the 10K bps data channel used by the current system to establish communications and perform the control functions. The newer systems will digitize the voice signal and transmit it as a digital bit stream, which will modulate a carrier using BPSK or some variation thereof.

The advantage of an all-digital system is a significant improvement in noise rejection and reliability of communications. Further, digital techniques facilitate the use of multiplexing methods which will increase the number of telephone calls that can be handled within a given spectrum space.

17-5 PAGING SYSTEMS

Paging is a radio communications system designed to signal individuals wherever they may be. Paging systems operate in the simplex mode, for they broadcast signals or messages to individuals who carry small battery-operated receivers. Millions of people carry paging receivers. Typically they work in jobs that require maintaining constant communications with their employer and/or customers. The paging receiver operates continuously. To contact an individual with a pager, all you need to do is make a telephone call. A paging company will send a radio signal that will be received by the pager. The paging receiver usually has a built-in audible signaling device that informs the person that he or she is being paged. The signal may be as simple as an audio tone that indicates that the individual should call a telephone number to receive a message or otherwise make contact. Alternatively, the paging company can transmit a short printed message to the paging receiver. Some paging receivers have a small LCD screen on which a telephone number is displayed. This tells the paged individual which number to call. Some paging receivers have larger LCD screens that are capable of displaying several lines of alphanumeric text information.

PAGING SYSTEM OPERATION

Although the paging business is not owned by telephone companies, it is closely allied with the telephone business, because the telephone system provides the initial and final communications process. The most common paging process is described below (see Fig. 17-37).

To contact a person who has a pager, an individual dials the telephone number assigned to that person. The call is received at the office of the paging company. The paging company responds with one or more signaling tones that tell the caller to enter the telephone number which the paged person should call. Once the number is entered, the caller presses the pound sign key to signal the end of the telephone entry. The calling party then hangs up.

The paging system records the telephone number in a computer and translates this number into a serial binary-coded message. A unique protocol is used. The message is transmitted as a data bit stream to the paging receiver. The serial binary-coded message modulates the carrier of a radio transmitter. Paging systems usually operate in the VHF and UHF frequency ranges. A variety of bands have been assigned by the FCC specifically for paging purposes.

The paging company usually has a large antenna system mounted on a tower or the top of a tall building so that its communications range is considerable. Most paging systems can locate an individual within a 30-mi radius.

When the signal is transmitted, the paging receiver picks it up on its assigned frequency. The paged individual receives the message, as described above, and responds.

As indicated earlier, modern paging systems can transmit complete messages to an individual. These messages are usually entered on a personal computer and transmitted through a modem over the telephone system to the paging company. In this case the message is received, stored in a computer, and formulated into the correct protocol for transmission to the pager. The length of the message is typically restricted to several lines of text. A typical paging receiver may be able to display four lines of 20 characters each.

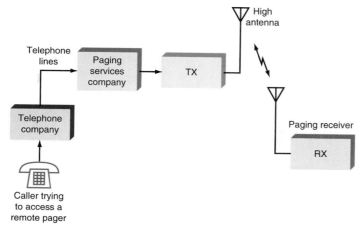

FIG. 17-37 The paging system.

The earliest paging systems used a tone signaling system. Each paging receiver is assigned a special code called a *cap code,* which is a sequence of numbers or a combination of letters and numbers. The code is broadcast over the paging region. If the pager is within the region, it will pick up and recognize its unique code. The cap code is encoded using audio tones. Early systems used a two-tone system. The typical tone sequence is shown in Fig. 17-38(*a*). A typical cap code might be C432. This is coded using several groups of audio tones.

The two-tone system limited the number of users; therefore, a five-tone code has been developed. It permits a system to accommodate up to about 100,000 users. The five-tone coding format is shown in Fig. 17-38(*b*). The cap code, which might be something like 2-53617, is encoded into a five-tone sequence. The transmission begins with a preamble that signals all receivers that a paging alert is coming. After a time gap, the five-tone sequence is transmitted.

The tones frequency-modulate the transmitter carrier in a fixed protocol or sequence. All paging receivers pick up every transmission, but they recognize only their own code. When a code is recognized, the beeper in the receiver goes off, informing the user that a call has been received.

Figure 17-39 shows a typical digital protocol. It begins with a preamble packet of bits that is the dotting sequence of clock pulses that help establish synchronization in the receiver. Motorola, the leading pager manufacturer, calls this sequence *dotting comma.* A unique preamble binary word is also transmitted multiple times. This transmission is followed by a sync packet that provides further dotting and a sequence of sync words.

Digital systems use FSK at a slow bit rate, usually in the 200- to 600-bps range. The transmission protocol starts with a bit stream of alternate 1s and 0s, called *dotting,* that helps the receiver synchronize itself to the transmitter. Once sync has been established, the cap code is transmitted. If it is recognized, the receiver reads the digitized number or message being transmitted to it and displays it on a small LCD.

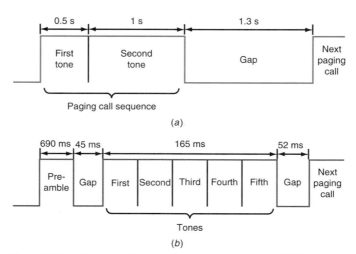

FIG. 17-38   Tone code formats for paging receivers. (*a*) Two-tone format. (*b*) A five-tone format.

Preamble	Sync. packet	Address packet	Display data	Data

Fig. 17-39   A digital protocol for pagers.

A Paging Receiver

A *paging receiver* is a small battery-powered superheterodyne receiver. Most pagers use a single-chip IC receiver similar to the one described in Chap. 8. Both single- and double-conversion models are available. Most basic paging systems use some form of frequency modulation although the more advanced digital message sending pagers use FSK or PSK.

A paging receiver is shown in Fig. 17-40. This is a simple single-conversion receiver. The local oscillator is crystal controlled and, therefore, sets the paging receiver to a specific frequency.

The output of the demodulator circuit is a series of tones that is sent to the tone decoder, which recognizes the receiver's unique code. If recognition occurs, the piezo-electric sounder is activated to provide the familiar beeping sound.

On the more advanced digital receivers, a built-in microcontroller provides the necessary message decoding and display operation. The microcontroller stores the message in its memory and then displays it on the LCD. A control program handles all protocol disassembly and LCD display formatting and operation.

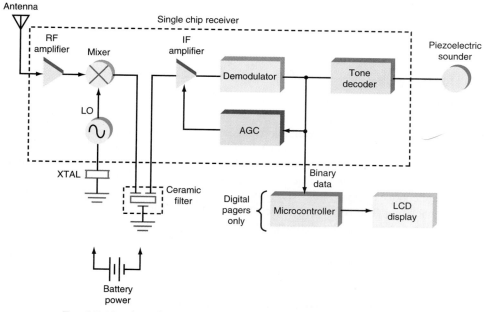

Fig. 17-40   A paging receiver.

Integrated services digital network (ISDN) is a digital communications interface designed to replace the local analog loop now used in the public switched network. It is designed to support digital voice telephones as well as fax machines, computers, video, and other digital data sources. It will allow computers to access online services directly without a modem. And it will permit remote LANs to be easily interconnected without a modem or other expensive interface. ISDN is a telephone company service that anyone can order and use just like standard telephone services. It was created so that the installed base of twisted-pair cable used for analog local loops can be used for digital transmission.

Digital techniques have already been widely adopted in the telephone system. An example is the T1 service described in Chap. 11. This service and higher-speed multiplexed digital systems are used between central offices and for long-distance communications. Higher-speed systems use fiber-optic cable.

ISDN is an international standard that is expected to be as widely adopted as current world telephone standards, thus permitting full systems interoperability across international borders. ISDN was introduced in the 1980s. It is more accepted in Europe and Japan than in the United States. During the 1990s, however, ISDN has had steadily increasing acceptance and adoption in the United States.

It was expected that the initial application for ISDN would be digital voice communications. But the use of ISDN for computer data transmission has turned out to be the primary application. It is widely used to interconnect remote LANs and to replace modems for higher-speed data transmission over long distances. In the future, the ISDN system is expected to be used to transmit compressed digital video, thus permitting inexpensive video telephones and video services such as video conferencing.

DID YOU KNOW?

ISDN is expected to provide inexpensive video telephones and other services such as video conferencing.

THE ISDN INTERFACE

Although ISDN service is available from local telephone companies, all-new equipment or special interfaces are required to use it. Once in place, ISDN can be used for voice telephone communications, fax, digital video, and all types of data transmissions. Figure 17-41 shows the interface connections to an ISDN line. The NT_1 box connects to the telephone company lines via twisted-pair, called the *U interface,* which uses multilevel encoding. The NT_1 box connects to the NT_2 box which provides the interface to the user's equipment. The circuitry for NT_1 and NT_2 is usually in the same enclosure, and this piece of equipment is owned by the customer. The NT_1 interface provides encoding and decoding, multiplexing, and timing. The NT_2 interface provides for connection to LANA, PBX, and other devices.

The communications devices connect to the NT_2 circuits over what is called the *S bus.* The S bus is twisted pair using special three-level modified AMI encoding. ISDN–compatible devices are called *terminal equipment.* Examples are special ISDN telephones, digital fax machines, or personal computer ISDN interfaces. These are des-

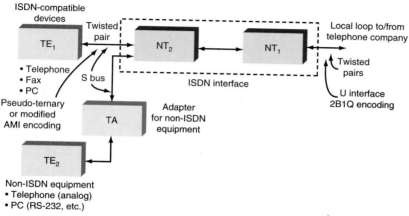

FIG. 17-41 The ISDN interface.

ignated as TE_1 devices. Equipment that is not ISDN compatible is designated TE_2. To use such equipment on an ISDN line, a *terminal adapter (TA)* is required. Different types of TAs are used depending upon the type of device. Standard analog telephones and computers with non-ISDN I/O interfaces are instances of such equipment.

BASIC RATE INTERFACE

The two basic types of ISDN connections are the *basic rate interface (BRI)* and the *primary rate interface (PRI)*. The basic rate interface is made up of a single twisted pair. Two bearer (B) channels for voice and one data control (D) channel are time multiplexed on the line. Its designation is 2B + D. The B channels are capable of 64K bps data transmission. The D channel can operate at 16K bps. The two B channels are used for 64K bps baseband transmissions of any kind of data. One channel is normally used for transmitting, and the other for receiving. The 16K bps channel is used for signaling and control functions between the user and the telephone company such as dialing and busy signals.

Because multiple channels are provided in an ISDN cable, various configurations of data transmissions are possible: simplex, half duplex, or full duplex. ISDN interfaces permit reconfiguration "on the fly," depending upon the application. For instance, the data to be transmitted can be partitioned into 2-bit words which can be transmitted simultaneously at 64K bps. This is the equivalent of transmitting at a rate of 128K bps. The total information bandwidth is the sum of the individual line data rates, in this case, 64 + 64 + 16 = 144K bps. Considering the framing and auxiliary bits also used, the total data rate is 192K bps. Most ISDN interfaces will be of this type.

PRIMARY RATE INTERFACE

The primary rate interface is called 23B + D and is made up of 23 64K bps B channels and a 64K bps D channel. Again, a single twisted pair is used with all channels being time multiplexed. The total data bandwidth is 24 × 64K bps, or 1536K bps, or

1.536M bps. Special designations are given to the various popular combinations such as H0 for 6 64K bps channels used at 384K bps, H10 for 23 64K bps channels used for 1472K bps transmission, and H11 for 24 channels at 1536K bps.

Primary rate ISDN is primarily for larger companies that need to accommodate a PBX system or needs internetwork connections to remote points. The primary rate service is T_1 compatible.

ISDN SIGNALS

Since the standard twisted pair used in the local loop is bandwidth limited to about 4 kHz, special techniques are needed to increase the speed of data transmission. ISDN uses a special multilevel digital signal on the U interface to the central office.

ISDN uses baseband communications methods; that is, the digital signal is connected directly to the line without modulation. However, the standard serial two-level binary signal is not used. A special multivoltage signal is used. Known as 2B1Q, this signal uses four voltage levels (-3, -1, $+1$, and $+3$ V) to represent pairs of bits according to the scheme listed below:

$$+1 \text{ V} = 11$$
$$+3 \text{ V} = 10$$
$$-1 \text{ V} = 01$$
$$-3 \text{ V} = 00$$

The data to be transmitted is split up into 2-bit sections and encoded as the appropriate voltage level. This is the same technique used in modems to increase the data rate. Each level represents 1 symbol or baud. And since 2 bits per baud are being encoded, the data rate is increased over the standard binary rate possible. Figure 17-42 shows a 2B1Q or 2 bits per quaternary symbol.

Figure 17-43 shows the frame format for the basic rate ISDN signal. This is the format of the data at the S bus. There are two different but similar formats, one for network to subscriber transmission and the other for subscriber to network transmission. Each frame contains 48 bits that occur in 250 μs. This gives a bit time of 250/48, or 5.208 μs. This translates to a bit rate of $1/5.208 \times 10^{-6}$, or 192,000 bps, or 192K bps.

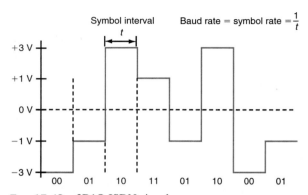

FIG. 17-42 2B1Q ISDN signal.

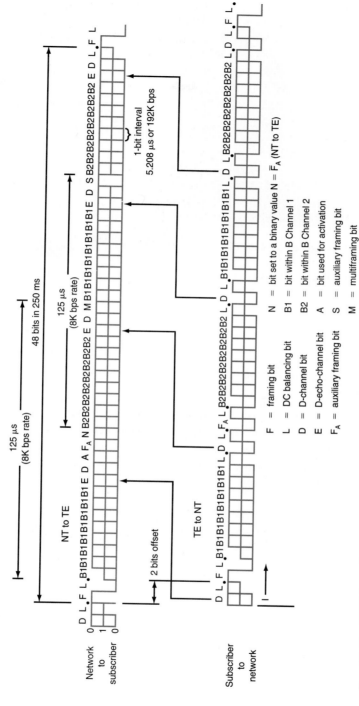

Fig. 17-43 ISDN frame formats.

F = framing bit
L = DC balancing bit
D = D-channel bit
E = D-echo-channel bit
F_A = auxiliary framing bit

N = bit set to a binary value N = $\overline{F}_A$ (NT to TE)
B1 = bit within B Channel 1
B2 = bit within B Channel 2
A = bit used for activation
S = auxiliary framing bit
M = multiframing bit

With access to the company's catalog inventories and the customer data base, this telemarketer can receive a customer's order, close the sale, and request express shipment with a single phone call.

In each frame there are 2 bytes (also called *octets*) of data for the B_1 bearer channel and 2 bytes of data for the B_2 bearer channel. The spacing between bytes is 125 μs, which translates to an 8-kHz sampling rate, which is standard for digital conversion of voice in the telephone system. There are 4 D channel bits per frame, but they are distributed rather than clustered. The other bits are used for framing, control, and signaling functions. DC balancing bits are added to ensure that the number of positive bits equals the number of negative bits so that the line remains DC balanced, that is, that there is no net positive or negative charge over one frame. Special multiplexing and demultiplexing circuits at each end of the system format the data into frames or extract it for use.

The encoding method used for this S-bus signal is a special pseudoternary, or three-level, binary signal as shown in Fig. 17-44. A binary 1 or mark is represented as a 0-V level, and a binary 0 or space is represented as a positive or negative voltage pulse (approximately 2 V). This is a modified form of the *alternate mark inversion (AMI) format* used in the T1 system. Consecutive binary 0s are transmitted as alternating polarity pulses to ensure that no DC offset occurs on the line. A 0 followed by another 0 of the same polarity is a code violation. Code violations are used to identify the start and stopping points in a frame.

FIG. 17-44 S-bus line encoding. Pseudoternary or modified AMI format. Number of pulses equals number of negative pulses per frame to maintain 0 Vdc average on the line.

SUMMARY

This chapter introduced the parts of the telecommunications systems most commonly in use and accessible. The basic operation of standard and electronic telephones was discussed, as were the basic parts and operations of cordless telephones, facsimile machines, cellular telephones, and pagers. The chapter also covered telecommunications systems, including the ISDN, for telephones, facsimile machines, cellular telephones, and pagers.

The standard telephone was invented in 1876, long before electronic components and circuits had been discovered. For that reason, the telephone was basically an electrical device, and all functions were handled by simple passive components. The original telephone system was designed for full-duplex analog communications of voice signals. Today the telephone system is still primarily analog in nature, although it employs a considerable number of digital techniques not only in signal transmission but also in control mechanisms. Even after the discovery of the vacuum tube and the creation of the electronics industry as we know it today, most telephones retained their simple passive component design, which continues to be in wide use to this day because of its simplicity and reliability. Through this basic system, any telephone can connect with any other telephone in the world.

A telephone incorporating vacuum tubes was never contemplated simply because of the high cost, large size, and heavy power consumption that this type of telephone would require. However, when solid-state circuits came along in the late 1950s, an electronic telephone became possible and practical. Today most new telephones are electronic, and they use IC technology. The development of the microprocessor has also affected telephone design. A built-in microprocessor permits automatic control of a multiple-line and full-featured telephone's functions. A microprocessor provides such features as number storage, automatic dialing, and re-dialing that are not possible with standard telephones.

The ultimate convenience in the home or office is the cordless telephone, which uses two-way radio transmission and provides total portability. An advanced version of the cordless telephone is the digital spread spectrum cordless telephone. Because of the use of digital techniques, signal quality is excellent with virtually no noise. These advanced telephones cost significantly more than the standard cordless telephones but are more desirable for critical applications. As the technology further develops and demand increases, prices of these phones will continue to drop.

The telephone itself is a small but relatively complex entity, but it is nothing compared to the massive system that backs it up. The telephone system can interconnect any two telephones in the world, and most people can only speculate on the method by which this interconnection takes place. In reality, it takes place on many levels and involves an incredible array of systems and technology managed by an equally complex array of businesses. Through the telephone system, facsimile machines can send printed information by wire or radio. The system can also accommodate cellular and paging telephone communications, which are actually two-way radio transmission systems.

The integrated services digital network (ISDN) is a telephone service that has been designed to eventually replace the local analog loop now used. The ISDN was created so that the installed base of twisted-pair cable used for analog local loops can be used for digital transmission such as digital voice telephones, fax machines, computers, video, and other digital data sources. It will allow computers to access online services directly without a modem. And it will permit remote LANs to be easily interconnected without a modem or other expensive interface.

Advanced mobile phone service (AMPS)
American Telephone and Telegraph (AT&T)
Automatic voice level adjustment
Base station
Basic rate interface (BRI)
Battery
BORSCHT
Cap code
Cellular telephone systems
Cellular telephone unit
Central office (local exchange)
Charged coupled device (CCD)
Coding
Computer branch exchange (CPX)
Cordless telephone
Data compression techniques
Dialing circuits
Dual-tone multifrequency (DTMF) system
Facsimile
Facsimile operation
Facsimile standards
Frequency allocations
Handset
Hybrid circuit

IC electronic telephone
Image processing
Integrated services digital network (ISDN)
Interexchange carriers (IXCs)
International Telecommunications Union (ITU-T)
International Telegraph and Telephone Consultative Committee (CCITT)
Key systems
Local access and transport areas (LATAS)
Local exchange companies (or carriers; LECs)
Local loop
Logic unit
Long-distance operation
MCI
Mobile identification number (MIN)
Mobile telephone switching office (MTSO)
Modem
Number assignment module (NAM)
Operational procedure with cellular telephone
Overvoltage protection
Paging receiver

Paging systems
Personal communications network (PSN)
Personal communications service (PCS)
Point of presence (POP) office
Primary rate interface (PRI)
Private automatic branch exchange (PBAX)
Private branch exchange (PBX)
Private telephone systems
Pulse dialing
Receiver
Ringer
Ringing
Subscriber interface
Subscriber line interface circuit (SLIC)
Supervision
Switch hook
Telephone hierarchy
Telephone set
Telephone system
Telephones
Test signals
Tone dialing
Transmitter
Varistor

REVIEW

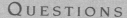

QUESTIONS

1. Define specifically what is meant by the *local loop*.
2. What type of power supply is used to power a standard telephone, what are its specifications, and where is it located?
3. State the characteristics of the ringing signal supplied by the telephone company.
4. What is a hybrid?
5. True or false. Most telephone companies can still accommodate pulse dial telephones.
6. What type of transmitter (microphone) is used in a standard telephone, and how does it work?
7. Define what is meant by *tip and ring* and state what colors are used to represent them.
8. What is the name of the TouchTone dialing system?
9. What two tone frequencies are generated when you press the pound key (#) on a TouchTone phone?
10. What is the name of the building or facility to which every telephone is connected?
11. What kind of microphone is used in an electronic telephone?
12. What is the purpose of the bridge rectifier circuit at the input to the connection of the telephone to the line to the telephone company?
13. Name one type of low-cost sounding device used to implement the bell or ringer in an electronic telephone.
14. True or false. In an electronic telephone, the hybrid is a special type of transformer.
15. Give two names or designations for the standard connector used on telephones.
16. State the two frequency ranges used by cordless telephones, and tell which is used for transmission and reception for the handheld and base units. How many channels are available?
17. What do you call the circuits that make up the connections to each telephone at the telephone office?
18. What is a key telephone system?
19. Does your local telephone company supply long-distance service?
20. True or false. A PBX may be used as a LAN under some conditions.
21. True or false. Fax can transmit photos and drawings as well as printed text.
22. What is the most common transmission medium for fax signals? What other medium is commonly used?
23. True or false. Facsimile was invented before radio.
24. Who sets the standards for fax transmission?
25. Vestigial sideband AM is used in what group type fax machines?
26. What is the name of the semiconductor photosensitive device used in most modern fax machines to convert a scanned line into an analog signal?
27. What is the group designation given to most modern fax machines?
28. To ensure compatibility between sending and receiving fax machines, the control logic carries out a procedure using audio tones to establish communications. What is this process called?
29. What circuit in the fax machine makes the fax signal compatible with the telephone line?
30. What is the upper speed limit of a G3 fax machine over the telephone lines?
31. What is the resolution of a G3 fax machine in lines per inch?

R
E
V
I
E
W

32. Fax transmission is usually:
    a. Full duplex
    b. Half duplex
33. Fax signals representing the image to be transmitted before they are prepared for the telephone lines are:
    a. Analog
    b. Digital
34. True or false. Group 4 transmissions do not use the standard telephone lines.
35. What are the speed and resolution of Group 4 fax transmissions?
36. What do you call the small zones into which the area to be served by a cellular telephone system is divided?
37. What is the master control station for a group of cells called?
38. How many cellular channels were initially available, and how many are available now?
39. True or false. Cellular telephone radios operate full duplex.
40. What type of modulation is used in a cellular radio?
41. What is the maximum power output of a large mobile cellular transmitter? A handheld unit?
42. What is the transmit frequency range?
43. What is the receive frequency range?
44. What is the channel spacing between cellular stations?
45. What is the frequency separation between send and receive frequencies?
46. Name common values for first IFs and second IFs in a cellular receiver.
47. What circuit in the cellular telephone allows a transmitter and receiver to share an antenna?
48. What is the source of the frequency divider ratios in the frequency synthesizer PLLs in the cellular telephone?
49. Name two signals that are transmitted back to the cell site and monitored by the MTSO.
50. Name three conditions in the transceiver controlled by the MTSO.
51. What is the name of the section of the cellular transceiver that interprets the serial digital data from the cell site and MTSO?
52. What is the NAM, and where is it kept?
53. What is AMPS?
54. Name the major parts of a paging system, and briefly explain its operation.
55. Name the two main code signaling methods used in paging systems.
56. What is dotting, and why is it needed?
57. What is ISDN?
58. What type of wiring does ISDN use?
59. What type of signals can an ISDN system carry?

PROBLEMS

1. Describe how one cordless telephone is prevented from interfering with another nearby cordless telephone. ◆
2. State the basic specifications and benefits of the newer class of cordless telephone.
3. List the basic BORSCHT functions that are performed by the telephone company for every telephone. ◆
4. Explain what each of the number groups in a 10-digit telephone number mean.
5. Describe the types of possible links between telephone exchanges. Are they two line or four line? ◆
6. Briefly define the terms *LATA, LEC, POP,* and *IXC*. Which one means "long-distance carrier"?

7. Define *PBX*, and explain its needs and benefits. Describe briefly how a PBX works.

8. What is Centrex?

9. Explain the process and hardware used to convert images to be transmitted into electrical signals in a fax machine.

10. Describe two methods of scanning used in modern fax machines.

11. What is the most commonly used type of printer in a fax machine? How does it work?

12. Describe how fax signals are processed to speed up transmission.

13. Give two reasons why there are more channels available with cellular systems than in older mobile telephone systems.

14. How are the transmit and receive frequencies used by a cellular telephone determined?

15. What is the purpose of the APC circuit? What operates this circuit?

16. A receiver has a first IF of 82.2 MHz. The second IF is 456 kHz. What is the local oscillator frequency on the second mixer?

17. Explain the concept of a communications channel using TDMA.

18. State the specifications for basic rate ISDN.

19. State the specifications for primary rate ISDN.

20. Define the designations TE_1, TE_2, TA, NT_1, and NT_2.

21. How can data rates exceeding the bit rate capacity of a single cable be achieved with ISDN?

22. How can non-ISDN devices be used with an ISDN system?

23. List five possible applications for ISDN.

24. What is the name of the signal encoding used in ISDN to achieve high data rates on a low-data-rate line? State the specifications of the coding method.

25. What is the basic frequency response of the telephone local loop? Can it carry digital as well as analog signals?

CRITICAL THINKING

1. What must be done to a primary rate ISDN signal to make it compatible with a T1 system?

2. Discuss how it would be possible to use the standard AC power lines for telephone voice transmission. What circuits might be needed, and what would be the limitations of this system?

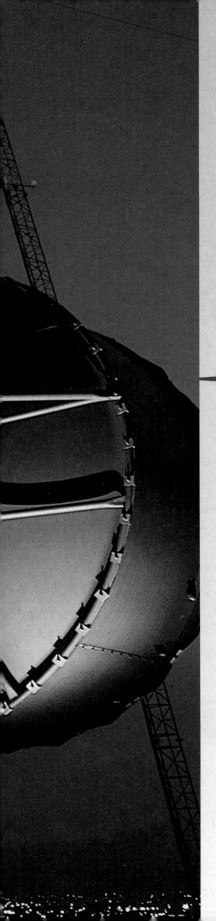

CHAPTER 18

OPTICAL COMMUNICATIONS

Objectives

After completing this chapter, you will be able to:

◆ *Define* the terms *optical* and *light*. Name the three main bands of the optical spectrum, and state their wavelength ranges.

◆ *State* eight benefits of fiber-optic cables over electrical cables for communications. Name six typical communications applications for fiber-optic cable.

◆ *Explain* how light is propagated through a fiber-optic cable. Name the three basic types of fiber-optic cables, and state the two materials from which they are made. Calculate the transmission loss in decibels of fiber-optic cable and connectors over a distance.

◆ *Name* the two types of optical transmitter components and their main operating range.

◆ *Explain* the operation of an optical detector and receiver.

◆ *State* the nature and frequency range of the infrared band, and name natural and artificial sources of infrared light.

Optical communications systems use light to transmit information from one place to another. Light is a type of electromagnetic radiation like radio waves, but it occupies that part of the frequency spectrum lower in frequency than x-rays but higher than the highest microwaves known as *millimeter waves.* We normally associate the term *light* only with visible light as we know it. But it also consists of infrared and ultraviolet waves which we cannot see.

Today, light is being used increasingly as the carrier for information in a communications system. Although visible light is used to some extent, most optical communications systems are infrared. The transmission medium is either free space or a special light-carrying cable called a *fiber-optic cable.* The light is modulated by the information to be transmitted. Because the frequency of light is extremely high, it can accommodate very wide bandwidths of baseband information and/or extremely high rates of data transmission with excellent reliability. Optical communications offers other benefits as well. This chapter introduces the basic concepts and circuits used in optical communications.

18-1 OPTICAL PRINCIPLES

Optics is a major field of study in itself and far beyond the scope of this book. Most courses in physics will introduce the basic principles of light and optics. We will review briefly those principles that relate directly to optical communications systems and components.

LIGHT

Light, radio waves, and microwaves are all forms of electromagnetic radiation. Light frequencies fall between microwaves and x-rays, as shown in Fig. 18-1(*a*). Radio frequencies range from approximately 10 kHz to 300 GHz. Microwaves extend from 1 to 300 GHz. The range of about 30 to 300 GHz is generally defined as *millimeter waves.*

Further up the scale is the optical spectrum, made up of infrared, visible, and ultraviolet light. The frequency of the optical spectrum is in the range of 3×10^{11} to 3×10^{16} Hz. This includes both the infrared and ultraviolet bands as well as the visible parts of the spectrum. The visible spectrum is from 4.3×10^{14} to 7.5×10^{14} Hz.

We rarely refer to the "frequency of light." Light is expressed in terms of wavelength. Recall that *wavelength* is a distance measured in meters between peaks of a wave. It is calculated with the familiar expression:

$$\lambda = \frac{300,000,000}{f}$$

where λ (lowercase lambda) is the wavelength in meters, 300,000,000 is the speed of light in meters per second, and f is the frequency in Hz. The details of the optical spectrum are shown in Fig. 18-1(*b*).

Light waves are very short and usually expressed in nanometers (nm, one-billionth of a meter) or micrometers (μm, one-millionth of a meter). An older term for *micrometer* is *micron*. Visible light is in the

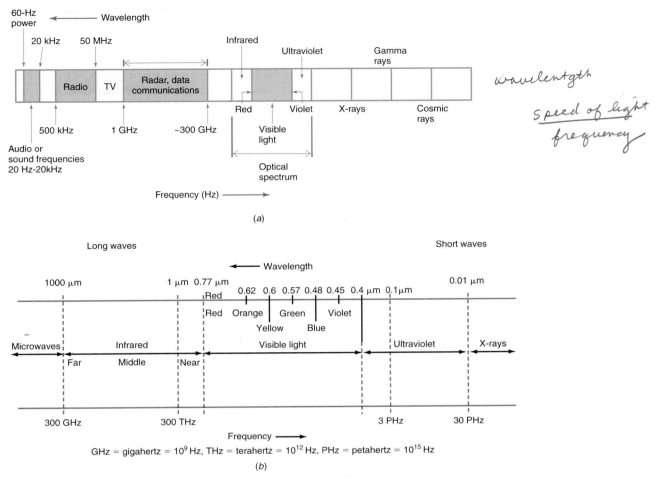

wavelength

$$\frac{speed\ of\ light}{frequency}$$

FIG. 18-1 The optical spectrum. (*a*) Electromagnetic frequency spectrum showing the optical spectrum. (*b*) Optical spectrum details.

400- to 700-nm range or 0.4 to 0.7 μm depending upon the color of the light. Short-wavelength light is violet (400 nm) while red (700 nm) is long-wavelength light.

Another unit of measure for light wavelength is the angstrom. One angstrom (Å) is equal to 10^{-10} m or 10^{-4} μm. To say it the other way, 1 μm equals 10,000 Å.

Right below visible light is a region known as *infrared*. Its spectrum is from 0.7 to 1000 μm. Sometimes you will hear infrared referred to as *near* or *far infrared*. Near infrared are those frequencies near the optical spectrum while far infrared is lower in frequency near the upper microwave region [see Fig. 18-1(*b*)].

Right above the visible spectrum is the ultraviolet range. The ultraviolet range is above that of violet visible light which has a wavelength of 400 nm or 0.4 μm up to about 10^{-8} m or 0.01 μm or 10 nm. The higher the frequency of the light, the shorter the wavelength. The primary source of ultraviolet light is the sun. Infrared and ultraviolet are included in what we call the *optical spectrum*.

SPEED OF LIGHT. Light waves travel in a straight line as microwaves do. Light rays emitted by a candle, light bulb, or other light source move out in a straight line in all directions. Light waves are assumed to have a spherical wavefront like radio waves.

The speed of light is approximately 300,000,000 m/s, or about 186,000 mi/s, in free space. These are the values normally used in calculation, but for a more accurate outcome, the actual values are closer to 2.998×10^8 m/s, or 186,280 mi/s.

The speed of light depends upon the medium through which the light passes. The figures given above are correct for light traveling in *free space,* that is, for light traveling in air or a vacuum. When light passes through another material such as glass, water, or plastic, its speed is slower.

PHYSICAL OPTICS

Physical optics refers to the ways that light can be processed. Light can be processed or manipulated in many ways. For example, lenses are widely used to focus, enlarge, or decrease the size of light waves from some source.

REFLECTION. The simplest way of manipulating light is to reflect it. When light rays strike a reflective surface, such as a mirror, the light waves are thrown back or reflected. By using mirrors, the direction of a light beam can be changed.

The reflection of light from a mirror follows a simple physical law. That is, the direction of the reflected light wave can be easily predicted if the angle of the light beam striking the mirror is known (refer to Fig. 18-2). Assume an imaginary line that is perpendicular with the flat mirror surface. A perpendicular line, of course, makes a right angle with the surface, as shown. This imaginary perpendicular line is referred to as the *normal.* The normal is usually drawn at the point where the mirror reflects the light beam.

If the light beam follows the normal, the reflection will simply go back along the same path. The reflected light ray will exactly coincide with the original light ray.

If the light ray strikes the mirror at some angle *A* from the normal, the reflected light ray will leave the mirror at the same angle *B* to the normal. This principle is known as the *law of reflection.* It is usually expressed in the following form: *The angle of incidence is equal to the angle of reflection.*

The light ray from the light source is normally called the *incident ray.* It makes an angle *A* with the normal at the reflecting surface called the *angle of incidence.* The reflected ray is the light wave that leaves the mirror surface. Its direction is determined by the angle of reflection *B,* which is exactly equal to the angle of incidence.

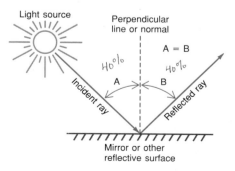

FIG. 18-2 The law of light reflection. The angle of reflection equals the angle of incidence.

REFRACTION. The direction of the light ray can also be changed by *refraction,* which is the bending of a light ray that occurs when the light rays pass from one medium to another. In reflection, the light ray bounces away from the reflecting surface rather than being absorbed by or passing through the mirror. Refraction occurs only when light passes through transparent material such as air, water, and glass. This refraction takes place at the point where two different substances come together. For example, where air and water come together, refraction will occur. The dividing line between the two different substances or media is known as the *boundary,* or *interface.*

To visualize refraction, place a spoon or straw into a glass of water as shown in Fig. 18-3(*a*). If you observe the glass of water from the side, it will look as if the spoon or straw is bent or offset at the surface of the water.

Another phenomenon caused by refraction occurs whenever you observe an object under water. You may be standing in a clear stream and observing a stone at the bottom. The rock is in a different position from where it appears to be from your observation [see Fig. 18-3(*b*)].

The refraction occurs because light travels at different speeds in different materials. The speed of light in free space is typically much higher than the speed of light in water, glass, or other materials. The amount of refraction of the light of a material is usually expressed in terms of the index of refraction n. This is the ratio of the speed of light in air to the speed of light in the substance.

Naturally, the index of refraction of air is 1, simply because 1 divided by itself is 1. The refractive index of water is approximately 1.3, and that of glass is 1.5.

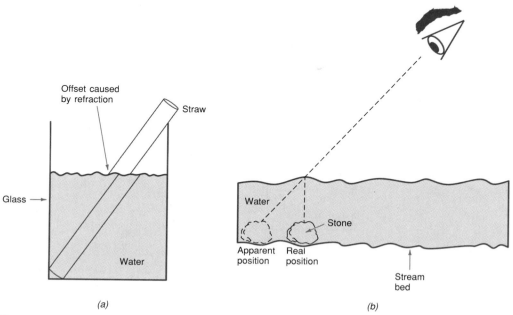

FIG. 18-3 Examples of the effect of refraction.

To get a better understanding of this idea, consider a piece of glass with a refractive index of 1.5. This means that light will travel through 1.5 ft of air, but during that same time, the light will travel only 1 ft through the glass. The glass slows down the light wave considerably. The index of the refraction is important because it tells exactly how much a light wave will be bent in various substances.

When a light ray passes from one medium to another, the light wave will be bent depending upon the index of refraction. In Fig. 18-4, a light ray is passing through air. It makes an angle A with the normal. At the interface between the air and the glass, the direction of the light ray is changed. The speed of light is slower; therefore, the angle that the light beam makes to the normal B is different from the incident angle. If the index of refraction is known, the exact angle can be determined.

If the light ray passes from the glass back into air, it will again change direction as Fig. 18-4 shows. The important point to note is that the angle of the refracted ray B is not equal to the angle of incidence A.

If the angle of incidence is increased, at some point the angle of refraction will equal 90° to the normal as shown in Fig. 18-5(a). When this happens, the refracted light ray travels along the interface between the air and glass. In this case, the angle of incidence A is said to be the *critical angle*. The critical angle value depends upon the index of refraction of the glass.

If you make the angle of incidence greater than the critical angle, the light ray will be reflected from the interface [see Fig. 18-5(b)]. When the light ray strikes the interface at an angle greater than the critical angle, the light ray does not pass through the interface into the glass. The effect is as if a mirror existed at the interface. When this occurs, the angle of reflection B is equal to the angle of incidence A as if a real mirror were used. This action is known as *total internal reflection,* which occurs only in materials in which the velocity of light is slower than in air. This is the basic principle that allows a fiber-optic cable to work.

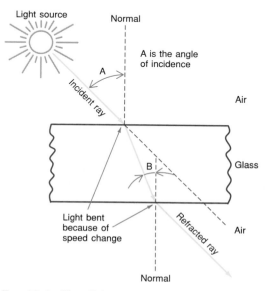

Fig. 18-4 How light rays are bent when passing from one medium to another.

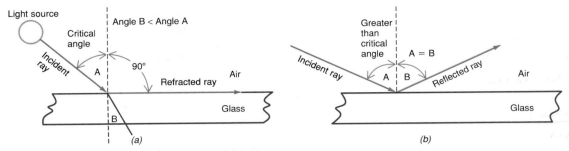

FIG. 18-5 Special cases of refraction. (*a*) Along the surface. (*b*) Reflection.

18-2 OPTICAL COMMUNICATIONS SYSTEMS

Optical communications systems use light as the carrier of the information to be transmitted. As indicated earlier, the medium may be free space as with radio waves or a special light "pipe" or *waveguide* known as *fiber-optic cable*. Both media are used, although the fiber-optic cable is far more practical and more widely used. This chapter focuses on fiber-optic cable.

RATIONALE FOR LIGHT WAVE COMMUNICATIONS

The main limitation of communications systems is their restricted information-carrying capabilities. In more specific terms, this means that the communications medium can carry only just so many messages. This information-handling ability is directly proportional to the bandwidth of the communications channel. Using light as the transmission medium provides vastly increased bandwidths. Instead of using an electrical signal traveling over a cable or electromagnetic waves traveling through space, the information is put on a light beam and transmitted through space or through a special fiber-optic waveguide.

LIGHT WAVE COMMUNICATIONS IN FREE SPACE

Figure 18-6 shows the elements of an optical communications system using free space. It consists of a light source modulated by the signal to be transmitted, a photodetector to pick up the light and convert it back into an electrical signal, an amplifier, and a demodulator to recover the original information signal.

LIGHT SOURCES.   The transmitter is a light source. Common light sources are light-emitting diodes (LEDs) and lasers. These sources can follow electrical signal changes as fast as 1 GHz or more.

Lasers generate *monochromatic,* or single-frequency, light that is fully *coherent;* that is, all the light waves are lined up in sync with one another and as a result produce a very narrow and intense light beam.

MODULATOR. A modulator is used to vary the intensity of the light beam in accordance with the modulating baseband signal. Amplitude modulation, also referred to as *intensity modulation,* is used where the information or intelligence signal controls the brightness of the light. Analog signals vary the brightness continuously over a specified range. This technique is used in some cable TV systems. Digital signals simply turn the light beam off and on at the data rate.

A modulator for analog signals can be a power transistor in series with the light source and its DC power supply (see Fig. 18-7). The voice, video, or other information signal is applied to an amplifier that drives the class A modulator transistor. As the analog signal goes positive, the base drive on the transistor increases, turning the transistor on harder and decreasing its collector-to-emitter base voltage. This applies more of the supply voltage to the light source making it brighter. A negative-going or decreasing signal amplitude drives the transistor toward cutoff, thereby reducing its collector current and increasing the voltage drop across the transistor. This decreases the voltage to the light source.

Frequency modulation is not used in light communication. There is no way to vary the frequency of the light source, even a monochromatic source like an LED or laser. Amplitude modulation is used with analog signals, but otherwise most light wave communication is accomplished by pulse modulation.

Pulse modulation refers to turning the light source off and on in accordance with some serial binary signal. The most common type of pulse modulation is *PCM,* which is serial binary data organized into bytes or longer words. RZ and Manchester formats are common.

RECEIVER. The modulated light wave is picked up by a photodetector. This is usually a photodiode or transistor whose conduction is varied by the light. The small signal is amplified and then demodulated to recover the originally transmitted signal. Digital processing may be necessary. For example, if the original signal is voice that was digitized by an A/D converter before being transmitted as a PCM signal, then a D/A converter will be needed at the receiver to recover the voice signal.

Communications by light beam in free space is impractical over very long distances because of the great attenuation of the light due to atmospheric effects. Fog, haze, smog, rain, snow, and other conditions absorb, reflect, refract, and disperse the light, greatly attenuating it and thereby limiting the transmission distance. Artificial light beams used to carry information are also obliterated during daylight hours by the sun. And they can be interfered with by any other light source that points in the direction of the receiver. Distances are normally limited to several hundred feet with low-power LEDs and lasers. If high-powered lasers are used, a distance of several miles may be possible.

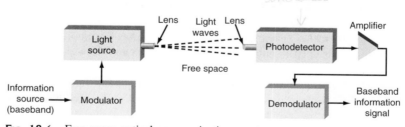

FIG. 18-6 Free-space optical communications system.

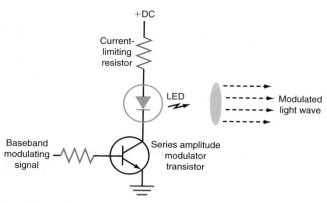

FIG. 18-7 A simple light transmitter with series amplitude modulator. Analog signals: transistor varies its conduction and acts like a variable resistance. Pulse signals: Transistor acts like a saturated ON-OFF switch.

Light beam communication has become far more practical with the invention of the *laser,* a special high-intensity, single-frequency light source. It produces a very narrow beam of brilliant light of a specific wavelength (color). Because of its great intensity, the laser beam can penetrate through atmospheric obstacles better than other types of light can, thereby making light beam communications more reliable over longer distances. When lasers are used, the light beam is so narrow that the transmitter and receiver must be perfectly aligned with one another in order for communications to occur. Although this causes initial installation alignment problems, it also helps to eliminate external interfering light sources.

A FIBER-OPTIC COMMUNICATIONS SYSTEM

Instead of using free space, some type of light-carrying cable can also be used. Today, fiber-optic cables have been highly refined. Cables many miles long can be constructed and then interconnected for the purpose of transmitting information on a light beam over very long distances. Thanks to these fiber-optic cables, a new transmission medium is now available. Its great advantage is its immense information-carrying capacity (wide bandwidth). Whereas hundreds of telephone conversations may be transmitted simultaneously at microwave frequencies, many thousands of signals can be carried on a light beam through a fiber-optic cable. Using multiplexing techniques similar to those used in telephone and radio systems, fiber-optic communications systems have an almost limitless capacity for information transfer.

The components of a typical fiber-optic communications system are shown in Fig. 18-8. The information signal to be transmitted may be voice, video, or computer data. The first step is to convert the information into a form compatible with the communications medium, usually by converting continuous analog signals such as voice and video (TV) signals into a series of digital pulses. An A/D converter is used for this purpose.

These digital pulses are then used to flash a powerful light source off and on very rapidly. In simple low-cost systems that transmit over short distances, the light source is usually a light-emitting diode (LED) that emits a low-intensity infrared light beam. Infrared beams like those used in TV remote controls are also used in transmission.

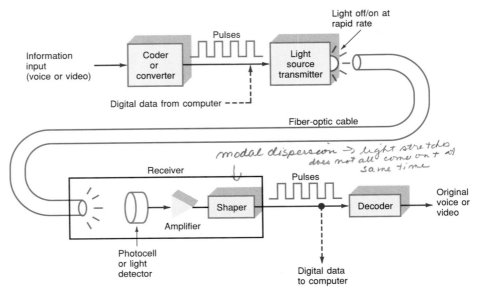

Fig. 18-8 Basic elements of a fiber-optic communications system.

The light beam pulses are then fed into a fiber-optic cable, which can transmit them over long distances. At the receiving end, a light-sensitive device known as a *photocell,* or *light detector,* is used to detect the light pulses. This photocell, or photodetector, converts the light pulses into an electrical signal. The electrical pulses are amplified and reshaped back into digital form. They are fed to a decoder, such as a D/A converter, where the original voice or video is recovered.

In very long transmission systems, repeater units must be used along the way. Since the light is greatly attenuated when it travels over long runs of cable, at some point it may be too weak to be received reliably. To overcome this problem, special relay stations are used to pick up the light beam, convert it back into electrical pulses that are amplified, and then retransmitted on another light beam.

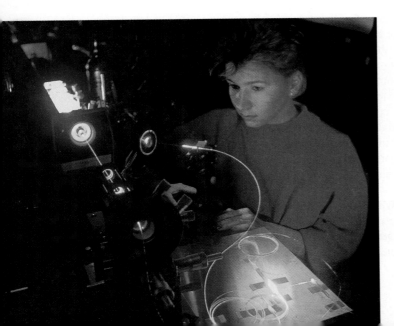

Continuous optical waveguide and laser research created a whole new generation of communications technology. This photograph represents on-going fiber-optic research and development.

1. TV studio to transmitter interconnection eliminating a microwave radio link

2. Closed-circuit TV systems used in buildings for security

3. Secure communications systems at military bases

4. Computer networks, wide area and local area

5. Shipboard communications

6. Aircraft communications

7. Aircraft controls

8. Interconnection of measuring and monitoring instruments in plants and laboratories

9. Data acquisition and control signal communications in industrial process control systems

10. Nuclear plant instrumentation

11. College campus communication

12. Utilities (electric, gas, and so on) station communications

13. Cable TV systems replacing coaxial cable.

Fig. 18-9 Applications of fiber-optic cables.

APPLICATIONS OF FIBER OPTICS

Fiber-optic communications systems are being used more and more each day. Their primary use is in long-distance telephone systems and cable TV systems. Fiber-optic cables are no more expensive or complex to install than standard electrical cables, yet their information-carrying capacity is many times greater.

Fiber-optic communications systems are being used in other applications. For example, they are being used to interconnect computers in networks within a large building, to carry control signals in airplanes and in ships, and in TV systems because of the wide bandwidth. In all cases, the fiber-optic cables replace conventional coaxial or twisted-pair cables. Figure 18-9 lists some of the applications in which fiber-optic cables are being used.

BENEFITS OF FIBER OPTICS

The main benefit of fiber-optic cables is their enormous information-carrying capability. This capacity is dependent upon the bandwidth of the cable. *Bandwidth* refers to the

1. *Wider bandwidth.* Fiber-optic cables have high information-carrying capability.

2. *Low loss.* Fiber-optic cables have less signal attenuation over a given distance than an equivalent length of coaxial cable.

3. *Lightweight.* Glass or plastic cables are much lighter than copper cables and offer benefits in those areas where low weight is critical (for example, aircraft).

4. *Small size.* Practical fiber-optic cables are much smaller in diameter than electrical cables and thus can be contained in a relatively small space.

5. *Security.* Fiber-optic cables cannot be as easily "tapped" as electrical cables, and they do not radiate signals that can be picked up for eavesdropping purposes. There is less need for complex and expensive encryption techniques.

6. *Interference immunity.* Fiber-optic cables do not radiate signals, as some electrical cables do, and cause interference to other cables. They are immune to the picking up of interference from other sources.

7. *Greater safety.* Fiber-optic cables do not carry electricity. Therefore, there is no shock hazard. They are also insulators and thus not susceptible to lightning strikes as electrical cables are. They can be used in corrosive and/or explosive environments without danger of sparks.

Fig. 18-10 Benefits of fiber-optic cables over conventional electrical cables.

Did You Know?

The small size and brittleness of fiber-optic cable along with the special tools and techniques required make it difficult to work with.

range of frequencies which a cable will carry. Electrical cables such as coaxial have a wide bandwidth (up to about 500 MHz), but the bandwidth of fiber-optic cable is much greater. Fiber-optic cable has many other benefits, summarized in Fig. 18-10.

There are some disadvantages to fiber-optic cable. For example, its small size and brittleness make it difficult to work with. Special, expensive tools and techniques are required.

18-3 FIBER-OPTIC CABLES

A *fiber-optic cable* is thin glass or plastic cable that acts as a light "pipe." It is not really a hollow tube carrying light but a long thin strand of glass or plastic fiber. Fiber cables have a circular cross section with a diameter of only a fraction of an inch. Some fiber-optic cables are only the size of a human hair. A light source is placed at the end of the fiber, and light passes through it and exits at the other end of the cable. How the light actually propagates through the fiber depends upon the laws of optics.

PRINCIPLES OF FIBER-OPTIC CABLE

Fiber-optic cables operate on the optical principles of total internal reflection as described earlier in this chapter. Figure 18-11(a) shows a thin fiber-optic cable. A beam of light is focused on the end of the cable. It can be positioned a number of different ways so that the light enters the fiber at different angles. For example, light ray *A* enters the cable perpendicular to the end surface. Therefore, the light beam travels straight down the fiber and exits the other end. This is the most desirable condition.

CRITICAL ANGLE. The angle of light beam *B* is such that its angle of incidence is less than the critical angle, and, therefore, refraction takes place. The light wave passes through the fiber and exits the edge into the air at a different angle.

The angle of incidence of light beams *C* and *D* is greater than the critical angle. Therefore, total internal reflection takes place, and the light beams are reflected off the surface of the fiber cable. The light beam bounces back and forth between the surfaces until it exits at the other end of the cable.

When the light beam reflects off the inner surface, the angle of incidence is equal to the angle of reflection. Because of this principle, light rays entering at different angles will take different paths through the cable, and some paths will be longer than others. Therefore, if multiple light rays enter the end of the cable, they will take different paths, and, therefore, some will exit sooner and some later than others.

In practice, the light source is placed so that the angle is such that the light beam passes directly down the center axis of the cable so that the reflection angles are great. This prevents the light from being lost due to refraction at the interface. Because of

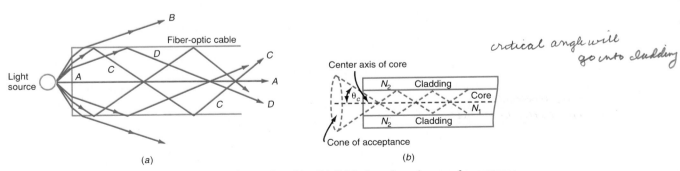

FIG. 18-11 (*a*) Light rays in a fiber-optic cable. (*b*) Critical angle and cone of acceptance.

total internal reflection, the light beam will continue to propagate through the fiber even though it is bent. With long slow bends, the light will stay within the cable.

Figure 18-11(*b*) shows the critical angle of the cable θ_c, sometimes referred to as the *acceptance angle* θ_A. This angle is formed between the center axis line of the core and the line that defines the maximum point where light entering the cable will undergo total internal reflection. If the light beam entering the end of the cable has an angle less than the critical angle, it will be internally reflected and propagated down the cable.

NUMERICAL APERTURE. Refer to the angles in Fig. 18-11(*b*). External to the end of the cable is what is called a *cone of acceptance;* it is defined by the critical angle. Any light beam outside the cone will not be internally reflected and transmitted down the cable.

The cone of acceptance defines the *numerical aperture* (NA) of the cable. This is a number less than 1 that gives some indication of the range of angles over which a particular cable will work. The NA can be calculated with the expression:

$$NA = \sin \theta_c$$

For example, if the critical angle is 20°, the NA is:

$$NA = \sin 20° = 0.342$$

The NA can also be determined from the indices of refraction of the core and cladding:

$$NA = \sqrt{[(N_1)^2 - (N_2)^2]}$$

If $N_1 = 1.5$ and $N_2 = 1.4$, the numerical aperture is:

$$NA = \sqrt{(1.5)^2 - (1.4)^2} = \sqrt{(2.25 - 1.96)} = \sqrt{0.29} = 0.5385$$

Typical numerical apertures for common cables are 0.275 and 0.29.

Example 18-1

The numerical aperture (NA) of a fiber-optic cable is 0.29. What is the critical angle?

$$NA = \sin \theta_c$$
$$0.29 = \sin \theta_c$$
$$\theta_c = \sin^{-1} (0.29) = \arcsin (0.29)$$
$$\theta_c = 16.86°$$

FIBER-OPTIC CABLE CONSTRUCTION

Fiber-optic cables come in a variety of sizes, shapes, and types. The simplest cable contains a single strand of fiber; a complex cable is made up of multiple fibers with different layers and other elements.

The portion of a fiber-optic cable that carries the light is made from either glass or plastic. Another name for glass is *silica.* The optical characteristics of glass are superior to those of plastic. However, glass is far more expensive and more fragile than plastic. Although plastic is less expensive and more flexible, its attenuation of light is greater. For a given intensity, light will travel a farther distance in glass than in plastic.

The construction of a fiber-optic cable is shown in Fig. 18-11(*b*). The glass or plastic optical fiber is contained within an outer cladding. The index of refraction of the outer cladding N_2 is slightly less than the index of refraction N_1 of the core. Typical values for N_1 and N_2 are 1.5 and 1.4. Over the cladding is a plastic jacket similar to the outer insulation on an electrical cable.

The fiber, which is called the *core,* is usually surrounded by a protective cladding (see Fig. 18-12). The cladding is also made of glass or plastic but has a lower index of refraction. This ensures that the proper interface is achieved so that the light waves remain within the core. In addition to protecting the fiber core from nicks and scratches, the cladding gives strength. Some fiber-optic cable has a glass core with a glass cladding. Other cables have a plastic core with a plastic cladding. Another arrangement, *plastic-clad silica (PCS) cable,* is a glass core with a plastic cladding.

Typically the division between the core and the cladding cannot be seen, they are usually made of the same types of material. Over the cladding is usually a plastic jacket similar to the outer insulation on an electrical cable.

In addition to the core, its cladding, and a jacket, fiber-optic cable usually contains one or more additional elements to form a complete cable. The simplest cable is the core with its cladding surrounded by a protective jacket. As in electrical cables, this outer jacket, or insulation, is made of some type of plastic, typically polyethylene, polyurethane, or polyvinyl chloride (PVC). The main purpose of this outer jacket is to protect the core and cladding from damage. Usually fiber-optic cables are buried underground or strung between supports. Therefore, the fibers must be protected from moisture, dirt, temperature variations, and other conditions. The outer jacket also helps minimize physical damage such as cuts, nicks, and crushing.

The more complex cables may contain two or more fiber-optic elements. Typical cables are available with 2, 6, 12, 18, and 24 optical fiber cores.

There are many different types of cable configurations. Many have several layers of protective jackets. Some cables incorporate a flexible strength or tension element that helps minimize damage to the fiber-optic elements when the cable is being pulled or must support its own weight. Typically, this strength member is made up of a stranded steel or a special yarn known as *Kevlar,* which is strong and preferred over the steel because it is an insulator. In some cables, the Kevlar forms a protective sleeve or jacket over the cladding. Most claddings are covered with a clear protective coating for added strength and resistance to moisture and damage (see Fig. 18-13).

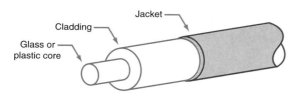

FIG. 18-12 Basic construction of a fiber-optic cable.

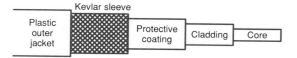

FIG. 18-13 Typical layers in a fiber-optic cable.

Fiber-optic cables are also available in a flat ribbon form (see Fig. 18-14). Flat ribbon cable works well with multiple fibers. Handling and identification of individual fibers are easy. The flat cable is also more space efficient for some applications.

TYPES OF FIBER-OPTIC CABLES

There are two basic ways of classifying fiber-optic cables: The first method is by the index of refraction, which varies across the cross section of the cable. The second method of classification is by *mode,* which refers to the various paths the light rays can take in passing through the fiber. Usually these two methods of classification are combined to define the types of cable.

STEP INDEX CABLE.   The two ways to define the index of refraction variation across a cable are step index and graded index. *Step index* refers to the fact that there is a sharply defined step in the index of refraction where the fiber core and the cladding interface. It means that the core has one constant index of refraction N_1 and the cladding has another constant index of refraction N_2. When the two come together, there is a distinct step (see Fig. 18-15). If you were to plot a curve showing the index of refraction as it varies vertically from left to right across the cross section of the cable, you would see a sharp increase in the index of refraction as the core is encountered and then a sharp decline in index refraction as the cladding is encountered.

GRADED INDEX CABLE.   The other type of cable has a graded index. Here, the index of refraction of the core is not constant. Instead, it varies smoothly and continuously over the diameter of the core (see Fig. 18-16). As you get closer to the center of

center of core greater density → graded index

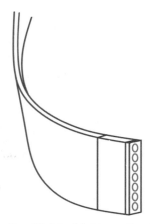

FIG. 18-14 Multicore flat ribbon cable.

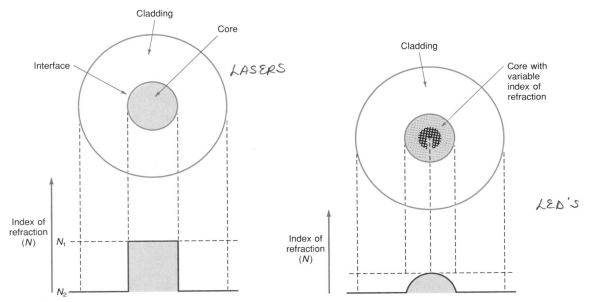

LASERS

LED'S

FIG. 18-15 A step index cable cross section.

FIG. 18-16 Graded index cable cross section.

the core, the index of refraction gradually increases, reaching a peak at the center and then declining as the other outer edge of the core is reached. The index of refraction of the cladding is constant.

CABLE MODE. *Mode* refers to the number of paths for the light rays in the cable. There are two classifications: single mode and multimode. In *single mode,* light follows a single path through the core; in *multimode,* the light takes many paths.

Each type of fiber-optic cable uses one of these methods of rating the index or mode. In practice, there are three commonly used types of fiber-optic cable: *multimode step index, single-mode step index,* and *multimode graded index.*

MULTIMODE STEP INDEX CABLE. The multimode step index fiber cable is probably the most common and widely used type. It is also the easiest to make and therefore the least expensive. It is widely used for short to medium distances at relatively low pulse frequencies.

The main advantage of a multimode stepped index fiber is its large size. Typical core diameters are in the 50- to 1000-μm range. Such large-diameter cores are excellent at gathering light and transmitting it efficiently. This means that an inexpensive light source such as an LED can be used to produce the light pulses. The light takes many hundreds or even thousands of paths through the core before exiting (refer to Fig. 18-11). Because of the different lengths of these paths, some of the light rays take longer to reach the other end of the cable than others. The problem with this is that it stretches the light pulses.

For example, in Fig. 18-17, a short light pulse is applied to the end of the cable by the source. Light rays from the source travel in multiple paths. At the end of the cable,

do not have single mode graded index

HINTS AND HELPS

The multimode step index fiber cable is widely used at short to medium distances at relatively low pulse frequencies. This cable is also the easiest to make and the least expensive of the fiber-optic cables.

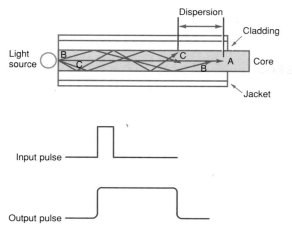

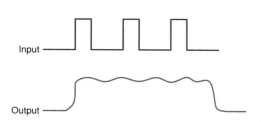

Fig. 18-18 The effect of modal dispersion on pulses occurring too rapidly in a multimode step index cable.

Fig. 18-17 A multimode step index cable.

the rays that travel the shortest distance reach the end first. Other rays begin to reach the end of the cable later, until the light ray with the longest path finally reaches the end, concluding the pulse. In Fig. 18-17, ray A reaches the end first, then B, and then C. The result is a pulse at the other end of the cable that is lower in amplitude because of the attenuation of the light in the cable and increased in duration because of the different arrival times of the various light rays. This stretching of the pulse is referred to as *modal dispersion.*

Because the pulse has been stretched, input pulses at the input cannot occur at a rate faster than the output pulse duration permits. Otherwise, the pulses will essentially merge (see Fig. 18-18). At the output, one long pulse will occur and will be indistinguishable from the three separate pulses originally transmitted. This means that incorrect information will be received. The only cure for this problem is to reduce the pulse repetition rate or the frequency of the pulses. When this is done, proper operation occurs. But with pulses at a lower frequency, less information can be handled. The key point here is that the type of cable selected must have a modal dispersion sufficiently low to handle the desired upper frequency of operation.

SINGLE-MODE STEP INDEX. A single-mode or monomode step index fiber cable essentially eliminates modal dispersion by making the core so small that the total number of modes or paths through the core are minimized (see Fig. 18-19). Typical core sizes are 2 to 15 μm. The only path through the core is down the center. With minimum refraction, little pulse stretching occurs. The output pulse has essentially the same duration as the input pulse.

Single-mode step index fibers are by far the best since the pulse repetition rate can be high and the maximum amount of information can be carried. For very long dis-

not very efficient
will LED better with laser

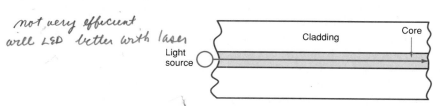

Fig. 18-19 Single-mode step index cable.

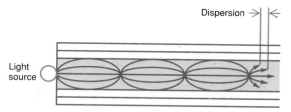

Fig. 18-20 A multimode graded index cable.

best with lasers

tance transmission and maximum information content, single-mode step index fiber cables should be used.

The main problem with this type of cable is that it is extremely small and difficult to make and therefore very expensive. It is also more difficult to handle. Splicing and making interconnections are also more difficult. Finally, for proper operation, an expensive, superintense light source such as a laser must be used. For long distances, however, this is the type of cable preferred.

MULTIMODE GRADED INDEX. Multimode graded index fiber cables have several modes, or paths, of transmission through the cable, but they are much more orderly and predictable. Figure 18-20 shows the typical paths of the light beams. Because of the continuously varying index of refraction across the core, the light rays are bent smoothly and repeatedly converge at points along the cable. The light rays near the edge of the core take a longer path but travel faster because the index of refraction is lower. All the modes or light paths tend to arrive at one point simultaneously. The result is less modal dispersion. As a result, this cable can be used at very high pulse rates, and, therefore, a considerable amount of information can be carried. This type of cable is also much wider in diameter, with core sizes in the 50- to 100-m range. Therefore, it is easier to splice and interconnect, and cheaper, less intense light sources can be used.

FIBER-OPTIC CABLE SPECIFICATIONS

The most important specifications of a fiber-optic cable are size, attenuation, and bandwidth. The NA is also sometimes given, although this specification is needed only when connectors are being designed—a rare occurrence, because standard connectors are available.

CABLE SIZE. Fiber-optic cable comes in a variety of sizes and configurations as previously indicated. Size is normally specified as the diameter of the core, and cladding is given in microns (μm). For example, a common size is 62.5/125, where 62.5 is the diameter of the core and 125 is the diameter of the cladding, both in μm (see Fig. 18-21). A PVC or polyurethane plastic jacket over the outer cladding gives a total outside diameter of about 3 mm or 0.118 in. Other common cable sizes are 50/125 and 100/140, although the 62.5/125 is by far the most widely used.

Cables come in two common varieties, simplex and duplex. Simplex cable, as the name implies, is just a single fiber core cable as shown in Fig. 18-21. In a common duplex cable, as shown in Fig. 18-22, two cables are combined within a single outer cladding.

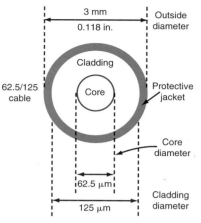

FIG. 18-21   Fiber-optic cable dimensions.

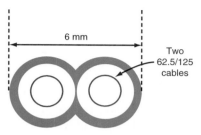

FIG. 18-22   Cross section of a duplex cable.

ATTENUATION.   The most important specification of a fiber-optic cable is its attenuation. *Attenuation* refers to the loss of light energy as the light pulse travels from one end of the cable to the other. The light pulse of a specific amplitude or brilliance is applied to one end of the cable, but the light pulse output at the other end of the cable will be much lower in amplitude. The intensity of the light at the output is lower because of various losses in the cable. The main reason for the loss in light intensity over the length of the cable is light absorption, scattering, and dispersion.

Absorption refers to how light energy is converted to heat in the core material due to the impurity of the glass or plastic. This phenomenon is similar to electrical resistance. *Scattering* refers to the light lost due to light waves entering at the wrong angle and being lost in the cladding due to refraction. *Dispersion,* as mentioned, refers to the pulse stretching caused by the many different paths through the cable. While no light is lost as such in dispersion, the output is still lower in amplitude than the input, although the length of the light pulse has increased in duration.

The amount of attenuation varies with the type of cable and its size. Glass has less attenuation than plastic. Wider cores have less attenuation than narrower cores of the same material. Wider cores have less absorption and much more dispersion. Wider cores are also usually plastic, which has more absorption capacity than glass.

More important, attenuation is directly proportional to the length of the cable. It is obvious that the longer the distance the light has to travel, the greater the loss due to absorption, scattering, and dispersion.

The attenuation of a fiber-optic cable is expressed in decibels (dB) per unit of length. The standard is decibels per kilometer. The standard decibel formula is used:

$$dB = 10 \log \left(\frac{P_{out}}{P_{in}} \right)$$

where P_{out} is the power out and P_{in} is the power in. Because light intensity is a type of electromagnetic radiation, it is normally expressed and measured in power units, watts, or some fraction thereof. Figure 18-23 shows the percentage of output power for various decibel losses.

For example, 3 dB represents half power. In other words, a 3-dB loss means that only 50 percent of the input appears at the output. The

HINTS AND HELPS

For fiber optics, attenuation is directly proportional to the length of the cable. For instance, doubling the length of a cable doubles the attenuation.

Loss, dB	Power Output, %
1	79
2	63
3	50
4	40
5	31
6	25
7	20
8	14
9	12
10	10
20	1
30	0.1
40	0.01
50	0.001

Fig. 18-23 Decibel (dB) loss table.

other 50 percent of the power is lost in the cable. The higher the decibel figure, the greater the attenuation and loss. A 30-dB loss means that only one-thousandth of the input power appears at the end. The standard specification for fiber-optic cable is attenuation expressed in terms of decibels per kilometer (dB/km).

The attenuation ratings of fiber-optic cables vary over a considerable range. The finest single-mode step index cables have an attenuation of only 1 dB/km. However, very large core plastic fiber cables can have an attenuation of several thousand decibels per kilometer. A typical 62.5/125 cable has a loss in the 3- to 5-dB/km range.

Telecommunications equipment requires normal servicing and proper maintenance. Preventative maintenance is constantly needed, as are upgrading and updating equipment. Many new installations like this use fiber-optic links.

Typically, fibers with an attenuation of less than 10 dB/km are called *low-loss fibers* and those with an attenuation of between 10 and 100 dB/km are *medium-loss fibers*. *High-loss fibers* have over 100 dB/km ratings. Naturally, the smaller the decibel number, the less the attenuation and the better the cable.

The total attenuation for a particular cable can be determined from the attenuation rating of the cable. For example, if a cable has an attenuation of 3.75 dB/km, a 5-km long cable has a total attenuation of 3.75 × 15, or 56.25 dB. If two cables are spliced together and one has an attenuation of 17 dB and the other 24 dB, the total attenuation is simply the sum, or 17 + 24, or 41 dB.

BANDWIDTH. The bandwidth of a fiber-optic cable determines the maximum speed of the data pulses the cable can handle. The bandwidth is normally stated in terms of megahertz-kilometers (MHz-km). A common 62.5/125-μm cable has a bandwidth in the 100- to 300-MHz-km range. Cables with 500 and 600 MHz-km are also common. Even higher-bandwidth cables are available to carry gigahertz-range signals.

As the length of the cable is increased, the bandwidth decreases in proportion. If a 160-MHz-km cable length is doubled from 1 to 2 km, its bandwidth is halved to 80 MHz-km.

Example 18-2

A fiber-optic cable has a bandwidth rating of 600 MHz-km. What is the bandwidth of a 500-foot segment of cable?

$$1 \text{ km} = 0.62 \text{ mi}$$
$$1 \text{ mi} = 5280 \text{ ft}$$
$$1 \text{ km} = 0.62(5280) = 3274 \text{ ft}$$
$$500 \text{ ft} = \frac{500}{3274} = 0.153 \text{ km}$$
$$600 \text{ MHz-km} = 600 \text{ MHz} (1 \text{ km})$$
$$600 \text{ MHz-km} = x_{\text{MHz}}(0.153 \text{ km})$$
$$x_{\text{MHz}} = \frac{600 \text{ MHz-km}}{0.153 \text{ km}} = 3928.3 \text{ MHz, or } 3.93 \text{ GHz}$$

FREQUENCY RANGE. Most fiber-optic cable operates over a relatively wide light frequency range, although it is normally optimized for a narrow range of light frequencies. The most commonly used light frequencies are 860, 1300, and 1550 nm (or 0.86, 1.3, and 1.55 μm). The cable has minimum attenuation to these frequencies.

Figure 18-24 shows attenuation versus wavelength for a typical cable. Note that there is a peak in the attenuation at approximately 1.4 μm. This is caused by hydroxyl ions or undesired hydrogen oxygen ions produced during the manufacturing process. These so-called water losses are avoided by carefully selecting the frequency of light transmission. Minimal loss occurs at 1.3 and 1.55 μm. Most LED or laser light sources operate at one of these wavelengths.

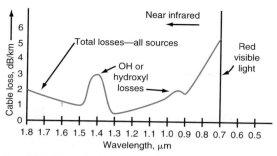

Fig. 18-24 Attenuation versus wavelength of a typical fiber-optic cable.

CONNECTORS AND SPLICING

When long fiber-optic cables are needed, two or more cables can be spliced together. The ends of the cable are perfectly aligned and then fused together with heat. A variety of connectors are available that provide a convenient way to splice cables and attach them to transmitters, receivers, and repeaters.

CONNECTORS. Connectors are special mechanical assemblies that allow fiber-optic cables to be connected to one another. Fiber-optic connectors are the optical equivalent of electrical plugs and sockets. They are mechanical assemblies that hold the ends of a cable and cause them to be accurately aligned with the ends of another cable. Most fiber-optic connectors either snap or twist together or have threads that allow the two pieces to be screwed together.

Connectors ensure precise alignment of the cables. The ends of the cables must be aligned with precision so that maximum light from one cable is transferred to another. A poor splice or connection will introduce excessive attenuation as the light exits one cable and enters the other. Figure 18-25 shows several ways that cores can be misaligned. A connector can correct these problems.

A typical fiber-optic connector is shown in Fig. 18-26(*a*). One end of the connector, called the *ferrule,* holds the fiber securely in place. A matching fitting holds the other fiber securely in place. When the two are screwed together, the ends of the fibers touch, thereby establishing a low-loss coupling. Figure 18-26(*b*) shows in more detail how the connector aligns the fibers.

Dozens of different kinds of connectors are available for different applications. They are used in the parts of the system where it may be desirable to occasionally dis-

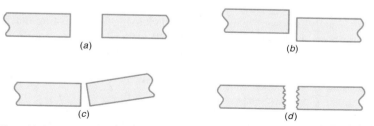

Fig. 18-25 Misalignment and rough end surfaces cause loss of light and high attenuation. (*a*) Too much end separation. (*b*) Axial misalignment. (*c*) Angular misalignment. (*d*) Rough, uneven surfaces.

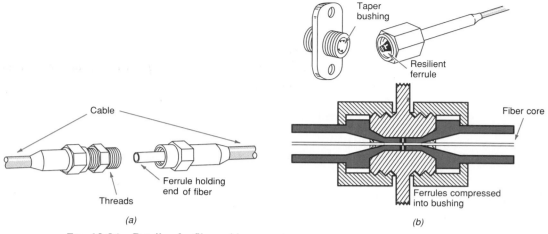

Fig. 18-26 Details of a fiber cable connector.

connect the fiber-optic cable for making tests or repairs. Connectors are normally used at the end of the cable applied to the light source or the end of the cable connected to the photodetector.

Connectors are also used at the repeater units where the light is picked up, converted into an electrical pulse, amplified and reshaped, and then used to create a new pulse to continue the transmission over a long line. Connectors are used on the back of interface adapters that plug into computers.

The two most common connector designations are ST and SMA. The *ST connectors,* also referred to as *bayonet connectors,* use a half-twist cam-type arrangement like that used in BNC coaxial connectors. They are convenient for quick connect and disconnect. The *SMA connectors* are about the same size but have threaded connections.

SPLICING. *Splicing fiber-optic cable* means permanently attaching the end of one cable to another. This is usually done without a connector. The first step is cutting the cable, called cleaving the cable, so that it is perfectly square on the end. Cleaving is so important to minimizing light loss that special tools have been developed to ensure perfect cuts.

The two cables to be spliced are then permanently bonded together by heating them instantaneously to high temperatures so that they fuse or melt together. Special tools and splicing machines must be used to ensure perfect alignment.

Installing a connector begins with cleaving the fiber so that it is perfectly square. Polishing usually follows. Again, the special cleaving and polishing machines devised for this purpose must always be used. Poorly spliced cable or poorly installed connectors create an enormous loss.

18-4 OPTICAL TRANSMITTERS AND RECEIVERS

In an optical communications system, transmission begins with the transmitter, which consists of a carrier generator and a modulator. The carrier is a light beam that is usually modulated by turning it on and off with digital pulses. The basic transmitter is essentially a light source.

The receiver is a light or photodetector that converts the received light back into an electrical signal. In this section the types of light sources used in fiber-optic systems and the transmitter circuitry, as well as the various light detectors and the related receiver circuits, will be discussed.

LIGHT SOURCES

Conventional light sources such as incandescent lamps cannot be used in fiber-optic systems because they are too slow. In order to transmit high-speed digital pulses, a very fast light source must be used. The two most commonly used light sources are *light-emitting diodes (LEDs)* and *semiconductor lasers.*

LIGHT-EMITTING DIODES.

A *light-emitting diode* is a PN-junction semiconductor device that emits light when forward-biased. When a free electron encounters a hole in the semiconductor structure, the two combine, and in the process they give up energy in the form of light. Semiconductors such as gallium arsenide (GaAs) are superior to silicon in light emission. Most LEDs are GaAs devices optimized for producing red light. LEDs are widely used for displays indicating whether a circuit is off or on, or for displaying decimal and binary data. However, because an LED is a fast semiconductor device, it can be turned off and on very quickly and is capable of transmitting the narrow light pulses required in a digital fiber-optics system.

LEDs can be designed to emit virtually any color light desired. The LEDs used for fiber-optic transmission are usually in the red and near-infrared ranges. Typical wavelengths of LED light commonly used are 0.82, 0.94, 1.3, and 1.55 μm. These frequencies are all in the near-infrared range just below red light, which is not visible to the naked eye. These frequencies have been chosen primarily because most fiber-optic cables have the lowest losses in these frequency ranges. The most commonly used wavelength is 1.3 μm.

One physical arrangement of the LED is shown in Fig. 18-27(*a*). A p-type material is diffused into the n-type substrate creating a diode. Radiation occurs from the p-type material and around the junction. Figure 18-27(*b*) shows a common light radiation pattern.

The light output from an LED is expressed in terms of power. Typical light output levels are in the 10- to 50-μW range. Sometimes the light output will be expressed

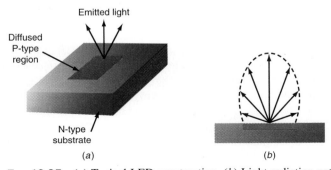

FIG. 18-27 (*a*) Typical LED construction. (*b*) Light radiation pattern.

in dBm or dB referenced of 1 mW (milliwatt). Common levels are -15 to -30 dBm. Forward-bias current levels to achieve this power level are in the 50- to 200-mA range. High-output LEDs with output ratings in the 600- to 2500-μW range are also available.

A typical LED used for lighting is relatively slow to turn on and off. A typical turn-on/turn-off time is about 150 ns. This is too slow for most data communications applications by fiber optics. Faster LEDs capable of data rates up to 50 MHz are available. For faster data rates, a laser diode must be used.

Special LEDs are made just for fiber-optic applications. These units are made of *gallium arsenide* or *indium phosphide (GaAs or InP)* and emit light at 1.3 μm. Other LEDs with as many as six multiple layers of semiconductor material are used to optimize the device for a particular frequency and light output. Figure 18-28 shows some LED assemblies made to accept ST bayonet and SMA fiber-optic cable connectors. These are made for PC board mounting.

LASER DIODES. The other commonly used light transmitter is a *laser,* which is a light source that emits coherent monochromatic light. *Monochromatic light* is a pure single-frequency light. While an LED emits red light, that light covers a narrow spectrum around the red frequencies. *Coherent* refers to the fact that all the light waves emitted are in phase with one another. Coherence produces a focusing effect on a beam so that it is narrow and, as a result, extremely intense. The effect is somewhat similar to that of using a highly directional antenna to focus radio waves into a narrow beam that also increases the intensity of the signal.

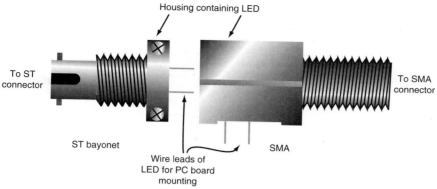

Housing containing LED

To ST connector

ST bayonet

Wire leads of LED for PC board mounting

To SMA connector

SMA

Fig. 18-28 LED assemblies with connectors.

Lasers can be made in a variety of ways. Some are made from a solid rod of special material which emits light when properly stimulated. Certain types of gases are also used. Lasers used in fiber-optic systems are usually specially made LEDs that can operate as lasers.

The most widely used light source in fiber-optic systems is an *injection laser diode (ILD)*. Like the LED, it is a PN-junction diode usually made with GaAs. With a low-level forward-bias current, the ILD operates like a standard LED, producing a low-level light over a relatively broad frequency range. At a higher forward current level known as the *threshold,* the ILD operates as a laser and emits a brilliant light over a much narrower frequency range.

ILDs are capable of developing light power up to several watts. They are far more powerful than LEDs and therefore capable of transmitting over much longer distances. Another advantage over LEDs is their ability to turn off and on at a faster rate. High-speed laser diodes are capable of gigabit-per-second digital data rates.

Injection lasers dissipate a tremendous amount of heat and, therefore, must be connected to a heat sink for proper operation. Because their operation is heat sensitive, most injection lasers are used in a circuit that provides some feedback for temperature control. This not only protects the laser but also ensures proper light intensity and frequency.

LIGHT TRANSMITTERS

LED TRANSMITTER. A light transmitter consists of the LED and its associated driving circuitry. A typical circuit is shown in Fig. 18-29. The binary data pulses are applied to a logic gate which, in turn, operates a transistor switch Q_1 that turns the LED off and on. A positive pulse at the NAND gate input causes the NAND output to go to zero. This turns off Q_1, so the LED is then forward-biased through R_2 and turns on. With zero input, the NAND output is high, so that Q_1 turns on and shunts current away from the LED. Very high current pulses are used to ensure very bright light. High intensity is required if data is to be transmitted reliably over long distances. Most LEDs are capable of generating power levels up to approximately several thousand microwatts. With such low intensity, LED transmitters are good for only short distances. Further, the speed of the LED is limited. Turn-off and turn-on times are no faster than tens of nanoseconds, and, therefore, transmission rates are limited. Most LED-like transmitters are used for short-distance, low-speed, digital fiber-optic systems.

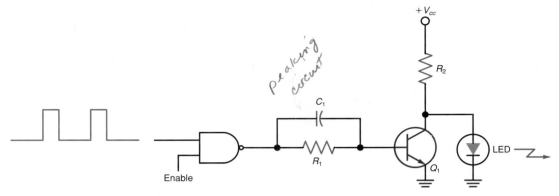

FIG. 18-29 Optical transmitter circuit using an LED.

LASER TRANSMITTER. A typical injection laser transmitter circuit is shown in Fig. 18-30. When the input is zero, the AND gate output is zero, so Q_1 is off, as is the ILD. C_2 charges through R_3 to the high voltage. When a binary 1 input occurs, Q_1 conducts, connecting C_2 to the ILD. C_2 discharges a very high current pulse into the laser, turning it on briefly and creating an intense light pulse.

Laser transmitters can develop an output power from a few milliwatts to several watts. And ILDs can turn on and off at very high speeds. Picosecond speeds permit rates of up to several gigahertz.

DATA FORMATS. Digital data is formatted in a number of ways in fiber-optic systems. To transmit information by fiber-optic cable, data is usually converted into a serial digital data stream. A common NRZ serial format is shown in Fig. 18-31(a). The NRZ format at A was discussed in Chap. 11. Each bit occupies a separate time slot and is either a binary 1 or binary 0 during that time period.

In the RZ format of Fig. 18-31(b), the same time period is allotted for each bit, but each bit is transmitted as a very narrow pulse (usually 50 percent of the bit time) or as an absence of a pulse. In some systems the Manchester or biphase code is used, as shown in Fig. 18-31(c). While the NRZ code can be used, the RZ and Manchester codes are easier to detect reliably at the higher data rates and are thus preferred.

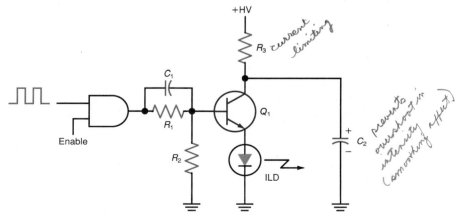

FIG. 18-30 An ILD transmitter.

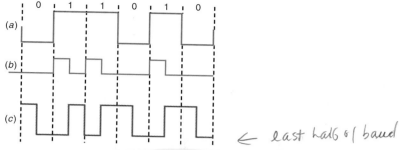

0 1 1 0 1 0

(a)

(b)

(c)   ← *last half of baud*

Fig. 18-31   Data formats used in fiber-optic transmission. (*a*) NRZ (nonreturn to zero). (*b*) RZ (return to zero). (*c*) Biphase (Manchester).

LIGHT DETECTORS

The receiver part of the optical communications system is relatively simple. It consists of a detector that senses the light pulses and converts them into an electrical signal. This signal is amplified and shaped into the original serial digital data. The most critical component is the light sensor.

PHOTODIODE.   The most widely used light sensor is a *photodiode*. This is a silicon PN-junction diode that is sensitive to light. This diode is normally reverse-biased, as shown in Fig. 18-32. The only current that flows through it is an extremely small reverse leakage current. When light strikes the diode, this leakage current increases significantly. This current flows through a resistor and develops a voltage drop across it. The result is an output voltage pulse.

PHOTOTRANSISTOR.   The reverse current in a diode is extremely small even when exposed to light. The resulting voltage pulse is very small, and so must be amplified. The base-collector junction is exposed to light. The base leakage current produced causes a larger emitter-to-collector current to flow. Thus the transistor amplifies the small leakage current into a larger, more useful output (see Fig. 18-33). Phototransistor circuits are far more sensitive to small light levels, but they are relatively slow. Thus further amplification and pulse shaping are normally used.

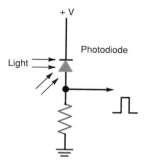

Fig. 18-32   The conversion of light, by a photodiode, into voltage pulses.

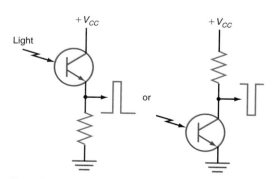

Fig. 18-33   A phototransistor is more sensitive to the light level.

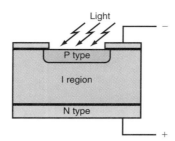

FIG. 18-34 Structure of a PIN photodiode.

PIN DIODE. The sensitivity of a standard PN-junction photodiode can be increased and the response time decreased by creating a new device that adds an undoped or *intrinsic (I) layer* between the P and N semiconductors. The result is a PIN diode (Fig. 18-34). The thin P layer is exposed to the light which penetrates to the junction, causing electron flow proportional to the amount of light. The diode is reverse-biased, and the current is very low until light strikes the diode, which significantly increases the current.

PIN diodes are significantly faster in response to rapid light pulses of high frequency. And their light sensitivity is far greater than that of an ordinary photodiode.

AVALANCHE DIODE. The *avalanche photodiode (APD)* is a more widely used photosensor. It is the fastest and most sensitive photodiode available, but it is expensive, and its circuitry is complex. Like the standard photodiode, the APD is reverse-biased. However, the operation is different. The APD uses the reverse breakdown mode of operation that is commonly found in zener and IMPATT microwave diodes. When a sufficient amount of reverse voltage is applied, an extremely high current flows because of the avalanche effect. Normally, several hundred volts of reverse bias, just below the avalanche threshold, are applied. When light strikes the junction, breakdown occurs, and a large current flows. This high reverse current requires less amplification than the small current in a standard photodiode. Germanium APDs are also significantly faster than the other photodiodes and are capable of handling the very high gigabit-per-second data rates possible in some systems.

LIGHT RECEIVERS

Figure 18-35 shows the basic circuit used in most receivers. The current through the photodiode generated when light is sensed produces a current which is amplified in an op amp. The op amp output voltage is the diode current multiplied by the feedback resistor R_f. Very high gain is used. Following the amplifier is an op amp comparator used as a shaping circuit that squares the pulses to ensure fast rise and fall times. The output is passed through a logic gate so that the correct binary voltage levels are produced.

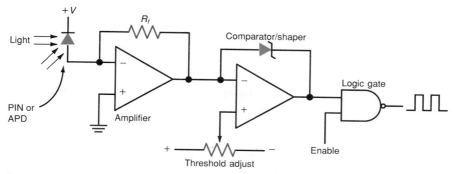

Fig. 18-35 A photodiode fiber-optic receiver.

PERFORMANCE CONSIDERATIONS

The most important specification in a fiber-optic communications system is the data rate, that is, the speed of the optical pulses. The very best systems use high-power injection laser diodes and APD detectors. This combination can produce data rates of several billion (giga) bits per second (bps). The rate is known as a *gigabit rate*. Most systems have a data rate of 10 to 500 Mbps. The popular FDDI fiber-optic LAN configuration use 100 Mbps.

The performance of a fiber-optic cable system is usually indicated by the *bit rate-distance product*. This rating tells the fastest bit rate that can be achieved over a 1-km cable. Assume a system with a 100 Mbps-km rating. This is a constant figure and is a product of the megabits per second and the kilometer values. If the distance increases, the bit rate decreases in proportion. In the above system, at 2 km the rate would drop to 50 Mbps. At 4 km, the rate is 25 Mbps and so on.

The upper data rate is also limited by the dispersion factor. The rise and fall times of the received pulse will be increased by an amount equal to the dispersion value. If the dispersion factor is 10 ns/km, over a 2-km distance the rise and fall times will be increased by 20 ns each. The data rate can never be more than the frequency corresponding to the sum of the rise and fall times at the receiver.

A handy formula for determining the maximum data rate R in megabits per second (Mbps) for a given distance D in kilometers of cable with a dispersion factor of d given in microseconds per kilometer (μs/km) is:

$$R = \frac{1}{5dD}$$

Assume a cable length of 8 km and a dispersion factor of 10 ns/km, or 0.01 μs/km:

$$R = \frac{1}{5(0.01)(8)} = \frac{1}{0.4} = 2.5 \text{ Mbps}$$

This relationship is only an approximation, but it is a handy way to predict system limitations.

Example 18-3

A measurement is made on a fiber-optic cable 1200 ft long. Its upper frequency limit is determined to be 43 Mbps. What is the dispersion factor d?

$$1 \text{ km} = 3274 \text{ ft}$$

$$D = 1200 \text{ ft} = \frac{1200}{3274} = 0.367 \text{ km}$$

$$R = \frac{1}{5dD}$$

$$d = \frac{1}{5RD}$$

$$= \frac{1}{5(43 \times 10^6)(0.367)} = 12.7 \text{ ns/km}$$

POWER BUDGET

A *power budget,* sometimes called a *flux budget,* is an accounting of all the attenuation and gains in a fiber-optic system. Gains must be greater than losses in order for the system to work. A designer must determine whether an adequate amount of light power reaches the receiver. Does the receiver get as much input light as its sensitivity dictates, given the power output of the transmitter and all the cable and other losses?

LOSSES.   There are numerous sources of losses along a fiber-optic cable system:

1. *Cable losses.*   These vary with the type of cable and its length. The range is from less than 1 dB/km up to tens of dB/km.
2. *Connections between cable and light source and photodetector.*   These vary widely depending upon how they are made. Today, the attachment is made by the manufacturer who specifies a loss factor. The resulting assembly has a connector to be attached to the cable. Such terminations can produce attenuations of 1 to 6 dB.
3. *Connectors.*   Despite their precision, connectors still introduce losses. These typically run from 0.5 to 2 dB each.
4. *Splices.*   If done properly, the splice may introduce an attenuation of only a few tenths of a decibel. But the amount can be much more, up to several decibels if splicing is done incorrectly.
5. *Cable bends.*   If a fiber-optic cable is bent at too great an angle, the light rays will come under a radically changed set of internal conditions. The total internal reflection will no longer be effective, for angles have changed due to the bend. The result is that some of the light will be lost by refraction in the cladding. If bend radii are made 1000 times more than the diameter of the cable, the losses are

These fiber-optic bundles are capable of carrying millions of telephone signals simultaneously.

minimal. Decreasing the bend radius to less than about 100 times the cable diameter will increase the attenuation. A bend radius approaching 100 to 200 times the cable diameter could cause cable breakage or internal damage.

All the above losses will vary widely depending upon the hardware used. Check the manufacturer's specifications for every component to be sure that you have the correct information. Then, as a safety factor, add 5 to 10 dB of loss. This contingency factor will cover incidental losses that cannot be predicted.

CALCULATING THE BUDGET. The losses work against the light generated by the LED or ILD. The idea is to use a light power sufficient to give an amount of received power in excess of the minimum receiver sensitivity with the losses in the system.

Assume a system with the following specifications:

1. Light transmitter LED output power: 30 μW
2. Light receiver sensitivity: 1 μW
3. Cable length: 6 km
4. Cable attenuation: 3 dB/km, $3 \times 6 = 18$ dB, total
5. Four connectors: attenuation 0.8 dB each, $4 \times 0.8 = 3.2$ dB total
6. LED to connector loss: 2 dB
7. Connector to photodetector loss: 2 dB
8. Cable dispersion: 8 ns/km
9. Data rate: 3 Mbps

First, calculate all the losses; add up all the decibel loss factors.

$$\text{Total loss, dB} = 18 + 3.2 + 2 + 2 = 25.2 \text{ dB}$$

Also add in a 4-dB contingency factor, making the total loss $25.2 + 4$, or 29.2 dB. What power gain is needed to overcome this loss?

$$dB = 10 \log \left(\frac{P_t}{P_r} \right)$$

P_t is the transmitted power, and P_r is the received power:

$$29.2 \text{ dB} = 10 \log \left(\frac{P_t}{P_r} \right)$$

Therefore:

$$\frac{P_t}{P_r} = 10^{db/10} = 10^{2.92} = 831.8$$

If P_t is 30 μW, then:

$$P_r = \frac{30}{831.8} = 0.036 \ \mu\text{W}$$

Accordingly, the received power will be 0.036 μW. The sensitivity of the receiver is only 1 μW. The received signal is below the threshold of the receiver. This problem may be solved in one of three ways:

1. Increase transmitter power.
2. Get a more sensitive receiver.
3. Add a repeater.

In an initial design, the problem would be solved by increasing transmitter power and/or increasing receiver sensitivity. Theoretically a lower-loss cable could also be used. Over short distances, a repeater is an unnecessary expense; therefore using a repeater is not a good option.

Assume that the transmitter output power is increased to 1 mW or 1000 μW. The new received power then is:

$$P_r = \frac{1000}{831.8} = 1.2 \ \mu\text{W}$$

This is just over the threshold of the receiver sensitivity. Now, we can determine the upper frequency or data rate.

$$R = \frac{1}{5dD} = \frac{1}{5(0.008)(6)} = 4.1666 \text{ Mbps}$$

This is higher than the proposed data rate of 3 Mbps so the system should work.

18-5 INFRARED COMMUNICATIONS

Fiber-optic communications, discussed in detail in the foregoing sections, is a form of infrared communications. There are other infrared applications, and in the next section, we will summarize how infrared is used in communications and discuss two common applications, namely, TV remote controls and infrared wireless LANs.

The source of infrared in a fiber-optic cable system is an LED or a laser diode. These same sources are also used in all other wireless infrared communications systems. However, there are many natural sources of infrared. In fact, all objects emit infrared radiation! Infrared radiation comes from heat. And all objects have and emit heat.

Infrared is generated by the vibration of the atoms in a substance when it is heated. As the temperature of an object is increased, the atoms are further agitated, causing them to vibrate. As a result they emit infrared rays. The higher the temperature, the greater the amount of infrared energy emitted.

As an object is cooled, it emits less heat. The colder the temperature, the less the atomic and molecular vibration and the smaller the amount of infrared emitted. At absolute zero, the lowest possible temperature ($-273°C$), all molecular motion stops, and no heat or infrared is emitted. This is a hypothetical state, for it is impossible to achieve absolute zero.

The color of an object also affects how much infrared is emitted. The best emitter color of infrared is black. It absorbs and radiates heat and infrared better than any other color. In the visible light spectrum, colors at the low end of the frequency range near infrared emit the greatest amount of heat and infrared. Red is the greatest emitter; the amount of infrared radiation decreases as the colors fade from red to orange to yellow to green to blue and then to violet, which is the "coolest" color. Distinguishing temperature by color provides a way to analyze infrared images and photographs to determine what they mean and represent.

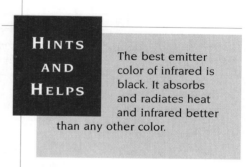

HINTS AND HELPS

The best emitter color of infrared is black. It absorbs and radiates heat and infrared better than any other color.

INFRARED COMMUNICATIONS SYSTEMS

This section describes two of the most widely used infrared communications systems: the ubiquitous TV remote control and the infrared wireless LAN.

TV REMOTE CONTROL. Almost every TV set sold these days, regardless of size or cost, has a wireless remote control. Other consumer electronic products have remote controls including VCRs, cable TV converters, CD players, stereo audio systems, and some ordinary radios. Generic remote controls are available to hook up to any device that you wish to control remotely.

All these devices work on the same basic principle. A small handheld battery-powered unit transmits a serial digital code via an infrared beam to a receiver that decodes it and carries out the specific action defined by the code. A TV remote control is one of the more sophisticated of these controls, for it requires many codes to perform volume control, channel selection, and other functions.

Figure 18-36 is a general block diagram of a remote control transmitter. In most modern units, all the circuitry, except perhaps for the IR LED driver transistors, is contained within a single IC. The purpose of the transmitter is to convert a keyboard entry into a serial binary code that is transmitted by IR to the receiver.

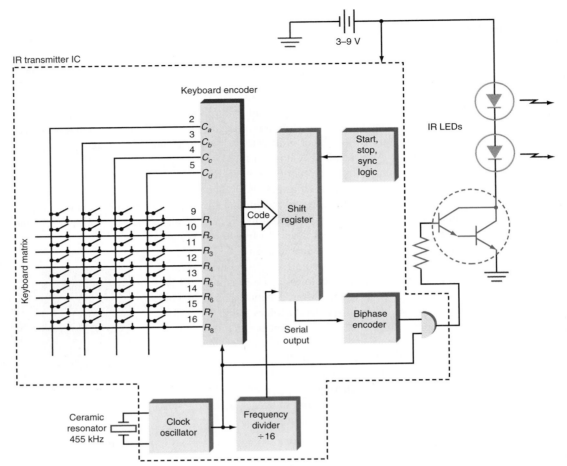

Fig. 18-36 IR TV remote control transmitter.

The keyboard is a matrix of momentary contact SPST pushbuttons. The arrangement shown is organized as 8 rows and 4 columns. The row and column connections are made to a keyboard encoder circuit inside the IC. Pulses generated internally are applied to the column lines. When a key is depressed, the pulses from one of the column outputs are connected to one of the row inputs. The encoder circuit converts this input into a unique binary code representing a number for channel selection or some function such as volume control. Some encoders generate a few as 6 bits and others generate up to 32 bit codes. Nine- and ten-bit codes are very common.

The serial output is generated by the shift register as data is shifted out. A standard NRZ serial code is generated. This is usually applied to a serial encoder to generate a standard biphase or Manchester code. Recall that the biphase code provides more reliable transmission and reception because there is a signal change for every 0-to-1 or 1-to-0 transition. The actual bit rate is usually in the 30 to 70 kbps range.

The serial bit stream turns a higher-frequency pulse source off and on according to the code's binary 1s and 0s. The transmitter IC contains a clock oscillator that runs at a frequency in the 445- to 510-kHZ range. A typical unit runs at 455 kHz using an external ceramic resonator to set the frequency. The serial data turns the 455-kHz pulses off and on. For example, a binary 1 generates a burst of 16 455-kHz pulses as shown

Modern communication research includes the use of argon lasers for light wave data links.

in Fig. 18-37. When a binary 0 occurs in the data train, no pulses are transmitted. The figure shows a 6-bit code (011001) with a start pulse. The period T of the 455-kHz pulses is 2.2 μs. The pulse width is set for a duty cycle of about 25 percent, or $T/4$.

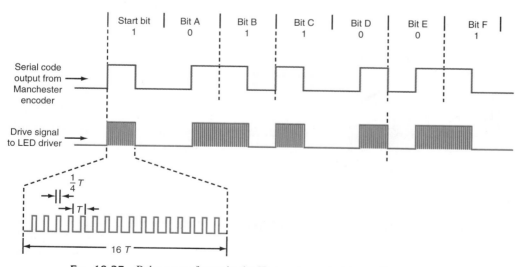

FIG. 18-37 Pulse wave forms in the IR remote control transmitter.

If 16 pulses make up a binary 1 interval, that duration is 16×2.2, or 35.2 μs. This translates into a code bit rate of $1/35.2 \times 10^{-6}$, or 28.410 kHz.

The 455-kHz pulses modulate the infrared light source by turning it off and on. The IR source is usually one or more IR LEDs. These are driven by a Darlington transistor pair external to the IC as shown in Fig. 18-36. Two or more LEDs are used to ensure a sufficient level of IR radiation to the receiver in the TV set. The LED current is usually very high, giving high IR output levels for reliable transmission to the receiver. Some remote units use three LEDs for a wide-angle transmission signal so that a high-amplitude signal will be received despite the direction in which the remote is pointed.

An IR receiver is shown in Fig. 18-38. The PIN IR photodiode is mounted on the front of the TV set, where it will pick up the IR signal from the transmitter. The received signal is very small despite the fact that the distance between the transmitter and receiver is only 6 to 15 ft on average. Two or more high-gain amplifiers boost the signal level. Most circuits have some form of AGC. The incoming pulses are then detected, shaped, and converted into the original serial data train. This serial data is then read by the control microcomputer that is usually part of the TV receiver.

The microcontroller is a dedicated microcomputer built into every TV set. A master control program is stored in a ROM. The microcomputer converts inputs from the remote control and front panel controls into output signals that control the various functions in a TV set such as channel selection and volume control.

The microcontroller inputs and decodes the incoming signal and then issues output control signals to all other circuits: the PLL frequency synthesizer that controls the TV tuner, the volume control circuits in the audio section, and in the more advanced receivers, chroma and video such as hue, saturation, brightness, and contrast. The microcontroller also generates, sometimes with the help of an external IC, the characters and simple graphics that can be displayed on the screen. Most microcontrollers also contain a built-in clock.

INFRARED LANS. Most LANs are implemented with cable, either coaxial or twisted pair. However, wireless LANs are growing in popularity because they do not have the

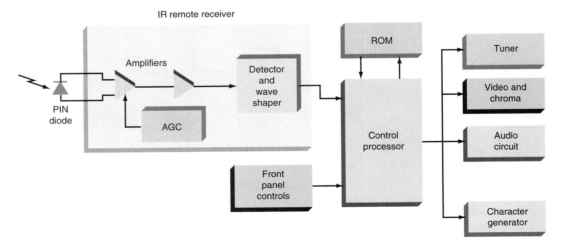

FIG. 18-38 The IR receiver and control microprocessor.

This fiber amplifier research and development resulted in improved delivery of video signals on optical fiber.

restriction and cost associated with laying cable and maintaining it. Most wireless LANs use radio. Spread spectrum modulation in the 902- to 928-MHz range is the most popular, but higher-frequency LANs in the 2.4- to 18-GHz range are available. Radio waves penetrate walls and ceilings to a significant distance, making wireless LANs very popular in offices where many moves occur. However, a growing alternative is the IR wireless LAN.

When infrared is used in a wireless LAN, each PC in the network has a special IR transceiver (see Fig. 18-39). A network adapter interface board in the PC performs all the packet formation and decoding normally associated with transmission and reception of data over the LAN. The serial data to be transmitted is fed to a transistor driver that operates one or more IR LEDs.

The interface also contains an IR receiver, which consists of an infrared PIN diode and the related circuitry. The received IR pulses are converted by the diode into electrical signals which are amplified, detected, shaped, and delivered to the interface that recovers the original serial data. The serial data is then translated back into parallel data that the computer expects to see. The PC and its network operating system software deal with the data as if it were received by a cable.

In order for the IR wireless LAN to work properly, all nodes must be within the line of sight of one another. IR waves travel in straight lines and cannot penetrate walls. As a result, the IR LEDs and PIN photodiodes on each unit must be within sight of one another. Transmission distances are usually limited to several hundred feet, and even that must be open free space with no obstructions. The IR module is normally mounted high on the wall and is connected to the PC via cable. Large high-power, ceiling-mounted units are also available.

Since IR wireless LANs demand line-of-sight conditions, the LED transmitter in one PC must be pointed directly at the PIN photodetector in the receiving PC, usually the LAN server. There must be a clear, unobstructed path between the two. IR LED transmitters may have a 30 to 60°

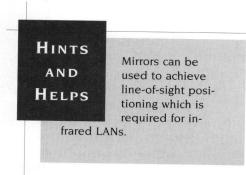

HINTS AND HELPS

Mirrors can be used to achieve line-of-sight positioning which is required for infrared LANs.

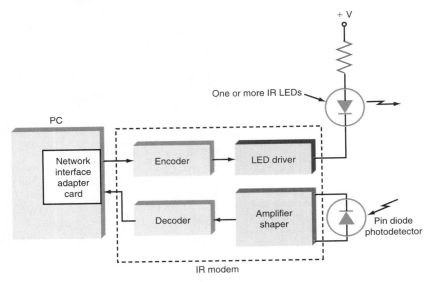

Fig. 18-39 IR wireless LAN transceiver.

radiation angle, which means that the receiver must be positioned to pick up optimum radiation.

As for speed, inexpensive low-end IR LANs have a speed of 115.2 kbps. Newer IR LANs permit speeds of up to 20 Mbps.

Although IR LANs do not have the range and "through-wall" flexibility of a radio wireless LAN, they are useful when office moves and rearrangements are common or anticipated. Rerouting of LAN cabling is expensive and causes a considerable amount of LAN downtime. With an IR LAN, moves are fast and easy with little disruption of normal work. IR wireless LAN equipment is also considerably less expensive than the radio modem units.

SUMMARY

Today, light is being used increasingly as the carrier for information in a communications system. Although visible light is used to some extent, most optical communications systems are infrared. The transmission medium is either free space or a special light-carrying vacuum called a *fiber-optic cable*. The light is modulated by the information to be transmitted. Because the frequency of light is extremely high, it can accommodate very wide bandwidths of baseband information and/or extremely high rates of data transmission with excellent reliability.

Light beam communication has become practical with the invention of the laser. The *laser* is a special high-intensity, single-frequency light source. It produces a very narrow beam of brilliant light of a specific wavelength (color). Because of its great intensity, the laser beam can penetrate through atmospheric obstacles better than other types of light can, thereby making light beam communications more reliable over longer distances. Instead of using free space, some type of light-carrying cable can also be used. Since the 1950s, first glass fibers, and later on low-cost plastic fibers, have been used in cable construction because they permit miles-long, light-carrying cables to be constructed. These fiber-optic cables have been refined to a high state, and they have an almost limitless capacity for information transfer.

Fiber-optic communications is only one form of infrared communications. Infrared is also used in military, scientific, medicinal, and industrial applications. It is the principal component of TV remote controls and other system controls. There are many sources of infrared, including natural sources. Infrared is generated by the vibration of the atoms in a substance when it is heated. As the temperature of an object is increased, the atoms are further agitated, causing them to vibrate. As a result, they emit infrared rays. The higher the temperature, the greater the amount of infrared energy emitted. As an object is cooled, it emits less heat. Any object that generates heat emits infrared. Automobiles generate a great deal of heat, and therefore infrared. The human body is an enormous generator of heat and infrared. All heat sources are infrared transmitters and can form the transmitter end of an infrared communications system.

S
U
M
M
A
R
Y

KEY TERMS

Absorption
Amplitude modulation
Angle of incidence
Angstrom (Å)
Attenuation
Avalanche photodiode (APD)
Bandwidth
Cable size
Cladding
Cone of acceptance
Connectors

Core
Critical angle
Data format
Data rate
Fiber-optic cable
Fiber-optic communications system
Frequency range
Gallium arsenide (GaAs) semiconductor
Glass
Graded index cable

Incident ray
Index of refraction
Infrared communications
Infrared spectrum
Injection laser diode (ILD)
Intensity modulation
Jacket
Laser
Laser diode
Laser transmitter
Law of reflection
LED transmitter

Light
Light detector
Light receiver
Light transmitter
Light-emitting diode
    (LED)
Local area network
    (LAN)
Losses in fiber-optic cable
    system
Micrometer (μm) or
    micron
Millimeter waves
Mode of cable
Modulator
Multimode graded index
    cable

Multimode step index
    cable
Nanometer
    (nm)
Numerical aperture
    (NA)
Optical communications
    systems
Optical spectrum
Optics
Photodiode
Phototransistor
Physical optics
PIN diode
Power (flux) budget
Pulse modulation
Receiver

Reflection
Refraction
Silica
Single-mode step index
    cable
Speed of light
Splicing
Step index cable
Threshold
TV remote control
Ultraviolet range
Visible light

QUESTIONS

1. True or false. Light is electromagnetic radiation.
2. The optical spectrum is made up of three parts. Name them.
3. Which segment of the optical spectrum has the highest frequencies? The lowest frequencies?
4. Light waves travel in a:
    a. Circle
    b. Straight line
    c. Curve
    d. Random path
5. What units are used to express the wavelength of light?
6. State the lowest wavelength and highest wavelength of visible light, and state the color of each.
7. What is the wavelength range of infrared light?
8. True or false. The speed of light is faster in glass or plastic than it is in air.
9. What is the name of the number that tells how fast light travels in a medium compared to air?
10. What device can be used to bounce or change the direction of a light wave?
11. What is the term used to describe the bending of light rays due to speed changes when moving from one medium to another?
12. When the angle of refraction is 90° to the normal, describe how the ray travels with relationship between the two media involved.

13. What special effect occurs when the incident ray strikes the interface between two media at an angle greater than the critical angle?
14. What factor determines the critical angle in a medium?
15. Of what value are ultraviolet rays in communications?
16. In an optical communications system, name the two most common transmission media.
17. Describe how a light source is modulated. What type of modulation is the most common?
18. Describe how both analog and digital signals may modulate a light transmitter.
19. What are the two main types of light sources used in optical communications systems? Which is the preferred and why?
20. What factors limit the free-space transmission of information on a light beam?
21. What type of signals do fiber-optic cables carry?
22. What optical principle makes fiber-optic cable possible?
23. What two materials are used to make fiber-optic cable?
24. Name the three main types of information carried by fiber-optic cables.
25. What is the major application of fiber-optic cable?
26. State the main benefit of fiber-optic cable over electrical cable.
27. True or false. Fiber-optic cable has more loss than electrical cable over long distances.
28. True or false. Fiber-optic cable is smaller, lighter, and stronger than electrical cable.
29. State the two main disadvantages of fiber-optic cable.
30. What is the name of the device that converts the light pulses into an electrical signal?
31. What is the name given to the regenerative units used to compensate for signal attenuation over long distances?

32. Which material has the best optical characteristics and lowest loss?
    a. Plastic
    b. Glass
    c. They are equal.
33. What covers and protects the core in a fiber-optic cable?
34. In PCS–type cable, what materials are used to make the core and the cladding?
35. The index of refraction is highest in the:
    a. Core
    b. Cladding
36. List the three main types of fiber-optic cable.
37. What is the name given to the phenomenon of stretching of the light pulse by the fiber-optic cable? What actually causes it?
38. Light pulse stretching occurs in what two types of cable? What type of fiber-optic cable does not cause any significant degree of pulse stretching?
39. What type of cable is the best to use for very high frequency pulses?
40. Pulse stretching causes the information capacity of a cable to:
    a. Increase
    b. Decrease
41. Define what is meant by *graded index fiber-optic cable.*
42. What is the typical core diameter range of a single-mode step index cable?
43. What types of covering are applied over the cladding in a fiber-optic cable to protect against moisture and damage?
44. True or false. Fiber-optic cables are available with multiple cores.
45. What term is used to refer to light loss in a cable?
46. What three major factors cause light loss in a fiber-optic cable?
47. How is the amount of light loss in a fiber-optic cable expressed and measured?
48. True or false. Fiber-optic cables may be spliced.
49. State how fiber-optic cables are conveniently linked and attached to one another and to related equipment.

50. Name the designations of the two most common types of fiber-optic cable connectors, and state the primary difference between them.
51. Name two common and popular sizes of fiber-optic cable.
52. Define what is meant by *bandwidth* as it applies to fiber-optic cable. What units are used to express bandwidth?
53. What causes the greatest loss peaks in a fiber-optic cable. At what frequencies do these peak losses occur?
54. What are the two most common light sources used in fiber-optic transmitters?
55. True or false. Visible light is the most common type of light used in fiber-optic systems.
56. Name the three most common light frequencies used in fiber-optic cable systems. Why are these frequencies used?
57. True or false. The light from a 1.55-μm LED is visible.
58. What semiconductor material are LEDs usually made of?
59. What term is used to describe a single light frequency?
60. What do you call the condition of all emitted light waves being in phase?
61. What light source produces light like that described in questions 59 and 60 above?
62. What special structure is created when reflective surfaces are added to a laser diode? How does it affect the light waves generated and emitted?
63. For normal operation, LEDs and ILDs are:
    a. Reverse-biased
    b. Forward-biased
64. Which is fastest?
    a. LED
    b. ILD
65. Which produces the brightest light?
    a. LED
    b. ILD
66. During normal operation, all photodiodes are:
    a. Reverse-biased
    b. Forward-biased
67. Name the two most sensitive and fastest light detectors.
68. Name the two main circuits in a fiber-optic receiver.
69. Besides the reason stated in the text, name one key reason that Manchester encoding might be preferred.
70. What popular local area network system uses fiber-optic cable?
71. What electrical unit is used to state the output of a light transmitter and the sensitivity of a light receiver?
72. What causes infrared to be radiated?
73. Explain why and how IR is radiated.
74. What colors radiate the most IR? The least?
75. Explain how pressing a button on a TV remote control generates an IR signal.
76. What is the typical data rate of a TV remote control?
77. What frequency is used to modulate the IR beam?
78. What encoding method is commonly used in IR data transmission?
79. What component in the TV receiver actually decodes the serial data code sent by the transmitter and initiates the indented action?

PROBLEMS

1. State the speed of light in air in meters per second and miles per second.
2. If a light ray strikes a mirror at an angle of 28.4 degrees from the normal, at what angle is it reflected from the normal?
3. Draw a simple block diagram of an optical communications system, and explain the purpose and operation of each element.
4. Explain the process by which voice and video signals are transmitted by light beam.
5. A cable has a loss of 9 dB. What percentage of its input power will appear at the output?
6. Express 1.2 km in terms of miles and feet.
7. How many kilometers are there in 6 mi?
8. Four cables with attenuations of 7, 16, 29, and 34 dB are spliced together. What is the total attenuation in decibels?
9. A fiber-optic cable has a bandwidth of 160 MHz-km. What is the bandwidth of a 1-mi segment? A 0.5-mi segment?
10. Explain briefly how light falling on the PN junction of a photodiode causes the diode's conductance to change.
11. The bit rate-distance product of a system is 600 Mbps-km. What is the speed rating at 8 km?
12. What is the average maximum distance between repeaters in a fiber-optic system?
13. Explain how analog signals are transmitted over a fiber-optic cable.
14. A fiber-optic cable system has a dispersion factor of 33 ns/km. The length of the system is 0.8 km. What is the highest data rate that can be achieved on this link?
15. Calculate the minimum receiver sensitivity to reliably detect a signal transmitted with a power of 0.7 mW over a link that is 3.5 km long with attenuation of 2.8 dB/km. It has two splices with attenuation of 0.3 dB each and four connectors with attenuation of 1 dB each. The losses in the connections to the IPD and APD devices are 2 dB each. Assume a contingency factor of 5 dB. The dispersion factor is 18 ns/km. What is the maximum data rate that this system can achieve?

CRITICAL THINKING

1. Name three potential new applications for fiber-optic communications not on the list in Fig. 18-9 but that take advantage of the benefits listed in Fig. 18-10.
2. Explain how a single fiber-optic cable can handle two-way communications, both half and full duplex.
3. Could an incandescent light be used for a fiber-optic transmitter? Explain its possible benefits and disadvantages.
4. Compare a wireless radio system with a fiber-optic communications system for digital data communications over a distance of 1 km. Assume a desired data rate of 75M bps. Give pros and cons, advantages and disadvantages, of each. Which one would be best, all factors being considered?
5. Describe how a simple IR system could be used to replace a cable between a personal computer and a laser printer. Give typical data rates and distance limitations as well as the pros and cons of an IR wireless system.

CHAPTER 19

TELEVISION

Objectives

After completing this chapter, you will be able to:

◆ *Describe* and give specifications for a complete television signal including all of its individual components.

◆ *Explain* the process used by a television camera to convert a visual scene into a video signal.

◆ *Draw* a simplified block diagram, showing main components, of a television transmitter, a television receiver and the signal flow, a television tuner, an IF amplifier and video detector section of a TV receiver, an audio and sweep section of a TV receiver, a picture tube, and a satellite TV receiver.

◆ *Draw* a block diagram of a cable TV system.

◆ *Name* all the elements of and explain the operation of a cable TV system.

◆ *Explain* the difference between a conventional satellite receiver and a DBS TV receiver.

◆ *Define* *high-definition television* (HDTV), and state the basic specifications of proposed HDTV receivers.

To say that the technological developments within the field of electronic communications have greatly affected our lives would be a gross understatement. And perhaps none of these developments has had more influence than TV. Recent census data indicates that 99 percent of all U.S. households have TV sets, which means that more U.S. households have TV sets than bathtubs. Estimates indicate that the average person watches TV 3 to 7 hours per day. It is our primary source of entertainment and the major method by which we are informed. Because of TV's great impact upon our lives, it is worthwhile to understand this technology.

This chapter introduces you to the subject of TV technology. It embodies almost all the principles and circuits covered elsewhere in this book. Studying TV is an excellent review of the communications fundamentals including modulation and multiplexing, transmitters and receivers, antennas and transmission lines, and even digital techniques.

In addition to standard audio transmission, TV systems use a TV camera to convert a visual scene into a voltage known as the *video signal.* The video signal represents the picture information and is used to modulate a transmitter. Both the picture and sound signals are transmitted to the receiver, which demodulates the signals and presents the information to the user. The TV receiver is a special superheterodyne that recovers both the sound and the picture information. The picture is displayed on a picture tube.

19-1 TV SIGNAL

A considerable amount of intelligence is contained in a complete TV signal. As a result, the signal occupies a significant amount of spectrum space. As indicated earlier, the TV signal consists of two main parts: the sound and the picture. But it is far more complex than that. The sound today is usually stereo, and the picture carries color information as well as the synchronizing signals that keep the receiver in step with the transmitter. In the next section the complete TV signal and the position it occupies in the electromagnetic spectrum will be discussed.

SIGNAL BANDWIDTH

The complete signal bandwidth of a TV signal is shown in Fig. 19-1. The entire TV signal occupies a channel in the spectrum with a bandwidth of 6 MHz. There are two carriers, one each for the picture and the sound.

AUDIO SIGNAL.   The sound carrier is at the upper end of the spectrum. Frequency modulation is used to impress the sound signal on the carrier. The audio bandwidth of the signal is 50 Hz to 15 kHz. The maximum permitted frequency deviation is 25 kHz, considerably less than the deviation permitted by conventional FM broadcasting. As a result, a TV sound signal occupies somewhat less bandwidth in the spectrum than a

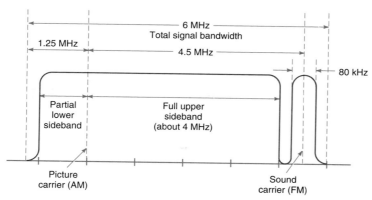

FIG. 19-1 Spectrum of a broadcast TV signal.

standard FM broadcast station. Stereo sound is also available in TV, and the multiplexing method used to transmit two channels of sound information is virtually identical to that used in stereo transmission for FM broadcasting.

VIDEO SIGNAL. The picture information is transmitted on a separate carrier located 4.5 MHz lower in frequency than the sound carrier (refer again to Fig. 19-1). The video signal derived from a camera is used to amplitude modulate the picture carrier. Different methods of modulation are used for both sound and picture information so that there is less interference between the picture and sound signals. Further, amplitude modulation of the carrier takes up less bandwidth in the spectrum, and this is important when a high-frequency, content-modulating signal such as video is to be transmitted.

Note in Fig. 19-1 that vestigial sideband AM is used. The full upper sidebands of the picture information are transmitted, but a major portion of the lower sidebands is suppressed to conserve spectrum space. Only a vestige of the lower sibeband is transmitted.

The color information in a picture is transmitted by way of frequency division multiplexing techniques. Two color signals derived from the camera are used to modulate a 3.85-MHz subcarrier which, in turn, modulates the picture carrier along with the main video information. The color subcarriers use double-sideband suppressed carrier AM.

The video signal can contain frequency components up to about 4.2 MHz. Therefore, if both sidebands were transmitted simultaneously, the picture signal would occupy 8.4 MHz. The vestigial sideband transmission reduces this excessive bandwidth.

TV SPECTRUM ALLOCATION. Because a TV signal occupies so much bandwidth, it must be transmitted in a very high frequency portion of the spectrum. TV signals are assigned to frequencies in the VHF and UHF range. U.S. TV stations use the frequency range between 54 and 806 MHz. This portion of the spectrum is divided into 68 6-MHz channels which are assigned frequencies (see Fig. 19-2). Channels 2 through 7 occupy the frequency range from 54 to 88 MHz. The standard FM radio broadcast band occupies the

CHANNEL	FREQUENCY, MHz	CHANNEL	FREQUENCY, MHz
Low-band VHF		**UHF (cont.)**	
2	54–60	32	578–584
3	60–66	33	584–590
4	66–72	34	590–596
5	76–82	35	596–602
6	82–88	36	602–608
FM broadcast	88–108	37	608–614
Aircraft	118–135	38	614–620
Ham radio	144–148	39	620–626
Mobile or marine	150–173	40	626–632
		41	632–638
		42	638–644
High-band VHF		43	644–650
7	174–180	44	650–656
8	180–186	45	656–662
9	186–192	46	662–668
10	192–198	47	668–674
11	198–204	48	674–680
12	204–210	49	680–686
13	210–216	50	686–692
UHF		51	692–698
		52	698–704
14	470–476	53	704–710
15	476–482	54	710–716
16	482–488	55	716–722
17	488–494	56	722–728
18	494–500	57	728–734
19	500–506	58	734–740
20	506–512	59	740–746
21	512–518	60	746–752
22	518–524	61	752–758
23	524–530	62	758–764
24	530–536	63	764–770
25	536–542	64	770–776
26	542–548	65	776–782
27	548–554	66	782–788
28	554–560	67	788–794
29	560–566	68	794–800
30	566–572	69	800–806
31	572–578	Cellular telephone	806–902

FIG. 19-2 VHF and UHF TV channel frequency assignments.

88- to 108-MHz range. Aircraft, amateur radio, and marine and mobile radio communications services occupy the frequency spectrum from approximately 118 to 173 MHz. Additional TV channels occupy the space between 470 and 806 MHz. Figure 19-2 shows the frequency range of each TV channel.

To find the exact frequencies of the transmitter and sound carriers, use Fig. 19-2 and the spectrum outline in Fig. 19-1. To compute the picture carrier, add 1.25 MHz to the lower frequency of range given in Fig. 19-2. For example, for channel 6, the lower frequency is 82 MHz. The picture carrier is 82 + 1.25, or 83.25 MHz. The sound carrier is 4.5 MHz higher, or 83.25 + 4.5, that is, 87.75 MHz.

Example 19-1

Compute the picture and sound carrier frequencies for UHF TV channel 39.

1. From Fig. 19-2, channel 39 extends from 620 to 626 MHz.
2. The picture carrier is 1.25 MHz above the lower band limit, or
$$1.25 + 620 = 621.25 \text{ MHz}$$
3. The sound carrier is 4.5 MHz above the picture carrier:
$$4.5 + 621.25 = 625.75 \text{ MHz}$$

GENERATING THE VIDEO SIGNAL

The video signal is most often generated by a TV camera, a very sophisticated electronic device that incorporates lenses and light-sensitive transducers to convert the scene or object to be viewed into an electrical signal that can be used to modulate a carrier. All visible scenes and objects are simply light that has been reflected and absorbed and then transmitted to our eyes. It is the purpose of the camera to take the light intensity and color details in a scene and convert them into an electrical signal.

To do this, the scene to be transmitted is collected and focused by a lens upon a light-sensitive imaging device. Both vacuum tube and semiconductor devices are used for converting the light information in the scene into an electrical signal. Some examples are the *vidicon tube* and the *charge coupled device (CCD)* so widely used in camcorders and all modern TV cameras.

The scene is divided into smaller segments that can be transmitted serially over a period of time. Again, it is the job of the camera to subdivide the scene in an orderly manner so that an acceptable signal is developed. This process is known as *scanning.*

DID YOU KNOW?

A given scene is divided into segments that can be transmitted serially over a period of time, because any scene contains so much light information that it would be impossible for an electronic device to perform a simultaneous conversion of all of it.

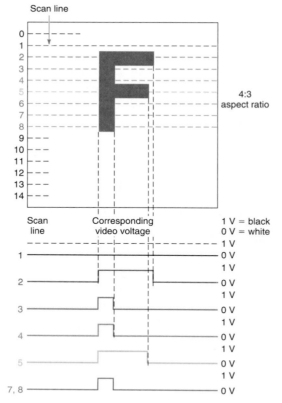

Fig. 19-3 Simplified explanation of scanning.

PRINCIPLES OF SCANNING. Scanning is a technique that divides a rectangular scene up into individual lines. The standard TV scene dimensions have an aspect ratio of 4:3; that is, the scene width is 4 units for every 3 units of height. To create a picture, the scene is subdivided into many fine horizontal lines called *scan lines.* Each line represents a very narrow portion of light variations in the scene. The greater the number of scan lines, the higher the resolution and the greater the detail that can be observed. U.S. TV standards call for the scene to be divided into a maximum of 525 horizontal lines.

Figure 19-3 is a simplified drawing of the scanning process. In this example, the scene is a large black letter F on a white background. The task of the TV camera is to convert this scene into an electrical signal. The camera accomplishes this by transmitting a voltage of 1 V for black and 0 V for white. The scene is divided into 15 scan lines numbered 0 through 14. The scene is focused on the light-sensitive area of a vidicon tube or CCD imaging device which scans the scene 1 line at a time, transmitting the light variations along that line as voltage levels. Figure 19-3 shows the light variations along several of the lines. Where the white background is being scanned, a 0-V signal occurs. When a black picture element is encountered, a 1-V level is transmitted. The electrical signals derived from each scan line are referred to as the *video signal.* They are transmitted serially one after the other until the entire scene has been sent (see Fig. 19-4). This is exactly how a standard TV picture is developed and transmitted.

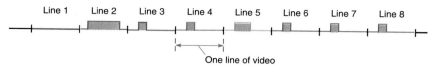

FIG. 19-4 The scan line voltages are transmitted serially. These correspond to the scanned letter F in Fig. 19-3.

Since the scene contains colors, there are different levels of light along each scan line. This information is transmitted as different shades of gray between black and white. Shades of gray are represented by some voltage level between the 0- and 1-V extremes represented by white and black. The resulting signal is known as the *brightness,* or *luminance, signal* and is usually designated by the letter *Y.*

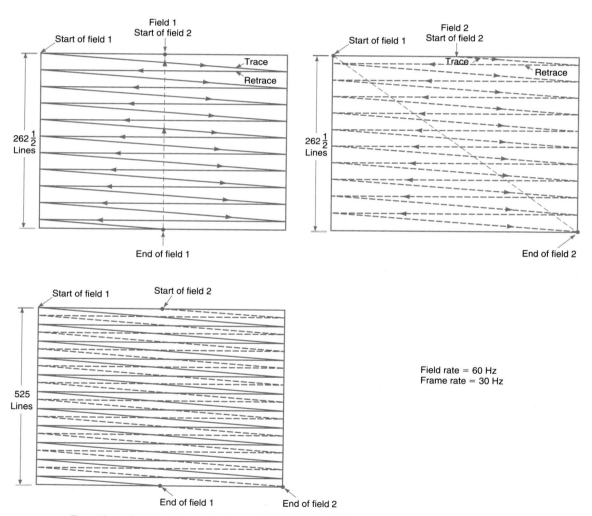

Field rate = 60 Hz
Frame rate = 30 Hz

FIG. 19-5 Interlaced scanning is used to minimize flicker.

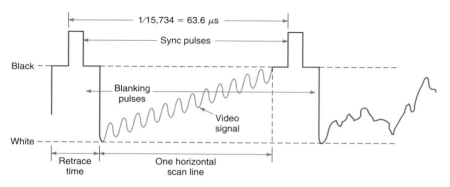

FIG. 19-6 Sync pulses are used to keep the receiver in step with the transmitter.

A more detailed illustration of the scanning process is given in Fig. 19-5. The scene is scanned twice. One complete scanning of the scene is called a *field* and contains 262½ lines. The entire field is scanned in 1/60 of a second for a 60-Hz field rate. In color TV the field rate is 59.94 Hz. Then the scene is scanned a second time, again using 262½ lines. This second field is scanned in such a way that its scan lines fall between those of the first field. This produces what is known as *interlaced scanning,* with a total of $2 \times 262\frac{1}{2} = 525$ lines. In practice, only about 480 lines show on the picture tube screen. Two interlaced fields produce a complete frame of video. With the field rate being 1/60 of a second, two fields produce a frame rate of 1/30 of a second, or 30 Hz. The frame rate in color TV is one-half the field rate, or 29.97 Hz. Interlaced scanning is used to reduce flicker, which is annoying to the eye. This rate is also fast enough that the human eye cannot detect individual scan lines, which results in a stable picture.

The rate of occurrence of the horizontal scan lines is 15,750 Hz for monochrome, or black and white, TV and 15,734 Hz for color TV. This means that it takes about 1/15,734 s, or 63.6 μs, to trace out 1 horizontal scan line.

At the TV receiver, the picture tube is scanned in step with the transmitter to accurately reproduce the picture. To ensure that the receiver stays exactly in synchronization with the transmitter, special horizontal and vertical sync pulses are added to and transmitted with the video signal (see Fig. 19-6). After 1 line has been scanned, a horizontal blanking pulse comes along. At the receiver, the blanking pulse is used to cut off the electron beam in the picture tube during the time the beam must retrace from right to left to get ready for the next left to right scan line. The horizontal sync pulse is used at the receiver to keep the sweep circuits that drive the picture tube in step with the transmitted signal. The width of the horizontal blanking pulse is about 10 μs. Since the total horizontal period is 63.6 μs, only about 53.5μs is devoted to the video signal.

At the end of each field, the scanning must retrace from bottom to top of the scene so that the next field can be scanned. This is initiated by the vertical blanking and sync pulses. The entire vertical pulse blanks the picture tube during the vertical retrace. The pulses on top of the vertical blanking pulse are the horizontal sync pulses that must continue to keep the horizontal sweep in sync during the vertical retrace. The equalizing pulses help synchronize the half scan lines in each field. Approximately 30 to 40 scan lines are used up during the vertical blanking interval. Therefore, only 480 to 495 lines of actual video are shown on the screen.

**RELATIONSHIP BETWEEN RESOLUTION AND BAND-
WIDTH.** Scanning a scene or picture is a kind of sampling
process. Consider the scene to be a continuous variation of
light intensities and colors. To capture this scene and transmit it electronically, the light intensity and color variations
must be converted into electrical signals. This conversion is
accomplished through a process called *scanning,* whereby
the picture is divided into many fine horizontal lines next to
one another.

The *resolution* of the picture refers to the amount of detail that can be shown. Pictures with high resolution have excellent *definition,* or distinction of detail, and the pictures ap-
pear to be clearly focused. A picture lacking detail looks *softer,* or somewhat out of focus. The bandwidth of a video system determines the resolution. The greater the bandwidth, the greater the amount of definition and detail.

Resolution in a video system is measured in terms of the number of lines defined
within the bounds of the picture. For example, the horizontal resolution (R_H) is given as
the maximum number of alternating black-and-white vertical lines that can be distinguished. Assume closely spaced vertical black-and-white lines of the same width. When
such lines are scanned, they will be converted into a square wave (50 percent duty cycle). One cycle, or period, t of this square wave is the time for 1 black and one white
line. If the lines are very thin, the resulting period will be short and the frequency will
be high. If the lines are wide, the period will be longer and the resulting frequency lower.

The *National Television Standards Committee* (*NTSC*) system restricts the bandwidth in the United States to 4.2 MHz. This translates into a period of 0.238 μs, or
238 ns. The width of a line is one-half this value, or 0.238/2 μs, or 0.119 μs. Remember that the horizontal sweep interval is about 63.6 μs. About 10 μs of this interval is taken up by the horizontal blanking interval, leaving 53.5 μs for the video. The
displayed scan line takes 53.5 μs. With 0.119 μs per line, 1 horizontal scan line can
resolve, or contain, up to 53.5/0.119, or 449.5, vertical lines. Therefore, the approximate horizontal resolution R_H is about 450 lines.

The vertical resolution R_V is the number of horizontal lines that can be distinguished. Only about 480 to 495 horizontal lines are shown on the screen. The vertical
resolution is about 0.7 times the number of actual lines N_L:

$$R_V = 0.7 \, N_L$$

If 485 lines are shown, the vertical resolution is 0.7 × 485, or 340, lines.

COLOR SIGNAL GENERATION. The video signal as described so far contains the
video or luminance information, which is a black-and-white version of the scene. This
is combined with the sync pulses. Now the color detail in the scene must somehow be
represented by an electrical signal. This is done by dividing the light in each scan line
into three separate signals, each representing one of the three basic colors, red, green,
or blue. It is a principle of physics that any color can be made by mixing some combination of the three primary light colors (see Fig. 19-7).

In the same way, the light in any scene can be divided into its three basic color
components by passing the light through red, green, and blue filters. This is done in a
color TV camera, which is really three cameras in one (see Fig. 19-8). The lens focuses the scene on three separate light-sensitive devices such as a vidicon tube or a
CCD imaging device by way of a series of mirrors and beam splitters. The red light in
the scene passes through the red filter, the green through the green filter, and the blue

Example 19-2

The European PAL TV system uses 625 interlaced scan lines occurring at a rate of 25 frames per second. The horizontal scanning rate is 15,625 Hz. About 80 percent of 1 complete horizontal scan is devoted to the displayed video, and 20 percent to the horizontal blanking. Assume that the horizontal resolution R_H is about 512 lines. Only about 580 horizontal scan lines are displayed on the screen. Calculate:

1. Bandwidth of the system
2. The vertical resolution

1. The time for 1 horizontal scan is;
$$\frac{1}{15,625} = 64 \ \mu s$$
About 80 percent of this is devoted to the video or:
$$0.8 \times 64 = 51.2 \ \mu s$$
If the horizontal resolution is 512 lines, then the time for 1 line is:
$$\frac{51.2}{512} = 0.1 \ \mu s$$
Two lines equals 1 period or:
$$2 \times 0.1 = 0.2 \ \mu s$$
Converting this to frequency gives the approximate bandwidth:
$$BW = \frac{1}{0.2 \times 10^{-6}} = 5 \ MHz$$

2. $R_V = 0.8 N_L = 0.8 \times 580 \ \text{lines} = 464 \ \text{lines}$

through the blue filter. The result is the generation of three simultaneous signals (R, G, and B) during the scanning process by the light-sensitive imaging devices.

The R, G, and B signals also contain the basic brightness or luminance information. If the color signals are mixed in the correct proportion, the result is the standard B&W video or luminance Y signal. The Y signal is generated by scaling each color

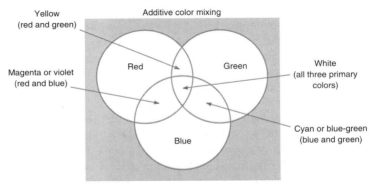

Fig. 19-7 Creating other colors with red, green, and blue light.

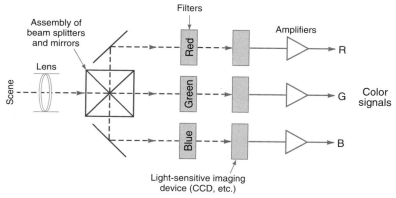

FIG. 19-8   How the camera generates the color signals.

signal with a tapped voltage divider and adding the signals together as shown in Fig. 19-9(a). Note that the Y signal is made up of 30 percent red, 59 percent green, and 11 percent blue. The resulting Y signal is what a B&W TV set will see.

The color signals must also be transmitted along with the luminance information in the same bandwidth allotted to the TV signal. This is done by a frequency division multiplexing technique shown in Fig. 19-9(a). Instead of all three color signals being transmitted, they are combined into color signals referred to as the I and Q signals. These signals are made up of different proportions of the R, G, and B signals according to the following specifications:

$$I = 60 \text{ percent red, } 28 \text{ percent green, } -32 \text{ percent blue}$$
$$Q = 21 \text{ percent red, } -52 \text{ percent green, } 31 \text{ percent blue}$$

The minus signs in the above expressions mean that the color signal has been phase-inverted before the mixing process.

The I and Q signals are referred to as the *chrominance signals.* To transmit them, they are *phase-encoded;* that is, they are used to modulate a subcarrier which is in turn mixed with the luminance signal to form a complete, or composite, video signal. These I and Q signals are fed to balanced modulators along with 3.58-MHz (actually 3.579545-MHz) subcarrier signals that are 90° out of phase [again refer to Fig. 19-9(a)]. This type of modulation is referred to as a *quadrature modulation,* where *quadrature* means a 90° phase shift. The output of each balanced modulator is a double-sideband suppressed carrier AM signal. The resulting two signals are added to the Y signal to create the composite video signal. The combined signal modulates the picture carrier. The resulting signal is called the *NTSC composite video signal.* This signal and its sidebands are within the 6-MHz TV signal bandwidth.

The I and Q color signals are also called the $R - Y$ and the $B - Y$ *signals* as the combination of the three color signals produces the effect of subtracting Y from the R or B signals. The phase of these signals with respect to the original 3.58-MHz subcarrier signal determines the color to be seen. The color tint can be varied at the receiver so that the viewer sees the correct colors. In many TV sets an extra phase shift of 57° is inserted to ensure that maximum color detail is seen. The resulting I and Q signals are shown as phasors in Fig. 19-9(b). There is still 90° between the I and Q signals, but their position is moved 57°. The reason for this extra phase shift is that the eye is more sensitive to the color orange. If the I signal is adjusted to the orange phase

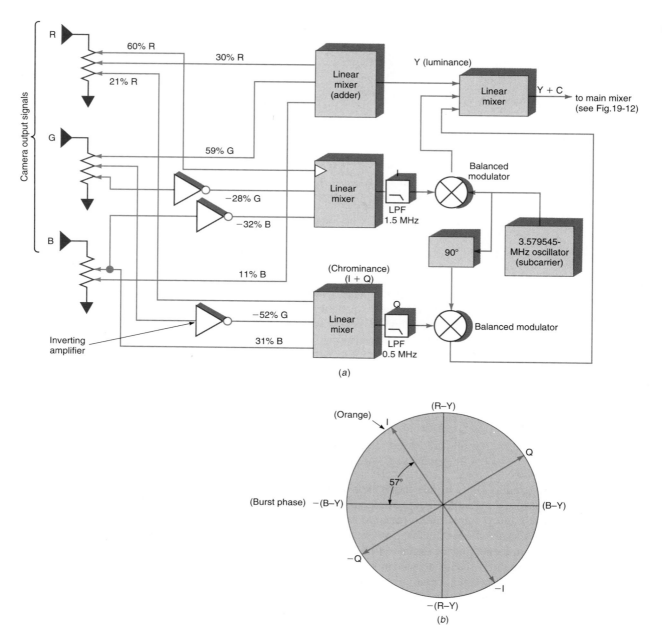

FIG. 19-9 (*a*) How the NTSC composite video signal is generated. (*b*) The chrominance signals are phase-encoded.

position, better detail will be seen. The *I* signal is transmitted with more bandwidth than the *Q* signal, as can be seen by the response of the low-pass filters at the outputs of the *I* and *Q* mixers in Fig. 19-9(*a*).

The complete spectrum of the transmitted color signal is shown in Fig. 19-10. Note the color portion of the signal. Because of the frequency of the subcarrier, the sidebands produced during amplitude modulation occur in clusters that are interleaved between the other sidebands produced by the video modulation.

Remember that the 3.58-MHz subcarrier is suppressed by the balanced modulators and therefore is not transmitted. Only the filtered upper and lower sidebands of

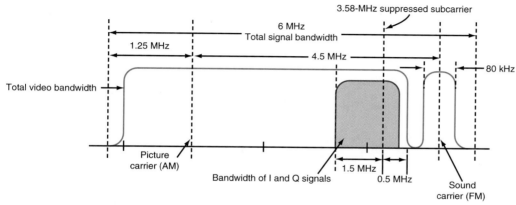

FIG. 19-10 The transmitted video and color signal spectrum.

the color signals are transmitted. To demodulate these *double-sided (DSB) AM signals,* the carrier must be reinserted at the receiver. A 3.58-MHz oscillator in the receiver generates the subcarrier for the balanced modulator-demodulator circuits.

For the color signals to be accurately recovered, the subcarrier at the receiver must have a phase related to the subcarrier at the transmitter. To ensure the proper conditions at the receiver, a sample of the 3.58-MHz subcarrier signal developed at the transmitter is added to the composite video signal. This is done by gating 8 to 12 cycles of the 3.58-MHz subcarrier and adding it to the horizontal sync and blanking pulse as shown in Fig. 19-11. This is called the *color burst,* and it rides on what is called the *back porch* of the horizontal sync pulse. The receiver uses this signal to phase-synchronize the internally generated subcarrier before it is used in the demodulation process.

A block diagram of the TV transmitter is shown in Fig. 19-12. Note the sweep and sync circuits that create the scanning signals for the vidicons or CCDs as well as generate the sync pulses that are transmitted along with the video and color signals. The sync signals, luminance Y, and the color signals are added to form the final video signal that is used to modulate the carrier. Low-level AM is used. The final AM signal is amplified by very high power linear amplifiers and sent to the antenna via a diplexer, which is a set of sharp bandpass filters that pass the transmitter signal to the antenna but prevent signals from getting back into the sound transmitter.

At the same time, the voice or sound signals frequency-modulate a carrier that is amplified by class C amplifiers and fed to the same antenna by way of the diplexer. The resulting VHF or UHF TV signal travels by line-of-sight propagation to the antenna and receiver.

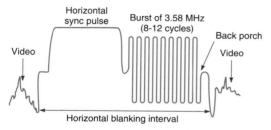

FIG. 19-11 The 3.58-MHz color subcarrier burst used to synchronize color demodulation at the receiver.

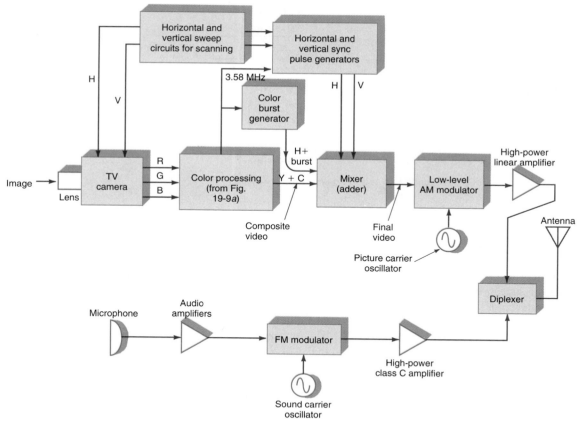

FIG. 19-12 Complete TV transmitter.

19-2 TV RECEIVER

The process involved in receiving a TV signal and recovering it to present the picture and sound outputs in a high-quality manner is complex. Over the course of the past 50 years since its invention, the TV set has evolved from a large vacuum tube unit into a smaller and more reliable solid-state unit made mostly with ICs.

A block diagram of a TV receiver is shown in Fig. 19-13. Although it is basically a superheterodyne receiver, it is one of the most sophisticated and complex electronic devices ever developed. Today, most of the circuitry is incorporated in large-scale ICs. Yet the typical TV receiver still uses many discrete component circuits.

THE TUNER

The signal from the antenna or the cable is connected to the tuner, which consists of an RF amplifier, mixer, and local oscillator. The tuner is used to select which TV channel is to be viewed and to convert the picture and sound carriers plus their modulation

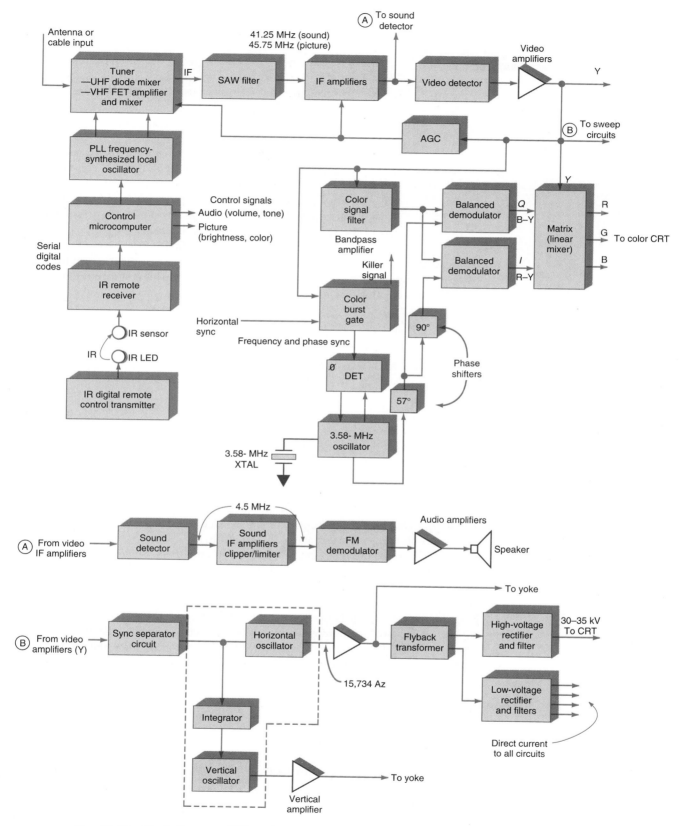

Fig. 19-13 Block diagram of TV receiver.

to an *intermediate frequency* (*IF*). As in most superheterodyne receivers, the local oscillator frequency is set higher than the incoming signal by the IF value.

Most TV set tuners are prepackaged in sealed and shielded enclosures. They are two tuners in one, one for the VHF signals and another for the UHF signals. The VHF tuner usually uses low-noise FETs for the RF amplifier and the mixer. UHF tuners use a diode mixer with no RF amplifier or a GaAs FET RF amplifier and mixer.

TUNING SYNTHESIZER. The local oscillators are *phase-locked loop* (*PLL*) *frequency synthesizers* set to frequencies that will convert the TV signals to the IF. Tuning of the local oscillator is typically done digitally. The PLL synthesizer is tuned by setting the feedback frequency division ratio. In a TV set this is changed by a microprocessor which is part of the master control system. The interstage LC–resonant circuits in the tuner are controlled by varactor diodes. By varying the DC bias on the varactors, their capacitance is changed, thereby changing the resonant frequency of the tuned circuits. The bias control signals also come from the control microprocessor. Most TV sets are also tuned by IR remote control.

VIDEO IF AND DEMODULATION

The standard TV receiver IFs are 41.25 MHz for the sound and 45.75 MHz for the picture. For example, if a receiver is tuned to channel 4, the picture carrier is 67.25 MHz, and the sound carrier is 71.75 MHz (the difference is 4.5 MHz). The synthesizer local oscillator is set to 113 MHz. The tuner produces an output that is the difference between the incoming signal and local oscillator frequencies, or $113 - 67.25$ MHz, or 45.75 MHz, for the picture and $113 - 71.75$ MHz, or 41.25 MHz, for the sound. Because the local oscillator frequency is above the frequency of incoming signals, the relationship of the picture and sound carriers is reversed at the intermediate frequencies, the picture IF being 4.5 MHz above the sound IF.

The IF signals are then sent to the video IF amplifiers. Selectivity is usually obtained with a *surface acoustic wave* (*SAW*) *filter*. This fixed tuned filter is designed to provide the exact selectivity required to pass both of the IF signals with the correct response to match the vestigial sideband signal transmitted. Figure 19-14(*a*) is a block diagram of the filter. It is made on a piezoelectric ceramic substrate such as lithium niobate. A pattern of interdigital fingers on the surface convert the IF signals into acoustic waves that travel across the filter surface. By controlling the shapes, sizes, and spacings of the interdigital filters, the response can be tailored to any application. Interdigital fingers at the output convert the acoustic waves into electrical signals at the IF.

The response of the SAW IF filter is shown in Fig. 19-14(*b*). Note that the filter greatly attenuates the sound IF to prevent it from getting into the video circuits. The maximum response occurs in the 43- to 44-MHz range. The picture carrier IF is down 50 percent on the curve.

Continue to refer to Fig. 19-13. The IF signals are next amplified by IC amplifiers. The video (luminance, or *Y*) signal is then recovered by an AM demodulator. In older sets, a simple diode detector was used for video detection. In most modern sets

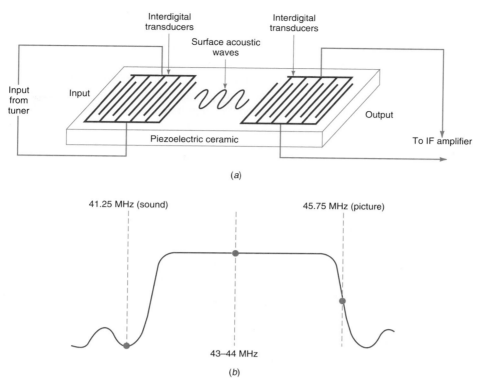

FIG. 19-14 (*a*) Surface acoustic wave (SAW) filter. (*b*) Typical IF response curve.

a synchronous balanced modulator type of synchronous demodulator is used. It is part of the IF amplifier IC.

The output of the video detector is the *Y* signal and the composite color signals, which are amplified by the video amplifiers. The *Y* signal is used to create an AGC voltage for controlling the gain of the IF amplifiers and the tuner amplifiers and mixers.

The composite color signal is taken from the video amplifier output by a filter and fed to color-balanced demodulator circuits. The color burst signal is also picked up by a gating circuit and sent to a phase detector (Φ DET) whose output is used to synchronize an oscillator that produces a 3.58-MHz subcarrier signal of the correct frequency and phase. The output of this oscillator is fed to two balanced demodulators that recover the *I* and *Q* signals. The carriers fed to the two balanced modulators are 90° out of phase. Note the 57° phase shifter used to correctly position the color phase for maximum recovery of color detail. The *Q* and *I* signals are combined in matrix with the *Y* signal, and out comes the three *R, G,* and *B* color signals. These are amplified and sent to the picture tube, which reproduces the picture.

SOUND IF AND DEMODULATION

To recover the sound part of the TV signal, a separate sound IF and detector section are used. Continuing to refer to Fig. 19-13, note that the 41.25- and 45.75-MHz sound and picture IF signals are fed to a sound detector circuit. This is a nonlinear circuit that heterodynes the two IFs and generates the sum and difference frequencies. The result is a 4.5-MHz difference signal which contains both the AM picture and the FM sound modulation. This is the sound IF signal. It is passed to the sound IF amplifiers which also perform a clipping-limiting function which removes the AM, leaving only the FM sound. The audio is recovered with a quadrature detector or differential peak detector as described in Chap. 6. The audio is amplified by one or more audio stages and sent to the speaker. If stereo is used, the appropriate demultiplexing is done by an IC, and the left and right channel audio signals are amplified.

Synchronizing Circuits

A major part of the TV receiver is dedicated to the sweep and synchronizing functions that are unique to TV receivers. In other words, the receiver's job does not end with demodulation and recovery of the picture and sound. To display the picture on a picture tube, special sweep circuits are needed to generate the voltages and currents to operate the picture tube, and sync circuits are needed to keep the sweep in step with the transmitted signal.

The sweep and sync operations begin in the video amplifier. The demodulated video includes the vertical and horizontal blanking and sync pulses. The sync pulses are stripped off the video signal with a sync separator circuit and fed to the sweep circuits (refer to the lower part of Fig. 19-13). The horizontal sync pulses are used to synchronize a horizontal oscillator to 15,734 Hz. This oscillator drives a horizontal output stage that develops a sawtooth of current that drives magnetic deflection coils in the picture tube yoke that sweep the electron beams in the picture tube.

The horizontal output stage, which is a high-power transistor switch, is also part of a switching power supply. The horizontal output transistor drives a step-up–step-down transformer called the *flyback*. The 15.734-kHz pulses developed are stepped up, rectified, and filtered to develop the 30- to 35-kV-high direct current required to operate the picture tube. Step-down windings on the flyback produce lower-voltage pulses that are rectified and filtered into low voltages which are used as power supplies for most of the circuits in the receiver.

The sync pulses are also fed to an IC that takes the horizontal sync pulses during the vertical blanking interval and integrates them into a 60-Hz sync pulse that is used to synchronize a vertical sweep oscillator. The output from this oscillator is a sawtooth sweep voltage at the field rate of 60 Hz (actually 59.94 Hz). This output is amplified and converted into a linear sweep current that drives the magnetic coils in the picture tube yoke. These coils produce vertical deflection of the electron beams in the picture tube.

In most modern TV sets, the horizontal and vertical oscillators are replaced by digital sync circuits (see Fig. 19-15). The horizontal sync pulses from the sync separator are normally used to phase-lock a 31.468-kHz *voltage controlled oscillator* (*VCO*) that runs at two times the normal horizontal rate of 15.734 kHz. Dividing this by 2 in a flip-flop gives the horizontal pulses that are amplified and shaped in the horizontal output stage to drive the deflection coils on the picture tube. A digital frequency divider divides the 31.468-kHz signal by 525 to get a 59.94-Hz signal for vertical sync. This signal is shaped into a current sawtooth and amplified by the vertical output stage which drives the deflection coils on the picture tube.

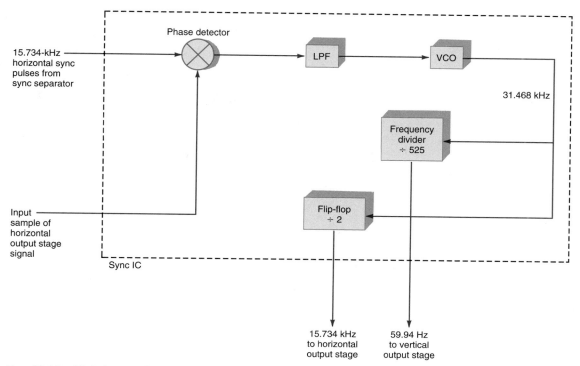

Fig. 19-15 Digital generation of horizontal and vertical sync pulses.

COLOR PICTURE TUBE

A picture tube is a vacuum tube called a *cathode-ray tube* (*CRT*). Both monochrome (B&W) and color picture tubes are available. The CRT used in computer video monitors works like the TV picture tube described here.

MONOCHROME CRT. The basic operation of a CRT is illustrated with a monochrome tube as shown in Fig. 19-16(*a*). The tube is housed in a bell-shaped glass enclosure. A filament heats a cathode which emits electrons. The negatively charged electrons are attracted and accelerated by positive-bias voltages on the elements in an electron gun assembly. The electron gun also focuses the electrons into a very narrow beam. A control grid that is made negative with respect to the cathode controls the intensity of the electron beam and the brightness of the spot it makes.

The beam is accelerated forward by a very high voltage applied to an internal metallic coating called *aquadag*. The *face*, or front, of the picture tube is coated internally with a phosphor that glows and produces white light when it is struck by the electron beam.

Around the neck of the picture tube is a structure of magnetic coils called the *deflection yoke*. The horizontal and vertical current linear sawtooth waves generated by the sweep and synchronizing circuits are applied to the yoke

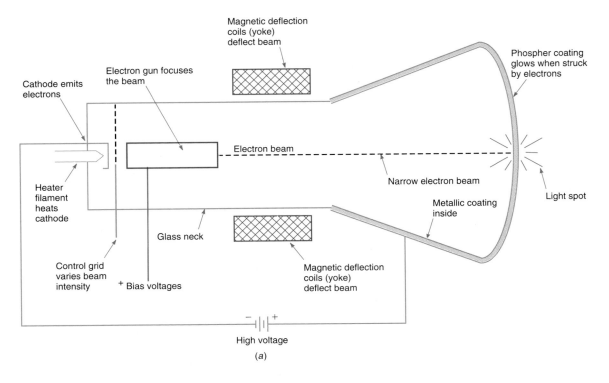

Magnetic deflection coils (yoke) deflect beam

Electron gun focuses the beam

Cathode emits electrons

Phospher coating glows when struck by electrons

Electron beam

Heater filament heats cathode

Narrow electron beam

Light spot

Glass neck

Metallic coating inside

Control grid varies beam intensity

+ Bias voltages

Magnetic deflection coils (yoke) deflect beam

High voltage

(a)

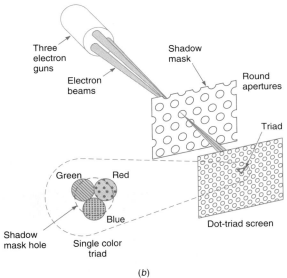

Three electron guns

Shadow mask

Round apertures

Electron beams

Triad

Green   Red

Shadow mask hole

Blue

Single color triad

Dot-triad screen

(b)

FIG. 19-16 (a) Basic construction and operation of a black and white (monochrome) cathode-ray tube C. (b) Details of color picture tube.

coils, which produce magnetic fields inside the tube that influence the position of the electron beam. When electrons flow, a magnetic field is produced around the conductor through which the current flows. The magnetic field that occurs around the electron beam is moved or deflected by the magnetic field produced by the deflection coils in the yoke. Thus the electron beam is swept across the face of the picture tube in the interlaced manner described earlier.

As the beam is being swept across the face of the tube to trace out the scene, the intensity of the electron beam is varied by the luminance, or *Y,* signal, which is applied to the cathode or in some cases to the control grid. The *control grid* is an element in the electron gun that is negatively biased with respect to the cathode. By varying the grid voltage, the beam can be made stronger or weaker, thereby varying the intensity of the light spot produced by the beam when it strikes the phosphor. Any shade of gray, from white to black, can be reproduced in this way.

COLOR CRT.   The operation of a color picture tube is similar to that just described. To produce color, the inside of the picture tube is coated with many tiny red, green, and blue phosphor dots arranged in groups of three called *triads.* Some tubes use a pattern of red, green, and blue stripes. These dots or stripes are energized by three separate cathodes and electron guns driven by the red, green, and blue color signals. Figure 19-16(*b*) shows how the three electron guns are focused so that they strike only the red, green, and blue dots as they are swept across the screen. A metallic plate with holes for each dot triad called a *shadow mask* is between the guns and the phosphor dots to ensure that the correct beam strikes the correct color dot. By varying the intensity of the color beams, the dot triads can be made to produce any color. The dots are small enough so that the eye cannot see them individually at a distance. What the eye sees is a color picture swept out on the face of the tube.

Figure 19-17 shows how all the signals come together at the picture tube to produce the color picture. The *R, G,* and *B* signals are mixed with the *Y* signal to control the cathodes of the CRT. Thus the beams are properly modulated to reproduce the color picture. Note the various controls associated with the picture tube. The *R-G-B* screen, brightness, focus, and centering controls vary the DC voltages that set the levels as desired. The convergence controls and assembly are used to control the positioning of the three electron beams so that they are centered on the holes in the shadow mask and the electron beams strike the color dots dead center. The deflection yoke over the neck of the tube deflects all three electron beams simultaneously.

19-3 CABLE TV

Cable TV, sometimes called *CATV,* is a system of delivering the TV signal to home receivers by way of a coaxial cable rather than over the air by radio wave propagation. A cable TV company collects all the available signals and programs and frequency multiplexes them on a single coaxial cable that is fed to the homes of subscribers. A special cable decoder box is used to receive the cable signals, select the desired channel, and feed a signal to the TV set.

CATV BACKGROUND

Many companies were established to offer TV signals by cable. They put up very tall high-gain TV antennas. The resulting signals were amplified and fed to the subscribers by cable. Similar systems were developed for apartments and condos. A single master antenna system was installed at a building, and the signals were amplified and distributed to each apartment or unit by cable.

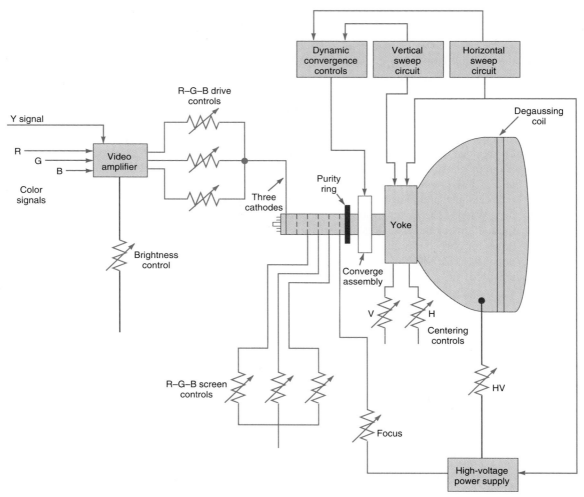

Fig. 19-17 Color picture tube circuits.

Modern Cable TV Systems

Today, cable TV companies collect signals and programs from many sources, multiplex them, and distribute them to subscribers (see Fig. 19-18). The main building or facility is called the *headend*. The antennas receive local TV stations and other nearby stations plus the special cable channel signals distributed by satellite. The cable companies use parabolic dishes to pick up the so-called premium cable channels. A cable TV company uses many TV antennas and receivers to pick up the stations whose programming it will redistribute. These signals are then processed and then combined or frequency multiplexed onto a single cable.

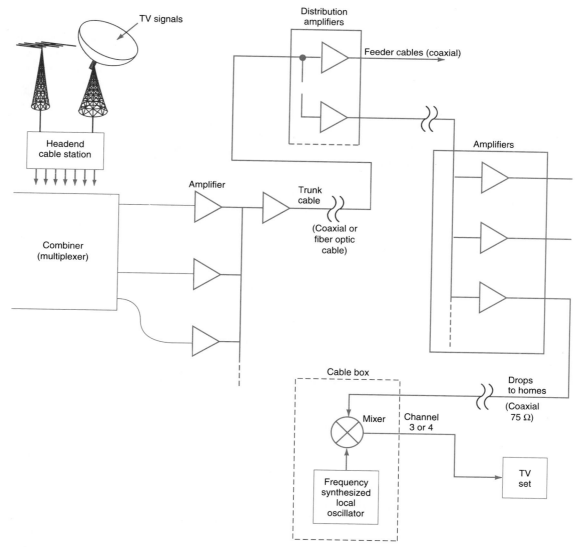

FIG. 19-18 The modern cable TV system.

The main output cable is called the *trunk cable.* It is usually a large, low-loss coaxial cable, although newer cable TV systems are using fiber-optic cable. The trunk cable is usually buried and extended to surrounding areas. A junction box containing amplifiers takes the signal and redistributes it to smaller cables called *feeders,* which go to specific areas and neighborhoods. From there the signals are again rejuvenated with amplifiers and then sent to individual homes by coaxial cables called *drops.*

The coaxial cable (usually 75 Ω RG-59/U) comes into a home and is connected to a cable decoder box, which is essentially a special TV tuner that picks up the cable channels and provides a frequency synthesizer and mixer to select the desired channel. The mixer output is heterodyned to TV channel 3 or 4 and then fed to the TV set antenna terminals. The desired signal is frequency translated by the cable box to channel 3 or 4 that the TV set can receive.

Cable TV is a popular and widely used service in the United States. More than 70 percent of U.S. homes have cable TV service. This service eliminates the need for antennas. And because of the direct connection of amplified signals, there is no such thing as poor, weak, noisy, or snowy signals. In addition, many TV programs are available only via cable, for example, the premium movie channels. The only downside to cable TV is that it is more expensive than connecting a TV to a standard antenna.

SIGNAL PROCESSING

The TV signals to be redistributed by the cable company usually undergo some kind of processing before they are put on the cable to the TV set. Amplification and impedance matching are the main processes involved in sending the signal to remote locations over what is sometimes many miles of coaxial cable. However, at the headend, other types of processes are involved.

STRAIGHT-THROUGH PROCESSORS. In early cable systems, the TV signals from local stations were picked up with antennas and the signal was amplified before being multiplexed onto the main cable. This is called *straight-through processing.* Amplifiers called *strip amplifiers* and tuned to the received channels pass the desired TV signal to the combiner. Most of these amplifiers include some kind of gain control or attenuators that can reduce the signal level to prevent distortion of strong local signals. This process can still be used with local VHF TV stations, but today heterodyne processing is used instead.

HETERODYNE PROCESSORS. *Heterodyne processing* translates the incoming TV signal to a different frequency. This is necessary when satellite signals are involved. Microwave carriers cannot be put on the cable, so they are down-converted to some available 6-MHz TV channel. In addition, heterodyne processing gives the cable companies the flexibility of putting the signals on any channel they want to use.

The cable TV industry has created a special set of nonbroadcast TV channels, as shown in Fig. 19-19. Some of the frequency assignments correspond to standard TV channels, but others do not. Since all these frequencies are confined to a cable, there can be duplication of any frequency that might be used in radio or TV broadcasting. Note that the spacing between the channels is 6 MHz.

The cable company uses modules called *heterodyne processors* to translate the received signals to the desired channel (see Fig. 19-20). The processor is a small TV receiver. It has a tuner set to pick up the desired over-the-air channel. The output of the mixer is the normal TV IFs of 45.75 and 41.25 MHz. These picture and sound IF signals are usually separated by filters, and they incorporate AGC and provide for individual gain control to make fine-tuning adjustments. These signals are then sent to a mixer where they are combined with a local oscillator signal to up-convert them to the final output frequency. A switch is usually provided to connect the input local oscillator to the output mixer. This puts the received signal back on the same frequency. In some cases that is done. However, setting the switch to the other position selects a different local oscillator frequency that will up-convert the signal to another desired channel frequency.

Channel	Frequency, video carrier, MHz	Channel	Frequency, video carrier, MHz
Low-Band VHF		**Superband (cont.)**	
2	55.25	N	241.25
3	61.25	O	247.25
4	67.25	P	253.25
5	77.25	Q	259.25
6	83.25	R	265.25
MidBand VHF		S	271.25
A-2	109.25	T	277.25
A-1	115.25	U	283.25
A	121.25	V	289.25
B	127.25	W	295.25
C	133.25		
D	139.25	**Hyperband**	
E	145.25	AA	301.25
F	151.25	BB	307.25
G	157.25	CC	313.25
H	163.25	DD	319.25
I	169.25	EE	325.25
High-Band VHF		FF	331.25
7	175.25	GG	337.25
8	181.25	HH	343.25
9	187.25	II	349.25
10	193.25	JJ	355.25
11	199.25	KK	361.25
12	205.25	LL	367.25
13	211.25	MM	373.25
Superband		NN	379.25
J	217.25	OO	385.25
K	223.25	PP	391.25
L	229.25	QQ	397.25
M	235.25	RR	403.25

Fig. 19-19 Special cable TV channels. Note that the video or picture carrier frequency is given.

Some heterodyne processors completely demodulate the received signal into its individual audio and video components. This gives the cable company full control over signal quality by making it adjustable. In this way, the cable company could also employ scrambling methods if desired. The signals are then sent to a modulator unit that puts the signals on carrier frequencies. The resulting signal is then up-converted to the desired output channel frequency.

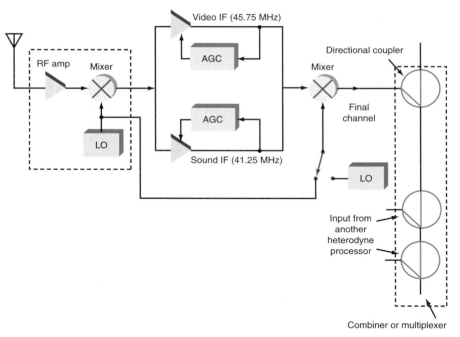

Fig. 19-20 A heterodyne processor.

All the signals on their final channel assignments are sent to a *combiner*, which is a large special-purpose linear mixer. Normally, directional couplers are used for the combining operation. Figure 19-20 shows how multiple directional couplers are connected to form the combiner or multiplexer. The result is that all of the signals are frequency-multiplexed into a composite signal that is put on the trunk cable.

CABLE TV CONVERTER

The receiving end of the cable TV system at the customer's home is a box of electronics that selects the desired channel signal from those on the cable and translates it to channel 3 or 4, where it is connected to the host TV receiver through the antenna input terminals. The *cable TV box* is thus a tuner that can select the special cable TV channels and convert them to a frequency that any TV set can pick up.

Figure 19-21 shows a basic block diagram of a cable TV converter. The 75-Ω RG-59/U cable connects to a tuner made up of a mixer and a frequency synthesizer local oscillator capable of selecting any of the desired channels. The synthesizer is phase-locked and microprocessor controlled. Most control processors provide for remote control with a digital infrared remote control like that used on virtually every modern TV set.

The output of the mixer is sent to a modulator that puts the signal on either channel 3 or 4. The output of the modulator connects to the TV set antenna input. The TV set is then set to the selected channel and left there. All channel changing is done with the cable converter remote control.

Today, the cable converters have many advanced features, among them automatic identification and remote control by the cable company. Each processor contains a

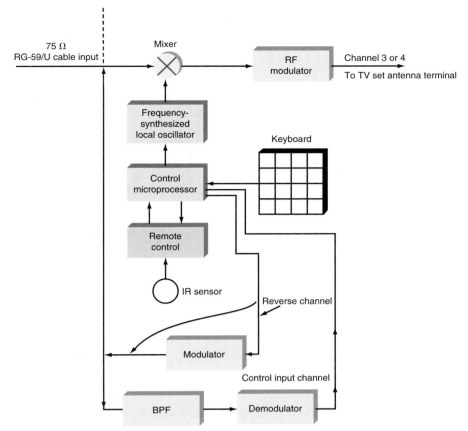

FIG. 19-21 Cable TV converter.

unique ID code that the cable company uses to identify the customer. This digital code is transmitted back to the cable company over a special reverse channel. There are several 6-MHz channels below channel 2 which can be used to transmit special signals to or from the cable converter. The digital ID modulates one of these special reverse channels. These low channels can also be used by the cable company to turn on or disable a cable converter box remotely. A digital signal is modulated onto a special channel and sent to the cable converter. It is picked off by a special tuner or with a bandpass filter as shown in Fig. 19-21. The signal is demodulated, and the recovered signal is sent to the microprocessor for control purposes. It can be used to lock out access to any special channels for which the customer has not subscribed. The reverse channels can also be used for simple troubleshooting.

19-4 SATELLITE TV

One of the most common methods of TV signal distribution is via communications satellite. A communications satellite orbits around the equator about 22,300 mi out in space. It rotates in synchronism with the earth and therefore appears to be stationary. The satellite is used as a radio relay station (refer to Fig. 19-22). The TV signal to be

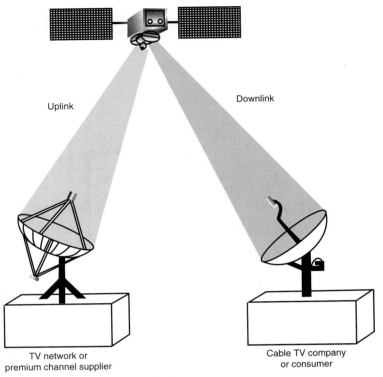

Uplink

Downlink

TV network or
premium channel supplier

Cable TV company
or consumer

Fig. 19-22 Satellite TV distribution.

distributed is used to frequency-modulate a microwave carrier, and then it is transmitted to the satellite. The path from earth to the satellite is called the *uplink*. The satellite translates the signal to another frequency and then retransmits it back to earth. This is called the *downlink*. A receive site on earth picks up the signal. The receive site may be a cable TV company or an individual consumer. Satellites are widely used by the TV networks, the premium channel companies, and the cable TV industry for distributing their signals nationally.

A newer form of consumer satellite TV is *direct broadcast satellite (DBS) TV*. The DBS systems are designed specifically for consumer reception directly from the satellite. The new DBS systems feature digitally encoded video and audio signals, which make transmission and reception more reliable and provide outstanding picture and sound quality. By using higher-frequency microwaves, higher-powered satellite transponders, and very low noise GaAsFETs in the receiver, the customer's satellite dish can be made very small. These systems typically use an 18-in dish as opposed to the 5- to 12-ft-diameter dishes still used in many satellite TV systems.

SATELLITE TRANSMISSION

The TV signal to be uplinked to the satellite from a ground station is used to modulate a carrier in one of several available microwave satellite bands. The C band between approximately 3.7 to 4.2 GHz is the most commonly used. The video signal frequency modulates the microwave carrier on one of 24 channel frequencies. The C-band channel frequencies and their corresponding downlinked transponder number are shown in Fig. 19-23.

The audio accompanying the video frequency modulates a subcarrier in the 5- to 8-MHz range. The 6.2- and 6.8-MHz subcarriers are the most common. Stereo sound is used to modulate the two subcarriers on 5.58 and 5.76 MHz. The video occupies the spectrum of approximately 0 to 5 MHz. The composite spectrum of the video and audio subcarrier signals used to frequency-modulate the uplink transmitter is illustrated in Fig. 19-24. This is the signal that must be recovered by the satellite receiver.

The signal is received by the satellite, and filters pass the signal through the selected transponder. In the transponder the signal is down converted to a lower frequency, amplified, and retransmitted.

SATELLITE RECEIVERS

A *satellite receiver* is a special subsystem designed to work with a consumer TV set. It consists of a parabolic dish antenna, a low-noise amplifier and down converter, an IF section with appropriate demodulators for both video and sound, and a method of

Satellite Channel Center Frequencies

DOWNLINK TRANSPONDER NUMBER	FREQUENCY, MHz	DOWNLINK TRANSPONDER NUMBER	FREQUENCY, MHz
1	3720	13	3960
2	3740	14	3980
3	3760	15	4000
4	3780	16	4020
5	3800	17	4040
6	3820	18	4060
7	3840	19	4080
8	3860	20	4100
9	3880	21	4120
10	3900	22	4140
11	3920	23	4160
12	3940	24	4180

FIG. 19-23 Satellite C-band downlink frequencies used for TV distribution. Odd channels: vertical polarization; even channels: horizontal polarization.

Television broadcasts from this Los Angeles mobile news unit can transmit local news nation-wide. Microwave communications are used to relay signals to the parent station.

interconnecting it to the conventional TV set. In addition, most satellite receivers contain circuitry for controlling the positioning of the satellite dish antenna. You will sometimes hear the satellite receiver referred to as *TVRO,* or *TV receive-only, system.* The following section describes the basic organization and operation of typical TVRO satellite receivers.

ANTENNA. The antenna is more critical in a satellite TV receiver than in any other kind of receiver. The signal from the satellite 22,300 mi away is extremely weak.

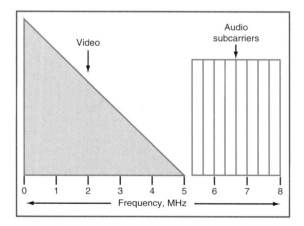

FIG. 19-24 Composite video-audio signal used to modulate the C-band uplink transmitter.

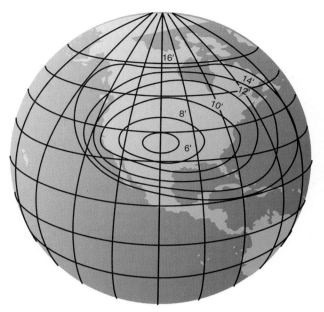

FIG. 19-25 Satellite "footprint" showing desired disk sizes for best reception.

In addition, there are hundreds of satellites in orbit above the earth, and their spacing is getting closer each year as the number of satellites continues to increase. A high-gain, highly directional parabolic dish antenna is used to select only the signal from the desired satellite and provide very high gain. Most satellite TV antennas range in size from approximately 6 to 15 ft. Over the years, lower-noise, higher-gain amplifiers have been developed using *gallium arsenide field effect* (*GaAs*) *transistors*. This has permitted dish antenna sizes to be reduced in size. In some of the newer systems, antennas as small as 3 to 4 ft in diameter are available. However, in most cases, the larger the dish, the higher the gain and the better the performance.

A part of determining antenna size is based on the location of the receiver in the United States. The signal strength of the satellite downlink signal varies considerably over the United States. In most cases, the signal strength is higher in the center of the country and considerably lower on the coasts. As a result, if the receiver is located on the east or west coast, higher gains and larger antennas are required for satisfactory reception. Figure 19-25 shows the "footprint" of the satellite antenna on earth. The contour lines indicate different signal levels, the strongest being in the center and the weakest being on the outside. Since the signal is the strongest in the center, smaller antennas can be used. Larger dishes must be used on the outer areas to receive an adequate signal level.

The antenna is a horn located at the focal point of a parabolic reflector (see Fig. 19-26). Signals picked up by the dish are focused on the horn, giving very high gain and exceptionally narrow directional characteristics. The antenna is built so that it can receive both horizontal and vertically polarized signals. The horn is usually coupled by a short piece of coaxial cable to the receiver input.

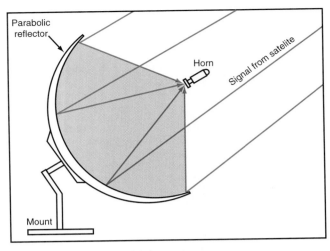

FIG. 19-26 Parabolic disk-horn antenna.

DIRECT BROADCAST SATELLITE SYSTEMS

The *direct broadcast satellite (DBS) system* is the newest form of satellite TV available to consumers. It was designed specifically to be an all-digital system—in contrast to the current analog systems in use. Data compression techniques are used to reduce the data rate required to produce high-quality picture and sound.

The DBS system features entirely new digital uplink ground stations and satellites. Since the satellites are designed to transmit directly to the home, extra high power transponders are used to ensure a satisfactorysignal level.

To receive the digital video from the satellite, a consumer must purchase a satellite TV receiver and antenna. These are similar to the satellite receivers just described; however, they work with digital signals and operate in the K_u rather than the C band. By using higher frequencies as well as higher-power satellite transponders, the necessary dish antenna can be extremely small. The new satellite DBS system antennas are only 18 in in diameter. There are several special digital broadcast satellites in orbit, and some of the direct satellite TV sources include DirecTV, USSB, and PrimeStar. All provide full coverage of the major cable networks and the premium channels usually distributed to homes by cable TV, and all can be received directly. In addition to purchasing the receiver and antenna, the consumer must also subscribe to one of the services supplying the desired channels.

This father and son can easily install their RCA direct digital TV satellite dish antenna.

SATELLITE TRANSMISSION. The video to be transmitted must first be placed into digital form. To digitize an analog signal, it must be sampled a minimum of two times per cycle in order for sufficient digital data to be developed for reconstruction of the signal. Assuming that video frequencies of up to 4.2 Mbps are used, the minimum sampling rate is twice this, or 8.4 Mbps. For each sample, a binary number proportional to the light amplitude is developed. This is done by an A/D converter, usually with an 8-bit output. The resulting video signal, therefore, has a data rate of 8 bits times 8.4 Mbps, or 67.2 Mbps. This is an extremely high data rate. However, for a color TV signal to be transmitted in this way, there must be a separate signal for each of the red, green, and blue components making up the video. This translates to a total data rate of 3×67.2, or 202 Mbps. Even with today's technology, this is an extremely high data rate that is hard to achieve reliably.

In order to lower the data rate and improve the reliability of transmission, the new DBS system uses compressed digital video. Once the video signals have been put into digital form, they are processed by (*digital signal processing*) (*DSP*) *circuits* to minimize the full amount of data to be transmitted. Digital compression greatly reduces the actual transmitting speed to somewhere in the 20- to 30-Mbps range. The compressed serial digital signal is then used to modulate the uplinked carrier using BPSK.

The DBS satellite uses the K_u band with a frequency range of 11 to 14 GHz. Uplink signals are usually in the 14- to 14.5-GHz range while the downlink usually covers the range of 10.95 to 12.75 GHz.

The primary advantage of using the K_u band rather than the C band is that the receiving antennas may be made much smaller for a given amount of gain. However, these higher frequencies are more affected by atmospheric conditions than the lower microwave frequencies. The biggest problem is the increased attenuation of the downlink signal caused by rain. Any type of weather involving rain or water vapor, such as fog, can seriously reduce the received signal. This is because the wavelength of K_u band signals is near that of water vapor. Therefore, the water vapor absorbs the signal. Although the power of the satellite transponder and the gain of the receiving antenna are typically sufficient to provide solid reception, there can be fadeout under heavy downpour conditions.

Finally, the digital signal is transmitted from the satellite to the receiver using circular polarization. The DBS satellites have *right-hand* and *left-hand circularly polarized* (*RHCP* and *LHCP*) *helical antennas.* By transmitting both polarities of signal, frequency reuse can be incorporated to double the channel capacity.

DBS RECEIVER. A block diagram of a typical DBS digital receiver is shown in Fig. 19-27. The receiver subsystem begins with the antenna and its low-noise block converter. The horn antenna picks up the K_u band signal and translates the entire 500-MHz band used by the signal down to the 950- to 1450-MHz range, as explained earlier. Control signals from the receiver to the antenna select between RHCP and LHCP. The RF signal from the antenna is sent by coaxial cable to the receiver.

A typical DBS downlink signal occurs in the 12.2- to 12.7-GHz portion of the K_u band. Each transponder has a bandwidth of approximately 24 MHz. The digital signal is usually occurring at a rate of approximately 27 Mbps.

Figure 19-28 shows how the digital signal is transmitted. The digital audio and video signals are organized into data packets. Each packet consists of a total of 147 bytes. The first 2 bytes (16 bits) contain the *service channel identification* (*SCID*) *number.* This is a 12-bit number that identifies the video program being carried by the packet. The 4 additional bits are used to indicate whether the packet is encrypted and

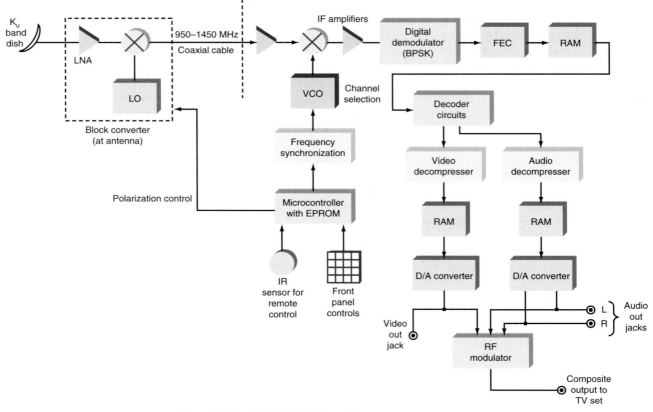

Fig. 19-27 Digital DBS TV receiver.

if so, which decoding key to use. One additional byte contains the packet type and a continuity counter.

The data block consists of 127 bytes, either 8-bit video signals or 16-bit audio signals. It may also contain digital data used for control purposes in the receiver. Finally, the last 17 bytes are the error-detection check codes. These 17 bytes are developed by an error-checking circuit at the transmitter. The appended bytes are checked at the receiver to detect any errors and correct them.

The received signal is passed through another mixer with a variable-frequency local oscillator to provide channel selection. The digital signal at the second IF is then demodulated to recover the originally transmitted digital signal, which is passed

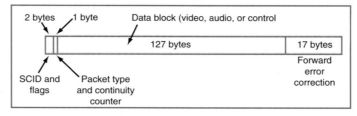

Fig. 19-28 Digital data packet format used in DBS TV.

through a *forward-error correction (FEC) circuit.* This circuit is designed to detect bit errors in the transmission and to correct them on the fly. Any bits lost or obscured by noise during the transmission process are usually caught and corrected to ensure a near-perfect digital signal.

The resulting error-corrected signal is stored in *random access memory (RAM),* after which the signal is decoded to separate it into both the video and the audio portions. The resulting signals are then sent to the audio and video decompression circuits. The DBS TV system uses digital compression-decompression standards referred to as *MPEG2. MPEG* means Motion Picture Experts Group, which is a standards organization that establishes technical standards for movies and video. MPEG2 is its latest and best compression method for video.

Although the new DBS digital systems will not replace cable TV, they provide the consumer with the capability of receiving a wide range of TV channels. The use of digital techniques provides an unusually high-quality signal.

19-5 HIGH-DEFINITION TELEVISION

High-definition television (HDTV) is the result of an attempt to improve the picture and sound quality of TV transmissions. The goal is to transmit finer picture detail and to minimize the effect of having to transmit pictures in the form of multiple horizontal scan lines. Improvements in sound transmission are also an objective.

After many years of evaluating proposals for HDTV systems, the Federal Communications Commission (FCC) is nearing a decision with regard to the specific details of a system. As this is being written, the final choice has not been made, but the options have been greatly reduced, and it is possible to discuss briefly the specifications of a possible forthcoming HDTV system. The proposed new formats are incompatible with computer video standards, which are becoming more widely used in communications, such as those over the Internet. This incompatibility may further delay the introduction of an HDTV system.

HDTV STANDARDS

The FCC first asked for proposals for HDTV systems in the late 1980s. As many as 23 different system proposals were submitted by 14 independent companies. During the early 1990s, the FCC narrowed the choices down by way of a series of actual hardware tests made on prototypes submitted by the competitors. Final specifications have not been chosen as of this date. Therefore, it would be presumptuous to attempt to guess what the final decision will be. However, sufficient data is available to provide a general idea as to the specifications of the final system.

PICTURE DETAILS. The proposed HDTV system will provide higher resolution and greater detail by offering more horizontal scan lines per frame. As indicated earlier, the current standard is 525 lines using interlaced scanning. The proposed new systems will use either 787.5 or 1050 horizontal scan lines. If the 1050-scan-line sys-

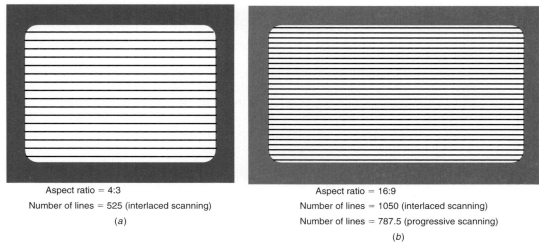

Aspect ratio = 4:3

Number of lines = 525 (interlaced scanning)

(a)

Aspect ratio = 16:9

Number of lines = 1050 (interlaced scanning)

Number of lines = 787.5 (progressive scanning)

(b)

FIG. 19-29   TV picture standards. (a) Current standard. (b) HDTV standard.

tem is selected, interlaced scanning will be used. If the 787.5-scan-line system is chosen, the scanning will be *progressive;* that is, 1 line will be scanned after the other at a 59.94-Hz frame rate. The progressive scanning system produces less picture flicker. The aspect ratio of the picture will change from the current 4:3 to 16:9 (see Fig. 19-29).

TRANSMISSION DETAILS.   Since the proposed HDTV system will be fully digital, the video is digitized and transmitted as a multibit binary number. The basic video sampling frequency is approximately 53 to 56 MHz for the 1050-line system and approximately 75.75 MHz for the 787.5-line system. The number of digitized points per scan line are known as *pixels* (i.e., *picture elements*). There are 1280 pixels per horizontal line for the 787.5-line system and 1440 pixels for the 1050-line system. The number of vertical pixels corresponds to the number of horizontal scan lines allocated to the picture. There are 960 lines in the 1050-line system and 720 lines in the 787.5-line system. The other horizontal lines are not shown on the screen, but digital data is encoded into these lines. An example is the closed-captioning text used by the hearing impaired.

The color signals are also digitized. The horizontal chrominance information produces 640 pixels in the 787.5-line system and either 352 or 720 chrominance pixels in the 1050-line system. There are 360 vertical pixels in the 787.5-line system or 480 vertical pixels in the 1050-line system.

The high sampling rate produces a very wide bandwidth of digital signal. For the 1050-line system, the luminance bandwidth is 21.5 to 23.6 MHz and the chrominance bandwidth is 4.5 or 11.8 MHz. The bandwidth requirement for the 787.5-line system is even greater, 34 MHz for the luminance bandwidth and 17 MHz for the chrominance. Obviously, it does not appear that such wide bandwidth signals will fit within the standard 6-MHz-wide TV channel.

DID YOU KNOW?

The new high-definition television system will utilize all-digital transmission techniques which provide higher-quality pictures.

SPECTRUM ALLOCATIONS. A primary requirement of HDTV is that the final standard make use of the available 6-MHz TV channels currently allocated by the FCC. Currently there are 68 6-MHz TV channels. The VHF channels (2 through 13) are fully occupied with standard TV transmissions, and many UHF channels are used.

OTHER CONSIDERATIONS. The final standards to be selected by the FCC are expected to take into consideration the compatibility with computer displays. The hope is that HDTV signals might be displayed on future computer screens while computer data would be compatible with the new HDTV sets. This would help bring together the computer and TV worlds and expedite the development of the so-called information superhighway. Such a system would allow TV signals to be transmitted over LANs and larger networking systems such as the new digital telephone systems and the more advanced fiber-optic cable TV systems being installed.

Excellent technical careers are available in television broadcasting. This technician is adjusting color patterns, adjusting and maintaining strong signal quality, and routing programming.

Television is radio communications with both pictures and sound. In addition to standard audio transmission, TV systems use a camera to convert a visual scene into a voltage known as the *video signal*. This signal represents the picture information and modulates a transmitter. The picture and the sound signals are transmitted to the receiver. The receiver demodulates the signals and presents the information to the user. The TV receiver is a special superheterodyne that recovers the sound and picture information. The picture is displayed on a picture tube.

As common as a TV set is, we usually take for granted the extremely complex process involved in receiving a TV signal and recovering it to present the picture and sound outputs in a high-quality manner. During the 50 years since its invention, the TV set has evolved from a large vacuum tube unit into a smaller, more reliable solid-state unit made mostly with ICs.

One of the most common means of TV signal distribution is via communications satellite. A communications satellite orbits around the equator about 22,300 mi out in space. In this position it rotates in synchronism with the earth and therefore appears to be stationary. The satellite is used as a radio relay station. Satellites are widely used by the TV networks, the premium channel companies, and the cable TV industry for distributing their signals nationally. A newer form of consumer satellite is *direct broadcast satellite (DBS) TV*. The DBS systems are designed for consumer reception directly from the satellite. These systems typically use an 18-in dish as opposed to the 5- to 12-ft-diameter dishes that many satellite TV systems still use.

A major change is in the works in the TV industry: *high-definition television (HDTV)*. When the specifications are approved for mass production, consumers will be able to buy a TV set with greatly improved picture and sound and, it is hoped, computer compatibility.

KEY TERMS

Antenna
Audio signal
$B - Y$ signal
CATV (cable TV)
CATV converter
Chrominance signal
Coaxial cable
Color CRT
Color picture tube
Color signal generation
Definition
Direct broadcast satellite (DBS) TV
Dish antennas
Downlink

Dual-conversion front-end
Heterodyne processors
High-definition TV (HDTV)
I color signal
Interlaced scanning
Luminance (Y) signal
Monochrome CRT
Motion Picture Experts Group (MPEG)
Q color signal
$R - Y$ signal
Resolution
Satellite receivers
Satellite transmission

Scanning
Signal bandwidth
Sound IF
Straight-through processors
Strip amplifiers
Synchronizing circuits
Trunk cable
Tuner
Tuning synthesizer
TV receiver
TV signal
TV spectrum allocation
Video signal

REVIEW

QUESTIONS

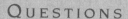

1. What is the bandwidth of a standard TV signal?
2. State the kind of modulation used on the video carrier and the sound carrier in a TV signal.
3. What is the frequency spacing between the sound and picture carriers?
4. What do you call the brightness signal produced by a video monochrome video camera?
5. Name two widely used electronic imaging devices in TV cameras that convert light variations into a video signal.
6. What is the name of the process that breaks up a picture or scene into serially transmitted signals?
7. What are the three basic colors that can be used to produce any other color light?
8. What is the maximum number of scan lines used in an NTSC TV picture or frame of video?
9. How many scan lines make up one field of a TV picture?
10. What are the field and frame rates in NTSC color TV?
11. What is the rate of scanning 1 horizontal line in a color TV set?
12. What is the name of the circuit that lets the picture and sound transmitters use the same antenna?
13. The color camera signals are combined in a resistive matrix to produce the two composite color signals. What are they called?
14. How are the two color signals multiplexed and modulated onto the main video carrier?
15. What is the frequency of the subcarrier that the color signals modulate?
16. What characteristic of the composite color signal tells the receiver what the transmitted color is?
17. What type of modulation is used in the generation of the chrominance signals?
18. What portion of a modern TV set uses digitally coded infrared signals to control channel selection, and volume level?
19. What is the name of the special filter that provides most of the selectivity for the TV receiver?
20. The picture and sound IFs are heterodyned together to form the sound IF. What is its frequency?
21. Name two common sound demodulators in TV receivers.
22. What is meant by a *quadrature 3.58-MHz subcarrier* as used in demodulation?
23. What circuit strips the horizontal sync pulses from the video detector output?
24. The horizontal sync pulses synchronize an internal sweep oscillator to what frequency in a color TV receiver?
25. What is the shape of the horizontal and vertical sweep signals which are currents applied to the horizontal and vertical deflection coils?
26. What is the name of the assembly around the neck of the picture tube to which the sweep signals are applied?
27. What element in the picture tube generates the electrons? What element in the picture tube focuses the electrons into a narrow beam?
28. By what process is the electron beam deflected and swept across the face of the picture tube?
29. How many electron guns are used in a color CRT to excite the color dot triads on the face of the tube?
30. What is the name of the circuits that ensure the electron beams strike the correct color dots?

31. What stage in the TV receiver is used as switching power supply to develop the HV required to operate the picture tube?
32. What is the name of the transformer used to step up and step down the horizontal sync pulses to produce horizontal sweep as well as most DC power supply voltages in a TV set?
33. What is the name given to the cable TV station that collects and distributes the cable signals?
34. What are the names of the main cables used to distribute the TV signals to subscribers?
35. What two types of cables are used for the main distribution of signals in a cable TV system?
36. Name the coaxial cable that feeds individual houses in a cable TV system. What is the type designation and impedance of this cable?
37. What is the name of the equipment at the cable station used to change the TV signal frequency to another frequency? Why is this done?
38. What local oscillator frequency would you use to translate the TV signal at its normal IF values to cable channel J?
39. What is the name of the circuit that is used to assemble or mix all of the different cable channels together to form a single signal that is distributed by the cable?
40. Name the two main sections of a cable converter box used by the subscriber.
41. What is a reverse channel on a cable TV converter box?
42. Describe the nature of the output signal developed by the cable converter box and where it is connected.
43. What is the name of the unit in a satellite that receives and then rebroadcasts the TV signal to be distributed?
44. What is the frequency range of standard satellites used for TV distribution, and what is the letter designation of that band?
45. In the United States, what direction must you point an antenna to "see" a satellite?
46. What is the approximate distance of the satellite from the earth?
47. What is the name given to a consumer satellite receiving station?
48. What kind of amplifiers are used in satellite TV receiver front-ends?
49. What allows an antenna to receive two different signals on the same frequency?
50. What is the channel number and antenna polarity of the downlink frequency 4120 MHz?
51. What kind of modulation is used to apply the video to the microwave uplink carrier?
52. What is a common intermediate frequency in a single-conversion satellite receiver?
53. What are two common first IFs used in double-conversion satellite receivers?
54. According to the satellite "footprint" in Fig. 19-25, what is the size of the antenna needed for good reception in south Florida?
55. What feature of the DBS system makes digital transmission and reception possible?
56. What natural occurrence can sometimes prohibit the reception of DBS signals?
57. Name two ways that a satellite receiver, either conventional or DBS, is connected to a conventional TV set.
58. Describe the proposed HDTV screen size and format.
59. True or false. HDTV is a full digital system.
60. Where in the spectrum will HDTV be transmitted?
61. True or false. HDTV is expected to fully replace conventional TV.

PROBLEMS

1. Compute the exact video and sound carriers for a channel 12 TV station.
2. What is the approximate upper frequency response of the video signal transmitted?

3. Using Carson's rule, calculate the approximate bandwidth of the sound spectrum of a TV signal. ➤

4. How is the bandwidth of the video portion of a TV signal restricted to minimize spectrum bandwidth?

5. Describe the process by which the picture at the receiver is kept in step with the transmitted signal? What component of the TV signal performs this function? ➤

6. Describe the process by which the color in a scene is converted into video signals.

7. What signal is formed by adding the color signals in the following proportion: $0.11B + 0.59G + 0.3R$? ➤

8. How is channel selection in the tuner accomplished? What type of local oscillator is used, and how does it work?

9. A channel 33 UHF TV station has a picture carrier frequency of 585.25 MHz. What is the sound carrier frequency? ➤

10. What are the TV receiver sound IF and the picture IF values?

11. What would be the local oscillator frequency to receive a channel 10 signal?

12. How is the spectrum space conserved in transmitting the video in a TV signal?

13. How is the composite chrominance signal demodulated? What type of demodulator circuits are used?

14. How is the 3.85-MHz subcarrier oscillator in the receiver phase and frequency synchronized to the transmitted signal?

15. How is the vertical sweep oscillator synchronized to a sync pulse derived from the horizontal sync pulses occurring during the vertical blanking interval? What is the vertical sync frequency in a color TV set?

16. How is cable attenuation occurring during distribution overcome?

17. Explain briefly how the cable company can remotely control a subscriber's cable converter.

18. Describe the antennas used in satellite transmission and reception.

19. Why are the amplifiers and mixers in a satellite receiver usually located at the antenna?

20. How is the audio signal modulated onto the microwave uplink carrier?

21. How is the tuning of a satellite receiver accomplished if the local oscillator is at the antenna?

22. Why are smaller antennas satisfactory in the central United States but not in Canada or Mexico?

23. Why are two demodulators needed in a satellite receiver? What form of modulation must be detected?

24. Describe how the antenna is oriented and positioned in a consumer satellite installation.

25. What is the main difference between conventional and DBS satellite TV systems?

26. What is the operational frequency range and band designation of a DBS receiver?

27. What is the rationale for HDTV?

28. Describe the audio format used in HDTV.

29. What will be the total number of pixels used to make up the visible portion of the HDTV system? Calculate this number for both the 1050- and 787.5-line system.

30. Describe briefly how the audio and video data are transmitted in both the DBS and HDTV systems. What is the packaging, or format, of the data?

CRITICAL THINKING

1. Name three reasons why digital TV transmission is preferred over analog TV on a satellite.

2. What limits the transmission of video over the telephone lines? Explain how it might be possible to transmit video over the conventional telephone network.

3. Given the nature of the proposed new HDTV standards, describe what you think a modern HDTV set will be like in terms of circuitry.

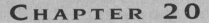

CHAPTER 20

COMMUNICATIONS
TESTS AND
MEASUREMENTS

Objectives

After completing this chapter, you will be able to:

◆ *List* 10 common test instruments used in testing communications equipment and describe the basic operation of each.

◆ *Test* common communications equipment tests on transmitters, receivers, and antennas including frequency measurements, power measurements, SWR measurements, sensitivity, and spectrum analyses.

◆ *Describe* the basic troubleshooting procedures used for locating problems in transmitters and receivers.

CHAPTER TWENTY

This book was written primarily to educate those of you seeking to become communications electronics technicians. The previous chapters provided a broad general overview of the circuits, systems, and methods used in electronic communications. The work you will perform in the specific field of concentration in communications will depend upon your employer. However, one thing is generally constant if you are employed as a communications electronics technician: Your work will involve some form of testing and measurement. All electronic equipment technicians, regardless of their field of specialization, perform the basic functions of equipment installation, operation, servicing, repair, and maintenance. Most communications electronics technicians are responsible for testing and troubleshooting communications equipment. Engineering and manufacturing technicians also perform a wide range of tests and measurements in their work. And those who troubleshoot and repair communications equipment will use tests and measurements to isolate the problems and perform necessary adjustments and final verifications of specifications.

The purpose of this chapter is to introduce you to the wide range of special testing equipment used in communications electronics. You will, of course, use standard test instruments such as a multimeter for measuring DC and AC voltage, current, and resistance and an oscilloscope. However, because of the high-frequency nature of communications equipment, special test instruments have been created to perform a wide range of RF tests and measurements. The chapter also includes a section that outlines some of the more common communications tests including the measurement of frequency, modulation, power, and *standing wave ratio* (*SWR*). The chapter concludes with a special section on troubleshooting techniques.

Because of the wide range of communications equipment available and the many differences from manufacturer to manufacturer, it is difficult to be specific. Only general test and measurement procedures are given. Just keep in mind that when you are performing real tests and measurements, you must familiarize yourself not only with the test equipment used but also the specific transmitter, receiver, or other communications device being tested. Manuals for both the test instruments and the communications equipment should be available for reference.

20-1 COMMUNICATIONS TEST EQUIPMENT

This section gives a broad overview of the many different types of test instruments available for use with communications equipment. It is assumed here that you already know basic test and measurement techniques used with conventional low-frequency test equipment such as multimeters, signal generators, and oscilloscopes. In your communications work, you will continue to use standard oscilloscopes and multimeters for measuring voltages, currents, and resistance. Coverage of these basic test instruments will not be repeated here. We will, however, draw upon your knowledge of the principles of those instruments as they apply to the test equipment discussed in this section.

VOLTAGE MEASUREMENTS

The most common measurement obtained for most electronic equipment is voltage. This is particularly true for DC and low-frequency AC applications. In RF applications, voltage measurements may be important under some conditions, but power measurements are far more common at higher frequencies, particularly microwave. In testing and troubleshooting communications equipment, you will still use a DC voltmeter to check power supplies and other DC conditions. There are also occasions when measurement of RF, that is, AC, voltage must be made.

There are two basic ways to make AC voltage measurements in electronic equipment. One is to use an AC voltmeter. Most conventional voltmeters can measure AC voltages from a few millivolts to several hundred volts. Typical bench or portable AC multimeters are restricted in their frequency range to a maximum of several thousand kilohertz. Higher-frequency AC voltmeters are available for measuring audio voltages up to several hundred thousand kilohertz. For higher frequencies, special RF voltmeters must be used.

RF Voltmeters. An *RF voltmeter* is a special piece of test equipment optimized for measuring the voltage of high-frequency signals. Typical units are available for making measurements up to 10 MHz. There are special units capable of measuring voltages from microvolts to hundreds of volts at frequencies up to 1 to 2 GHz.

RF voltmeters are made to measure sine-wave voltages, with the readout given in *root-mean-square (rms)*. Most RF voltmeters are of the analog variety with a moving pointer on a background scale. Measurement accuracy is within the 1 to 5 percent range depending upon the specific instrument. Accuracy is usually quoted as a percentage of the reading or as a percentage of the full-scale value of the voltage range selected. RF voltmeters with digital readout probes are also available with somewhat improved measurement accuracy.

RF Probes. One way to measure RF voltage is to use an RF probe with a standard DC multimeter. RF probes are sometimes referred to as *detector probes.* An RF probe is basically a rectifier with a filter capacitor that stores the peak value of the sine-wave RF voltage. The external DC voltmeter reads the capacitor voltage. The result is a peak value that can easily be converted to root-mean-square by multiplying it by 0.707.

Most RF probes are good for RF voltage measurements to about 250 MHz. The accuracy is about 5 percent, but that is usually very good for RF measurements.

Oscilloscopes. Two basic types of oscilloscopes are used in RF measurements: the *analog oscilloscope* and the *digital storage oscilloscope (DSO).*

Analog oscilloscopes amplify the signal to be measured and display it on the face of a CRT at a specific sweep rate. They are available for displaying and measuring RF voltages to about 500 MHz. As a rule, an analog oscilloscope should have a bandwidth of three or more times the highest-frequency component (a carrier, a harmonic, or a sideband) to be displayed.

Digital storage oscilloscopes, also known as *digital,* or *sampling oscilloscopes,* are growing in popularity and rapidly replacing analog oscilloscopes. DSOs use high-speed sampling or A/D techniques to convert the signal to be measured into a series of digital words that are stored in an internal memory. Sampling rates vary depending upon the oscilloscope but can range from approximately 20 million samples per second to more than 8 billion samples per second. Each measurement sample is usually converted into an 8- or 10-bit parallel binary number which is stored in an internal memory. Oscilloscopes with approximately 4 to 256K bytes are available depending upon the product.

Hints and Helps

The main disadvantage of equivalent time sampling is the time required to acquire enough samples to start the display.

Today, more than 50 percent of oscilloscope sales are of the digital sampling type. DSOs are very popular for high-frequency measurements because they provide the means to display signals with frequencies up to about 30 GHz. This means that complex modulated microwave signals can be readily viewed, measured, and analyzed.

POWER METERS

As indicated earlier, it is far more common to measure RF power than it is to measure RF voltage or current. This is particularly true in testing and adjusting transmitters that typically develop significant output power. One of the most commonly used RF test instruments is the *power meter*.

Power meters come in a variety of sizes and configurations. One of the most popular is a small in-line power meter designed to be inserted into the coaxial cable between a transmitter and an antenna. The meter is used to measure the transmitter output power supply to the antenna. A short coaxial cable connects the transmitter output to the power meter, and the output of the power meter is connected to the antenna or dummy load.

A more sophisticated power meter is the bench unit designed for laboratory or production line testing. The output of the transmitter or other device whose power is to be measured is connected by a short coaxial cable to the power meter input.

Power meters may have either an analog readout meter or a digital display. The dial or display is calibrated in milliwatts, watts, or kilowatts. The dial can also be calibrated in terms of dBm. This is the decibel power reference to 1 milliwatt (mW). In the smaller, handheld type of power meter, an SWR measurement capability is also usually included.

The operation of a power meter is generally based on converting signal power into heat. Whenever current flows through a resistance, power is dissipated in the form of

Example 20-1

An RF voltmeter with a detector probe is used to measure the voltage across a 75-Ω resistive load. The frequency is 137.5 MHz. The measured voltage is 8 V. What power is dissipated in the load?

An RF power meter with a detector probe produces a peak reading of voltage V_P:

$$V_P = 8 \text{ V}$$
$$V_{rms} = 0.707 \, V_P$$
$$= 0.707(8) = 5.656 \text{ V}$$

$$\text{Power} = \frac{V^2}{R} = \frac{(5.656)^2}{75}$$

$$= \frac{32}{75} = 0.4265 \text{ W}$$

$$= 426.5 \text{ mW}$$

heat. If the heat can be accurately measured, it can usually be converted into an electrical signal which can be displayed on a meter.

Power can also be measured indirectly. If the load impedance is known and resistive, you can measure the voltage across the load and then calculate the power with the formula $P = V^2/R$.

Dummy Loads

A *dummy load* is a resistor that is connected to the transmission line in place of the antenna to absorb the transmitter output power. When measuring the power or making other transmitter tests, it is usually desirable to disconnect the antenna so that the transmitter does not radiate and interfere with other stations on the same frequency. In addition, it is best that no radiation be released if the transmitter has a problem or does not meet frequency or emission standards. The dummy load meets this requirement. The dummy load may be connected directly to the transmitter coaxial output connector, or it may be connected to it by a small piece of coaxial cable.

The load is a resistor whose value is equal to the output impedance of the transmitter and that has sufficient power rating. For example, a CB transmitter has an output impedance of 50 Ω and a power rating of about 4 W. The resistor dummy load must be capable of dissipating that amount of power or more. For example, you could use three 150-Ω, 2-W resistors in parallel to give a load of 150/3, or 50 Ω and 3 $\times$ 2, or 6 W. Standard composition carbon resistors can be used. The tolerance is not critical, and resistors with 5 or 10 percent tolerance will work well.

For low-power transmitters such as CBs and amateur radios, an incandescent light bulb makes a reasonably good load. A type 47 pilot light is widely used for transmitter outputs of several watts. An ordinary light bulb of 75 or 100 W, or higher, can also be used for higher-power transmitters.

The best dummy load is a commercial unit designed for that purpose. These units are usually designed for some upper power limit such as 200 W or 1 kW. The higher-power units are made with a resistor immersed in oil to improve its heat dissipation capability without burning up. A typical unit is a resistor installed in a 1-gallon can filled with insulating oil. A coaxial connector on top is used to attach the unit to the transmitter. Other units are mounted in an aluminum housing with heat fins to improve heat dissipation. The resistors are noncritical, but they must be noninductive. The resistor should be as close to pure resistance at the operating frequency as possible.

Standing Wave Ratio Meters

A *standing wave ratio (SWR)* meter measures forward and reflected power and therefore can display SWR as well. In addition, most power meters contain built-in circuitry for measuring SWR. There are two basic ways to measure SWR. One is to use a bridge circuit, and the other is to use a short segment of a transmission line on which to detect reflections.

SWR Bridge Meter. Figure 20-1(*a*) shows a bridge SWR meter. A bridge is formed of precision, noninductive resistors and the antenna radiation resistance. In some SWR meters, resistors are replaced with a capacitive voltage divider. The meter is connected to measure the unbalance of the bridge. The transmitter is the AC power source.

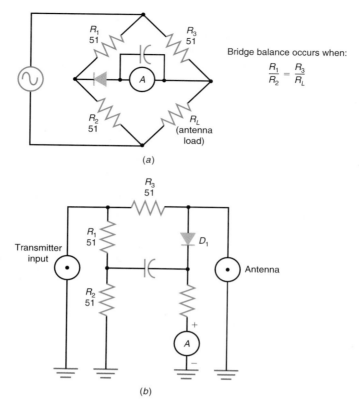

Bridge balance occurs when:

$$\frac{R_1}{R_2} = \frac{R_3}{R_L}$$

(a)

(b)

Fig. 20-1 Bridge SWR meter.

Figure 20-1(b) shows the circuit rearranged so that the meter and one side of the bridge are grounded, thereby creating a better match to unbalanced coaxial transmission lines. Note the use of coaxial connectors for the transmitter input and the antenna and transmission line. The meter is a basic DC microammeter. The diode D_1 rectifies the RF signal into a proportional direct current. If the radiation resistance of the antenna is 50 Ω, the bridge will be balanced and the meter reading will be zero. The meter is calibrated to display an SWR of 1. If the antenna radiation resistance is not 50 Ω, the bridge will be unbalanced and the meter will display a reading that is proportional to the degree of unbalance. The meter is calibrated in SWR values.

SIGNAL GENERATORS

A *signal generator* is one of the most often needed pieces of equipment in communications equipment servicing. As its name implies, a signal generator is a device that produces an output signal of a specific shape at a specific frequency and, in communications applications, usually with some form of modulation. The heart of all signal generators is a variable-frequency oscillator that generates a signal, which is usually a sine wave. Audio frequency sine waves are required for the testing of audio circuits in communications equipment, and sine waves in the radio frequency range from approximately 500 kHz to 30 GHz are required to test all types of RF amplifiers, filters, and other circuits. This section provides a general overview of the most common types of signal generators used in communications testing and servicing.

FUNCTION GENERATORS. A *function generator* is a signal generator designed to generate sine waves, square waves, and triangular waves over a frequency range of approximately 0.001 Hz to about 2 MHz. By changing capacitor values and varying the charging current with a variable resistance, a wide range of frequencies can be developed. The sine, square, and triangular waves are also available simultaneously at individual output jacks.

A function generator is one of the most flexible signal generators available. It covers all the frequencies needed for audio testing and also provides signals in the low RF range. The precision of the frequency setting is accurate for most testing purposes. A frequency counter can be used for precise frequency measurement if needed.

The output impedance of a function generator is typically 50 Ω. The output jacks are BNC connectors which are used with 50- or 75-Ω coaxial cable. The output amplitude is continuously adjustable with a potentiometer. Some function generators include a switched resistive attenuator that allows the output voltage to be reduced to the millivolt and microvolt level.

Because of the very low cost and flexibility of a function generator, it is the most popular bench instrument in use for general testing of radio amplifiers, filters, and low-frequency RF circuits. The square-wave output signals also make it useful in testing digital circuits.

RF SIGNAL GENERATORS. Two basic types of RF signal generators are in use. The first is a simple, inexpensive type that uses a variable-frequency oscillator to generate RF signals in the 100-kHz to 500-MHz range. The second type is frequency-synthesized.

These simple RF signal generators contain an output level control that can be used to adjust the signal to the desired level, from a few volts down to several millivolts. Some units contain built-in resistive step attenuators to reduce the signal level even further. The more sophisticated generators have built-in level control or *automatic gain control (AGC)*. This ensures that the output signal remains constant while it is tuned over a broad frequency range.

Most low-cost signal generators allow the RF signal being generated to be amplitude-modulated. Normally, a built-in audio oscillator with a fixed frequency somewhere in the 400- to 1000-Hz range is included. A modulation level control is provided to

Versatile RF equipment permits vector network analysis, spectrum analysis, and optional instrument controller testing. Simulated vector spectrum analysis is used in the laboratory manual accompaning this textbook.

adjust the modulation from 0 to 100 percent. Some RF generators have a built-in frequency modulator.

Such low-cost signal generators are useful in testing and troubleshooting communications receivers. They can provide an RF signal at the signal frequency for injection into the antenna terminals of the receiver. The generator can produce signals that can substitute for local oscillators or can be set to the intermediate frequencies for testing IF amplifiers.

The output frequency is usually set by a large calibrated dial. The precision of calibration is only a few percent, but more precise settings can be obtained by monitoring the signal output on a frequency counter.

When any type of generator based on LC or RC oscillators is being used, it is best to turn the generator on and let it warm up for several hours before it is used. When a generator is first turned on, its output frequency will drift because of changes in capacitance, inductance, and resistance values. Once the circuit has warmed up to its operating temperature, these variations cease or drop to a negligible amount.

Newer generators use *frequency synthesis* techniques. These generators include one or more mixer circuits that allow the generator to cover an extremely wide range of frequencies. The great value of a frequency-synthesized signal generator is its excellent frequency stability and precision of frequency setting.

Most frequency-synthesized signal generators use a front panel keyboard. The desired frequency is entered with the keypad and displayed on a digital readout. As with other signal generators, the output level is fully variable. The output impedance is typically 50 Ω, with both BNC- and N-type coaxial connectors being required.

Frequency-synthesized generators are available for frequencies into the 20- to 30-GHz range. Such generators are extremely expensive, but may be required if precision measurement and testing are necessary.

Sweep Generators. A *sweep generator* is a signal generator whose output frequency can be linearly varied over some specific range. Sweep generators have oscillators that can be frequency modulated, after which a linear sawtooth voltage can be used as a modulating signal. The resulting output waveform is a constant-amplitude sine wave whose frequency increases from some lower limit to some upper limit (see Fig. 20-2).

Sweep generators are normally used to provide a means of automatically varying the frequency over a narrow range to plot the frequency response of a filter or amplifier or to show the bandpass response curve of the tuned circuits in equipment such as a receiver. The sweep generator is connected to the input of the circuit, and the upper and lower frequencies are determined by adjustments on the generator. The generator then automatically sweeps over the desired frequency range.

At the same time, the output of the circuit being tested is monitored. The amplitude of the output will vary according to the frequency, depending on the type of circuit being tested. The output of the circuit is connected to an RF detector probe. The resulting signal is the envelope of the RF signal as determined by the output variation

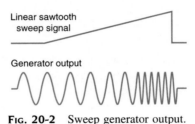

Fig. 20-2 Sweep generator output.

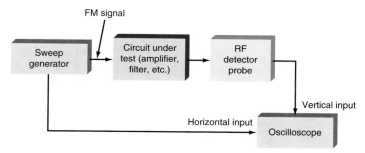

FIG. 20-3 Testing frequency response with a sweep generator.

of the circuit being tested. The signal displayed on the oscilloscope is an amplitude plot of the frequency response curve. The horizontal axis represents the frequency being varied with time, and the output represents the amplitude of the circuit output at each of the frequencies.

Figure 20-3 shows the general test setup. The linear sweep from the sweep generator is used in place of the oscilloscope's internal sweep so that the displayed response curve is perfectly synchronized with the generator.

Most sweep generators also have *marker capability;* that is, one or more reference oscillators are included to provide frequency markers at selected points so that the response curve can be actively interpreted. Marker increments may be 100 kHz or 1 MHz. They are added to (linearly mixed with) the output of the RF detector probe, and the composite signal is amplified and sent to the vertical input of the oscilloscope. Sweep generators can save a considerable amount of time in testing and adjusting complex tuned circuits in receivers and other equipment.

Some function generators have built-in sweep capability. If sweep capability is not built in, often an input jack is provided so that an external sawtooth wave can be connected to the generator for sweep purposes.

ARBITRARY WAVEFORM GENERATORS.   A newer type of signal generator is the *arbitrary waveform generator.* It uses digital techniques to generate almost any waveform. If many points or samples of the waveform and a binary word with many bits are used, the output wave with additional filtering is a very close approximation to the desired wave. Most arbitrary waveform generators come with preprogrammed standard waves like sine, rectangular, sawtooth, and triangular waves, and amplitude modulation. These generators are set up so that you can program a waveform. The arbitrary waveform generator provides a fast and easy way to generate almost any signal shape. Because digital sampling techniques are used, the upper frequency limit of the output is usually below 20 MHz.

FREQUENCY COUNTERS

One of the most widely used communications test instruments is the *frequency counter.* It measures the frequency of transmitters, local and carrier oscillators, frequency synthesizers, and any other signal-generating circuit or equipment. It is imperative that the frequency counter operate on its assigned frequency to ensure compliance with the rules and regulations and to avoid interference with other services.

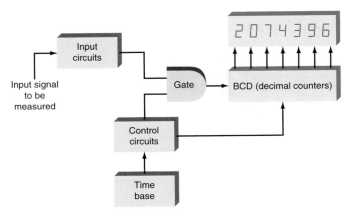

Fig. 20-4 Block diagram of a frequency counter.

A frequency counter displays the frequency of a signal on a decimal readout. Counters are available as bench instruments or portable battery-powered units. A block diagram of a frequency counter is shown in Fig. 20-4. Almost all digital counters are made up of six basic components: the input circuit, the gate, the decimal counter, the display, the control circuits, and the time base. In various combinations, these circuits permit the counter to make time and frequency measurements.

FREQUENCY MEASUREMENT. *Frequency* is a measure of the number of events or cycles of a signal that occur in a given period of time. The usual unit of frequency measurement is hertz (Hz), or cycles per second. The time base generates a very precise signal that is used to open, or enable, the main gate for an accurate period of time to allow the input pulses to pass through to the counter. The time base accuracy is the most critical specification of the counter. The counter accumulates the number of input cycles that occur during that 1-s interval. The display then shows the frequency in cycles per second, or hertz.

The number of decade counters and display digits also determines the resolution of the frequency measurement. The greater the number of digits, the better the resolution will be. Most low-cost counters have at least 5 digits of display. This provides reasonably good resolution on most frequency measurements. For very high frequency measurements, resolution is more limited. However, good resolution can still be obtained with a minimum number of digits by optimum selection of the time base frequency.

In most counters that have a selectable time base, the position of the display decimal point is automatically adjusted as the time base signal is adjusted. In this way, the display always shows the frequency in units of hertz, kilohertz, or megahertz. Some of the more sophisticated counters have an automatic time base selection feature called *autoranging*. Special autoranging circuitry in the counter automatically selects the best time base frequency for maximum measurement resolution without overranging. *Overranging* is the condition that occurs when the count capability of the counter is exceeded during the count interval. The number of counters and display digits determines the count capability and thus the overrange point for a given time base.

PRESCALING. All the techniques for measuring high frequencies involve a process that converts the high frequency into a proportional lower frequency that can be measured with conventional counting circuitry. This translation of the high frequency into the lower frequency is called *down conversion*.

Prescaling is a down-conversion technique that involves the division of the input frequency by a factor that puts the resulting signal into the normal frequency range of the counter. It is important to realize that although prescaling permits the measurement of higher frequencies, it is not without its disadvantages, one of which is loss of resolution. One digit of resolution is lost for each decade of prescaling incorporated.

The prescaling technique for extending the frequency-measuring capability of a counter is widely used. It is simple to implement with modern, high-speed ICs. It is also the most economical method of extending the counting range. Prescalers can be built into the counter and switched in when necessary. Alternatively, external prescalers, which are widely available for low-cost counters, can be used. Most prescalers operate in the range of 200 MHz to 20 GHz. For frequencies beyond 20 GHz, more sophisticated down-conversion techniques must be used.

DEVIATION METERS

A *deviation meter* is designed to measure the amount of carrier deviation of an FM/PM transmitter. The maximum amount of deviation and the maximum modulating frequency determine the modulation index, which, in turn, determines the number of significant sidebands and the bandwidth of the signal. The bandwidth is usually specified by the FCC and, to avoid adjacent channel interference, the signal may not deviate outside the channel.

The deviation meter is usually an FM demodulator that has been calibrated to provide an output indication of the amount of deviation on a meter scale or on a digital readout. The readout may be indicated in percent of total deviation or in frequency deviation from hertz to kilohertz.

Although separate deviation meters are made, they usually built into another piece of equipment containing signal generators, power meters, and other instruments which together form a complete transmitter test set.

DID YOU KNOW?

Deviation meters are usually built into another piece of test equipment containing signal generators, power meters, and other instruments which together form a complete transmitter test set.

SPECTRUM ANALYZERS

The *spectrum analyzer* is one of the most useful and popular communications test instruments. Its basic function is to display received signals in the frequency domain. Oscilloscopes are used primarily to display signals in the time domain. The sweep circuits in the oscilloscope deflect the electron beam in the CRT across the screen horizontally. This represents units of time. The input signal to be displayed is applied to deflect the electron beam vertically. Thus, electronic signals which are voltages occurring with respect to time are displayed on the oscilloscope screen.

The spectrum analyzer combines the display of an oscilloscope with circuits that convert the signal into the individual frequency components dictated by the Fourier analysis of the signal. Signals applied to the input of the spectrum analyzer are shown as vertical lines or narrow pulses at their frequency of operation.

Figure 20-5 shows the display of a spectrum analyzer. The horizontal display is calibrated in frequency units, and the vertical part of the display is calibrated in voltage, power, or decibels. The spectrum analyzer display shows three signals at fre-

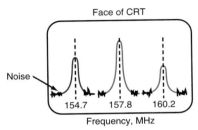

Face of CRT

Noise

154.7   157.8   160.2

Frequency, MHz

Fig. 20-5   A frequency display of a spectrum analyzer.

quencies of 154.7, 157.8, and 160.2 MHz. The vertical height represents the relative strength of the amplitude of each signal. Each signal might represent the carrier of a radio transmitter. A spectrum analysis of any noise between the signals is shown.

The spectrum analyzer can be used to view a complex signal in terms of its frequency components. A graticule on the face of the CRT allows the frequency spacing between adjacent frequency components to be determined. The vertical amplitude of the scale is calibrated in decibels. The spectrum analyzer is extremely useful in analyzing complex signals which may be difficult to analyze or whose content may be unrecognizable.

The four basic techniques of spectrum analysis are bank of filters, swept filter, swept spectrum superheterodyne, and fast Fourier transform (FFT). All these methods do the same thing, that is, decompose the input signal into its individual sine-wave frequency components. Both analog and digital methods are used to implement each type. The superheterodyne and FFT methods are the most widely used and are discussed below.

SUPERHETERODYNES.   Perhaps the most widely used RF spectrum analyzer is the *superheterodyne* type (see Fig. 20-6). It consists of a broadband front end, a mixer, and a tunable local oscillator. The frequency range of the input is restricted to some upper limit by a low-pass filter in the input. The mixer output is the difference between the input signal frequency component and the local oscillator frequency. This is the *intermediate frequency (IF)*.

As the local oscillator frequency increases, the output of the mixer stays at the IF. Each frequency component of the input signal is converted into the IF value by the varying local oscillator signal. If the frequency components are very close to one another and the bandwidth of the IF bandpass filter (BPF) is broad, the display will be just one broad pulse. Narrowing the bandwidth of the IF BPF allows more closely spaced components to be detected. Most spectrum analyzers have several switchable selectivity ranges for the IF.

Spectrum analyzers are available in many configurations with different specifications, and they are designed to display signals from approximately 100 kHz to approximately 30 GHz. Most RF and microwave spectrum analyzers are superheterodynes. Spectrum analyzers are usually calibrated to provide relatively good measurement accuracy of the signal. Most signals are displayed as power or decibel measurements, although some analyzers provide for voltage level displays. The input is usually applied through a 50-Ω coaxial cable and connector. On some analyzers, a 75-Ω cable input is provided.

FFT SPECTRUM ANALYZERS.   The *fast Fourier transform (FFT)* method of spectrum analysis relies on the FFT mathematical analysis described in Chap. 9. FFT spectrum analyzers give a high-resolution display and are generally superior to all other

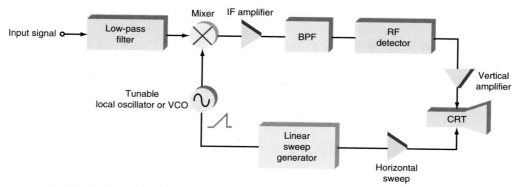

Fig. 20-6 Superheterodyne spectrum analyzer.

types of spectrum analyzers. However, the upper frequency of the input signal is limited to frequencies in the tens of megahertz range.

In addition to measuring the spectrum of a signal, spectrum analyzers are useful in detecting harmonics and other spurious signals generated unintentionally. Spectrum analyzers can be used to display the relative signal-to-noise ratio, and they are ideal for analyzing modulation components and displaying the harmonic spectrum of a rectangular pulse train.

Because of their extremely high prices (usually $10,000 to $50,000), spectrum analyzers are not widely used. However, many critical testing and measurement applications demand their use, especially in developing new RF equipment and in making final tests and measurements of manufactured units. Spectrum analyzers are also used in the field for testing cable TV systems, cellular telephone systems, and other complex communications systems.

FIELD STRENGTH METERS

One of the least expensive pieces of RF test equipment is the *field strength meter (FSM)*, a portable device used for detecting the presence of RF signals near an antenna. The FSM is a sensitive detector for RF energy being radiated by a transmitter into an antenna. It provides a relative indication of the strength of the electromagnetic waves reaching the meter.

The field strength meter is a vertical whip antenna, usually of the telescoping type, connected to a simple diode detector. The diode detector is exactly like the circuit of a simple crystal radio or a detector probe, as described earlier, but without any tuned circuits so that the unit will pick up signals on any frequency.

The field strength meter does not give an accurate measurement of signal strength. In fact, its only purpose is to detect the presence of a nearby signal (within about 100 ft or less). Its primary purpose is to determine whether a given transmitter and antenna system are working. The closer the meter is moved to the transmitter and antenna, the higher the signal level.

A useful function of the meter is in determining the radiation pattern of an antenna. The field strength meter is adjusted to give the maximum reading in the direction of the most radiation from the antenna. The meter is moved in a constant radius circle around the antenna for 360°. Every 5° or 10°, a field strength reading is taken from the meter. The resulting set of readings can be plotted on polar graph paper to reveal the horizontal radiation pattern of the antenna.

There are other types of field strength meters. A simple meter may be built to incorporate a resonant circuit to tune the input to a specific transmitter frequency. This makes the meter more sensitive. Some meters may have a built-in amplifier to make the meter even more sensitive and useful at greater distances from the antenna.

An *absolute* (rather than *relative*) *field strength meter* is available for accurate measurements of signal strength. The strength of the radiated signal is usually measured in microvolts per meter, or μV/m. This is the amount of voltage the signal will induce into an antenna that is 1 m long. An absolute field strength meter is calibrated in units of μV/m. Highly accurate signal measurements can be made.

OTHER TEST INSTRUMENTS

There are hundreds of types of communications test instruments, most of which are very specialized. The ones described previously in this chapter, plus those listed in Table 20-1, are the most common, but there are many others including the many special test instruments designed by equipment manufacturers for testing their production units or servicing customers' equipment.

TABLE 20-1   DESCRIPTION OF ABSORPTION WAVE METER, IMPEDANCE METER, DIP OSCILLATOR, AND NOISE BRIDGE

Instrument	Purpose
Absorption wave meter	Variable tuned circuit with an indicator that tells when the tuned circuit is resonant to a signal coupled to the meter by a transmitter; provides a rough indication of frequency.
Impedance meter	An instrument, usually of the bridge type, that accurately measures the impedance of a circuit, a component, or even an antenna at RF frequencies.
Dip oscillator	A tunable oscillator used to determine the approximate resonant frequency of any de-energized LC–resonant circuit. The oscillator inductor is coupled to the tuned circuit inductor, and the oscillator is tuned until its feedback is reduced as energy being taken by the tuned circuit, which is indicated by a "dip," or reduction, in current on a built-in meter. The approximate frequency is read from a calibrated dial.
Noise bridge	Bridge circuit driven by a random noise voltage source (usually a reversed-biased zener diode that generates random "white" or pseudo random "pink" noise) that has an antenna or coaxial cable as one leg of the bridge; used to make antenna characteristic impedance measurements and measurements of coaxial cable velocity factor and length.

There are many hundreds, even many thousands, of different tests that are made on communications equipment. There are, however, some common tests widely used on all types of communications equipment. This section summarizes the most common tests and measurements made in the servicing of communications equipment.

Most of the tests described here focus on standard radio communications equipment. Tests for transmitters, receivers, and antennas will be described, as well as special microwave tests, fiber-optics cable tests, and tests on data communications equipment. Keep in mind that the test procedures described are general. To make specific tests, follow the test setups recommended by the test equipment manufacturer. Always have the manuals for the test equipment and the equipment being serviced on hand for reference.

TRANSMITTER TESTS

Four main tests are made on most transmitters: tests of frequency, modulation, and power, and tests for any undesired output signal component such as harmonics and parasitic radiations. These tests and measurements are made for several reasons. First, any equipment that radiates a radio signal is governed by FCC rules and regulations. For a transmitter to meet its intended purpose, the FCC specifies frequency, power, and other measurements to which the equipment must comply. Second, the tests are nor-

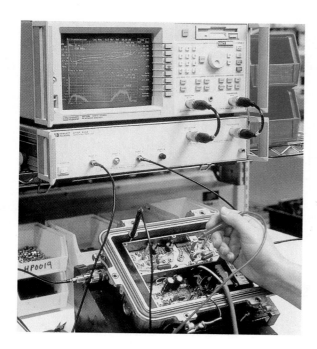

This network analyzer helps designers predict how multiple components will behave together and helps manufacturers know whether final products will meet performance specifications.

mally made when the equipment is first installed to be sure everything is working correctly. Third, such tests may be made to troubleshoot equipment. If the equipment is not working properly, these tests are some of the first that should be made to help identify the trouble.

FREQUENCY MEASUREMENT.   Regardless of the method of carrier generation, the frequency of the transmitter is important. The transmitter must operate on the assigned frequency to comply with FCC regulations and to ensure that the signal can be picked up by a receiver that is tuned to that frequency.

The output of a transmitter is measured directly to determine its frequency. The transmitted signal is independently picked up and its frequency measured on a frequency counter. Figure 20-7 shows several methods of picking up the signal. Many frequency counters designed for communications work come with an antenna that picks up the signal directly from the transmitter [see Fig. 20-7(a)]. The transmitter is keyed up (turned on) while it is connected to its regular antenna, and the antenna on the counter picks up the signal and translates it into one that can be measured by the counter circuitry. No modulation should be applied, especially if FM is used.

The transmitter output can be connected to a dummy load [see Fig. 20-7(b)]. This will ensure that no signal is radiated, but that there will be sufficient signal pickup to make a frequency measurement if the counter and its antenna are placed near the transmitter.

Another method of connecting the counter to the transmitter is to use a small coil, as shown in Fig. 20-7(c). A small pickup coil can be made of stiff copper wire. Enamel copper wire, size AWG 12 or 14 wire, is formed into a loop of two to four turns. The ends of the loop are connected to a coaxial cable with a BNC connector for attachment to the frequency counter. The loop can be placed near the transmitter circuits. This method of pickup is used when the transmitter has been opened and its circuits ex-

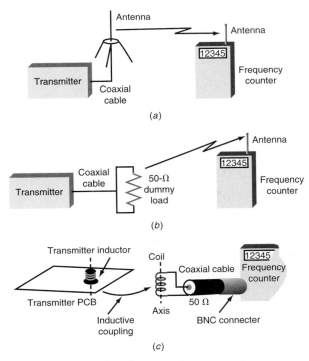

FIG. 20-7   Transmitter frequency measurement.

posed. For most transmitters, the loop has been placed only in the general vicinity of the circuitry. Normally the loop picks up radiation from one of the inductors in the final stage of the transmitter. Maximum coupling is achieved when the axis of the turns of the loop is parallel to the axis of one of the inductors in the final output stage.

Once the signal from the transmitter is coupled to the counter, the counter sensitivity is adjusted, and the counter is set to the desired range for displaying the frequency. The greater the number of digits the counter can display, the more accurate the measurement.

Normally this test is made without modulation. If only the carrier is transmitted, any modulation effects can be ignored. Modulation must not be applied to an FM transmitter, because the carrier frequency will be varied by the modulation, resulting in an inaccurate frequency measurement.

The quality of crystals today is excellent; thus, off-frequency operation is not common. If the transmitter is not within specifications, the crystal can be replaced. In some critical pieces of equipment, the crystal may be in an oven. If the oven temperature control circuits are not working correctly, the crystal may have drifted off frequency. This calls for repair of the oven circuitry or replacement of the entire unit.

If the signal source is a frequency synthesizer, the precision of the reference crystal can be checked. If it is within specifications, perhaps an off-frequency operation is being caused by a digital problem in the phase-locked loop. An incorrect frequency division ratio, faulty phase detector, or poorly tacking VCO may be the problem.

MODULATION TESTS. If amplitude modulation is being used in the transmitter, you should measure the percentage of modulation. It is best to keep the percentage of modulation as close to 100 as possible to ensure maximum output power, below 100 to prevent signal distortion and harmonic radiation. In FM or PM transmitters, you should measure the frequency deviation with modulation. The goal with FM is to keep the deviation within the specific range to prevent adjacent channel interference.

The best way to measure AM is to use an oscilloscope and display the AM signal directly. To do this, you must have an oscilloscope whose vertical amplifier bandwidth is sufficient to cover the transmitter frequency. Figure 20-8(*a*) shows the basic test setup. An audio signal generator is used to amplitude-modulate the transmitter. An audio signal of 400–1000 Hz is applied in place of the microphone signal.

The transmitter is then keyed up, and the oscilloscope attached to the output load. It is best to perform this test with a dummy load to prevent radiation of the signal. The oscilloscope is then adjusted to display the AM signal. The display will appear as shown in Fig. 20-8(*b*).

POWER MEASUREMENTS. Most transmitters have a tune-up procedure recommended by the manufacturer for adjusting each stage to produce maximum output power. In older transmitters, tuned circuits between stages have to be precisely adjusted in the correct sequence. In modern solid-state transmitters, there are fewer adjustments, but in most cases there are some adjustments in the driver and frequency multiplier stages as well as tuning adjustments for resonance at the operating frequency to the final amplifier. There may be impedance-matching adjustments in the final amplifier to ensure full coupling of the power to the antenna. The process is essentially that of adjusting the tuned circuits to resonance. These measurements are generally made while monitoring the output power of the transmitter.

The procedure for measuring the output power is to connect the transmitter output to an RF power meter and the dummy load as shown in Fig. 20-9. The transmitter is keyed up without modulation, and adjustments are made on the transmitter circuit

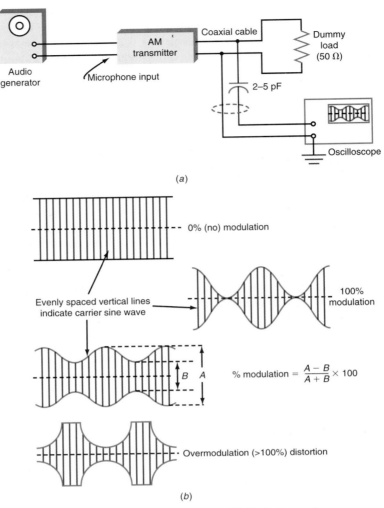

Fig. 20-8 AM measurement. (*a*) Test setup. (*b*) Typical waveforms.

to tune for maximum power output. With the test arrangement shown, the power meter will display the output power reading.

Once the transmitter is properly tuned up, it can be connected to the antenna. The power into the antenna will then be indicated. If the antenna is properly matched to the transmission line, the amount of output power will be the same as that in the dummy load. If not, SWR measurements should be made. It may be necessary to adjust the antenna or a matching circuit to ensure maximum output power with minimum SWR.

HARMONICS AND SPURIOUS OUTPUT MEASUREMENTS. A common problem in transmitters is the radiation of undesirable harmonics or spurious signals. Ideally, the output of the transmitter should be a pure signal at the carrier frequency with only those sideband components produced by the modulating signal. However, most transmitters will generate some harmonics and spurious signals. Transmitters that use class C, class D, and class E amplifiers generate a high har-

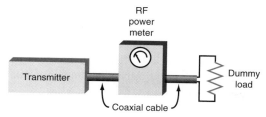

Fig. 20-9 Power measurement.

monic content. If the tuned circuits in the transmitter are properly designed, the Qs will be high enough to reduce the harmonic content level sufficiently. However, this is not always the case.

Another problem is that other spurious signals can be generated by transmitters. In high-power transmitters particularly, parasitic oscillations can occur. These are caused by the excitation of small tuned circuits whose components are the stray inductances and capacitances in the circuits or the transistors of the tubes involved. Parasitic oscillations can reach high levels and cause radiation on undesired frequencies.

For most transmitters, the FCC specifies maximum levels of harmonic and spurious radiations. Normally they must be at least 30 or 40 dB down from the main carrier signal. The best way to measure harmonics and spurious signals is to use a spectrum analyzer. The transmitter output is modulated with an audio tone, and its output is monitored directly on the spectrum analyzer. It is usually best to feed the transmitter output into a dummy load for this measurement. The spectrum analyzer is then adjusted to display the normal carrier and sideband pattern. The search for high-level signals can begin for spurious outputs by tuning the spectrum analyzer above and below the operating frequency. The spectrum analyzer can be tuned to search for signals at the second, third, and higher harmonics of the carrier frequency. If signals are detected, they can be measured to ensure that they are sufficiently low in power to meet FCC regulations and/or the manufacturer's specifications. The spectrum analyzer is then tuned over a broad range to ensure that no other spurious nonharmonic signals are present.

It may be possible to reduce the harmonic and spurious output content by making a minor transmitter tuning adjustment. If not, to meet specifications, it is often necessary to place a low-pass filter in the transmission line.

ANTENNA AND TRANSMISSION LINE TESTS

If the transmitter is working correctly and the antenna has been properly designed, about the only test that needs to be made on the transmission line and antenna is for standing waves. It will tell you whether any further adjustments are necessary. If the SWR is high, you can usually tune the antenna to reduce it.

You may also run into a transmission line problem. It may be open or short-circuited, which will show up on an SWR test as infinite SWR. But there may be other problems such as a cable that has been cut, short-circuited, or crushed between the transmitter and receiver. These kinds of problems can be located with a time domain reflectometer test.

SWR Tests. The test procedure for SWR is shown in Fig. 20-10. The SWR meter is connected between the transmission line and the antenna. Check with the manufacturer of the SWR meter to determine whether any specific connection location is required or whether other conditions must be met. Some of the lower-cost SWR meters must be connected directly at the antenna or a specific number of half wavelengths from the antenna back to the transmitter.

Once the meter is properly connected, key up the transmitter without modulation. The transmitter should have been previously tuned and adjusted for maximum output power. The SWR can be read directly from the instrument's meter. In some cases, the meter will give measurements for the relative amount of incident or forward and reflected power or will read out in terms of the reflection coefficient, in which case you must calculate the SWR as described earlier. Other meters will read out directly in SWR. The maximum range is usually 3:1.

The ideal SWR is 1 or 1:1, which means that all the power generated by the transmitter is absorbed by the antenna load. Nevertheless, in even the best systems, perfect matching is rarely achieved. Any mismatch will produce reflected power and standing waves. If the SWR is less than 2:1, the amount of power that will be lost or reflected will be minimal.

The primary procedure for reducing the SWR is to make antenna adjustments, usually in the form of modifying the element lengths to more closely tune the antenna to the frequency of operation. Using many antennas makes it possible to adjust their length over a narrow range to fine-tune the SWR. Other antennas permit adjustment to provide a better match of the transmission line to the driven element of the antenna. These adjustments can be made one at a time, and the SWR monitored.

TDR Tests. *Time domain reflectometry (TDR)* is a pulse for cables and transmission lines of all types. It is widely used in finding faults in cables used for digital data transmission, but it can be used for RF transmission lines. (Refer to the section Data Communications Tests later in this chapter for details.)

RECEIVER TESTS

The primary tests for receivers involve sensitivity and noise level. The greater the sensitivity of the receiver, the higher its gain and the better job it does of receiving very small signals.

As part of the sensitivity testing, signal-to-noise (S/N) ratio is also usually measured indirectly. The ability of a receiver to pick up weak signals is just as much a function of the receiver noise level as it is of overall receiver gain. The lower the noise level, the greater the ability of the receiver to detect weak signals.

As part of an overall sensitivity check, some receiver manufacturers specify an audio power output level. Since receiver sensitivity measurements are usually made by measuring the speaker output voltage, power output can also be checked if desired.

In this section, some of the common tests made on receivers are described. The information is generic in nature, and the actual testing procedures are often different from one receiver manufacturer to another.

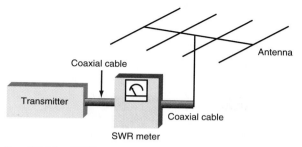

FIG. 20-10 SWR measurement.

EQUIPMENT REQUIRED. To make the sensitivity and noise measurements, the following equipment is necessary:

1. *Dual Trace Oscilloscope.* The vertical frequency response is not too critical, for you will be viewing noise and audio frequency signals.
2. *RF Signal Generator.* This generator will provide an RF signal at the receiver operating frequency. It should have an output attenuator so that signals as low as 1 μV (microvolt) or less can be set. This may indicate the need for an external attenuator if the generator does not have built-in attenuators or output level adjustments. The generator must also have modulation capability, either AM or FM, depending upon the type of receiver to be tested.
3. *RF Voltmeter.* The RF voltmeter is needed to measure the RF generator output voltage in some tests. Some higher-quality RF generators have an RF voltmeter built in to aid in setting the output attenuator and level controls.
4. *Frequency Counter.* A frequency counter capable of measuring the F generator output frequency is also needed.
5. *Multimeter.* A multimeter capable of measuring audio frequency voltage levels is needed. Any analog or digital multimeter with AC measurement capability in the AF range can be used.
6. *Dummy Loads.* A dummy load will be needed for the receiver antenna input for the noise test. This can be either a 50- or a 75-Ω resistor attached to the appropriate coaxial input connector. A dummy load is needed for the speaker. Since most noise and sensitivity tests are made with maximum receiver gain including audio gain, it is not practical or desirable to leave the speaker connected. Most communications receivers have an audio output power capability of 2 to 10 W, which is sufficiently high to make the output signal level too high for comfort. A speaker dummy load of 4, 8, or 16 Ω, depending upon the speaker impedance, is needed. Be sure that the dummy load can withstand the maximum audio power output of the receiver. Do not use wire-wound resistors for this application, for they have too much inductance. Check the receiver's specifications for both the impedance and the maximum power level specification.

NOISE TESTS. *Noise* consists of random signal variations picked up by the receiver or caused by thermal agitation and other conditions inside the receiver circuitry. External noise cannot be controlled or eliminated. However, noise contributed by the receiver can be controlled. Every effort is made during the design to minimize internally generated noise and thus to improve the ability of the receiver to pick up weak signals.

Most noise generated by the receiver occurs in the receiver's front end, primarily the RF amplifier and the mixer. Careful attention is given to the design of both these circuits so that they contribute minimum noise.

Because noise is a totally random signal that is a composite of varying frequency and varying amplitude signals, it is somewhat difficult to measure. However, the fol-

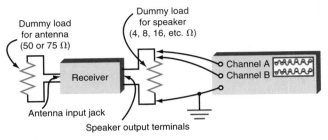

Fig. 20-11 Noise test setup.

lowing procedure has become a common and popular method that is easy to implement.

Refer to the test setup shown in Fig. 20-11. The antenna is removed from the receiver, and a dummy load of the correct impedance is used to replace it. A carbon composition resistor of 50 or 75 Ω can be used. The idea is to prevent the receiver from picking up any signals while maintaining the correct impedance.

At the output of the receiver, the speaker is replaced with a dummy load. Most speakers have an output impedance of 4 or 8 Ω. Check the receiver's specifications, and connect an appropriate value of resistor in place of the speaker. Be sure the dummy load resistors can withstand the output power.

Finally, connect the dual-trace oscilloscope across the dummy speaker load. The same signal should be displayed on both channels of the oscilloscope. The displayed signal will be an amplified version of the noise produced by the receiver and amplified by all of the stages between the antenna input and the speaker. Follow this step-by-step procedure:

1. Turn on the receiver, and tune it to a channel where no signal will be received.
2. Set the receiver volume control to maximum. If the receiver has any type of RF or IF gain control, it too should be set to its maximum setting.
3. Set the oscilloscope input for the lower trace (channel B) to ground. Most oscilloscopes have a switch that allows the input to be set for AC measurements, DC measurements, or ground. Grounding the channel B input will prevent any signal from being displayed. At this time you will see a straight horizontal line for the lower trace. Adjust that lower trace so that it lines up with one of the horizontal graticule lines near the bottom of the oscilloscope screen. This will provide a voltage measurement reference.
4. Set the channel B input and the channel A input to alternating current. Adjust the vertical sensitivities of channels A and B so that they are on the same range. Make the adjustments so that the signal is displayed something like that shown in Fig. 20-12(*a*). You should see exactly the same noise pattern on both channels.
5. Using the vertical position control on the upper or A channel, move the upper noise trace downward so that it begins to merge with the noise signal on channel B. The correct adjustment for the position of the channel A noise signal is such that the peaks of the upper and lower signals just barely merge. This will generally be indicated at the point where there is no blank space between upper and lower traces. The signal should look something like that shown in Fig. 20-12(*b*).
6. Now set both oscilloscope inputs to ground, thereby preventing the noise signal from being displayed. You will see two straight horizontal lines. The distance between the two lines is a measure of the noise voltage. The value indicated is two times the root-mean-square noise voltage [see Fig. 20-12(*c*)].

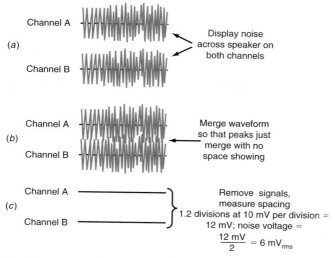

FIG. 20-12   Noise measurement procedure.

Assume that the adjustments described above were made and the separation between the two horizontal traces is 1.2 vertical divisions. If the vertical gains of both channels are set to the 10 mV per division range, the noise reading is 1.2 × 10 mV = 12 mV. The root-mean-square noise voltage is one-half of this figure, or 12 mV/2 = 6 mV.

POWER OUTPUT TESTS.   Sometimes it is necessary to measure the receiver's total power output capability. This is a good general test of all the receiver circuits. If the receiver can supply the manufacturer's specified maximum output power into the speaker with a given low RF signal level input, the receiver is operating correctly.

The test setup for the power output test is shown in Fig. 20-13. An RF signal generator for the correct frequency is used as the primary signal source. It must also be possible to modulate this generator with either AM or FM depending upon the type of receiver.

It is also desirable to connect a frequency counter to the signal generator output to provide an accurate measure of the receiver input frequency. Most communications

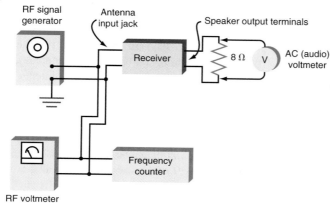

FIG. 20-13   Power output test.

receivers operate on specific frequency channels. In order for the test to be valid, the generator output frequency must be set to the center of the receiver frequency channel. This will usually be known from the receiver's specifications. The signal generator is tuned, and the frequency is set by monitoring the digital readout on the frequency counter.

Be sure to replace the speaker with a resistive load of an impedance equal to that of the speaker, such as 8 Ω. The dummy speaker load should also be able to carry the maximum rated output power of the receiver. Finally, connect an AC voltmeter across the speaker dummy load.

Turn on the receiver, and set the volume control to the maximum. If the receiver has a variable RF or IF gain control, you must set it to the maximum setting also. At this time, any other receiver features should be disabled. For example, in an FM receiver, the squelch should be turned off or disabled. In an AM receiver, if a noise limiter is used, it too should be turned off.

To begin the test, follow this procedure:

1. Set the RF generator output level to 1 mV. If the RF signal generator has a built-in RF voltmeter, use it to make this setting. Otherwise, an external RF voltmeter may be needed, as shown in Fig. 20-13.
2. Set the signal generator for modulation of the appropriate type. If amplitude modulation is used, set the percentage of modulation for 30. If FM is used, set the deviation for ±3 kHz. In most signal generators, the percentage of AM and the frequency deviation for FM is fixed. Refer to the signal generator specifications to find out what these values are.
3. With everything appropriately adjusted, measure the AC voltage across the speaker dummy load. This will be a root-mean-square reading.
4. To determine the receiver power output, use the standard power formula $P = V^2/R$.

Assume that you measure a root-mean-square voltage of 4 V across an 8-Ω speaker. The power output will be:

$$P = \frac{V^2}{R} = \frac{4^2}{8} = \frac{16}{8} = 2 \text{ W}$$

An optional test is to observe the signal across the speaker load with an oscilloscope. Most RF generators modulate the input signal with a sine wave of 400 Hz or 1 kHz. If an oscilloscope is placed across the dummy speaker load, the sine wave will be seen. This will indicate whether the receiver is distorting. The oscilloscope can also be used in place of the audio voltmeter making the voltage measurement across the dummy speaker load. Remember that oscilloscope measurements are peak to peak. The peak-to-peak value must be converted to root-mean-square to make the power output calculation.

20-dB Quieting Sensitivity Tests. In most cases, the sensitivity of a receiver is expressed in terms of the minimum RF voltage at the antenna terminals that will produce a specific audio output power level. Most measurements factor in the effect of noise.

The method of sensitivity measurement is determined according to whether AM or FM is used. Since most modern radio communications equipment uses frequency modulation, measuring FM receiver sensitivity will be illustrated. There are two basic methods, quieting and SINAD. The *quieting method* measures the amount of signal needed to reduce the output noise to 20 dB. As the signal level increases, the noise level decreases until the limiters in the IF section begin to start their clipping action. When this happens, the receiver output "quiets"; that is, its output is silent and blanks out the noise.

The *SINAD test* is a measure of the input signal voltage that will produce at least a 12-dB signal-to-noise ratio. The noise value includes any harmonics that are produced by the receiver circuits due to distortion.

The test setup for receiver sensitivity measurements is shown in Fig. 20-14(*a*). It consists of an RF signal generator, RF voltmeter, frequency counter, the receiver to be tested, and a voltmeter to measure the output across the speaker dummy load.

It is often necessary to provide an impedance-matching network between the generator and the receiver antenna input terminals. Most RF generators have a 50-Ω output impedance. This may match the receiver input impedance exactly. However, some receiver input impedances may be different. For example, if a receiver has a 75-Ω input impedance, some form of impedance matching will be required. This is usually handled by a resistive attenuator known as an *impedance-matching pad,* which is a resistive T network that provides the correct match between the receiver input and the generator. A typical impedance-matching pad is shown in Fig. 20-14(*b*). It matches the 50-Ω generator output to the 50-Ω receiver antenna input. Since resistors are used, the impedance-matching circuit is also an attenuator. With the values given in Fig. 20-14(*b*), the signal attenuation is 10 dB. This must be factored into all signal generator measurements to obtain the correct sensitivity figure. Different values of resistors can be used to create a pad with the correct impedance-matching qualities but with lower or higher values of attenuation.

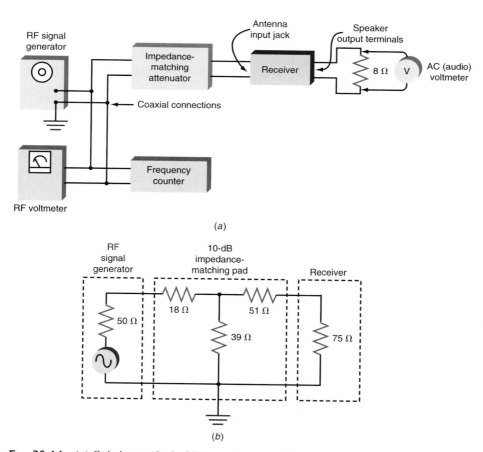

FIG. 20-14 (*a*) Quieting method of FM receiver sensitivity measurement. (*b*) T-type impedance-matching attenuator.

Some manufacturers specify a special input network made up of resistors, inductors, and/or capacitors for this or other sensitivity tests. This network helps simulate the antenna accurately in equipment that uses special antennas.

Follow this procedure to make the 20-dB quieting measurement:

1. Turn the receiver on, and set it to an unused channel.
2. Leave the signal generator off so that no signal is applied.
3. Set the receiver gain to maximum with any RF or IF gain control, if available.
4. Adjust the volume control of the receiver so that you read some convenient value of noise voltage on the meter connected across the speaker. One volt rms is a good value if you can achieve it, but if not, any other convenient value will do.
5. Turn on the signal generator, but set the output level to zero or some very low value. Adjust the generator frequency to the center of the receiver's channel setting. Turn off the modulation so that the generator supplies carrier only.
6. Increase the signal generator output signal level a little at a time, and observe the voltage across the speaker. The noise voltage level will decrease as the carrier signal gets strong enough to overpower the noise. Increase the signal level until the noise voltage drops to one-tenth of its previous value.
7. Measure the generator output voltage on the generator meter or the external RF voltmeter.
8. If an attenuator pad or other impedance-matching network was used, subtract the loss it introduces. The resulting value is the voltage level that produces 20 dB of quieting in the receiver.

Assume that you measured a generator output of 5 μV that produces the 20-dB noise decrease. This is applied across a 50-Ω load producing an input power of $P^2 = V^2/R = (5 \times 10^{-6})^2/50 = 0.5$ pW. This is attenuated further by the 10-dB matching pad to a level of one-tenth, or 0.05, pW. This translates to a voltage level across 50 Ω as

$$V = \sqrt{PR} = \sqrt{(0.05 \times 10^{-12} \times 50)} = 1.58 \times 10^{-6} = 1.58 \; \mu V$$

This is the receiver sensitivity. It takes 1.58 μV of a signal to produce 20 dB of quieting in the receiver.

For a good communications receiver, the 20-dB quieting value should be under 1 μV. A typical value is in the 0.2- to 0.5-μV range. The lower the value, the better the sensitivity.

MICROWAVE TESTS

Microwave tests are generally similar to those performed on standard transmitters and receivers. Transmitter measurements include output power, deviation, harmonics, and spurious signals as well as modulation. The techniques are similar but require the use of only those test instruments whose frequency response is in the desired microwave region. The same goes for receiver measurements and antenna-transmission line tests. The procedures are generally the same, but the equipment is different. For example, with power measurements, a directional coupler is normally used as the transmitter output to reduce the signal to a proper level for measurement with the power meter.

The tests for wireless data communications equipment are essentially the same as those for standard RF communications as described above. The only difference is the type of modulation used to apply the binary signal to the carrier. FSK and its many variants as well as PSK and spread spectrum are the most widely used. Special FSK/PSK deviation and modulation meters are available to make these measurements.

For data communications applications where binary signals are baseband on coaxial and twisted-pair cables, such as in LANs, more conventional testing methods may be used. For example, binary test patterns may be initiated in the transmitting equipment, and the signal viewed on the oscilloscope at the receiving end. Tests of signal attenuation and wave shape can then be made.

EYE DIAGRAMS. A common method of analyzing the quality of binary data transmitted on a cable is to display what is known as the *eye diagram* on a common oscilloscope. The eye diagram, or pattern, is a display of the individual bits overlapped with one another. The resulting output looks like an open eye. The shape of the pattern and the degree that the eye is "open" can be used to determine many things about the quality of transmission.

Eye diagrams are used for testing because it is difficult to display long streams of random serial bits on an oscilloscope. The randomness of the data prevents good synchronization of the oscilloscope with the data, and thus the display jitters and changes continuously. Sending the same pattern of bits such as repeating the ASCII code for the letter U (alternating 1s and 0s) may help the synchronization process, but the display of one whole word on the screen usually does not provide sufficient detail to determine the nature of the signal. The eye diagram solves these problems.

Figure 20-15(*a*) shows a serial pulse train of alternating binary 0s and 1s that is applied to a transmission line. The transmission line, either coaxial or twisted pair, is a low-pass filter, and therefore it eliminates or at least greatly attenuates the higher-frequency components in the pulse train. It also delays the signal. The result is that the signal is rounded and distorted at the end of the cable and the input to the receiver [see Fig. 20-15(*b*)].

> **HINTS AND HELPS**
>
> The eye diagram is an excellent way to get a nonprecise, quick, qualitative check of a signal. The eye pattern tells at a glance the degree of bandwidth limitation, signal distortion, jitter, and noise margin.

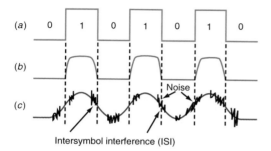

FIG. 20-15 (*a*) Ideal signal before transmission. (*b*) Signal attenuated, distorted, and delayed by medium (coaxial or twisted pair). (*c*) Severely attenuated signal with noise.

The longer the cable and/or the higher the bit rate, the greater the distortion. The pulses tend to blur into one another, causing what is called *intersymbol interference (ISI)*. ISI makes the voltage levels for binary 1 and 0 closer together with 1 bit smearing or overlapping into the other. This makes the receiver's job of clearly distinguishing a binary 1 from a binary 0 more difficult. Further, noise is usually picked up along the transmission path, making the received signal a poor representation of the original data signal. Too much signal rounding or ISI introduces bit errors. Figure 20-15(*c*) is a severely distorted noisy data signal.

The eye pattern provides a way to view a serial data signal and to make a determination about its quality. To display the eye diagram, you need an oscilloscope with a bandwidth at least five times the maximum bit rate. For example, if the bit rate is 10 Mbps, the oscilloscope bandwidth should be at least 10×5 MHz, or 50 MHz. The higher the better. The oscilloscope should have triggered sweep. Either a conventional analog or a digital oscilloscope can be used.

Apply the baseband binary signal at the end of the cable and the receiver input to the vertical input. Adjust the sweep rate of the scope so that one bit interval takes up the entire horizontal width of the screen. Use the variable sweep control to fine-tune the display and use the trigger controls to stabilize the display. The result will be an eye diagram.

Several different eye patterns are shown in Fig. 20-16. The multiple lines represent the overlapping pulses occurring over time. Their amplitude and phase shift are slightly shifted from sweep to sweep, thereby giving the kind of pattern shown. If the signal has not been severely rounded, delayed, or distorted, it might appear as shown in Fig. 20-16(*a*). The eye is "wide open" and has a trapezoidal shape. This is a composite display of the rise and fall times of the pulses overlapping one another. The steeper the sides, the less the distortion. The eye pattern in Fig. 20-16(*a*) indicates wide bandwidth of the medium.

In Fig. 20-16(*b*), the pattern looks more like an open eye. The pulses are rounded, indicating that the bandwidth is limited. In fact, the pulses approach the shape of a sine wave. The pattern shown in Fig. 20-16(*c*) indicates more severe bandwidth limiting. This reduces the amplitude of the rounded pulses, resulting in a pattern that appears to be an eye that is closing. The more the eye closes, the narrower the bandwidth, the greater the distortion, and the greater the intersymbol interference. The difference between the binary 0 and 1 levels is less, and the chance is greater for the receiver to misinterpret the level and create a bit error.

Note further in Fig. 20-16(*c*) that the amplitudes of some of the traces are different from others. This is caused by noise varying the amplitude of the signal. The noise can further confuse the receiver, thus producing bit errors. The amount of voltage between the lowest of the upper patterns and the highest of the lower patterns as shown is called the *noise margin*. The smaller this value, the greater the noise and the greater the bit error rate. Noise margin is sometimes expressed as a percentage based upon the ratio of the noise margin level *a* in Fig. 20-16(*c*) and the maximum peak-to-peak value of the eye *A*.

The eye diagram is not a precise measurement method. But it is an excellent way to get a quick qualitative check of the signal. The eye pattern tells at a glance the degree of bandwidth limitation, signal distortion, jitter, and noise margin.

PATTERN GENERATORS. A *pattern generator* is a device that produces fixed binary bit patterns in serial form to use as test signals in data communications systems. The pattern generator may generate a repeating ASCII code or any desired stream of 1s and 0s. Pattern generators are used to replace the actual source of data such as the

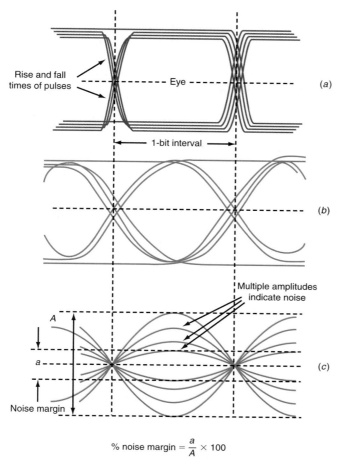

FIG. 20-16 Eye diagrams. (*a*) Good bandwidth. (*b*) Limited bandwidth. (*c*) Severe bandwidth limitation with noise.

computer. Their output patterns can be changed to standard codes or messages or may be programmable to some desired sequence. A pattern generator may be implemented in software at the sending computer.

BIT ERROR RATE TESTS. At the other end of the link the pattern generator sequence is detected and compared to the known actual pattern or message sent. Any errors in the comparison indicate errors. The instrument that detects the pulse's pattern and compares it is called a *bit error rate (BER) analyzer*. It compares on a bit-by-bit basis the transmitted and received data to point out every bit error made. It keeps track of the total number of bits sent and the number of errors that occur and then computes the BER by dividing the number of errors by the number of bits sent. The BER tester must of course know the exact pattern or message sent by the pattern generator.

$$\text{Percent BER} = \frac{\text{number of errors detected}}{\text{numbers of bits sent}} \times 100$$

TDR TESTS WITH AN ANALYZER. Special data communications test instruments are also available. A popular instrument is the *TDR tester*, also known as a *cable* *analyzer* or *LAN meter*. This instrument, often hand held, is connected to coaxial or twisted-pair cable and is able to make tests and measurements, some of which are:

1. Tests for open or short circuits and impedance anomalies on coaxial or twisted-pair cables.
2. Measurements of cable length, capacitance, and loop resistance.
3. Measurements of cable attenuation.
4. Tests for cable miswiring such as so-called split pairs.

A *split pair* is a wiring error often made when a cable contains multiple twisted-pair lines. One of the wires from one twisted pair is wrongly paired with one wire from another twisted pair.

Many tests are based upon what is called *time domain reflectometry (TDR)*. The TDR technique can be used on any cable or transmission line, such as antenna lines or LAN cables, to determine SWR, short and open circuits, and characteristic impedance mismatches between cable and load. It can even determine the distance to the short or open circuit or to any other glitch anywhere along the line. TDR testing is based upon the presence of standing waves on the line if impedances are not matched.

The basic TDR process is to apply a rectangular pulse to the cable input and to monitor the signal at the input. The test setup is shown in Fig. 20-17. If the load impedance is matched, the pulse will be absorbed by the load and no reflections will occur. However, if there is a short or open circuit or impedance mismatch, a pulse will be reflected from the point of the mismatch.

PROTOCOL ANALYZERS.   The most sophisticated data communications test equipment is the *protocol analyzer*. Its purpose is to capture and analyze the data transmitted in a particular system. Most data communications systems transmit data in frames or packets that include preamble information such as sync bits or frames, start of header codes, addresses of source and destination, and a finite block of data, followed by error-detection codes. A protocol analyzer can capture these frames, analyze them, and tell you whether the system is operating properly. The analyzer will determine the specific protocol being captured and then indicate whether the data transmitted is following the protocol or whether there are errors in transmission or in formatting the frame.

Protocol analyzers contain complex microcomputer circuitry programmed to recognize a wide range of data communications protocols such as Bisync, SDLC, HDLC, Ethernet, and other LAN and network protocols. These instruments normally cost tens of thousands of dollars and may, in fact, be primarily a computer containing the stored protocols and software that read and store and then compare and analyze the received data so that it can report any differences or errors. Most protocol analyzers have a video display if it is not a separate computer.

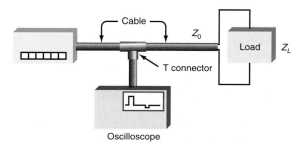

FIG. 20-17   TDR test setup.

A variety of special instruments are available for testing and measuring fiber-optic systems. The most widely used fiber-optic instruments are the *automatic splicer* and the *optical time domain reflectometer (OTDR)*.

Automatic Splicers.   Splicing fiber-optic cable is a common occurrence in installing and maintaining fiber-optic systems. This operation can be accomplished with hand tools especially made for cutting, polishing, and splicing the cable. However, as the cable thickness has gotten finer, hand splicing has become more difficult than ever. It is very difficult to align the two cable ends perfectly before the splice is made.

To overcome this problem, a special splicer has been developed by several manufacturers. It provides a way to automatically align the cable ends and splice them. The two cables to be spliced are stripped and cleaved by hand and then placed in the unit. A special mechanism holds the two cable ends close together. Then an optical system with a light source, lenses, and light sensors detects the physical alignment of the two cables, and a servofeedback mechanism drives a motor so that the two cable ends are perfectly centered on one another. An optical viewing screen is provided so that the operator can view the alignment from two directions at 90° to one another.

Once the alignment is perfect, the splicer is activated. The splicer is a pair of probes centered over the junction of the two cable ends. Pressing the "splice" button causes the probes to generate an electrical arc hot enough to fuse the two glass cable ends together.

The automatic splicer is very expensive, but it must be used because it is not possible for humans to make good splices visually and by hand. Handmade splices have high attenuation, whereas minimum attenuation is best achieved with the automatic splicer.

OPTICAL TIME DOMAIN REFLECTOMETER.   Another essential instrument for fiber-optic work is the optical TDR, or OTDR. It is an oscilloscope-like device with a CRT display and a built-in microcomputer.

The OTDR works like a standard TDR in that it generates a pluse, in this case, a light pulse, and sends it down a cable to be tested. If there is a break or defect, there will be a light reflection just as there is a reflection on an electric transmission line. The reflection is detected. Internal circuitry measures the time between the transmitted and reflected pulses so that the location of the break or other fault can be calculated and displayed. The OTDR also detects splices, connectors, and other anomalies such as dents in the cable. The attenuation of each of these irregularities can be determined and displayed.

20-3 TROUBLESHOOTING TECHNIQUES

Some of the main duties of a communications technician are troubleshooting, servicing, and maintaining communications equipment. Most communications equipment is relatively reliable, and most of it requires little maintenance. However, equipment does fail. Most electronic communications equipment fails because of on-the-job wear and tear. Of course, it is still possible for equipment to fail as the result of component defects. The equipment might fail eventually because of poor design, exceeding of product capabilities, or misapplication. In any case, you must locate such failures and repair them. This is where troubleshooting techniques will be valuable. The goal is to find the trouble quickly, solve the problem, and put the equipment back into use as economically as possible.

GENERAL SERVICING ADVICE

One of the main decisions you must make in dealing with any kind of electronic equipment is to repair or not to repair. Because of the nature of electronic equipment today, repairing it may not be the fastest and most economical approach.

Assume that you have a defective radio transceiver. One of your options is to send the unit out for repair. Repair rates run anywhere from $25 an hour to over $100 an hour depending upon the equipment, the manufacturer, and other factors. If the problem is a difficult one, it may take several hours to locate and repair it. Many communications transceivers are inexpensive units that may in some cases cost less to buy new than to repair.

There are two types of repair approaches: (1) to replace modules, or (2) to troubleshoot to the component level and replace individual components. Some electronic equipment is built in sections or modules. The module is, in most cases, a separate PCB containing a portion of the circuitry inside the unit. The typical arrangement might be for the receiver to be on one PCB, the transmitter on another, and the power supply on another, with another unit such as a tuner or frequency synthesizer also separate. A fast and easy way to troubleshoot and repair a unit is to replace the entire defective module. If you are a manufacturer repairing your own units in volume for customers, or if your organization does many repairs of a similar nature on a particular brand or model of equipment, the repair at the component level is the best approach.

COMMON PROBLEMS

Many repairs can be made quickly and easily because they result from problems that occur on a regular basis. Some of the most common problems in communications equipment are power supply failures, cable and connector failures, and antenna troubles.

POWER SUPPLIES.   All equipment is powered by some type of DC power supply. If the power supply doesn't work, the equipment is completely inoperable. Therefore, one of the first things you should do is to check that the power supply is working.

If the unit is used in a fixed location and operates from standard AC power lines, the first test should be to check for AC power and the availability of the correct DC power supply voltages. Is the unit plugged in, and if so, does AC power actually get to the outlet? If AC power is indeed available, check the power supply inside the unit next. These power supplies convert AC power into one or more DC voltages to operate the equipment. Open the equipment and, using the manufacturer's service information, determine the power supply voltages. Then use a multimeter to verify that they are at the correct levels. Most power supplies these days are regulated and, therefore, the voltages should be very close to those specified, at least within ±5 percent. Anything outside that range should be suspect. Any voltages that are obviously quite different from the specified value indicate a power supply problem.

Another common power supply problem is bad batteries. With continuous usage, batteries quickly run down. If primary batteries are used, the batteries must be replaced with new ones. If secondary or rechargeable batteries continue to fail even after short periods of usage after charging, it means that they too should be replaced. Most rechargeable batteries can be charged and discharged only so many times before they are no longer effective.

CABLES AND CONNECTORS. Perhaps the most common failure points in any electronic system or equipment are the mechanical components. Connectors and cables are mechanical in nature and can be a weak link in electronic equipment. Once it has been confirmed that the power supplies are operating correctly in the equipment, the next step is to check the cables and connectors. Start by verifying that the connectors are correctly attached. Another common problem is for the cable attached to the connector to break internally. Most of the time the cable will not break completely, but one or more wires in the cable may be broken while others remain attached.

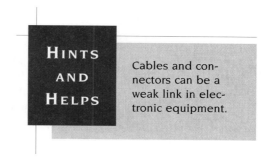

HINTS AND HELPS

Cables and connectors can be a weak link in electronic equipment.

Occasionally connectors get dirty. Removing the connector and cleaning the connections often solve the problem. It may be necessary to replace the connectors to ensure a reliable physical connection, however.

ANTENNAS. Another common failure in communications systems is the antenna. In most cases, antennas on portable equipment are fragile. A bad antenna is a common problem on handheld transceivers, cordless telephones, cellular telephones, and similar equipment.

DOCUMENTATION

Before you begin any serious detailed troubleshooting and repair of communications equipment, be sure that you have all the necessary documentation. This includes the manufacturer's user operation manual and any technical service manuals that you can acquire. Nothing speeds up troubleshooting and repair faster than having all the technical information before you begin. Manufacturers often regularly identify common problems and suggest troubleshooting approaches. But perhaps most important, manufacturers provide specifications as well as measurement data and procedures that are critical to the operation of the equipment. By having this information, you will be able to make the necessary tests, measurements, and adjustments to ensure that the equipment complies.

TROUBLESHOOTING METHODS

There are two basic approaches to troubleshooting transmitters, receivers, and other equipment: signal tracing and signal injection. Both methods work equally well and may often be used together to isolate a difficult problem.

SIGNAL TRACING. A commonly used technique in troubleshooting communications equipment is called *signal tracing*. The idea is to use an oscilloscope or other signal detection device to follow a signal through the various stages of the equipment. As long as the signal is present and of the correct amplitude, the circuits are good. The point at which you lose the signal in the equipment or at which the signal no longer conforms to specifications is the location of the problem.

To perform signal tracing in a transmitter, you need some type of monitoring or measuring instrument: an RF voltmeter or an oscilloscope, an RF detector probe on an oscilloscope, a spectrum analyzer, and power meters and frequency counters.

Figure 20-18 is a block diagram of a generic FM communications transmitter. It is assumed that the power supply is working and that it is supplying the correct voltage to all the circuits. Further, it is assumed that the transmitter has been declared inoperable. A quick verification of this should be made by testing the transmitter to see whether it is putting out a signal. A good overall check is to connect a dummy load, key up the transmitter, and attempt to pick up the signal on a nearby frequency counter or field strength meter with antenna. If no signal is indicated, troubleshooting can begin.

To troubleshoot the transmitter, connect a dummy load to the antenna jack and turn on the power. Using the signal-tracing method, start with the carrier signal source. Locate the carrier crystal oscillator or the frequency synthesizer responsible for generating the basic carrier signal. Monitor the output of this circuit on the oscilloscope to see that it is generating a sine-wave carrier of the correct amplitude and frequency. The frequency of the signal can be checked with a frequency counter. For the circuit in Fig. 20-18, you should measure and observe a 13.5-MHz sine wave.

If the carrier oscillator or frequency synthesizer is working correctly, you can go on to the next stages. Most transmitters have additional buffer amplifiers, frequency multipliers, and power amplifiers. Each should be checked in sequence from one step to the other. In Fig. 20-18, the carrier signal passes through a phase modulator to a sequence of three frequency multipliers. The output of the phase modulator should be the carrier frequency. The output of the first multiplier, a doubler, should be 27 MHz. The output of the second multiplier should be 3×27, or 81, MHz. The output of the third multiplier should be 2×81, or 162, MHz. This is also the output of the final power amplifier stage.

The output of every individual stage or circuit should be verified. The schematics or service manual will usually provide typical values of signal level at the output of each stage or at selected points in the transmitter. When discrete component circuits

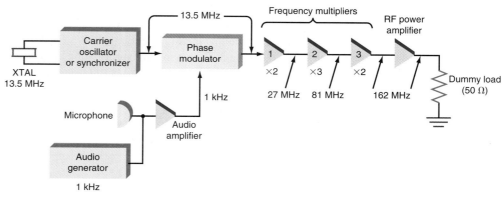

FIG. 20-18 FM transmitter.

are used, it is easy to check each transistor stage. In modern equipment each stage is implemented with one or more ICs. In such cases you can verify the input and then determine what outputs are available. Then you can determine what you should expect at the outputs.

You may lose the signal at some point as you follow it through the stages. For example, you may be tracing the signal through a sequence of frequency multipliers and power amplifiers as shown in Fig. 20-18. Assume that the carrier oscillator is working correctly, as was the first multiplier. You measure the correct frequency at the output of the first multiplier. This signal also appears at the input to the next stage. But suppose you can measure no signal at the output of the second multiplier. If this is the case, you have effectively isolated the problem to the second frequency multiplier circuit. It has an input but no output.

The most likely problem is a bad transistor or IC. Active components such as transistors and ICs fail more often than passive components like resistors, capacitors, and inductors or transformers. You should verify bias voltages to determine that they are correct, but if they are and the circuit still does not operate, it usually means that there is an open or short-circuited transistor.

Other components may also fail. Capacitors fail more often than any other type of component except the semiconductor devices. Resistors are less likely to fail but can open or change in value. Inductors rarely fail. Delicate, sensitive components like crystal, ceramic, and SAW filters can also break. Once you isolate the problem to a particular component, turn off the power, replace the suspected part, and repeat the tests. Continue testing until the transmitter delivers a signal to the dummy load.

If the carrier circuits are working but the unit is not receiving modulation, verify that the microphone is operating correctly. In portable and mobile operation, the microphone is a separate unit attached by a cable to the transmitter. Microphone cables and connectors frequently fail because of the continuous stress they receive. Microphones are also subject to physical damage because they are banged around and almost universally abused in a mobile operating environment.

If the microphone is OK, start checking the audio amplifier and modulator circuits. You can use a simple audio signal generator to inject the signal into the audio amplifiers and follow it through with an oscilloscope. A common test is to use a 1-kHz sine wave at a low level to prevent distortion. Then follow the signal through the circuitry with an oscilloscope, noting both the amplitude and any distortion that may occur. Distortion is really the result of applying too much signal to the input. Remember, microphones are very low level voltage-generating devices. The audio circuits are designed to provide a substantial amount of gain even with signal input voice signals that are in the microvolt or millivolt range.

Using higher output voltage from a signal generator will overload the circuitry and cause slipping and distortion. If you see that the audio signals you are tracing are square or otherwise distorted, back off on the signal generator output voltage until the distortion disappears. Then continue with the signal tracing. The point at which you lose the signal or at which the signal becomes distorted or attenuated will be the location of the problem. Once the transmitter is delivering power to the dummy load, you will want to run frequency, power output, and deviation tests to ensure that everything is working correctly and the unit meets manufacturer's specifications and FCC regulations.

You can also perform signal tracing on a receiver. You will need an RF signal generator with appropriate modulation and an oscilloscope, RF voltmeter, or other signal-measuring instrument to trace the signal with. You may prefer to use a speaker dummy load rather than the speaker itself.

SIGNAL INJECTION.   *Signal injection* is somewhat similar to signal tracing. It is normally used with receivers. The process is to use signal generators of the correct output frequency to inject a signal into the various stages of the receiver and to check for the appropriate output response, usually a correct signal in the speaker.

Signal injection is the opposite of signal tracing as it starts at the speaker output and works backward through the receiver from speaker to antenna. The signal injection would begin by testing the audio power output amplifier. You would inject a 1-kHz sine wave from an audio oscillator or function generator into the input of the amplifier. Follow the receiver documentation's information with regard to how much signal should be present at the outputs of each stage. If an audio signal is heard, the speaker and power amplifier are OK. The audio signal is then injected into any other audio amplifier stages back to the demodulator circuit output.

The next injection will take place at the input to the second mixer. Set the RF signal generator to 4.5 MHz with appropriate audio modulation. You should hear the audio tone in the speaker. If not, check the local oscillator and IF stages as described before. Do not overlook the demodulator. Keep in mind that you may not actually be able to get to the inputs and outputs of some circuits as they may be contained within an IC. If this is the case, you can restrict the injection to the IC input while monitoring the output. If you get no output, replace the IC.

Next, inject a signal of 45 MHz with modulation at the input to the first mixer. The tone should be heard, but if not, inspect and test the first IF stages and local oscillator. Finally, test the RF amplifier with a signal at the receive frequency of 478 MHz.

This spectrum analyzer can provide measurement speed up to four times faster than conventional swept-tuned analyzers with outstanding frequency and resolution.

SUMMARY

The work performed by communications electronics technicians will involve some form of testing and measurement over the course of equipment installation, operation, servicing, repair, and maintenance. Engineering and manufacturing technicians will also perform a wide range of tests and measurements in their work, including verifying and making final adjustments to specifications. It is important, therefore, for technicians to be familiar with the types of malfunctions that can occur and with the factors to consider in repair and maintenance of equipment.

There exists a wide range of choices in electronic communications testing equipment such as oscilloscopes and multimeters for measuring AC and DC voltages, current, resistance, and frequency. Measurements are routinely made of frequency, power, modulation, and standing wave ratios. Some equipment is manufactured with its own testing microcomputers built in; other equipment must be tested with separate equipment. Because electronic communications equipment is so complex and varied in makeup, the manufacturers' own manuals are the best resource in diagnosing and repairing their equipment.

KEY TERMS

Absolute field strength meter
Analog oscilloscope
Arbitrary waveform generator
Automatic gain control (AGC)
Automatic splicer
Autoranging
Bit error rate (BER) analyzer
Detector probes
Deviation meter
Digital (sampling) oscilloscope
Digital storage oscilloscope (DSO)
Down conversion
Dummy load

Eye diagram
Fast Fourier transform (FFT)
Field strength meter (FSM)
Frequency
Frequency counter
Frequency synthesis
Function generator
Impedance-matching pad
Intermediate frequency (IF)
Intersymbol interference (ISI)
Marker capability
Noise
Noise margin
Optical time domain relectometer (OTDR)
Pattern generator
Power meter

Prescaling
Protocol analyzer
Quieting
RF voltmeter
Root-mean-square (rms)
Signal generator
Signal injection
Signal tracing
SINAD test
Spectrum analyzer
Standing wave ratio (SWR)
Superheterodyne RF spectrum analyzer
Sweep generator
TDR tester (cable analyzer or LAN meter)
Time domain reflectometry (TDR)

1. What is the most important specification of an AC voltmeter for measuring radio frequency voltages?
2. What is a digital oscilloscope? Name the major sections of a DSO, and describe how they work.
3. What kind of resistors must be used in a dummy load?
4. Name two common types of SWR meters.
5. What is the name of the versatile generator that generates sine, square, and triangular waves?
6. How is the output level of an RF generator controlled?
7. What is the typical output impedance of an RF generator?
8. What cautions should be taken and what guidelines should be followed when connecting measuring instruments to the equipment being tested? Why should you allow an RF generator to warm up before using it?
9. What is a sweep generator, and how does it work?
10. Name two applications for a sweep generator.
11. How is the output of the circuit being tested with a sweep generator detected and displayed?
12. Name the six main sections of a frequency counter.
13. What is a prescaler? Name two types.
14. What is a deviation meter?
15. What is displayed on the CRT of a spectrum analyzer? Explain.
16. State the function of a field strength meter. What is its circuit?
17. Name the four most common transmitter tests.
18. What test instrument is used to measure the modulation of an AM transmitter?
19. What kind of test instrument is best for detecting the presence of harmonics and spurious radiations from a transmitter?
20. Generally, what level of SWR should not be exceeded in normal operation?
21. What does *SINAD* mean?
22. What is jitter? How does it show up on an eye pattern?
23. What does a "closing eye" usually indicate about the medium or circuit being tested?
24. Explain how an excessively noisy binary bit stream appears in an eye diagram.
25. What kind of instrument is used for bit error rate testing?
26. Describe the basic process of bit error rate testing.
27. What is a protocol analyzer? How does it display its results?
28. What type of components are most likely to fail in communications equipment?
29. What circuits should be tested first before any troubleshooting is initiated?
30. Besides the right test equipment, what should you have on hand before you attempt to service or troubleshoot any kind of communications equipment?
31. Name two ways that an antenna may be defective.
32. Name the two basic types of troubleshooting methods used with communications equipment. Which is most likely to be used on receivers? Transmitters?
33. What test instruments are needed to carry out the two troubleshooting methods given above?

R E V I E W

PROBLEMS

1. Explain the theory behind the operation of an arbitrary waveform generator. What is its main limitation? ◆

2. Explain the operation of a superheterodyne spectrum analyzer. What determines the resolution of the analyzer?

3. A noise test is performed on a receiver. Where is the noise measured? Using the procedure described in the text, the spacing between the two horizontal lines on the oscillope screen is 2.3 divisions. The vertical sensitivity calibration is 20 millivolts per division. What is the noise level? ◆

4. Refer to the transmitter diagram in Fig. 20-18. No modulation is applied to the circuit. The transmitter output power across the dummy load is zero. A measurement of the signal at the output of the phase modulator reveals a sine wave of 14.8 MHz. What is the problem? Explain why there is no output. Solve this problem two ways: (*a*) Assume that the carrier frequency is as shown in Fig. 20-18. (*b*) Assume a carrier crystal of 14.8 MHz.

5. In Fig. 20-18, the output of the third multiplier is OK, but there is no output across the dummy load. What component is most likely defective? ◆

CRITICAL THINKING

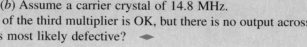

1. State two benefits of DSOs over analog scopes.

2. How do you lower SWR if it is too high?

3. A handheld cellular telephone worth approximately $175, consumer retail price, was obtained free when the cellular service was first initiated. It does not work. A call to the dealer indicates that this model can be repaired, but the labor rate is $75 per hour, and most repairs require a minimum of 2 h to fix. Parts charges are extra. Should you have the telephone fixed, or should you buy a new one? Give your reasons for the decision you would make.

ANSWERS TO SELECTED PROBLEMS

CHAPTER 1

1. 7.5 MHz, 60 MHz, 3750 MHz, or 3.75 GHz

3. 2200 MHz, 8.333 MHz, 27.875 MHz, 17.5 GHz

CHAPTER 2

1. 50,000

3. 30,357

5. 5.4, 4074

7. 14 dB

9. 37 dBm

CHAPTER 3

1. $m = (V_{max} - V_{min})/(V_{max} + V_{min})$

3. 100 percent

5. 80 percent

7. 3896 kHz, 3904 kHz; BW = 8 kHz

9. 800 W

CHAPTER 4

1. 28.8 W, 14.4 W

3. 200 μV

CHAPTER 5

1. $m_f = 12$ kHz/2 kHz $= 6$

5. -0.1

7. 64 percent

9. $\delta = 149.5$ Hz, S/N $= 26.76{:}1$

CHAPTER 6

1. 1.29 MHz

3. 3141.4 Hz or ± 1570.7 Hz at 4000-Hz modulating frequency

5. $f_0 = 446.43$ kHz, $f_L = 71.43$ kHz

CHAPTER 7

1. 206.4 MHz

3. 25.005 MHz

5. 1627

7. 132 MHz, 50 kHz

9. 72 W

CHAPTER 8

1. 6 kHz

3. 18.06 and 17.94 MHz

5. 2.4

7. 46 MHz

9. 27, 162, 189, and 351 MHz

CHAPTER 9

1. 7 MHz

CHAPTER 10

1. 66

3. Bit rate 1.544 MHz, 24 channels

CHAPTER 11

1. EBCDIC

3. 69.44 μS

5. 60 kbps

7. 278.95 Mbps

9. 8×10^{-6}

CHAPTER 12

1. 10 Mbps; $t = 1/10 \times 10^6 = 0.1 \times 10^{-6} = 0.1$ μS $= 100$ nS.

CHAPTER 13

1. 0.857 m

3. 51 Ω

5. VF $= 0.6324$

7. 4.02 ft

9. 1.325 nS

CHAPTER 14

1. 0.1 wavelength $= 20$ Ω; 0.3 wavelength $= 90$ Ω (see Fig. 14-11)

3. $468/27 = 17.333$ mHz, 520 kHz to 1.04 MHz

5. 75-Ω coaxial cable

7. 1.822 ft

9. 1.65 ft

CHAPTER 15

1. 36 percent, 0.36 percent

3. 0.632 in

5. 50 mW

7. 8.43 in

9. Gain = 960,000 or 59.82 dB; beamwidth = 0.145°

CHAPTER 16

1. Zero degrees (equator), 90° (polar)

3. The TC&C system is used to fire small hydrazine thrusters that move the satellite in one of several directions.

5. Satellite position information is transmitted as part of the NAV-msg 1500-bit message containing ephemeris and timing information. The data rate is 50 bps.

7. Differential GPS uses a precisely placed earth station that receives the L1 GPS signals and computes the error between them and its own precisely known location. The error is transmitted to special DGPS receivers that can decipher the error data and correct the output. Accuracies to within 3 to 6 ft are possible.

CHAPTER 17

1. By the use of tone-signaling codes unique to a particular telephone.

3. BORSCHT functions are battery power, overvoltage protection, ringing, supervision, coding, hybrid, test.

5. Links between exchanges are four line rather than two line; that is, the send and receive signals are separated before transmission between exchanges. These links may be twisted pair, fiber-optic cable, or microwave radio and may be analog or digital, in either case with multiplexing.

7. PBX means *private branch exchange.* It is a telephone system for medium-to-large organizations that require internal communications by telephone as well as connection to one of several lines to a local central office. The PBX provides switching and related functions to connect any internal phone to any other internal telephone or any internal telephone to an outside line. The PBX is beneficial as it minimizes the number of central office lines needed and eliminates the cost of having the telephone handle internal communications.

9. The document to be transmitted is divided into many fine horizontal lines as it is scanned a line at a time by a transducer that converts the dark and light portions into an electrical signal that is used to modulate a carrier.

CHAPTER 18

1. 300,000,000 m/s, 186,000 mi/s

3. See Fig. 18-8 and the related text.

5. 12 percent

7. 9.6 km

9. 100 MHz, 200 MHz

CHAPTER 19

1. Picture carrier = 205.25 MHz; sound carrier = 209.75 MHz

3. 180 kHz

5. Horizontal sync pulses are transmitted before and after each line of video. Vertical sync pulses are transmitted before and after each field of lines. The sync pulses are used at the receiver to perform the sweep operations that display the picture on the CRT.

7. The Y or luminance signal.

9. 589.75 MHz

CHAPTER 20

1. An arbitrary waveform generator uses digital techniques to generate a wave of any shape by storing binary samples in a memory and outputting them through a digital-to-analog converter and filter.

2. 23 mV rms

3. The RF power amplifier transistor is probably defective.

INDEX

INDEX

I
N
D
E
X